MW01628132

Red-cockaded Woodpecker by John James Audubon, (1821), watercolor, graphite gouche, black crayon, accession number 1863.17.389. Collection of the New York Historical Society

Red-cockaded Woodpecker

Recovery

Ecology

and

Management

Red-cockaded Woodpecker:

Recovery, Ecology and Management

Edited by:

David L. Kulhavy
Piper Professor,
Forest Entomology and Landscape Ecology
College of Forestry
Stephen F. Austin State University
Nacogdoches, Texas

Robert G. Hooper
Research Wildlife Biologist
Southern Research Station
USDA Forest Service
Charleston, South Carolina

Ralph Costa
Red-cockaded Woodpecker Recovery Coordinator
USDI Fish and Wildlife Service
College of Agriculture, Forestry and Life Sciences
Clemson University
Clemson, South Carolina

A Center for Applied Studies in Forestry Publication

The views expressed are not necessarily those of the Center for Applied Studies, College of Forestry, Stephen F. Austin State University

Printed in the United States of America

First Edition
10 9 8 7 6 5 4 3 2 1

ISBN 0-938361-12-0

Cover text and composition by P. R. Blackwell and David L. Kulhavy

DEDICATION

This volume is dedicated to two biologists whose inventions have greatly improved our ability to manage the red-cockaded woodpecker. The significance of their contribution is only partially reflected by the numerous chapters in this volume that report the successful application of their new technology. Looking into the future, it is clear that recovery of the red-cockaded woodpecker would be much less certain, and perhaps impossible, without the artificial cavities that they developed.

Carole K. Copeyon

David H. Allen

FOREWORD

A decade has elapsed since the second red-cockaded woodpecker symposium in 1983. Perhaps more significant things happened in the world of the red-cockaded woodpecker between 1983-1992 than in the previous 2 decades combined. Following is a list of some of the more important events:

(1) Second Red-cockaded Woodpecker Symposium is held, January 1983.
(2) 1985 Recovery Plan defines recovery objectives.
(3) USDA Forest Service adopts new guidelines in 1985.
(4) Detailed habitat requirements are published (Henry 1989).
(5) Large-scale studies on forest stands selected for foraging are published (Porter and Labisky 1986, Hooper and Harlow 1986, DeLotelle et al. 1987).
(6) Report of the American Ornithologists' Union Committee for the Conservation of the Red-cockaded Woodpecker is published (Ligon 1986).
(7) First successful translocations of red-cockaded woodpeckers are made (DeFazio et al. 1987).
(8) US Forest Service found in violation of the Endangered Species Act in Texas in 1989.
(9) Costa and Escano (1989) report and threat of additional lawsuits lead to Forest Service interim guidelines on National Forests.
(10) First increase in a population of red-cockaded woodpeckers reported (Hooper et al. 1991).
(11) Hurricane Hugo devastates Francis Marion National Forest, 28 September, 1989 (Hooper et al. 1990).
(12) The Red-cockaded Woodpecker Summit is held, March 1990.
(13) Studies confirm that red-cockaded woodpeckers select trees with heart rot for cavity excavation (Hooper 1988, Hooper et al. 1991).
(14) First indepth studies into the sociobiology of the red-cockaded woodpecker are published Lennartz et al. 1987, Walters et al. 1988, Walters 1990).
(15) First effective artificial cavities are invented (Copeyon 1990, Allen 1991).
(16) Studies confirm the impact of midstory development on the abandonment of cavity tree clusters (Conner and Rudolph 1989, Kalisz and Boettcher 1991, Loeb et al. 1991).
(17) Impacts of forest fragmentation are studied for the first time (Conner and Rudolph 1989, Conner and Rudolph 1991).
(18) First investigations into the genetic health of red-cockaded woodpecker populations are made (Stangel et al. 1992).
(19) First estimate of viable population size is made (Reed et al. 1988).
(20) First study on growth of heartwood is made (Clark 1992).
(21) Southern pine beetles recognized as major threat to red-cockaded woodpecker cavity trees (Conner et al. 1991).
(22) Declines in red-cockaded woodpecker populations continue (Ortego and Lay 1988, Ortego et al. 1988, Masters et al. 1989, Conner and Rudolph 1989, Costa and Escano 1989, James 1991).
(23) Release of the Final Environmental Impact Statement for the Management of the Red-cockaded Woodpecker and its Habitat on National Forests in the Southern Region in June, 1995.

The second symposium published 33 papers, many of which are still important references. Given the activity of the past decade, it is not surprising the response the third symposium has generated. The 65 papers in this volume touch nearly every aspect of the biology and management of the red-cockaded woodpecker. At least during the next decade this volume will be a beginning place for those striving to understand this endangered species and many of its papers will be important references for years to come.

R. Scott Beasley, Dean, College of Forestry
Stephen F. Austin State University

Preface and Acknowledgments

Red-cockaded Woodpecker Symposium III: Species Recovery, Ecology and Management was held 24-28 January 1993 at the Marriott in North Charleston, South Carolina. Primary organizers and cosponsors of the symposium were the U.S. Fish and Wildlife Service; Center for Applied Studies, College of Forestry, Stephen F. Austin State University; and the U.S. Forest Service, Southern Research Station. Additional cosponsors, contributing funding for the symposium and publication of this volume, included the Center for Applied Studies, College of Forestry, Stephen F. Austin State University; U.S. Forest Service, Region 8; National Fish and Wildlife Foundation; American Forest and Paper Association; Georgia-Pacific Corporation; and the Southern Timber Purchasers Council.

Registration totaled 310; all 13 southeastern states harboring red-cockaded woodpeckers were represented. Most federal and state organizations responsible for red-cockaded woodpecker recovery and management were in attendance. Additionally, many academic, conservation, and forestry organizations and interests were represented.

Opening remarks were provided by Ralph Costa, Red-cockaded Woodpecker Recovery Coordinator, U. S. Fish and Wildlife Service and David Kulhavy, Stephen F. Austin State University. Dave Wilson, Forest Supervisor, Francis Marion and Sumter National Forests, provided the welcoming remarks. Invited keynote presentations provided a broad overview of Federal agencies' roles and responsibilities regarding red-cockaded woodpecker recovery; John F. Turner, Director, U.S. Fish and Wildlife Service, Marv Meier, Deputy Regional Forester, Resources, Region 8, U.S. Forest Service, and Lieutenant General Charles E. Dominy, U. S. Army, The Pentagon.

A total of 11 technical sessions were conducted. Session I covered federal agency conservation strategies and management responsibilities chaired by Ralph Costa; sessions II and III incorporated status of, and management needs for, federal populations chaired by Ron Escano and Jeff Hardesty; sessions IV and V focused on disturbances, primarily hurricanes and southern pine beetles, in red-cockaded woodpecker clusters moderated by Phil Doerr and Robert Hooper; sessions VI and VII investigated red-cockaded woodpecker cavities use, construction, and competition chaired by Richard N. Conner and Frances James; sessions VIII and IX concentrated on forest management in red-cockaded woodpecker habitat chaired by Lonnette Edwards and Jerome Jackson; sessions X and XI covered red-cockaded woodpecker status, distribution, and management on private lands chaired by Roy DeLotelle and Emlyn Smith. In all, 60 technical papers were presented; several posters and displays were also exhibited.

Seven Red-cockaded Woodpecker Working Group Meetings were held on Sunday, 24 January, these included: (1) USDI Fish and Wildlife Service Field Office Biologists and State Agency Biologists, (2) Education Newsletter Committee, (3) USDI Fish and Wildlife Service Refuge Biologists, (4) Department of Defense Biologists and Military Personnel, (5) Symposium Proceedings Technical Review Committee, (6) USDA Forest Service Employees, and (7) Private Industry Biologists. An all day field trip was conducted on, and hosted by, the Francis Marion National Forest on February 27; Craig Watson, Eddie Taylor and Danny Carlson led the primary portions of the trip and discussions at stops. The field trip supplemented the "Restoration of the Red-cockaded Woodpecker Population on the Francis Marion National Forest: Three Years Post Hugo" paper/presentation by Watson, Hooper, Carlson, Taylor and Milling. Special thanks to Jerome Jackson, Mississippi State University, for bringing "Lady Red," our only captive red-cockaded woodpecker, to the conference; and to "Lady Red" for vocalizing her concerns and interest as appropriate!

This volume is organized into 6 sections:

(1) Outlook for Recovery of the Red-cockaded Woodpecker;

(2) Strategies for Recovery of the Red-cockaded Woodpecker;

(3) Recovery of the Red-cockaded Woodpecker in the Face of Disturbances;

(4) New Insights into the Biology of the Red-cockaded Woodpecker;
(5) Cavity Trees: the Most Critical Resource; and
(6) Status of Red-cockaded Woodpecker Populations and Habitat.

Individuals responsible for organizing and convening the conference included Ralph Costa and David Kulhavy assisted by Dennis Krusac, USDA Forest Service, and Jerry McIlwain, Director, Fisheries, Wildlife and Range, Region 8, USDA Forest Service. The excellent hotel accommodations were provided by Peggy Bauder, Sales Manager, Marriott.

Center For Applied Studies individuals assisting with this volume included: Graeme Seibel and William G. Ross technical assistance; Jennifer Pitarresi, John Tull, Joy Attebury, Janet Musgraves, Michelle Clendenin, Jay Mac Donald, James Matthews, Steve Croft, Nadir Erbilgin and Judy Haney, provided needed assistance. Debbie Corbin, Joleen Briggs and Robin Miller provided assistance within the College of Forestry with registration and records. Joe Gage, Gloria Chrismer, Michelle Clendenin, William G. Ross and Bobby Cahal provided assistance at the meeting.

Ultimately, the success of the conference and the proceedings would not have been possible without the participation of the 300+ attendees and the 114 authors and co-authors who contributed their time and expertise. We would like to acknowledge the support of Dave Flemming, Chief, Division of Endangered Species, U. S. Fish and Wildlife Service; and to R. Scott Beasley, Dean, College of Forestry.

DLK Nacogdoches, TX
RGH Charleston, SC
RC Clemson, SC

CONTENTS

SECTION 2 Strategies for Recovery

SECTION 3 Natural Disturbances

SECTION 4 New Insights on the Biology

SECTION 5 Cavity Trees as a Resource

SECTION 6 Status

SECTION 1

Outlook for Recovery

Outlook for Recovery of the Red-cockaded Woodpecker

Ralph Costa

Until recently, recovery of the red-cockaded woodpecker (RCW) seemed an insurmountable challenge. Faced with widespread and significant RCW population declines (Costa and Escano 1989), most biologists, managers, and researchers have been generally pessimistic regarding the long term survival and recovery of the RCW (Jackson 1978; Ligon et al. 1986). Additionally, even where populations were increasing or considered stable, managers were continuously challenged in their recovery efforts by other factors related to their agencies or organization's structure and/or mission. These factors included, inflexible or inadequate budgets, political ideologies and agendas, a multitude of land management and resource allocations alternatives, and some degree of scientific uncertainty. Although these and other issues will remain challenges to RCW recovery into the future, renewed optimism for recovery of the RCW has recently flourished. This new hope is a direct result of:

(1) technological advances,
(2) new directions in ecosystem management,
(3) significant population increases in both small and large populations, and, perhaps most important,
(4) new attitudes about, and emphasis on, cooperative ventures and innovative partnerships.

The papers presented in this first section provide valuable insights into the agency and individual accomplishments and philosophies that have renewed our hopes, and bolstered our spirits, regarding RCW recovery.

In his opening remarks, John Turner, U.S. Fish and Wildlife Service Director, said "we've reached a turning point, and we have great expectations for recovery. We're seeing a new spirit of cooperation, the forging of new partnerships, and the willingness of all concerned parties to work together toward mutually beneficial solutions. We're starting to see some tangible results of this cooperation, in the form of working plans to conserve red-cockaded woodpeckers on both public and private land." He was referring to the draft management plans being developed for National Forests, National Wildlife Refuges, and Army installations. At the time of the symposium, numerous private land partnerships were also in progress. In conclusion, Director Turner stated "Part of the challenge we all face in recovering this species is getting the word out to people affected by our actions. The southern pine ecosystem is unique—not only for its biological diversity but for its historic importance to local economies."

Building on Director Turner's introduction, Marvin C. Meier, Deputy Regional Forester, Region 8, USDA Forest Service, elaborated on the new directions for RCW recovery on southern National Forests. Highlights of this ecosystem management strategy include, habitat management at a landscape scale, restoration of native plant communities, and implementation of more natural disturbance regimes. Additionally, more intensive management will be practiced in smaller, and thus vulnerable populations. He concluded, "The key to recovery is cooperative efforts...", and "Together we can get the job done."

Lieutenant General Charles E. Dominy, Director of the Army staff, provided insights into the unique challenges faced by the Army as they carry out their primary mission to stay trained and ready to win the nation's wars, while caring for 25 million acres of public land and its associated biological diversity, including RCWs. However, he was positive about the dual missions and stated "...I am confident that we are up to the challenge." Further, Lt. General Dominy acknowledged that "...we must be active participants in the processes that restore and preserve these resources...we have also learned that, by integrating military and ecological planning, we can minimize restrictions on military training..." Like the previous speakers, he emphasized that cooperative efforts will be critical to future environmental programs.

In his paper, "Red-cockaded Woodpecker Species Recovery: Agencies' and Individuals' Challenges, Responsibilities and Opportunities", Ralph Costa, RCW Recovery Coordinator, U.S. Fish and Wildlife Service, highlighted the challenges to, and strategies for, RCW recovery. Regarding challenges, he focused on the importance of both individual and group efforts, imagination, and scientific credibility. He pointed out that our success will be based on our individual and group skills at negotiation, communication and cooperation, and our ability to provide credible conservation and management proposals founded in sound ecological principles. Following the theme of the invited keynote speakers, Costa also recognized that "Shared responsibilities and cooperative ventures will be a key to our success." Costa provided an optimistic outlook for recovery, as he concluded with, "We have the wisdom and experience of the veteran to guide us and the enthusiasm and vitality of the newcomers to stimulate us."

In "Red-cockaded Woodpecker Extinction or Recovery: Summary of Status and Management on Our National Forests", Ron Escano, National Program Manager, Threatened and Endangered Species, USDA Forest Service, provided a positive outlook for RCW populations on National Forests. He presented population status and trend information along with management summaries and accomplishments (i.e., monitoring, midstory treatments, artificial cavities, etc.) for 1986-1992. Based on these data, he concluded "...most of the populations are responding to the changes in RCW management that have taken place since 1988." This response is reflected in the improved population conditions (status and trend estimates) apparent since 1990. Escano's outlook for recovery was expressed in the following excerpt from his paper: "Two years ago, if you asked me the question, 'What do you foresee, RCW extinction or recovery', my answer would have been extinction. Today, after looking at these data, I see recovery within our grasp."

Bruce A. Sneddon, Office of the Deputy Chief of Staff for Operations and Plans, Department of the Army, made numerous references to the Army's commitment to environmental stewardship in his paper, "Trained and Ready While Protecting our Environment." He acknowledged that, "The challenge facing the army is to develop strategies which protect and conserve these resources while retaining the capability to meet the Army's missions." Regarding the RCW, he stated that the Army "... is a key player to the success of the species' recovery." The Army is demonstrating its commitment to the conservation of endangered species via several programs, including the establishment of: **(1)** a Directorate for Environmental Programs, **(2)** an Endangered Species Team, and **(3)** RCW Management Guidelines. Based on Sneddon's concluding remarks, "The Army will be a leader in environmental policy and management", the outlook for RCW recovery on Army lands is positive.

Jerome A. Jackson, in his chapter "The Red-cockaded Woodpecker: Two Hundred Years of Knowledge, Twenty Years Under the Endangered Species Act", outlined (and in some cases, detailed) our past knowledge about RCWs, in reference to its naming, listing, management/recovery programs and philosophies, and involvement in legal/court battles. Of particular interest was Jackson's analysis of the numbers of publications dealing specifically with RCWs in selected journals between 1950 and 1992. These data, and the continuing wealth of information being "produced" on RCWs (this volume for example), suggests that the outlook for RCW recovery, assuming knowledge provides a significant pillar in our recovery foundation, must be considered very positive. Jackson agrees; he states, "We can save the red-cockaded woodpecker. Understanding of the species' ecosystem and wise management must be our tools for the future beyond recovery."

In the paper, "Ecological Conscience, Economics, and Ethics: Exploring Leopoldian Thinking", Gene Wood provides a thought provoking and philosophical treatise on Leopoldian thinking which suggests "That conservation policy must be formulated within a paradigm that is both ecologically and economically sound." While Wood does not specifically address RCW recovery issues, his conclusion suggests that the outlook for recovery should be bright if we do not "... charge toward an uncertain future with imprudent disregard for ecosystem 'stability, integrity and beauty'."

The papers presented in this section provide information that has to instill, in all but the most pessimistic, a belief that RCW recovery is achievable. Some in our ranks have fought the RCW conservation/recovery battle for decades and are understandably, at times, short on hope. However, in recent years, it has not been difficult to find a renewed spirit of optimism among all concerned parties. We are at a crossroads in RCW recovery. Carole Copeyon and Dave Allen, to whom this volume is dedicated, provided techniques that rejuvenated the majority of us struggling with declining populations; artificial cavities will see us through the next couple of decades. Ecosystem management strategies, soon to be in place, will prepare and maintain the current and future habitat for RCWs, primarily on public lands. Our cooperators and partners in the private sector are contributing significant resources and populations toward RCW conservation. From my perspective, I see a coalition of individuals, representing all aspects of RCW conservation, who, as a team, are providing credible research information, management implementation, and policy formulation that will ensure RCW recovery in an expeditious, efficient, and professional manner. The outlook for RCW recovery is excellent; our wisdom and knowledge, gathered individually, and collectively shared and put into action, will take us to recovery. Our cooperative spirit is focused and strong and will be the basis of our success.

Director Turner's Remarks

Red-cockaded Symposium III:
Species Recovery, Ecology, and Management
Monday, January 25, 1993

- I would like to thank Region 4 of the U.S. Fish and Wildlife Service, the College of Forestry at Stephen F. Austin State University in Texas, the Southeast Forest Experiment Station of the U.S. Forest Service and all the cooperators who helped organize and fund this symposium. I'm delighted to be here with Lt. General Dominy, Director of Army Staff, Marv Meier, Deputy Regional Forester in Region 8 of the Forest Service and the other speakers and participants attending this important and timely conference.

- This is an exciting time for those of us involved in efforts to recover the red-cockaded woodpecker. The last time such a group gather was 10 years ago. Now we've reached a turning point, and we have great expectations for recovery. We're seeing a new spirit of cooperation, the forging of new partnerships, and the willingness of all concerned parties to work together toward mutually beneficial solutions.

- It is that spirit of cooperation I would like to address today, because that is what will help us meet the challenges we are facing in the next decade and beyond.

- It takes real commitment to get a job done, and that's what we have here. We're starting to see some tangible results of this cooperation, in the form of working plans to conserve red-cockaded woodpeckers on both public and private land.

- One development I'm very excited about, is the creation of the Red-cockaded Woodpecker Conservation Committee. Shortly after this symposium ends, a group of scientist will gather to form this committee. They will represent all factions involved in recovery of the woodpecker, including environmental, military, timber, development, and conservation interests. All members of the committee will be recognized experts with extensive experience with, and working knowledge of, red-cockaded woodpecker management and conservation.

- One task of the committee will be to review and comment on a red-cockaded woodpecker recovery plan, revision, scheduled for 1994/1995; the plan was updated last in 1985. But the scope of the team's responsibility is to be broader than that. We see the group addressing issues on both a local and regional level, and seeking solutions to specific management problems and challenges.

- We will rely on the committee's collective expertise to review guidelines, address private lands issues, and explore cooperative opportunities such as habitat conservation plans and conservation agreements. The committee, in my view, will function in the role of advisor, providing the Service with input to help in development of conservation and recovery strategies.

- This type of cooperative working group, which gives everybody a place at the table, is without question the best way to address complex resource issues. Hand-in-hand with that partnership must come commitment from each of the players, and we are seeing great strides in that arena.

- I point to the Forest Service's efforts in development of their Environmental Impact Statement (EIS) and Red-cockaded Woodpecker Management Plan. National forests harbor 12 of the 15 woodpecker populations needed for the species' recover, and the Forest Service's recently drafted EIS, which addresses red-cockaded woodpecker conservation in their Southeast Region, is a landmark document.

- The plan addresses changes in existing timber management programs, silvicultural practices, and harvest rotations. But what I find so exciting about it is that the Forest Service has taken the "ecosystem-based" approach to woodpecker conservation. This concept is one that just plain makes sense for endangered species management. I applaud the Forest

Service's vision -- its recognition of the long-term benefits of this kind of plan.

- Under General Sullivan's strong leadership, the Army is carrying forth his commitment to responsible resource management. His "can-do" attitude is infectious. For the first time, the Army has a full-time, Pentagon-level Endangered Species Task Force charged with drafting the Army's own guidelines for red-cockaded woodpecker management.

- Army lands harbor all or parts of 4 of the 15 woodpecker populations needed for recovery in addition to several other significant populations, so this conservation plan is critical. It outlines the Army's strategy for management and conservation of red-cockaded woodpeckers on their land. The Fish and Wildlife Service strongly supported this effort from the start, and I know this partnership will continue. Similarly, the military installations at Elgin Air Force Base in Florida and the Marine Corps' Camp LeJeune in North Carolina harbor recover red-cockaded woodpecker populations and are drafting long-range conservation plans.

- The Fish and Wildlife Service, like the Forest Service and the Defense Department, faces the challenge of managing public lands for the red-cockaded woodpecker. Ten national wildlife refuges support about 200 groups of woodpeckers. After this symposium ends, the Service will begin to revise its plan for conservation of the species on refuge lands. The agency will draw on new technologies for woodpecker recovery and incorporate into its plan public information and education strategies. The goal is to develop a plan that could triple the number of woodpecker groups that are using national wildlife refuges.

- I mentioned the use of new technologies a minute ago. Recovery of the red-cockaded woodpecker on Federal lands will depend at least in part on our use of new techniques, or new ways of thinking about our current practices. Currently artificial cavities are being used extensively on some lands. Translocation of birds is playing an increasingly important role in our recovery efforts. Likewise, growing season prescribed burning must be a significant component in any ecosystem based management plan.

- These practices will be at the forefront of our efforts on public land, but one of the greatest challenged we face is on private lands. Of the nearly 4,000 existing groups of red-cockaded woodpeckers, well over 500 are on private land. If we're to make real progress in recovery, which in large measure depends on public support, we've got to make sure the role of private landowners is identified and integrated into the process.

- We're making some headway along these lines, and are developing a private lands strategy for the woodpecker. The strategy embraces elements important to recovery and focuses on helping landowners understand and work within the Endangered Species Act (Act). Once the strategy is in place, we hope to maintain, or indeed in some critically important increase woodpecker populations on private lands. This will be accomplished by fostering partnerships with major landowners harboring (small 400 groups, but) significant populations. Perhaps more importantly the strategy is also designed to minimize the economic impacts of red-cockaded woodpecker management, especially to small property landowners and those with non-recovery populations.

- We are putting together a procedures manual that will provide habitat guidelines designed to protect landowners from a habitat-related "take" of red-cockaded woodpeckers. We owe it to private landowners to increase their understanding of the take prohibition and suggest ways to avoid it. Land holders willing to go a step further will be able to look to the procedures manual for ideas on how they can enhance woodpecker populations on their land.

- We would also like to see development of a range-wide habitat conservation plan that allows for incidental take of woodpeckers on small private lands, under Section 10 of the Act. This plan, which could cover the 13-state range of the species, would be designed to help small property landowners who may have trouble meeting the habitat standards in the procedures manual.

- Along the same lines, but better suited to large landowners, is the concept of memorandums of understanding (MOU). MOU's can lead to efforts over and above simply meeting the requirements of the law. The cooperative spirit needed to forge an MOU often leads to extra conservation measures to enhance local and often relatively large populations, such as those found on industrial forest land or southern quail plantations.

- This approach has the potential to conserve the remaining large populations of red-cockaded woodpeckers on private lands because MOU's can help link habitats and populations, especially where private lands adjoin state or Federal property. On Thursday, Ralph Costa, the Service's Red Cockaded Woodpecker Recovery Coordinator, will discuss the background and details of the private lands strategy.

- As previously mentioned, the success of these programs depends directly on public support. Part of the challenge we all face in recovering this species is getting the word out to people affected by our actions. I see the need for outreach programs that not only raise awareness about the red-cockaded woodpecker but show the interdependency of the whole ecosystem. The southern pine ecosystem is unique—not only for its biological diversity but for its historic importance to local economies.

- I've often said that we do our best work for wildlife when we work together. I can't think of a better example than this symposium I'm very optimistic that the fruits of your labors this week will be seen for years to come.

- Thank you for allowing me to be a part of this symposium. I know great things will come of this effort.

Red-cockaded Woodpecker Recovery: an Emerging Success Story

Marvin C. Meier, USDA Forest Service, Atlanta, GA

ABSTRACT: The red-cockaded woodpecker (*Picoides borealis*) was listed as endangered in 1970. In the 23 years since its listing, Forest Service management of the red-cockaded woodpecker (RCW) has evolved as our knowledge increased. A brief history of events and changes in RCW management direction from 1968 to present is discussed. The Forest Service is currently preparing an Environmental Impact Statement (EIS) detailing a recovery strategy for the RCW on National Forest System lands. The strategy is based on an ecological approach to management at the landscape scale. Three broad areas that are critical to RCW recovery are discussed, including habitat management, research and development, and cooperative relations.
KEYWORDS: Ecosystem management, red-cockaded woodpecker, recovery, cooperative relations, National Forests

The 23 years since the red-cockaded woodpecker was listed have been a learning experience for the Forest Service, and I expect others. We've learned that management, active ecologically based management, is absolutely essential to recovery of the RCW.

Is it premature to title this talk "An Emerging Success Story?" I think not and here's why. There are 4 main reasons for the emerging success story.

Cooperative Relations

We have enjoyed an excellent working relationship with the US Fish and Wildlife Service Field Supervisors and their Regional Office, including Gary Henry, Tom Olds, and Jim Pulliam. We are currently working very closely with Ralph Costa and Dave Flemming as we develop our new RCW management strategy. Maintaining this excellent working relationship has not always been easy -- reasonable people can disagree; the key is to maintain a professional commitment to working out solutions based on current knowledge!

Research Support

Another reason for the emerging success is the outstanding support from our research branch and significant contributions from a much broader array of research scientist.

Dedicated Employees

Our employees, from the biologists, silviculturists, and foresters at the district level, to the Chief of the Forest Service are dedicated to RCW recovery. In the past, we have not fully understood the magnitude of this recovery task. Our nearly completed EIS documents our recovery obligations and provides a recovery management strategy. Our employees are eagerly awaiting the opportunity to implement this new strategy.

The RCW's Resiliency

And finally, the resilience of the red-cockaded woodpecker itself has added to this emerging success story. The bird has come through times when habitat management knowledge was limited and in retrospect, inadequate to provide for recovery. Realizing the inadequacies of previous management direction, the Forest Service implemented emergency management actions to halt the declines in our smaller populations. These actions were successful, and in some cases we are seeing increases in populations. As stated previously, we are developing a RCW recovery strategy for National Forest System lands. We believe this strategy will yield significant gains toward recovery in the next 10-20 years. Before I discuss some of the key elements of the new strategy, I would like to provide a little history:

What we've done in the past, what we're doing now, and then what we see in the future.

FOREST SERVICE RED-COCKADED WOODPECKER HISTORICAL REVIEW

For discussion, the following dates correspond to past, present, and future time frames.

Past: 1968 thru 1988
Present: 1989 thru today (January 1993)
Future: Tomorrow and beyond

Past

In the past, as new knowledge became available, National Forest RCW management evolved accordingly. What follows is a chronology of events leading up to the present.

•**1968:** The RCW was identified by the USDI Fish and Wildlife Service as rare and endangered. The Forest Service policy was to protect cavity trees and a 200' buffer around them.

•**1970:** The RCW was listed as an endangered species.

•**1971:** The first RCW symposium. This was basically the foundation of RCW management.

•**1973:** The passage of the Endangered Species Act providing protection to the RCW.

•**1975:** The Forest Service amended its Wildlife Habitat Management Handbook to include a RCW chapter. Management recommendations included protecting cavity trees and a 200' buffer, providing 40 acres of pine > 20 years old adjacent to colony sites, and controlling midstory.

•**1979:** The red-cockaded woodpecker recovery plan was approved.

Following approval of the RCW recovery plan, the Forest Service amended its RCW Handbook chapter. Guidelines included protecting cavity trees and a 200' buffer, providing 100-250 acres of foraging habitat > 20 years old adjacent to colony sites, identification of recruitment stands, midstory control, and recommended rotations of at least 80 years for longleaf pine and 70 years for other yellow pines.

•**1980-82:** Baseline RCW surveys were conducted on 17 National Forests.

•**1983:** The second RCW symposium was held.

•**1985:** The RCW Recovery Plan was revised.

The Forest Service revised its RCW Handbook chapter. Guidelines included protecting cavity trees and a 200' buffer, establishing replacement stands for all active colonies and recruitment stands for all areas below their population objective, establishing foraging habitat requirements of 125 acres of pine and pine/hardwood ≥ 30 years old with at least 40% ≥ 60 years old or 6,350 pine stems ≥ 10" DBH, and 8490 sq ft of basal area (required for active colonies and recruitment stands and had to be within 1/2 mile of colony sites or recruitment stands), midstory control, and rotations of at least 80 years for longleaf pine and 70 years for other yellow pines.

•**1988:** Environmental groups in Texas initiate litigation on Forest Service management of RCWs in Texas. The Forest Service begins implementing Court Ordered RCW management plan in Texas.

Present

•**1989:** The Costa and Escano (1989) report was published indicating 67% of Forest Service RCW populations were declining, some drastically.

The Regional Forester established a 3-phase process to immediately protect existing RCW colonies and to develop new management direction.

The Regional Forester issued the March 27, 1989 Policy. This was Phase I of the process. The March 27 policy halted all active timber sales involving suitable habitat within 3/4 mile of active and inactive RCW colonies. All regeneration units within 1/4 mile of colonies were deleted or changed to thinnings if possible. Harvest units that isolated colonies from foraging habitat or replacement/ recruitment stands were modified. All advertised sales were withdrawn if they included any regeneration cutting within 3/4 mile of a colony. Proposed sales only included thinning within 3/4 mile of colonies.

Foraging guidelines were changed to delete

the 125 acre option. Foraging habitat had to include 6,350 stems ≥ 10" DBH that were ≥ 30 years old and comprised 8,490 sq. ft. of basal area, all within 1/2 mile of the colony site.

The Forest Service revisited proposed sale plans to ensure they met the new foraging guidelines.

The Forest Service was sued by the Southern Timber Purchasers Council over the March 27, 1989 Policy.

Hurricane Hugo devastated the Francis Marion National Forest. Tremendous efforts were made to stabilize the RCW population. In the process, artificial cavity technology was refined and implemented on a massive scale.

•**1990:** Phase II of Regional Forester's strategy began. Interim RCW management direction was issued for all RCW Forests except the National Forests in Texas, Apalachicola National Forest, and Vernon-Kisatchie-Evangeline population in Louisiana. Texas was excluded because of the court ordered plan. The Apalachicola and Vernon-Kisatchie-Evangeline populations were excluded because they contained more than 250 active colonies.

The intent of interim was to stop the declines that were occurring in most of our RCW populations. Interim direction was very similar to the March 27 policy, except it allowed limited regeneration under very restricted conditions.

•**1991:** Interim RCW management direction was issued for the Apalachicola National Forest and Vernon-Kisatchie-Evangiline populations. Apalachicola National Forest was included because further analysis indicated the Wakulla District was declining and the Apalachicola District may not have been as stable as once thought. Vernon-Kisatchie-Evangiline was originally treated as one population, but USFWS in consultation with the Forest Service determined that they were 3 demographically isolated populations.

Interim RCW management direction for Texas was submitted to the court for approval. To date, no action has been taken by the Court.

The EIS process began for the National Forest RCW recovery strategy. This began Phase III of Regional Forester's RCW strategy.

•**1992:** Preliminary analysis of RCW data from all Forests indicates that interim RCW direction was successful. We have stopped or slowed population declines on most Forests (Escano 1995).

Future: Where Are We Going From Here

The future looks bright. In the next 6 - 8 weeks we hope to have our Draft Environmental Impact Statement for the RCW recovery Strategy available for public review. By late summer we hope to be implementing this new RCW management strategy (Krusac et al 1995).

I will now discuss 3 broad areas that are critical to RCW recovery. These areas include habitat management, research and development, and cooperative efforts.

HABITAT MANAGEMENT

Unless the RCW recovery plan changes drastically, National Forest system lands will be responsible for 80% of the identified recovery populations. We do not anticipate major changes in the recovery plan because National Forest System lands provide most of the only suitable habitat or potential habitat for recovery. As such, habitat management on National Forests is critical to recovery. The RCW recovery strategy soon to be released will ensure suitable habitat is provided through time. This strategy will be another step in improved management, with recognizable differences on the ground.

Habitat will be managed at the landscape scale. Recovery populations will have adequate contiguous suitable habitat available to ensure successful dispersal within populations, and long - term viability. We will emphasize restoration of the native pine communities and growing season prescribed burns, imitating the natural fire disturbance regimes. We will practice more intensive management in our smaller populations because the conservation biology literature indicates smaller populations are more vulnerable to extinction (Shaffer 1981, 1987, Gilpin and Soulé 1986, Goodman 1987).

We believe our proposal is an excellent example of ecosystem management. We see tremendous benefits from this ecological approach to management. There are numerous federally listed species and others of federal concern that are associated with the ecosystems used by RCW. These include the gopher tortoise (*Gopherus polyphemus*), eastern indigo snake (*Drymarchon corais_couperi*), American burying beetle (*Nicrophorus americanus*), flatwoods salamander (*Ambystoma cingulatum*), Bachman's sparrow (*Aimophila aestivalis*), 3 sub-species of gopher frog (*Rana areolata* spp), 4 sub-species pine snake (*Pituophis melanoleucus* spp), and many plant species. Add to this list the numerous species of State concern and the benefits to all species from an ecosystem approach to RCW recovery become obvious. Ecosystem management is a proactive means of addressing viability concerns for a multitude of species.

Basically, this new management approach consists of 3 areas: habitat management at a landscape scale, restoration of the natural pine communities to the extent feasible, and a more natural disturbance regime. At the same time, we will continue our augmentation, artificial cavity, and midstory control programs. Proactive, hand-on management is essential to provide the critical needs of the RCW while more "natural habitat" is developed.

RESEARCH AND DEVELOPMENT

As I mentioned earlier, we've had excellent help and support from research scientists. Artificial cavities are an excellent example of research, development, and technology transfer in regards to RCW management. Carole Copeyon, while at North Carolina State University, developed a technique for drilling artificial cavities (Copeyon 1990). Shortly after Copeyon's work, David Allen with the Southeastern Forest Experiment Station developed a technology for cavity inserts (Allen 1991). Both of these techniques were utilized in stabilizing the Francis Marion population immediately after hurricane Hugo. In the 3 nesting seasons since Hugo, there has been a 34% increase in the number of clans, largely due to artificial cavities. Since the development of these techniques, Forest Service personnel have provided artificial cavity construction training to numerous State and Federal Agencies. Artificial cavities will be a critical element in our recovery efforts.

Another vital componant of the recovery strategy is augmentation. This involves the translocation of sub-adult RCWs from one location to another to either provide a single bird mate or to create a new breeding pair. These techniques help stabilize or expand populations. Research continues as does augmentation on many National Forests. Augmentation will be used hand in hand with artificial cavities to help recover the RCW.

We are comfortable with our knowledge of, and success with, the two above techniques. However, several area need additional investigation. These include:

RCW Southern Pine Beetle Interactions

In parts of the RCWs range, we are having serious problems with southern pine beetle (SPB) infestations. In Texas, SPB appear to prefer RCW cavity trees. We need to examine the relationships between SPB and cavity trees and determine what management techniques might be useful in decreasing the likelihood of SPB infestation.

Long Term Impact of Hurricanes

Any RCW population in the lower coastal plain is susceptible to hurricane damage. Continued monitoring of the RCWs response on the Francis Marion is important. We also need to investigate silvicultural options that may make forests less susceptible to hurricane damage.

Red-heart

We know there is a relationship between RCW cavity trees and red-heart fungus. The question is how do we manage for red-heart?

Foraging Habitat

Currently we are providing 6,350 pine stems > 10" DBH within 1/2 mile of colony sites. In many cases, these conditions are not met because the stems don't exist, but clans appear healthy. Examples include Francis Marion (pre and post Hugo), Apalachicola Ranger District, Vernon Ranger District. These are our largest populations. There may be other factors more critical than foraging substrate, but providing 6,350 stems where available may be precluding other management options. A good example is unevenaged management. Unevenaged management will result in less trees per acre. Providing 6,350 stems may limit this management option in some situations. Other factors that may be more critical than foraging habitat include habitat fragmentation, demographic isolation, cavity tree competition, and lack of suitable trees for cavity development.

Effects of RCW Management on Other Species

With the Forest Service commitment to

ecosystem management and conserving biological diversity, it will be important to quantify the effects of RCW management on other plant and animal species. You will hear some preliminary results of a study conducted by Mississippi State University on Thursday morning that addresses this issue.

As the answers to the above questions are determined, we will amend our RCW management direction to incorporate the latest research findings. We will strive to implement the best strategy possible. This strategy will effectively and efficiently increase the population and lead to recovery, while allowing for a subsidiary flow of products, uses, and values.

COOPERATIVE RELATIONS

As I mentioned in my introduction, we have enjoyed excellent cooperation with those involved in RCW research and management. Cooperative efforts are probably the most critical element to RCW recovery. None of us can do it alone. If we all go our own way trying to recover the RCW, it won't happen. The job is too large, complex, and interrelated for anyone or any agency to accomplish alone. It will take a cooperative effort from all involved parties including Federal and State agencies, private organizations and individuals, and the timber industry. Some examples of existing cooperative efforts include:

Oconee National Forest/Hitchiti Experimental Forest/Piedmont National Wildlife Refuge. This effort involves managing the area as one population under a memorandum of understanding.

Francis Marion National Forest/Fort Jackson and Charleston Naval Weapons Station and Fort Jackson are providing financial support.

Some examples where cooperative efforts could be initiated include:

Conecuh National Forest/ Blackwater River State Forest/Elgin Air Force Base: An opportunity exists to enter into a cooperative agreement to mange these lands as one population.

Tombigbee National Forest/ Noxubee National Wildlife Refuge: Although there are currently no known RCW on the Tombigbee National Forest, if a suitable habitat link exists between the two areas, there may be an opportunity to enter into a cooperative agreement. The Noxubee National Wildlife Refuge has had tremendous success in recent years with their RCW management programs. You will hear about their success this afternoon in a paper presented by Dave Richardson. Continued analysis will help us explore the opportunity to mange our habitats as one unit.

The USFWS is currently developing a private lands strategy for RCW management. If this proposed strategy involves moving sub-adult RCWs from small private landholdings that are demographically isolated, the US Forest Service will be a recipient for these sub-adults. This proposal appears to have merit. It will prevent the likely loss of sub-adults that could disperse into unsuitable habitat. These sub-adults could be used to bolster our smaller more vulnerable populations.

Our cooperative efforts with research need to be maintained and strengthened. As stated earlier, North Carolina State University (NCSU) developed the drilling technique for creating artificial cavities. Shortly after Hurricane Hugo hit the Francis Marion, a crew from NCSU came down to the forest and trained Forest Service personnel in the use of their technique. This cooperative effort went a long way toward stabilizing the situation after Hugo.

SUMMARY

In closing, I want to reiterate our commitment to RCW recovery. The Southern Region will be initiating new management direction in the near future. This will be an ecosystem approach to management dedsigned to benefit many species of concern, not just the RCW. Some questions remain unanswered, however, we must move ahead while research continues to investigate the unknowns. The job of RCW recovery is too large for any one of us to tackle alone. The key to recovery is cooperative efforts, and the Forest Service looks forward to continued cooperation efforts with all of you and your respective organizations. Together we can get the job done!

Lieutenant General Charles E. Dominy
Director of the Army Staff

Red-cockaded Woodpecker:
Species Recovery, Ecology, and Management
February 25, 1993, Charleston, SC

Thank you, John, for that kind introduction. And thanks to **Region Four of the Fish and Wildlife Service, Stephen F. Austin State University, and the Southeastern Forest Experiment Station of the Forest Service** who organized and, along with other federal and private parties, are graciously co-sponsoring this gathering. I'm honored to be here with you this morning to help kick-off this worthwhile symposium dedicated to the preservation of a national asset—the red-cockaded woodpecker (RCW). It is heartening to see so many people here sharing the concern for this species, working together to overcome the effects of nature's caprice and man's past negligence.

I am here this morning representing the Army's commitment to the RCW and other environmental issues. As the Defense Department's **Executive Agent for the Red-cockaded Woodpecker Management Improvement Initiative**, the Army will continue to work with concerned, responsible groups like this one to preserve not only the RCW, but also the other threatened and endangered plant and animal species found on the federal lands for which we provide stewardship.

The Army faces a unique challenge. We are the stewards for about **25 million acres of public lands**, and as such, we are critical to the protection and conservation of many plant and animal species. We take this responsibility seriously. At the same time, we must never forget that our primary mission is to stay trained and ready to fight and win the nation's wars. Balancing environmental concerns with the Army's Training needs has presented us with some significant challenges in recent years, but I am confident that we are up to the challenge.

Training is the glue which binds our Army together and yields the trained and ready force that achieved decisive victories in Panama and Desert Storm. Training remains the key to combat-ready units, and training combat-ready units requires land. Modernization of weapons systems and war-fighting doctrine place further demands on land. The same systems that allow us to engage the enemy at extended ranges with great accuracy and limited collateral damage require larger tracts of land for realistic, integrated training.

A growing number of endangered species compete for the same land the Army uses for training. As we transition from forward-deployed to a CONUS-based Army, the competition for land becomes even more fierce. The challenge is to balance environmental needs with training requirements. We do this through innovative training methods—such as increased use of simulators and simulations for training and testing weapon systems; we educate our soldiers, civilians, and family members to be more environmentally aware and active; and we integrate environmental factors into our long and short range training plan processes.

Why do endangered species like Army land so much? Over half of the Army's major installations have at least one federally protected endangered species. Over 100 different endangered species are found on our installations, and that number is estimated to double by 2006. As natural habitats around our installations are destroyed by urban expansion or commercial and agricultural activities, many wildlife species migrate to Army installations.

The RCW is one of these species. Since it was first listed as an endangered species in 1968, the Army has learned a lot about this small bird. some of the lessons have been painful, but we (and the woodpeckers) have benefited in the long run.

Our experience with the RCW at Fort Bragg has taught us that all of us in the Army need to be sensitive to the value of ecological resources, and that we must be active participants in the processes that restore and preserve these resources. We have also learned that, by integrating military and ecological planning, we can minimize restrictions on military training - we can optimize the use of both our military and environmental resources.

We host the RCW at 6 Army installations in North Carolina, South Carolina, Georgia, and Louisiana. There are about **760 active colonies** throughout these locations. We can point with pride to the Fort Bragg community and their sustained efforts to help the North Carolina Sandhills population achieve recovery status. Working with wildlife biologists at Fort Bragg and other Army installations, we have implemented strategies for maximizing growth of existing colonies.

The strategies include:

- Installation of artificial cavities.
- Protecting the RCW from predators.
- Controlling mid- and understory growth through an effective burn policy.
- Translocating juvenile birds to unoccupied colony trees in attempts to facilitate pair bonding.
- We've implemented a standard RCW cavity tree marking system - two white bands - for all Army installations.
- We've restricted access to RCW colonies, allowing only transient foot traffic and vehicular traffic on existing roads.
- We've posted signs with maps, "the Woodpecker Special", throughout the posts to educate the community.

We're also exploring other innovative methods for RCW recovery. In cooperation with the **Nature Conservancy**, we have begun to research the benefits of artificial cavities on RCW populations We are incorporating new technologies, such as the Global Positioning System, to precisely locate cavity trees and prevent damage to active colonies. This is just one of many examples of technology—produced for military use—that has far-reaching benefits for mankind and our environment.

We are involved with **Project Reliance, a tri-service** research and development program with focus on research for endangered species and land use management practices. It is **cooperative efforts** like those I've just mentioned that will be critical to future environmental programs. We are looking at partnering with other federal, state, and DOD agencies to maximize the effectiveness of our research and development dollars.

We are establishing a Threatened and Endangered Species Research Development Plan to study the impacts of military training on specific plant and animal species. We are also committed to research and management which investigates and promotes **biological diversity**. Preserving native species in numbers and distributions that provide a high likelihood of continued existence is a crucial element in habitat management. Conserving biological diversity will minimize the number of species that might otherwise be added to the threatened and endangered list.

By **working with the Fish and Wildlife Service and the National Marine Fisheries Service**, the Army's Endangered Species Team has developed a comprehensive threatened- and endangered-species management policy. The team is currently developing the **Army's Red-cockaded Woodpecker Management Guidelines** which will provide common rules, methodologies, and management techniques for the RCW on Army installations.

Yes, we have come a long way with the RCW. But, it's important to understand that the Army's RCW efforts are part of a larger environmental program. I want to tell you a little about the Army's larger role in environmental stewardship.

On November 19, 1992, the Secretary of the Army and the Chief of Staff of the Army General Gordon Sullivan signed the **U. S. Army Environmental Strategy—Into the 21st Century**. This Strategy defines our leadership commitment to the environmental challenges of the present as well as the future.

We are moving out with four simultaneous efforts: giving immediate priority to sustained compliance with all environmental laws; continuing to restore previously contaminated sites as quickly as funds permit; focusing efforts on pollution prevention to reduce or eliminate pollution at the source; and conserving and preserving natural resources so they will be available for present and future generations to use.

With the Strategy firmly in place, we've quickly moved on to building the framework to ensure success in our environmental efforts. As of January 1st, 1993, we have established a new Directorate for Environmental Programs. Now, all environmental programs within the Army are managed within one organization. This centralization will ensure that we integrate our programs focus our environmental efforts, and succeed in our conservation and preservation missions.

Having mentioned some recent achievements, I must point out that the Army has a solid record of environmental accomplishments. We have been actively involved with the environment and land-use management for many, many years, and our history is replete with environmental awards and recognition for past actions.

In 1988, Fort Carson and the 4th Infantry Division (Mechanized) received the National Wildlife Federation's prestigious National Conservation Achievement Award, with the notation that Fort Carson's natural resources are in better condition today than when the land was in private hands.

Army installations across the country can boast of environmental awards: Fort Sill, Oklahoma for conservation; Fort Benning, Georgia for environmental protection and enhancement; Fort Belvoir, Virginia for erosion prevention, restoration, forestry, fish and wildlife conservation, and cultural resources management. The list goes on. We take quiet pride in all these achievements, but we do not rest on our laurels.

The Army continues aggressive actions to clean up contaminated sites at its installations. As part of the Chesapeake Bay Initiative, we have established projects to extract and treat contaminated ground water, remove underground storage tanks, remediate soil, create and restore mudflats, wetlands, and wildlife habitat, and prevent erosion.

Just three weeks ago the Chief of Staff of the Army, representing the uniformed Army, signed a memorandum of understanding with the Department of the Interior to transfer management of Rocky Mountain Arsenal, near Denver, Colorado. This agreement will make the 27 square miles of land at this newly designated national wildlife refuge a valuable habitat for such animals as the endangered bald eagle, many species of hawks, prairie dogs, deer, coyote, owls, and migratory waterfowl.

We take our responsibilities seriously. It is incumbent upon us to raise the consciousness not only of our military members and communities, but also all other Americans that share the ecosystems of the RCW and other threatened and endangered species. This is a **shared responsibility**—not solely on the Army, not solely on this group, but shared with all who inhabit this planet.

The Army will be a national leader in environmental issues as an integral part of our mission. We will continue to exercise responsible stewardship of our federal lands. We are Americans—we love this country, and we've vowed to protect it. We will do so by integrating training and environmental requirements. We realize that environmental factors weigh heavily in protecting our nation.

We are America's Army —Your Army—soldiers, civilians, families, active and reserve. If you wonder how I can be so sure of success in our environmental leadership role, I have only to point with great pride to the men and women of our Army. We are the premiere Army in the world—highly trained, equipped with modern systems, guided by well-developed, confident leaders executing missions based on solid doctrine.

The key to all of that is quality soldiers. We have the finest quality soldiers of any Army, at any time, in the world. We are committed to maintaining quality soldiers through realistic, intensive training. We are also committed to integrating environmental requirements with this training mission. We are focused—and when today's Army focuses on a mission—whether it's the combat of Desert Storm, hurricane relief in Florida and Hawaii, or restoring hope in Somalia—success is just a matter of time. Thank you.

Red-cockaded Woodpecker Species Recovery: Agencies' and Individuals' Challenges, Responsibilities and Opportunities

Ralph Costa, USDI, Fish and Wildlife Service,
Red-cockaded Woodpecker Field Office, Clemson, SC.

ABSTRACT: Red-cockaded woodpecker recovery progress has been limited at best during the past 2 decades; only one of the required 15 populations is considered "recovered", and most others are a long way from recovery status. Challenges facing individuals and managers responsible for, or involved in, recovery are formidable and include, management and resource allocation alternatives, biopolitical ideologies and agendas, scientific uncertainty, inflexible budget and monetary systems, social responsibility questions, and a host of others frequently inherent to endangered species issues. In spite of the challenges, meaningful progress is being made in the form of new, comprehensive Federal management plans, private land conservation strategies, and population stabilization/growth programs. Continued success will be dependent upon building additional partnerships, focusing on a short-term attainable objective of downlisting to threatened, increasing artificial cavity programs, fully implementing the new Federal management plans, addressing and resolving private land/red-cockaded woodpecker issues, and sharing information via a red-cockaded woodpecker newsletter.
KEYWORDS: red-cockaded woodpecker, recovery, Endangered Species Act

On January 27, 1983, a year ago to the day on this coming Wednesday, some of you in this room assembled at Red-cockaded Woodpecker Symposium II. Indeed, a smaller group of you are veterans of Red-cockaded Woodpecker Symposium I, held in May of 1971, 22 years ago! I have to wonder how the diary would read if each participant of the two previous Red-cockaded Woodpecker Symposia wrote a couple of pages providing his/her opinions and insights concerning our progress toward red-cockaded woodpecker (RCW) recovery during the last 22 years. At Symposium II, Richard Thompson, of the U.S. Fish and Wildlife Service (Service), in his overview suggested that "much of what we may have labeled as progress [he was speaking about the past 12 years since Red-cockaded Woodpecker Symposium I] may be attributed to the efforts of those confronted with keeping those conflicts [he was referring to the conflicts between humans and birds] from becoming a contest of public emotions as has happened with some other endangered species" (Thompson 1983). In other words, progress was seen as avoidance of conflicts and not as achievement of specific recovery objectives. Lawrence Givens (1971), also a U.S. Fish and Wildlife Service employee, in his Introduction to Red-cockaded Woodpecker Symposium I took a more philosophical note and suggested that "the symposium was or should be a small step toward helping establish an acceptable land ethic." He also said, and I quote, "he [referring to RCWs] is a part of the whole web of life and unless we can better understand how he is to survive, then the understanding of our own survival becomes less clear."

Building on their thoughts, I would suggest that species extinctions and the associated destruction of naturally functioning ecosystems are not realities to be proud of as we pass them on to future generations, and thus, deny them opportunities and options for a healthy and secure future. That said, where are we with RCW recovery today?

Our recovery progress, as measured by numbers of increasing populations, has been poor. However, we have made significant progress in understanding the bird's ecology.

This increased knowledge has, in recent years, led to the development of technological and management strategies that promise to provide a catalyst for renewed recovery progress.

Within this room, we have the collective wisdom of two decades of RCW research, management, and conservation philosophy. That wisdom and knowledge, gathered individually and collectively shared, is our recovery foundation. The Red-cockaded Woodpecker Recovery Plan (U.S. Fish and Wildlife Service 1985), specifies that we need 15 populations in 5 physiographic provinces in 8 States to delist the RCW. RCW recovery involves a multitude of Federal, state and private interests. Shared responsibilities and cooperative ventures will be a key to our success. We need a comprehensive and cohesive regional plan, that specifies the who, where, when, and how of RCW recovery. Many components of this plan are in place or soon will be. The U.S. Forest Service has prepared a Draft Environmental Impact Statement for the Management of the Red-cockaded Woodpecker and its Habitat on National Forests in the Southern Region (U.S. Forest Service 1993); the U.S. Army has drafted Management Guidelines for the Red-cockaded Woodpecker on Army Installations (U.S. Army 1993); and the U.S. Fish and Wildlife Service has issued a draft Red-cockaded Woodpecker Procedures Manual for Private Lands (Costa 1992). In spite of these mileposts, RCW recovery will remain a significant challenge well into the future. It is a concept fraught with management and resource allocation alternatives, biopolitical ideologies, some degree of scientific uncertainty, economic unknowns, social responsibility questions, and a host of other variables inherent to endangered species issues.

Identifying and resolving these issues can only be successful if we identify clearly defined population objectives and time frames for achievement of stated goals. The identification of an attainable, measurable goal for the coming years is paramount to our success. For 20 years the goal has been recovery, period. No time frame, to my knowledge, was proposed nor were any milestones or other points of measurable success identified. We have simply had this awesome challenge called recovery. It is very intimidating to focus on recovery, or 15 populations, when we have only one recovered population and the majority of others are seemingly far away from recovery status. Acknowledging that RCW recovery will happen incrementally, I propose today that we as a group establish a 17-year objective of downlisting the species to threatened status by the year 2010. This would require six populations with the following distributions:

1. Coastal Plain of North Carolina or South Carolina.

2. Sandhills of North Carolina or South Carolina.

3. Coastal Plain of Georgia or the Florida peninsula.

4. Coastal Plain of Alabama or the Florida panhandle.

5. Coastal Plain of Mississippi.

6. Coastal Plain of Louisiana or Texas.

I believe reaching that objective within the first decade of the next century is an attainable goal. Achieving threatened status will provide us the experience and knowledge required to head toward recovery within another specified time frame. In fact, if all 12 populations potentially involved in downlisting would work toward that goal simultaneously, when we reached 6 recovered populations, 6 other populations (5 of which are required for delisting) could also be well on their way to viability.

There are additional challenges related to RCW recovery. We certainly have challenges to overcome within our local, state, and Federal governmental organizations. Many of the potential barriers are related to agency traditions, inflexible budget and monetary systems, and the biopolitical realities of administration and agency policies. Thus, RCW recovery will require an understanding of, and the ability to operate within, the bureaucratic framework of our governments. Success will be dependent upon our individual and group skills of negotiation, communication, and cooperation, and our ability to provide credible RCW conservation and management proposals founded in sound ecological principles.

Many of you, regardless of your role in RCW conservation, face challenging issues on a frequent basis. These include, developing RCW management strategies for industrial forest landowners; designing urban landscapes in and around RCW habitat; changing state laws to permit RCW management activities in state wilderness areas; dealing with generational changes in family ownership and how that affects land management decisions;

drafting biological opinions for Federal projects; hurricanes; salvaging small, seemingly doomed populations; integrating military training and national defense needs with RCW habitat requirements, and the list goes on. How these challenges are approached and eventually resolved is due in large measure to your individual efforts, imagination, and scientific credibility. It is these experiences and efforts that provide the success stories from which other projects can be designed and successfully implemented.

I hope I have made the point that RCW recovery is a mission faced with many challenges at the local, state, regional and indeed national levels. Individual efforts during the past 20 years, by many in this audience, have built the foundation from which RCW recovery must proceed. Again, I propose that we as individuals and a coalition, focus our attention during the next 17 years on downlisting the RCW to threatened status. If all populations work with equal vigor on this effort we could, in that time frame, have 50 percent or more of our required recovery populations in place by the second decade of the next century.

I would like to briefly cover our responsibilities under the Endangered Species Act of 1973 as amended (Act) (U.S. Fish and Wildlife Service 1988). The Act forms the legal portion of our recovery foundation. The Act, along with Federal agency management plans, state conservation programs, and our RCW Recovery Plan provides the legal framework within which all recovery actions should proceed. In a sense, the Act and its successful implementation by the Service is an insurance policy that not only guarantees alternatives and options for future generations, but also measures and evaluates the condition of today's natural resource base, that ultimately we depend upon for our own continued well-being and survival. Although some suggest otherwise, I have found during the past 18 months, since accepting this position, that the Act is a flexible piece of legislation.

In a sense, it is a large scale conservation plan. Historically, I think it has traditionally been viewed only as a law enforcement tool and not as a carefully developed program for long-term human sustainability through conservation of our limited natural resources. However, the law enforcement and compliance section is only one small part, albeit an important one, of a very thoughtful and comprehensive document whose full implementation would benefit our society and future generations.

What are our individual and agency responsibilities under the Act? Federal agency cooperation under Section 7(a)(1) of the Act states, "Federal agencies shall utilize their authorities in furtherance of the purposes of this Act by carrying out programs for the conservation of endangered species." This affirmative conservation mandate is further clarified in the definition section of the Act. The term "conservation" means "to use and the use of all methods and procedures which are necessary to bring an endangered species or threatened species to the point at which the measures provided pursuant to this Act are no longer necessary. Such methods and procedures include, but are not limited to, all activities associated with scientific resources management such as research, census, law enforcement, habitat acquisition and maintenance, propagation, live trapping, and transplantation,..." The undisputed success stories that you will hear this week concerning artificial cavity and translocation technologies should send a very clear and loud message to Federal agencies concerning their obligations under Section 7(a)(1) of the Act. Under Section 7(a)(2) of the Act (interagency cooperation), each Federal agency must ensure that any action authorized, funded, or carried out by them is not likely to jeopardize the continued existence of a listed species. These determinations are made in consultation with the Service.

Under Section 6 of the Act, (cooperation with the states), state agencies can enter into cooperative agreements with the Service. Through this process, the states receive funds to assist them in development of programs for the conservation of Federally listed species. The cooperative agreement and issuance of funds is contingent upon the state having established acceptable conservation programs, consistent with the purposes of the Act, for all resident, federally listed species occurring in the state. This Section 6 nexus or link with the states provides an opportunity for the Service to examine the states' RCW plans and programs, and recommend changes or improvements where necessary.

Under Section 10 of the Act (permits and exemptions), a process is available that allows "incidental take" of listed species. Such taking is incidental to, and not the purpose of, the carrying out of an otherwise lawful activity. For the RCW, this could mean the translocation of birds from point A to point B. I see this tool as one way of dealing with small, demographically isolated RCW populations on private lands. I think the responsibility for dealing with the issue of RCWs on private lands is a coalition challenge involving Service leadership. National Forests, Department of Defense lands and National Wildlife Refuges would be the primary recipients of private land birds. State natural resource agencies could serve as liaisons between the private landowners and the Federal organizations.

The Act contains the flexibility and mechanisms necessary to achieve RCW recovery. The Service, under the Act, has the responsibility to enforce compliance with its requirements. When it becomes necessary to initiate law enforcement investigations it is generally a sign that individuals or organizations are ignoring their obligations under the Act or worse, intentionally violating the law. While successful prosecutions for known violations are important, it is perhaps more important that law enforcement actions are followed by implementation of positive conservation measures by the agencies and individuals involved. In other words, Act enforcement actions should provide a catalyst for improvements in RCW conservation. Indeed this has been the case with the two most notable RCW legal issues to date; the Florida developer case and the National Forest in Texas litigation. The $300,000 of fine money from the Florida developer case was made available, through the National Fish and Wildlife Foundation, for RCW conservation programs in the state of Florida where the violation occurred; the funds being equitably distributed between the Service, U.S. Forest Service and Florida Game and Fresh Water Fish Commission.

The National Forests in Texas litigation was the catalyst that initiated the process to develop new regional direction for RCW management on Forest Service lands. The Service has demonstrated that it will respond to its responsibility of enforcing compliance with Act provisions when RCWs are concerned. In concert with these actions, we will also make sure that positive conservation measures are one result of the process.

A successful RCW recovery program requires building partnerships and sharing information. We are going to discuss many of the potential and existing partnerships in the coming days. Today, I would like to present a new RCW partnership concept involving the formation of a RCW Conservation Committee.

I have been intensively and extensively involved with the RCW since 1985. From the start, I have always sensed some degree of divisiveness within the ranks of researchers, managers, academicians, and administrators. This is neither bad nor necessarily alarming. Constructive discourse on our differing opinions concerning RCW conservation and management is one of the pillars of our solid foundation. That pillar represents our diversity of experiences and our individual personalities and values, all vital components of our collective knowledge and influence. Given that we should retain this pillar of diversity for the support it provides, what can we add to our foundation to increase our opportunities for success? I believe we need a scientifically based partnership that speaks with a common voice on major RCW conservation and management issues.

We need a coalition which can address the complex issues by providing constructive analyses and scientifically based recommendations. In a sense, I am looking for a team. As RCW Coordinator I have incredible access to a lot of information concerning RCW activities and projects. It is time to begin sharing this information. Also, as your RCW Coordinator, I have some ideas and plans that I believe are responsive to the issues at hand. The private lands RCW conservation strategy (Costa 1994) is an example.

Selling recovery strategies to the individuals, agencies, and citizens affected or involved requires marketing. Marketing not only requires advertising (we will come back to that point in a bit), but also endorsement. The simple fact is, we have a product to sell, RCW recovery. We will be most successful in this endeavor if we, as individuals, recognize what others have contributed scientifically and philosophically to our common goal and then through a RCW Conservation Committee, present unified, not necessarily unanimous, opinions on the major RCW issues facing us today. We need a coalition of individuals that is recognized for its scientific credibility, political savvy, and logical approach to complex endangered species issues.

Specifically, I see a committee that reaches decisions through informed consent, a majority opinion if you will. Because of Service responsibilities under the Act, the Service has to remain neutral and also retain the right to disagree with committee findings, hopefully a rare occurrence. I can imagine the RCW Conservation Committee providing opinions and analyses of current foraging habitat standards for Federal lands, and the recently proposed standards for private lands. It could serve as a recovery team during the revision of our current RCW Recovery plan, scheduled for 1994/1995. I can see the committee, or parts of the committee along with other parties, serving as the steering committee for statewide private land habitat conservation plans. They could serve as an oversight committee, periodically examining Federal and state RCW management plan implementation and progress. The list goes on. Challenges to the formation and operation of such a committee include:

(1) 13 state responsibilities,
(2) travel and associated funding,
(3) committee member selection,
(4) prioritization of issues,
(5) communication network development, and,
(6) other unforeseen issues.

Some of these challenges are already being addressed. Our challenge grant with the National Fish and Wildlife Foundation could provide up to $12,000 towards the operation of this committee. I believe the formation of a RCW Conservation Committee has merit and I would enjoy talking to any of you about it this week and in the coming weeks.

Species recovery requires building partnerships and sharing information. As the group of individuals and agencies primarily responsible for, and interested in, the survival of this species, it behooves us to establish a better communication network. I believe our lack of a reliable method to share and exchange information has slowed down our potential progress toward recovery. We have our journals and telephones, but these do not meet our need to keep all concerned parties abreast of the "business" end of recovery. In short, we need an RCW newsletter, the contents of which would only be limited by our imagination. Some ideas that come to mind immediately are: summaries/progress reports of Federal recovery programs; RCW Conservation Committee updates; jobs available; editorials and essays; new technologies; recovery progress chart (perhaps by population); recent publications, etc. It could also be self-sustaining and perhaps a money-making endeavor if it was used to solicit contributions towards our recovery fund.

Successful recovery depends in large measure on us as a coalition sharing information among ourselves, and then individually taking that information to others who are uninformed. This newsletter concept is not a new idea. The "Red-cockaded Woodpecker News" (Number 1 Winter 1982) was the first and last attempt at such an effort. In fact, as part of our first issue at this second attempt, we could include the original edition. It would not only be interesting historical reading, but also provide ideas for our new version.

On a similar note concerning information sharing, I think we should pursue one or two major public educational initiatives on RCWs. One idea that I have thought about is getting the RCW story, and perhaps other endangered species' stories on McDonalds' bags. Can you imagine how many thousands of people in a month could be potentially educated by such a simple technique. We need a group of folks willing to make it happen.

How many of you flew in on Delta yesterday? Did you see the cover article on eagles? Did you wonder why it was not a RCW on the cover? Why? Because we have not pursued the idea. It is a great one. This is the second time in 12 months that Delta has had a 3-month series on endangered communities and endangered species. The next time, we need to be there with the longleaf pine ecosystem and RCW story. We have got to take advantage of these opportunities and create our own initiatives for selling and marketing our story. It is in the public's best interest and no other group of individuals is better suited to accomplish the tasks.

As a government employee with a very specific job description, I have been entrusted by the public to coordinate efforts relating to RCW recovery. This morning, I have shared with you some ideas that I believe will help us realize our mutual goal. My perception, which in part comes from extensive communication with many of you in this room, is that after 18 months in this position we need to:

(1) focus on the attainable short-term objective of downlisting to threatened;
(2) convene a RCW Conservation

Committee;

(3) significantly increase our artificial cavity programs;

(4) address and resolve the private land issues;

(5) publish a newsletter; and,

(6) implement public education campaigns.

I believe that if we accomplish these tasks in the coming years, we will make significant progress toward our goal of RCW recovery.

I strongly encourage each of you, during your time at this meeting, to seek out someone whose values, philosophical make-up, and/or conservation principles may differ from yours, perhaps an old acquaintance. Spend some time searching for the common ground you share concerning RCWs. Seek out new faces and begin building additional partnerships. The more frequently those concerned with the survival of this species can speak with a unified voice, the faster we will complete our mission. I am ready and willing to do my part. Are each of you? The challenge is yours, as an individual and as a coalition.

Our track record as a species in caring for our co-inhabitants and their associated ecosystems is pretty dismal. One need only examine what remains of our Nation's old-growth forests and tall grass prairies to understand the potential of human impacts to our planet. Since the arrival of European colonists in 1620 we have lost 500 species and subspecies from the North America continent, including such notable birds as the ivory-billed woodpecker, carolina parakeet, great auk, passenger pigeon (15 billion) and more recently, the dusky seaside sparrow.

'The survival of the RCW is by no means secure either. Security will come with 15 viable populations. Today, we have one. However, I can honestly say that I have nothing but optimism about the future. The challenges and opportunities are many and varied. There are many key people in the right place at the right time. We have the wisdom and experience of the veterans to guide us and the enthusiasm and vitality of the newcomers to stimulate us. It is an opportune time for some important changes in our recovery program. I challenge all of you to discard any excess baggage, renew old friendships and join the coalition focused on a vision and a plan for RCW recovery. I thank all of you for the opportunity to work with you. You have taught me a lot in the past 18 months, not the least of which is that when we decide, as a group, to do something we can and do accomplish it. This symposium proves that point quite well. Thank you all for your help, your papers, and your continued involvement after the conference. You made it all possible. Enjoy the symposium!

Red-cockaded Woodpecker Extinction or Recovery: Summary of Status and Management on Our National Forests

Ronald E. F. Escano, USDA Forest Service, Washington, DC 20090-6090

ABSTRACT: This paper describes the population status and summarizes management for the red-cockaded woodpecker on lands administered by the Forest Service for the period 1986 to 1992. Baseline information for this species on National Forest lands was established in 1986 and annual monitoring of population status and management accomplishments has occurred since 1988. While 76 percent of the populations are at extreme risk of extirpation, data suggests improved population conditions, especially since 1990. It appears that the red-cockaded woodpecker is responding positively to the changes in management for this species that have taken place since 1988.
KEYWORDS: red-cockaded woodpecker, endangered species, southeastern pine, habitat management

The success of red-cockaded woodpecker (RCW) recovery efforts and the ability to properly manage for this species is predicated by knowledge of population status and trends and good accounting of management accomplishments. The first Forest Service (FS) attempt to describe the status of RCW's across the Region was the multi-agency rangewide survey conducted between 1980 and 1982. As part of this survey, 17 National Forests were systematically sampled to determine an RCW population estimate (4 National Forests were not included). The FS regional population estimate based on this survey was 2121±405 active colonies (Lennartz, et al. 1983). In 1986, there was a general confusion of the overall status and health of FS RCW populations across the Region and there was a lack of information on management accomplishments. In an effort to create an information baseline on RCW population and management status, a survey of FS District RCW information was completed in 1987. This survey identified that the FS District records showed that 3,357 RCW colonies were known to exist in 1986 and that 2,115 of these were considered active (Costa and Escano, 1989). Starting in 1988, the FS has monitored RCW population status and management accomplishments annually across the Region. This paper will summarize the results of this monitoring, and where possible, compare the results with the 1986 baseline condition described by Costa and Escano, 1989.

FINDINGS

Population Size and Trends

The number of total RCW colonies known to occur on FS-administered lands has increased approximately 10% from 1986 to 1992 (**Table 1**). This is a result of the increased survey efforts implemented since 1986. The number of active colonies reported has decreased between 1986 and 1992 but has been increasing since 1990 (**Table 2**). The number of populations aggregated by size (number of active colonies) is shown on **Figure 1**. In 1986,

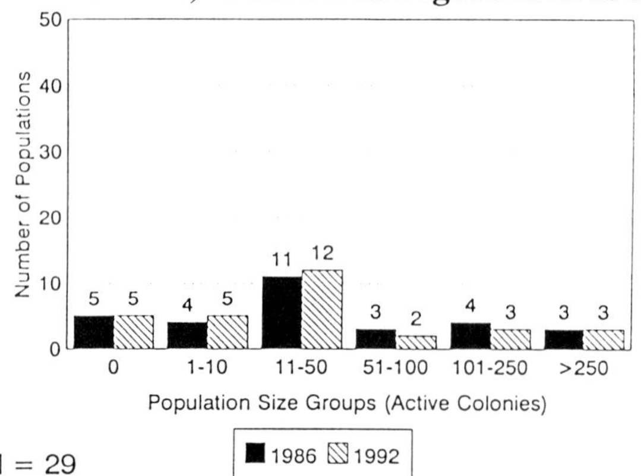

Figure 1. Number of populations by size 1986-1992

Table 1. Total red-cockaded woodpecker colonies on U.S. Forest Service administered lands 1986-1992.

POPULATION	1986[1]	1988	1989	1990	1991	1992
Angelina-Sabine NF's	124	121	127	125	125	128
Apalachicola NF	662	881	899	899	893	891
Bankhead NF	9	8	8	7	7	7
Bienville NF	181	170	193	196	193	193
Caney RD	3	3	3	3	3	3
Catahoula-Winn RD's	127	123	145	160	171	171
Cherokee NF	1	1	1	2	2	2
Conecuh NF	51	55	52	49	50	52
Croatan NF	67	61	73	76	76	70
Daniel Boone NF	28	18	24	28	33	34
Davy Crockett NF	114	98	106	109	113	118
Desoto NF	75	63	114	108	111	108
Evangeline RD	52	57	73	78	86	89
Francis Marion NF	500	518	HUGO	369	369	376
Homochitto NF	65	67	61	61	61	65
Kisatchie RD	104	118	121	134	143	146
Oakmulgee RD	325	309	301	299	303	304
Ocala NF	61	60	56	60	61	61
Oconee-Hitchiti	23	23	25	25	27	32
Osceola NF	98	90	102	105	127	132
Ouachita NF	20	21	25	27	29	32
Sam Houston NF	241	226	231	247	254	260
Savannah River Plant	21	18	21	27	32	32
Sumter NF	10	10	10	10	10	10
Talladega NF	145	125	161	173	168	153
Tombigbee NF	3	0	0	0	0	0
Tuskegee RD	2	2	3	3	3	3
Uwharrie NF	25	2	2	2	2	2
Vernon RD	210	224	231	231	236	242
Regional total	**3357**	**3434**	**3156[2]**	**3611**	**3686**	**3716**

[1] **Baseline**

[2] **Total with pre-HUGO Francis Marion 518 colonies included is 3674**

69% of the red-cockaded woodpecker populations had less than 50 active colonies. In 1992, the proportion of red-cockaded woodpecker populations with less than 50 active colonies has increased to 76% and 2 populations are down to 1 single bird clan each. A comparison of population trend estimations between 1986 and 1992 are shown on **Table 3**. The 1986 trend estimates are from the Costa and Escano (1989) report and are based on District survey data through 1986. The 1992 trend estimate is based on District survey data from 1988 through 1992. Both estimates are based on the same considerations; number of active colonies, proportion of colonies checked in a survey year found to be active and the change in total number of colonies as a measure of survey effort. It is not just the plot of the number of active colonies, which may reflect changes in survey effort more than actual population trends. In 1986, 82% of the populations were decreasing, while in 1992, it had dropped to 50% (**Figure 2**). There are four populations that have lost half or more of their

Table 2. Active red-cockaded woodpecker colonies on U.S. Forest Service administered lands 1986-1992.

POPULATION	1986[1]	1988	1989	1990	1991	1992
Angelina-Sabine NF's	46	34	34	33	31	32
Apalachicola NF	487	705	693	684	689	685
Bankhead NF	1	0	0	0	0	1
Bienville NF	127	101	88	86	88	91
Caney RD	0	0	0	0	0	0
Catahoula-Winn RD's	82	46	50	50	49	49
Cherokee NF	1	1	1	1	1	1
Conecuh NF	32	14	16	13	12	11
Croatan NF	50	43	45	48	48	48
Daniel Boone NF	7	4	6	4	5	5
Davy Crockett NF	30	24	27	29	30	36
Desoto NF	25	22	18	15	16	12
Evangeline RD	50	31	41	43	46	46
Francis Marion NF	483	487	HUGO	318	327	346
Homochitto NF	40	35	26	25	25	22
Kisatchie RD	71	75	63	68	54	59
Oakmulgee RD	195	166	157	120	121	122
Ocala NF	22	23	14	11	12	11
Oconee-Hitchiti	12	12	11	11	16	16
Osceola NF	55	56	50	44	44	43
Ouachita NF	20	19	16	13	15	16
Sam Houston NF	112	122	133	135	132	132
Savannah River Plant	3	5	5	7	9	10
Sumter NF	0	0	0	0	0	0
Talladega NF	11	5	5	7	6	5
Tombigbee NF	0	0	0	0	0	0
Tuskegee RD	0	1	1	1	0	0
Uwharrie NF	0	0	0	0	0	0
Vernon RD	153	177	170	169	174	186
REGIONAL TOTAL	**2115**	**2214**	**1670[2]**	**1936**	**1949**	**1985**

[1]Baseline

[2] Total with pre-Hugo Francis Marion 487 colonies included is 2157

active colonies since 1986 and are still showing a decreasing trend (Conecuh, Desoto, Homochitto and Ocala). These populations are at a high risk of extirpation. In 1986, only one population was showing signs of an increasing population trend (Vernon Ranger District), while in 1992, nine populations appear to be increasing (Bienville, Croatan, Davy Crockett, Francis Marion, Oakmulgee, Oconee-Hitchiti, Ouachita, Savannah River Plant, and the Vernon).

Monitoring

Monitoring intensity has greatly increased since 1986. The percent of colonies checked for status per year has almost tripled since 1986 (**Figure 3**). The number of acres surveyed for "new" colonies has gone from 0 in 1986 to over 200,000 acres in 1992 (**Table 4**).

Midstory Treatments

The Regional summaries for colony midstory treatments and burning

Table 3. Red-cockaded woodpecker population trend comparison 1986 and 1992

POPULATION	1986[1] TREND ESTIMATION	1988-1992[2] TREND ESTIMATION
Angelina-Sabine NF's	Decreasing	Decreasing
Apalachicola NF	Stable	Stable
Bankhead NF	Decreasing	Decreasing
Bienville NF	Decreasing	Stable/Increasing
Caney RD	Extirpated 1986	
Catahoula-Winn RD's	Decreasing	Stable/Decreasing
Cherokee NF	Decreasing	Decreasing
Coneuch NF	Stable/Decreasing	Decreasing
Croatan NF	Stable/Decreasing	Stable/Increasing
Daniel Boone NF	Decreasing	Stable/Decreasing
Davy Crockett NF	Decreasing	Increasing
Desoto NF	Stable/Decreasing	Decreasing
Evangeline RD		Stable/Decreasing
Francis Marion NF	Stable	Increasing
Homochitto NF	Stable/Decreasing	Decreasing
Kisatchie RD		Stable/Decreasing
Oakmulgee RD	Stable/Decreasing	Stable/Increasing
Ocala NF	Stable/Decreasing	Decreasing
Oconee-Hitchiti	Stable/Decreasing	Increasing
Osceola NF	Stable	Decreasing
Ouachita NF	Stable/Decreasing	Stable/Increasing
Sam Houston NF	Decreasing	Stable
Savannah River Plant	Decreasing	Increasing
Sumter NF	Extirpated 1985	
Talladega	Decreasing	Stable/Decreasing
Tombigbee NF	Extirpated 1970's	
Tuskegee RD	Extirpated 1985	Extirpated 1991
Uwharrie RD	Extirpated 1970's	
Vernon RD	Stable/Increasing	Increasing

[1]From Costa and Escano, 1989 [2]Based on percent of colonies checked for status that are active, total number of active colonies and total number of colonies for the period 1988 through 1992.

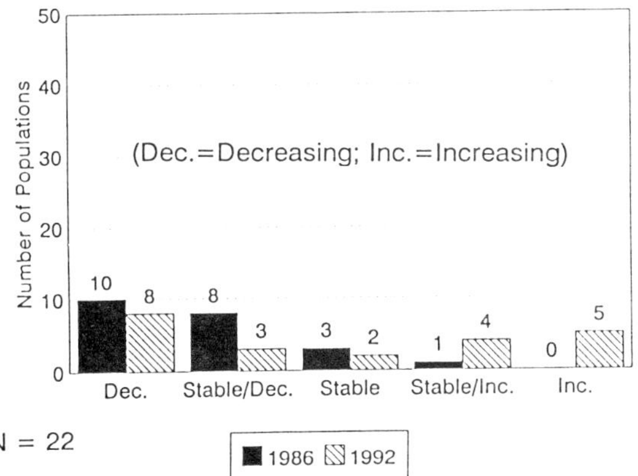

Figure 2. Number of populations by trend 1986-1992.

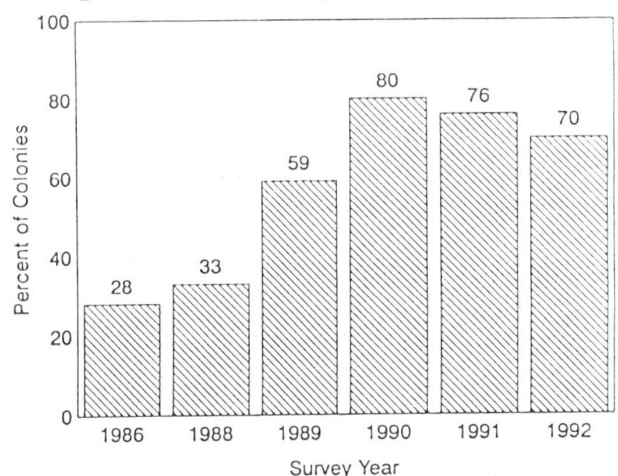

Figure 3. Percent of colonies checked for status 1986-1992.

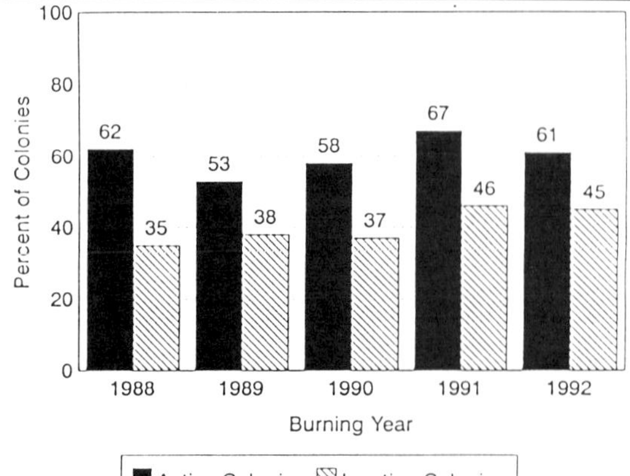

Figure 4. Percent of colonies needing burning-successfully burned.

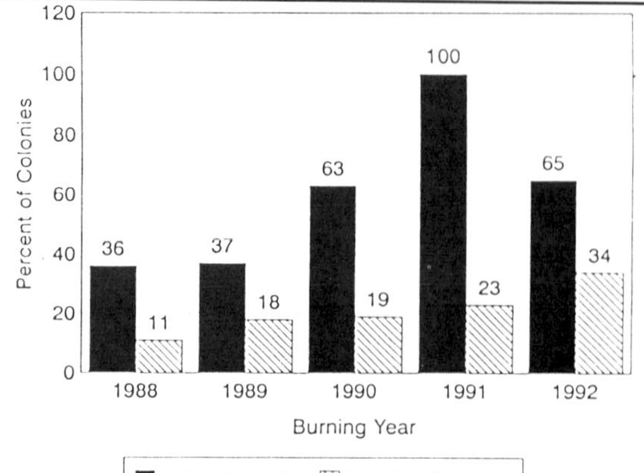

Figure 5. Percent of colonies needing midstory treatment successfully treated.

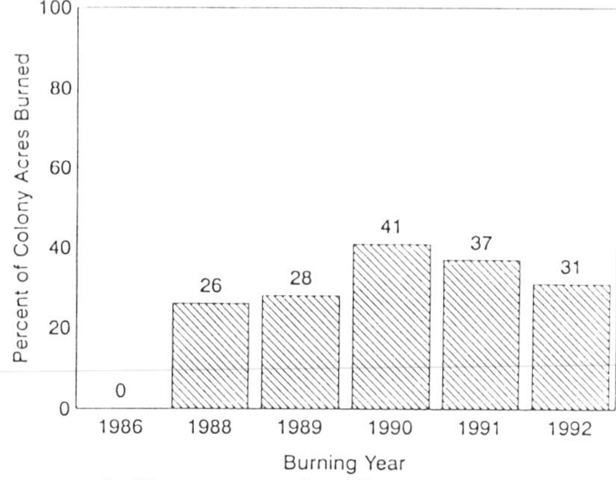

Figure 6. Percent of colony acres burned during the summer.

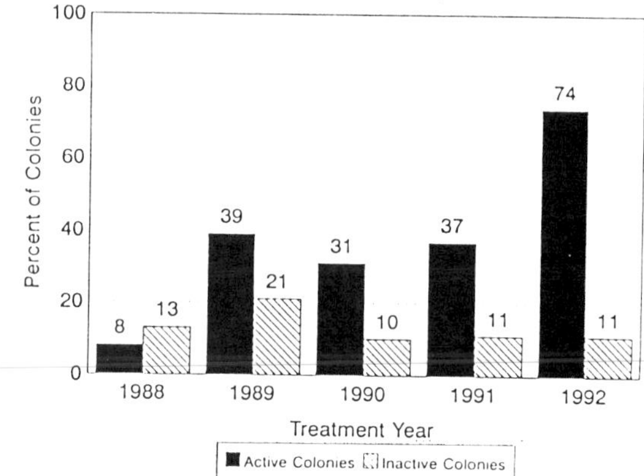

Figure 7. Percent of colonies needing restrictors receiving restrictors.

accomplishments are shown on **Table 4**. The percent of active colonies needing burning to maintain suitable midstory conditions that were actually burned in a given year range from 53% to 67% since 1988 (**Figure 4**). Since 1988, 424 active colonies have been burned on the average per year. This equates to about a 5-year burning cycle. The percent of high priority inactive colonies, those within 3 miles of an active colony, needing burning that were actually burned in a given year has increased from 35% in 1988 to 45% in 1992 (**Figure 4**). The proportion of active colonies needing midstory treatment that were actually treated has almost doubled since 1988, while the proportion of high priority inactive colonies treated has tripled (**Figure 5**). The average number of active colonies treated for midstory control per year is 253, and the backlog of active colonies needing midstory treatment has decreased from 548 colonies in 1988 to less than 100 in 1992 (**Table 4**). In 1986, none of the colony treatments were summer burns, while in 1992, almost 1/3 of all the colony acres burned were accomplished in the growing season (**Figure 6**).

Special Techniques

Several habitat improvement techniques have had widespread FS use that were not available in 1986. These include use of cavity restrictors, artificial cavities and augmentation. Since 1988, 564 active colonies have been treated with restrictors, and the backlog of active colonies needing restrictors has been reduced to less than 100 in 1992 (**Table 4**). The percent of active colonies needing restrictors that have been treated has increased dramatically since 1988 (**Figure 7**).

Since 1990, 653 active colonies and 97 inactive colonies have received artificial cavities or inserts (**Table 4**). In 1992, 45% of the active colonies needing artificial cavities or inserts were treated (**Figure 8**), leaving about 250 active colonies still needing treatment. The 1990 artificial cavity and insert data is dominated by the Hurricane Hugo colony restoration efforts on the Francis Marion

Table 4. Regional management summaries.

	1986	1988	1989	1990	1991	1992
MONITORING						
1. Number of colonies checked for status	941	1132	1861	2940	2813	2605
2. Acres surveyed for "new" colonies	0	0	85423	15178	20806	20761
MIDSTORY TREATMENTS						
1. Active colonies needing burning		807	655	742	547	777
2. Active colonies burned		502	347	428	368	473
3. Inactive colonies <3 miles from active needing burning		374	455	559	491	565
4. Inactive colonies burned		130	174	208	228	257
5. Active colonies needing midstory treatment		548	528	546	350	226
6. Active colonies treated		199	196	342	381	146
7. Inactive colonies <3 miles from active colony needing midstory treatment		389	575	614	535	354
8. Inactive colonies treated		44	106	117	121	122
9. Colony acres winter burned	21944	8900	12456	12818	17342	16685
10. Colony acres summer burned	0	3142	4819	8806	9608	7644
11. Colony acres intensive treatment	1054	3927	6748	9587	11513	7820
RESTRICTORS						
1. Number of active colonies needing restrictors		146	241	344	391	284
2. Number of active colonies treated	0	11	93	107	144	209
3. Number of inactive colonies needing restrictors		206	310	445	506	271
4. Number of inactive colonies treated	0	26	64	43	55	31
ARTIFICIAL CAVITIES						
1. Number of active colonies needing cavities				347	448	482
2. Number of active colonies treated	0	0	0	286	142	225
3. Number treated colonies with use of artificial cavity				213	127	204
4. Number of inactive colonies needing cavities				363	666	1272
5. Number of inactive colonies treated	0	0	0	39	10	48
6. Number of treated colonies with use of artificial cavity				5	5	6
AUGMENTATION						
1. Number of single bird clans				67	118	56
2. Number of clans augmented	0	0	0	16	12	10
3. Number of augmented clans which paired				10	11	9
4. Number of augmented clans which fledged young				5	7	6

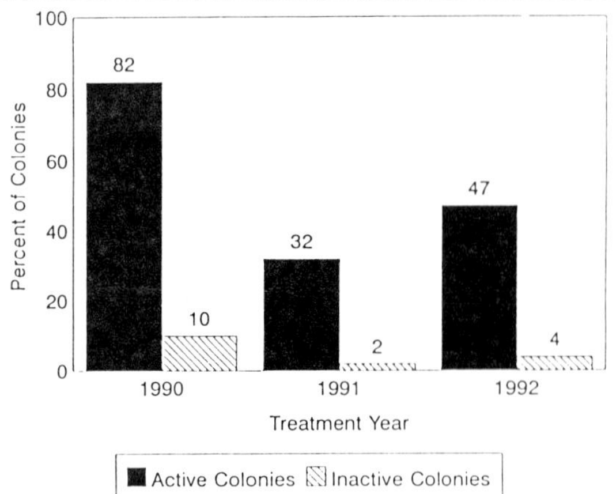

Figure 8. Percent of colonies needing cavities receiving artificial cavities.

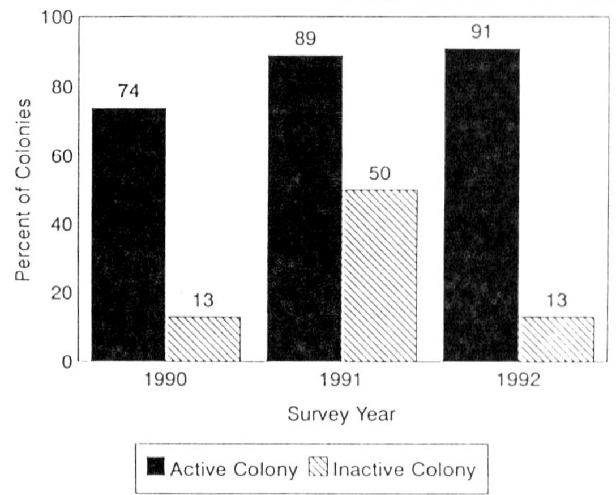

Figure 9. Percent of colonies with artificial cavities with use of artificial cavity.

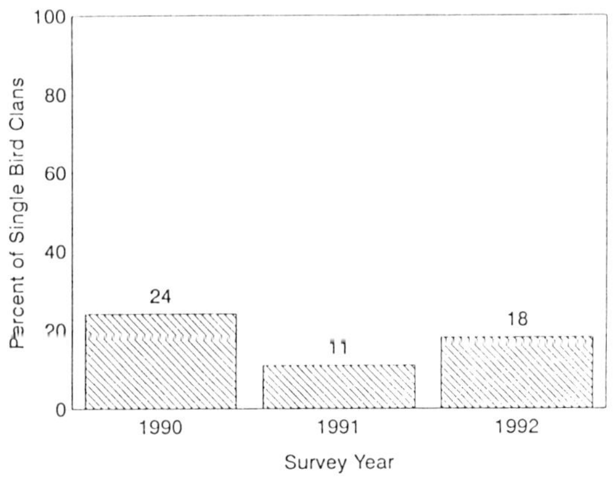

Figure 10. Percent of single bird clans augmented.

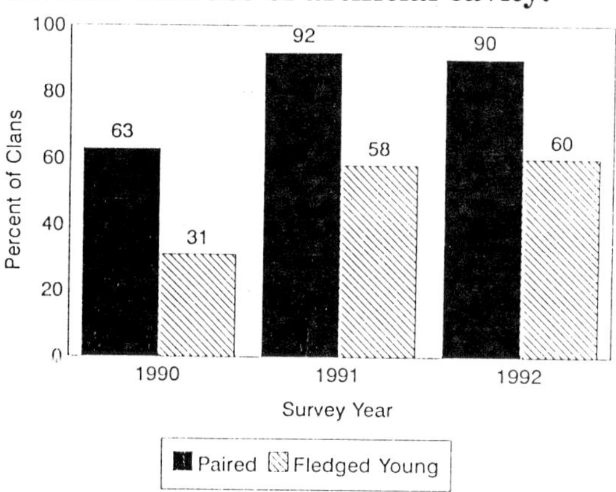

Figure 11. Augmented clan success

National Forest. The use of artificial cavities or inserts by RCW in previously active colonies is running close to 90% and there is about 13% use in previously inactive colonies **(Figure 9)**. Less than 20% of the identified single bird clans are being augmented in a given year **(Figure 10)**. Overall, 38 clans on National Forest lands have been augmented since 1990 **(Table 4)**. The success of the augmentation efforts has been running about 90% of the augmented clans actually showing a pair bond and about 60% of the augmented clans fledging young **(Figure 11)**.

DISCUSSION

There is no longer a void of information on how RCW populations are doing on our National Forests, and a system is in place to describe the management efforts taking place on an annual basis. This monitoring effort will be key for the future success of FS RCW recovery efforts.

Population Size and Trends

At first glance, the population size distribution changes that have occurred between 1986 and 1992, and the changes in population trend interpretations over this same time period appear to conflict. Based on the population size distributions, the number of FS RCW populations at risk of extirpation has increased since 1986, while the trend information indicates that fewer populations appear to be decreasing and more appear to be increasing in 1992 compared to 1986. The population size distribution changes I believe reflect the steep declines that were occurring in most of our populations until quite recently. The number of active colony data is not a very sensitive barometer and through more intensive monitoring efforts, the data has caught up with the population changes taking place. The population trend information is, I believe, correctly describing the improved population conditions since 1990 and could be a sign that most of the populations are responding to the changes in RCW management that have taken place since 1988.

Monitoring

Monitoring intensity has increased greatly since 1986. Most small populations are receiving annual 100% checks of all known colonies and single bird clans are being identified. Data are currently being collected to determine actual effective population sizes of all populations. This should eliminate the inherent shortcomings of tracking only active colony status.

Midstory Treatments

The backlog of active colonies needing midstory control has virtually been taken care of between 1986 and 1992. The midstory control work is now shifting to the high priority (within 3 miles of an active colony) inactive colonies. The greatest challenge in midstory treatments is burning enough colony acres to maintain suitable conditions. About 700 active colonies need to be burned annually to achieve a 3-year burning cycle, but we are burning only about 400. We run the risk of losing the tremendous investment in intensive midstory treatments because of the inability to follow up these treatments with the burning needed to maintain suitable conditions.

Special Techniques

RCW management has become a lot more complex since 1986. The use of cavity restrictors, artificial cavities and inserts and the translocation of birds between colonies (and populations) has become an integral part of the RCW recovery program. The habitat improvement accomplishments tend to follow (all except burning) a curve of increasing backlogs until the program builds momentum and then achieves a quick reduction in the backlog. Restrictor placement in active colonies has proceeded through this cycle, while artificial cavity placement and augmentation are still in the increasing backlog stage. Both placement of artificial cavities and augmentation require very specialized expertise, and it will take longer to build the technical support base needed for these programs than the normal habitat improvement techniques.

CONCLUSIONS

It was very interesting for me to look at the RCW data after being away from it for 2 years--a clear fresh look if you will. Two years ago, if you asked me the question, "What do you foresee RCW extinction or recovery," my answer would have been extinction. Today, after looking at this data, I see recovery within our grasp. RCW populations appear to be responding to management. Not that we are out-of-the-woods with 75% of the FS populations at risk, 2 populations virtually extirpated, and 4 more that we will likely lose in the near future. I see light at the end of the tunnel with a real turning point in 1990. I feel the aggressive recovery actions the FS has implemented since 1986 has made the difference. The important point is that recovery is achievable, but it will take a long-term investment. We cannot achieve RCW recovery with a short-term fix, but I think we are up for the challenge of a long-term program.

Trained and Ready While Protecting Our Environment

Bruce A. Sneddon, United States Army
Office of the Deputy Chief of Staff for Operations and Plans
Department of the Army Washington, D.C. 20310

"The Army will be a national leader in environmental and natural resource stewardship for present and future generations as an integral part of our mission." (Vision statement from the U.S. Army Environmental Strategy into the 21st Century).

The Army's leadership is committed to environmental stewardship, which includes the recovery of the many endangered species on our installations. The Secretary of the Army plays a vital role in environmental stewardship as one of the seven members on the Endangered Species Committee. Additionally, the Army is the executive agent for the Department of Defense for the RCW Management Improvement Initiative. We are also committed to a trained and ready force which will perform as well as our young men and women did in Desert Storm. Good training equals victory in combat and fewer casualties on the battlefield.

The focus of this paper is to outline the Army's commitment to environmental stewardship and to maintaining the Nation's Army trained and ready. The red-cockaded woodpecker (*Picoides borealis*) [RCW] and supporting ongoing Army initiatives will be used throughout the paper to illustrate some of the endangered species management challenges and initiatives.

ENVIRONMENTAL STEWARDSHIP

On November 19, 1992, the Secretary of the Army, the Honorable M.P.W. Stone, and the Army Chief of Staff, General Gordon R. Sullivan, signed the U.S. Army Environmental Strategy Into the 21st Century. This strategy defines the Army's leadership commitment and philosophy for meeting present and future environmental challenges. It provides a framework to ensure that environmental considerations are integral to the Army mission and that an environmental stewardship ethic governs all Army activities. The strategy provides a unity of direction and a cohesive framework for all Army activities associated with:

(1) Army installations, facilities and training areas,

(2) acquisition, manufacturing, industrial operations and activities; and,

(3) the Army's civil works mission. Through the strategy's framework, the Army strives to achieve environmentally sustainable operations at all of its military installations, to enhance national security and quality of life, and to meet similar environmental standards at all Army civil works facilities. This strategy entails four simultaneous efforts: giving immediate priority to sustained compliance, continuing to restore previously contaminated sites, focusing efforts on prevention, and conserving and preserving natural resources.

Army environmental stewardship is built around four pillars. These pillars represent the four major activity areas of compliance, restoration, prevention, and conservation. Maximum support for the Army mission is realized when all four pillars are strong and well-balanced. Most of the activities that the Army conducts regarding threatened and endangered species are in the areas of compliance and conservation.

ENVIRONMENTAL CHALLENGE

The Army faces multiple challenges in its effort to implement its environmental strategy and meet mission requirements. Successful resolution of natural resources, endangered species, and cultural resource issues are integral to continued sustained use of the Nation's land entrusted to the Army's care. All of these challenges impact, to some degree, on the Army's ability to train.

The following list is a sample of some of the environmental issues which impact on the Army's management and use of its more than 12 million acres of land under its control (figure does not include the land managed by the Army's Civil Works organization).

- **Maneuver suitability:**

Fort Bragg, NC: Soil erosion impacts on red-cockaded woodpecker habitat and aquatic environments. Training operations have been changed reducing erosion to acceptable levels.

Fort Stewart, GA: Topography of the land limits maneuver training for mechanized forces. Training impacts on the land have to be monitored to assure future use of good training land.

- **Endangered species:**

Fort Bragg, NC: Red-cockaded woodpecker nesting areas restrict operational use of many portions of the training land. Training and land management practices have been revised.

Pohakuloa Training Area, HI: Rare and endangered plants found in an area under construction for a large training range complex. Construction has been halted until workable alternatives are developed.

- **Wetlands:**

Fort Richardson, AK: The firing of artillery ammunition containing white phosphorous into wetlands in the installation's artillery impact area contributed to the death of waterfowl in the wetlands. Firing of white phosphorous munitions into wetlands have been suspended Army-wide.

Fort Bragg, NC: Wetlands in a range complex under construction have halted construction until alternatives are developed.

- **Cultural resources:**

Fort Bliss, TX: Historic gravesites and significant archaeological sites have been found in training areas. Training activities around these sites have been restricted or curtailed.

- **Pohakuloa Training Area:** Significant archaeological sites have been found in training areas. Training activities around these sites have been restricted or curtailed.

The above list is but a small representative list of the natural and cultural resource challenges facing the Army today. The challenge facing the Army is to develop strategies which protect and conserve these resources while retaining the capability to meet the Army's missions.

The Army faces a unique challenge. The 12 million acres which it controls supports its military mission to be trained and ready. At the same time, as one of the federal government's major landholders, the Army's land is, in many instances, critical for the protection and conservation of many plant and animal species.

Army installations are, in many cases, a safe haven for many endangered species. When the Army first acquired most of its current land, much of it was either overused, poorly situated for commercial or agricultural use, or just far removed from civilization at the time. Through well planned and executed long term land use management strategies, these lands today provide not only the platform on which to support intensive training, but also provide some of the best habitat for the Nation's natural and cultural treasures.

Herein lies the challenge facing today's Army: Meeting the land use requirements for two diverse interests. To maintain a trained and ready force, the Army relies on the large expanses of land available on its installations. These lands provide the large maneuver areas necessary to support various types of training. These dispersed lands also provide varied types of terrain which supports training units under various climatic and geographic conditions. By its very nature, training activities are intense uses of the land. Large numbers of tracked and wheeled vehicles, in addition to foot traffic, maneuver over large tracks of land, at times causing severe stress to the land's carry capacity. Additionally, large impact areas are used to contain the munitions resulting from the concentrated firing of various types and caliber of weapons. Land rehabilitation and maintenance programs have been implemented to maintain quality lands to support the training requirements for today and the future.

Endangered species compete for the same land on which the Army trains. Good land use management practices and large tracts of land provide excellent habitat for many plants and animals. As the natural habitat around installations is destroyed by the expansion of urbanization and commercial and agricultural uses, many of the wildlife species immigrate to Army installations. Here they find well maintained natural habitat and are buffered from daily contact with civilization.

In the past, the competition for the same land between Army mission requirements and the installation's wildlife and plants was not as severe as it is today. The converging trends for land use today to meet endangered species and training needs present a real challenge to the Army. Since enactment of the Endangered Species Act in 1973, the number of federally listed endangered species has dramatically increased. Today, over 50 per cent of the Army's major installations have at least one federally protected endangered species, and the number will undoubtedly increase in the future. Over 100 different endangered species are found on these installations, and the number is growing. Adding federal candidate species and state listed species further increases the number of species for which some conservation measures must be taken.

Concurrent with this increase in the number of threatened and endangered species, the Army is undergoing dramatic changes which also increase the pressure for land use. The end of the Cold War has resulted in an extensive reshaping of the Nation's army. Some forces are being withdrawn from overseas for restationing within the U.S. At the same time, the Army is also reducing the number of installations needed to support this reshaped army. Decreasing the number of installations and adding additional units at some of the installations increases the demand for use of existing training land.

Changes in the way the army fights also increases its demand for land. The warfighting doctrine of today and tomorrow require large tracts of land to practice for war. Training today's army far outstrips the land requirements needed to support the army of the 1970's. Modernized weapon systems with significantly increased range, lethality, and speed also require large amounts of land to support training requirements. Although additional lands may be needed, most likely the army will continue to use its current holdings. Land expansion, if any, will only be initiated for a few selected locations. This means that the lands which supported yesterday's force must also accommodate the modernized force of today and tomorrow. The increase in the number of endangered species and the increase in training density and intensity have the potential to adversely impact each other.

TRAINING—#1 PEACETIME MISSION

Training is the glue that binds the Army into a force capable of achieving decisive victory through land force dominance. Training is not a cost, but an investment in readiness and our National security. Training is the Army's most important peacetime mission. Through tough, realistic, battle-focused training, commanders build and confirm the ability of soldiers and leaders to fight and win as a combined arms team in both joint and combined operations.

Training remains the key to combat-ready units. In the future, commanders and leaders will remain the principal trainers and must ensure that the Army trains as it will fight, executes doctrine as envisioned, and builds troop and unit confidence in equipment. The Army trains to established standards. This provides the capability to evaluate training proficiency and identify training requirements. Training that is doctrinally correct, performance-oriented, challenging, and effective will endure as the main ingredient of sustained performance, success and victory. The success of Just Cause and Desert Storm were the result of this training focus.

TRAINING READINESS AND ENDANGERED SPECIES MANAGEMENT—CASE STUDY

Up to this point, a broad overview of both endangered species and training requirements has been presented. This section provides a case study to identify and discuss the potential conflicts and solutions to managing endangered species and the Army's training mission. The red-cockaded woodpecker will be used to illustrate the challenge of balancing training with endangered species protection and conservation.

The historical range of the red-cockaded woodpecker is the southeastern United States; from Texas, east to the Atlantic seaboard, and from Kentucky, south to Florida. It is estimated that 3500 active colonies are all that remain of this species. Approximately 760 active colonies, 22 percent of the total number of active colonies, are on Army installations.

The Red-cockaded Woodpecker Recovery Plan has designated 15 recovery areas in the woodpecker's range as critical to the species' long-term recovery. Most of the six Army installations which support the red-cockaded woodpecker are major components of one of the designated red-cockaded woodpecker recovery areas. It is estimated that 1.6 million acres of potential habitat are available to support the woodpecker's recovery. The Army's potential contribution to this habitat is one-half of a million acres.

Although the Army is not the major federal landholder of potential red-cockaded woodpecker habitat, it is a key player to the success of the species' recovery. The Endangered Species Act is very specific in delineating the responsibilities federal agencies have for protecting and conserving endangered species. Protection and conservation have primacy over all other agency actions which may or could adversely affect the endangered species or its habitat. For the Army, this could mean the curtailment or restriction of training activities in specific areas on the installation.

Failure to develop strategies which support both training needs and endangered species conservation requirements results in the loss or restriction of critical training land. This loss of land directly affects the Army's ability to maintain a trained and ready force. Some of the impacts to training readiness due to the loss of training land include:

• The inability of units to train to standard. Units are not able to conduct doctrinal training with the required amount of land, facilities, or ranges.

• Training realism is degraded because portions of operations are curtailed or administrative requirements are implemented when units encounter specific situations. Examples include not being able to conduct combat operations in nesting areas and the suspension of live fire operations on specific ranges due to the proximity of colony sites and habitat.

• Training scope is decreased due to restrictions on available training acreage. Battalion and brigade-size operations are reduced to company and platoon due to restrictions in maneuver areas. In some areas, training becomes limited to foot movement where before, tracked and wheeled vehicle movement had been possible.

• Combined arms training cannot be realistically conducted. An infantry brigade cannot train on all of its wartime missions with supporting engineers (no excavation activity), armor (no tracked vehicles in habitat areas), and aviation (gunnery ranges closed due to habitat in the impact area). Additionally, some training events must now be accomplished at other installations where previously they had been conducted at the home installation.

• Increased training costs as units conduct training at other locations. For example, it costs an additional $42,000 to transport an attack helicopter battalion from Fort Bragg, NC to Fort Stewart, GA to conduct required gunnery qualifications.

• Training is curtailed because no training facilities are within a reasonable distance or the unit cannot afford the additional transportation and training costs.

The training impacts presented above are real. They are part of the "cost of doing business" at some of the Army's installations today which have red-cockaded woodpeckers. The challenge is to identify and develop potential means to mitigate training impacts. Some of these means include:

• Developing standard guidelines for training activities. Today, each installation has different guidelines. This precludes training to standard by units which train at different installations. Training guidelines for all installations must be the same. This supports the Army's training tenant of training to standard.

• Improved leader and soldier education. Development of an environmental ethic provides the requisite foundation for executing the Army's vision of environmental stewardship. Specific education regarding training activities in and around red-cockaded woodpecker habitat areas significantly reduces the potential for conflict.

• Relocation of either the woodpecker or resiting training facilities. If well planned, this can be a cost effective alternative. Early planning can identify potential conflicts and provide sufficient time to develop other siting locations or alternatives without significant expenditure of resources.

• Land acquisition for either training activities or habitat support. Easement, lease, or out-right purchase are potential alternatives to support mission requirements and lessen the demand for the same piece of real estate.

• Focused research. Decisions regarding the impact of specific training activities on the red-cockaded woodpecker are based on the best available scientific and technical information. Prior to this year, little research had been conducted to determine the impact that certain training activities have on the woodpecker or other species. Development of sound research data will significantly assist the Army in developing supportable proposals for training and other activities.

• Improved land use planning coordination. Installations must develop a comprehensive land use master plan, to include an integrated natural resource plan. This plan, the environmental inventory of the installation, will provide the basis for developing plans to support Army missions and supporting endangered species requirements.

ARMY'S STEWARDSHIP COMMITMENT

The issues involving the protection and conservation of the red-cockaded woodpecker are not isolated to this species. Rather, all installations in the Army face the same issues albeit with different threatened and endangered species. Successful conflict resolution requires an army willing to meet the challenge. Today, the Army is meeting that challenge head-on. Under the mantle of the Army Environmental Strategy, the Army is demonstrating its commitment to the conservation of endangered species. Significant initiatives include:

• Establishment of the Directorate for Environmental Programs. Effective January 1, 1993, all environmental programs within the Army are managed within one organization. This assures proper integration of resources and focus of effort. All endangered species matters are managed by the Conservation Division of the directorate.

• Department of the Army Endangered Species Team. This team was established in May 1992 to develop Army guidelines for the red-cockaded woodpecker and to develop management strategies and policies for addressing endangered species requirements within the Army. This integrated team consists of an army trainer, an engineer, and an environmental lawyer. With the establishment of the Directorate of Environmental Programs, the team's functions will be performed by the Conservation Division.

• Army Red-cockaded Woodpecker Management Guidelines. These guidelines, currently being developed by the Department of the Army Endangered Species Team, will provide standard guidelines for managing the red-cockaded woodpecker on Army installations. These guidelines will establish common rules, methodologies, and techniques for RCW management. These guidelines will replace the current 1984 Red-cockaded Woodpecker Guidelines, after they are approved for implementation by the U.S. Fish and Wildlife Service.

• Threatened and endangered species management policy. A comprehensive document addressing the management policies and requirements for endangered species has been developed by the Department of the Army Endangered Species Team. This document provides army major commands and installations definitive guidance regarding endangered species management. The document addresses the development of endangered species management plans, biological assessments, environmental assessments, NEPA requirements, and the informal and formal consultation process with the U.S. Fish and Wildlife Service and the National Marine Fisheries Service. This is the first document to provide comprehensive management guidance for the Army.

• Increased threatened and endangered species research. The focus of army research will be on military impacts on various species and habitats, and on installation ecosystem and biodiversity management techniques. These research efforts are being done through a tri-service (Army, Navy, Air Force) research program and through existing Army programs. Efforts are ongoing to identify requirements and develop a comprehensive research plan. This plan will document the research requirements and provide the basis for conducting much needed research. This research program will also include cooperative efforts with other federal agencies, state agencies and academic institutions.

- Endangered Species Action Plan. This plan, developed by the Department of the Army Endangered Species Team, establishes goals and supporting objectives for implementing endangered species initiatives. The plan integrates current and new endangered species activities. It serves as the audit document to monitor the progress of the implementation of various initiatives. This plan Tough, realistic, battle-focused training forms the teamwork needed to win as a combined arms team. This training excellence is the foundation for the Army of the 21st Century--A Total Force trained and ready to fight, serving our nation at home and abroad, a strategic force capable of decisive victory. The Army is also committed to the environment. It will not wait for the future to take shape, rather, the Army will shape the future. The Army will be a leader in environmental policy and management. The initiatives presented above are part of that commitment.

Tough and realistic training will remain the cornerstone of readiness. Effective land management will be needed to continue providing this training at the highest standards. Preparing for war is inherently destructive; however, the Army's continued success requires an aggressive effort to minimize or avoid permanently destroying training lands. Anticipating and planning for the effects military activities have upon the environment and natural and cultural resources are highly effective investments for the Army of the future.[1]

---

[1] U.S. Army Environmental Strategy Into the 21st Century.

The Red-cockaded Woodpecker: Two Hundred Years of Knowledge, Twenty Years Under the Endangered Species Act

Jerome A. Jackson
Department of Biological Sciences
Mississippi State University, Mississippi State, Mississippi 39762

ABSTRACT: The red-cockaded woodpecker (*Picoides borealis*) was first described in 1807 by the amateur French ornithologist Louis Jean Pierre Vieillot. Its English common name and some basic understanding of its natural history were given by Alexander Wilson in 1810. John James Audubon contributed more to our knowledge of the species during the early 1800s, including a basic description of its range and association with the southern pine forest ecosystem. Further knowledge of the species came only in brief anecdotal accounts until the mid-20th century. By then the species was in obvious difficulty as a result of habitat losses. Named as an endangered species in 1968, knowledge of the bird has increased dramatically in the 20 years since passage of the Endangered Species Act of 1973, the infusion of federal dollars, and controversy over its protection and management.
KEYWORDS: red-cockaded woodpecker, *Picoides borealis*, history, Endangered Species Act, publications

The late 1700s were turbulent times in world history; colonists in North America had revolted and gained independence from Great Britain, wars were wracking Europe, and revolt by slaves was destroying European colonialism in the West Indies. Into this scene arrived a French businessman who was involved in the spice trade and who, on the side, made substantial contributions to our knowledge of New World birds. Louis Jean Pierre Vieillot apparently escaped the draft in France by retiring to the West Indies, only to have to flee from there to North America because of slave uprisings. He arrived in the U.S. about 1792 and sometime between then and his departure six years later, he discovered the red-cockaded woodpecker (RCW). In 1807 Vieillot, named it "*Picus borealis*" -- the northern woodpecker -- an apparent misnomer to us, but from his perspective as an individual who had resided in the West Indies, this was indeed a northern woodpecker (Vieillot 1807). The painting accompanying Vieillot's work included several errors (Jackson 1971), but was accurate enough to stand as documentation for the description of the species.

Three years later, Alexander Wilson, known as the Father of American Ornithology, described the RCW once again as a new species, *Picus querulus*, the name *querulus* referring to the constant chattering of these birds as they move through the southern pines (Wilson and Bonaparte 1831). Since Vieillot's scientific name came first, we know the bird as *Picoides borealis* today, but since Vieillot failed to give the species an English name, we have adopted Wilson's—the red-cockaded woodpecker. This name, which confuses some people today, was easily understood in the early 1800s. The American revolutionary army could not afford brass for badges of rank and often substituted a colored feather stuck in the corner of the hat to indicate rank—these feathers were called "cockades." Since the male RCW has only a few red feathers "stuck in the corner of his hat," he became the red-cockaded woodpecker. This use of cockades by the military also, incidentally, gave us the words to the song we know so well: "Yankee Doodle went to town a riding on a pony. He stuck a feather in his cap and called it macaroni." Macaroni was a slang term referring to the gold braid and other decorations of European military officers.

Alexander Wilson taught us much about RCWs, noting that they were found in the pine woods of the Carolinas and Georgia. John James Audubon (1849) credited Wilson with the discovery of the species and did a better job painting it than did either of his predecessors. He added extensively to our knowledge of its ecology. Thanks to Audubon, by the early 1840s, its range was fairly well known. Audubon reported it as associated with pine forests from Texas to New Jersey and as far inland as Tennessee. Curiously, none of these early naturalists mentioned the characteristic nest and roost cavities of the species. C.J. Maynard (1881:244) correctly noted that nests of the RCW are almost always excavated in living pines, and Phelps (1914) described the association with trees containing rotted heartwood and having a coating of gum on the nest tree above and below the cavity.

The beginning of problems for the RCW came with wholesale clearing of the virgin pine forests of the southeastern United States. This began in the 1880s when, following depletion of forest resources in the Great Lakes region, forest industry moved South to purchase newly available federal lands for as little as $1.25/acre (Lillard 1947). The virgin longleaf forests that survived this onslaught fell for patriotic reasons. Southern congressmen complained that the South was not getting its fair share of war industry money during World War I and succeeded in getting a bill passed in Congress providing for the construction of 1000 ships of southern pine (Fickle 1978). Placards went up around the South stating that "Every swing of an axe, every cut of a saw may score as heavily as a shot fired from the trenches" (Clark 1986). Only a little over 300 ships were built of southern pine and they never saw action in the war, but the forests were gone.

The slash left behind fueled devastating forest fires and southerners joined a national bandwagon in condemning fire as evil. Ignoring the wisdom of a few who saw the positive role of fire in southern pine ecosystems, the message of Smokey the Bear generated an era of hardwood replacement that further diminished RCW populations.

By 1919, Pangburn (1919:401) noted evidence that the species might be declining in Florida. Murphey (1939:73) was very early to put the demise of the RCW in blunt perspective: "From many sections of the South, where it was formerly common, the red-cockaded woodpecker has disappeared by reason of the ruthless destruction of pine forests by the lumberman. When the large timber is cut out, the birds leave the locality and apparently do not return." Although in Missouri, conservation interests were able to save a 61-m (200-ft)-wide buffer of virgin pine forest—including the last known RCW colony in the state—by the late 1940s, the species was gone from Missouri (McKinley, 1951).

Beginning in the late 1940s and continuing into the early 1970s, C.C. Steirly published a series of papers describing the basic ecology and decline of RCW in Virginia (e.g., Steirly 1949, 1950, 1957a, b). In 1960, acting on a request from the Mississippi Ornithological Society, the U.S. Forest Service began protecting RCW cavity trees on National Forests in Mississippi (Newcomb 1960). Dan Lay, in Texas, called attention to the plight of the species in an article titled "Destined for Oblivion" (Lay 1969).

It was in the late 1960s, that a committee within the U.S. Fish and Wildlife Service (FWS) began considering species that might be included on a federal endangered species list. The RCW was suggested as a possibility and several southern wildlife biologists were asked for their opinions as to the species' status. With no data, but several opinions that this was a bird in trouble, the RCW was listed as endangered (U.S.D.I. 1968b). Protection was extended to cavity trees on federal lands, but we soon learned that this was not enough.

In order to pull together what was known of this newly declared endangered species, in May 1971, the FWS and Tall Timbers Research Station hosted the first symposium on the species at Okefenokee National Wildlife Refuge in Folkston, Georgia. About 40 people attended and the resulting proceedings included 18 papers (Thompson 1971). One major result of the symposium was the realization of just how unusual the RCW is --even among woodpeckers. Manifestation of this understanding is the glossary of specialized terms published as an appendix to the symposium proceedings and used with regard to the species (Jackson and Thompson 1971). Another major result of the meeting was the clear message that this was a species that was suffering from loss of habitat— conversion of forest lands to non-forest uses, short rotation forestry, and exclusion of fire from those older southern pine forests that remained. Jackson (1988) elaborated on the nature and extent of these changes in southern pine ecosystems.

At the conclusion of the first symposium, those in attendance, representing various federal and state agencies, university researchers, conservation groups, and forest industry—were polled. Was the RCW truly an endangered species? The response was unanimous. Yes.

Two years later I happened to be in Washington doing some work at the Smithsonian when a colleague of mine there casually commented that my "favorite bird was being taken off of the endangered species list." I was dumb-founded. I said that that couldn't be, that at a symposium sponsored by the FWS, the participants had been queried as to what the status of the species should be and the support for its endangered status had been unanimous. In response, he obtained a copy of the proof for what was to become the 1973 red book—the RCW was not included. Immediately I went to the Office of Endangered Species to investigate further. When I arrived, no one except a secretary was in the office. I asked if minutes were kept of meetings at which endangered species decisions were made. The secretary said yes and showed me those relating to the RCW. There had been little said. A representative from the Endangered Species Office had said something to the effect: "What about the red-cockaded woodpecker? We're told that it is no longer endangered." To which an ornithologist from the Smithsonian replied: "Yes, there are plenty of red pines in the South."

I contacted Richard Thompson of the FWS, the organizer of the first symposium. He was equally upset by the news. The word was spread and a formal statement came from Thompson expressing the sentiments of the symposium participants. In short, the RCW was added back to the list. As evidence of its intended removal, you can note that it is the only species not listed in the index to the 1973 red book (Office of Endangered Species and International Activities 1973). Within a month, evidence as to the source of the delisting effort came in the form of a published, but premature, news release (Schenck 1973) from a prominent forest industry declaring that "Recently the red-headed nest builder [red-cockaded woodpecker!] (among 67 listed as endangered or rare by the U.S. Department of the Interior) was removed from the 'endangered' list to the more plentiful 'rare' category, coming on the heels of efforts like those of the forest products company in saving nesting trees".

Neither the listing nor the possible delisting of an endangered species would today be possible in such cavalier fashions. The novelty has worn off. The seriousness of listing and delisting is now well understood. The procedural mechanisms, while still subject to human frailties and perhaps even lobbying efforts, are now protected by greater understanding and professionalism as well as greater involvement of the public. The Endangered Species Act of 1973 has aged well, albeit with scars and bruises. It may well be the most significant piece of environmental legislation ever passed. Those implementing the Act, those enforcing it, and those subject to it have grown as the Act has been defined and interpreted by practitioners of endangered species science and the courts.

In 1975, a recovery team was appointed to draft a recovery plan for the species. Five individuals were appointed to the team: myself, W. Wilson Baker, Verlon Carter, Thad Cherry, and Melvin Hopkins. As Team Leader I invited others to attend recovery team meetings as consultants. The first team meeting was held in Charleston in November 1975, so that team members could see the status and characteristics of the then largest known population of the species, that on the Francis Marion National Forest. Work on the recovery plan was difficult to say the least. Team members came from a variety of backgrounds and represented a variety of interests. But in spite of very heated discussions and major disagreements (King et al. 1977), a plan was forged and finally approved in 1979 (Jackson et al. 1979a). In the meantime, considerable research was being conducted.

We learned, in part by holding recovery team meetings in several states, that RCW habitat varied greatly from region to region and from pine species to pine species. Recovery Team members and consultants brought together further knowledge of the range of habitat diversity in which the species could be found.

Lawsuits or threats of lawsuits by conservation groups brought public attention to the RCW and federal, state, and private land owners were forced to take the Endangered Species Act seriously. There were still problems. A barracks complex at Fort Benning, Georgia, replaced an active RCW colony in the mid-1970s, and it was local individuals who brought the problem to the public eye. It was too late to save the colony site, but an attempt was made to provide roost sites (Jackson et al. 1983). Further mitigation included protection and management of another colony, but there was no follow-up and that colony site became an obstacle course (Freeman 1984). When D'Arbonne National Wildlife Refuge was established as mitigation for a water project, the mineral rights to the land were not purchased. No consideration was given that an endangered species might be impacted by future exploitation of gas beneath the refuge. Legal action again brought the RCW to the public eye and the responsibility for managing the RCW to the attention of land managers (Turner 1987).

In January 1983, a second red-cockaded woodpecker Symposium was held, this time in Panama City, Florida. It was cosponsored by the FWS, the U.S. Forest Service, and the Florida Game and Fresh Water Fish Commission. About 150 people attended and the proceedings included 16 papers and 16 notes (Wood 1983c).

Interest in the species was growing dramatically, but the species continued to decline everywhere that populations were monitored (e.g., Costa and Escano 1989). Imminent extinction was feared for several populations (e.g., McFarlane 1992). Legal action to change management practices on the National Forests began in the 1970s. However populations on Texas National Forests plummeted by as much as 76% in a nine-year period before Judge Robert Parker, in June 1988, ruled that the Forest Service must alter management practices to favor the birds. The decision was appealed, but in March 1991 the Fifth Circuit Court of Appeals affirmed Judge Parker's ruling.

On another legal front, southern forest industry, through the Region 8 Forest Service Timber Purchasers Council, filed suit against the U.S. Forest Service challenging management efforts for the RCW (Baker 1989). Clearly, the Forest Service and the bird have been caught in the middle!

With all of this focus on RCWs, answers were needed for many management questions. Federal dollars employed many researchers and our knowledge of the RCW grew tremendously during the 1980s. A revised Recovery Plan (U.S. Fish and Wildlife Service 1985) brought further controversy, but helped further identify needs and bring focus to the challenges.

The species also continued to make headlines (**Figure 1**). Considerable negative publicity stirred further efforts to get the species delisted (John E. Wishart, pers. comm., 19 Nov. 1981), but those efforts were not successful. Polarization of parties has been common as conflict has dominated many situations. The problems have often been people problems and many of these have been described in McFarlane (1992).

Biological problems also continue to plague the species. Southern pine beetles continue to be a serious problem, especially in east Texas. Understory control and the potential of competition and predation by southern flying squirrels (*Glaucomys volans*) are viewed as problems range wide; however the full nature of the problems, and the severity and type of action needed to solve them, remain controversial.

Finally, by the late 1980s, we began to see more serious proactive steps being taken on behalf of RCWs by many affected and interested parties, including forest industry. Education was, and remains, a serious need. A training handbook from Camp Lejeune (**Figure 2**), a training video at Ft. Polk, a Scientific American Frontiers program ("Woodpecker Housing Crunch", 10 Oct. 1990) on public television, and other educational efforts have made a positive difference.

As a measure of our growth in knowledge about the RCW, I tallied articles that dealt specifically with the RCW, from five professional journals. We can see this growth in knowledge as indexed by this tally (**Figure 3**). I believe we can also see the influence of major events in the recent history of this species.

While there have been 77 articles published on this species in these professional journals since 1950, 74 of them have been published since the species was listed as endangered. Many more research articles have appeared in other journals, in symposia proceedings, and in government publications.

As a simple comparison, I tallied articles published during the same time period for

Figure 1. Representative headlines from news articles relating to the RCW.

another troubled species—one that has been even more in the news—the spotted owl (*Strix occidentalis*) (**Figure 4**). During the same time period there were only 30 publications on the spotted owl in these professional journals. Most only in very recent years, since the species has gained notoriety and following publication of the Report of the Advisory Panel on the Spotted Owl (Dawson et al. 1986).

I believe that this comparison shows several things:

(**1**) the value of endangered species listing in drawing attention to a species;

(**2**) the effect of federal research dollars on the growth of our knowledge about the species;

(**3**) the level of committment agencies and researchers have toward a species; and

(**4**) some basic differences in the biology of these two species.

The red-cockaded woodpecker has been referred to as the "spotted owl of the Southeast." There are obvious parallels—the most important of which is the need for old growth forest. However, from a research perspective, we have had it easy. The RCW makes conspicuous cavities, its colony sites are traditional sites, its foraging habitat is open, and its vocalizations and behavior make it easily observable. Much of the research on

Figure 2. Cover of a training manual from Camp Lejeune, NC. Cartoon caricatures of red-cockaded woodpeckers and training materials such as this have helped inform the public about this endangered species.

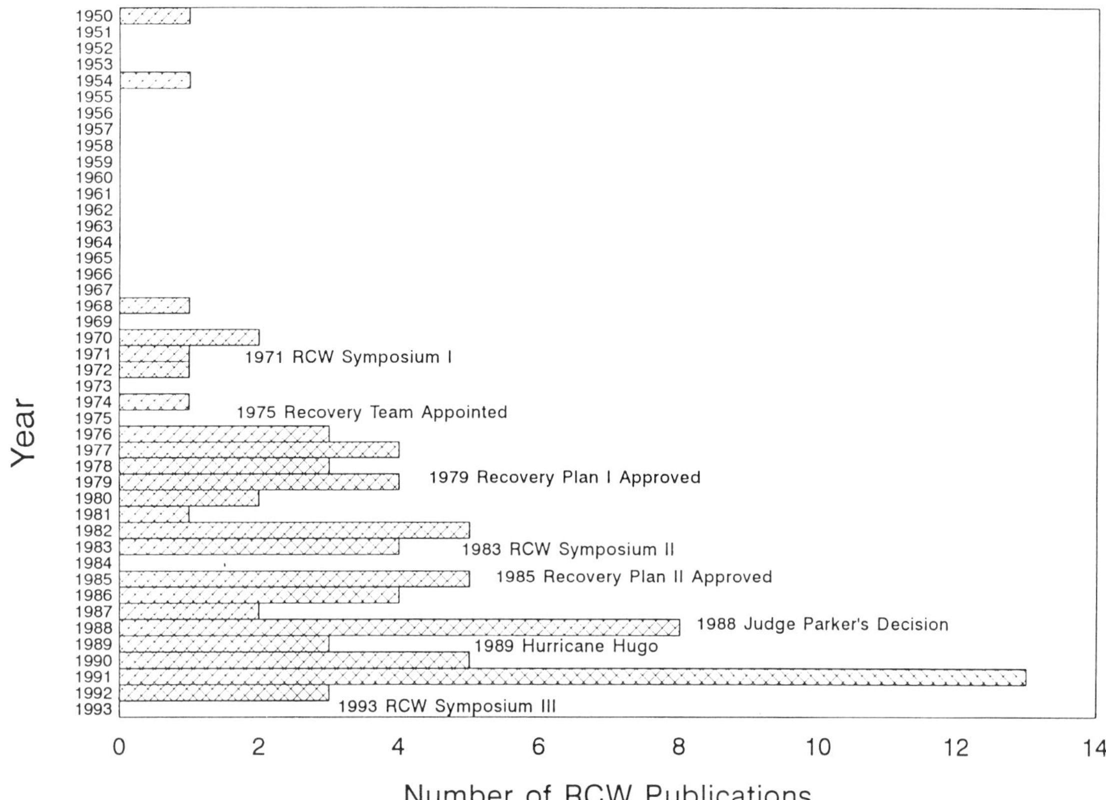

Figure 3. Numbers of publications dealing specifically with the red-cockaded woodpecker in selected professional journals between 1950 and 1992. The column for 1993 is present only to indicate the timing of the third RCW symposium. Journals surveyed included: The Auk, The Wilson Bulletin, The Condor, Journal of Field Ornithology, Journal of Wildlife Management and the Wildlife Society Bulletin. Superimposed on these data is the timing of major events associated with red-cockaded woodpecker conservation and management.

RCWs has been focused on these habitat and behavioral characteristics. We can easily locate the colony sites and birds and monitor them; we can also easily capture the birds from their roosts. I think it is no accident that the weak link in our chain of knowledge about the RCW is related to the foraging ecology and needs of the bird away from the colony site. It is something more difficult to study.

Spotted owl researchers have neither colony sites nor conspicuous cavity trees that they can focus their efforts on. They also have a bird whose behavior is difficult to study as a result of its nocturnal and more furtive nature.

There is no question that much of the management success that we have seen in recent years has been, in part, a result of land managers having the colony site as a focus for their efforts. These differences in the conspicuousness of the species have benefitted the RCW in other ways too. Birders and others can reliably find and see the RCW; its social behavior and constant vocalizations draw attention to it and make friends for it..

"Who gives a hoot?" was the message on a cover (25 June 1990) of Time Magazine focusing on the spotted owl. This is not a question that is relevant to the RCW. Successes of recent years have been a result of many people at many levels and from many walks of life who have cared.

Finally, technology has often saved the day. Cavity restrictors were an important innovation (Carter et al. 1989) -but only a tool, not an answer for every ill. Hurricane Hugo may have devastated the Francis Marion, but it may also have saved the RCW. Implementation of innovative techniques might have been slow to come had we not been faced with such a disaster. We were just fortunate that invention had preceded necessity and that agencies and biologists were ready to meet the need. The rapid response of all involved and the transference of cavity insert (Allen 1991) and

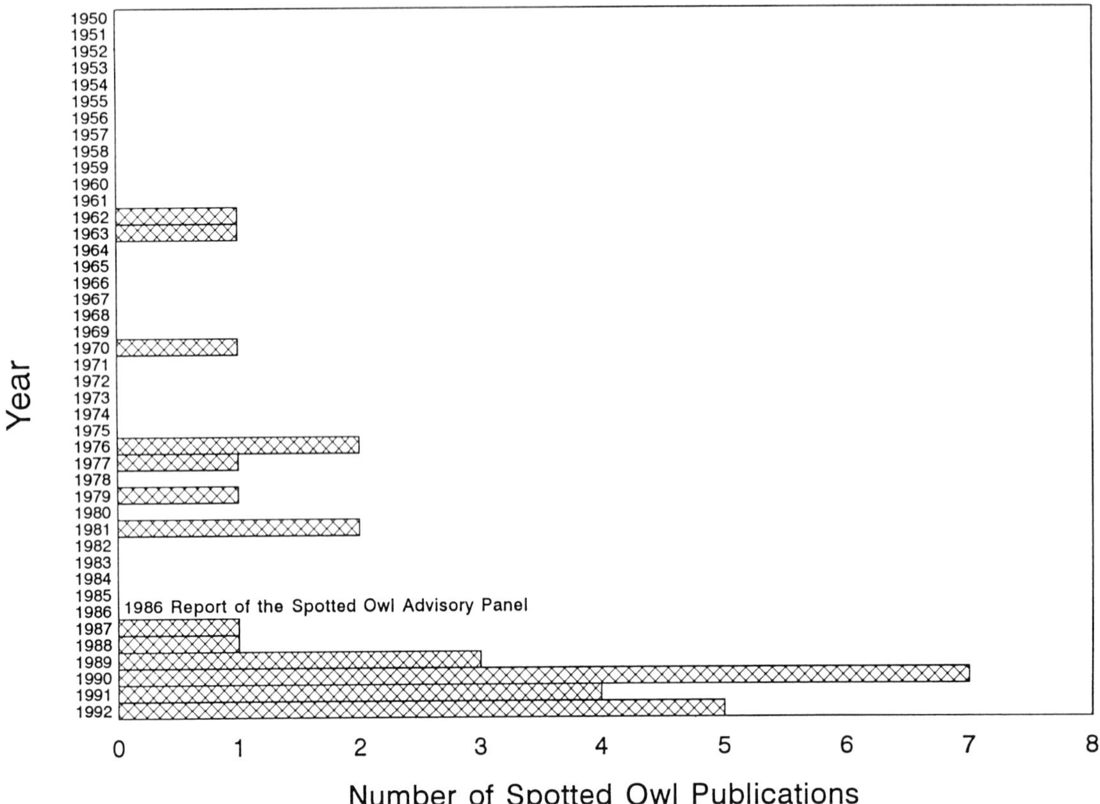

Figure 4. Numbers of publications dealing specifically with the spotted owl in selected professional journals between 1950 and 1992. Journals surveyed are the same as those included in Figure 3.

artificial cavity construction (Copeyon 1990) technology to general use as management tools is a wonderful success story.

But I am reminded of the TV series the "Six Million Dollar Man." The hero was involved in an accident and was incredibly injured. As a prelude to each episode the origin of the bionic six-million-dollar man is briefly retold. A voice-over tells the viewer: "We can save him. We have the technology."

We can save the RCW. We have the technology. But we can't afford - and would not want - a bionic woodpecker. Technology can get us over the hump. Understanding of the species' ecosystem and wise management must be our tools for the future, beyond recovery.

Ecological Conscience, Economics, and Ethics: Exploring Leopoldian Thinking

Gene W. Wood, Department of Aquaculture, Fisheries And Wildlife, Clemson University, Clemson, SC 29634

ABSTRACT: Leopoldian thinking suggests that conservation policy must be formulated within a paradigm that is both ecologically and economically sound. An ecological conscience promotes concerns for natural resources with commodity and intrinsic values. Economic concerns are prompted by the necessity to both maintain a stable society and to provide for the science and technology requisite to the advancement of civilization. Ethics constrain us to careful consideration of the importance of ecosystem integrity, stability and beauty.
KEYWORDS: Leopold, conservation policy, economics, ecological conscience, ethics

As outlined by Marx (1964), American society had a bipolar perspective on the relationship between progress and environmental quality from the colonial period through the early Twentieth Century. While environmental awareness is not a new phenomenon, within the last two decades it has risen to unprecedented levels. An environmentally conscious American public worries about the ecological condition of the landscape and the possibility that many species and biotic communities may be at risk (Times Mirror Magazines 1992).

It is logical that an individual or society should worry about the land base that supplies goods and services essential to basic life processes, societal development and progress, and that contains qualities of intrinsic value. However, concurrent with this newest culmination in American environmental concern is the very high level of separation of individual Americans from a direct economic dependency on land. Our society is about 75 percent urbanistic (Brown and Jacobson 1987:8), and it is substantially employed by the service industry. While Americans are increasingly landless and decreasingly experienced with land management, they are increasingly involved in political decisions regarding land-use.

In this paper, I offer pragmatic rationale that economic land-use is economically and sociologically necessary and ecologically legitimate. I also suggest that implicit in the term conservation is appropriate restraint of the consumptive use of natural resources. Such restraint is for the benefit of both human progeny and the sustainability of the ecological communities in which we presently live and that contain options for the future. I further suggest that the man-land economic relationship is a natural process, and that the land ethic (Leopold 1949:262) is intended to restrain and guide land-use practices, not to prevent them.

CONFUSION AND EMOTIONAL SIMPLIFICATION

Aldo Leopold noted that the forces that drive the practice of conservation consist of both intellectual and emotional components (Leopold 1949:263). Progress in scientific understanding of ecosystem composition, structure, and function continues to extend our awareness of the complexity of nature. Both the written and electronic media have appealed to our emotional sensitivity to the beauty and intrigue of nature. However, society seems to be failing to recognize that the complexities of conservation of an ecosystem are proportional to the complexity of that ecosystem, the nature of other ecosystems to which it is linked, the nature of the linkage, and the nature of human involvement in that system (Myers 1993).

Although simple solutions are frequently offered, realistic paths to resolution of most of our conservation issues usually are unclear. Perhaps our often contradictory scientific evidence mixed with our tendencies to be emotional and impatient with hard data have confused us even less than they should have. A case in point is the contradictory evidence and theories concerned with global warming, one of our most pervasive environmental issues (Solomon et al. 1991, George C. Marshall Institute 1992, Lacis and Carlson 1992, Scuderi 1993). But our society appears to be intolerant of complexities, and it often chooses arbitrarily among messages. The presumed objectivity of science is often subsumed by the subjectivity needed for clarification in the mind of society at large. Much of what society believes is what it wants to believe, and want in this case is an emotional force that is often guided by populist notions.

Intelligent decision-making does not discount the importance and validity of emotion. It does, however, recognize emotion as only one force causing debate. Rational decision making requires a platform more substantial than that planked primarily with notions heavily biased towards intrinsic values. To achieve objectivity, I suggest that the environmental quality - economic land-use debate should strive to maintain certain basic tenets about:

a) who we are in an ecological context;

b) why we practice conservation, and

c) how we practice conservation in the American politico-socio-economic format.

BASIC TENETS

1. The human species is a biotic component in ecosystems

a) The human species is the most powerful consumer in the history of the planet.

b) Human effectiveness as a competitor with other species is based on a superior intellectual capacity that has provided for the development of machines to greatly magnify human physical capability. This is in contrast with all other species that are constrained primarily to their physical attributes.

c) The human species is the only one to consciously consider the long-term well-being of its progeny.

d) The human species is the only one to consciously consider the well-being of its ecological associates.

e) The human species is the only one to consciously both restrain its consumptive activity and regulate its own population growth. These behaviors are in response to an " ecological conscience " (Leopold 1949:243-264).

2. The purpose of conservation

a) Conservation is aimed at prevention and treatment of over-exploitation. Over-exploitation is the result of inappropriate human behavior and is manifested in the deterioration of land health which thereby puts the well-being of human progeny at risk.

b) Conservation is a response to the land ethic which has the same generic goal as do social ethics in separating appropriate behavior from inappropriate behavior.

3. Conservation in the American politico-socio-economic format

a) American society is both democratic and capitalistic.

b) The success of American democracy is based on the existence of a free-enterprise system.

c) Rights of the private landowner, including the right to benefits derived from economic land-use, are essential to a free-enterprise system.

d) Certain elements of capitalism and certain elements of conservation are antagonistic.

e) A system that allows both the accrual of individual benefits from economic land-use and the freedom to argue the difference between rights and wrongs in land-use is the one most likely to have a sustainable conservation program.

In general, we develop and implement conservation to restrain our powerful inclination and ability to consume goods and services that come directly from land. The practice of conservation is prompted by our ecological conscience which is both unique to the human species and ultimately anthropocentric as it seeks to ensure environmental quality for human progeny. The practice of conservation implements the integration of the ecological values of "integrity, stability, and beauty" (Leopold 1949:262) with commodity resource production. In American society where both the rights of the individual are constitutionally protected while the goals of the public are served, a free enterprize system with an assurance of private landowner rights are considerations prerequisite to meaningful conservation programming.

ECOLOGICAL CONSCIENCE

While members of every species perceive themselves collectively as being at the center of the universe, only the human species has contemplated that perspective. This is an uncomfortable position for us since our ability to compete with other species for ecological niche space is without equal in the history of the planet. Intellectually we cannot resist introspection about our existence relative to that of other species. On the other hand, it is critically important that introspection not deteriorate to paranoia.

Because the powerfulness of the human species is intellectually based as opposed to physically based, as it is in most species, physical regulators in the form of biological feedback mechanisms and predator-prey strategies have not evolved to stabilize our population growth and consumptive activities. In our case, the feedback mechanism has had to be a manifestation of our intellectual capacity. This mechanism is called the ecological conscience.

The intellectual capacity of the human species fortunately is a double-edged sword. In one mode it fuels our basic tendencies to consume excessively by providing strategies for high rates of consumption. This "greed" mode can lead to inappropriate behavior in both social and ecological contexts as the strongest become wealthy at the expense of the weak and consumption of natural resources exceeds biological productivity. However, counterbalancing the greed mode is a second mode called "conscience." Intellect informs the conscience of what may be right and what may be wrong for the long-term good of the species. It thereby separates appropriate from inappropriate behavior.

Leopold suggested that the ecological conscience was an extension of the social conscience and might be a "community instinct in the making" (Leopold 1949:239,246,263). However, there is an important basic difference in the intellect that fuels the social conscience and that which fuels the ecological conscience. The information base for social conscience is based on millennia of experimentation with social codes of conduct in an array of politico-socio-economic systems (societies). Basic rules governing behavior essential to stability in social relationships, e.g. the Mosaic Decalogue, the Golden Rule, and democracy appear to be highly evolved and enduring, although they are constantly challenged and not universally accepted. Humans are constantly informed about appropriate and inappropriate behavior through their individual successes and failures in day-to-day relationships with other humans and society at large.

In contrast, a comparable data base cannot be claimed as the foundation for the intellect that informs the ecological conscience. The information fueling our ecological conscience is presumed to be based on scientific information. But ecological science is substantially less than one century old and it is variously interpreted by different scientists. In addition, to the extent that ecological information is science-based, a minute fraction of society (ecological scientists) is informing the whole of society on what is right and what is wrong (e.g., Lubchenco et al. 1991, and Hughes and Noss 1992:16).

Even at the level of the individual in contemporary society the assimilation of information is highly biased to the sociological. Almost every American daily experiences human-to-human relationships. But in modern America, relatively few people are intimately involved in a day-to-day direct dependency on land. Therefore the ecological conscience of American society at large is derived more in an abstract framework, i.e. perceptions, than it is in an empirical framework, i.e., personally experiential and scientifically experimental.

In an era in which the planet's ability to support life is for the first time ever being challenged by the activities of a single species, it is critically important that that species have an ecological conscience. It is equally important that that conscience be correctly informed. It must be aware that while unrestrained economic growth and development can lead to environmental degradation, economics is an essential component of human ecology. Furthermore, the economic component must be sufficiently robust as to be capable of ensuring a flow of goods and services beyond those required for minimal survival, i.e. some reasonable level of physical comfort, and capable of providing for scientific and social progress leading to the advancement of civilization.

In summary, the function of the ecological conscience is not to prevent economic land-use, but instead to guide it (Leopold 1949:239). It does not prompt the removal of humans from the landscape, but instead their appropriate integration into it. The human species has a biotic right to existence and an ecological responsibility to ensure an environment suitable for the well-being of its progeny. The ecological conscience prompts us to differentiate between right and wrong behaviors as we claim our biotic right to compete with other species for ecological niche space. Economic planning is inseparable from our ecological strategies for both natural resource consumption and conservation. Therefore, the purpose of the ecological conscience is to strive for a harmony between our socio-economic status and our ecological well-being. "Conservation is a state of harmony between men and land" (Leopold 1949:243).

ECONOMIC LAND-USE AND CONSERVATION

Thomas Jefferson's early Nineteenth Century socio-economic concept for the United States of America was a nation of farmers (Marx 1964). He would have found progress in the form of the industrial revolution to be considerably disturbing. A century later, Aldo Leopold found that inappropriate farming, ranching and logging practices had degraded the productivity and stability of many, if not most, of America's forest and prairie ecosystems (Flader 1974, Meine 1988). Interestingly, these two savants, living a century apart, shared two key perspectives critical to American conservation. First, both believed that the economic dependency of men on land was not only necessary, it was also desirable. And second, both had an unshakable faith in the American democratic system. Jefferson's record on this point is well known. Leopold biographer Curt Meine (1988:430) wrote "Leopold had a Jeffersonian faith in democracy: the only sure cure for democracy's ills was still more democracy." While Leopold wrestled with the apparent incongruities of implementing a strong conservation program in a democracy (Flader 1974), he was both optimistic and determined that the goal could be achieved. Evidence for this point can be found in a 1942 statement: "Hitler's taunt that no democracy uses its land decently, while true of our past must be proven untrue in the years to come" (Leopold 1942 in Meine 1988:430).

Both Jefferson and Leopold strongly believed in minimal federal government regulation and maximal freedom for the exercise of rights and responsibilities by individuals and communities. Leopold was convinced that the federal government could not shoulder the full responsibility for conservation in America (Leopold 1949:250). Evidence for his dislike for broad sweeping federal programs is abundant. While he was a U.S. Forest Service employee from 1909 to 1928 and a participant in the federal conservation programs of the "New Deal" of the 1930s, he was also often critical of politicized conservation. In addition to his technical and logical arguments, his cynicism in the often-used term "alphabetic conservation" is obvious.

Leopold's belief that, for the most part, any meaningful national conservation strategy would be dependent on the private landowner was first presented at the Seventeenth American Game Conference: "Only the landholder can practice management efficiently, because he is the only person who resides on the land and has complete authority over it" (Leopold 1930). His sense of the critical importance of private lands and rights of landowners in conservation was also obvious in the statement taken from his journal: "One of the curious evidences that 'conservation programs' are losing their grip is that they have seldom resorted to self-government as a cure for land abuse" (Leopold 1939 in Meine 1988:430).

Obviously, landscape ownership has to be a primary concern in conservation programming. Approximately 67% of the American land mass is privately owned. While the federal government owns 30% and state and local governments own 3% of the landscape, there is twice as much land managed for private citizen and corporate goals as there is public ownership (U.S. Department of Interior and U.S. Department of Agriculture 1992). Most of the federal ownership is in Alaska and the West; relatively minor amounts occur in the Mid-West and East. For instance, in the case of forest lands, federal ownership in many of the western states exceeds 50%, but in the East it accounts for only about 10% (Waddell et al. 1989). It should be of note that southern forests, which are rapidly becoming the fiber basket of the nation, are less than 9% federally owned.

When considering these data, three important questions emerge. First, can any meaningful conservation programming be done without substantial involvement of private landowners? Second, can private landowners be involved in meaningful conservation programs without sacrificing, or even jeopardizing, their constitutional rights to economic use of their lands? And third, is there critical importance to integrating conservation with economic land-use?

National conservation programming that is confined to federal ownership has little to no chance for success because most of the land in the 48 contiguous states is in private ownership. Organisms, both plant and animal, occur, or have the potential to occur, wherever there are habitats suitable for the species' growth, maintenance and reproduction. The organism cannot perceive any basic biological difference between public and private ownerships. Our current predicament in trying to implement the Endangered Species Act of 1973 (ESA) clearly demonstrates this point.

When the ESA was first enacted it was generally perceived as an instrument for protection of species in danger of extinction or threatened with reaching that condition, and to provide for the recovery of those species using federal agency programs on federal lands. However, in the last 20 years, we have seen the ESA broadened by both amendments and interpretations to the point that it is having substantial impact on private lands and their economic use. This current situation was inevitable. Obviously, all members of all listed species would not occur exclusively on federal lands. Furthermore, as the number of listed species has increased, thc number that do not occur at all on federal lands has increased. The federal government cannot possibly own all of the lands that support listed species. If the spirit of the ESA, purported to be the world's most sweeping environmental legislation, is to be served, private lands and landowners have to be a part of the program.

Can private landowners become involved directly with conservation programming without fear of loss or placing in jeopardy their constitutional rights to land-use? To paraphrase Leopold (1942) in the context of the current issue, while this may have been the case in the past, it must not be the case in the future. Conservation that is totally dependent upon legislative and administrative "thou shalt not's" is loaded with disincentives for the private landowner. His/her participation under this circumstance is under threat of penalty, and, in this situation, is usually both minimally and begrudgingly given. Anything that is as necessarily dynamically creative as conservation must be, cannot be progressive when key participants are contemptuous of the task.

The ecological conscience of society at large will prompt democratic legislation on environmental conservation matters. This is, and will continue to be, a majority rule for the common good process. Following Hardin's (1968) "Tragedy of the Commons" thesis, while this process is not perfect, it is necessary because humans are better at collectively making and implementing rules than they are at individual self-discipline, even when the logic against a certain behavior is obvious.

The problem with most of our conservation legislation in the past, particularly when it has focused on non-commodity resources at the expense of commodity resources, is that it has rarely taken into account the economic plight forced on those that own the land. The problem has been compounded as government agency rule-making and interpretations typically have had even less regard for economic impact or how impact might be ameliorated. A case in point can be drawn from the ESA. A 1981 opinion of a Department of Interior Solicitor stated "Under the definition of 'harm' a private landowner may be criminally liable for significant habitat modifications regardless of the reasonableness of his efforts to mitigate harm to individual wildlife" [46 Federal Register 29490 (June 2, 1981)]. A biological principle may have been served by this statement, but the approach is one of alienation and ultimately subsumes the larger principle of conservation of the species throughout its range. An example can be seen in the increase in negativism toward the red-cockaded woodpecker (*Picoides borealis*) and the concomitant decline in its numbers on private lands in the southern U.S. over the past 15 years.

Negative experiences of the past will jeopardize proposals for private landowner participation in conservation programs in the future. I offer two examples. First, in the negotiations with the U.S. Fish and Wildlife Service on a proactive conservation plan for a listed species, a major timber company recently offered to exempt from the plan its lands in Virginia because of that state's attitude toward listed species problems on private lands.

Because the Service did not require Virginia's approval of the plan, the issue was dropped and the negotiations were successfully completed to the benefit of the species. If the Service would have required Virginia's independent approval, a proactive conservation plan for the company's ownership in that state would have been lost.

The second example is the program to re-establish the red wolf (*Canis rufus*) in the wild. This program was initiated under Section 10(j) of the ESA which provides for experimental/introduced populations. ESA Section 7 and Section 9 requirements and standards have minimal effects on private landowners so long as such a population is classified as "non-essential" to the recovery of the species in the wild. However, if the Secretary of Interior reclassifies the population to "essential," the full force of the ESA will apply (Parker and Phillips 1991). The Service wants to substantially expand the introduced, non-essential red wolf population in northeastern North Carolina, and it needs voluntary participation of private landowners in this program. Even though no conflicts between the wolf program and the current land uses are anticipated, because of previous bad experiences with the ESA, private landowners are reluctant to join this conservation effort for fear of the potential impact of reclassification.

Private landowners can be brought into the mainstream of conservation if they do not perceive it as hostile territory. The process must not be economically debilitating or offensive to their values for their land. The "pitch" must be a solicitation on the grounds that a democracy only works when citizens cooperate to protect societal values that have been agreed upon in a democratic process. Concomitantly, the rights of the individual must be assured. Of ultimate importance will be recognition that in natural resource conservation, protection and enhancement of values of a society largely removed from a direct relationship with land may often depend substantially on the efforts of a small minority of its members, i.e. private landowners. To paraphrase Meine's (1988:430) characterization of Leopold on this type of issue, conservation in our democratic society can be improved upon by applying more democracy.

Finally, is it critically important to integrate conservation into economic land-use? The answer is emphatically "Yes." Furthermore, it is important on both public and private lands. Since the early 1960's American's have increasingly believed that the only meaningful conservation was that achieved through the "lock-up" process. Possibly the Wilderness Act of 1964 was the stimulus for this philosophy. Leopold, who shares with Bob Marshall the credit as father of wilderness conservation, was responsible for the nation's first designated wilderness area in 1924 (Meine 1988:205,226), and would have applauded the Wilderness Act. On the other hand, his position that national conservation programming that was totally dependent upon government programs and ownership was doomed to failure in the long run is abundantly clear.

According to a joint report by the U.S. Department of Interior and the U.S. Department of Agriculture (1992:5-1), 10% of the U.S. land mass is in national park, wilderness, and wildlife areas. Between 1959 and 1987 the amount of land area under this designation increased from 47 million to 225 million acres (4.8-fold). The total designated wilderness area alone now occupies 4% of the U.S. land mass. The value of maintaining some wilderness for a society that only a short time ago originated in a wilderness is important for establishment of a relative sense of progress in a landscape context. On the other hand, it is plain that we cannot afford, either economically, socially or ecologically, the wilderness-set-aside concept as the primary girder for mainstream conservation.

There are no reports suggesting that the demand for raw materials produced by economic land-use will decline in the future. In fact, every report indicates substantial increases in demands both in the national and global economy. Every acre that is permanently removed from commodity production or potential production, either has or will have a negative socio-economic impact. Whether or not the benefits from new set-aside lands will offset the commodity forfeitures must await the evidence of the future.

To illustrate the above point, the forest lock-up associated with protection of northern spotted owl (*Strix occidentalis caucina*) habitat in the Pacific Northwest had adverse consequences both for the economy of the region and for the ecological values of private lands not affected by the owl plan. When the federal timber was made unavailable the value of private lands timber soared and far more than normal amounts were cut in a short period of time because it became so valuable that the landowners could not afford to keep it. In addition, lumber prices in the South also increased greatly as the supply-and-demand principle remained operational. There is a general view that we may already be stressing a substantial portion of the southern pine timber land (U.S. Forest Service 1988:142-163). The need for forest products is predicted to increase, so where do we go from here? The U.S. already imports more wood than it exports, and forest products companies are already looking for land in foreign countries. Only someone who believes in an infinite supply of land, or an infinite capacity of land to produce raw material would not worry about the repercussions of locking up land.

As land is withdrawn from commodity production, a greater demand is placed on the remaining landscape to supply the raw materials needed to feed, clothe and shelter our species. This change is arithmetic, but when the change in demand for raw materials becomes part of the equation, the change in stress on the remaining landscape becomes exponential. Norman Myers (1993) essay "The Question of linkages in Environment and Development" speaks directly to this concept. We have so compartmentalized our conservation ideology, e.g. "the species comes first," that the impact that a particular conservation process may have on other societal and ecological processes has either been ignored or forgotten. In Myers'(1993:309) words ".... we will respond to linkages either by reactions of sufficient scope and character, or by salvage measures in a world impoverished by our disregard for linkages."

THE LAND ETHIC: LEOPOLDIAN THINKING

Nash (1987) pointed out that there has been a substantial number of land ethicists other than Aldo Leopold. But the thing that set Leopold apart was his ability to articulate an ethic to guide human behavior in an ecological context, i.e. an integration of ethics and science. Susan Flader, who was the first to chronicle the developmental history of Leopold's thinking, perceptively and correctly pointed out ".... many who fancy themselves disciples of Leopold shun the rigorous thinking and constructive openmindedness that to him were essential" (Flader 1974:xvi-xvii).

In this age of struggle between preservationists and utilitarianist arguing whose values are right for the land and ultimately for society at large, the land ethic is quoted out of context by the former and rejected from misunderstanding by the latter. The tragedy of this situation is that the land ethic as articulated by Leopold, while both misunderstood and misused, currently guides our national conservation programming (e.g. ESA, National Wild and Scenic Rivers Act, Wilderness Act, biodiversity and ecosystem management, neotropical migrant bird species programs, etc.) and will do so into the foreseeable future.

The underpinning of the "land" ethic lies in three basic ideas (Leopold 1949:255): "**(1)** That land is not merely soil. **(2)** That the native plants and animals kept the energy circuit open; others may not. **(3)** That man-made changes are of a different order than evolutionary changes, and have effects more comprehensive than is intended or foreseen."

In Leopold's mind, the term land meant ecosystem. He was familiar with and indeed intrigued by the relatively new science of ecology and the term ecosystem. But his basic utilitarian senses seemed to ground him to the term land. He recognized the farm as a "food factory," but it was also a habitat. Forests produced lumber for everything from barns to "the finest oak desks," but they were also biotic communities. To Leopold, ethical land management would involve answering the challenge to intelligently consider ecosystem composition, structure and function while knowing that an expectation for complete understanding would be an exercise in futility and could have dangerous consequences. According to Meine (1988:471) "Land health, in his thinking, and in his public actions, looked to the 'functional integrity' of land, though not to the exclusion of those who most closely lived and worked with land."

A sense of the necessity of economic land-use was not ignored in the development of the land ethic. "A land ethic of course cannot prevent the alteration, management and use of these 'resources,' but it does affirm their right to continued existence, and, at least in spots, their continued existence in a natural state" (Leopold 1949:240). And with respect to societal reimbursement of the private landowner for economic sacrifice, Leopold (1949:250) wrote "If the act costs him cash this is fair and proper, but when it costs only fore-thought, openmindedness, or time, the issue is at least debatable."

Leopold presented his articulation of the land ethic as "an evolutionary possibility and an ecological necessity" (Leopold 1949:239). While he used the evolution of social ethics as a model for reference, the land ethic that he concluded resulted from a life-time spent in intense thinking about land while working as a forester, game manager, public policy-maker, private landowner, consultant, teacher, and writer. He was not so naive as to think that the land ethic realistically could ever be more than a guideline. Implementation to perfection would be impossible and even dangerous to attempt.

While Leopold flirted with metaphysical thinking and moralistic views (Norton 1988), he was basically a pragmatist and anthropocentrist: man had inherited the earth for whatever length of time his species might be here; prolonging that span for as long as possible and ensuring an environment suitable to future human well-being was each generation's responsibility. He never gave up utilitarianism, but he found the pure utilitarian approach to land to be too narrow in scope to accommodate responsibilities to the future. The definition of utility, or commodity value, was too limited by current contexts of time and space to be the sole yardstick for value.

While subscribing to the general philosophy of Gifford Pinchot that conservation was "wise use" and wise meant for the greatest good, for the greatest number, for the longest time, the definition of usefulness was still too constraining. Usefulness was too easily limited to economic value. Most species and many biotic communities have no economic value to our current society. Evidence that Leopold correctly anticipated this problem is revealed in debates over endangered species protection with the question "What good is it (the species)?" and those concerning commercial development of unique biotic communities considered to be economic wastelands.

As was pointed out previously, Leopold wrestled with the incongruities of holistic conservation programming in a democratic, free-enterprize system. Yet his land ethic statement was the greatest incongruity of all in this context. He admitted that modifications to the landscape as a result of management aimed at feeding, clothing, and sheltering a society could not be avoided (Leopold 1949:240). He saw the human species as a member of a biotic team (op. cit.:241) and legitimate on the landscape. Hunting was a natural human instinct (op. cit.:184, 227). He chastised those who could only see the Kansas plains with endless corn fields as tedious (op. cit.:180). The ventures of Native Americans, Spanish explorers, trappers, ranchers, loggers were our heritage, including both the good and the bad. Yet this realist would write the ethical guideline "A thing is right when it tends to preserve the integrity, stability, and beauty of the biotic community. It is wrong when it tends otherwise."

Obviously, we cannot farm or practice commercial forestry or even have normal home sites without interfering with normal functioning of the native ecosystems. Land management in the context of farming, ranching, and forestry channels energy and nutrients into ecosystem components, both wild and domestic, that are important to human growth, maintenance and reproduction, and we have to carry out the process in an economically viable manner. The value of the land ethic is that it prompts us to pursue our ecological and economic well-being with intelligent forethought. Like most of our social ethics, it is an impossible standard to live to perfection. It does not intend to stimulate paranoia although the evidence is that in some cases it may have done so. Leopold was prompted on this potential by Albert Hochbaum's review of an earlier essay, "The Green Lagoons." Hochbaum reminded his former teacher "If we regret [by dwelling on] what we have done, we must regret that we are men. It is only by accepting ourselves for what we are, the best of us and the worst of us, that we can hold any hope for the future" (in Meine 1988:455). Only three paragraphs after the land ethic statement, Leopold wrote in retrospect "Conservation is paved with good intentions which prove futile, or even dangerous, because they are devoid of critical understanding either of the land, or of economic land-use" (Leopold 1949:263).

Hochbaum saw Leopold as probably less a person than he was a "standard" (Meine 1988:456). Norton's deeply philosophical essay "The Constancy of Leopold's Land Ethic" saw him in the same way. It has been my intention to offer a similar perspective. It is not Leopold the person that is important in this context. The focus should be Leopoldian thinking as a model for those who create, orchestrate and implement land-use policies.

Perhaps the immense complexities of our biosphere are beyond the comprehension of the human mind. While we cannot suspend attempts at societal progress and the advancement of civilization until we reach general agreement on "best guesses," no person should be so callous and reckless as to charge toward an uncertain future with impudent disregard for ecosystem "stability, integrity and beauty."

SECTION 2

Strategies for Recovery

Strategies for Recovery of the Red-cockaded Woodpecker: An Introduction

Robert G. Hooper

The first management practices performed intentionally to benefit the red-cockaded woodpecker on public lands probably occurred on national forests and state-owned lands in the mid-1960's (Beland 1971, Shaw 1971). Initially, only cavity trees were protected during timber harvests. This practice was recognized by biologists and most managers from the beginning as primarily a public relations gesture that fell far short of meeting the needs of the species. Soon thereafter, managers were encouraged to leave replacement trees, then to leave uncut buffers of 200 feet around each cavity tree, then to leave foraging areas of 40 acres and then 125 acres, etc. The complexity of management guidelines for the red-cockaded woodpecker evolved steadily, but alas the species continued its overall decline (James 1995).

But we have learned from the failures of the past 2 decades. Yes, we really do have to control mid-story encroachment in cluster sites (Conner and Rudolph 1989, Costa and Escano 1989, Loeb et al. 1992). And yes, the bird really does need an ample supply of trees suitable for cavities (Hooper 1988, Costa and Escano 1989). Simple concepts, but when followed, populations do increase (Hooper et al. 1991). We have also learned that small populations have demographic problems that may preclude their recovery without direct assistance (Walters et al. 1988, Conner and Rudolph 1991, Hooper and Lennartz 1995). A final concept of importance is that it takes time to grow habitat and many areas are 20-40 years short of having an adequate supply of cavity trees (Costa and Escano 1989).

In their chapter, "An Ecological Approach to Recovering the Red-cockaded Woodpecker," Krusac et al. focus sharply on the knowledge gained from past failures and successes in recovery of the species, and offer a complete management strategy that will soon be implemented on the National Forests. The major features of their strategy include landscape level management, increased emphasis on prescribed fire, extended rotations, development of old growth characteristics in forest stands, use of timber harvest methods that mimic natural disturbances, the natural regeneration of forest stands, and direct aid to small populations via augmentation and artificial cavities. Their plan is a giant step towards ecosystem management.

Already we can see the encouraging results of some of these new practices. In the chapter, "Red-cockaded Woodpecker Management on the Savannah River Site: A Management/Research Success Story," Gaines et al. outline the intensive management that resulted in the remarkable turnaround in that population. Only 4 woodpeckers inhabited the area in 1985. Through an innovative program of augmentation of birds from other areas, the translocation of resident birds, and the installation of cavity inserts, the number of birds increased to 39 at 12 clusters in 1992.

There were 26 active clusters in Noxubee National Wildlife Refuge in 1971, but only 16 clusters in 1987. Richardson and Stockie describe how the population increased to 32 clusters in just 6 years in their chapter, "Response of a Small Red-cockaded Woodpecker Population to Intensive Management at Noxubee National Wildlife Refuge".

In the chapter, "Red-cockaded Woodpecker Management Initiatives at Fort Bragg Military Installation," Cantrell et al. outline recent management initiatives at that military base. Fort Bragg is the most intensively used military installation in the world, yet has one of the most significant populations of red-cockaded woodpeckers. Consequently, there are many management challenges.

Reinman discusses the potential value of mixed-aged stand management in "Population Status and Management of Red-cockaded Woodpeckers on St. Marks National Wildlife Refuge, 1980-1992." It is hoped that this approach to forest management along with augmentation and artificial cavities will increase this small population.

In a landmark paper, "Potential Red-cockaded Woodpecker Habitat Produced on a Sustained Basis Under Different Silvicultural Systems" Walker models an array of management strategies for providing cavity trees and foraging habitat. It is clear that the land manager has several options for providing abundant habitat for the red-cockaded woodpecker. These options need to be considered carefully because some strategies are implemented more readily than others, and some provide much more of a harvestable surplus of timber (over and above the needs of the woodpecker) than others.

The issue of red-cockaded woodpeckers on private lands is biologically, politically, economically, sociologically and legally complex. Two seminal chapters in this section discuss these complexities in detail. More importantly, they offer solutions to some of the problems associated with the red-cockaded woodpecker on private lands.

Costa outlines the US Fish and Wildlife Service conservation strategy for private lands in "Red-Cockaded Woodpecker Recovery and Private Lands: A Conservation Strategy Responsive to the Issues." The objectives of the private lands strategy are to:

(1) minimize the loss of genetic diversity for the species;

(2) increase populations on public lands;

(3) develop cooperative relationships between government and private landowners; and

(4) minimize the economic impact of the red-cockaded woodpecker to small private landowners.

These objectives will be met through state-wide habitat conservation plans for small landowners and memoranda of agreements with larger landowners.

In "Integrating Timber Management and Red-cockaded Woodpecker Conservation: The Georgia Pacific Plan," Wood and Kleinhofs describe the first memorandum of agreement with the US Fish and Wildlife Service for management of an endangered species on private lands. This pioneer effort resulted in a totally new way for private landowners and the Service to cooperate to conserve endangered species. Under the agreement, Georgia Pacific Corporation is able to continue its economic activities effectively, while at the same time provide essential habitat for the red-cockaded woodpeckers living on its lands.

The final chapter in the section, "Use of Geographic Information Systems for Determination of Red-cockaded Woodpecker Management Areas," by Lipscomb and Williams, examines the use of GIS for determining habitat allocation around woodpecker clusters. Their strategy for handling allocation problems has special application to private lands where land values can be high and several landowners may be involved.

Various chapters throughout this volume make it very clear that if the red-cockaded woodpecker is to be recovered it will be through active management practices such as those discussed in this section. Indeed it is difficult to foresee a future time when active management in some form is not necessary to ensure the continued existence of the species. However, it is naive to think that 50 years from now and perhaps even 20 years from now, that there will not have been major changes in the management of both public and private lands from what we now envision. There is some truth in the statement made by wildlife biologist and land manager Mickey Beland at the first red-cockaded symposium (Beland 1971):

> Managers usually think of past management practices as being inadequate, poorly conceived, and generally inferior to what is being practiced at the moment. They usually believe that present management reflects good judgment and is adequate to get the desired job accomplished.

Part of the manager's dilemma is that new (and hopefully better) ways of doing the same job are advanced. However, the biggest problem facing the manager is that the job changes—or more specifically the goals of management change. Even though changes will occur in the future, I am convinced the strategies discussed in this section will recover the red-cockaded woodpecker in the context that recovery is currently defined.

An Ecological Approach to Recovering the Red-cockaded Woodpecker on Southern National Forests

Dennis L. Krusac, Joseph M. Dabney and John J. Petrick
USDA Forest Service, Atlanta, Georgia 30367

ABSTRACT: The USDA Forest Service is currently preparing an environmental impact statement (EIS) detailing new management direction for the red-cockaded woodpecker (*Picoides borealis*) on southern National Forests. Three broad areas are discussed: **(1)** the EIS process, including the implications of the National Environmental Policy Act, National Forest Management Act, and the Endangered Species Act; **(2)** key management activities that are critical to recovery of the red-cockaded woodpecker; and **(3)** how these key management activities comprise an ecological approach to recovery of the red-cockaded woodpecker.
KEYWORDS: red-cockaded woodpecker, recovery, National Forest, ecosystem management, restoration, ecological approach, biological diversity

Over the past year the USDA Forest Service made a commitment to manage the National Forests with an ecosystem approach. Three years ago the Southern Region of the Forest Service began developing new management direction for recovering the endangered red-cockaded woodpecker (*Picoides borealis*) (RCW) on National Forest System lands. The Region quickly recognized the best approach to recovering the RCW was through restoring, to the degree practicable, the fire-dependent pine ecosystems in which the RCW evolved.

This paper covers three major topics. The first is an overview of the EIS process. This has little to do with an ecological approach to recovering the RCW, but such a discussion may be appropriate for those unfamiliar with the intricacies of addressing the three major laws with which the Forest Service must comply. The second is a discussion of several key RCW management activities which the Forest Service feels are critical to recovery of the species on its lands. The third is a discussion of how these key management activities comprise an ecological approach to recovering the RCW.

THE EIS PROCESS

The Forest Service has been criticized for not simply "getting on with the program" when it comes to changing it's management direction for the RCW. Unfortunately for everyone involved, including the RCW, it is not that simple. The Forest Service must comply with 3 laws anytime major actions are proposed. These are the National Environmental Policy Act (NEPA), the National Forest Management Act (NFMA) and the Endangered Species Act (ESA). All 3 provide very specific guidelines which must be followed. In addition to complying with each law individually, the Forest Service must ensure there are no conflicts between them. The inherent complexity of these laws, in conjunction with the complexities associated with RCW management, causes the mandatory NEPA document to be relatively complex.

The first step of the NEPA process is the publishing of a Notice of Intent in the Federal Register. This is a summary of the action proposed by the Forest Service. The initial Notice of Intent for our current EIS was published in May 1989. This notice was subsequently revised in October of 1991.

Part of the notice is a request for comments on the proposed action. This is referred to as scoping. The current notice allowed the public 30 days in which to submit comments. In addition to this 30 day formal comment period, comments received relative to the May 1989 notice, as well as those received in conjunction with preparation of the Interim Standards and Guidelines, were also used to help identify significant issues.

The next step of the process is the preparation of a draft environmental impact statement (DEIS). The DEIS identifies the issues and alternatives developed in response to those issues, and describes the environmental and socio-economic effects of each alternative. The DEIS also discloses the preferred alternative.

The DEIS is distributed for a 90 day public review period. It is sent to those individuals, agencies, and groups who have indicated an interest in the proposed action throughout the scoping process. In addition, a notice of availability is published in the Federal Register allowing others to request the document for review and comment.

At the end of the public review period all comments are reviewed and analyzed. Pertinent comments are then incorporated into the preparation of the final environmental impact statement (FEIS). A Record of Decision (ROD) is then prepared. This is the document in which the deciding official selects an alternative and lays the frame work to implement the action. The decision is subject to administrative appeal pursuant to 36 CFR 217. The decision may not be implemented sooner than 30 days after the notice of availability for the FEIS is published in the Federal Register.

KEYS TO RCW RECOVERY

A significant volume of pertinent information has become available since the 1985 revision of the US Fish and Wildlife Service's RCW Recovery Plan and the Forest Service's Wildlife Habitat Management Handbook. The following is a discussion of several aspects of RCW management which the Forest Service feels are key to recovery of the species. Most of these activities are based on knowledge which has become available since 1985.

Artificial Cavities

In 1989 Hurricane Hugo slammed into the Francis Marion National Forest in the coastal plain of South Carolina. The hurricane destroyed 87% of the known cavity trees (Hooper et al. 1990) of what had been described as the healthiest RCW population in existence, and the only population to have a documented population increase (Hooper et al. 1991a). Many people mourned the loss of this population, however, their remorse proved to be premature. The year before Hugo, Carol Copeyon, then a graduate student at North Carolina State University, developed a technique to drill artificial cavities suitable for the RCW (Copeyon 1990, Copeyon et al. 1991). David Allen, a Forest Service biologist, had just developed a different artificial cavity technique that used a prefabricated cavity or insert (Allen 1991). Prior to Hugo, these techniques had been used experimentally. The Hugo catastrophe catapulted this technology to the forefront. The role it played in the miraculous restoration of the Francis Marion's RCW population is described in detail by Watson et al. (1995). In addition, the use of artificial cavities has resulted in increases in the RCW in struggling populations throughout the range of the species (Gaines et al. 1995, Reinman 1995, Richardson and Stockie 1995).

The lack of trees old enough to provide natural cavities for RCW is a significant limiting factor on many National Forests. This shortage was inherited by the Forest Service. The majority of National Forests were heavily cut over prior to their acquisition by the federal government. Therefore, the newly established forests are simply not old enough to provide the necessary older aged trees (Costa and Escano 1989). Artificial cavities will play a major role in sustaining and expanding RCW populations until existing forest stands reach an age to provide adequate natural cavities. These techniques are seen as a short-term measure, as the Forest Service's long range goal is to provide the RCW with habitat in which it can thrive without such artificial measures.

Augmentation

Augmentation is a term used to describe the trapping and translocation of RCW from one locale to another. To date, the translocation of both juvenile males and females has been successful (DeFazio et al. 1987, Allen et al. 1993, Rudolph et al. 1995). A technique to move adult birds is being evaluated by the Southeastern Forest Experiment Station.

Each year the juvenile female RCWs (and less frequently the juvenile males) disperse from their natal cluster in search of a mate (Lennartz et al. 1987, Walters 1989). In large RCW populations with high bird densities, the probability of these young birds finding a mate is high. However, in smaller populations with very low bird density, the probability of a successful dispersal is very low or nonexistent. It is in these situations that augmentation can be used to give mother nature a hand. By trapping juvenile birds of the appropriate sex and releasing them at known single bird clusters, often many miles away, new breeding units can be created, bolstering existing populations.

A new augmentation strategy involves the translocation of both a juvenile male and juvenile female from separate groups, to a previously prepared cluster site. This approach allows the reintroduction of RCW into habitats where the species has been extirpated (Rudolph et al. 1995). As with artificial cavities, the Forest Service sees augmentation as a short term measure until successful natural dispersals can sustain a population.

Habitat Management Areas (HMA)

In the past, Forest Service management of RCW centered around protection of the cluster and an appropriate amount of foraging habitat, usually 125 acres of pine or pine/hardwood forest meeting specific criteria. In large, high density RCW populations, the impact to RCW by other resource management activities was often minimal because these foraging areas frequently overlapped. However, in the smaller populations where active clusters may be several miles apart, other management activities occurred in the areas outside the foraging habitat with little consideration for the RCW. These activities were not malicious acts by Forest Service managers. They were well removed from clusters and should therefore have had no impact on them. The full impact of such activities and the subsequent habitat fragmentation on RCW populations was not fully understood until research conducted in North Carolina (Walters 1989) and Texas (Conner and Rudolph 1991b) was published. This research pointed out the need to manage relatively large contiguous blocks of land (HMAs) in a manner conducive to providing good RCW habitat, if recovery of the species is the objective.

Midstory Control

The impact of midstory on RCW and subsequent cluster abandonment is well documented (USDI 1985, Conner and Rudolph 1989, Costa and Escano 1989, Hooper et al. 1991b, Loeb et al. 1992). Without some form of midstory control, an RCW population is doomed. This critical element of RCW recovery is best achieved and maintained with a good prescribed burning program. Historically, periodic fires caused by nature and Native Americans were the primary midstory control agents (Christensen 1977, Landers 1987, Landers et al. 1989, Foti and Glenn 1991, Runkle 1991). On many National Forests, fire has been excluded from the ecosystem for extended periods of time, allowing tall and dense midstories to develop. In such situations, where fire in effective, an initial treatment using manual, mechanical or chemical means will be necessary.

Management Intensity Levels (MIL)

Large, high density RCW populations are much less susceptible to disturbance within the immediate vicinity of individual clusters than are smaller, widely scattered populations (Conner and Rudolph 1991b, Hooper and Lennartz 1995). This revelation indicated that the degree of management and protection provided RCW should vary depending on the size and trend of the population in question. More intensive management will occur in our smaller populations because smaller populations in general are more vulnerable to extinction (Shaffer 1981, 1987, Gilpin and Soulé 1986, Goodman 1987). The following is an example of how different management intensity levels (MIL) may be prescribed for different size populations:

MIL 1:>250 actively breeding groups with a stable or increasing population trend.
MIL 2:>250 actively breeding groups with a decreasing trend; or 125 to 149 actively breeding groups with a stable or increasing population trend.
MIL 3: 50 to 249 actively breeding groups with a decreasing trend; or 25 to 124 actively breeding groups with a stable or increasing population trend.
MIL 4:<25 actively breeding groups, regardless of trend; or 25 to 49 actively breeding groups with a decreasing population trend.

In this example MIL 1 represents a recovered population. Therefore, restrictions on other resource management within an RCW habitat management area would be minimal. Conversely MIL 4 is a very small population that could disappear at any time. In this situation restrictions on other resource management and the amount of direct RCW management would be greater.

Longer Rotations

The RCW is not an old-growth obligate. However, it does need trees with old-growth characteristics in which to efficiently excavate its cavities. These characteristics include relatively little sapwood and the presence of heart rot, both of which reduce excavation time, and sufficient heartwood to contain the cavity chamber. The occurrence of these characteristics increases with tree age (Conner and O'Halloran 1987, Hooper 1988, Hooper and Lennartz 1991, Clark 1992, Conner and Rudolph 1995a, Rudolph et al. 1995). Intensive timber harvest on most National Forests prior to their acquisition by the federal government, has resulted in an average stand age of 60-70 years. As a rule, these stands are too young to provide suitable cavity trees. Most existing RCW cavities are in relict trees, the trees that were left during the massive timber harvest of the 1920's and 30's. These relicts are often 30-40 or more years older than the surrounding stands. Unfortunately, mortality of these older trees often exceeds recruitment, resulting in a net loss of RCW cavity trees or potential cavity trees (Conner et al. 1991a).

The RCW Recovery Plan (USDI 1985) offered 4 options for providing suitable cavity trees:

(1) lengthened rotations,

(2) by leaving old-growth remnant trees (relicts) well distributed throughout younger stands,

(3) by perpetuating small remnant stands of old growth (recruitment stands) throughout the forest area, or

(4) by a combination of these methods.

The Forest Service, in its 1985 Wildlife Habitat Management Handbook offered its land managers 2 of these options. Longer rotations, 100 years for longleaf pine and 80 years for other yellow pines, could be used. The other option was the perpetuation of small remnant stands of old-growth or recruitment stands. If the latter option had been selected, rotations in the general forest area would have been 80 years for longleaf and 70 years for other yellow pines. The majority of the National Forests chose the recruitment stand option. Recruitment stands were usually the oldest available pine stand within 3/4 mile of active RCW clusters, but few had trees suitable for cavity excavation. These stands were set aside and protected. Not surprisingly this strategy has not been effective in the short time it was used. To date, there has not been a single documented case of a recruitment stand being colonized by RCW, unless enhanced by artificial cavities.

The average age of trees in which RCW have initiated cavity excavation is about 95 years for longleaf and 74-77 years for loblolly (Jackson et al. 1979b). If one considers the ineffectiveness of the recruitment stand concept under current forest conditions, and the fact that most trees in the general forest area were being harvested before reaching these average ages for cavity initiation, it is easy to see the potential for RCW to find potential cavity trees was very low to nonexistent.

Therefore, a major key to recovery of the RCW is the extension of rotations in HMAs, to an age that will ensure an appropriate percentage of the area will be occupied by trees old enough to provide adequate potential RCW nesting habitat at any given time. This also includes setting diameter limits large enough in areas being managed by uneven-age methods to ensure an adequate supply of nesting habitat. Unfortunately, the extension of rotations on paper (as with the earlier designations of recruitment stands) does not result in the immediate presence of older trees on the ground. This is why the use of artificial cavities, previously discussed, will be so critical to recovery of the RCW in the short term.

AN ECOLOGICAL APPROACH TO RCW RECOVERY

The RCW evolved in the fire-dependent pine ecosystems of the southeastern United States. There were millions of acres associated with these ecosystems, which were maintained by periodic fires caused by nature and Native Americans. By the early 1900's, practically all forest lands of the southeast had been cleared for agriculture or had the virgin timber harvested from it. The RCW's once vast habitat had now shrunk to a few scattered islands of suitable habitat. In the 1930's, well intentioned but mis-informed natural resource managers decided that fire in the forest was bad. In the 1940's, Smokey Bear waged a successful campaign to prevent forest fires. This included a significant reduction in prescribed burning as well. The importance of fire in the maintenance of these forests was poorly understood and the subsequent exclusion of fire further contributed to the decline of RCW habitat. Today the role of fire in sustaining these ecosystems is much better understood, and is accepted as a valuable tool by much of the environmental community.

The role of fire in providing RCW habitat is also accepted and considered desirable. The logical next step up from using fire on a regular basis is to use an ecological approach to managing the forest to provide not only habitat for the RCW, but for the many other species of plants and animals associated with the fire-dependent pine ecosystems.

Management at the Landscape Level

A prerequisite to any attempt at ecosystem management is the designation of relatively large areas of land on which this management will be implemented. Margules et al. (1988) and Saunders et al. (1991) discussed how to manage biodiversity in an already fragmented system. The minimum subset of existing remnants required to represent the diversity of a given area must be determined. The HMAs being designated for RCW management provide such areas. Depending on the population objectives, HMAs will range in size from a minimum of 6,150 to 100,000 acres or more. In several situations entire ranger districts and even entire national forests will be included in HMAs. These HMAs will indeed offer the opportunity to conserve biological diversity through ecosystem management, at the landscape level.

The Reintroduction of Fire

As previously stated, fire was the dominant force which sustained the ecosystems in which the RCW evolved (Christens 1977, Landers 1987, Landers et al. 1989, Foti and Glenn 1991, Runkle 1991). Analysis of various historic information indicates fires in the pine and pine-hardwood ecosystems occurred every 3 to 5 years. Over most of the RCW's range, fires could occur at any time of the year, however the period of most frequent occurrence was during the growing season. The goal of reintroducing fire, and specifically growing season fire where appropriate, is the creation of open-park like stands of pine with a lush herbaceous understory. Such stands were often described by early explorers as pine barrens. Ironically, botanists today consider these to be some of the most diverse ecosystems in the southeast. Fire must once again become the dominant force in the maintenance of these ecosystems.

Natural Regeneration

The pre-Columbian forest was regenerated in a variety of ways. Some of the frequent fires were probably of a catastrophic nature and possibly killed mature pines. Hurricanes, tornados and other less severe wind storms have blown down huge acreages of pine forest. At the other extreme is the death of one or a few trees due to insect attack, old age, etc. In all of these situations, the new forest regenerated naturally from residual seed on the forest floor or from seed produced by surrounding trees (Wahlenberg 1946). Within the HMAs, harvest methods which mimic the natural forces will be utilized and natural regeneration will be emphasized.

Older Trees

As previously stated, the RCW is not an old-growth obligate, but does need older trees in which to excavate its cavities. Within the RCW HMAs, rotations will be extended and diameter limits will be increased to ensure an appropriate portion of the general forest area will be occupied by trees old enough to provide adequate numbers of potential cavity trees. In addition, RCW cluster sites, recruitment stands and/or replacement stands will allow a significant portion, approximately 10%, of each HMA to develop old growth characteristics. In many HMAs, scattered or clumped residual trees will be left in areas where timber is being harvested. These trees will reach an age at least twice the rotation for that area, ie., if the rotation is 120 years, these trees will remain in place for at least 240 years. Even in an unmanaged old-growth forest, such as the Wade Tract, a relatively small percentage of the trees reach truly old ages, therefore, the proposed management of the RCW HMAs will closely mimic such systems.

Manage to Mimic Natural Disturbances

The question is frequently ask, "Why do we need to manage the forest for the RCW anyway?." After all, the pre-Columbian forest was not managed. The answer lies in the amount of available suitable habitat. In pre-

Columbian times, a hurricane wiping out 200,000 acres of prime RCW habitat was no big deal. There were still millions and millions of acres available. Today this is not the case. Only small islands of suitable RCW habitat remain. When one of these is hit by a catastrophic storm, the consequences can be severe. In addition, the probability of the coastal plain RCW populations being impacted by storms is quite high (Hooper and McAdie 1995).

Based on the Hurricane Hugo experience on the Francis Marion National Forest, a key to making a forest "hurricane proof" is to ensure a good distribution of age classes (Hooper and McAdie 1995), whether through even or uneven age management systems. If we are to manage the forest and our goal is the restoration of the ecosystem as well as recovery of the RCW, it is logical to utilize harvest methods which mimic natural disturbances. In order to do this, a full range of vegetation management options is needed. The harvest methods which utilize natural regeneration will be emphasized in HMAs, but there are also situations where clearcutting may be desirable. Examples are to restore a more desirable native pine species or to provide habitat for another threatened, endangered or sensitive species. In situations where clearcutting may be desirable, low intensity site preparation methods, such as prescribed burning, will be encouraged. We are fortunate indeed that the RCW is compatible with a wide range of resource management activities and forest uses, as long as its habitat needs are considered.

CONCLUSIONS

The long range management goal within the HMAs is recovery of the RCW, and ultimately the restoration of the associated ecosystems, to the highest degree practicable. This will be accomplished through management at a landscape level, the reintroduction of fire as a management tool, by extending rotations and allowing old growth characteristics to develop, by utilizing harvest methods which mimic natural disturbances, and emphasizing natural regeneration of the forest stands.

It will not be easy to achieve this goal, but it is attainable. There will be costs and there will be trade-offs, both biological and economic. The Forest Service is committed to the recovery of the RCW and it is committed to ecosystem management. Such management may not be a panacea, but in the long term it will provide the most good for the largest number of species, including human-kind.

Red-cockaded Woodpecker Recovery and Private Lands: a Conservation Strategy Responsive to the Issues

Ralph Costa, USDI, Fish and Wildlife Service,
Red-cockaded Woodpecker Field Office, Clemson, SC

ABSTRACT: Although recovery of the red-cockaded woodpecker will occur on public land, significant economic, social, political, legal and biological issues not only surround its presence on private land, but also potentially influence the success of Federal recovery programs. The U.S. Fish and Wildlife Service has developed a conservation strategy responsive to the specific issues related to red-cockaded woodpecker private land losses, economic impacts to private landowners of providing woodpecker habitat, and cooperative conservation efforts between the public and private sectors. A private lands habitat manual, statewide habitat conservation plans, and memorandums of agreement are covered in detail.
KEYWORDS: red-cockaded woodpecker, private lands, habitat conservation plans

Although the presence and treatment of endangered species on private lands has been an issue since the passage of the Endangered Species Act of 1973, as amended (Act), the level of intensity and controversy surrounding this issue has increased significantly since 1991. This growing controversy can be attributed to the national awareness of the northern spotted owl situation, the pending reauthorization of the Act, and the increasing concern over private property rights.

Because of the wide distribution of the RCW (13 southern States) and its significant presence on private lands, there is potential for conflict between endangered species conservation and private property rights. Thus, how to deal with RCWs on private lands has reached a new level of concern for the U.S. Fish and Wildlife Service (Service), private landowners, and other interested parties. Additionally, the Forest Service, through their State and Private Forestry Program, and the southeastern State agencies with wildlife and forestry responsibilities, are directly involved with the issue because of their on-the-ground contacts and land management support programs with private landowners.

Because economic, social, political, legal, and biological issues are commonly involved, resolution of conflicts between landowners' objectives for their land and their legal responsibilities under the Act has been difficult. Specifically, private landowners fall into several different categories of land ownership in relation to their economic resources and the size, location, and management objective of their land. Not surprisingly, existing or "potential" RCW populations on a particular property are also related to the size, location, and management objectives for that land. Recognition and acceptance of these two points formed much of the foundation for the RCW private lands strategy as currently envisioned.

It is the Services' intent to design a RCW strategy for private lands that will address the following specific issues:

(1) the continuing decline and permanent loss of RCWs from small, demographically isolated populations;

(2) the significant economic impacts to small property landowners trying to meet minimum habitat standards for RCWs; and,

(3) the lack of an organized effort to bring private, state, and Federal entities together in the furtherance of RCW conservation and recovery efforts.

The objectives of the RCW private lands strategy are to:

(1) minimize loss of genetic diversity at the species level;

(2) increase RCW populations on public lands, particularly in recovery populations, and on appropriate non-Federal lands;

(3) foster and develop cooperative relationships between and among Federal, state, and private parties with the goal of maintaining and/or establishing additional, relatively large populations beyond those targeted as recovery populations; and,

(4) minimize the economic impacts of RCW habitat retention to small private property landowners.

BACKGROUND

Population Status

Although fairly reliable population estimates exist for Federal and state lands, determining the number of RCWs on private lands remains problematic, particularly for small property landowners. James (1995) conducted a range-wide mail survey on RCW population changes between the early 1980's and 1990. State natural heritage programs, Federal and state RCW biologists, and other knowledgeable individuals, reported a total of 1,017 and 672 active clusters on private lands in 1980 and 1990, respectively, indicating a 34 percent decrease. Eliminating industrial forest populations and South Carolina and Georgia quail plantation populations, drops the totals for 1980 and 1990 to 464 and 350, respectively, a 25 percent decrease. These adjusted totals generally reflect the clusters reported under the "scattered additional records" category and probably represent RCWs located on small, nonindustrial owned, forested tracts.

Several locations only reported data for 1990, undoubtedly due to poor or nonexistent records for 1980. The above totals include all reported clusters for both survey periods, regardless of whether or not a particular location only reported 1990 information. If only those locations in the "scattered additional records" category reporting numbers for both 1980 and 1990 are totaled, the results are 454 and 264, respectively, a 42 percent decrease.

These data do not accurately reflect population trends because data collection efforts and areas covered were not the same during the two periods. However, they do suggest a disturbing and alarming situation for RCWs on private lands. Other biologists (Baker 1995, Carter, et al. 1995, Cely and Ferral 1995, Cox, et al. 1995) investigating long-term RCW population information by state, have also documented declining populations on private land. There is no reason to doubt that the same scenario exists for the other southern states with relict RCW clusters scattered across private lands.

Although the loss of RCWs on larger tracts of private land, like Georgia and South Carolina quail plantations and some industrial forest properties, has not been as rapid as on small parcels, these populations are by no means secure over the long-term. Without comprehensive plans to:

(1) supply future cavity trees and foraging habitat;

(2) maintain open park-like clusters free of hardwood midstory; and,

(3) prevent or mitigate demographic and genetic problems, even relatively large populations on private lands will remain in jeopardy. However, past and present management on some of these tracts, particularly quail plantations, has been and continues to be generally beneficial to RCW conservation. The frequent prescribed burning and the current availability of old trees on these properties has definitely benefited the RCW during the past several decades (Engstrom 1995). Hopefully, these lands will remain under this management scenario and thus provide RCW habitat well into the future.

Reasons for Decline

The decline of RCWs on private lands is primarily a result of habitat loss, degradation, and fragmentation. Habitat loss occurs when tracts are cut for timber management purposes or converted to other land uses. In some cases, landowners are conducting these activities without the knowledge that RCWs occupy the property.

Habitat degradation, while not resulting in immediate loss of habitat or RCWs, ultimately leads to the same conclusion, cluster abandonment and population decline. Degradation of habitat generally takes the form of hardwood midstory development, a result of fire suppression efforts and/or an inadequate or nonexistent prescribed burning program. Suboptimal habitat also occurs when clusters are isolated or fragmented by fields, pastures, timber management practices, or urban development. These situations lead to increased cavity competition (Dr. J.H. Carter III, pers. comm.), increased exposure to predation, and direct habitat loss/fragmentation which may reduce foraging availability and group size in small populations (Conner and Rudolph

1991b). Cumulatively, these impacts may affect reproductive success and, ultimately, the groups' chances for survival.

Even if habitat loss and degradation could be halted and reversed on the remaining private properties containing RCWs, their future survival remains extremely tenuous. The majority of RCWs remaining on private lands today are geographically and demographically isolated from larger, more stable populations. This isolation makes it virtually impossible for individual groups to replace group members, when mates or helpers die or disperse. The result is a spiralling cycle of group deterioration leading to a single bird, generally a male with strong territory fidelity, and then with his death or dispersal, an abandoned cluster and another locality without RCWs. Additionally, small, demographically isolated populations may be susceptible to the loss of genetic variability (Stangel 1992).

PRIVATE LANDS STRATEGY

Assumptions

The following assumptions were made during the development of the RCW private lands strategy:

1. Delisting the RCW will be primarily dependent upon recovering it on Federal lands, including national forests, military installations and national wildlife refuges. While some key state and private lands may play an important role in this species recovery, most private lands will not have a significant role in the future survival of RCWs. The role of private lands in RCW recovery was suggested in the 1985 RCW Recovery Plan (U.S. Fish and Wildlife Service 1985b) which states, "habitat trends and landowner objectives indicate that the most realistic opportunity for managing RCWs will be on Federal and other public forest properties," and "consequently, whatever RCW populations exist on private lands must be considered in peril. The prospective future is extirpation of the species on most private properties unless efforts are implemented to reverse current habitat trends."

2. The genetic variability contained in the small, isolated populations on private lands has unknown value, and thus merits saving and reintroduction into remaining larger populations.

3. There is a minimum effective population size (25 successfully breeding pairs) necessary to maintain short-term (100's of years) genetic viability (U.S. Forest Service 1993).

4. The impacts of providing nesting and foraging habitat to the small, private property landowner can:

(a) be economically prohibitive;

(b) appear socially unacceptable; and,

(c) justify generating tremendous amounts of both anti-RCW and anti-Act publicity. Unfortunately and ironically, in a vast majority of cases, the particular RCW group in question is doomed to extirpation anyway for the reasons previously discussed.

5. Landowners with large (1,000's of acres), contiguous blocks of mature forested land and relatively large (10+ groups) RCW populations are more likely to cooperate with the Service on RCW conservation if they do not have to adhere to what they perceive as overly restrictive minimum habitat standards.

6. Management flexibility is a key component and must be considered throughout the process when dealing with individual and corporate private landowners. Because of differences in economic resources of landowners, RCW population size and location, property value and size, and landowner objectives, treatment of RCW/private land situations must, to some degree, be dealt with on a case-by-case basis within established guidelines.

7. Although most private landowners believe that the Act and its implementation are necessary, they do not believe it is their personal responsibility to provide this social good on their land without just compensation. At the same time, most citizens want to do the right thing and are not interested in intentionally violating the Act.

Components

The private lands strategy contains three programs including:

(1) the Draft Red-cockaded Woodpecker Procedures Manual for Private Lands (Manual) (Costa 1992);

(2) Statewide Habitat Conservation Plans (HCP); and,

(3) Memorandum of Agreement (MOA) and individual HCPs. The different programs have been designed around several variable factors, including tract size, RCW population size, distance of subject population from a recovery population, and landowner objectives. Understanding the relationship of these factors to one another helps direct both the landowner and the Service to acceptable solutions on a case-by-case basis. This system provides landowners some flexibility in deciding which program best satisfies their objective(s), while at the same time safeguarding the well-being of the species.

Red-cockaded Woodpecker Procedures Manual for Private Lands

The Manual's objective is to provide private landowners with descriptions and explanations of the Services' policies regarding RCWs on private lands. It's purpose is to identify and clarify procedures that private landowners should follow to be in compliance with applicable provisions of the Act. It provides RCW habitat management guidance in an understandable, manual-type format.

The Manual includes recommendations for habitat retention standards that if followed will provide adequate habitat for RCW groups and thus, minimize landowners exposure to a potential Section 9 "take" violation under the Act.

The Manual is applicable to all landowners, regardless of acres owned, number of RCW groups on the property, or the location of the RCW population. Its most important feature is the identification of the type and amount of habitat necessary to maintain a RCW group, i.e., cavity trees and buffer, plus the recommended foraging substrate. If these recommendations are ignored, cluster abandonment following habitat alteration could be evidence of a "take" violation, if the abandonment can be connected to habitat modifications that altered cluster integrity or reduced foraging substrate below the standards suggested in the guidelines. The Manual provides baseline habitat guidelines from which all parties should operate. The habitat recommendations provided in the Manual establish the foundation from which other programs, including development and approval of HCPs and MOAs, are built.

The need for this Manual has been, and continues to be, clearly stated by many individuals and organizations involved with, or representing, private timber and development interests. There are numerous reasons why the public has requested information concerning their responsibilities for RCWs under the Act. They include:

1. The belief that the government has the responsibility to administer, enforce, and interpret the Act for the public, while at the same time addressing the rights and concerns of private property landowners.

2. The importance of reliability, predictability, and consistency when competing interests (e.g., timber sale buyers) are bidding on timber sales or designing development projects. If competing interests do not have the same RCW guidance from which to help formulate their projects and identify their associated economic investments and returns, unnecessary legal and perhaps ethical problems could occur. Additionally, and perhaps more importantly, the RCW groups in question may be adversely impacted by implementation of an inadequate habitat retention plan.

3. The fact that the majority of private landowners with RCWs on their property, even if they disagree with the concept of retaining some habitat for RCWs on private lands, are law abiding citizens and are not interested in violating the Act and possibly incurring potential legal problems. However, to avoid a violation they have to know what the Services' RCW guidelines are and thus, they continue to contact the Service for direction and guidance.

The habitat guidelines outlined in the current Manual were developed using:

(1) selected input from 9 RCW biologists;

(2) personal knowledge and experience;

(3) the latest research information available that examines the relationships between RCWs and their habitat.

Since the "Guidelines for Preparation of Biological Assessments and Evaluations for the Red-cockaded Woodpecker" (Bluebook) (V.G. Henry 1989) were issued, the standards for habitat retention on private lands have been drawn from the Bluebook. In the absence of other information, this was a conservative (i.e., "error in favor of the species") and thus, acceptable approach. However, it is generally accepted that the recommendations presented in the Bluebook are adequate to achieve RCW recovery when applied to large populations and/or land areas, i.e., national forests or military installations. Recognizing that private landowners do not specifically have recovery responsibilities for the RCW, but only an obligation to avoid a "take," the Manual attempts to differentiate between habitat related responsibilities for the public sector (i.e., achieve recovery) and private sector (i.e., avoid "take").

Table 1: Comparisons of the Foraging Habitat Standards Provided in the Guidelines for Preparation of Biological Assessments and Evaluations for the Red-cockaded Woodpecker (Bluebook) and the Draft Red-cockaded Woodpecker Procedures Manual for Private Lands (Manual).

	Bluebook	Manual
Acres/RCW Group	125	60-300
Sq.ft. BA/Territory (10" DBH)	6,100	3,000
Sq.ft. BA/Acre	70	10-80
10" DBH Stems/Territory	6,350	2,000-5,000 (16"-10")

Table 1 illustrates the differences between the two standards. If the Manual receives general acceptance by the private sector, it will not only be due to the habitat differences between the Bluebook and it, but also to the flexibility built into the Manual standards. The success of endangered species programs involving private lands will in large measure be dependent upon the ability of the private landowner to integrate his/her land management objectives with the endangered species' needs. Management flexibility and the associated timber management options are key features of the RCW Manual.

Statewide Habitat Conservation Plan

The objectives of statewide HCPs are to

(1) minimize and mitigate the permanent loss of RCWs and their associated genetic diversity from small, demographically isolated populations usually associated with small landholdings;

(2) provide a limited supply of RCWs for designated recovery populations or other public/private lands with commitment to, and management programs for, RCW conservation; and,

(3) relieve the private landowner of the potential economic hardship of managing minor populations of RCWs on small, isolated properties.

The development of statewide HCPs would be a cooperative effort. As a minimum, in most states, the Forest Service, Fish and Wildlife Service, and appropriate applicants (state wildlife and forestry agencies) would be involved in statewide HCPs. Other organizations or agencies, would also be encouraged and invited to participate. These may include state affiliates of national wildlife groups (e.g., The Wildlife Society, National Wildlife Federation), conservation groups (e.g., National Audubon Society, Sierra Club) and forestry groups (e.g., Society of American Foresters); the state forestry association and farm bureau; and other Federal parties (e.g., Department of Defense). Upon completion and approval of a HCP and issuance of the incidental take permit, this program might involve the following actions:

1. Identification of qualified RCW populations (small and demographically isolated) and the associated landowners.

2. Determination of the number of groups, group size and distance from other populations for each affected property.

3. Translocation of RCW offspring (juveniles) to appropriate Federal or non-Federal (state or private) populations. Non-Federal populations would have to be approved under a MOA or some other type of cooperative agreement with the Service before birds could be placed there.

4. Translocation of adults to an appropriate Federal or non-Federal population.

The continuing documented loss of RCWs on small, isolated private tracts is an unfortunate, but inevitable result of:

(1) human population growth and associated development;

(2) inadequate economic resources to implement conservation and management programs; and,

(3) an insufficient land base to support even a short-term viable RCW population. Quite simply, attempting to manage and conserve RCW groups on small, geographically isolated parcels of private land, neither fits into many landowners objectives or makes biological sense, at least from a species recovery perspective.

The birds life history and ecological requirements, including its:

(1) relatively large home range;

(2) dependence on old growth stand components; and,

(3) sociality and cooperative breeding habits, do not lend themselves to small tracts of private land. Saving RCWs and their genetic diversity by translocating juveniles and adults (technology and biology permitting) from small, isolated private tracts to Federal and other non-Federal approved properties is a biologically responsible course to pursue. Fortunately, and perhaps ironically, it happens to coincide with most small property landowners' objectives. The bottom line is, the ecosystem upon which this species depends can neither be found nor created on small private tracts of land.

Hopefully of course, the removal of RCWs from these private lands will result in a net gain on Federal recovery areas or establishment of larger, more secure private populations. Today's RCW technology already permits successful augmentation of single RCW (primarily male) groups with juvenile or adult female birds from other groups (Allen, et al. 1993).

There have also been successful attempts at establishing new groups by moving juvenile males and females from different natal territories to new, unoccupied habitat, provisioned with artificial cavities (Rudolph et al. 1992). Artificial cavity technology (Copeyon 1990, Allen 1991, Taylor and Hooper 1991) provides the opportunity to create new clusters and repair inactive and older abandoned sites. Ongoing research at the Southeast Forest Experiment Station in Clemson, South Carolina is investigating the possibility and feasibility of relocating entire groups from an established territory to a new territory (Dr. Kay Franzreb, pers. comm.).

Although relatively effective, all of these technologies will probably not result directly in the creation of 1 RCW group on Federal land for each group covered under a statewide HCP on private land. However, the demographic and genetic contribution to a small population on Federal land could be significantly greater than 1:1. Knowing that a substantial percentage of private RCW groups are probably single birds or non-breeding groups, we cannot expect success with every translocation effort. On the other hand, permanent loss of RCW groups on private lands is guaranteed if no action is taken. Remaining on an isolated, small parcel of land within a highly fragmented landscape, miles from other RCWs, is a biologically poor choice compared to translocation to a recovery area where RCW management and conservation programs are in place and populations are larger; thus providing opportunities for the social interaction critical to this birds survival.

Statewide HCPs are not biologically appropriate for aggregates of private property landowners who:

(1) support at least a short-term viable population, or

(2) comprise a significant portion of a subpopulation, within a recovery population. The Southern Pines and Pinehurst area in the Sandhills of North Carolina is an example of this situation.

Statewide HCPs are also not appropriate for private property landowners who, individually or collectively, harbor a relatively large RCW population. The quail plantations of southern Georgia, South Carolina, and coastal North Carolina, and some industrial forest lands are good examples of this category. RCW populations in these situations are perhaps best managed via MOAs or with individual HCPs.

Memorandum of Agreement

The objectives of a MOA are to:

(1) increase the acreage of private land habitat under RCW management;

(2) maintain (or in some cases increase) the larger existing RCW populations on private lands; and,

(3) foster and develop cooperative relationships for RCW management and conservation between and among the Federal, state, and private sectors.

The format of a MOA, although somewhat standard, must be flexible enough to cover all aspects of the conservation program for the particular population involved. Similarly, MOA content is dependent upon various factors, which usually differ by population. Some of these factors are ownership, population size and objective, available economic resources, legal requirements, and acreage available. Specifically, a MOA should contain:

1. Purpose and Background: This section covers the rationale behind the MOA and any necessary legal aspects of the agreement. It also identifies the parties involved, their RCW population size, and the acreage available for habitat management.

2. Agreement: This section has several subparts including:

A. The cooperative program which outlines and details the specific responsibilities for each cooperator in relation to population objectives, habitat management, and population monitoring. Other items are covered as needed.

B. Duration of agreement and effective date.

C. Amendment procedures.

D. Termination procedures.

E. Implementation disclaimer and list of responsible officials for implementation of the agreement.

3. Authorities: If appropriate, this section is included to cover the legal background for the agreement and any other legal documentation relative to the particular situation.

4. Signatures: Landowners and responsible officials (i.e., Regional Director, Service) who will approve the MOA.

MOAs will be used to establish cooperative agreements between private landowners and the Service for habitat management and population monitoring for private RCW populations. In this context, MOAs could serve to increase the Service's knowledge about potentially significant (large) private populations, while at the same time allowing the landowner to continue with their existing management program, albeit slightly modified to accommodate RCW habitat needs. The Georgia-Pacific (G-P) (Wood and Kleinhofs 1995) initiative is an example of this type of MOA. Specifically, the G-P/Service MOA will establish a sound management program for the largest remaining Arkansas RCW population, located on G-P's Crossett Forest, in southern Arkansas and northern Louisiana.

All major industrial forest corporations will be invited to join in this private lands RCW management strategy. The Service is confident other industrial forest landowners will participate in MOAs after realizing that:

(1) the commitments for RCW habitat management are not economically unreasonable;

(2) there are marketing and "green" image benefits to participating in a MOA; and,

(3) a MOA provides a legal and economic framework within which operational flexibility and predictability exists.

MOAs are probably the best hope of maintaining the remaining relatively large RCW populations on private lands. Working with cooperative landowners can frequently lead to a management plan that not only satisfies Service recommendations, but also goes a step further by adding additional, or modifying existing, practices or programs that increase the population's survival probability. Extra measures might include increased foraging substrate, establishment of replacement/recruitment stands, and aggressive thinning and/or prescribed burning needed to control hardwood midstory and provide quality foraging habitat.

In some cases, landowners of large private properties will not be able to participate in a MOA because proposed timber management or land development projects would either remove existing RCW habitat or not satisfy the Services' minimum habitat recommendations. Either of these situations could result in some level of incidental take. In these cases, the development of an individualized HCP and the issuance of an incidental take permit by the Service, under Section 10(A)(1)(b) of the Act, could allow the landowner to proceed with his/her proposed project. Currently, the Service is discussing HCPs with several large private property landowners in South Carolina and Alabama.

Relevance to Recovery Plan

The Manual and the other components of the private lands strategy respond specifically to recovery activity number 3, as listed in the RCW Recovery Plan: "encourage protection and management on private lands," as listed in the RCW Recovery Plan. The step-down outline for recovery activity number 3 is presented below. The private lands strategy has been, and continues to be, designed and developed in response to this recovery activity.

3.1. Provide information on management and legal requirements to private landowners and managers. [Manual].

3.1.1. Develop information articles and management guidelines oriented to private lands. [Manual and existing pamphlets].

3.1.2. Distribute information to private landowners and managers through professional and industrial associations. [RCW Symposium III proceedings and ongoing negotiations to publish Manual direction in state and industrial publications when it is issued in final form].

3.2. Develop model cooperative agreements between Federal agencies and private landowners and implement where feasible. [MOA and individual HCPs being discussed].

3.3. Protect RCW habitat on private lands

through easements, acquisitions, and donations. [Tall Timbers Research Station initiative on southern Georgia quail plantations, Environmental Defense Fund initiative in the Sandhills of North Carolina, and others].

3.4. Recognize or reward protection and management efforts. [This will be accomplished upon approval of the G-P MOA and with other individuals and organizations as additional MOA, HCPs, and cooperative management agreements are implemented].

3.4.1. Provide favorable publicity through news media. [same as 3.4]

3.4.2. Provide certificates of recognition.

3.4.3. Explore and implement, if feasible, tax incentive programs. [This concept is frequently raised by the public at RCW meetings; it deserves serious consideration].

CONCLUSION

Given that we will neither find nor create sufficient quantities of the habitat necessary to perpetuate the RCW on most private land, we need a private lands strategy that compliments, and thus contributes to, the recovery program. Building partnerships with the private sector will help accomplish this. Some agreements will involve private land management plans in the form of a HCP or a MOA. Other partnerships will provide a limited supply of juvenile birds for recovery efforts on Federal lands, with eventual removal of the adult RCWs from private lands.

Recovery of the RCW is dependent upon building conservation programs on selected Federal lands that have as their foundation ecosystem scale thinking and planning. The role of private lands in this mission must be kept in perspective biologically and socially if the efforts on Federal lands are to be successful. The private lands strategy after all, is part of the recovery strategy. They are inextricably tied together; dependent on one another for success. RCW recovery on Federal properties is dependent upon broad based public support, which in turn is influenced to some degree, by the Services' policies regarding RCW management on private lands. The private lands strategy presented in this paper, provides a biologically and socially acceptable framework for addressing the RCW/private lands issue, and ultimately recovery of the species.

Integrating Timber Management and Red-cockaded Woodpecker Conservation: The Georgia-Pacific Plan

Gene W. Wood, Consulting Forest Wildlife Ecologist,
Georgia-Pacific Corporation, Seneca, SC 29678
John Kleinhofs, Forest Manager, West-Central Region,
Georgia-Pacific Corporation, Crossett, AR 71635

ABSTRACT: A plan was devised to integrate the conservation of the red-cockaded woodpecker (*Picoides borealis*) RCW into commercial timber management on industrial forest lands managed on a 40-year rotation. The U.S. Fish and Wildlife Service has agreed that while this plan is not identical to the federal guidelines for RCW management on private lands, it is equivalent in its conservation effect. The Memorandum of Agreement on a no-take plan for private lands is a new paradigm in endangered species conservation.
KEYWORDS: Georgia Pacific, no-take plan, endangered species management, private lands, memorandum of agreement

Georgia-Pacific Corporation (G-P) owns approximately 3.5 million ac. (acres) of pine (*Pinus* sp.) and pine-hardwood timberlands in the geographic range of the red-cockaded woodpecker [RCW] (*Picoides borealis*). Timber management consists of both pulpwood and sawtimber rotations with even-aged management and intermediate thinnings prescribed according to forest productivity and economic return assessments. The relative importance of G-P's conservation programs on the southern landscape is demonstrated by recognition that the company owns 10% of the total industrial forest ownership in the South and 5% of the southern pine forest (U.S.D.A.-Forest Service 1988:12-13).

From the early 1970s until the late 1980s, G-P's timber management in habitat occupied by the RCW consisted primarily of protection of cavity trees. In October 1989, the company decided to prepare a proactive plan to integrate timber management and RCW conservation. The first field work began in mid-December 1989 on the Crossett Forest, Crossett, Arkansas. The Crossett Forest was chosen as the focal point for G-P's RCW work because it was known to have more active colonies than the rest of the company's lands combined. Intensive field inventorying and some reproductive success investigation began in early May 1990 and was completed by mid-August 1990. Using these initial field data as background information, the first draft of the plan was presented to the company director of forestry in November 1990.

The G-P plan was presented to the Regional Director, Region 4, U.S. Fish and Wildlife Service (USFWS) in April 1991. Our first conference with the USFWS was in late July 1991. The second and final conference was in late April 1992. The final draft of the plan was agreed upon by the company and the USFWS RCW Coordinator on October 9, 1992. A memorandum of agreement (MOA) has been prepared by G-P attorneys and U.S. Department of Interior solicitors with expectations of final agreement signing by the appropriate officials in April 1993. The MOA states " While the G-P plan is not identical to the USFWS plan for RCW management on private lands, it is equivalent in its conversation effect."

The G-P plan was the first of its kind to be presented to the USFWS. Because this plan was a different approach, the Service had to make sure that no legal or biological considerations were being ignored, and that this new approach was at least as beneficial to the conservation of the species as traditional methodologies, such as habitat conservation planning as provided for under Section 10(a) of the Endangered Species Act of 1973 (ESA). A second problem was the transition of personnel in the USFWS RCW Coordinator position which required the new coordinator to familiarize himself both with a new agency and a new proposal that could have far reaching consequences for RCW issues on private lands.

At this point, however, both G-P and the USFWS believe that the trail has been broken for government consideration of innovative approaches to listed species management on private lands. In the case of the RCW, the USFWS believes that the G-P plan is a generalized model that other forest industry corporations may want to consider when attempting to blend RCW conservation into timber management (R. Costa, pers. commun.).

PLAN PURPOSE

While the ESA does not require the private landowner to participate in recovery efforts for a listed species, and the Red-cockaded Woodpecker Recovery Plan (USFWS 1985:26-27) does not expect such actions, "taking" of a listed species is prohibited under ESA Section 9. Where timber management is concerned, prohibition focuses on the "harm" aspect of the definition of take [ESA Section 3 (19)]. Quarles et al. (1991:14-25) explored the definition of harm as given in 50 C.F.R. Section 1532(19) and related court cases, and warned that any habitat modifications that significantly impair essential behavioral patterns, "including breeding, feeding, or sheltering," of a listed species might be interpreted as constituting harm and thus violate the take prohibition under Section 9.

Consideration of these points of law led G-P to the conclusion that the formulation of guidelines to integrate listed species conservation into timber management would be a prudent measure. In 1989 the federal government's guidelines for RCW management on private lands were limited to Guidelines For Preparation Of Biological Assessments And Evaluations For The Red-cockaded Woodpecker (Henry 1989), including the warning "If this recommendation is ignored, colony abandonment following habitat alteration would be strong evidence of a 'take' violation" (Henry 1989:12).

Examination of the 1989 Guidelines revealed that they were based on the 1985 Recovery Plan. However, the private landowner is not required to implement recovery efforts. Furthermore, the guidelines were not compatible with management of private commercial timberlands. Therefore, if G-P was going to not only abide by the law, but also become proactive in RCW conservation while continuing industrial forest management, a new strategy had to be devised and receive approval from the USFWS.

PLAN OBJECTIVES

With the development and implementation of the RCW plan, it is G-P's intention to demonstrate the company's desire to comply with the spirit of the ESA. Furthermore, G-P expects to demonstrate that the integration of timber management and RCW conservation practices is a proposition that can be both ecologically and economically sound. The biological objective for this demonstration will be no net losses in numbers of active colonies attributable to future timber management in habitats supporting currently demographically viable populations.

HABITAT MANAGEMENT PROCEDURES

The G-P plan was developed as a procedures manual to be used by foresters managing fee lands (company-owned lands) and long-term (≥10 years) lease lands. The plan is composed of five basic topics:

a) definitions of terminology,
b) colony area protection and management,
c) foraging habitat management,
d) management of abandoned colonies and associated foraging habitat,
e) demographic isolation, and
f) multiple-ownerships.

Definitions

The definitions used in the G-P plan are those common to the recent RCW literature with a few notable exceptions. Definitions of active and inactive cavities and cavity-starts are those in normal use, although inactive cavity-starts have not been previously defined in the literature. In the context of the G-P plan they are defined as start-holes with no evidence of recent tree bark scaling or excavation activity, i.e., the walls of the start-hole are "sooty" colored and there has been no resin flow from the bottom of the hole in several years. Another term used in the plan and not common to most of the literature is inactive, unaltered cavity tree. This is a tree that has an unaltered entrance tunnel and the cavity is neither filled with water nor occupied by a RCW cavity competitor species.

The colony area is the area over which active cavity trees, active cavity-start trees, and inactive, unaltered cavity trees are grouped and surrounded by a 3-ch (chains) (198 ft) buffer zone. Minimum size of the colony area is 10 ac. The term "zone of influence" is not standard in the RCW literature, but is used here to mean the area within 0.5 mi (miles) of the geometric center of the colony area. Where the Rectangular Land Survey is in effect, it means the 13 40-ac blocks (forties) closest to the geometric center of the colony area.

Foraging habitat consists of pine and pine-hardwood stands that are at least 25 years old and that have been thinned at least once to a stocking $\leq$100 ft^2 total BA (basal area)/ac (target= 70 ft^2 BA/ac), and burned or otherwise treated to promote a generally open stand structure. Foraging substrate consists of pine trees $\geq$10 inches DBH (diameter 4.5 ft. aboveground).

Demographic isolation describes a situation in which a RCW clan is highly unlikely to exchange genetic material with a definable RCW population. This situation will be constituted when a RCW colony is located more than 3 mi. from the next nearest colony with a breeding pair and the intervening space is fragmented by four or more breaks in foraging habitat that are each >10 ch. wide. If the habitat is continuous, the spatial separation required to constitute demographic isolation is 5 mi. We believe that these definitions are conservative on the side of the species.

Colony Areas

Colony area protection and management begins with discovery of at least one active cavity tree. All trees in the colony are marked with two painted bands that constitute a color code identifying each tree according to its current condition: active, recently abandoned, long-abandoned, or active start. Inactive start-trees are coded the same as those classified as long-abandoned. The boundary of the colony area is marked with flagging and a plowed fire line.

Activities prohibited in colony areas include the following:

1. Entry with logging equipment or felling of trees during the breeding season except for emergency reasons, such as southern pine beetle (*Dendroctonus frontalis*) outbreak, and then only after conference with the USFWS.

2. Removal of trees at any time unless that removal will result in colony habitat improvement.

3. Construction of roads.

4. Skidding or hauling unless no viable alternative exists.

5. Skidding or hauling within 1 chain. of an active cavity or active cavity-start tree.

Proactive management of colony habitat is guided by the following standards:

1. Maintain pine overstory stocking at 50-60 ft^2 BA/ac. Removals that result in a lower stocking may be made only for the purpose of SPB control or risk reduction. The forester is to use discretion in balancing SPB risks and risks of logging damage to existing and potential cavity trees.

2. Remove pine and hardwood mid-stories for colony habitat structure improvement.

3. Use fire and USFWS approved herbicides, singly or in combination, to prevent mid-story development.

4. Rake fuel from the bases of cavity trees whenever resin flow is sufficiently low on the bole to have a reasonable probability of ignition by a prescribed fire.

5. Where active cavity trees are seedtrees in regeneration areas, prevent regeneration development within 50 ft of each cavity tree.

6. Orient management planning to provide a 10-ac replacement colony area when the current colony area phases out due to natural mortality.

Colony management is aimed at creating and maintaining habitat conditions considered optimal according to descriptive data in the 1985 Recovery Plan (USFWS 1985), Conner and Rudolph (1989), and Loeb et al.(1992). Silvicultural practices aimed at SPB control and risk management within colony areas are based on both experience and the data of Conner et al. (1991b). These practices are guided by the recommendations of Belanger and Malac (1980) and Belanger et al. (1988).

Foraging Habitat

Based on numerous observations of RCW foraging in well managed, open (50-70 ft^2 BA/ac) pine stands that were less than 30 years old, and because it was normal to sawlog rotation management to produce such stands by age 25 years, the G-P plan departed from the 1985 Recovery Plan in defining foraging habitat as being ≥ 30 years old. In addition, the current proposal by the USFWS RCW Coordinator for the minimum age standard for foraging habitat is 25 years (R. Costa, pers. commun.).

A second notable departure from both the 1985 Recovery Plan and the currently proposed guidelines for private lands is our maximum distance for separation of foraging stands. The Recovery Plan recommended 5 ch. The new private lands guidelines prescribes 300 ft. The G-P plan standard is 10 ch. While Conner and Rudolph (1989, 1991b) suggested that foraging habitat removal might adversely affect probability of successful dispersal by females, we do not believe that the G-P plan will cause such a problem. Our position is based on numerous observations made of RCWs flying across large open areas (marsh fingers, fields and clearcuts) in the studies reported by Hooper et al. (1982), and Wood et al. (1985a) and continuing research observations made by R.G. Hooper (pers. commun.). While there must be some limit to the open space distance that the RCW either will attempt to cross or is capable of crossing, based on both research observations and other experiences, we feel certain that that distance is >10 ch. The 5 ch standard recommended in the 1985 Recovery Plan was an arbitrarily chosen value, and we feel that it was unnecessarily restrictive.

The second issue raised by Conner and Rudolph (1989, 1991b) concerning foraging habitat removal was forage insufficiency. However, experimental research aimed at testing forage insufficiency caused by clearcutting failed to do so (Wood et al. 1985a, 1985b; Hooper and Lennartz 1995). Furthermore, even though Conner and Rudolph interpreted their own data as suggesting that cutting was causing forage insufficiency problems, forage availability appeared to be lower in RCW populations presumed to be healthy than in those presumed to be unhealthy. The healthy population areas had higher densities of active colonies and more cutting (Hooper and Lennartz 1995).

Conner and Rudolph (1989, 1991b) conjectured that it might take 10 years for the effects of cutting to become apparent. In contrast, Wood et al. (1985b) reasoned that if forage availability became insufficient at some level of cutting, then the problem should be greatest during the nesting season immediately following habitat removal. During the nesting season foraging efficiency must be greater than at any time of the year as the birds have to procure and metabolize energy and nutrients in amounts sufficient to support their own maintenance, breeding activity, egg production, incubation, and growth of nestlings. Therefore, it seems logical to conclude that a forage insufficiency problem should be immediately detectable by lowered reproductive success. Wood et al. (1985b) were unable to demonstrate that reproductive success was lower in the 3 years following clearcutting up to 37% of the foraging habitat than it had been in the 2 years prior to cutting. Hooper and Lennartz (1995) tested the Wood et al. (1985a, 1985b) findings with a far more stringent design, but arrived at the same general conclusions.

Based on the existing biological data on red-cockaded woodpecker foraging habitat and its use, and the need to blend foraging habitat management with timber management on a 40-year rotation, the G-P plan contains the following standards:

1. Foraging habitat stands will be stocked at 50-100 ft^2 BA/ac (target maximum = 70 ft^2 BA/ac) of foraging substrate.

2. At least 10 ac of foraging habitat will be adjacent to the colony area at all times.

3. Total foraging substrate minimum will be 6000 ft^2 BA.

4. At least 4000 ft^2 BA of forage substrate will be ≥30 years old.

5. No more than 2000 ft^2 BA of forage may be 25-29 years old.

6. Distances between foraging habitat stands will not exceed 10 ch.

7. The minimum substrate standards will be maintained within the zone of influence of each colony if adequate amounts of habitat are available at the time of management initiation. If they are not available at that time, the zone of influence will be enlarged until the minimum available standard is met. In addition, no regeneration cutting will be allowed within the zone of influence until the normal standards can be met and maintained.

So long as the standards described above are met, there are no constraints on silvicultural applications by method or timing.

Abandoned Colonies

Management of abandoned colony areas and their associated foraging habitat is dealt with three different ways depending upon the circumstances of the situation. Abandoned colonies that are already abandoned at the time of discovery, are not treated any differently than any other timber area unless, in the judgment of the forester, it would be prudent to hold that habitat for replacement colony habitat. If that is the case, then management aimed at meeting colony habitat standards would be begun.

If a demographically isolated colony becomes abandoned, the colony area and its associated foraging habitat immediately return to management normal to the rest of that particular forest. However, if a colony that is not demographically isolated is abandoned subsequent to discovery, then based on the probability that it might be reoccupied (Doerr et al. 1989), the colony and foraging habitat standards are maintained for 5 consecutive years. If the colony is not reoccupied within this time, management reverts to that normal to the rest of that particular forest.

Multiple Ownership

The G-P plan follows the principle of proportional responsibility where there is multiple ownership of RCW habitat and as described in the USFWS 1989 Guidelines (Henry 1989:12). However, the company does not accept the responsibility for informing neighbors of their proportional responsibility for habitat to support an RCW clan that is roosting in a colony on G-P land. In such cases, the company informs the nearest USFWS field office as to the legal description of the location of the particular colony area and that G-P is proceeding with its proportional share of responsibility to provide habitat.

G-P does not accept responsibility for determining the presence of RCW colonies on neighboring lands and that may be located within 0.5 mi of the common boundary. However, if the company is informed of such a situation, it will proceed with meeting its proportional responsibility.

RCW STATUS ON G-P LANDS

Our best estimate of the number of currently active RCW colonies on fee lands is 112 (**Table 1**). The number on the Crossett Forest is the hardest to track because both discoveries of new colonies and colony abandonments are occurring. The minimum estimate for the Crossett based on current district forester reports is 85 active colonies. Another 11 colonies are known to exist on the Fordyce Forest also located in Arkansas. There are nine active colonies in central Mississippi and four in the southern portion of that state. There are three active colonies in the upper coastal plain of South Carolina.

The total acreage currently involved in G-P's RCW conservation program is 56,806 ac (**Table 1**). Within this area, 1,576 ac is currently managed to meet colony habitat standards.

MONITORING PROCEDURES

Each district forester is responsible for locating all RCW colonies on his district, implementing appropriate management procedures, and maintaining appropriate file data. While annual checks of the colony status are not required, this procedure is encouraged. Colony status must be checked prior to prescribing regeneration cutting in a zone of influence. In addition, the file showing current amounts of forage substrate within each zone of influence must be updated annually. The district foresters are also encouraged to record observations of numbers of adult birds seen in colonies and signs of nesting activity.

A 4-year study of nesting success on the Crossett is currently in progress. Procedures involve checking colonies to determine the number of adult RCWs present after March 15 and before April 15. From the population of colonies with at least two adults present, 25-35 colonies are monitored for nesting activity beginning about April 15. Monitoring entails determinations of nesting attempts, clutch size, hatching success, and fledging success. The current study will be completed in 1994.

PERMITS AND AGREEMENTS

In addition to the USFWS-G-P MOU on the RCW plan, several other agreements are either already in effect or they are being developed. An informal conference with the USFWS in 1992 led to clearance for the company to use the herbicides Arsenal, Accord, and Escort for vegetation management in colony areas in Arkansas, Mississippi, and Louisiana. These herbicides may be used singly or in combination and applied according to label

Table 1. Red-cockaded woodpecker status on Georgia-Pacific lands in fall 1992.

Location	Number Active Colonies	Habitat Management Area(ac)	
		Colony	Foraging
Crossett, Ark.	85	1,249	43,890
Fordyce, Ark.	11	167	5,890
Gloster, Miss.	3	30	1,550
Taylorsville, Miss.	1	10	520
Louisville, Miss.	9	90	2,100
Alcolu, SC.	3	30	1,270
Total	112	1,576	55,220

restrictions.

We are developing a MOA for a cooperative effort with Noxubee National Wildlife Refuge in Mississippi. Five individually demographically isolated colonies are located on G-P lands in Noxubee County. The company has agreed to allow the refuge biologist to translocate juvenile birds from these colonies to augment the Refuge population. The refuge biologist may also experiment with translocation of adult birds under USFWS authority and responsibility and G-P agreement. G-P will continue to manage these colonies and their associated foraging habitats so long as they remain active.

A cooperative study is being developed with the USFWS as part of the G-P RCW MOA. The study will be on the effects of foraging habitat removals to the maximum levels allowed by the new USFWS guidelines proposed for private lands. Under this cooperative study agreement, G-P will give the USFWS a 5-year grant to study the population dynamics in a demographically isolated population consisting of four active colonies on a 600-ac area following shelterwood regeneration cutting. The study will be in Winston County, Mississippi.

PERSONNEL TRAINING

In order to proceed with the integration of RCW conservation and timber management, G-P timber managers are being trained in RCW biology, terminology, management concepts, habitat components recognition, and how to implement the G-P plan. Training sessions involve one full day of which one-half is spent in the classroom and the other half in the field. In addition to training focused on the G-P plan, the foresters are also familiarized with the ESA and Federal guidelines for private lands. Training in the latter area is of particular importance for procurement foresters. Approximately 400 G-P personnel have been trained in 17 of these training sessions.

SUMMARY

In Fall 1989 G-P initiated development of a strategy aimed at blending RCW conservation with industrial forest management. The magnitude of this effort represented a pioneering venture for an industrial forest landowner as well as a major shift in company philosophy about non-commodity resources on industrial forest lands.

Two major benefits have accrued from the development of the G-P RCW plan. First, a generalized model that demonstrates the feasibility of blending RCW conservation and industrial forest management, in a manner that is both ecologically and environmentally sound, has been developed and deployed. Second, extensive collaboration with the USFWS during the development of this strategy has resulted in a totally new means for cooperation between private landowners and the federal government for threatened and endangered species conservation on lands primarily devoted to economic use.

Hopefully, the G-P experience will encourage other forest industry companies to explore this type of conservation effort. G-P believes that it is providing evidence that profitable economic land-use need not preclude meaningful conservation measures.

Red-cockaded Woodpecker Management on the Savannah River Site: A Management/Research Success Story

Glen D. Gaines, USDA Forest Service, Savannah River Forest Station, New Ellenton, SC 29809
Kathleen E. Franzreb, USDA Forest Service, Southern Research Station, Clemson University, Clemson, SC 29634
David H. Allen, North Carolina Wildlife Resource Commission, New Bern, NC 28560
Kevin S. Laves, USDA Forest Service, Southern Research Station, Clemson University, Clemson, SC 29634
William L. Jarvis, USDA Forest Service, Savannah River Forest Station, New Ellenton, SC 29809

ABSTRACT: In December, 1985 the RCW (*Picoides borealis*) population on the Savannah River Site, South Carolina had declined to a low of 4 birds (1 breeding pair and 2 single males) comprising 3 clusters. At that time, the Department of Energy, Southeastern Forest Experiment Station, and Savannah River Forest Station began a concentrated management/research partnership to establish a viable population of RCWs on the Savannah River Site. In December 1992, the population had increased to 39 birds (7 breeding pairs) in 12 clusters. The increase is attributed to an extensive program of nesting habitat restoration, translocations, artificial cavity installation, cavity competitor control, and monitoring. The Savannah River Site RCW Management Plan outlines a short-term strategy to achieve 30 active clusters by the year 2000 and a long-term objective of 400 clusters.
KEYWORDS: RCW, viable population, Savannah River Site, management, research, *Picoides borealis*

The Savannah River Site (SRS) comprises 80,269 ha (198,344 acres) in Aiken, Barnwell, and Allendale Counties in South Carolina. The SRS produces nuclear materials for the national defense of the United States. It has been administered by the Department of Energy (DOE) since 1951. Through an Interagency Agreement with DOE, the USDA Forest Service - Savannah River Forest Station (SRFS), has managed the natural resources on the SRS since 1952. The SRFS administers and implements the SRS Wildlife, Fisheries, and Botany Management Program (DOE 1991), a comprehensive program designed to address both traditional wildlife values as well as current issues in conservation biology. The management of threatened and endangered species is a major component of the program. Management efforts have concentrated on the red-cockaded woodpecker (*Picoides borealis*), (RCW) 1 of the 6 federally listed threatened and endangered species on the SRS.

In December 1985, the SRS RCW population had declined to a low of 4 birds, a breeding pair and 2 single males. Under the guidance of the Southeastern Forest Experiment Station (SEFES), a research and management program was developed to establish a viable RCW population on the SRS. The program was approved by DOE in 1985. The initial Interagency Agreement, identified major activities that would reverse the population decline. Among these were extensive monitoring, translocation of birds, genetic surveys, flying squirrel (*Glaucomys volans*) monitoring and control, and nesting habitat improvement (DeFazio and Lennartz 1987). Artificial cavity technology emerged as the program proceeded.

This is a cooperative project of the

Department of Energy, USDA Forest Service - Southeastern Forest Experiment Station, and the USDA Forest Service - Savannah River Forest Station. Early genetic survey data was collected and analyzed by the Savannah River Ecology Laboratory (SREL) - University of Georgia. Currently the USDI Fish and Wildlife Service - Cooperative Fish and Wildlife Research Unit at Clemson University is providing the genetic analysis and related population management strategies.

MANAGEMENT/RESEARCH PROGRAM (1985-1992)

Habitat Improvement Program

Suitable red-cockaded woodpecker nesting habitat is found in mature stands of southern pine (*Pinus* sp.), preferably longleaf pine (*P. palustris*) forest, with certain characteristics (Hooper et al. 1980). Pine basal areas should be relatively low (11.5 to 18.4 m^2 per ha {50 to 80 square feet per acre}). There must be little or no mid-story, but a herbaceous and/or shrub understory is common in good red-cockaded habitat. To create or maintain these conditions requires an extensive habitat improvement and/or management program that includes frequent prescribed burning. Since 1985, enhancement activities have concentrated on active and inactive clusters and recruitment habitat within 4.8 km (3 miles) of active clusters. During the first 4 years (1985-1988), the restoration and maintenance of suitable nesting habitat in the active and inactive clusters was emphasized. Over the past 4 years (1989-1992) nesting habitat restoration shifted to previously unoccupied recruitment stands within 4.8 km (3 miles) of the active clusters. Improvements since 1985 have included mid-story treatments, commercial thinnings, and prescribed burning.

Since 1985, 21 active and inactive clusters containing 190 ha (470 acres), and 46 recruitment stands containing 589 ha (1,456 acres), have been treated to control midstory vegetation (**Figure 1**). While the amount of treated acreage totals 779 ha (1,926 acres), some sites required repeated treatments so that a total of 1,254 ha (3,099 acres) has actually been subject to midstory control treatments. Midstory control included a combination of mechanical and herbicide treatments, designed to remove encroaching, mid-canopy hardwood species. Specific treatments included broadcast applications of hexazinone (pronone granular and Velpar L liquid), cutting stems with chainsaw or brushsaw, and cutting stems followed by herbicide stump treatment with triclopyr amine (Garlon 3A).

Commercial thinning was used primarily to reduce intermediate and co-dominant pine and hardwood in the overstory, so that the remaining pine basal area comprises 13.8 to 18.3 m^2 per ha (60 and 80 square feet per acre). The pines that were cut were thought to be unsuitable as future cavity trees.

About 2,428 ha (6,000 acres) a year were prescribed burned to improve and maintain nesting and foraging habitat for the RCW. From 1985 to 1990 the emphasis was on burning small areas to improve nesting habitat. Since 1991, larger areas have been burned to improve nesting, as well as foraging habitat.

Population Monitoring Program

Since 1985, the SEFES has monitored population size and makeup, reproductive success, mortality rates, and factors limiting population expansion. All adults and nestlings receive a U.S. Fish and Wildlife Service aluminum leg band with a unique color leg band combination to ease identification in the field. All woodpeckers translocated from other populations to the SRS were banded prior to release.

Group checks are conducted monthly to determine the identity and location of all individuals. More intense weekly checks occur during the breeding season (April through July) to determine the location of nests; the identity of breeding adults; the status of helpers, auxiliaries, and aliens; and the number and sex of nestlings. All inactive cluster sites are visited at least once a year for evidence of activity (Allen et al. 1992).

Using the extensive monitoring data for the SRS population and blood samples collected by Peter Stangel, of the SREL, Haig et al. (1992) conducted a series of population and pedigree analysis to examine the viability of the SRS population. They used the OGENES gene drop pedigree analysis to measure genetic diversity of the 1990 population. Population viability over the next 200 years was simulated with the VORTEX Monte Carlo model. The model predicted that the SRS population has a low probability of persistence. The chance for extinction under various scenarios was estimated between 68 and 100 percent during the next 200 years, if birds from other donor

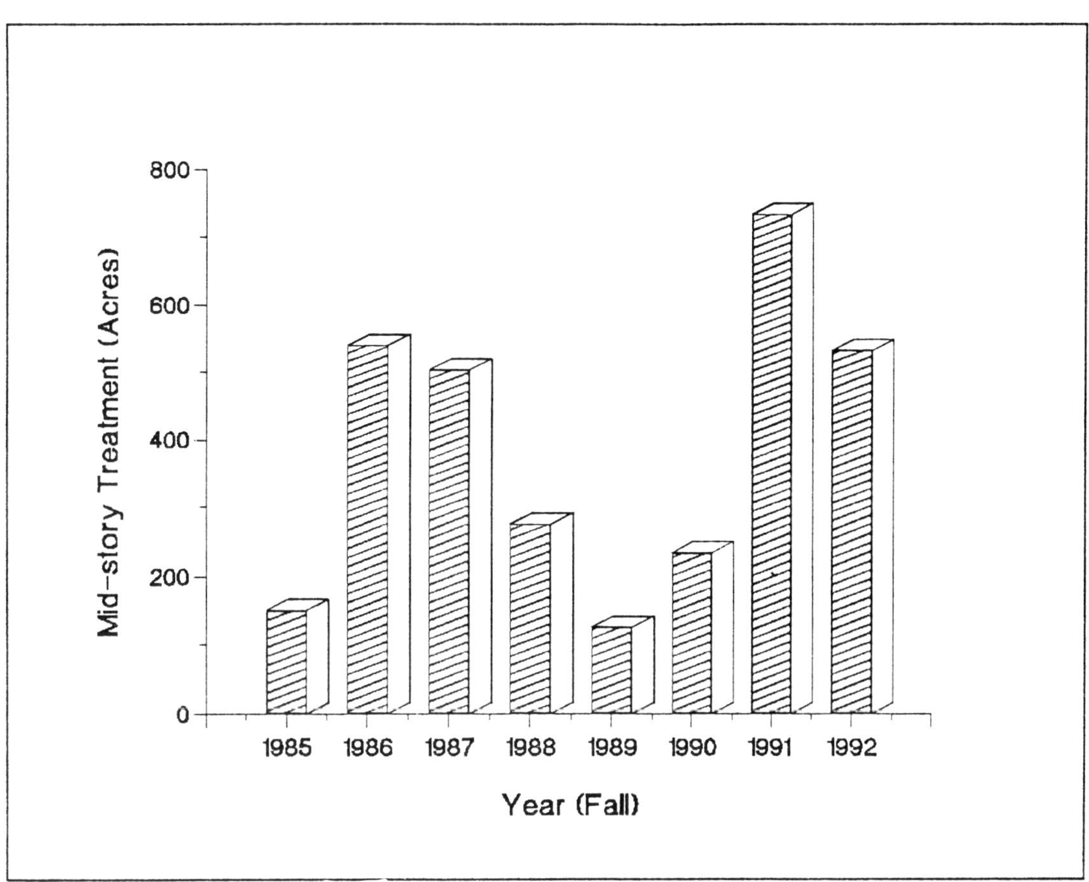

Figure 1. Acres of red-cockaded woodpecker nesting habitat improvement (mid-story removal) from 1985-1992, for the Savannah River Site, South Carolina.

populations are not translocated to the SRS in proper quantity (Haig et al. 1993b).

Translocation Program

It was apparent in early 1986 that without some intervention, the SRS population was moving rapidly toward extinction. In 1986, the SEFES began to translocate birds, as an emergency measure, to reverse the SRS population decline. They tried various strategies, including moving yearling females to bachelor males, moving a female and male to unoccupied habitat, moving a family unit (mated pair and nestlings) to unoccupied habitat, and cross-fostering nestlings (**Figure. 2**).

From 1986 to 1992, a total of 19 RCWs were translocated to the SRS from other populations. These individuals included 13 (7 females, 1 male, and 5 nestlings) from the Francis Marion National Forest in South Carolina, 5 females from the Apalachicola National Forest in Florida, and 1 female from Fort Bragg, North Carolina. Of the 19 individuals translocated to the SRS, 8 (7 females and 1 male) became part of the SRS population (Allen et al. 1991, Haig et al. 1993b, Laves 1992). This total does not include the 3 females moved to the SRS from the Apalachicola National Forest in the Fall, 1992, because the success of these translocations has yet to be determined.

In addition, 16 individuals (7 males and 9 females) from the SRS have been translocated within the SRS population. These moves were designed to provide a mate for an unpaired bird, to establish a new breeding pair, or to create favorable situations for translocations from other donor populations.

Artificial Cavity Inserts

The presence of cavities has been viewed as the single most important component of RCW territories (Ligon 1970). Because natural excavation of cavities is a slow process, requiring one to several years to complete

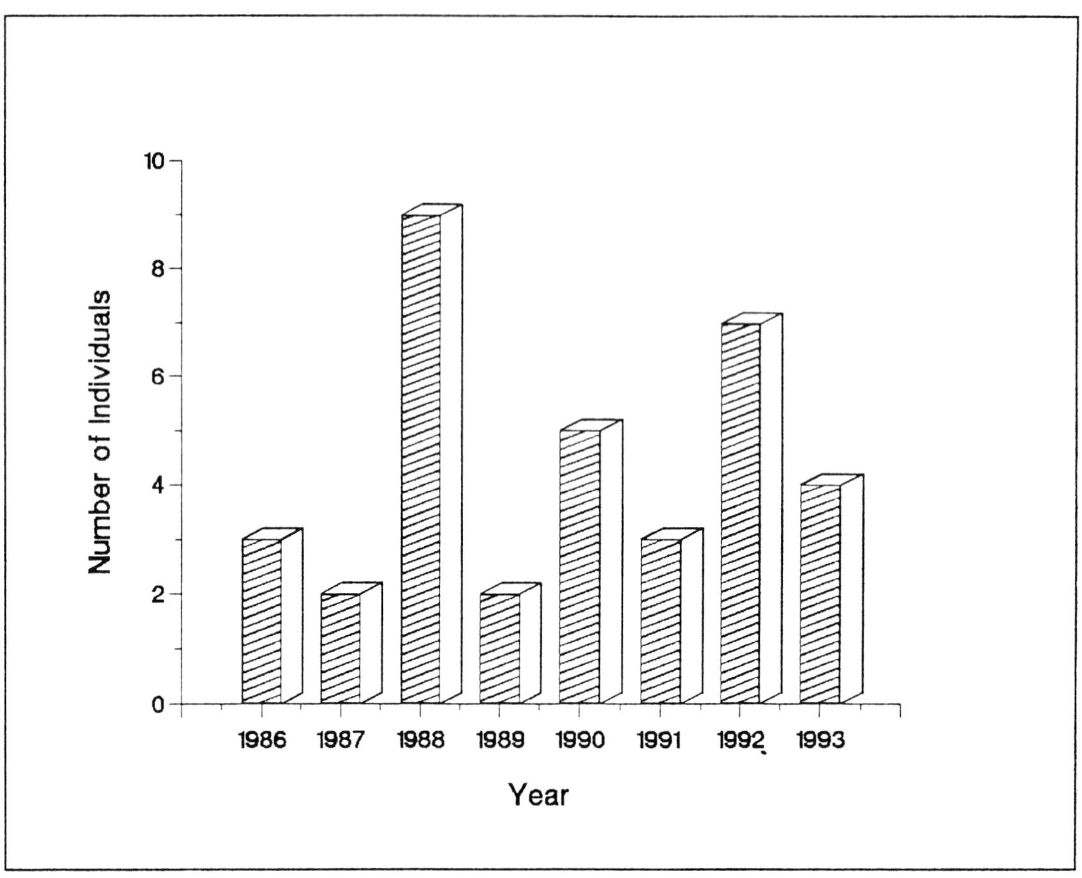

Figure 2. The number of red-cockaded individuals translocated from donor populations and from within the population from 1986 to 1993 on the Savannah River Site, South Carolina.

(Hooper et al. 1980), relying on it to expand small populations may not be practical. Further, with the limited number of potential cavity trees and lack of cavities at the SRS, recovery efforts would be greatly limited.

To compensate for this limiting resource, Allen (1991) developed a technique for constructing and installing artificial cavities (cavity inserts). The technique involves cutting out a rectangular hole and placing a prefabricated cavity insert in the hole. Details of the technique are found in Allen (1991). The advantage of using this technique is that the entire insert can be placed in the sapwood of a tree -- an important consideration on the SRS where old trees containing enough heartwood to drill cavities (Copeyon 1990) are limited. This technique was implemented on the Francis Marion National Forest beginning in 1989 after Hurricane Hugo. To date, 828 artificial cavities (356 inserts, 280 drilled cavities, and 192 drilled start holes) have been installed on the Francis Marion National Forest. The artificial cavities contained 43 to 64 percent of the RCW nests from 1990 to 1992 (Watson et al. 1993).

On the SRS, 101 cavity inserts were installed by personnel from SRFS and SEFES since 1990. By the end of 1992, 18 inserts were being utilized as nest or roost trees by RCWs. In comparison, 17 natural cavities were utilized as nest or roost trees **(Figure. 3)**. The priority for locations to install cavity inserts was

(1) active clusters containing few natural cavities to support existing woodpeckers or offspring,

(2) recruitment stands, within 3 miles of active clusters, identified as eminent translocation sites, and

(3) other recruitment stands within 4.8 km (3 miles) of active clusters.

Competitor Control

Because of the small number of available cavities, it was important to minimize use of the cavities by competitors. Biologists from SEFES regularly climbed RCW cavity trees to

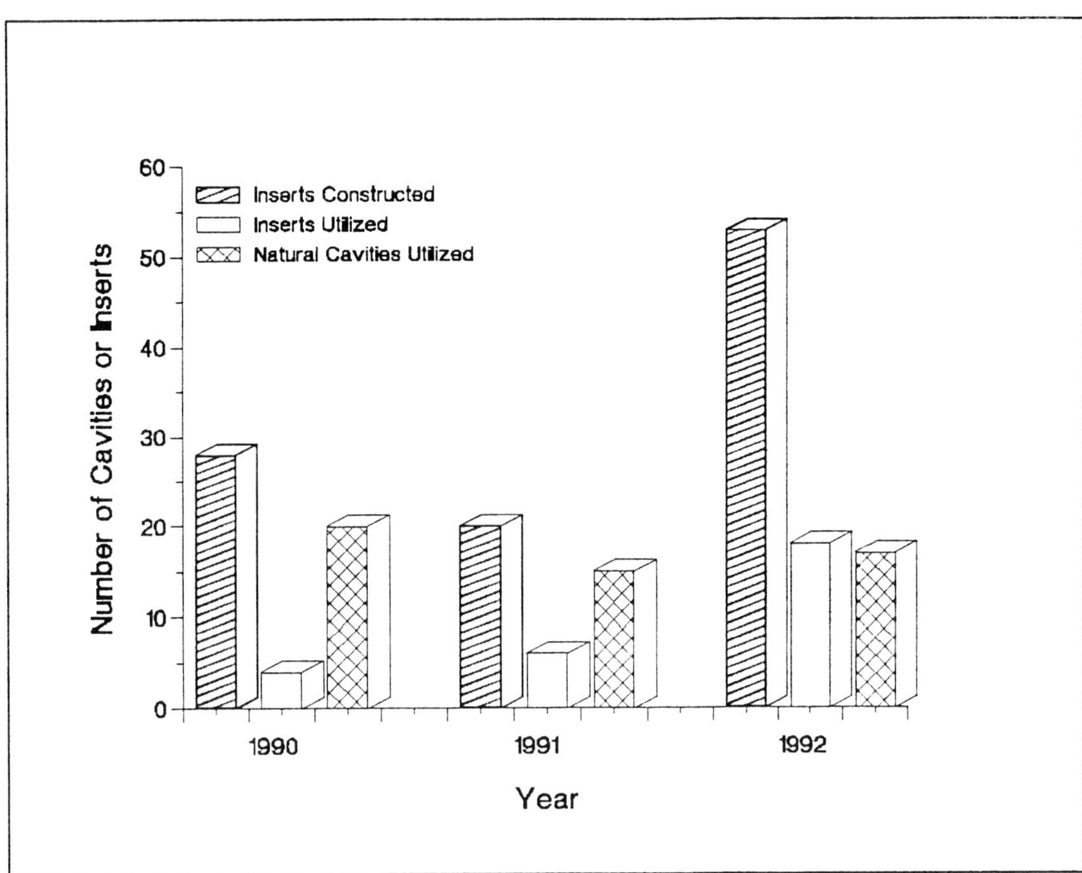

Figure 3. The number of artificial cavity inserts installed, along with the utilization of both artificial and natural cavities from 1990 to 1992 on the Savannah River Site, South Carolina.

determine the presence of flying squirrels. To discourage use of woodpecker cavities, squirrel nest boxes were placed at the base of some cavity trees in the hope a squirrel will enter the first suitable cavity it encountered. Squirrels found in the cavities or boxes during regularly scheduled checks were removed and killed.

Existing RCW cavities were subject to appropriation by the pileated woodpecker (*Dryocopus pileatus*). Pileated woodpeckers tended to enlarge the entrances to both natural and artificial cavities, making them unsuitable for future use by RCWs. As a preventive measure, each cavity on the SRS was fitted with a cavity restrictor (Carter et al. 1989; Allen 1991), a metal plate that fits around the cavity entrance to prevent enlargement by other woodpeckers.

PROGRAM RESULTS

The SRS RCW population responded well to the management and research actions, increasing from 4 birds in December 1985 to 39 in December 1992 (**Figure. 4**), an 875 percent increase in population size. During the same period the number of breeding pairs increased from 1 to 7, and the number of active clusters from 3 to 12 (**Figure. 5**). The 12 clusters were located in 2 subpopulations separated by approximately 16 km (10 miles). Each contained 6 active clusters.

The positive response of the SRS RCW population can be directly attributed to the management and research strategies that have been described. Although all elements of the program were necessary for its success, translocation and artificial cavity inserts were the key components responsible for the increase in this small population. Presently 87 percent of the birds (34 of 39) present on the SRS have been translocated (8) or are descendents (26) of translocated birds (Laves 1992). 5 of the 12 active clusters rely totally on artificial cavity inserts. 2 of the 12 active clusters are formally unoccupied recruitment stands that became active after the installation of cavity inserts.

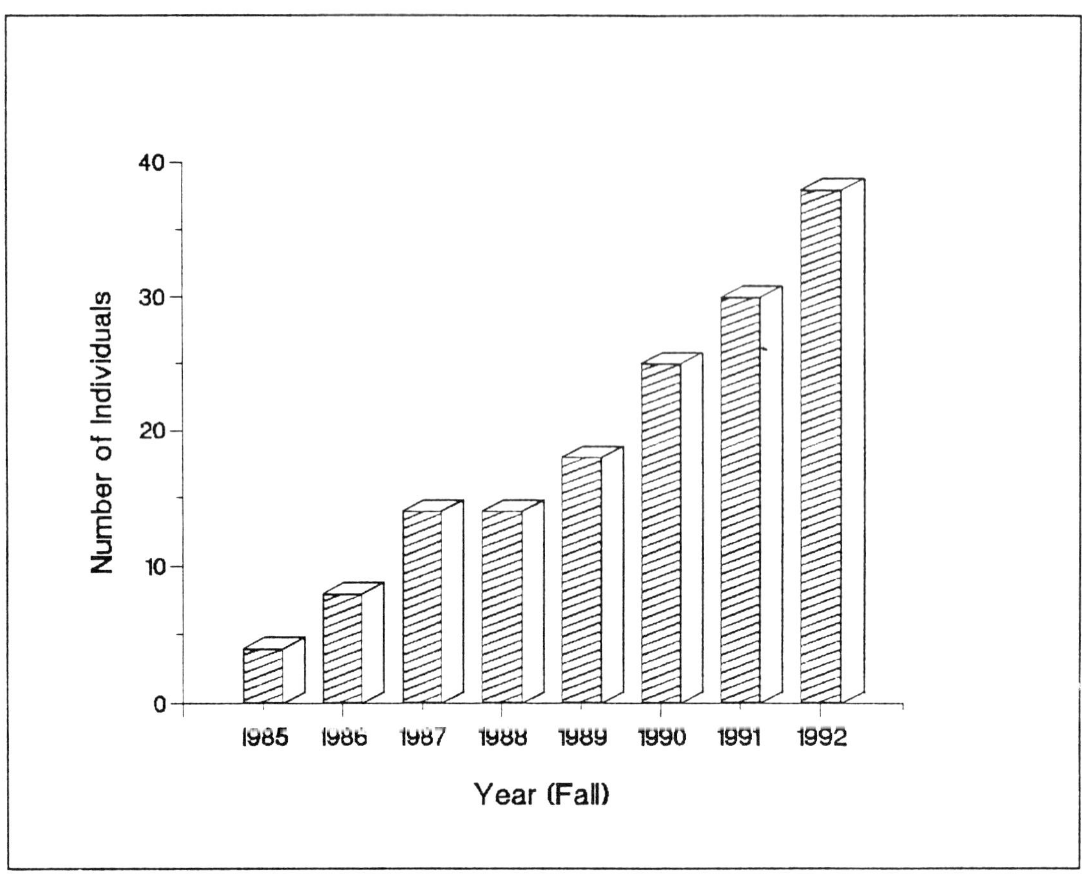

Figure 4. Red-cockaded woodpecker population trend shown as number of individuals from December, 1985 to December, 1992 on the Savannah River Site, South Carolina.

FUTURE DIRECTION

In 1990, in response to continued evidence of regionwide declines in RCW numbers and growing sentiment from managers and scientists that new management strategies needed to be developed, the DOE requested the SRFS tó develop a new management plan to recover the RCW on the SRS. In November, 1991, the Savannah River Site RCW Management Plan (USDA Forest Service 1991) was completed, submitted for formal review, and approved by the U.S. Fish and Wildlife Service. In this plan, a management area was identified that contained approximately 45,326 ha (112,000 acres) of forest habitat, of which approximately 31,971 ha (79,000 acres) were pine forest. The strategy provides management direction for the short-term (next 10 - 20 years) as well as for developing a forest that will support the long-term population objective for the SRS. The strategy was incorporated into the Wildlife, Fisheries, and Botany Operations Plan (USDA Forest Service 1992) that provided specific direction for managing the wildlife management program on the SRS.

Short-Term Direction

The short-term goal is to achieve 30 active RCW clusters by the year 2000. Accomplishment of this goal will require extensive population monitoring, translocations, installation of artificial cavities, competitor control, and nesting habitat improvements within 4.8 km (3 miles) of active clusters.

A minimum of 5 individuals (3 females and 2 males) per year will be translocated from outside donor populations, as recommended by Haig et al. (1992). These annual additions will increase the probability of the long-term survival of the SRS population to 96 percent. Continued within population translocations will occur whenever appropriate opportunities arise.

Artificial cavities will be installed at a rate of 3 to 4 per 4 ha (10 acres) of recruitment area within 4.8 km (3 miles) of active clusters and in

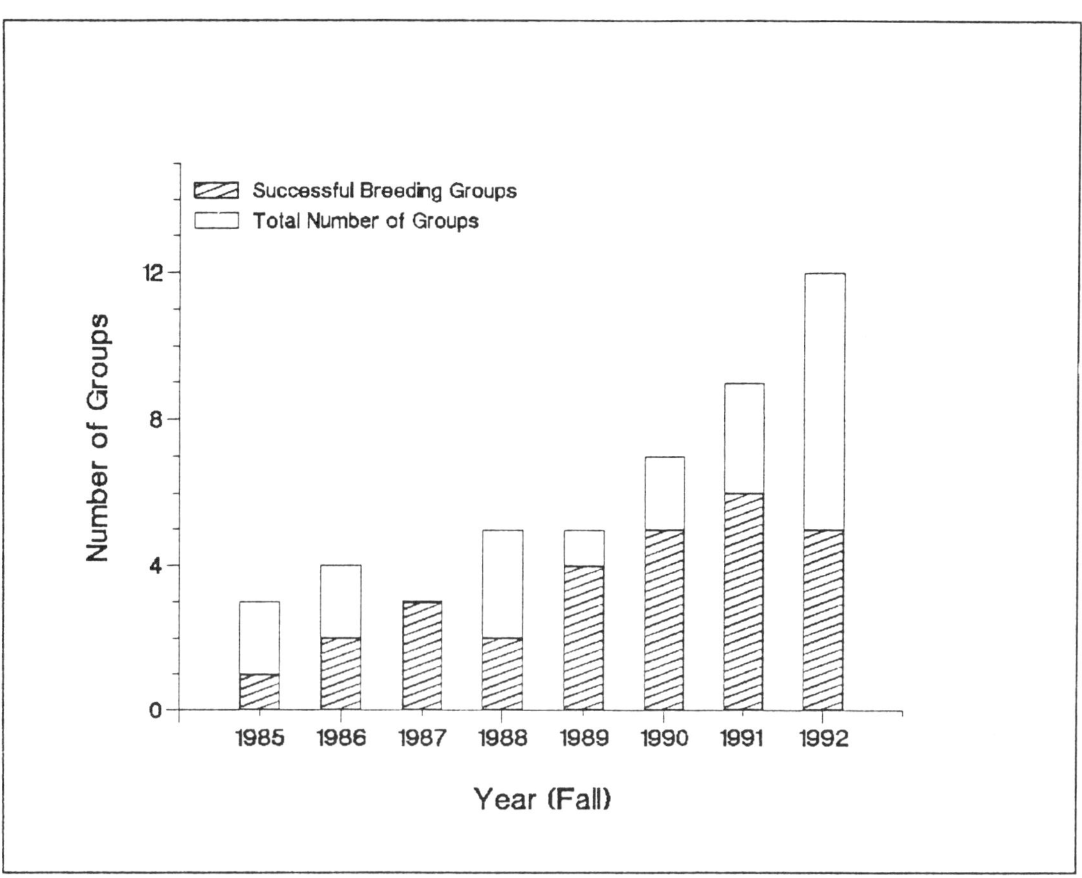

Figure 5. The Red-cockaded woodpecker population trend, shown as number of groups and breeding groups from December, 1985 to December, 1992 for the Savannah River Site, South Carolina.

other areas if needed for translocations. Installing cavities in recruitment areas will increase the likelihood of recruiting dispersing individuals from active clusters. It will also increase and accelerate options for establishing new clusters in unoccupied recruitment stands.

Recruitment stands are being selected at a rate of one per 81 ha (200 acres) of pine habitat, from the oldest available stands, with longleaf as the preferred species. Although the minimum size for recruitment stands is 4 ha (10 acres), the emphasis is on selecting larger areas from 10.1 to 16.2 ha (25 to 40 acres). Foraging habitat will be provided within 0.8 km (one-half mile) of clusters or recruitment stands. This habitat will consist of 6,350 pine stems that are 25.4 cm (10 inches) or greater in diameter at breast height and 790 m^2 (8,500 square feet) of pine basal area in stands greater than 30 years old.

Long-Term Direction

The long-term goal at the SRS is to have 400 active RCW clusters. This objective was based on one cluster per 81 ha (200 acres) of pine habitat in the RCW Management Area. The SRS Management Plan describes forest management standards and guidelines that will provide a forest landscape capable of sustaining a minimum of 400 groups of woodpeckers. Highlights of selected standards and guidelines are:

- A maximum of 9.5 percent of the regulated pine forest in early successional habitats (0-10 years old) will be provided across the RCWmanagement area. Rotation lengths will be 120 years for longleaf pine and 80 years for loblolly pine.

- Between 4 and 14 percent of the regulated pine forest may be in early successional habitat (0-10 years old) at the compartment level (i.e.

project planning level).

- Areas for regeneration harvests will range from 4 to 16 ha (10 and 40 acres), and will average less than 10 ha (25 acres).
- Regeneration harvest will be designed to minimize fragmentation of foraging habitat;
- Priorities for regeneration will be to convert off-site slash pine to longleaf pine and to retain existing longleaf pine stands.
- Relict and potential cavity trees will be retained in regeneration and intermediate harvest cuts.

If these strategies are implemented as planned, at least 20 percent of the RCW management area will be in suitable nesting habitat (longleaf pine greater than 80 years old).

CONCLUSION

Since 1985, the SRS RCW population has increased from 4 birds in 3 clusters to 39 birds in 12 clusters. All components of the management/research program have been essential for success. Vital program components include habitat improvement, translocation, artificial cavity installation, population monitoring, and competitor control. Of these, the most crucial elements have been the translocation program and the development of artificial cavity technology. It is apparent that recovery of small populations depends on translocation and artificial cavity installation, along with a well supported monitoring and habitat improvement program.

Our challenge now on the SRS is to continue the RCW program in the direction it is headed. By aggressively continuing the described program, by providing a management/research setting that is open to new ideas, and by implementing a forest landscape management strategy now that will support viable populations in the future, the RCW success story on the Savannah River Site will continue.

ACKNOWLEDGEMENTS

Funding for this program has been provided by the Department of Energy, whose interest and support has been vital to the programs success. Much credit for the success of this program goes to Dr. Michael R. Lennartz, who developed and played a leading role in the implementation of the program through 1990. We are also indebted to John Irwin for his long-standing support and encouragement. Thanks should go to the dedicated research biologists, wildlife biologists, foresters, and technicians, not only at the SRS but also at the donor population forests (Francis Marion National Forest and Apalachicola National Forest) whose work has made the program go forward. Thanks also goes to Dr. John Blake, SRFS for advice in putting this manuscript together and Iona Hauser for her tireless effort and assistance in putting together the oral presentation for this paper.

Red-cockaded Woodpecker Management Initiatives at Fort Bragg Military Installation

Mark A. Cantrell, Jacqueline J. Britcher and Erich L. Hoffman, Endangered Species Branch, XVIII Airborne Corps and Fort Bragg HQ, Fort Bragg, NC 28307

ABSTRACT: We describe some of the recent red-cockaded woodpecker (*Picoides borealis*) management initiatives at Fort Bragg Military Installation, a 63,143 ha military training facility in the Sandhills physiographic region of North Carolina. These initiatives resulted from consultations with the United States Fish and Wildlife Service (USFWS) for military activities and include surveys and monitoring of red-cockaded woodpecker groups and cavity trees, cavity enhancement, and demographic linkage with corridors. Management practices also include growing-season burning, midstory control of hardwoods, reforestation, and public education. When fully implemented, these initiatives will contribute to recovery of the red-cockaded woodpecker population in the Sandhills of North Carolina.
KEYWORDS: Fort Bragg Military Installation, military training, Sandhills, cavity enhancement, demographic linkage

The primary mission of Fort Bragg Military Installation (Fort Bragg) is training, logistical support, and mobilization of the U.S. Army's XVIII Airborne Corps (**Figure 1**). Fort Bragg is the most intensely utilized military installation in the world. It is also the largest tract of Federal land in the Sandhills physiographic region of North Carolina, and is the core location of the second largest of 15 designated recovery populations (USFWS 1985b).

The Endangered Species Act mandates recovery of endangered species by Federal agencies (Section 2), and provides for a review of their actions (Section 7). Our objectives here are to describe some recent initiatives at Fort Bragg to recover 1 of the 4 endangered species at Fort Bragg, the red-cockaded woodpecker.

We are grateful to those at North Carolina State University who have provided guidance and discussion during development of the endangered species programs at Fort Bragg: J.H. Carter, III, P.D. Doerr, K.H. Pollock, and J.R. Walters. We thank Fort Bragg personnel who have been involved with these programs: E.S. Bebb, R.A. Chisholm, C.A. George, W.M. Hunnicutt, T.M. Myers, and J.M. Patten, and D.L. Sewell. D.L. Sewell provided information on past harvests and current forest conditions. Personnel of the U.S. Fish and Wildlife Service, R. Costa, L.K. Gantt, and D. Mignogno have encouraged these initiatives. B. Gorsira and C.R. Rossell, Jr. provided helpful comments on an earlier draft of this manuscript.

STUDY AREA

Fort Bragg Military Installation is located in the Sandhills physiographic province of North Carolina. The Installation contains 63,143 ha, of which 64% are forested. An extensive system of firebreaks, parachute-drop zones, and artillery-impact areas are interspersed throughout the forest (**Figure 2**). The upland soils are mostly classified as sandy loams or loamy sands, which are low in organic matter and low in fertility. Lowland soils are typically loamy. The Sandhills area, like most of the southeastern coastal plain, was once forested with extensive stands of longleaf pine (*Pinus palustris*) (Wahlenburg 1946). Naval stores, livestock grazing, and lumber operations influenced the forest prior to the establishment of Fort Bragg in 1917. What remains are second-growth and typically understocked stands (**Figure 3**). The oldest stands

Figure 1. Fort Bragg soldiers utilize a small-arms qualification range adjacent to red-cockaded woodpecker habitat. Fort Bragg regulations describe the training activities allowed in endangered species habitat.

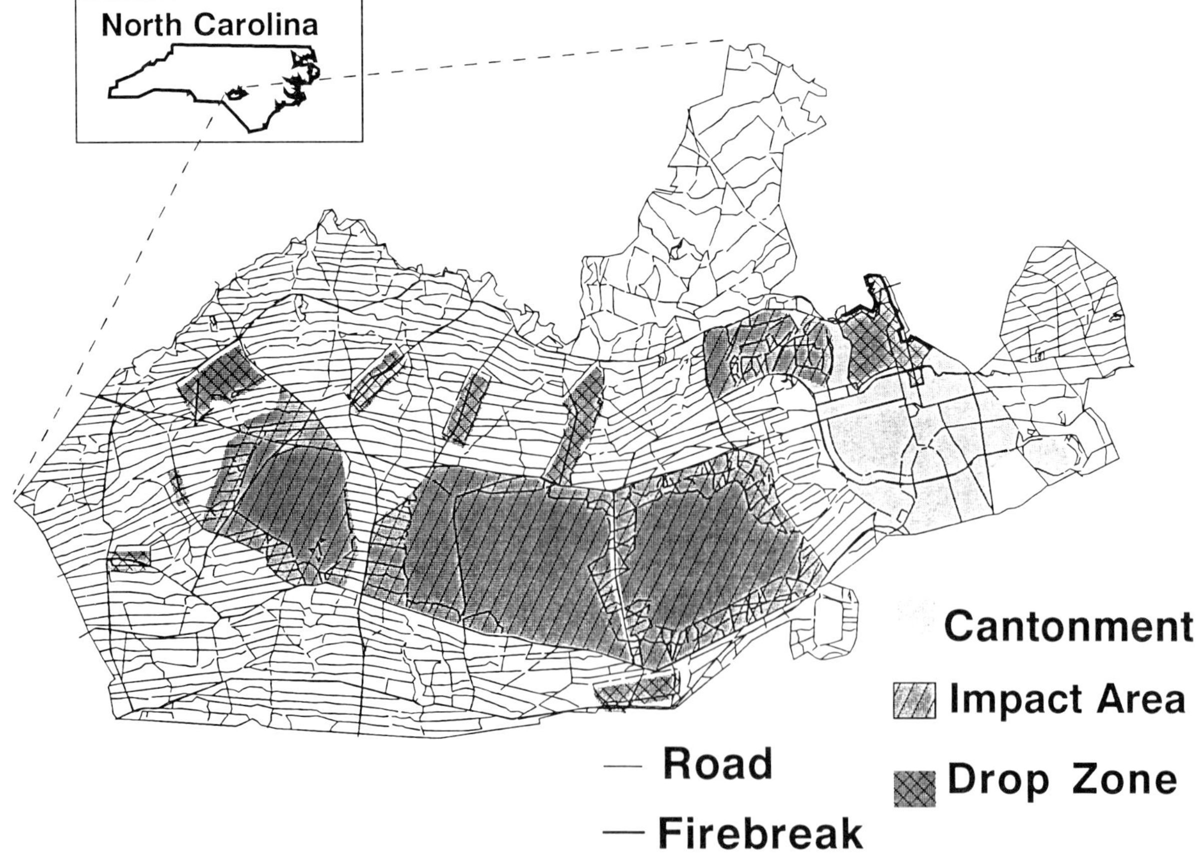

Figure 2. Fort Bragg Military Installation (63,143 ha) is located in the Sandhills physiographic province in eastern North Carolina. An extensive firebreak system, parachute-drop zone, and artillery-impact areas punctuate the forested landscape.

Table 1. Harvest volumes of some forest products, 1947 - 1990, on Fort Bragg, N.C. Records were not maintained during the first Army harvests, 1942 - 1947, nor prior to establishment of Fort Bragg. The period 1947 - 1950 represents the first contract by the Army-Navy Lumber Agency which distributed lumber to all military installations; the value of the timber harvested during that period was $661,396. Small volumes of cypress, juniper, yellow-poplar, and oak sawtimber, as well as other categories of forest products were also harvested intermittently.

Period	Sawtimber (board feet)	Pine Pulpwood (cords)	Resinous Stumps (tons)	Pine Straw (tons)
1947 - 1950	30,506,000	--	--	--
FY 55-65	54,663,476	109,669	--	--
FY 66-75	70,017,382	125,274	106,116	6,223
FY 76-80	15,639,704	10,915	5,698	9,236
FY 81-90	24,147,855	22,019	1,484	59,889

of longleaf pine on Fort Bragg are approximately 90 years old **(Figure 4).** Remnant, flat-top trees are scattered throughout the installation, most spared from past timber harvests because of poor form and/or red-cockaded woodpecker occupation.

Understory vegetation consists of turkey oak (*Quercus laevis*) on xeric sites, with other oaks (*Q. marilandica* and *Q. incana*) on less xeric sites, and all especially dense in areas of fire exclusion. Wiregrass (*Aristida stricta*) dominates the herb layer **(Figure 5)** with other common species (*Schizachyrum scoparium, Andropogon gyrans, Sporobolus junceus, Euphorbia ipecacuan, E. exserta, Tephrosia virginiana, Baptisia cinerea, Stylisma patens*). Pond pines (*P. serotina*) occur along stream ecotones, hillside seepages, and pocosins, along with a dense shrub layer of fetterbush (*Leucothoe racemosa and Lyonia lucida*), sweetpepper bush (*Clethra alnifolia*), blueberries (*Vaccinium spp.*), and gallberries (*Ilex glabra, I. coriacea*). Old field sites are dominated by loblolly pine (*P. taeda*), often mixed with longleaf. Slash pine (*Pinus elliottii*) occurs in off-site, stagnant plantations.

MANAGEMENT PRACTICES

The forests are managed by civilian Army personnel of the Natural Resources Branch with primary objectives of endangered species enhancement - a significant undertaking given the history of timber production **(Table 1)** and the resulting young, third-growth stands.

In 1989, a new emphasis was placed on warm-season prescribed fires to restore and maintain an open forest midstory; approximately 14,575 ha are targeted annually, for a 3-year rotation. Prior to this, dormant season burns were conducted on a 5-year rotation. Wildfires have occurred during all seasons incidental to military training activities. Harvest rotations are 120 years for longleaf pine stands, 100 years for other pine stands. Natural regeneration methods with minimal site preparation are favored. The revised Fort Bragg Forest Management Plan (1992) details the integration of endangered species, soil conservation, and wildlife concerns into woodland management through a Geographic Information System (GIS)-based inventory.

The Endangered Species Branch is tasked with the management of protected species at Fort Bragg. The civilian staff of this branch manage the significant populations of the endangered red-cockaded woodpecker, Michaux's sumac (*Rhus michauxii*), rough-leaf loosestrife (*Lysimachia asperulifolia*), and American chaffseed (*Schwalbea americana*).

Red-cockaded woodpecker management on Fort Bragg has been guided by Army policy and guidelines (1984), USFWS Biological Opinions concerning woodland management (1980), range modernization and construction (1985a), USFWS guidelines (1989), military training (1990), and specific construction projects (1992b). Guidance from the Opinions has been in the form of "reasonable and prudent alternatives," "conservation enhancement measures," "conservation measures," and "conservation recommendations." This guidance has required monitoring and surveys, habitat improvement (midstory hardwood removal, prescribed burning, cavity creation, and soil conservation), and restrictions on activities within clusters (e.g., military training, harvest of forest products).

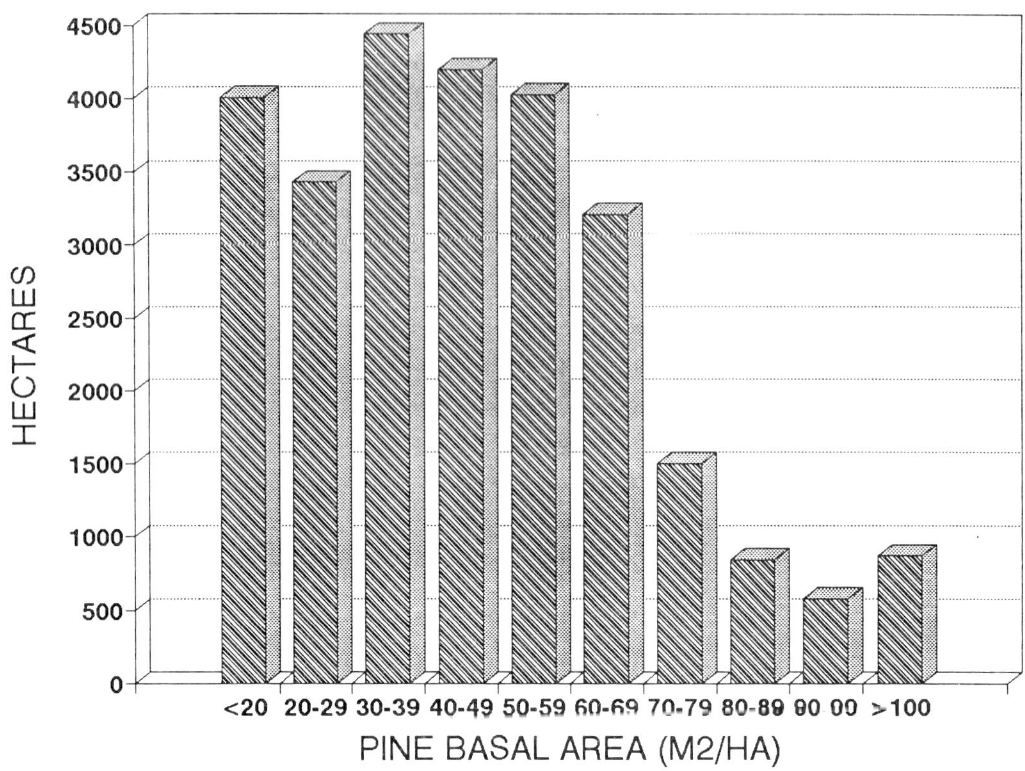

Figure 3. There was a wide range of stocking densities in pine and pine-hardwood stands on Fort Bragg in 1992. Some densely stocked stands (>90 m^2/ha) will be thinned in the future.

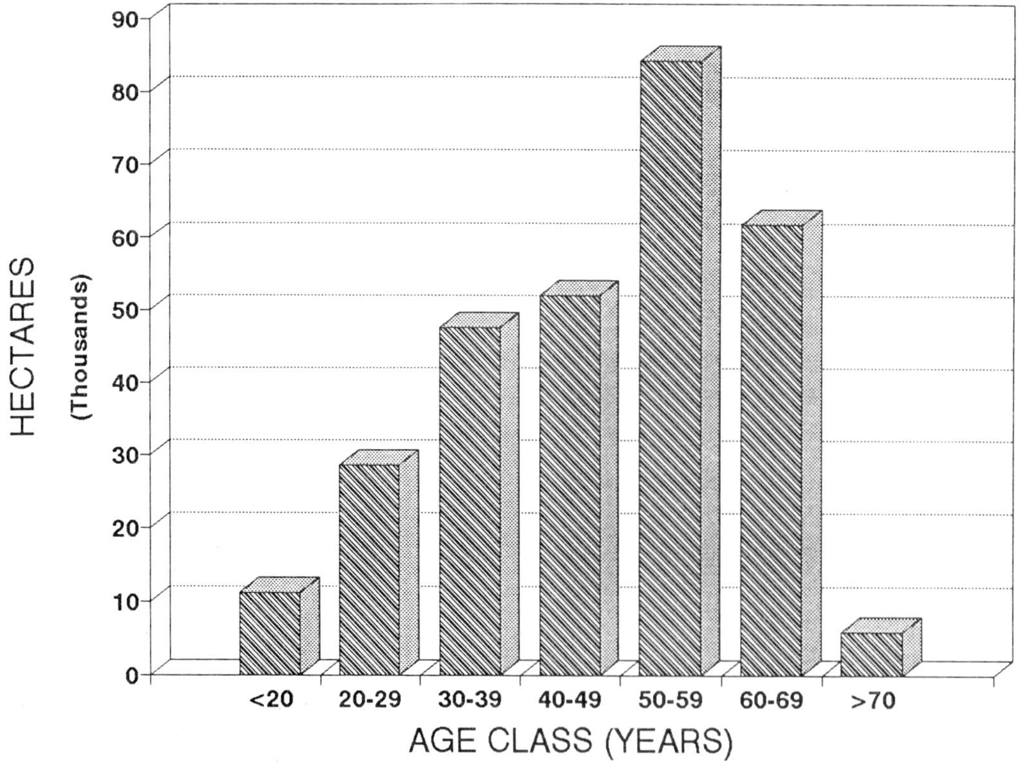

Figure 4. Ages of pine and pine-hardwood stands were relatively young during a 1992 forest inventory. Most stands were established after Fort Bragg was purchased in 1917.

Figure 5. Longleaf pine - wiregrass community at Fort Bragg Military Installation, North Carolina. The vegetative diversity of Fort Bragg includes more than 1,100 species and 33 natural community types.

Reproductive Success and Group Composition

The red-cockaded woodpecker monitoring program was developed in 1992 to improve estimates of reproductive success and group composition on Fort Bragg. The Installation was divided into 20 sample blocks based on geographic location and general habitat differences (**Figure 6**). Sample size was derived from previous observations of demographic variability. Sample clusters (n = 134) were selected randomly, except for those clusters located in artillery impact areas which were chosen based on accessibility. Within each block, 30% of both the active and inactive clusters were selected. A systematic survey for red-cockaded woodpecker cavity trees was conducted in 1992. Surveys are scheduled every 5 years to locate and map new cavity trees. All clusters are known.

During the breeding season, each of the sample clusters are intensively monitored. This includes collection of nest data on a 7-11 day cycle, identification of previously banded birds, banding of nestlings, and determination of sex at fledging. Data are also collected on each cavity tree at sample and non-sample clusters. Clusters within the Western Study Area (n = 157) are intensively monitored by North Carolina State University **(Figure 6).**

Cavity Enhancement

Past management practices, the extensive system of fire breaks, and large training areas have created a young, understocked, and fragmented forest on Fort Bragg with a limited availability of suitable trees for natural cavity excavation. Coupled with the time needed for birds to create a cavity (Baker 1971, Jackson et al. 1979b, Hooper et al. 1980), the lack of potential cavity trees (Jackson et al. 1979, Hooper 1988) likely influences fledgling and adult survival, reproductive success, group stability, and colonization. The USFWS included management recommendations in several Opinions (1992a, 1992b) to increase the number of usable cavities in cavity-limited sites. To accomplish this, we utilize the drilled-cavity (Copeyon 1990, Taylor and Hooper 1991) and the cavity-insert (Allen 1991) techniques. Additionally, existing cavities are improved by installation of entrance restrictor plates (Carter et al. 1989) and refurbishing of degraded cavities by vacuuming to remove usurper nest materials. Because the lack of suitable cavities is so widespread, we prioritized clusters for treatment.

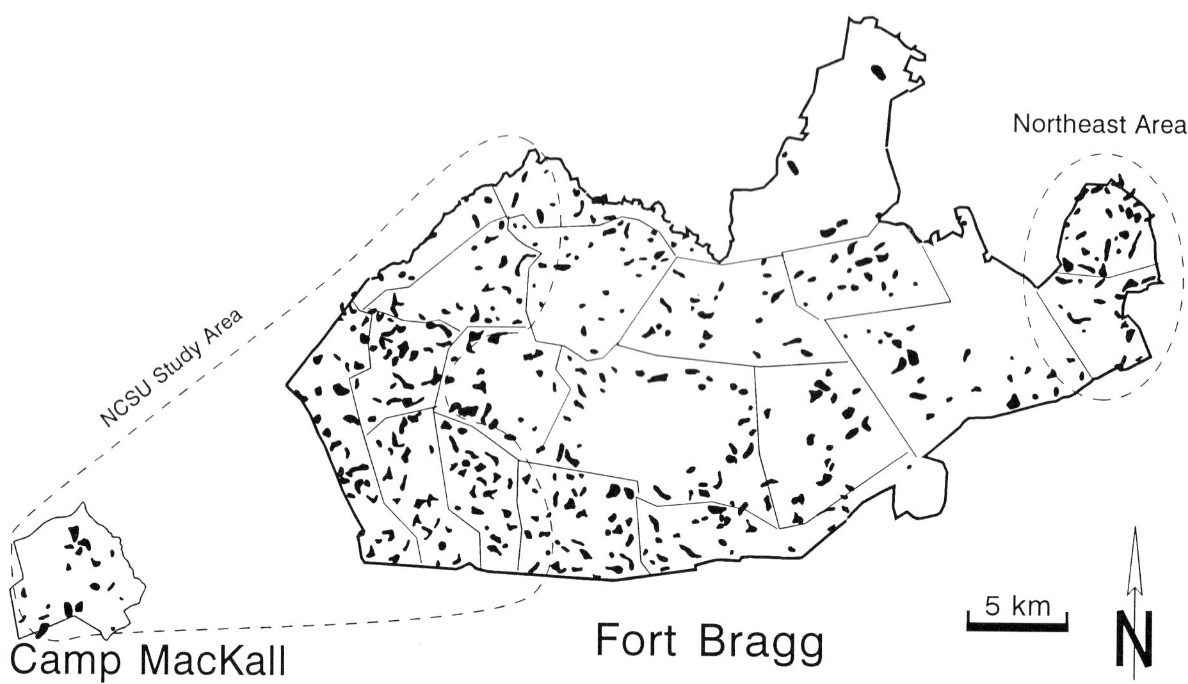

Figure 6. Red-cockaded woodpecker clusters on Fort Bragg Military Installation (n = 425) are divided into 20 sample blocks. A random sample of each block includes 30% of the active clusters and 30% of the inactive clusters. At occupied clusters in the sample, all red-cockaded woodpeckers are banded and monitored.

Demographic Linkage Through Corridors

The Fort Bragg Greenbelt is a 2,712 ha special management area established to provide a corridor between demes within the North Carolina Sandhills red-cockaded woodpecker population. An August 1991 biological assessment found that a proposed Army construction project in the Greenbelt "may affect" the red-cockaded woodpecker by genetic isolation of the eastern 10% of the Sandhills population (**Figure 7**). Further, the project would reduce the available forage substrate for one active cluster below the USFWS minimum requirements for basal area and stems ≥10 in. dbh. During the Section 7 Formal Consultation, a plan was developed to offset the projected habitat losses and to reestablish demographic and population contiguity. We developed the plan to provide habitat for red-cockaded woodpeckers in a corridor (Beier and Loe 1992) using some theoretical and practical direction (MacArthur and Wilson 1967, Harris 1984, Noss 1987, Simberloff and Cox 1987, Soulé 1987, Adams and Dove 1989, Harris and Gallagher 1989, Hudson 1991). We felt that reoccupation and enhancement of the Greenbelt area would encourage red-cockaded woodpecker stability at both the demographic and population level. In addition, the plan provides an overall strategy for urban development in this area. The proposal called for extensive reforestation, soil stabilization, fire management, cavity enhancement, public awareness, and project management.

ACCOMPLISHMENTS

Reproductive Success and Clan Composition Monitoring

Our sample clusters were occupied (n = 207) and unoccupied (n = 87); we located 156 nests in the occupied clusters. We banded nestlings (n = 115) and adults (**Figure 8**) (n = 73), exclusive of the western study area. NCSU banding records for the western study area will be available in the future for comparison. Fledging success was 51%. All red-cockaded woodpecker cavity trees (n = 3048) on Fort Bragg were inspected. We classified 51% as active, having fresh resin wells and chipping (Hooper et al. 1980). We observed

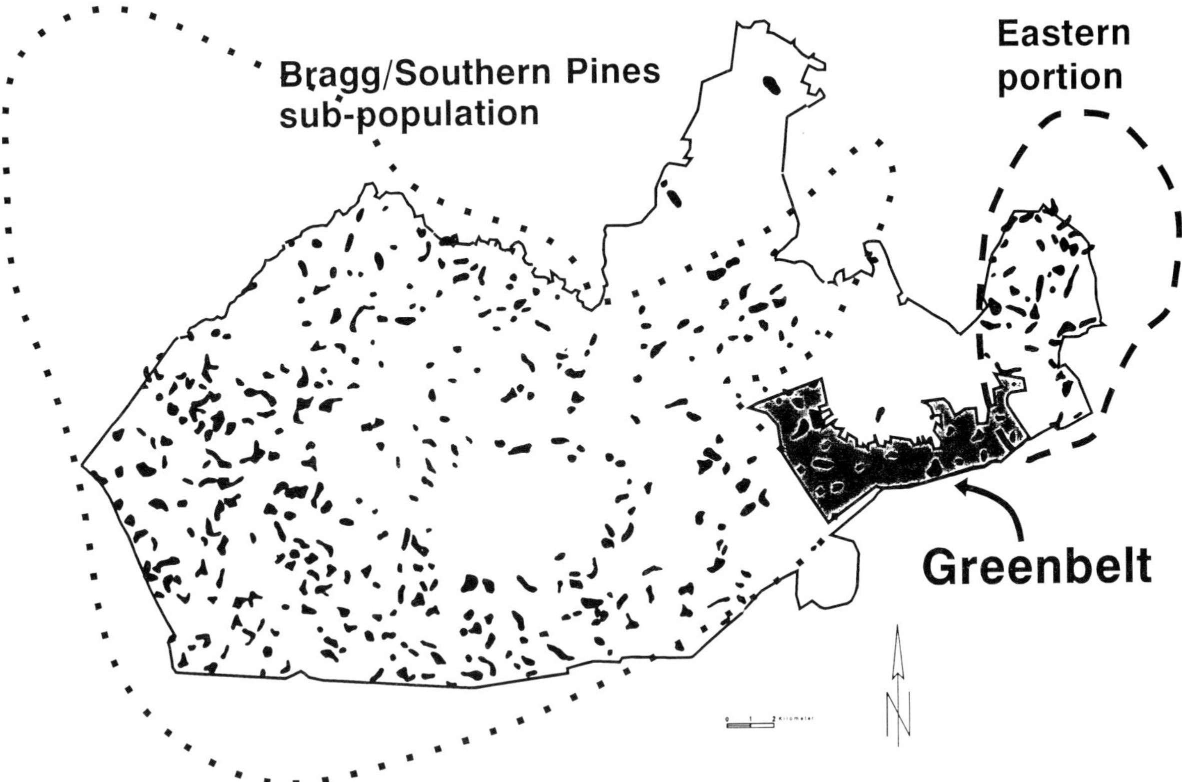

Figure 7. The Greenbelt habitat management area joins the eastern 10% of the North Carolina Sandhills red-cockaded woodpecker population. The dashed line represents the eastern-most of 2 subpopulations that occur in the Sandhills, including habitat on Fort Bragg Military Installation and the Southern Pines area. The sixth-largest metropolitan area of North Carolina is bridged by the forested Greenbelt area on Fort Bragg.

red-cockaded woodpeckers during fledge and nest checks, and made positive identifications (n = 695) of color-banded individuals (**Figure 9**).

Since our sampling has only begun, we have not analyzed our data, however, we hope to observe changes in some population parameters from habitat improvements throughout the Installation. The effects of prescribed burning, midstory-hardwood control, and cavity enhancement will likely be measured following this next breeding season; we hope to see a reduction in the number of unoccupied red-cockaded woodpecker clusters, an increase in nests success and fledging success at occupied clusters.

Cavity Enhancement

The cavity enhancement program was started in 1991. The objectives of our efforts over the next 5 years are to stabilize existing groups, encourage reoccupation of unoccupied clusters, and to create new groups. Long-term, we seek to increase the number of helpers within a group, the number of breeding groups, and thereby, the population size and stability. We have prioritized cavity-limited clusters where survival of group members is presumed threatened. Cluster sites are also prioritized based on USFWS consultations, demographic importance and reproductive status. We have "created" clusters in unoccupied habitat. We band all red-cockaded woodpeckers occupying adjacent clusters to monitor demographic responses to newly created cavities.

We have created 64 cavities and 17 starts in clusters (n = 23) prioritized for first treatment. An example of how our cavity enhancement program has benefited these priority sites is illustrated at cluster 134: the red-cockaded woodpecker group lost the nest cavity tree to a lightning strike in 1990, and failed to nest in 1991. Two cavities were drilled in January and March of 1992, the breeding pair began

Figure 8. Susan Ladd Brown of the Fort Bragg Endangered Species Branch bands an adult red-cockaded woodpecker.

Figure 9. Red-cockaded woodpeckers are banded and marked with color-leg bands at sample clusters. Later observations with spotting scopes allow descriptions of group composition, dispersal and survival.

pecking resin wells within days, and the pair nested in June, fledging 2 young in July. An insert box placed in the cluster (**Figure 10**) in September was occupied by one of the fledglings within 2 days. Other examples of the immediate success of artificial cavity creation occurred in clusters containing cavity trees lost due to siltation and fire. The specialized equipment and intensive training requisite to the cavity enhancement program is more than compensated by the potential for group stabilization, reproductive success, and population increase.

Demographic Linkage Through Corridors

During 1992, 863 ha were prescribe burned during the growing season. Due to heavy fuel loads 482 ha were burned in winter, to be followed in 1993 with a growing season burn. Smoke management and safety were especially critical in 930 ha of developed residential and industrial areas.

We contracted for reforestation of 41 ha with longleaf pine. Designs were completed for

Figure 10. Cavity insert in a longleaf pine at Fort Bragg Military Installation.

Figure 11. Biologists collect locational data at a red-cockaded woodpecker cavity tree using global positioning system technology. These data are incorporated into a Geographic Information System for mapping and analysis.

reclamation of 2 abandoned borrow pits, including provisions for experimental plots of seed mixtures for soil stabilization and future prescribed burning of reclaimed forest. These areas, 11 ha and 2.5 ha, will be reclaimed and reforested by March 1994. Non-essential roads in red-cockaded woodpecker clusters were closed. We drilled (n = 18) and/or inserted artificial cavities (n = 5) and drilled start-holes (n = 8) in 8 red-cockaded woodpecker clusters in the Greenbelt during 1992; trees (n = 41) were selected in 8 other red-cockaded woodpecker clusters to receive artificial cavities in 1993. Endangered species educational programs were initiated for residents of Army housing areas in and adjacent to the Greenbelt. Releases in local newspapers highlighted the establishment of the Greenbelt program.

FUTURE MANAGEMENT INITIATIVES

Our management efforts for the red-cockaded woodpecker demonstrate how a Federal agency has responded to mandates of the Endangered Species Act by hiring a professional staff and supporting an ambitious program to recover the red-cockaded woodpecker. Future programs include:

1) development and participation in a regional red-cockaded woodpecker conservation plan with North Carolina Wildlife Resources Commission, municipal, corporate, and private landowners,

2) utilization of Geographic Information System to catalog and analyze spatial data (**Figure 11**) concerning the Sandhills red-cockaded woodpecker population, and

3) public awareness and education through workshops, demonstrations, and mass media presentations.

Response of a Small Red-cockaded Woodpecker Population to Intensive Management at Noxubee National Wildlife Refuge

David M. Richardson and James M. Stockie, Noxubee National Wildlife Refuge, U.S. Fish & Wildlife Service, Route 1, Box 142, Brooksville, MS 39739

ABSTRACT: We discuss population status and management activities of red-cockaded woodpeckers (*Picoides borealis*) on Noxubee National Wildlife Refuge from 1972 to 1992. In 1972 there was a minimum of 26 active clusters but only 16 in 1987. Since 1990, intensive management increased the number of active clusters to 32. Abandonment of clusters in the past was primarily attributed to hardwood encroachment, cutting of old-growth pines, and tornados. Cavity inserts (n = 80) were installed to augment active clusters having <4 suitable cavities. Seventeen artificial clusters were also created. Selective removal of gray rat snakes (*Elaphe obsoleta spiloides*) and southern flying squirrels (*Glaucomys volans*) in clusters was done to reduce predation and competition with red-cockaded woodpeckers. Hardwood midstory within 19 active and 5 inactive clusters was removed using a shear V-blade, hand clearing, pulpwood sales, and a public firewood cutting program. Red-cockaded woodpecker conservation has been integrated with forestry management, and now is the principal force for the refuge's pine management program. The recent case history of RCW management at Noxubee NWR illustrates that habitat enhancement and installing artificial cavities facilitated increases of the red-cockaded populations.
KEYWORDS: Cavity insert, prescribed fire, red-cockaded woodpecker, flying squirrel, gray rat snake, artificial cluster, habitat enhancement.

Until recently, red-cockaded woodpecker (RCW) management on federal lands was based primarily on maintaining open, park-like, mature stands of pine required for nesting and roosting by the birds (U.S. Fish & Wildlife Service, 1985, U.S. Forest Service 1985, U.S. Fish and Wildlife Service 1987). Nonetheless, most populations on federal and private lands continued to decrease (Ortego and Lay 1988, Conner and Rudolph 1989, Costa and Escano 1989, James 1991). Major reasons for this decline were attributed to hardwood midstory encroachment around cavity trees (Van Balen and Doerr 1978, Locke et al. 1983, Costa and Escano 1989, Conner and Rudolph 1989, Loeb et al 1992); shortage of potential cavity trees (Hooper 1988, Costa and Escano 1989); demographic isolation (Costa and Escano 1989); and population density (Hooper and Lennartz 1995). Copeyon et al. (1991) and Walters et al. (1992a,b) have suggested that cavity limitation and competition for existing territories (i.e., clusters) were important factors affecting population growth of this species. This information, combined with techniques to create artificial cavities (Copeyon 1990, Taylor and Hooper 1991, Allen 1991), provided resource managers with new conservation and recovery possibilities.

In this paper, we present 6 years (i.e., 1986-1992) of intensive RCW management results on Noxubee National Wildlife Refuge (NWR). RCW management was based on the most current management technology and research. Particular emphasis was placed on the population growth we attributed to artificial cavities and habitat enhancement.

STUDY AREA

Noxubee NWR is in Oktibbeha, Noxubee and Winston Counties of east-central Mississippi. The refuge is located on the Upper Coastal Plain in three distinct regional physiographic zones: the north central plateau, the black prairie belt and the flatwoods (Rafferty, 1979). The forested portion of the refuge encompasses 42,725 acres; 13,204 and 8,820 acres were typed as pine, and pine-hardwood, respectively. Clusters were dominated by second-growth loblolly pine (*Pinus taeda*) and scattered shortleaf pine (*P. echinata*). The most common understory and midstory deciduous trees were sweetgum (*Liquidambar styraciflua*), oaks (*Quercus* spp.), and hickories (*Carya* spp.). A more extensive vegetative description of the clusters was provided by Nebeker et al. (1995).

The primary forest management practice in the pine and pine-hardwood during the last 50 years, has been even-aged stand regeneration on a 70-year rotation. Each refuge compartment was inventoried once every 10 years. Stands within each compartment were generally thinned to a basal area of 80 sq. ft. Forest regeneration cuts (primarily seed tree method) became a regular practice in the mid-1980's. Southern pine beetle (SPB) (*Dendroctonus frontalis*) infestations occurred regularly but were aggressively controlled. SPB spots were generally <5 acres in size. Nongrowing season prescribed burning in the pine type and in a portion of the pine-hardwood type was done since 1960 at 3-4 year intervals.

The RCW population on the refuge was small and demographically isolated from other populations. With the exception of 2 clusters, all known clusters (n = 10) within the refuge counties were >10 miles from refuge boundaries. Fragmented forests around the refuge, interspersed with agriculture, may have limited immigration of birds from these clusters. Abandoned clusters (n ≥ 30) occurred on the refuge as a result of extensive hardwood midstory and overstory development.

METHODS

Cluster Census

Cluster status was determined using criteria described by Jackson (1977, 1978b) and via RCW observations at clusters during the breeding season. Beginning in 1990, spring (March-May) and fall (September-November) clan size at each cluster was determined. This was accomplished by stationing 1-6 observers in clusters and recording the presence or absence of birds roosting in trees. In spring, most group sizes were determined by observing adults feeding young. All spring surveys were completed prior to young fledging.

Habitat Modification

Since 1986, active and inactive clusters on the refuge were treated for hardwood midstory encroachment using either roller-chopping, pulpwood sales, timber sales, public cutting of firewood, chainsaw crews, and/or shearing (Richardson and Smith 1992) (**Figure 1**). All hardwood control was done outside the breeding season. During 1990-1992, midstory control in conjunction with artificial cavity installation were used to create 17 artificial clusters (i.e., sites lacking red-cockaded woodpecker cavities) through reduction of pine and hardwood basal area. Artificial cavities were then added to these sites. Minimally, the area encompassing cavities in active and inactive clusters was modified to meet the Red-cockaded Woodpecker Recovery Plan Guidelines (U.S. Fish and Wildlife Service 1985) of <20 ft^2/acre of hardwoods. In general, clusters were about 10 acres in size, except for artificial clusters created through timber sales which were approximately 20 acres. Inactive clusters and artificial clusters controlled for hardwood were <0.75 to 2.5 miles from an active cluster.

Artificial Cavities

Our trees did not meet the criteria of <3.5 inches of sapwood needed for drilling artificial cavities (Copeyon 1990, Taylor and Hooper 1991) so we used prefabricated artificial cavities (i.e., inserts) (Allen 1991). The first 30 inserts were installed using Swedish climbing ladders; and thereafter, a customized, climbing tree stand was used (**Figure 2**). The climbing tree stand made installation both easier and safer than the climbing ladders. An attempt was made to orient the entrance of the inserts south-southwest to simulate the natural orientation of cavity entrances (Locke and Conner 1983). However, inserts were always placed on the backside lean of a tree to prevent water from entering the cavity which prevented many from being placed in that orientation. Inserts were placed at a height of 18-24 ft. A minimum of 2 inserts was placed at inactive and artificial sites. If these sites subsequently became active, 2 additional inserts were installed. Active clusters with <4 suitable cavities were provisioned with artificial cavities.

Predator/Competitor Control

A southern flying squirrel removal program was initiated in 1992. This program was

Figure 1. D-7 tractor fitted with a V-blade that was being used to shear hardwoods in red-cockaded woodpecker clusters on Noxubee National Wildlife Refuge, Mississippi.

Figure 2. Installation of a cavity insert on Noxubee National Wildlife Refuge, Mississippi utilizing an oversize, custom, climbing tree stand.

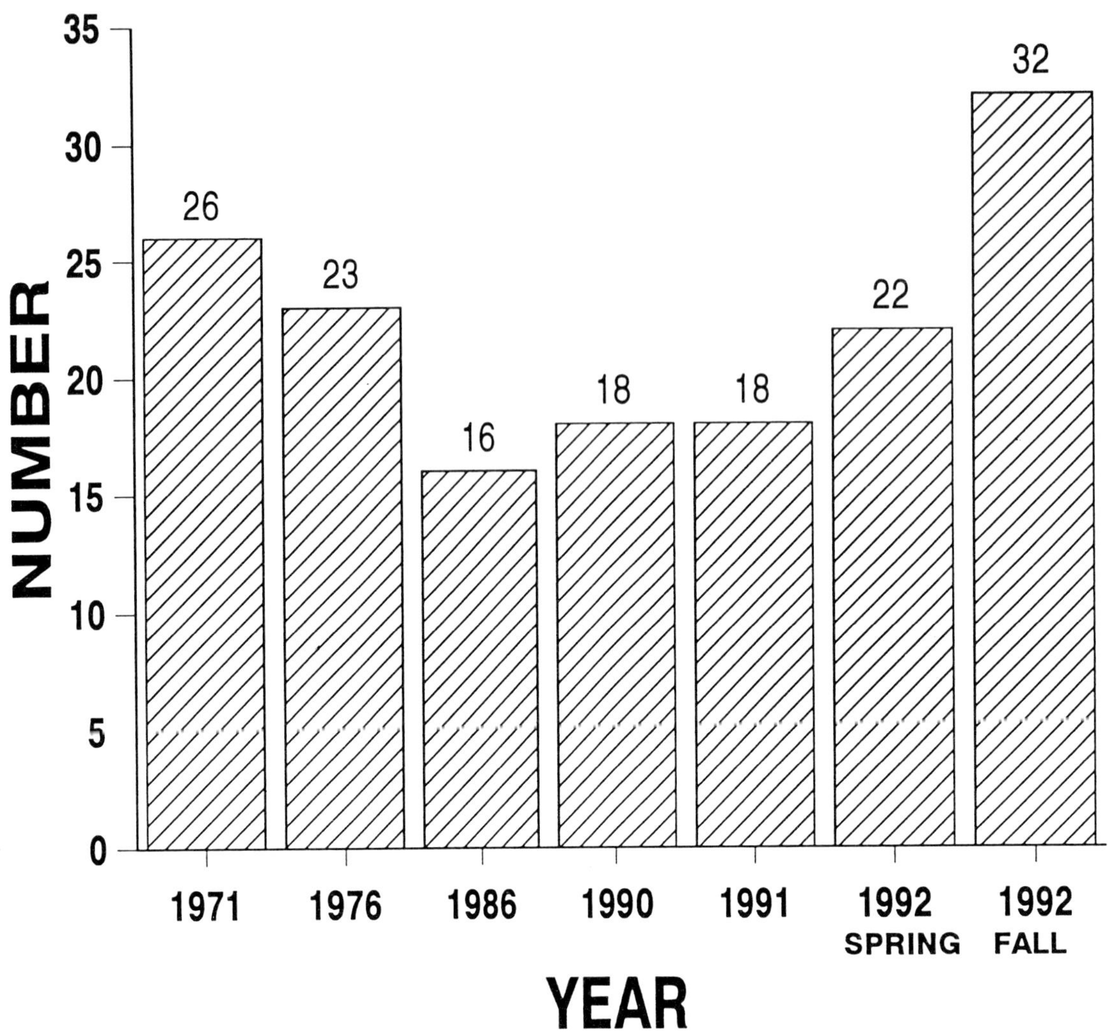

Figure 3. Number of known occupied red-cockaded woodpecker clusters on Noxubee National Wildlife Refuge in Mississippi.

deemed necessary because of the suspected flying squirrel related failure of 3 RCW nests in 1990 and 1991, and abundance of flying squirrels in RCW cavities. Beginning in March, in an attempt to provide ≥4 suitable cavities per cluster, flying squirrels were removed from active cavities and all inserts. Flying squirrels were shot with a .22 cal. rifle or caught in a net as they exited cavities.

To prevent nest destruction by gray rat snakes, snake traps were installed on nest trees in spring 1992 (Neal 1992). A 24-inch piece of 3/4-inch nylon mesh, folded in half, was stapled at breast height around the tree; 21 nest trees and 1 roost tree were treated. Snakes were entangled when they attempted to climb through the mesh. Trapped snakes were relocated.

RESULTS

Population Status and Habitat Improvement

From 1972, active clusters on the refuge decreased from a minimum of 26 to a low of 16 in 1986 and was attributed to hardwood encroachment, cutting of old-growth pines, and tornados (J.A. Jackson, Mississippi State Univ., pers. commun.) (**Figure 3**). No clusters were abandoned after 1986. By November 1992, there were 32 occupied clusters (**Figure 3**). Previously unknown "natural" clusters (n = 4) were found during 1990 and 1992. One was a reoccupied inactive cluster which had received no habitat enhancement or inserts.

Two new territories were believed to have been formed by pioneering and budding (Hooper 1983). The budded cluster consisted of a single nest cavity found in 1992; it was

Table 1. Number of sites treated and method of midstory hardwood removal and pine basal area reduction in active, inactive, and artificial colony sites at Noxubee National Wildlife Refuge during 1986-1992.

	Method of Treatment					
Cluster type	Firewood cutting	Pulpwood sale	Chainsaw[1]	Roller-chopping	Shear V-blade[2]	FCP[3]
Active	3	3	4	2	7	--
Inactive	--	--	--	--	5	--
Artificial	2	3	3	--	7	2
Total	**5**	**6**	**7**	**2**	**19**	**2**

[1]Felling of trees was done by refuge employees with chainsaws.

[2]Following felling of trees, sites were opened to the public for firewood cutting.

[3]Forestry compartment prescription; hardwood and pine BA reduced through a timber sale.

Table 2. Type of cluster and year of installation of cavity inserts at Noxubee National Wildlife Refuge during 1990-1992.

	Cluster Type		
Year	Active[1]	Inactive[1]	Artificial[2]
1990	5	2	4
1991	10	8	10
1992	15	2	24
Total	**30**	**12**	**38**

[1]Status of the cluster when inserts were installed.

[2]Site modified to the Red-cockaded Woodpecker Recovery Plan guidelines (U.S. Fish and Wildl. Serv. 1985) for a cluster, but the site lacked natural cavities.

occupied by a breeding pair. An adjacent group nested <125 yards from this site, and the female from the budded cluster was roosting in a cavity at this site. The pioneered cluster was found in 1990 and consisted of a cavity under construction where the bird was roosting in the cavity. Placement of an insert resulted in establishment of a breeding pair. The nearest active cluster to this site was 0.75 miles away. The fourth cluster had been occupied for several years but was previously unknown. The estimated fall RCW population increased from 64 to 90 birds in 1990 and 1992, respectively. Twenty-six of the 32 occupied sites contained ≥2 birds in 1992.

Since 1986, 41 sites were enhanced to meet recovery guidelines (U.S. Fish and Wildlife Service 1985) for a cluster (**Table 1**). Hardwood midstory was removed in 19 active and 5 inactive clusters. Abandonment of a cluster did not occur following midstory hardwood removal. Seventeen artificial clusters were created in 1990-1992; seven of these sites have been occupied. Three of these sites contained a pair, 2 had solitary males, and 2 were captured by adjacent groups. Four of 5 inactive clusters were occupied. Two of these sites had a pair, 1 had a solitary male, and 1 was captured (see Walters et al. 1992a).

Cavity Inserts

Cavity inserts (n = 80) were placed at 39 sites on the refuge during 1990-1992 (**Table 2**). In spring 1992, 5 of 22 RCW nests were in inserts and 4 produced young. Inserts were used for roosting by 43 RCW in fall 1992. However, 22 of the 37 unused inserts were at vacant clusters.

Competitor/Predator Control

Gray rat snakes (n =15) where caught in traps at 10 clusters. Rat snakes attempted to climb 9 nest trees with young and 1 nest with eggs. At one cluster, 3 different snakes were caught climbing a nest tree with young during a 7-day period. Four snakes were caught after young had fledged or the nest failed. A snake was also caught at a roost cavity tree.

Flying squirrels (n = 65) were removed from 19 trees within 13 clusters. With the exception of 2 squirrels, all were removed from occupied sites. In 3 instances, squirrels were removed from non-nest cavities within the RCW nest trees while young or eggs were present. One nest failure in 1992 was suspected to have been caused by flying squirrels.

DISCUSSION

Six years of intensive RCW management from 1986-1992 resulted in a 2-fold increase in the number of occupied RCW clusters. We attribute this increase to the provision of cavity inserts at inactive and artificial clusters. This increase was important given that he rate of new group formation was only 0.75 groups per year in a population of >200 groups (Walters 1990), and 6.1 groups per year in a population of >400 groups (Hooper et al. 1991a). Fortunately, 2 other small RCW populations recently increased following intensive management (Gaines et al 1995, Reinman 1995). thus it now seems possible to recover many of the small declining RCW populations that until now had faced extirpation.

We observed an increase from 1990-1992 of 16 occupied sites. This represents at least 13 new groups. The remaining 3 occupied sites were believed to be captured. However, Doerr et al. (1989) indicated that occupation of clusters was dynamic. We suggest that captured sites might be an intermediate step to new group formation. Nonetheless, our results reaffirm those of Copeyon et al. (1991) that the formation of new groups can be achieved through a combination of habitat enhancement and provision of artificial cavities. Moreover, the establishment of new groups can occur within small populations through natural dispersal.

The reoccupation of 4 out of 5 inactive clusters following midstory removal and installation of inserts was expected. Doerr et al. (1989) reported that without habitat enhancement the rate of reoccupation of abandoned clusters was 8.7% annually with nearly a 60% chance of being occupied during a 10-year period. The habitat enhancement and placement of inserts within inactive clusters at Noxubee undoubtedly contributed to an increased rate of occupancy at these sites.

Because midstory encroachment around cavities is believed to be a major cause of cavity and cluster abandonment (Van Balen and Doerr 1978, Hovis and Labisky 1985, Loeb et al. 1992), controlling hardwood encroachment in clusters was essential and should be a high priority. Hardwood midstory control methods employed at Noxubee NWR represented a variety of techniques that all afforded the same result without cluster abandonment.

The capture of 14 rat snakes attempting to climb nest trees illustrates the evolutionary importance of resin wells as a defense mechanism against snake predation. Jackson (1974), and Rudolph et al. (1990b) experimentally showed the resin barrier was an effective deterrent for preventing snake predation on the eggs or young. However, Rudolph et al. (1990b) reported that 3 of 18 climbs by snakes were successful. Moreover, Jackson (1978c) and (Neal 1992) discovered rat snakes in RCW nests; the young had been eaten by the snakes in both cases. Though the experimental rate of successful climbs was only 17% (Rudolph et al. 1990b), this could severely affect reproduction in small populations and make them more susceptible to extirpation. The use of newly-installed artificial cavities for nesting by red-cockaded woodpeckers could predispose the nests to predation because these cavities lacked a substantial resin barrier. Therefore, we believe that snake predation on red-cockaded woodpecker nests should be prevented, and suggest that the non-destructive method described by Withgott et al. (1995) be used on each cavity tree along with the snake netting above.

Utilization of RCW cavities by other vertebrate species has been well documented (Jackson 1978d, Harlow and Lennartz 1983, Rudolph et al. 1990a, Loeb 1993, Loeb and Stevens 1995). The extensive use by flying squirrels of suitable cavities in our study suggests that exploitive competition for cavities occurs. In most cases, the removal of flying squirrels from cavities or inserts resulted in the reoccupation of the cavity by RCW (D. Richardson, unpubl. data). Rudolph et al. (1990a) also observed reoccupation of a cavity by a red-cockaded woodpecker following the absence of a flying squirrel. However, we repeated checks of cavities at some clusters on a weekly basis to keep cavities absent of flying squirrels. At one cluster, the removal of 17 flying squirrels resulted in the sequential occupation by RCW's and squirrels. The nest at this site failed and flying squirrels were removed from the nest cavity. This cavity was reoccupied by a RCW. Although this was not direct evidence of competition between the species or competition by flying squirrels with RCW's, it clearly indicates a management concern at clusters with limited numbers of cavities.

DeFazio et al. (1987) indicated the need to control flying squirrels at clusters because it threatened translocation and nesting attempts at a small population in South Carolina. They concluded that use of nest boxes and removal of squirrels may have reduced competition for cavities. Another nondestructive method to prevent flying squirrel access to cavities has been developed (Montague et al. 1995). Regardless of the method used, we recommend that excluding or removing flying squirrels from cavities in small populations to reduce possible competition and predation.

MANAGEMENT IMPLICATIONS

Copeyon et al. (1991) showed that direct habitat enhancement and the installation of artificial cavities in mature pine stands induced the formation of new groups. However, their work was done within a population of >200 active clusters. The intensive management for RCW at Noxubee NWR followed Walter et al. (1992a) hypothesis that cavity limitation and habitat conditions limited population increase. Our results tend to support their hypothesis, though the availability of suitable cavities for natural cavity excavation can not be eliminated. Our results provide evidence that even a small, isolated RCW populations can increase substantially through natural dispersal. Population increases in other small RCW populations relied in part on augmentation of RCW's from other populations (Gaines et al. 1995, Reinman 1995). However, our increase was directly related to:

(1) creation of suitable nesting habitat,
(2) providing sufficient foraging habitat,
(3) control of RCW predators and cavity competitors, and,
(4) installing artificial cavities. Because small RCW populations can be induced to expand, efforts must be made to do so throughout the species distribution to lessen the probability of extinction. To increase small populations, resource managers should integrate management for RCW into a proactive forest management program which recognizes the short-term as well as long term objectives for species survival. Programs should focus on longer timber rotation, growing season burns, and current technology involving the use of artificial cavities.

ACKNOWLEDGMENTS

We thank the forestry and maintenance staff at Noxubee National Wildlife Refuge for their assistance in enhancement of clusters and creation of artificial clusters. Johnny Bradford provided field assistance throughout the project. Margaret Copeland provided assistance in data collection. The members of the Oktibbeha Audubon Society helped collected annual group size data. Comments by L. Brennan, R. Costa, B. Hooper, and J. Jackson improved earlier drafts of this manuscript.

Status and Management of Red-cockaded Woodpeckers on St. Marks National Wildlife Refuge, Florida, 1980-1992

Joseph P. Reinman
U. S. Fish and Wildlife Service
St. Marks National Wildlife Refuge, St. Marks, Florida 32355

ABSTRACT: Red-cockaded woodpecker (*Picoides borealis*) numbers on the St. Marks National Wildlife Refuge and the adjoining Ochlockonee River State Park declined dramatically in the early 1980's. In an attempt to supplement these declining numbers, two pairs and one male red-cockaded woodpeckers were experimentally translocated from private land to the refuge in 1984 and 1986. One female remained in the vicinity and nested successfully 4 consecutive years. An increase in red-cockaded numbers since 1983 is attributed to the successful translocation of the single female to the refuge.
KEYWORDS: Translocation/stocking, *Picoides borealis*, radio telemetry, mortality, nests/nesting, habitat alterations, private land

St. Marks National Wildlife Refuge contains a small number of red-cockaded woodpeckers that has occupied from 1 to 5 active clusters of cavity trees since 1978 (Baker et al. 1980, Lennartz et al. 1983a, Wood and Wenner 1983). Within 1 km of refuge clusters is an occupied cluster at Ochlockonee River State Park, and within 6 km are occupied clusters in the Apalachicola National Forest (**Figure 1**). Since the occupied clusters on the refuge and state park are within 29 km of occupied sites on the national forest and no significant barriers to dispersal exist, the refuge and state park red-cockadeds are likely part of the large Apalachicola National Forest population (see U. S. Fish and Wildlife Service 1985, Anon. 1990).

Despite this likely connection with the national forest population, refuge and state park numbers declined dramatically in the early 1980's, and by early 1984 had dwindled to 5 males occupying 3 clusters. The exact cause of this decline is unknown, although it may be related to the limited number of red-cockaded woodpeckers occupying the area and their relative isolation on the fringe of the population. Extirpation, similar to that observed by Baker (1983b) at another north Florida site, appeared likely.

Also in early 1984, red-cockaded woodpeckers were discovered on St. Joseph Land and Development Company lands scheduled for extensive clearcutting. These groups formed part of a small, presumably isolated population of red-cockadeds, entirely on St Joe lands managed for pine pulpwood production. St Joe's policy was to cut all but approximately 5 acres of the habitat containing the cluster of occupied cavity trees. Red-cockaded woodpecker abandonment of clusters after extensive clearcutting had already been observed in the area. Extensive clearcutting of large areas has also been implicated as a major cause of red-cockaded woodpecker declines in other populations (Jackson 1986, Ortego and Lay 1988, Conner and Rudolph 1989). Therefore, the abandonment of these newly discovered colonies seemed likely.

As a result of these developments, red-cockaded woodpeckers were experimentally translocated from St. Joe lands to St. Marks National Wildlife Refuge in 1984 and 1986. Objectives of the translocations were to:

1) perpetuate red-cockaded woodpeckers on the refuge by supplementing the declining numbers;

2) prevent the break-up and loss of certain red-cockaded woodpecker groups on St. Joe

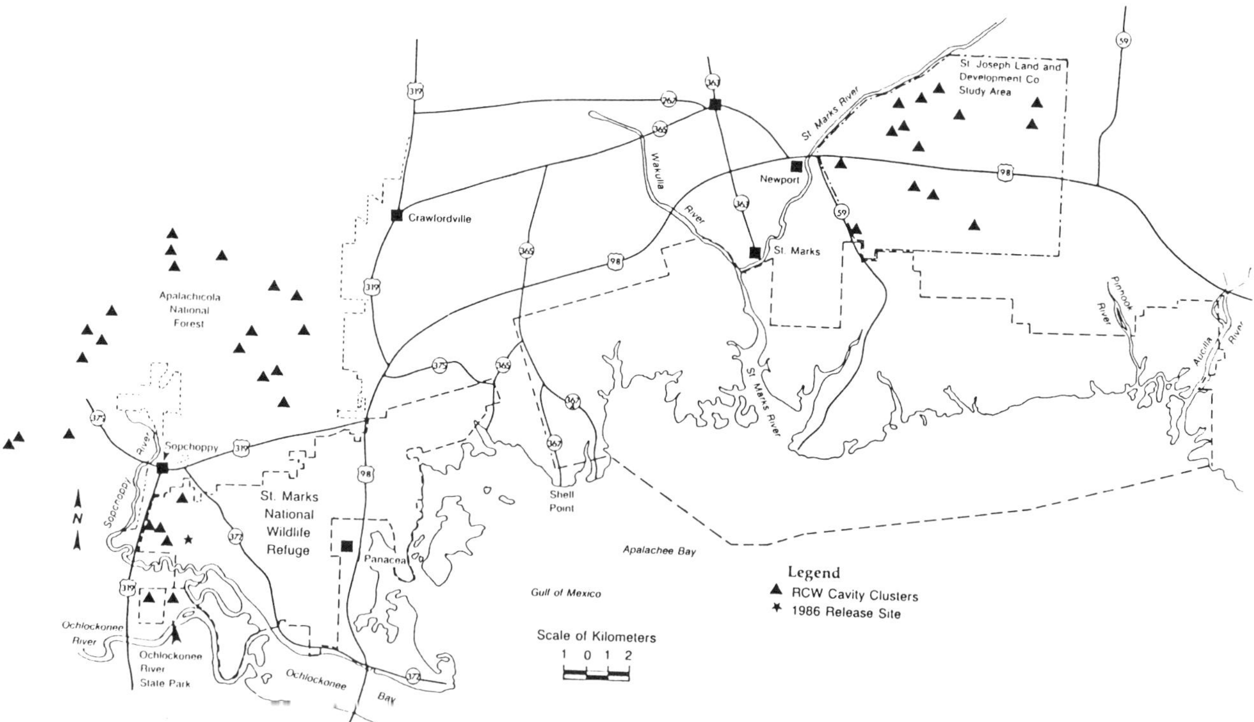

Figure 1. Red-cockaded woodpecker clusters on St. Marks National Wildlife Refuge, Ochlockonee River State Park, St. Joseph Land and Development Company study area, and adjacent lands of Apalachicola National Forest.

lands; and

3) contribute to the development and evaluation of translocation techniques.

This paper reports on the results of those translocations and examines the status of the red-cockaded woodpecker in the refuge/state park clusters and in the St. Joe study area.

STUDY AREAS

St. Marks National Wildlife Refuge encompasses over 26,000 ha of lower coastal plain along the Gulf coast of North Florida. Approximately one-half of the area is forested and 6,800 ha is considered pinelands (U.S. Fish and Wildlife Service, 1989). The red-cockaded woodpecker clusters occur on the western edge of the refuge in mixed-aged stands of longleaf pine (*Pinus palustris*) forest interspersed with depressions of titi (*Cyrilla racemiflora*), myrtle-leaved holly (*Ilex myrtifolia*), pond cypress (*Taxodium ascendens*), and blackgum (*Nyssa biflora*). The longleaf pines range in age up to 100 years or more. Cavities are frequently found in old relict trees that exhibit scars from turpentine operations conducted prior to refuge acquisition in 1938. Prescribed fires have been utilized in the pinelands since 1941, contributing to a generally low (under 1 m), diverse, grass-dominated understory of wiregrass (*Aristida stricta*), dropseed (*Sporobolus floridanus*), and gallberry (*Ilex glabra*) (Reinman 1985).

Ochlockonee River State Park is comprised of 159 ha of former refuge lands transferred to the State of Florida in 1968. Habitats are similar to those on the adjacent refuge.

The St. Joseph Land and Development Company study area comprises approximately 6,200 ha, representing only a small portion of the company's extensive landholdings in the area. Prior to recent clearcutting, the habitat generally consisted of old, relict longleaf pines interspersed in a younger, planted slash pine (*Pinus elliottii*) forest, with a dense gallberry-saw palmetto (*Serenoa repens*) understory. The understory was frequently 1-3 m high due to infrequent prescribed fires which usually occurred only in preparation for clearcutting. After clearcutting, the sites were double-chopped, bedded, and planted to slash pine. The resulting plantations are often a dense tangle of slash pine, wax myrtle (*Myrica cerifera*), titi, gallberry, and saw palmetto. Depressions of pond cypress and titi are also interspersed throughout the St. Joe study area.

METHODS

Monitoring of the St. Marks National Wildlife Refuge red-cockaded woodpeckers was initiated in 1980 and involved 3 known and 1 newly-discovered occupied clusters of cavity trees. In 1981, the monitoring program was expanded to an occupied refuge cluster discovered that year and an occupied cluster at Ochlockonee River State Park.

Color-marking red-cockaded woodpeckers was initiated in July 1981 on the refuge, March 1982 on the state park, and April 1984 on the St. Joe study area. Birds were flushed from roost cavities into nets mounted on wire frames and placed over cavity entrances. Each bird was uniquely banded with a U. S. Fish And Wildlife Service band and color leg-band(s). Initially, 21 red-cockadeds were also marked with patagial wing tags (Baker 1983a) to aid in identification.

Transmitters were attached to the translocated males by eyelash cement in 1984 (see Nesbitt et al. 1978), and by a harness made from surgical tubing in 1986 (see Nesbitt et al. 1982). A successful 3-day trial of the harness attachment was also conducted on a male red-cockaded woodpecker on the St. Joe study area in April 1985.

In 1984, 2 pairs (complete groups) of red-cockaded woodpeckers were translocated 37 km from 2 clusters on St. Joe land to 2 recently abandoned refuge clusters. Each of the abandoned refuge clusters contained at least 2 inactive cavities. These non-enlarged cavities were checked, cleaned if necessary, and plugged with wire screening a week or more prior to the translocations. One pair was moved on the evening of April 2 and the second on April 9. Each pair was captured soon after going to roost, placed in a separate holding box, transported to the refuge, and placed in the plugged cavities in 1 of the abandoned sites. Shortly after sunrise the next morning, the screen plugs were removed from the cavity entrances by pulling on an attached string. Monitoring of the birds began at that time.

On December 7, 1986, a single male was translocated from a cluster on St. Joe land to the refuge. The bird was captured upon leaving its roost cavity, transported to the refuge, and released on a relict longleaf pine in suitable nesting habitat. The release site contained numerous turpentine-scarred relict longleaf pines, and was located approximately 0.9 km from the nearest abandoned cluster and 1.3 km from the nearest active cluster (**Figure 1**). The bird had been radio-instrumented on the morning of December 4 to allow it to become accustomed to the radio transmitter. This also provided for a later check of the proper attachment of the harness. No problems with flight or feather wear were noted with this harnessed bird, or the bird harnessed in 1985.

Nesting activity was monitored by visiting occupied clusters at least once during the breeding season and flushing incubating or brooding adults and/or listening for nestling vocalizations. Fledgling success was determined by censusing nesting groups usually within one month after suspected fledging. Due to time constraints, a few occupied clusters on St. Joe were not visited during the nesting season but were generally censused in the early summer to determine nest success. Fledglings were identified by behavior and/or plumage characteristics (Jackson 1979a).

Censuses of red-cockaded woodpecker groups were generally conducted from the morning flush from cavities and continued for at least one hour. Additional observations at other times of the day were also used to supplement the censuses. Occupation of clusters was also determined by cavity tree checks for recent woodpecker activity (see Hooper et al. 1980).

RESULTS

By 1982, red-cockaded woodpecker numbers on the refuge and state park were declining seriously (**Figure 2**). Only 3 males and 1 female were present during the 1982 nesting season. By March 1984, only 5 males remained in 3 clusters.

The first translocated pair of red-cockaded woodpeckers were removed from a cluster on St. Joe lands where adjoining foraging habitat was mostly intact and clearcutting had just begun. The birds flushed from the unplugged refuge cavities at approximately 0710 on April 3, 1984, and moved through the surrounding stands erratically. By 1015 a light rain became a heavy downpour, which lasted the remainder of the day. During the storm, the red-cockadeds essentially stopped all movement. The following morning, only the radio-instrumented male was observed. At first his movements were even more erratic than the previous day, then the bird foraged slowly from 1030 to

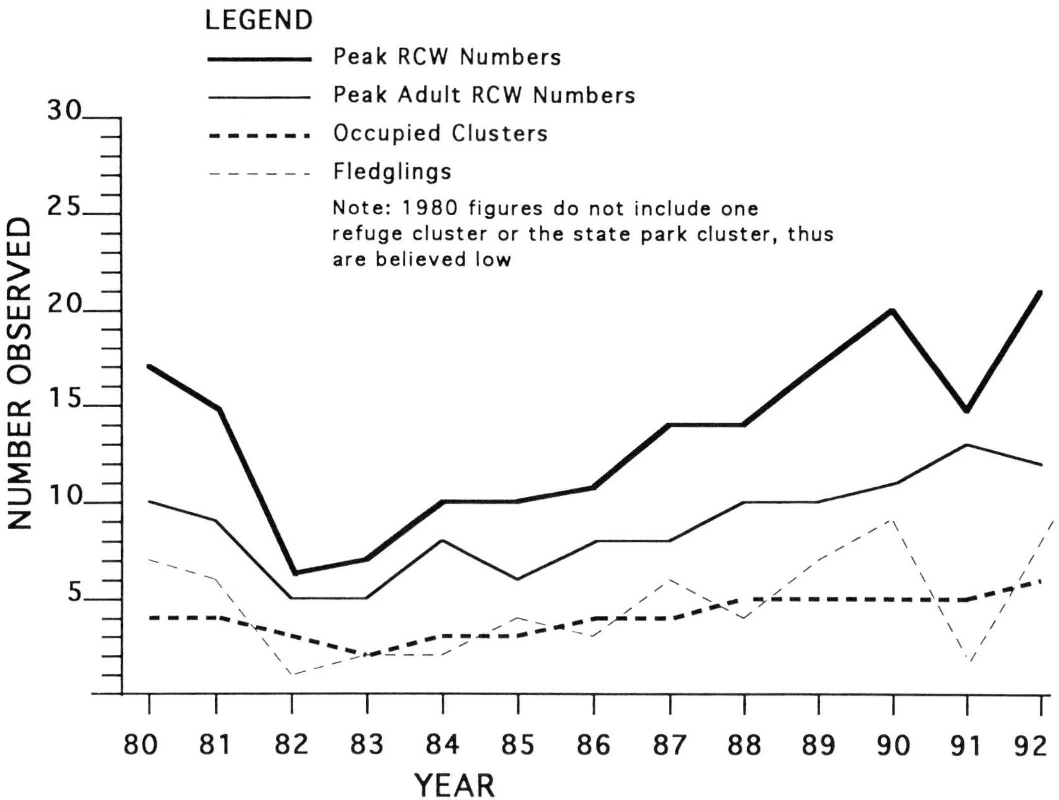

Figure 2. Status and nesting success of red-cockaded woodpeckers on St. Marks National Wildlife Refuge and Ochlockonee River State Park, 1980-1992.

1315. By 1415 the radio-transmitter had apparently detached and contact with the bird was lost.

On the morning of April 26 (3 weeks after the bird was last seen), the same male red-cockaded woodpecker flushed from the cavity in the St. Joe cluster where he was originally captured. Previous checks of the original site, including an April 22 evening roost check indicated that the male was absent and the site was occupied by other red-cockaded woodpeckers. The translocated male remained at the clearcut-isolated St. Joe cluster through July 1991. He successfully nested in 1984 and six of seven subsequent years, raising a total of 10 fledglings.

The second pair of red-cockaded woodpeckers were removed from a St. Joe cluster that had been isolated by clearcuts during the summer of 1983. After emerging from the refuge cavities at 0632 on April 10, 1984, the birds moved slowly and foraged normally. They exhibited none of the erratic flight movements of the previously translocated pair. At 0810 a male red-cockaded woodpecker, known to roost at the Ochlockonee River State Park (a distance of 2.2 km), joined the translocated pair. He began chasing the paired male, drumming, and foraging close to the female. By early afternoon the translocated female was foraging exclusively with the state park male. Both translocated birds roosted in the open that night in pine canopies, but 0.7 km apart. The translocated male returned to the release site, while the female roosted in the abandoned cluster where the first translocated pair had been released. Over the next two days only the state park male and translocated female foraged together. Contact with the translocated male was lost on April 12 after the radio transmitter fell off.

The translocated female eventually moved to the state park where she successfully nested with a second resident male from 1984 through 1987. Over the 4-year period 9 fledglings were produced.

Subsequent censuses of nearest known red-cockaded woodpecker clusters on Apalachicola National Forest and other clusters on the refuge failed to locate either the female from the first translocation or the male from the second translocation.

The single male red-cockaded woodpecker translocated on December 7, 1986, was removed from a cluster that had been isolated by clearcutting in 1984 and 1985. Upon release at 0900 the male moved and foraged normally but vocalized very little. In the afternoon the bird began to move about more rapidly and ranged farther in a large clockwise circle of the release site. The red-cockaded eventually roosted in a 25-year old slash pine plantation approximately 0.5 km from the release site. The following morning the radio receiver failed and the red-cockaded woodpecker was not located. Two days later, the bird was found dead approximately 0.2 km from the roost site. The carcass was maggot-infested and a necropsy revealed several lacerations and a fractured rib. The wounds suggest that the bird was killed by an avian predator.

The loss of 1 of the 3 radio-instrumented, translocated red-cockaded woodpeckers supports the observation by Odom et al. (1982) that translocated red-cockadeds are highly susceptible to avian predation if they have been radio-instrumented. Although activities and movements of the birds may appear normal, the radio transmitter may subtly alter behavior and/or provide a visual cue for avian predators, especially when the woodpeckers are unfamiliar with the habitat.

Even though only 1 in 5 translocated red-cockaded woodpeckers remained on the refuge or state park, that female was instrumental in halting the decline and fueling a rebound of those red-cockadeds (**Figure 2**). From 1984 through 1987 the translocated female and breeding male successfully produced 9 fledglings (4 pair-years) compared to 6 for all other pairs combined (at least 8 pair-years). With the addition of the successfully translocated female, nesting success increased 200 percent and total numbers climbed by 100 percent between 1983 and 1987.

In contrast to the increasing refuge and state park numbers, birds on the St. Joe study area have declined dramatically since 1985. Of the 10 known occupied clusters in 1985 only 3 were occupied in 1992, with total numbers declining from 28 birds to 6 (**Figure 3**). During this time period only 1 male red-cockaded woodpecker was lost due to translocation. The apparent reason for this decline is the extensive clearcutting that has occurred in the study area over the past 15 years. Nearly all primary foraging habitat outside the 5-acre clusters has been converted to unsuitable young slash pine plantations. Many inactive cavity trees and most potential replacement cavity trees have also been cut outside the small 5-acre "islands". Those clusters that have remained active are those that have inactive clusters and other suitable tracts of uncut pine forest nearby for foraging. Red-cockadeds occupying these sites have been observed foraging over 2 km from their cluster.

DISCUSSION

Translocations of red-cockaded woodpeckers to small declining populations can be a very effective tool in halting the decline and stimulating growth. This study and those of DeFazio et al. (1987) and Hess and Costa (1995) suggest that the most successful translocations may be those that introduce females to single males, a process which mimics the natural dynamics of pair formation (Lennartz et al. 1987, Walters et al. 1988). The technique is best applied when donor populations are large and stable or increasing, and limited to recipient populations with all-male clans and a shortage of females.

When the donor population is small and in danger of extirpation, removing females will only salvage a portion of the birds and genetic diversity of the population. The males, which often comprise a large proportion of a declining population (Baker 1983b, DeFazio et al. 1987, James 1991, this study), are soon lost as the population crashes. Therefore, moving pairs and even all-male clans may be important when the salvage of a small declining population is an objective.

In the 3 translocations described in this paper, there was a noticeable difference in the behavior of the translocated birds that may have been related to the condition of the habitat at the donor cluster. Although individual variation cannot be ruled out, those birds that resided in clusters where most of the foraging habitat had been adversely impacted by clearcutting appeared to respond much more positively to the translocation (based on behavior) than the pair that was removed from a site with most foraging habitat intact. Therefore, the success of translocating entire red-cockaded woodpecker groups may be enhanced if the birds are moved after donor habitats have been seriously degraded.

Considering the importance of cavities to

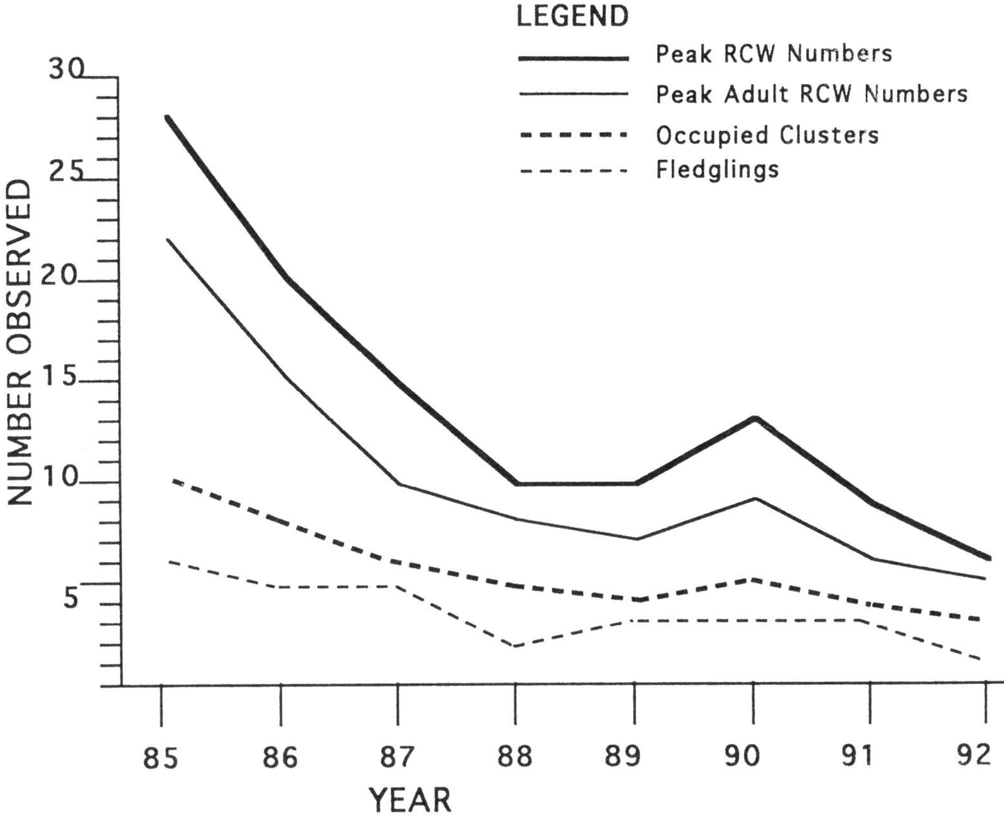

Figure 3. Status and nesting success of known 1985 red-cockaded woodpecker clusters on the St. Joseph Land and Development Company study area, 1985-1992.

the occupation of a site by red-cockaded woodpeckers (Doerr et al. 1989, Copeyon et al. 1991), release sites should contain at least as many suitable unoccupied cavities as translocated birds. Excess cavities would also be beneficial to help relieve interspecific competition for the cavities (see Kappes and Harris 1995). The artificial cavity construction techniques developed by Copeyon (1990), Allen (1991), and Taylor and Hooper (1991) have been widely adopted and highly successful (Copeyon et al. 1991, Carter and Engstrom 1995, Richardson and Stockie 1995, Watson et al. 1995).

Release sites for mixed-sex groups should be sufficient distances from resident groups containing unmated males to minimize the chance of disruption of the translocated group by roaming unmated males. This may only be a concern in spring translocations. However, the release sites should be near enough to occupied clusters to facilitate long-term interchange of dispersing birds.

ACKNOWLEDGMENTS

I am grateful to those contributing to this project over the years. Much of the equipment was supplied by R. Thompson, S. Nesbitt, or purchased via a grant from the National Fish and Wildlife Foundation. R. Matteson, U. S. Fish and Wildlife Service, loaned and assisted with the radio telemetry equipment. Field assistance was provided by G. Fishman, C. Gidden, S. Kelly, J. Hovis, G. Nelson, L. Williams, N. Brand, M. Porter, M. Balboni, D. Holle, M. Reeves, D. Bryan, S. Kirkman, R. Will, and other personnel and volunteers from St. Marks National Wildlife Refuge. I thank the Florida Park Service, especially C. Nichols, and the St. Joseph Land and Development Company, especially G. Forester and B. Newsom for access to study their birds. I gratefully acknowledge the support of D. Wood, Florida Game and Fresh Water Fish Commission, and R. Hooper, R. DeLotelle, D. Wood, and W. Baker for providing helpful suggestions on an earlier draft of this manuscript.

Potential Red-cockaded Woodpecker Habitat Produced on a Sustained Basis Under Different Silvicultural Systems

Jimmy S. Walker, Silviculturist
USDA Forest Service, Atlanta, Georgia 30367

ABSTRACT: Silviculture is the art and science of regenerating and tending a forest to meet specific objectives and is based on a knowledge of silvics. The underlying objective of nearly all forest management is the sustained yield of products and benefits (Smith 1986). Forests managed for red-cockaded woodpecker (*Picoides borealis*) (RCW) habitat or for timber production have many common silvicultural objectives. A sustained uniform flow of RCW habitat through time can be provided with a managed-forest approach. I modeled balanced uneven-aged, two-aged, and even-aged stands in order to compare the number of trees that meet size requirements for potential RCW habitat when grown under idealized conditions for loblolly pine (*Pinus taeda*) and longleaf pine (*P. palustris*) under different rotation lengths and regeneration methods. In addition, an irregular uneven-aged stand unmanaged for stand structure was modeled from an extant old-growth longleaf stand.
KEYWORDS: Forest management, sustained-yield, silviculture, regeneration, RCW habitat, timber production, models

In pre-Columbian times there may have been more than 60 million acres of prime RCW habitat across the range of the red-cockaded woodpecker (RCW). Naturally occurring wildfires and those set by the native Americans maintained the open park-like stands of pines desired by the RCW for colony sites and foraging areas. These stands were not static but ever changing with trees dying from lightning, insects, windthrow, and breakage. Sometimes a lone tree would die, sometimes a small patch of trees would be killed and sometimes large areas would be devastated by major hurricanes. The perpetuation of the forest resulted from those disturbances of from less than 1 acre to more than 1,000 acres (Wahlenberg 1946).

Today, the RCW is restricted to a fraction of its former habitat, and there are many demands and potential uses for that remaining habitat. But clearly, if the RCW is to survive and be recovered in the long-term, it is imperative that the remaining areas be properly managed to provide cavity trees and foraging habitat on a sustained basis.

As part of recent efforts by the Forest Service to develop a plan for providing sustained RCW habitat (Krusac et al. 1995), I analyzed different regeneration methods and rotation lengths for longleaf and loblolly pine, the 2 pine species most frequently associated with the RCW, to determine the quantity of suitable habitat provided. This paper is a brief synopsis of a portion of that analysis.

MANAGED FORESTS FOR SUSTAINED YIELD: AN OVERVIEW

A goal of forest management and the objective of forest regulation is to make a forest the source of an indefinitely sustained and uniform flow of benefits and services. Regulation is obtained at the forest (landscape level) and not at the stand level. The forest, not the stand, is the unit from which a sustained annual yield of products and benefits are provided. The longer the production cycle of the benefit or the age of the stand on which it is dependent, the more difficult the goal of sustained yield and uniform flow becomes because it takes decades to grow trees to a certain size or age (Smith 1986).

To provide a sustained uniform flow (yield)

of RCW habitat indefinitely requires that each annual age and/or size class of pine trees from Year 1 to recommended rotation age (which will be harvested and replaced each year) be equally represented in the areas managed for RCW. This balance of different stand structures and spatial patterns can provide habitats for a diversity of plants and animals (Oliver 1992).

Young pine stand management

The requirements for establishment, growth, and development of young pine trees (Oliver and Larson 1990) is the same whether for RCW habitat or for timber production. Stocking of seedlings, saplings, or pole-size trees should not be too dense or too sparse. Density should be low enough to maintain tree vigor but high enough to encourage natural pruning of limbs. The negative effects of residual older trees on establishment, growth and development of the young pine trees is the same whether for RCW habitat or for timber production, and must be taken into account.

Older pine stand management

The goal of providing potential cavity trees to meet the needs of a population of RCWs focuses on growing pine trees to a size and age where an adequate number will develop sufficient heartwood and red heart (*Phellinus pini*) for cavity excavation. In contrast, management for timber production focuses on growing healthy, sound trees (no decay) to a desired size. At our current level of understanding, the primary difference in providing potential cavity trees and prime sawtimber is the age to which trees are allowed to grow.

Rotation and Regulation

For the following discussion, rotation age refers to the oldest age class grown before an even-aged or two-aged regeneration method is used to reproduce a new age class and/or the maximum size class grown before an uneven-aged regeneration method is used to reproduce a new size class.

Rotation length affects the acres of forest in each age class (**Table 1**), the maximum size of trees, the number of trees, and the age of the oldest trees. The longer the rotation the fewer acres and trees there will be in each age class (Davis 1966). For a given rotation, soil properties, site quality, stocking, growth rate, disturbances, and tree species largely determine the number of trees in each age class (Smith 1986).

The successful regeneration, growth, and development of adequate numbers of pine trees is essential to providing RCW habitat in the long-term. It is only through the successful reestablishment of pine and pine-hardwood stands that a sustained uniform flow of RCW foraging and nesting habitat can be provided through time. Selection of appropriate rotation ages for a wide variety of sites allows planned regeneration where the probability that most stands have adequate tree vigor and health (except for catastrophic events) to maintain a reasonable level of stocking. Maintaining adequate stocking reduces the odds of isolation of colony sites and fragmentation of habitat. Poor sites may only produce foraging habitat. Those sites do not have the capability to produce potential nesting habitat or may not be able to sustain it for even a short period of time.

A forest is regulated when the area in each age class and/or size class is represented in equal proportion and consistently grown to provide a continual and approximately equal amount of outputs indefinitely (Davis 1966). Regulation applies whether these products are cavity trees or high quality pine sawtimber.

Tree Health and Longevity

Soil, site, and climate are the principal factors that alter the expected longevity of each tree species. Competition, senescence, and external factors such as insects, diseases, wind, and lightning are the main reasons that overstory trees die (Baker and Langdon 1990; Belanger 1981; Belanger et al. 1986; Boyer 1990; Bramlett and Kitchens 1983; Carter and Snow 1990; Guldin 1986; Hebb and Clewell 1976; Hepting 1971; Knight and Heikkenen 1980; Lawson 1990; Lohrey and Kossuth 1990; Mohr 1897; Oliver and Larson 1990; Platt et al. 1988; Spurr and Barnes 1980; Wahlenberg 1946, 1960).

One way of improving forest health and increasing tree longevity is to restore longleaf, shortleaf, slash, and loblolly pines to their natural sites, if they are currently occupied by another pine species. This would allow the site to grow trees that are more vigorous, less susceptible to insects and disease, and have the potential to live longer (Boyce 1961; Pritchett and Fisher 1987).

Intolerant, early successional species such as longleaf, slash, loblolly, shortleaf, and

Table 1. Numbers of acres containing each 10-year age class by rotation length. Values are acres per 10-year age class based on a 1,000 acre management compartment (does not include colony sites, replacement and recruitment stands).

	Rotation Length (years)						
	80	100	120	150	180	200	250
Age Class (years)				(acres)			
10	125	100	83	66	56	50	40
20	125	100	83	66	56	50	40
30	125	100	83	66	56	50	40
40	125	100	83	66	56	50	40
50	125	100	83	66	56	50	40
60	125	100	83	67	56	50	40
70	125	100	83	67	56	50	40
80	125	100	83	67	56	50	40
90		100	83	67	56	50	40
100		100	84	67	56	50	40
110			84	67	55	50	40
120			84	67	55	50	40
130				67	55	50	40
140				67	55	50	40
150				67	55	50	40
160					55	50	40
170					55	50	40
180					55	50	40
190						50	40
200						50	40
210							40
220							40
230							40
240							40
250							40

Virginia pine survive, grow, and develop best when free of competition and grown in full light (Baker 1987; Baker and Balmer 1983; Boyer and White 1990; Burns 1983; Zedaker et al. 1987). The relative intolerance of these species varies. For example, longleaf pine is especially intolerant of competition from any source, especially overtopping trees including parent trees. Over a 35 year period, basal area growth of the young longleaf pine trees was reduced about 55 percent under only 9 square feet of basal area per acre of parent trees and over 80 percent under 18 square feet as compared to the stand where the shelterwood seed trees were removed. Adequate longleaf regeneration was not retained (after 35 years) where parent tree density was greater than 27 square feet of basal area per acre (Boyer 1993).

Regeneration

The primary ecological and physiological factors that effect the establishment, survival, growth, and development of the desired tree species include percentage of canopy removed in the residual stand, root competition of residual trees, seed supply from residual trees, direct solar radiation, diffuse solar radiation, and competing vegetation (Smith 1986). Management can control some of these factors to enhance the establishment of new pine stands.

Smith (1986) states, "A **reproduction method** is a procedure by which a stand is established or renewed; the process is accomplished during the regeneration period by artificial or natural reproduction." Some species can only be regenerated by the clearcut method (e.g., sand and Virginia pine), while other species can be regenerated by one of several different regeneration methods (e.g., loblolly and longleaf pine).

Smith (1986) described the following

general reproduction methods:

Even-aged Stands.

The **clearcutting method** involves removal of the entire stand in one cutting operation. Regeneration of the new stand is obtained artificially by planting or seeding, or by natural seeding from adjacent stands or from trees cut in the clearing operation. The clearcutting method has been a successful way to regenerate loblolly, shortleaf, longleaf, slash, and Virginia pine (Baker 1987).

The **seed-tree method** involves removal of most of the stand in one cutting, except for a small number of seed trees left singly or in small groups (typically 6 to 12 square feet of basal area per acre). Well-distributed dominant and codominant trees of seed-bearing size on soils that do not cause the trees to be shallow-rooted which predisposes them to windthrow are necessary when using the seed-tree method. This method has been used successfully with loblolly, shortleaf, and slash pine (Baker 1987).

The **shelterwood method** involves removal of the trees in a stand in a series of cuttings (usually 2 or 3) over a relatively short portion of the rotation (Boyer 1963). Well-distributed dominant and codominant trees of seed bearing size on soils that do not cause the trees to be shallow-rooted which predisposes them to windthrow are necessary when using the shelterwood method. This method has been used successfully with loblolly, shortleaf, longleaf, and slash pine (Baker 1987).

Two-aged Stands.

The **irregular shelterwood method** involves the removal of trees in a stand in a series of cuttings (usually 2 or 3), similar to the even-aged shelterwood method, where establishment of essentially even-aged regeneration under the seed trees is encouraged. The irregular shelterwood method differs in that the final removal cut may take place much later or not at all. This creates a stand with two-age classes for long periods of time and then becomes an even-aged stand if the overstory is removed. Trees in this stand structure will usually be very irregular in height and diameter within the new age class as well as between the two age classes. This method can only be used in stands with sufficient, well-distributed dominant and codominant trees of seed-bearing size on sites that do not cause the trees to be shallow-rooted which predisposes them to windthrow.

The irregular shelterwood method is an untested method for loblolly, shortleaf, and slash pine. This method should be able to regenerate loblolly, shortleaf, and slash pine provided too many residual trees are not left (depends on growth and mortality rates as well as soil and site conditions). The irregular shelterwood method is not appropriate for longleaf pine because of longleaf pines's intolerance to competition from any source, especially overtopping trees including parent trees (Boyer 1993).

Balanced Uneven-aged Stands.

The **theoretical** balanced uneven-aged stand should have trees of each age and/or size class from seedlings to mature trees of rotation age and/or maximum tree size, with each age and/or size class occupying an equal area. Structure in the merchantable component (usually 6-inch and larger diameter classes) of the stand is best maintained by the BDQ (basal area, maximum diameter, and constant ratio of trees in successive diameter classes) method (Farrar 1984; Farrar and Murphy 1989). To maintain an adequate uneven-aged pine stand structure, establishment of pine regeneration is usually necessary at least once every 10-year period.

The **group selection method** involves removal of the trees, usually the oldest or largest trees, in scattered patches at relatively short intervals (about every 10 years), repeated indefinitely, to encourage the continuous establishment of regeneration and maintenance of a balanced uneven-aged stand (Smith 1986; Farrar 1984).

The group selection method should work to regenerate uneven-aged stands of loblolly, shortleaf, and longleaf pine on some sites (Baker 1987). Use of the group selection method to regenerate longleaf pine on medium sites has only been tested for about 15 years. Farrar and Boyer (1991) state: "A selection system may not work well for longleaf pine on very poor, dry, sandy sites, wet flatwoods sites with dense palmetto understories, or very good mesic sites because effective prescribed burning for competition control and/or seedbed preparation may be difficult to achieve".

The **single-tree selection method** involves removal of selected trees from all merchantable diameter classes (usually 6" DBH and larger) at relatively short intervals (every 3 to 15 years), repeated indefinitely, to encourage the continuous establishment of regeneration and maintenance of a balanced uneven-aged stand (Smith 1986; Farrar 1984; Farrar and Murphy 1989). Care must be taken not to reduce genetic quality and diversity by cutting only the best dominant individuals (high grading).

The single-tree selection method is best adapted to tolerant, late-successional species, but has been successfully used to regenerate loblolly and shortleaf pine in **uneven-aged** stands, provided hardwood competition is controlled on a regular basis (Baker 1987). This method is not appropriate to regenerate longleaf pine because of longleaf pine's intolerance to competition from any source, especially overtopping trees including parent trees (Boyer 1993).

Stand Initiation

Smith (1986) states, "Site preparation, whether deliberate or unintentional, may be more crucial in the establishment of regeneration than the method of reproduction cutting. Many species can be reproduced by several different methods of cutting but by only one general program of site preparation dictated by the characteristics of species and site. The important objective is to prescribe and create environmental conditions conducive to the establishment and growth of the desired species. Most of the treatments are applied during the period of establishment, but some are started well in advance of harvest cutting or applied occasionally throughout the rotation."

Site preparation methods that have been successful and practical are prescribed fire, mechanical, manual, herbicides, and biological.

Spatial and Temporal Arrangement of Age and/or Size Classes

Regeneration cuts and natural disturbances of similar magnitude (Hunter 1990; Spurr and Barnes 1980) determine the times when new trees (regeneration) appear or start active development on any unit of ground area. The new aggregation of trees produced are essentially the same age. Various spatial patterns of age classes are created by differences in the timing of regenerative events. The area occupied by a given age class can be of any size, provided that it is large enough that some new trees can continue to grow in height without being arrested in growth by expansion of crowns and/or root competition of adjacent older trees (Smith 1986).

There is a distinct difference between balanced and irregular uneven-aged stands. If the objective is to grow an intolerant southern yellow pine to a certain size and it takes 100 years to grow this size tree, a **balanced** uneven-aged stand should consist of ten different age and/or size classes, each of which occupy an approximately equal area. The age and/or size classes should also be spaced at uniform intervals from newly established reproduction to trees near rotation age. In contrast, **irregular** uneven-aged stands do not contain all the age and/or size classes necessary to ensure a continual supply of trees that will arrive at rotation age and/or size at short intervals indefinitely (Smith 1986; Farrar 1984). As a consequence of missing age classes, sooner or later there will be a shortage of what ever product you are trying to produce e.g., RCW cavity trees.

Stand Density

Uneven-aged stands for certain southern yellow pines usually have basal areas maintained between 45 to 75 square feet per acre (does not include smaller trees up to the 4-inch diameter class) while even-aged stands usually have basal areas maintained between 80 to 120 square feet per acre (in fully stocked stands of 6-inch or larger diameter classes).

Stand densities before cutting of uneven-aged stands on average sites should not exceed 75 square feet of basal area per acre and after cutting should not be less than 45 square feet of basal area per acre. Uneven-aged forests grow fewer trees in each age (size) class than do even-aged forests because of the ecological conditions needed to grow intolerant pine trees of all sizes intimately mixed throughout a stand. The frequency of cutting can range from every 3 years up to every 15 years depending on basal area growth rate and basal areas left after each cutting (Farrar 1984).

Role of Thinnings

After a new stand is established, a long period follows during which trees grow through various stages until they are ready for replacement by a succeeding generation. Various intermediate cuttings or tending operations may be made during the life of the stand from the regeneration stage to maturity. Thinning is an intermediate stand-tending practice that is used mostly to control the growth, health, and vigor of trees by reducing stand density (Smith 1986).

The main objective of thinning is to keep the most promising trees growing steadily by removing less desirable, neighboring trees before their competition becomes injurious to the health and vigor of the entire stand. The number of trees per acre is reduced as it would be under natural conditions, but at a substantially faster rate (Smith 1986).

METHODS

Modeling

Projected outputs are based on a **1,000-acre** unit of land. The 1,000-acre unit of land was used to compare even-aged, two-aged, and uneven-aged regeneration methods, silvicultural systems, and management strategies upon a forest-wide regulation basis. The 1,000 acre units did not include colony sites, or replacement and recruitment stands. Stated out-puts are post-thinning and pre-regeneration values for even-aged and two-aged stands and post-cutting cycle values (maximum diameter adjusted to retain rotation size and/or age trees) for uneven-aged stands.

The number of trees per acre used for the modeling and analysis were based on average diameter and square feet of basal area per acre for each age and/or size class (Farrar and Boyer 1991; Farrar and Murphy 1989; Forest Service 1976; Schumacher and Coile 1960, Wahlenberg 1946; Wahlenberg 1960). For even-aged and two-aged stands, growth rates and basal areas were selected for each diameter and age class to be representative of what could reasonably be assumed to occur under certain conditions. For example, the average diameter, basal area, and number of trees per acre for each longleaf pine age class from 10 to 80 are the same for an 80-year, 100-year, 120-year, 150-year, 200-year and 250-year rotation and then at the same rates for the next older age class where appropriate for longer rotation lengths. The relationships between rotation lengths and the other parameters should be comparable if other growth rates and basal areas are used. The uneven-aged stands were analyzed using computer programs developed by the Southern Forest Experiment Station for loblolly and longleaf pine. These evaluations assumed that age and/or size class distributions were balanced with one exception: The longleaf irregular uneven-aged stand was modeled after the Wade Tract (Landers et al. 1990; Platt et al. 1988). In addition, all sites were assumed to have the potential for an adequate number of trees to live to rotation age (except for catastrophic events) while providing at least a minimum stand structure level in most cases. It was also assumed that hardwoods in all pine stands were controlled at the understory level, and that all regeneration and tending operations were accomplished in a timely manner.

I considered pines ≥ 10 inches DBH suitable for foraging. This decision was based on Lennartz and Henry (1985). It is well documented that smaller trees are used readily as foraging substrate by the RCW (Hooper and Lennartz 1981; DeLotelle et al. 1983b) Thus, the 10 inch criteria does not represent the minimum tree suitable for foraging, but I ignored smaller trees in my modeling. Loblolly pines ≥ 18 inches DBH and ≥ 70 years old, and longleaf pines ≥ 15 inches DBH and ≥ 90 years old were considered potential cavity trees. Such pines would have heartwood columns of sufficient diameter (minimum of 5 inches) to contain a RCW cavity (Clark 1992). While trees younger than this are frequently used for cavities (Thompson and Baker 1971; Jackson et al. 1979b; Hooper 1988), there is reason to believe that even older trees have greater potential as cavity trees (Conner and O'Halloran 1987; DeLotelle and Epting 1988; Hooper et al. 1991b; Conner and Rudolph 1995a). In most of the scenarios projected here, many potential cavity trees are much older and larger than my minimum criteria.

The modeling represented the long term look at the number of trees that met size and age requirements for potential RCW habitat grown under **idealized** conditions on either a loblolly pine (site index of 90 feet in 50 years) or a longleaf pine site (site index of 70 feet in 50 years) for different rotation lengths and regeneration methods (except irregular uneven-aged stand). The models showed the expected number of trees occurring and the acres occupied by those trees when age and/or size class distribution was balanced (except Irregular Uneven-aged). In reality, the number of trees vary by growth rates, soil, site, and stand conditions; however, the relationships between the number (more or less) of trees occurring for other conditions will generally be similar to those predicted in these models.

Rotation length affects the number of acres

and trees in each age class and, to some degree, their distribution over the forest (**Table 1**). The regeneration method affects the stand structure and length of time for regeneration, survival, growth, and development of trees in the new stand. These models do not show the effects of going from the current conditions of the national forests to the desired future condition, but instead represents conditions after the forests are in a state of equilibrium (regulated).

The even-aged and uneven-aged stands were fairly easy to model. The two-aged stands were very complex stands to simulate. In two-aged stands, the growth and development of the older age class (parent trees) will pass through one of three major conditions depending on age (size):

(1) The basal area per acre in parent trees increase at various rates for many years before stabilizing and then declining.

(2) The basal area per acre in parent trees remains constant for many years because growth and mortality balance out before declining.

(3) The basal area per acre in parent trees decreases because mortality exceeds growth.

The effects on the new age class of pine and other vegetation depends on the number, size, and basal area increase or decrease of pine parent trees, soil and site conditions, and mortality patterns. The effects of root competition from the parent pine trees and other vegetation on the younger pine trees are greater on droughty sites and other sites during severe droughts. Only **one** condition was modeled for the loblolly pine parent trees in two-aged stands, where mortality exceeded growth at 90 years of age and all trees were dead by 160 years of age.

There are some rounding differences between the numbers in the Tables and Figures. For example in **Table 6** and **Figure 2** for an 80-year rotation, the actual number of trees is 7,525 but is rounded to 8,000 (**Table 6**) and adds up to 7,000 (**Figure 2**) because of the rounding for each age class.

Description of regeneration methods and rotation lengths that were modeled

Rotations. For the various methods 80, 100, and 120 year rotations were used for managing loblolly pine. Rotations used for longleaf pine were 80, 100, 120, 150, 200, and 250 years. Age and/or size class distribution was balanced. Table 1 shows the acres in each 10-year age class for the various rotation lengths.

Clearcut. Stands were clearcut at the end of the specified rotations. All stand establishment (site preparation and planting) and tending (thinnings) practices were accomplished in a timely manner.

Shelterwood. At the end of the specified rotation, stands to be regenerated were cut back to 20-30 square feet of basal area per acre. After the shelterwood seed cut was made, the remaining parent trees were not included as part of the RCW habitat figures. In 10 years or less the final shelterwood overstory removal cut was made.

Shelterwood With 20% Clumps. The shelterwood method was used on 80% of each pine stand scheduled to be regenerated. The remaining 20% of the stand would be left at its present stocking level unless a thinning was needed. A 1,000 acre unit of land on a 80-year rotation with balanced age classes would have 125 acres in each age class (**Table 1**). For example, the 80 year age class could be divided into 5 25-acre stands (or some other acreage combination) depending on existing size of stands. A shelterwood seed cut would be made on 20 acres of each 25-acre stand. The remaining 5 acres could be left as several small clumps (1-2 preferred) but usually not less than 1 acre in size. If needed, these clumps could be thinned to the desired stocking level for the existing site and stand conditions. In about 10 years, the final shelterwood overstory removal cut would be made on each 20 acres that had been regenerated. The 5 acres in clumps would not be regenerated until they were twice the rotation age if adequate stand structure was maintained or until stand structure was destroyed prior to rotation aged (remaining parent trees would usually not be cut). In the longer rotations for loblolly and longleaf pines, most of the trees left in the clumps would die from natural causes before 2 rotations had transpired. Nonetheless, many very old trees for cavity excavation would be provided across the forest in colony-sized clumps.

Irregular Shelterwood—40 BA Per Acre—Staged Removal-20% Clumps. This method is similar to the shelterwood with clumps except that 40 square feet of basal area per acre would be left initially. Then, in about 10 years, the parent trees would be cut back to about 20-30 BA per acre. The remainder of the parent trees would be removed (1 or more cuts) when the new stand was considered suitable for foraging (after another 10-20 years). The trees left in the uncut clumps would be treated the same as the above method. This was only modeled for loblolly pine: longleaf is too intolerant of competition for this method. This method is untested and has been proposed because it maintains continuous forest coverage.

Irregular Shelterwood—40 BA Per Acre —No Removal—No Clumps. This is also an untested method. The shelterwood seed cut would leave about 40 BA per acre. There would be no additional removal of parent trees. Only **one** condition was modeled for the loblolly pine parent trees in two-aged stands, where mortality exceeded growth at 90 years of age and all trees were dead by 160 years of age. Longleaf is too intolerant of competition for this method.

Irregular Shelterwood—30 BA Per Acre—No Removal—No Clumps. This method is similar to the above method except leave BA would be 30 square feet per acre. It is not suitable for longleaf because of intolerance. Again, this is an untested method.

Group Selection. This method has had limited research and field testing with southern pines. Two uneven-aged models were used, 1 for loblolly and 1 for longleaf pine. Structure in the merchantable component (usually 6-inch and larger diameter classes) of the stand was maintained by the BDQ (basal area, maximum diameter, and constant ratio of trees in successive diameter classes) method (Farrar 1984; Farrar and Murphy 1989). Maximum diameter for loblolly pine was assumed to be the same size as the largest size class of trees in a stand managed using the shelterwood method **(Table 2)**. Maximum diameter used for longleaf pine was 1 inch smaller than the largest size class of trees in a stand managed using the shelterwood method **(Table 3)** because of the slower growth rates in small openings. A 1-inch Q of 1.1 was used for loblolly pine and a 1.2 was used for longleaf pine. Residual stand basal area averaged 60 square feet per acre for loblolly pine and ranged from 50 to 60 square feet depending on the rotation length for longleaf pine. During each cutting cycle, the overstocked groups were thinned and the groups where most of the trees exceed the maximum diameter or where there is a surplus of trees needed to maintain the desired stand structure were cut. This cutting would occur at relatively short intervals, repeated indefinitely, to encourage the continuous establishment of regeneration and maintenance of a balanced uneven-aged stand structure (Smith 1986; Farrar 1984).

Single-Tree Selection. The parameters for loblolly pine were the same as used with the above method. During each cutting cycle, selected trees are cut that exceed the maximum diameter in addition to some trees from most merchantable diameter classes. Cutting would occur at relatively short intervals, repeated indefinitely, to encourage the continuous establishment of regeneration and maintenance of a balanced uneven-aged stand structure (Smith 1986; Farrar 1984). This method is not suitable for longleaf because of its intolerance to competition.

Irregular Uneven-aged Stand. This condition was modeled for longleaf after the Wade Tract, a stand of old-growth longleaf pine in south Georgia (Landers et al. 1990; Platt et al. 1988). There was no planned regeneration method for insuring the establishment, survival, growth, and development of longleaf pine regeneration. Some regeneration would be established naturally. A 250-year rotation was used based on the maximum age shown in data from the Wade Tract. A few older trees existed on the Wade Tract (Platt et al. 1988) but I ignored them in the modeling. Nonetheless, a few trees older than 250 years would occur under the conditions modeled but they would make up a very small percentage of the trees, just as in the Wade Tract. The data were derived on a per acre basis and then expanded to 1,000 acres so as to be comparable to the other.

RESULTS

Potential cavity trees

Loblolly Pine. The spatial arrangement of potential cavity trees varies between regeneration methods and rotation ages primarily because of stand structure and tree longevity. The even-aged stands managed using

Table 2. Mean and maximum diameters of loblolly pine trees ≥ 10 inches DBH and ≥ 30 years old produced under different regeneration methods, management strategies) and rotation ages. Values are averages for a 1,000 acre management compartment (does not include colony sites, replacement and recruitment stands).

	Rotation Age (years)		
	80	100	120
Regeneration Methods and Management Strategies	Mean DBH (Max.)	Mean DBH (Max.)	Mean DBH (Max.)
	(inches)		
Clearcut	14.8 (23)	16.0 (26)	16.9 (28)
Shelterwood	14.9 (22)	16.1 (25)	16.9 (27)
Shelterwood with 20% of Area in Clumps	15.6 (28)	16.0 (28)	16.8 (28)
Irregular Shelterwood with 40 BA Retained, Staged Removal, 20% of Area in Clumps	16.0 (28)	16.9 (28)	17.3 (28)
Irregular Shelterwood with 40 BA Retained, No Removal, No Clumps	18.0 (28)	17.1 (28)	17.1 (28)
Irregular Shelterwood with 30 BA Retained, No Removal, No Clumps	17.0 (28)	17.0 (28)	16.9 (28)
Group Selection	14.7 (22)	15.5 (25)	16.0 (27)
Single-tree Selection	14.7 (22)	15.5 (25)	16.0 (27)

Table 3. Mean and maximum diameters of longleaf pine trees ≥ 10 inches DBH and ≥ 30 years old produced under different regeneration methods, management strategies and rotation ages. Values are averages for a 1,000 acre management compartment (does not include colony sites, replacement and recruitment stands).

	Rotation Age (years)					
	80	100	120	150	200	250
Regeneration Methods and Management Strategies	Mean DBH (Max.)	Mean DBH (Max.)	Mean DBH (Max.)	Mean DBH (Max.)	Mean DBH (Max.)	Mean DBH (Max.)
	(inches)					
Clearcut	13.1 (17)	14.0 (19)	14.6 (20)	15.4 (21)	16.3 (24)	17.0 (26)
Shelterwood	12.7 (16)	13.6 (18)	14.3 (19)	15.0 (20)	16.0 (23)	16.6 (25)
Shelterwood with 20% of Area in Clumps	13.6 (21)	14.4 (23)	15.0 (25)	15.6 (27)	16.0 (27)	16.6 (27)
Group Selection	12.0 (15)	12.6 (17)	12.8 (18)	13.1 (19)	13.7 (22)	14.0 (24)

the Clearcut and Shelterwood methods would have stand size areas (10+acres) of potential cavity trees scattered over the 1,000 acres. In addition, the Shelterwood With Clumps would also have 20% of each stand in scattered clumps (usually 1 acre or larger in size). All rotation age loblolly pine trees left in the clumps died by age 160 in this model. In addition, the Irregular Shelterwood With Clumps method would have potential cavity trees present in the regenerated areas until the final irregular shelterwood removal cut. The Irregular Shelterwood Without Clumps method would have varying numbers of potential cavity trees scattered over the stands until about age 160. Because all the parent trees died by age 160, there was a decrease in the number of potential cavity trees and the acres they occupied **(Tables 4, 5).** The uneven-aged stands managed using the Group Selection system would not have potential nesting trees on every acre (**Figure 3**). For example, loblolly pine where the objective was to grow trees of a certain size in 100 years could have about 400 1-acre groups containing potential cavity trees intimately scattered over the 1,000 acres. Those groups could average from 15-20 trees per acre. The Clearcut method and Shelterwood method also provide 400 acres of potential cavity trees (**Figure 3**) for a 100-year rotation. Uneven-aged stands managed using the Single-Tree Selection system would average about 5-9 potential cavity trees per acre intimately scattered over the 1,000 acres (**Figure 3**).

The retention of rotation age trees, whether in clumps (Shelterwood With Clumps) or scattered (Irregular Shelterwood Without Clumps), affects the relationships between the number of foraging **(Table 8)** and potential cavity trees **(Table 4)** and average stand DBH **(Table 2)** more than the other methods. As rotation age increases and more of this old age class dies, the distribution and numbers of foraging and potential cavity trees change as new age classes develop and grow. The Irregular Shelterwood With Clumps had a different pattern. These methods are the most unpredictable in terms of foraging and potential cavity trees provided because of the possible growth and mortality combinations that may occur.

In general, regeneration method had somewhat more influence on numbers of potential loblolly pine cavity trees than did rotation lengths modeled (**Table 4**). Group Selection and Single-Tree Selection methods provided the fewest potential cavity trees (5,000 to 9,000) for a given rotation. All the other methods provided from 10,000 to 15,000 potential cavity trees depending upon method and rotation age. The primary reason Group Selection and Single-Tree Selection methods provided fewer cavity trees resulted from the ecological conditions needed to grow intolerant pine trees of all sizes intimately mixed throughout a stand.

Rotation age did have a significant effect on the numbers of potential cavity trees but the effect was not consistent among regeneration methods (**Table 4**). With the Irregular Shelterwood -40BA-No Removal-No Clumps and the Irregular Shelterwood-30BA-No Removal-No Clumps, the number of potential cavity trees decreased with increasing rotation age. This decrease in potential cavity trees resulted from increasing mortality of the retained parent trees and decreasing acres in each age class as rotation age increased. With all other methods the number of potential cavity trees increased as rotation age increased.

Regardless of the effect of rotation age on numbers of potential cavity trees (**Table 4, Figure 1**), their is little doubt that the quality of such trees increases with increasing age. However, there is some possibility that beyond a certain (but unknown) age that the quality of loblolly pine as potential cavity trees decreases (Baker 1983b).

The area of forest with potential cavity trees increased with increasing rotation age with the Clearcut, Shelterwood, Shelterwood With Clumps, and Irregular Shelterwood With Clumps methods (**Table 5, Figure 3**). The Irregular Shelterwood Without Clumps methods had a decrease in area with cavity trees as rotation age increased.

The numbers of potential cavity trees per acre vary between regeneration methods and rotation ages (**Figure 3**). Even-aged methods (Clearcut, Shelterwood and Shelterwood With Clumps) produced more potential cavity trees per acre on more acres than the other methods. Two-aged methods (Irregular Shelterwood With Clumps and Irregular Shelterwood Without Clumps) had a much higher proportion of acres (32-62%) with 1-9 potential cavity trees per acre. Uneven-aged methods (Group Selection and Single-Tree Selection) aggregate the potential cavity trees differently across the

Table 4. Numbers of potential cavity trees produced under different regeneration methods, management strategies and rotation ages. Values are 1,000's of loblolly pine trees ≥ 18 inches DBH and ≥ 70 years old based on a 1,000 acre management compartment (does not include colony sites, replacement and recruitment stands).

Regeneration Methods and Management Strategies	Rotation Age (years) 80	100	120
	(thousands of trees)		
Clearcut	10	13	14
Shelterwood	11	14	15
Shelterwood with 20% of Area in Clumps	11	14	15
Irregular Shelterwood with 40 BA Retained, Staged Removal, 20% of Area in Clumps	11	12	13
Irregular Shelterwood with 40 BA Retained, No Removal, No Clumps	14	12	10
Irregular Shelterwood with 30 BA Retained, No Removal, No Clumps	12	12	11
Group Selection	5	8	9
Single-tree Selection	5	8	9

Table 5. Numbers of acres containing potential cavity trees produced under different regeneration methods, management strategies and rotation ages. Values are acres with loblolly pine trees ≥ 18 inches DBH and ≥ 70 years old based on a 1,000 acre management compartment (does not include colony sites, replacement and recruitment stands).

Regeneration Methods and Management Strategies	Rotation Age (years) 80	100	120
	(acres)		
Clearcut	250	400	498
Shelterwood	250	400	498
Shelterwood with 20% of Area in Clumps	375	440	532
Irregular Shelterwood with 40 BA Retained, Staged Removal, 20% of Area in Clumps	675	680	730
Irregular Shelterwood with 40 BA Retained, No Removal, No Clumps	1000	900	747
Irregular Shelterwood with 30 BA Retained, No Removal, No Clumps	1000	900	747
Group Selection	250	400	498
Single-tree Selection	1000	1000	1000

Table 6. Numbers of potential cavity trees produced under different regeneration methods, management strategies and rotation ages. Values are 1,000's of longleaf pine trees ≥ 15 inches DBH and ≥ 90 years old based on a 1,000 acre management compartment (does not include colony sites, replacement and recruitment stands).

	Rotation Age (years)					
	80	100	120	150	200	250
Regeneration Methods and Management Strategies			(thousands of trees)			
Clearcut	0	9	13	16	18	17
Shelterwood	0	10	15	18	19	18
Shelterwood with 20% of Area in Clumps	8	14	16	18	17	18
Group Selection	0	5	7	7	9	9

Table 7. Numbers of acres containing potential cavity trees produced under different regeneration methods, management strategies and rotation ages. Values are acres with longleaf pine trees ≥ 15 inches DBH and ≥ 90 years old based on a 1,000 acre management compartment (does not include colony sites, replacement and recruitment stands).

	Rotation Age (years)					
	80	100	120	150	200	250
Regeneration Methods and Management Strategies			(acres)			
Clearcut	0	200	332	469	600	680
Shelterwood	0	200	332	469	600	680
Shelterwood with 20% of Area in Clumps	200	360	468	573	610	696
Group Selection	0	200	332	469	600	680

1,000 acres (**Figure 3**).

Longleaf Pine. The spatial arrangement of the potential cavity trees would be similar to the discussion on loblolly pine for the Clearcut, Shelterwood, and Shelterwood With Clumps methods (**Tables 6, 7**). The Shelterwood With Clumps method would have clumps ranging in age from 80 years to 160 years for the 80-year rotation, from 100 years to 200 years for the 100-year rotation, from 120 years to 240 years for the 120-year rotation, and from 150 years to 300 years for the 150-year rotation. All longleaf pine trees left in the clumps in the stands managed by 200- and 250-year rotations died by age 330 in this model. At about age 165, these stands managed using the 200- and 250-year rotations would not have any clumps of old trees remaining.

The uneven-aged stands managed using the Group Selection system would not have potential nesting trees on every acre. For example, longleaf pine where the objective was to grow trees of a certain size in 120 years could have about 332 one-acre groups or 166 two-acre patches containing potential cavity trees intimately scattered over the 1,000 acres. Those groups could average about 21 trees per acre. Thus, the increase in the number of potential cavity trees with rotation ages shown in **Table 6**, was due primarily to the increase in acres with potential cavity trees (**Table 7**).

The retention of rotation age trees in clumps (Shelterwood With Clumps) affects the relationships between the number of foraging (**Table 9**) and potential cavity trees (**Table 6**) and average stand DBH (**Table 3**) more than the other methods. As rotation age increases and more of this old age class dies, the distribution and numbers of foraging and potential cavity trees change as new age classes

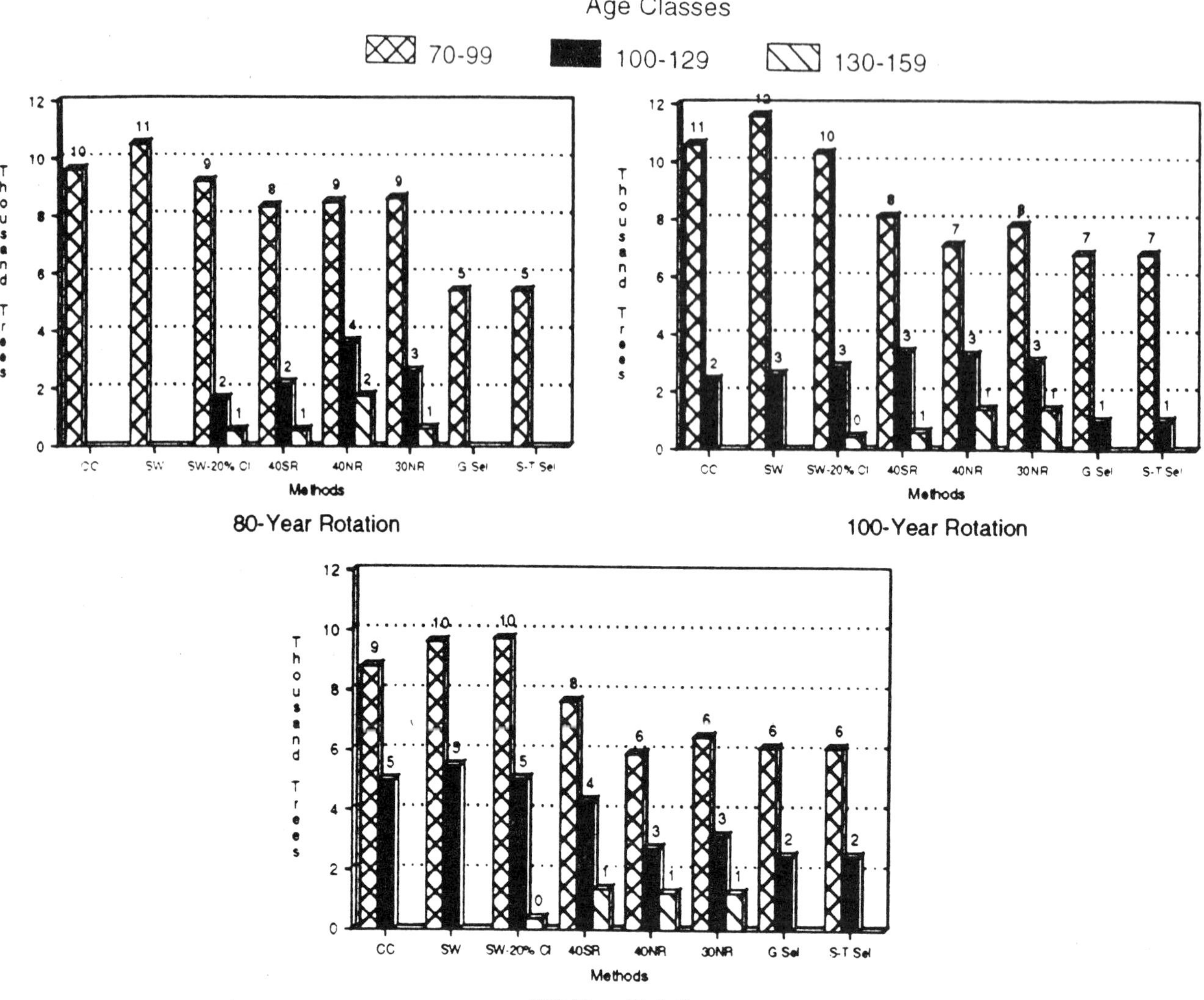

Figure 1. Number per 1,000 acres of potential loblolly pine cavity trees (≥18 inches dbh and ≥70 years old) by age classes for different regeneration methods and different rotation ages. CC = clearcut; SW = Shelterwood; SW-20% Cl = Shelterwood With 20% In Clumps; 40SR = Irregular Shelterwood - 40 Basal Area - Staged Removal - 20% In Clumps; 40NR = Irregular Shelterwood - 40 Basal Area - No Removal - No Clumps; 30NR = Irregular Shelterwood - 30 Basal Area - No Removal - No Clumps; G Sel = Group Selection; S-T Sel = Single-Tree Selection. See text for explanation of regeneration methods.

develop and grow. These methods are the most unpredictable because of the possible growth and mortality combinations that may occur.

If 80-year rotations are excluded, method of regeneration had a bigger influence on the number of potential longleaf pine cavity trees than rotation age (**Table 6, Figures 2, 5**). The even-aged methods (Clearcut, Shelterwood, and Shelterwood With Clumps) greatly out performed the uneven-aged methods (Group Selection and Irregular Uneven-aged) in producing numbers of potential cavity trees. For 80- and 100-year rotations, the Shelterwood With Clumps produced considerably more potential cavity trees than any of the other methods for those same rotations. It was the only method to provide potential cavity trees with 80-year rotations (**Table 6; Figure 2**). Also, the 80-year rotation provided nearly as many potential nesting trees as the Group Selection method with a 250-year rotation (**Table 6**) and the unmanaged Irregular Uneven-aged forest (**Figure 5**).

Rotation age had a significant effect on the number of potential cavity trees provided (**Table 6**), but this effect was mainly expressed

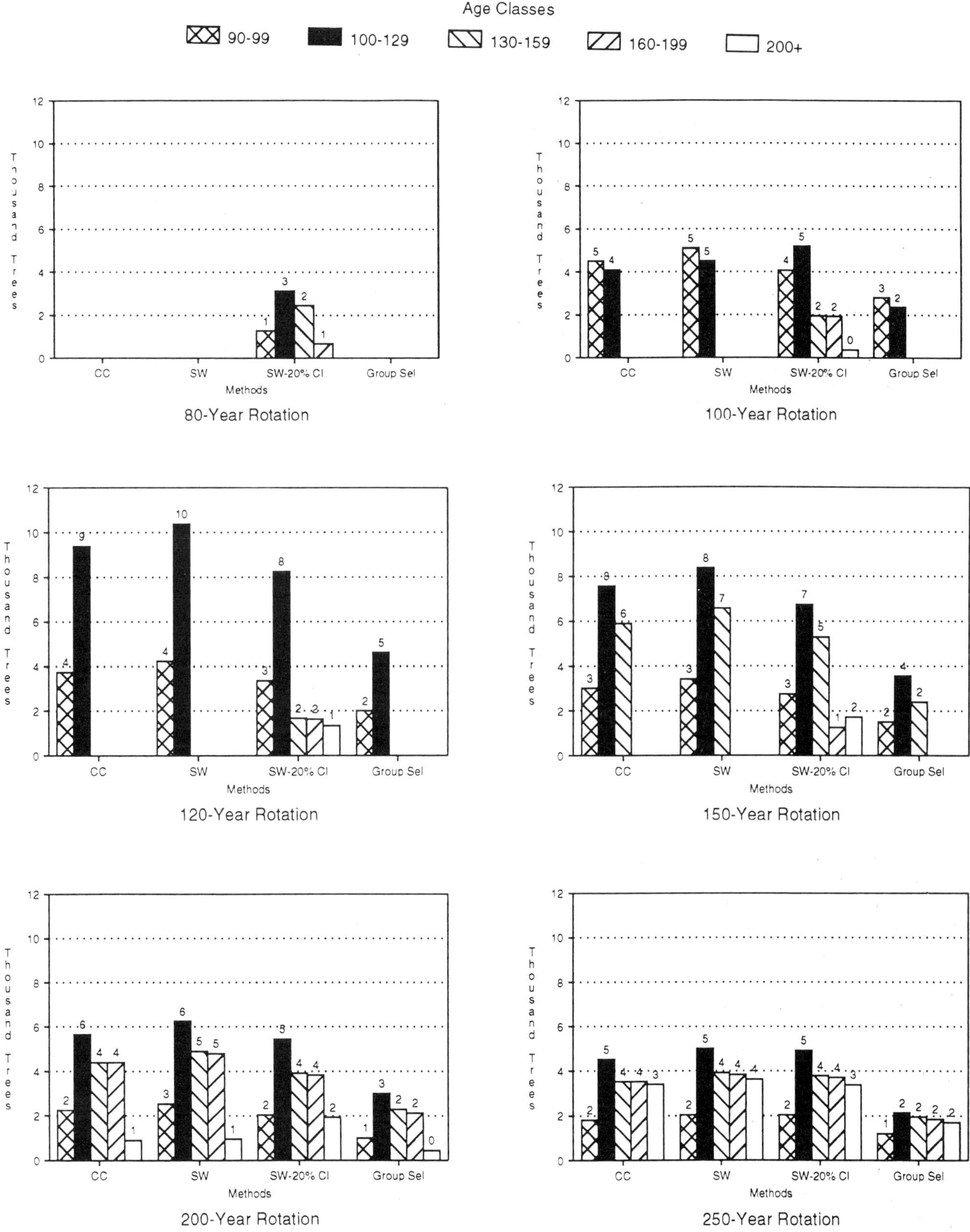

Figure 2. Number per 1,000 acres of potential longleaf pine cavity trees (≥15 inches DBH and ≥90 years old) by age classes for different regeneration methods and different rotation ages. CC = Clearcut; SW = Shelterwood; SW-20% Cl = Shelterwood With 20% In Clumps; Group Sel = Group Selection. See text for explanation of regeneration methods.

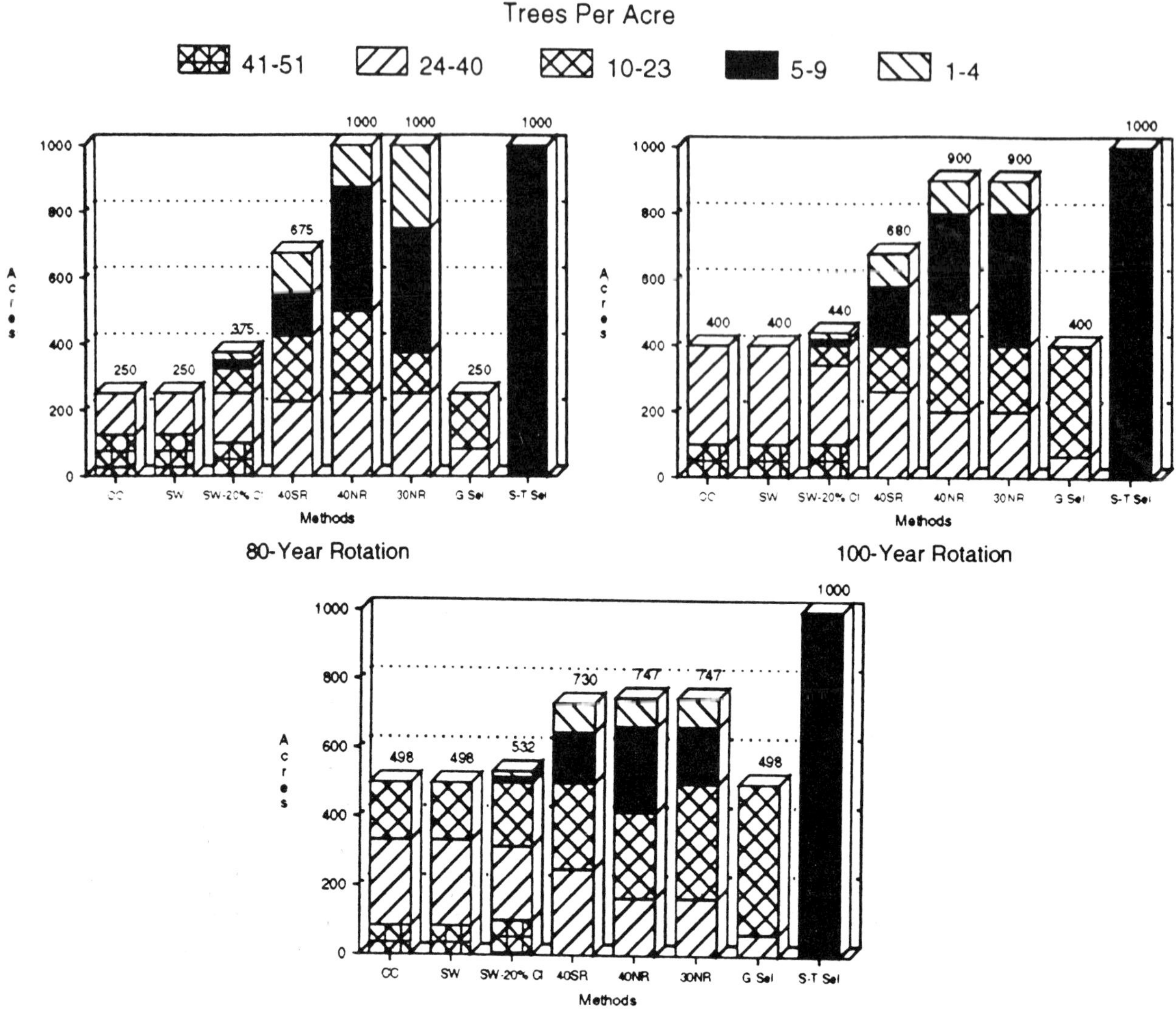

Figure 3. Acres of potential loblolly pine cavity trees (≥18 inches DBH and ≥70 years old) by trees per acre (average stand stocking levels) for different regeneration methods and different rotation ages. CC = Clearcut; SW = Shelterwood; SW-20% Cl = Shelterwood With 20% In Clumps; 40SR = Irregular Shelterwood - 40 Basal Area - Staged Removal - 20% In Clumps; 40NR = Irregular Shelterwood - 40 Basal Area - No Removal - No Clumps; 30NR = Irregular Shelterwood - 30 Basal Area - No Removal - No Clumps; G Sel = Group Selection; S-T Sel = Single-Tree Selection. See text for explanation of regeneration methods.

in rotations less than 150 years. With 150-year and longer rotations, there was little change in the number of potential cavity trees within any of the methods (**Table 6**). However, the longer the rotation, the older the potential cavity trees were (**Figure 2**).

The area of forest with potential cavity trees increased with increasing rotation age similarly with all regeneration methods (**Table 7; Figure 4**).

The numbers of potential cavity trees per acre vary between regeneration methods and rotation ages (**Figure 4**). For a 120-year rotation for longleaf pine, the Clearcut and Shelterwood methods ranged from 34-51 trees per acre, the Shelterwood With Clumps ranged from 10-51 trees per acre but with most in the 24-51 range, and the Group Selection method ranged from 15-24 trees per acre.

Foraging habitat

Method of regeneration had a greater effect on the number of trees suitable for foraging than did rotation length (**Tables 8, 9; Figure**

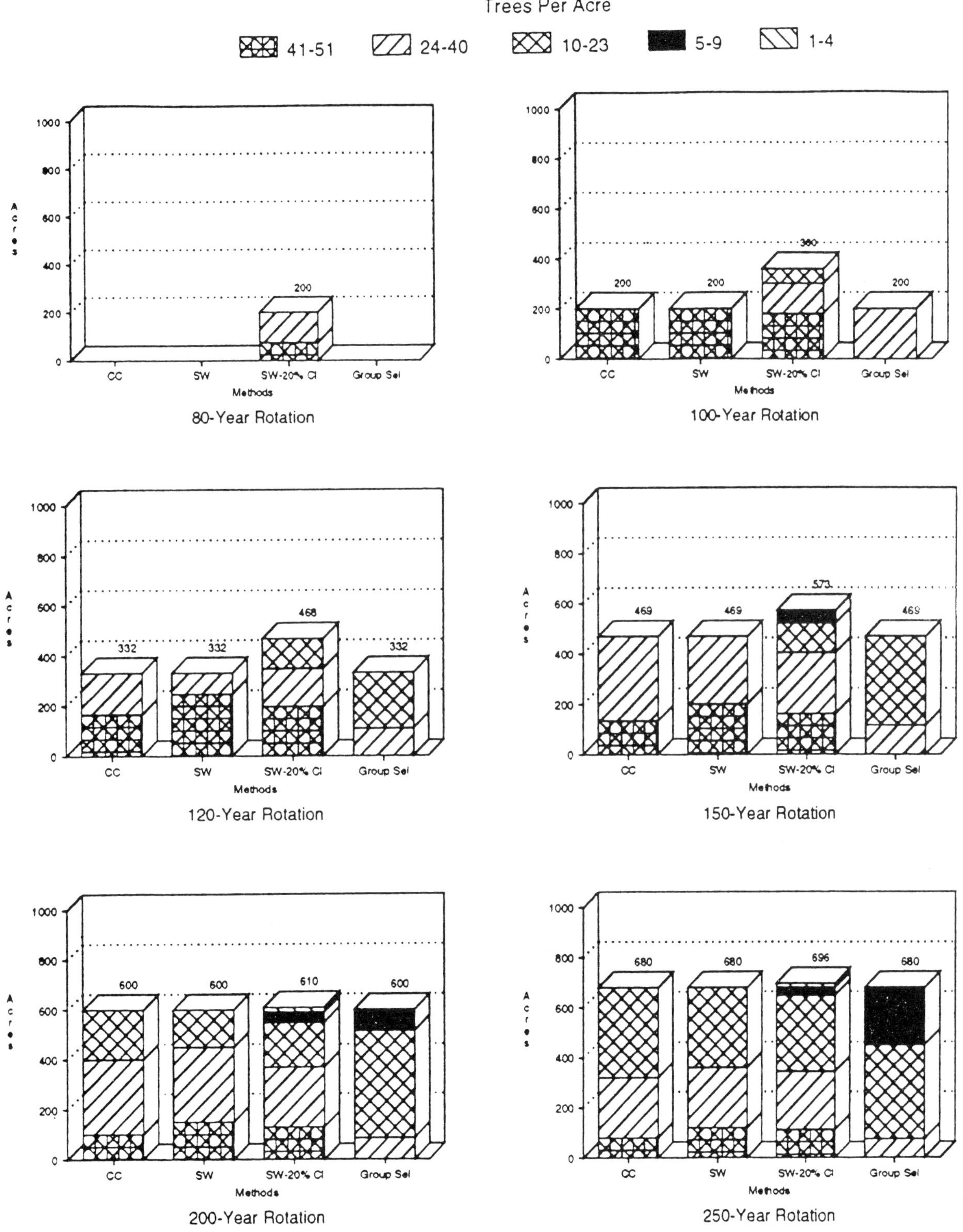

Figure 4. Acres of potential longleaf pine cavity trees (≥15 inches DBH and ≥90 years old) by trees per acre (average stand stocking levels) for different regeneration methods and different rotation ages. CC = Clearcut; SW = Shelterwood; SW-20% Cl = Shelterwood With 20% In Clumps; Group Sel = Group Selection. See text for Explanation of regeneration methods.

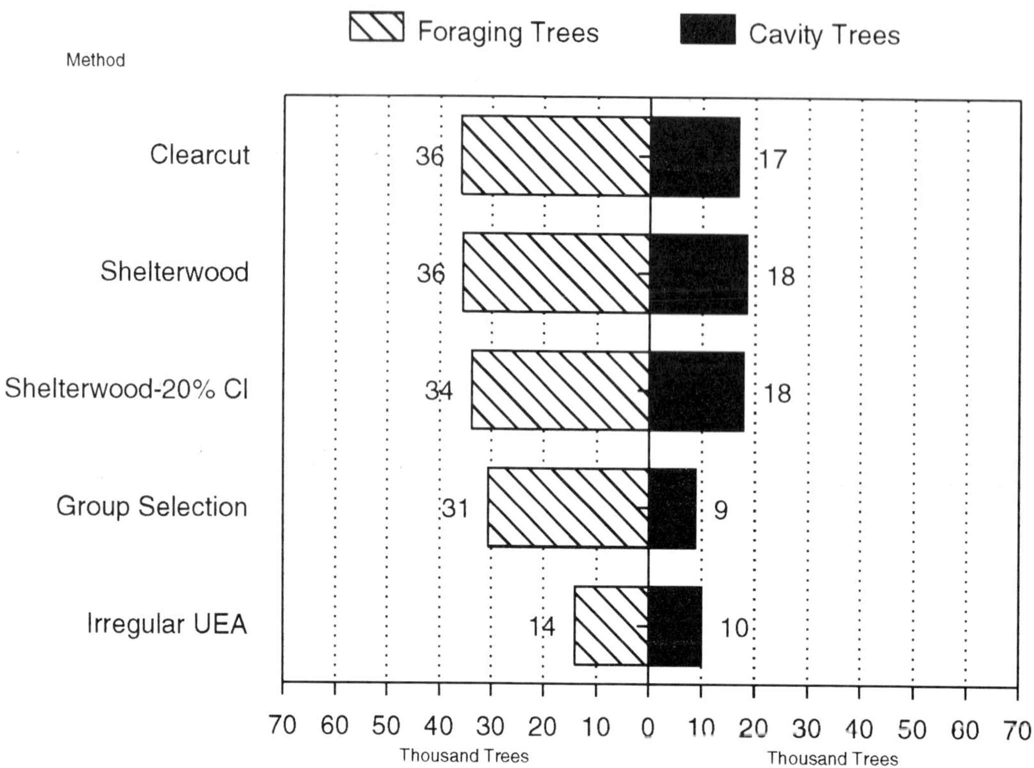

Figure 5. Numbers of potential longleaf pine cavity trees (≥15 inches DBH and ≥90 years old) and numbers of longleaf pines suitable for foraging (≥10 inches DBH and ≥30 years old) that would be provided by different regeneration methods and an irregular uneven-aged forest under a 250-year rotation. See text for explanation of regeneration methods and irregular uneven-aged.

5). With both loblolly and longleaf pine, the clearcut method provided the most foraging habitat for a given rotation age. Trees grown under the clearcut method were able to more fully achieve their growth potential than under any of the other methods. The Irregular Shelterwood-40 BA-No Removal-No Clumps provided the least amount of foraging habitat among the regeneration methods, but still enough to meet the foraging guidelines of 6350 pines ≥ 10 inches DBH and ≥ 30 years old (Lennartz and Henry 1985) for 4 clans per 1,000 acres. The Irregular Uneven-aged forest produced only enough foraging for 2 clans (**Figure 5**).

The number of pines ≥ 10 inches DBH and ≥ 30 years old declined as rotation age increased for both loblolly and longleaf pine with all regeneration methods except for Shelterwood With Clumps and Irregular Shelterwood Without Clumps for 100-year rotations (**Tables 8, 9**). Although increasing rotation age decreased the number of pines, it increased the mean diameter of pines ≥ 10 inches DBH (**Tables 2, 3**). The increase in pine tree diameters with increasing rotation age was simply because diameter continues to increase as tree age increases. In other words, with increasing rotation ages, there were fewer trees but they were bigger and older.

The retention of rotation age trees, whether in clumps (Shelterwood With Clumps) or scattered (Irregular Shelterwood Without Clumps), and its affects on foraging was previously discussed under the potential cavity trees section for both loblolly and longleaf pine (**Tables 2, 3, 4, 6, 8, 9**). These methods are the most unpredictable because of the possible growth and mortality combinations that may occur.

DISCUSSION

The numbers of trees for RCW cavities, foraging sites or sawtimber produced under various rotation lengths and regeneration

Table 8. Numbers of foraging size trees produced under different regeneration methods, management strategies and rotation ages. Values are 1,000's of loblolly pine trees ≥ 10 inches DBH and ≥ 30 years old based on a 1,000 acre management compartment (does not include colony sites, replacement and recruitment stands).

	Rotation Age (years)		
	80	100	120
Regeneration Methods and Management Strategies		(thousands of trees)	
Clearcut	55	49	44
Shelterwood	51	47	42
Shelterwood with 20% of Area in Clumps	44	45	41
Irregular Shelterwood with 40 BA Retained, Staged Removal, 20% of Area in Clumps	38	38	34
Irregular Shelterwood with 40 BA Retained, No Removal, No Clumps	31	32	28
Irregular Shelterwood with 30 BA Retained, No Removal, No Clumps	32	34	31
Group Selection	40	36	34
Single-tree Selection	40	36	34

Table 9. Numbers of foraging size trees produced under different regeneration methods, management strategies and rotation ages. Values are 1,000's of longleaf pine trees ≥ 10 inches DBH and ≥ 30 years old based on a 1,000 acre management compartment (does not include colony sites, replacement and recruitment stands).

	Rotation age (years)					
	80	100	120	150	200	250
Regeneration Methods and Management Strategies			(thousands of trees)			
Clearcut	59	56	53	49	42	36
Shelterwood	54	52	50	47	41	36
Shelterwood with 20% of Area in Clumps	50	48	44	41	38	34
Group Selection	40	39	38	36	33	31

methods vary considerably. The more complex the systems and methods used, the more costly and variable are the outputs. Soil and site conditions are the major factors in determining the health and longevity (except for catastrophic events) for the appropriate species on each site.

It is well beyond the scope of this study to determine which method is best for providing cavity trees and foraging habitat for the RCW, if indeed there is a best method. However, several generalities can be offered to those that ponder such questions.

First, it seems obvious that several regeneration methods and different rotation ages could provide very high quality potential cavity trees in considerable abundance as well as copious amounts of foraging habitat for the RCW.

Second, it is possible that any of the methods and rotations, except the 80-year rotation in longleaf, could provide adequate habitat for a RCW population in the long-term. And even the 80-year rotation in longleaf, if the Shelterwood With 20% Clumps was used, could sustain a healthy population. I base this conclusion on the fact that the numbers of cavity trees provided by this method would be about twice the number of potential cavity trees available in the Francis Marion National Forest prior to Hurricane Hugo (R.G. Hooper, pers. commun.). That population was essentially recovered and was increasing before Hugo (Hooper et al. 1991a).

Third, some scenarios seem more feasible than others to implement. For example, it is likely that the control of both hardwood and pine midstory in loblolly pine stands under Group Selection and especially under Single-Tree Selection is impractical, but essential (Loeb et al. 1992). Fire needed to control hardwoods would tend to destroy the loblolly pine regeneration and thus the future supply of cavity and foraging trees. Alternate methods of hardwood control may be costly (e.g., mechanical removal), or may not provide the ecological benefits of fire (e.g., botanical composition of the ground cover), or may be unacceptable to some publics (e.g., extensive use of herbicides). Longleaf pine stands under Group Selection would fare better than loblolly pine because of the possible fire regimes, but still it would require periodic use of herbicides to control hardwoods and would be more costly than the other methods.

Fourth, the irregular shelterwood methods have never been tested and it is not certain that they could supply a steady flow of RCW habitat in the long-term on many acres particularly where the parent trees (overwood) growth exceeds mortality for 10-20 or more years. It is my opinion that if there is not a staged reduction in the parent trees on most sites, many trees in the younger age class will die or will be severely suppressed. This would lead to a shortage of foraging and potential cavity trees, or at least numbers greatly reduced from those predicted here.

Fifth, some methods may grow a forest that is more vulnerable to wind damage than other methods. Wind has been identified as a major enemy of RCW cavity trees (Conner and Rudolph 1995b; Hooper and McAdie 1995). Trees grown in suppression when young or continually, such as in the irregular shelterwood, group selection and single-tree selection systems will usually fail to develop the bole taper and root mass necessary to survive strong winds (Savill 1983; Oliver and Larson 1990). Scattered tall large stem trees are proportionately more susceptible to stem breakage and windthrow (Oliver and Larson 1990). Given that on average, 45 hurricanes are expected to hit RCW recovery areas every 100 years (Hooper and McAdie 1995) and other catastrophic wind events occur periodically, some consideration should be given to using silvicultural systems that grow relatively wind resistant trees.

Sixth, forests managed with balanced even-aged stands provided more potential cavity trees and foraging habitat than two-aged and uneven-aged forests. The reason for this is that southern yellow pines grown under even aged methods and systems can do so with less competition and thus more fully realize their genetically determined growth potential. Compared to two-aged and uneven-aged, even-aged systems are more easily implemented, can grow more wind resistant trees if managed correctly, provide more wood products without having to reduce the amount of RCW habitat provided, allows essential control of midstory without damaging younger age classes of pines, and allows the use of fire to restore plant populations critical to an ecological approach to management.

Seventh, there must be a well-defined desired condition of the forest for RCW habitat or for any other output or benefit. This desired condition should be described at the stand level as well at the landscape level.

Interpretation of an Endangered Species: The Red-cockaded Woodpecker Story

David L. Kulhavy, College of Forestry
Stephen F. Austin State University, Nacogdoches, Texas 75962

ABSTRACT: The endangered red-cockaded woodpecker (RCW) has a unique life history and ecology that lends itself to interpretive techniques. This paper explores and explains: **1)** the reasons why the RCW has interpretive value, **2)** "tools" for successful RCW interpretation and **3)** various RCW interpretive success stories.
KEYWORDS: endangered species, interpretation, wildlife, environmental education

The red-cockaded woodpecker, *Picoides borealis*, has a unique biology that lends to its interpretive appeal. The red-cockaded woodpecker was once a common bird in the mature pine forests of the southeastern United States. Due primarily to habitat loss, the species is now on the federal endangered species list. This volume discusses the recovery, ecology and management of this endangered species and paramount to these efforts are effective and timely environmental education and interpretation.

Among woodpeckers, the red-cockaded woodpecker has an advanced social system. The birds live in a family unit called a group. The group may have 1 to 9 birds, but never more than one breeding pair. Appendix B, RCW Chronology of Significant Events, p. B-1, Record of Decision, Final Environmental Impact Statement for the Management of the Red-cockaded Woodpecker and Its Habitat on National Forests in the Southern Region, Volume 1, states "This document, as suggested by Walters et al. (1988b) [1988, this volume], will refer to the area (minimum of 10 acres) surrounding aggregates of cavity trees as 'clusters', and RCW family units as 'groups'. This is different from much of the literature, including the red-cockaded recovery plan and all existing National Forest System RCW management direction, which refers to the aggregate of cavity trees as 'colonies or colony sites' and the RCW family units as 'clans'." Hooper et al. (1980) document description, life history, habitat requirements and management suggestions; and provide a guide to appearance of red-cockaded woodpecker cavity trees. This article assists with interpretation of the red-cockaded woodpecker. Young birds remain with their parents through the summer and into the fall at which time the females disperse or are driven from the group. The male juveniles may stay on as helpers for several years. They assist the breeding pair with all aspects of group activity, including raising of future generations, construction of new cavities, and territory defense (see Cavity Trees as a Resource, Section 5).

The RCWs nest and roost in a group of cavity trees called a cluster. The cluster may contain one to more than 20 trees. A good cluster site is a mature, park-like stand (generally longleaf pine) with few or no hardwood trees. Cavities are excavated in live, mature pine trees. Studies throughout the southern United States found the average cavity tree age ranged from 62 to more than 176 years old. Conner and Rudolph (1995a) detail construction of new cavities in a 4-step process including: initial wounding of the pine; excavation of the entrance tube; penetration of the heartwood; and excavation of the full cavity chamber (see Figs. 1 and 2, pp. 346-347, this volume).

Typically, within any cluster, some cavities are finished and in use, others are under construction and some have been abandoned. Cavity construction can be completed in 4 to 5 months, but typically takes a year or more. Only the red-cockaded woodpecker constructs cavities in living pines in the southern pine ecosystems. However, 11 other birds, 5 mammals, 4 reptiles and bees are known to use the cavities. Some of the major competitors are the red-bellied woodpecker, pileated woodpecker, red-headed woodpecker, great crested flycatcher and flying squirrel (see Cavity Tree as a Resource section; and Kappes and Harris, this volume).

Scattered about the trunk near the cavity entrance are numerous small holes called resin wells, chipped through the bark. Resin flow from these wells eventually coats the trunk (see Cavity Trees as a Resource picture depicting the bole of a red-cockaded woodpecker cavity tree, the bark plate and resin wells; note placement of the cavity on the bole). The red-cockaded woodpecker regularly peck at resin wells to stimulate resin flow, and the resin-coated trunk acts as a predator deterrent, particularly for tree climbing snakes like red and black rat snakes (see Withgott et al., and Montague et al., this volume).

The red-cockaded woodpecker typically nests between late April and July. The female lays 2 to 4 eggs in her mate's roost cavity. Group members take turns incubating the eggs during the day, but the breeding male stays with the eggs at night. The eggs hatch in 10 to 12 days and nestlings are fed by the parents and the helpers. Young birds leave the nest (fledge) in about 26 days (see picture of fledging male before Outlook for Recovery section; this photograph tells its own interpretive story as the successful fledging of the young birds is critical for the long-term recovery of this species).

The group spends much of its time foraging for food as it travels about its territory (see picture before Section 4, New Insights on the Biology, depicting the feeding of the young by the parents; note also the copious resin and resin wells). Most of the searching is concentrated on the trunks and limbs of live pine trees. There, the birds scale the bark and dig into dead limbs for spiders, ants, cockroaches, centipedes and the eggs and larvae of various insects.

Adequate foraging habitat is vital to the red-cockaded woodpecker. The acreage of foraging habitat needed by a group varies with the quality of their habitat. One hundred twenty-five acres of pine is sufficient for some groups. In some areas, habitat conditions are not ideal or competition from other groups is low. In these situations, groups commonly forage over several hundred acres.

INTERPRETIVE VALUE OF THE RED-COCKADED WOODPECKER

The red-cockaded woodpecker is not only unique in its life history and ecology, but also in its "availability" and potential for endangered species interpretation. Unlike many other endangered species the red-cockaded woodpecker has several characteristics that lend to its interpretive appeal. These include:

1) a wide (Texas to North Carolina) distribution;

2) populations or individual cavity tree clusters close to urban centers;

3) ease of observation; and,

4) 100 percent probability of "hands on" (observation or trapping—with persons with the proper permits) experience for all participants, with a minimum of preparation by interpreters or group leaders. Logistically then, the red-cockaded woodpecker is an ideal endangered species to interpret.

A "day-in-the-life of a red-cockaded woodpecker cluster" can provide many simple lessons in the complexities of community ecology. Life in an active red-cockaded woodpecker cluster is very dynamic and educators and participants may view or discuss:

1) inter- and intraspecific competition for cavities and foraging substrates;

2) red-cockaded woodpecker predators including rat snakes, raptors and flying squirrels;

3) nesting and foraging habitat use and value; and,

4) relationships between ecosystem components such as longleaf pine-fire, tree age-red-cockaded woodpecker nesting or foraging use, red-cockaded woodpecker cavity trees-soil type, and red-cockaded woodpecker avian community diversity. The ability to illustrate these ecological principles and concepts in an understandable fashion and in a "field" setting close to communities is a definite bonus for interested educators and conservationists.

Another reason for the RCW's excellent interpretive appeal is that even locally (nearest National Forest, State Park or Forest), there are red-cockaded woodpecker conservation and management efforts being implemented and practiced. These activities provide opportunities for first-hand experience into one or more phases of an endangered species "recovery" program. Interpreters, educators and participants have opportunities to learn about, or participate in, various projects involving red-cockaded woodpecker recovery. Examples include assisting with artificial cavity installation. This technique has proved invaluable in recovery of the species, especially following the devastation from Hurricane Hugo. Hooper and McAdie (this volume) tell the story of the impact of hurricanes on RCW populations; and Watson et al. (this volume) detail the restoration efforts on the Francis Marion following Hurricane Hugo. Wind damage, bark beetles and lightning are impacts discussed in the Natural Disturbances section. Carter et al. and papers by Walters et al. (this volume) document artificial cavity use for recovery of RCW populations.

Coordination of recovery efforts with federal agencies, including the USDA Forest Service, and the USDI National Park Service and Fish and Wildlife Service, along with state and private agencies and universities, can lead to increased volunteer participation. Costa (this volume, pp. 22-27) stresses interagency agreements. Volunteer projects may include:

1) identification and protection of habitat;

2) inventorying suitable habitat for cavity trees;

3) monitoring colonies for group structure and composition, breeding success, and identification of foraging habitat;

4) assisting with trapping, banding and other research related activities; and,

5) hands-on habitat management work. Concurrent with these activities are award programs that dovetail with conservation efforts. Professional interpreters can increase the visibility of their efforts by coordinating programs and publicizing volunteer activities.

The value of the red-cockaded woodpecker and its associated habitat, as an environmental education "tool" includes examples of:

1) endangered species conservation principles;

2) community and landscape ecology concepts;

3) natural resource management alternatives and objectives; and,

4) recovery program processes.

The red-cockaded woodpecker story is successful because it is:

1) generally, logistically easy and efficient, available to all age groups, and to persons with disabilities;

2) always successful in the field (with a minimum of preparation), guaranteeing opportunities for close observation and/or actual bird handling, when working with qualified and "permitted" personnel;

3) easy to understand, yet illustrates and opens doors to many ecologically complex relationships; and,

4) available for hands-on habitat and study projects, enabling interested volunteers to become and remain interested in the species.

TOOLS FOR THE INTERPRETER

Basic to red-cockaded woodpecker interpretation are slide presentations. Whether illustrating life history and ecology, southern pine beetle-RCW relationships, effects of wildfires or hurricanes (Hugo) on RCWs, or habitat management programs, slide shows effectively convey information through the environmental educator to relatively large audiences in an indoor or outdoor setting. A logical follow-up to a slide show is a field trip or "discovery walk" through a red-cockaded woodpecker cluster.

Early mornings and evenings are the best times to view RCW activity. The red-cockaded woodpecker often forages and constructs cavities in open forest stands. Ease of access and visible evidence of the cavity trees lend themselves to interpretive success. Coordinating tours with local wildlife specialists, including federal and state wildlife biologists and resident ornithologists, will increase the success of the experience. Such excursions may or may not include banding birds. Field trips can also be designed around weekend projects to improve or maintain habitat. Care must be taken during the breeding season to minimize disturbance. Check local regulations on vehicle access and any other restrictions.

The above activities are especially suitable to youth groups, including Girl Scouts and Boy Scouts, Cub Scouts and Brownies, Webelos, 4H and other youth organizations. Outdoor education and landscape ecology courses can also be conducted in this environment. Field use of a slide projector with a built in screen coupled to a portable generator may enhance the experience by reinforcing the biology and ecology of the red-cockaded woodpecker and its role in the forest ecosystem. However, this technique must be weighed against the "teachability" of the moment and may be best left to a more formal lecture session. The ease of access and high visibility in many red-cockaded woodpecker colonies lend themselves to unique interpretive experiences. Elderhostels, youth groups, or wildlife ecology classes can easily traverse colonies.

Development of informational pamphlets, including self-guided auto tours through red-cockaded woodpecker habitat and adjacent red-cockaded woodpecker colonies, is a simple, yet effective interpretive method. Similarly, using the red-cockaded woodpecker as the symbol or logo of a program or project is beneficial.

Adopt-a-Cluster programs have been successful on National Forests. Individuals or small groups (Scout Troop) choose a red-cockaded woodpecker group to adopt. A group of banded birds may be monitored (weekly, monthly, or whenever possible) by the volunteers. Data may be gathered to:

1) identify habitat used for foraging,

2) record breeding and nesting activity and

3) investigate group dynamics. These programs have commonly fostered long-term volunteer commitments. The data collected can contribute significantly to the management program for a local population. Using the media to help interpret the red-cockaded woodpecker and its habitat can be very effective.

The manual by Allen (1991) is an excellent interpretative tool for use by either the volunteer leader and/or the wildlife biologist. With the precise instructions and diagrams, the ease of using this manual is increased as is the overall importance of this method as a valuable recovery tool for the red-cockaded woodpecker. Within Allen's manual are:

1) A brief review of Hurricane Hugo and the importance of the artificial cavity. Allen notes it takes about 45 minutes to construct each cavity (see picture before Strategies for Recovery section; this photograph depicts installation of an artificial cavity and is one of the main reasons for continued recovery of the red-cockaded woodpecker on the Francis Marion National Forest following Hurricane Hugo in 1989). The information passed on to youth (or adults) will increase the educational value of this technique and the importance of recovery of this endangered species.

2) Construction of cavity inserts. A demonstration of the insert and the importance of using either Douglas-fir (*Pseudotsuga menziesii*), western redcedar (*Thuja plicata*), basswood (*Tilia americana*), or southern yellow pine (*Pinus* spp.) are presented with the preferred use of western redcedar.

3) Selection of Trees and Cavity Placement on Tree. Allen notes that if two artificial cavities are inserted they should be within 100 yards of each other. Cavities are often installed at heights of 12, 22 or 32 feet and usually below the first live branch.

4) Cavity Excavation. Allen details the installation of artificial cavities from bark scraping; use of a chainsaw for preparation of the tree for the cavity; insertion of the cavity to adding "resin wells" and/or restrictor plates.

5) Use of Cavities. Allen describes the working of resin wells and the "acceptance" of these artificial cavities in the Department of Energy, Savannah River Site; and the USDA Forest Service Francis Marion National Forest following Hurricane Hugo where over 90 % of the natural cavity trees were destroyed in September of 1989. Over 60 % of the inserts are now being used for roosting or nesting by red-cockaded woodpeckers. (See Hooper and McAdie and Watson et al., this volume for more explanation. The picture before the Natural Disturbances section depicts the Francis Marion forest as it might have remained except for rapid response, installation of artificial cavities and use of remaining longleaf pine as potential red-cockaded woodpecker cavity trees).

6) Safety is stressed. Allen stresses the importance of the proper safety gear and the importance of the proper training in the use of both Swedish climbing ladders and chainsaw techniques. Safety equipment, including hardhats, hearing protection and gloves should be worn by crew members. The climber should also wear eye protection and chainsaw chaps (see Section 2, Strategies for Recovery).

7) Permits. Construction of artificial cavities for red-cockaded woodpeckers is subject to provisions in the Endangered Species Act. Care must be taken to work with the appropriate federal agency (e.g. USDA Forest Service) and compliance with the USDI Fish and Wildlife Service requirements. If the artificial cavity project is considered a federal action, Section 7 compliance (to ensure that any federally funded or approved program will not jeopardize an endangered species) is required. Section 9 compliance (prohibition against "take") is required on public and private lands.

The modification of the Copeyon method (Copeyon 1990, Taylor and Hooper 1991) may also be utilized as a successful interpretive and environmental education tool. These are specialized projects and must be undertaken with the supervision of the experienced biologist with the proper permits. In our projects, we worked directly with the USDA Forest Service as volunteers under their existing permits.

Gaining the public's consent and helping them make informed decisions concerning endangered species conservation requires increasing their ecological knowledge. This is the role of the professional interpreter (educator, biologist, naturalist).

RED-COCKADED WOODPECKER INTERPRETIVE SUCCESS STORIES

Although program success can be measured in several ways, awards presented to individuals or groups and sponsored by federal agencies responsible for species recovery, including the U.S. Forest Service, National Park Service and U.S. Fish and Wildlife Service, are commonly used as measures of successful projects. The Take Pride in America initiative was one such award program. This program is no longer in use, but local, regional and national awards were available for these programs.

This program enhanced the volunteer experience of individuals that worked to improve the environment. One example of this includes a Boy Scout troop in Nacogdoches, Texas, that cleared brush, small trees and other potential fuel from around the base of red-cockaded woodpecker cavity trees on the Angelina National Forest. This was done in preparation for a controlled burn conducted by the Forest Service. This project reduced fuel loading, complemented the District Ranger's work, and resulted in the award of a Take Pride in America plaque to both the individual Boy Scout who led the activity and to the troop.

Boy Scouts (and Explorers who have reached First Class) can earn merit badges, especially Fish and Wildlife Management for habitat restoration and examination of the habitat of this endangered species. Part of Soil and Water Conservation can be earned as cavity tree clusters may be charted using local soil surveys. With completion of these merit badges, and the Environmental Science and Citizenship in the World merit badge, the Scout or Explorer can earn the world conservation award.

In conjunction with habitat improvement was the interpretive value garnered during the project. This included, red-cockaded woodpecker observation, increased awareness and appreciation of the ecosystem's complexities, discussions about the variety of forest management options, and the camaraderie developed during the activity.

While stationed on the Kisatchie National Forest in Louisiana (1985-1988) Ralph Costa presented 28 red-cockaded woodpecker interpretive programs. These included, professional papers and poster displays, university seminars, school (K-12) talks and field trips, presentations at monthly meetings of conservation organizations, newspaper articles and a 10-minute television special.

On the Apalachicola National Forest in Florida, Costa presented over 40 red-cockaded woodpecker programs. In addition to the previously mentioned organizations, Girl and Boy Scout Troops, senior citizens homes, public radio and 4 television stations were contacted. The Apalachicola National Forest has also focused attention on the red-cockaded woodpecker by:

1) publishing a pamphlet on red-cockaded woodpecker life history and ecology, with a self guided auto tour;

2) featured the red-cockaded woodpecker on the logo for the Forest's first Natural Resources Career Camp; and,

3) initiated public awareness efforts via television specials and popular magazines. The Apalachicola is one of the targeted recovery populations.

The RCW Environmental Impact Statement will be a valuable tool for use in interpretation and environmental education programs. The increasing emphasis on environmental education and interpretation of our environment lends itself well to emphasis on the recovery, ecology and management of this endangered species.

Other successful programs include assistance with artificial cavity installation (completed in conjunction with the Wildlife Biologist on the Ranger district and with the proper authorization). Scouts and Explorers cannot use chainsaws (one of the tools for cavity excavation), however, they can assist with on-the-ground activities and the activity becomes an interpretive event. In conjunction with Mr. Alfredo Sanchez, Wildlife Biologist, on the Angelina Ranger District, two potential cavity trees were located in a longleaf pine stand on the southern portion of the ranger district in an active red-cockaded woodpecker cluster. With the proper signed volunteer agreement, activities include preparation of the cavity tree insert and assisting the Wildlife Biologist with passing up material via ropes and/or buckets for assistance with cavity installation.

In North Charleston at the Red-cockaded Woodpecker Symposium: Species Recovery, Ecology and Management, the symposium was coupled with a field expedition to the Francis Marion National Forest to view recovery procedures for the red-cockaded woodpecker coupled with a discussion of the effects of Hugo on this endangered species. Additional stops included longleaf pine regeneration; demonstration of the use of artificial cavities and their use in recovery; and a general discussion of birding and the importance of identification of this species. As part of the presentation, slides developed on site using Polaroid® film were utilized as an interpretative tool. Poster-size demonstrations can be used as well as literature on the recovery, ecology and management of the red-cockaded woodpecker.

SUMMARY

The unique ecology of the red-cockaded woodpecker lends itself to an interactive interpretive story that will attract the attention of people of all ages. It is essential that we as managers correctly and professionally interpret this endangered species. Part of the recovery effort will be based on our efforts as educators and on our role of telling the story of the red-cockaded woodpecker.

Use of Geographic Information Systems for Determination of Red-cockaded Woodpecker Management Areas[1]

Donald J. Lipscomb and Thomas M. Williams, The Belle W. Baruch Forest Science Institute[1] Clemson University, Georgetown, SC. 29442

ABSTRACT: Forest management practices in stands containing red-cockaded woodpeckers (*Picoides borealis*) (RCW) should follow habitat retention guidelines within designated distances of cavity trees. Geographic information systems (GIS) provide useful tools for combining map and tabular data into alternative management scenarios. However, these computerized systems require specific rules to define spatial relationships in colonies (clusters) and potential foraging habitat. Three alternatives for delineating colonies (clusters) and five alternatives for mapping potential foraging habitat were evaluated for agreement with existing guidelines and for application objectivity. To utilize GIS, consistent rules are needed which achieve mapped areas independent of operator bias. Differences in the affected area, basal area, and area locations of two documented colonies (clusters) were used to discuss the alternatives. Rules for GIS (spatial) location of colonies (clusters) and potential foraging habitat are recommended which comply with present guidelines, and produce consistent results.

KEYWORDS: red-cockaded woodpecker (*Picoides borealis*), geographic information systems, GIS, cluster (colony)[2], cavity trees, buffer zone, foraging area

Programs for the management of red-cockaded woodpecker habitat, such as the proposed strategy (Costa 1992) for private landowners or those already in use on federal land (Henry 1989, Lennartz and Henry 1985), describe the biological needs of the bird in detail and place constraints on habitat altering operations that may be detrimental to the birds in some way. These guidelines are based primarily on studies of the species biology (Loeb et al. 1992, Hooper et al. 1991b, Conner and O'Halloran 1987, Lennartz and Heckel 1987, Jackson 1977c, Van Balen and Doerr 1978, Hooper et al. 1980, Jackson et al. 1979b, Lennartz and Harlow 1979). More recent studies on the management implications of guidelines deal with entire forests and ignore spatial relationships of individual cavity trees or clusters (colonies) (Roise et al. 1990, Lancia et al. 1989). Timber management strategies developed under these guidelines will require a dynamic management planning process which can consistently recognize the location and area affected by the presence of RCW.

The spatial problems associated with providing adequate RCW habitat (Costa 1992, Henry 1989, Lennartz and Henry 1985, Shumway 1986) are readily solved with GIS technology. Management areas surrounding cavity trees must be mapped and related to the forest data base. GIS technology combined with global positioning systems (GPS) can be used to estimate new management requirements when new RCW cavity trees are found. Thus, GIS allows annual update of management plans for timber and other resources that change with movement of RCW affected areas.

To use GIS for RCW habitat management, consistent, objective and complete definitions of spatial relationships are needed. Also accurate location of cavity trees and forest

[1]Belle W. Baruch Forest Science Institute Technical Contribution: 93-06.

[2]Terminology for the aggregate of cavity trees (colony, cluster) and family of birds (clan, group) differs with authors. We will generally use the older terminology (colony, clan) except when literature is cited; in that case we will use the original author's terminology.

stand boundaries are critical. Sufficiently accurate points and distances can be obtained using standard cartographic coordinate systems and photogrammetric techniques (Light 1993).

Present definitions used in federal documents related to RCW habitat (Costa 1992, Henry 1989, Lennartz and Henry 1985), are inadequate to delineate potential foraging habitat and colony (cluster) boundaries in the GIS context. According to the latest (Costa 1992) definition, a 61 m buffer is required around the aggregate of all active and inactive cavity trees located within 403 m of one another. Foraging area for a RCW group (clan) must be within 804 m of the cluster (colony); contain a specified pine basal area with greater than 25 cm dbh stems; and contain non-foraging habitat gaps that are no more than 91 m wide. The foraging area must be located within a 804 m radius circle around the colony (cluster) center (Henry 1989). The GIS software will map and calculate basal areas of alternative foraging areas within this circle (potential foraging area) but requires a single coordinate for the center of the circle. Determination of a center coordinate depends on the geometry of the cavity trees and the rules used to relate cavity tree positions to the colony (cluster) area.

This study used GIS technology to determine the size and shape of mapped colony (cluster) area and potential foraging habitat resulting from different rules for defining these areas given the geometry of the cavity tree locations. Our choice of the best rules were those that:

(1) met the biological needs of the bird;

(2) did not conflict with the private landowner guidelines, and

(3) produced consistent results.

METHODS

Red-cockaded woodpecker cavity trees on Hobcaw experimental forest were mapped with traditional surveying techniques and included in a GIS data base (Lipscomb and Williams 1991 and Lipscomb and Williams 1995). Two colonies (clusters) were chosen for this study. Buffer algorithms in PC ARC/INFO were used for alternative rules to map a colony (cluster).

(1) The "point buffer" command creates a 61 m circular buffer around each cavity tree and aggregates circles around trees that are less than 122 m apart into a single polygon. If trees are located more than 122 m apart this command will create separate polygons.

(2) When separate polygons occur, a "line buffer" on a line connecting the separated cavity trees to the clump will produce a corridor.

(3) Alternatively, a polygon can be formed by connecting the outer-most trees. The "polygon buffer" command maps a larger polygon 61 m from the perimeter of the polygon formed by the line connecting the cavity trees. Areas and pine basal areas affected within the respective buffers were tabulated by overlay analysis.

The 804 meter radius zone for potential foraging was also determined in a variety of ways:

(1) It was mapped as a polygon 804 m from the perimeter of the colony (cluster) (polygon). It was mapped as an 804 m circle with the center of the colony defined as

(2) the center of a box formed by the Arc/Info command, "rebox" (box all). The rebox command returns the coordinates of a rectangle tangent to the outermost trees. The center could also be defined as

(3) the mean of the coordinates of all the cavity trees (mean all). With methods two and three, the greatest concentration (the clump formed by trees within 122 m of each other) of cavity trees

(4) could be used with the rebox command (box clump) or

(5) the mean of the coordinates of only these trees could be used (mean clump). This resulted in five alternative rules to determine the 804 meter zone for potential foraging areas.

The two colonies (clusters) chosen represent extremes of colony (cluster) types. Both colonies (clusters) have been active since, at least, 1971 in the same general locations (Dennis 1971). One colony (cluster), CL1, had ten trees, eight of which were clumped such that each tree was less than 122 m from at least one neighbor. The other two trees were 201 m to the northeast and 372 m to the southeast of the main clump (**Figure 1**). CL1 was about as large and disconnected as the definition of a colony (cluster) allows. The second colony (cluster), CL2 had two trees which were 15 m apart and a third that was 171 m to the east (**Figure 2**). We expected large variation around CL1 and little variation around CL2 for our different alternatives. The buffer and foraging polygons constructed by each of the

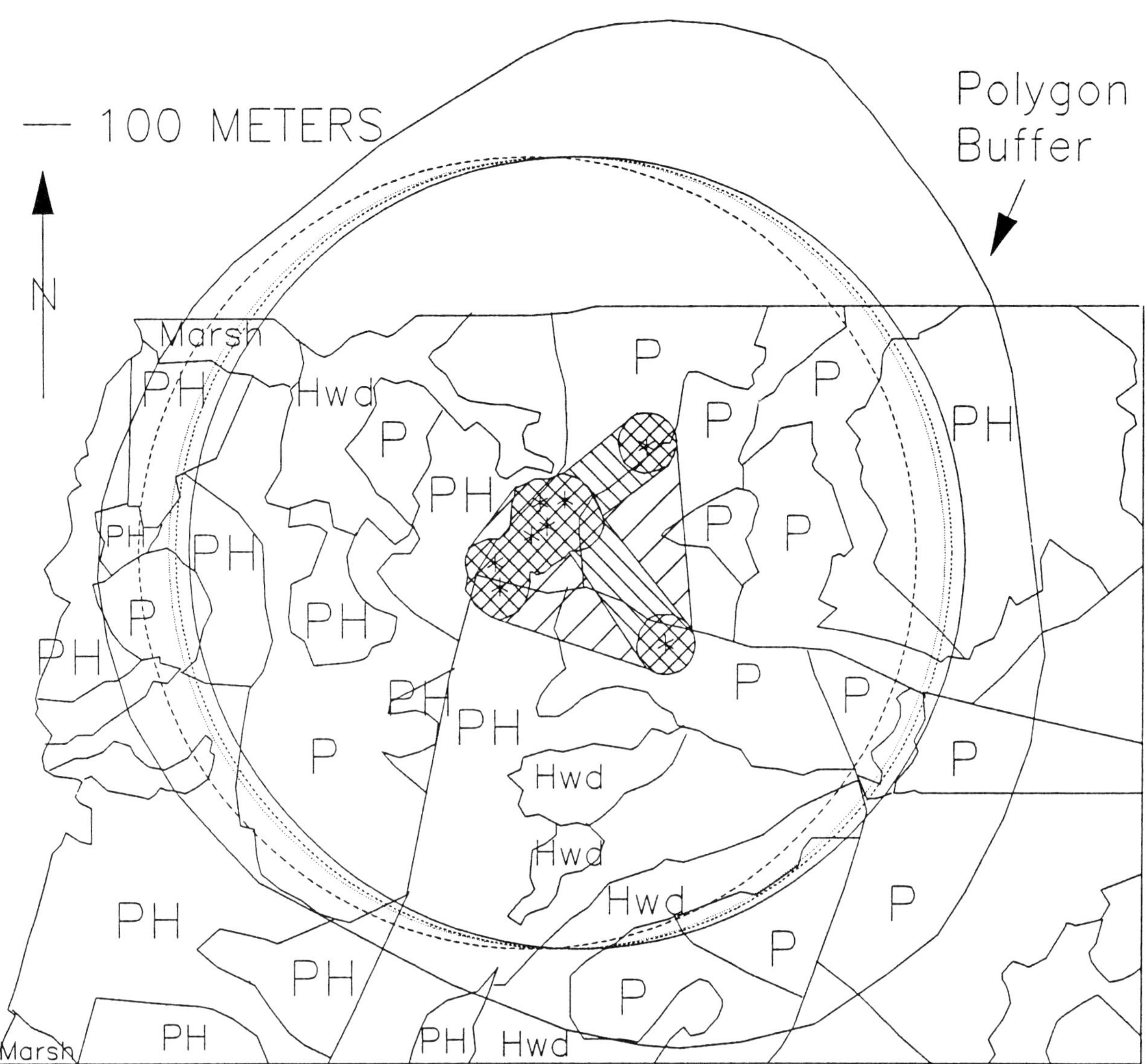

Figure 1. RCW Example CL1. Cavity trees (asterisks), three alternative colony (cluster) boundary rules, and potential foraging habitat on timber type map. Colony boundary: (1) separate zones - cross hatch; (2) corridors - add right single hatch; (3) polygon buffer - add left single hatch. Potential foraging habitat: (1) polygon buffer - labeled, (2) box all - solid circle, (3) mean all - short dash, (4) box clump - dots, (5) mean clump - long dash. Forest types: Pine - P, Pine-Hardwood - PH, Hardwood - Hwd

alternative definitions were used with the overlay module "clip" command and the forest stand map to create separate stand maps for each combination. The "clip" command is used as a "cookie cutter" to create new maps with only those portions of the stand map that fall within the borders of the buffer polygon. The function also attaches all tabular data associated with the new stand map allowing data base manipulation of the new units. Although the tabular data contains a variety of management information (Lipscomb and Williams 1991), only area, overstory type, and basal area of pines greater than 25 cm at dbh were used for comparison of the rules.

RESULTS AND DISCUSSION

Colony site rules

Total area in the colony (cluster) increased as the definition added either "corridors" or the polygon buffer **(Table 1)**. The area added by corridors increased as the number of, or

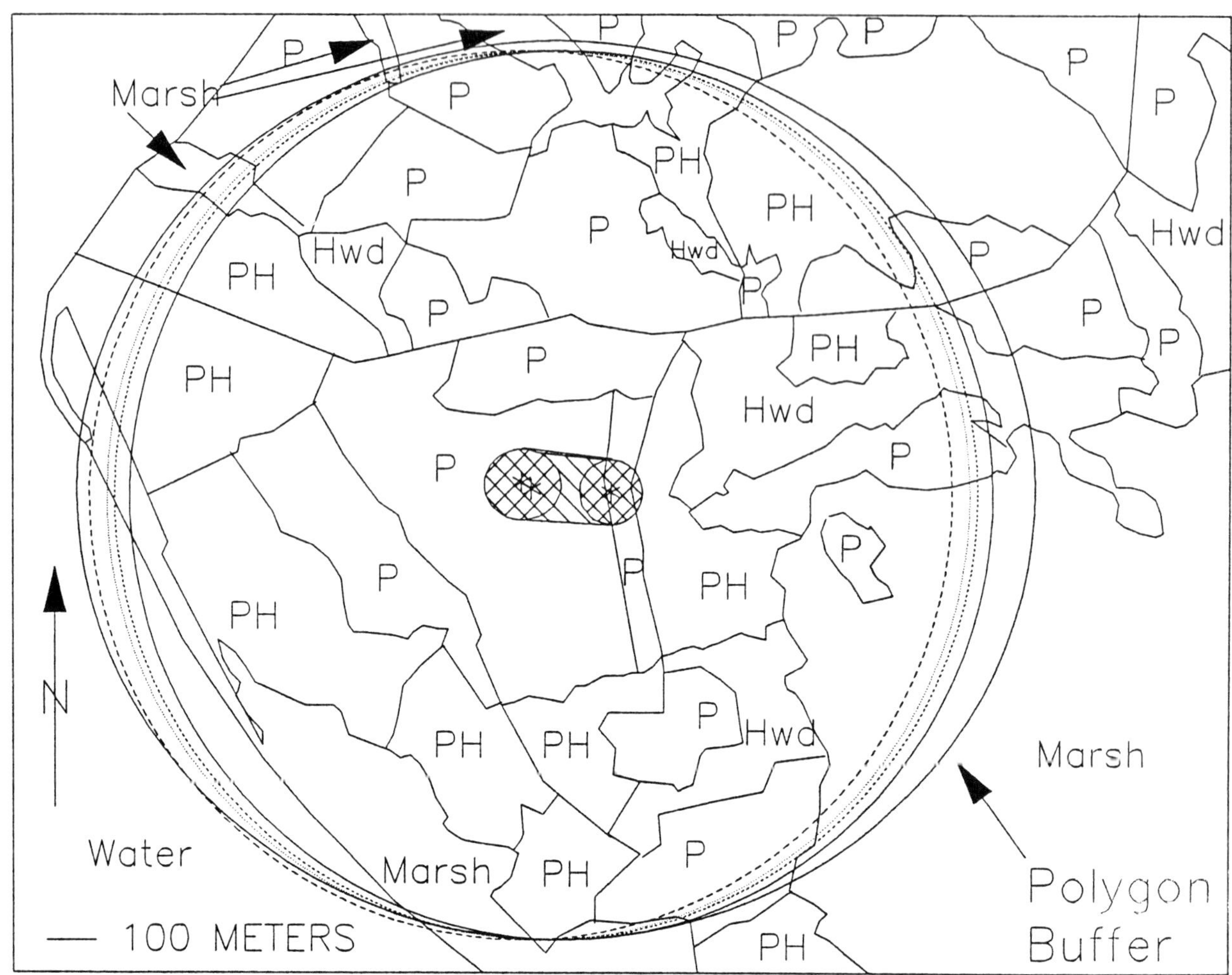

Figure 2. RCW Example CL2. Cavity trees (asterisks), three alternative colony (cluster) boundary rules, and potential foraging habitat on timber type map. Colony boundary: (1) separate zones - cross hatch; (2) corridors - add right single hatch; (3) polygon buffer - add left single hatch. Potential foraging habitat: (1) polygon buffer - labeled, (2) box all - solid circle, (3) mean all - short dash, (4) box clump - dots, (5) mean clump - long dash. Forest types: Pine - P, Pine-Hardwood - PH, Hardwood - Hwd.

distance to, outliers (cavity trees located more than 122 m from the clump) increased. The increase due to enclosing the trees in a polygon also depended on the geometric arrangement of the trees. In CL1, 4.1 ha were added by including corridors to the two outliers and since they were nearly perpendicular to the eight tree clump area, the polygon surrounding all trees added an additional 5.8 ha. Changes for CL2 were not as dramatic. The colony (cluster) area increased only 0.8 ha by adding a corridor and, since all of the trees were nearly on a line, the polygon added only an additional 0.1 ha.

The recommended minimum density of pines (> 25 cm dbh) in the colony (cluster) is 50 square feet per acre (11.5 m^2/ ha). Neither of the selected colonies (clusters) met that criteria. For CL1 the clumped trees were in a mature pine stand but the outliers were in younger stands. The rule that included corridors in the younger stands or the polygon which included more of the younger standand part of a pine-hardwood stand resulted in less pine density for these two alternatives. The polygon rule resulted in a pine density less than half the recommended density. For CL2 the corridors and polygon rules included more of the same stands where the cavity trees were located and resulted in no change in pine density.

Defining the colony (cluster) as buffers around individual trees connected with 122 meter wide buffer corridors for outliers best met all criteria for a GIS based definition. This definition will support the needs of the bird as

Table 1. Basal area of pine (>25 cm dbh) and area of colonies (clusters) for three alternative rules to determine colony.

	Pine		Pine-HWD		Hardwood		Totals	
	Area ha	BA m²	Area ha	BA m²	Area ha	BA m²	Area ha	BA m²
Cluster 1(CL1):								
Separate zones	5.5	41.4	1.4	8.4	0.0	0.0	6.9	49.8
Corridors	9.6	57.4	1.4	8.2	0.0	0.0	11.0	65.6
Polygon buffer	14.4	74.4	2.4	14.7	0.0	0.0	16.8	89.1
Cluster 2 (CL2):								
Separate zones	2.5	12.5	0.1	0.4	0.0	0.0	2.6	12.9
Corridors	3.3	16.5	0.1	0.4	0.0	0.0	3.4	16.9
Polygon buffer	3.4	17.0	0.1	0.4	0.0	0.0	3.5	17.4

Table 2. Basal area of pine (>25 cm dbh) and area for five alternative rules to determine potential foraging habitat boundaries with GIS.

	Pine		Pine-HWD		Hardwood		Totals	
	Area ha	BA m²	Area ha	BA m²	Area ha	BA m²	Area ha	BA m²
Cluster 1 (CL1):								
Polygon buffer	128.6	1123.0	86.8	531.7	28.2	79.1	243.6	1733.8
Box all	89.0	718.4	59.0	369.5	23.2	76.0	171.2	1163.9
Points all	89.1	752.7	59.3	398.9	21.4	75.4	169.8	1227.0
Box core	88.9	738.0	59.2	387.3	22.6	75.9	170.7	1201.2
Points core	89.4	786.7	59.9	424.1	19.7	74.4	169.0	1285.2
Cluster 2 (CL2):								
Polygon buffer	94.9	563.4	69.3	217.4	21.1	19.2	185.3	800.0
Box all	88.9	511.2	63.3	198.9	19.8	17.7	172.0	727.8
Points all	88.5	510.4	63.7	202.7	19.4	17.3	171.6	730.4
Box core	88.3	510.5	64.3	200.9	19.3	17.2	171.9	728.6
Points core	87.2	507.8	64.1	196.8	18.5	16.3	169.8	720.9

outlined in the guidelines in all cases . Also this definition is always determined completely by the geometry of the cavity trees and independent of decisions by the GIS operator. Neither of the other definitions meet these criteria in all instances.

The computer software will automatically create the buffers around trees for the first rule and, therefore, the location of the buffers are completely determined by the geometry of the cavity trees. Despite its ease of implementation this rule will only meet the latest definition of the cluster (colony) (Costa 1992) if the cavity trees are less than 122 meters apart. This was not true in either chosen case in spite of the small number of trees in CL2.

With only three trees in CL2, both the corridor and the polygon were uniquely defined so all three techniques produced buffer zones that were independent of operator decisions. For CL1 that was not the case. The polygon buffer technique was somewhat arbitrary because the two outliers and the three trees farthest to the west and south were used to form the polygon. Two other trees in the clump of eight were east and north of the trees chosen but could be included as outermost trees which would have resulted in a slightly smaller polygon. Also, for CL1 the polygon technique included more area with few large pines, resulting in a lower pine density for the colony defined by the polygon rule. Since the polygon rule always creates the largest colony it is most likely to include stands which are dissimilar to the stand where the cavity trees are located.

Foraging area rules

Application of the rules for determination of potential foraging habitat to the selected colonies produced some unexpected results

(**Table 2**). Four of the rules involve an 804 meter radius circle and should have identical areas of 205 ha. However, for both colonies the results are between 169 and 172 ha. These areas result from the intersection of the buffers with the forest stand map. CL1 is located near the northern property boundary and portions of the areas beyond the edge of the stand map are not included in the stand summaries (**Figure 1**). CL2 is near the southern edge of the forest and portions of the potential foraging areas extend into Winyah Bay on the west and a salt marsh on the east (**Figure 2**). As expected, the polygon area was larger than any of the circular areas. As with the colony area, the two distant and perpendicular outliers in CL1 produce a much larger area for the polygon rule. For CL2 the outlier simply made the area slightly oblong and extended it mostly into the marsh.

Potential foraging areas contained in the polygons contained more pine basal area than the potential foraging areas produced by other rules and much more in CL1 (**Table 2**). The changes in pine basal area among the various circular rules depended solely on the stands around the clusters. The arithmetic mean of the tree coordinates moved the center toward the largest number of cavity trees. In both cases this produced a center close to the grouped trees which were on the western side of the colony. A stand of thick loblolly pine (**P on Figure 1**) was located west of CL1 and rules that moved the center west increased the basal area of potential foraging available. Winyah Bay was at the western edge of the circles around CL2 (**Figure 2**). Therefore, definitions that moved the center west, decreased available forage.

A center described by the mean of the coordinates of all cavity trees in the colony (cluster);

(**1**) will always tend to be closest to the largest concentration of trees,

(**2**) is always uniquely defined by the location of the cavity trees, and

(**3**) will work with any mapping system. The box center rule does not bias the center toward the greatest concentration of cavity trees. However, the bias will tend to minimize flight distances for birds using the clumped trees. Elimination of outliers reinforces this bias, but determination of outliers may not be as obvious in other colonies as in the two cases considered in this example.

The choice of rule made little difference for CL2. Even the polygon technique added little to potential foraging area, primarily because most of the area added was in the salt marsh. The four circular centers differed by about 100 m in both colonies. The potential foraging habitat in CL2 changed by less than two percent, but for CL1 the potential foraging habitat changed by more than 11 percent. This was due primarily to the inclusion of larger portions of one thick stand. Since CL1 had two distant outliers, the polygon rule resulted in substantially more potential forage on a significantly larger area. However, portions of this potential forage were up to 1400 m from the most distant cavity tree. If this rule is used a distant stand might be included in the provided foraging habitat that is beyond the average foraging range of the RCW.

All techniques to define the potential foraging area, except the polygon buffer, appear to meet all criteria related to the biology of the woodpecker. The rules that exclude outliers may not always produce consistent results. In both the considered cases, there was a clear grouping of trees with obvious outliers. Trees may not always occur in such obvious groups. The center of a box that surrounds all trees will be more affected by outliers than the mean of all tree coordinates.

SUMMARY AND CONCLUSIONS

Management of forest land with RCW can be aided by GIS planning tools. To use these tools, management guidelines must contain complete, consistent, and objective rules to define all spatial relationships. We propose that the colony (cluster) be defined by a 61 m buffer around all cavity trees, with a 122 m wide corridor connecting to cavity trees more than 122 m from the clump; the center of the corridor being a straight line between the outlier and the nearest cavity tree in the clump. The potential foraging habitat should be defined as a 804 m radius circle centered at a point that is the mean of the x and y coordinates of all cavity trees in the colony. These two rules completely and uniquely define these areas and can be determined for any colony in which all cavity trees have been accurately mapped. If the woodpecker using an outlier is known to associate with a particular clan that cavity tree can easily be included into the proper colony.

The examples from Hobcaw forest

demonstrate compelling reasons for adoption of such unique definitions. The RCW populations had been well researched (Grimes 1977, Wood et al. 1985a), including determination of foraging ranges (Wood et al. 1985b). However, following elimination of 38% of the cavity trees by Hurricane Hugo, previous information was no longer adequate. The GIS system allowed rapid remapping of surviving and newly constructed cavity trees. With consistent spatial analysis, the response of each clan can be compared to patterns of forest damage and forest recovery in the colony or the potential foraging habitat.

One example, CL1, also demonstrates a future need. The potential foraging habitat extended beyond the Hobcaw forest boundary. The proposed private landowner guidelines anticipate landowner cooperation. Some of the definitions moved the potential foraging habitat about 100 m to the east or west. One hundred meters may not be significant to the biology of the bird or to management of a forest, but 100 m could encompass an entire ownership in a real estate development. The ability to determine exact locations of managed habitat will become more important as conservation efforts move to valuable private lands.

SECTION 3

Natural Disturbances

Natural Disturbances: Barriers to Recovery of the Red-cockaded Woodpecker

Robert G. Hooper and David L. Kulhavy

The red-cockaded woodpecker evolved with a recurring array of natural disturbances that affected the southern pine forest. Disturbances, such as fire, were a driving force in shaping plant and animal communities associated with the southern pine forests. Even today, fire is essential for perpetuating many of our southern ecosystems and it is universally accepted that the red-cockaded woodpecker cannot exist long-term without frequent fires within its habitat. Clearly fire is a disturbance that has benefits for the woodpecker. The chapters in this section on natural disturbance try to come to grips with perturbations that may or may not have a positive benefit for the red-cockaded woodpecker and that could indeed prevent its recovery.

Hurricanes and the southern pine beetle (*Dendroctonus frontalis*) have been around at least as long as the red-cockaded woodpecker. There is little doubt that hurricanes in pre-Columbian times had a significant impact on the habitat of the woodpecker. It does not take much imagination to envision relatively large areas where the old growth pine forests had been reduced to a few scattered trees. Populations of the woodpecker were probably greatly reduced in such areas. Indeed it is ludicrous to think the species existed at high population densities everywhere within its range in pre-Columbian times. However, in a relatively short time these areas were reforested and reinhabited by a dense population of woodpeckers, while other areas were beginning their recovery from a hurricane or beetle epidemic.

The extent of the pre-Columbian forest was such that disturbances we now judge to be destructive, such as hurricanes, down bursts, tornadoes, and pine beetle epidemics, were beneficial in the sense of regenerating forests and providing habitat for vertebrates dependent upon early successional forest stages. Both hurricanes and beetle epidemics are properly thought of on a regional scale. Coulson et al. discuss evidence from the literature that suggests initiation of southern pine beetle attacks in cavity trees is similar to the process as it occurs in the lightning-struck hosts and that growth of infestations in cluster sites is not likely because of the low basal area and wide spacing of trees. However, in the loblolly pine sites of eastern Texas, Rudolph and Conner note that the southern pine beetle was the major cause of red-cockaded woodpecker cavity tree mortality. Infestation of individual trees, primarily in the fall, resulted in annual mortality rates of active cavity trees exceeding 30%. Active cavity trees experienced higher mortality rates than inactive cavity trees. Chrismer et al. examined forest canopy gaps resulting from disturbances in red-cockaded woodpecker colonies. Canopy gaps in control stands (non-red-cockaded woodpecker stands) were more numerous but smaller than those in cavity tree clusters and replacement stands. Bark beetles and lightning were the most important disturbances in all stand types, both in terms of number of canopy gaps created and total area influenced. Nebeker et al. evaluated potential impact of southern pine beetle infestations on red-cockaded cavity tree clusters on the Noxubee National Wildlife Refuge. They concluded that there was little risk of infestations growing, but even small-scale infestations could expatriate the woodpecker. Nebeker et al. recommended the refuge continue an active southern pine beetle suppression program and maintenance of low pine basal area within the cavity tree clusters.

Wind as a factor in red-cockaded woodpecker management was examined on an individual tree and stand basis (Chrismer et al., Conner and Rudolph) and on an area wide basis (Hooper and McAdie, Lipscomb and Williams). Wind-related tree mortality was ranked third in occurrence behind lightning and bark beetles in red-cockaded woodpecker cavity tree clusters for all tree species (Chrismer et al.). Conner and Rudolph found wind damage was the second major cause of red-cockaded woodpecker cavity tree loss in eastern Texas and the distance between cavity trees and openings in the forest canopy affects the rate at which wind damage occurs. In Texas, most wind-damaged cavity trees were within 50 m of openings in the forest canopy.

According to Hooper and McAdie, 45 hurricanes are expected to hit the 11 most vulnerable recovery areas on average in 100 years. Six of these storms will be of near Hugo intensity. They predict that some recovery populations could be lost when they are struck by multiple hurricanes over a relatively short span of time. They suggest that through silviculture, it may be possible to grow a more wind resistant forest: the more trees that survive a hurricane, the shorter the restoration period.

Watson et al. relate the tragic impact of Hurricane Hugo **(Figures 1a, b)** upon the essentially recovered red-cockaded woodpecker population in the Francis Marion National Forest. Some 87%

Figure 1a. 100-year old longleaf pine stand, Francis Marion National Forest, SC, pre-Hugo photo (USDA Forest Service photo).

Figure 1b. Red-cockaded woodpecker cavity tree cluster site, Francis Marion National Forest after Hurricane Hugo. Note cavity tree on ground in foreground (USDA Forest Service).

of the cavity trees were destroyed and 63% of the birds killed in just a few hours. Fortunately new technology was available for construction of artificial cavities (**Figure 2**) and the population has responded remarkably. This chapter in effect provides a plan of action for managers of other recovery areas that will be hit by hurricanes in the future.

Lipscomb and Williams offer the hope that the red-cockaded woodpecker can recover under their own initiative from major loss of their cavity trees (38%) when potential cavity trees are available for them to do so. Their results may apply to serious loss of cavity trees from either wind or bark beetles (or a combination), assuming a healthy population of woodpeckers in the first place.

Catastrophes are costly. Reeves et al. look at economic implications of events such as Hurricane Hugo. They asked a random sample of United States citizens if they would be willing to pay more to increase the chances of an endangered species surviving. Preliminary results suggest the average person is willing to pay more for a higher probability that a species will survive.

This section speaks to the need for active forest management to recover from and reduce the impacts of natural disturbances, both before and after they occur. It is a certainty that natural disturbances will threaten red-cockaded woodpecker populations in the future (see Hooper and McAdie, **Table 1, p. 150**). How much of a barrier these potential catastrophes will be to recovery of the woodpecker depends almost entirely on us: how well we manage for beetle and wind resistant forests and how well we respond to the catastrophes when they occur. Dramatic recoveries following catastrophic events (i.e. Hurricane Hugo, Watson et al.) are possible with concerted efforts with artificial cavities, rapid response and intensive monitoring.

For recovery of the red-cockaded woodpecker to succeed, the influence of disturbances and their role in the long-term management of the red-cockaded woodpecker must be considered and integrated into long-term management goals for both recovery of the species and long-term forest health.

Figure 2. Tim Milling installing cavity insert, Francis Marion National Forest following Hurricane Hugo (USDA Forest Service).

Hurricanes and the Long-Term Management of the Red-cockaded Woodpecker

Robert G. Hooper, U.S.D.A., Forest Service,
Southern Research Station, Charleston, SC 29414
Colin J. McAdie, U.S.D.C., NOAA, National Weather Service,
National Hurricane Center, Coral Gables, FL 33146

ABSTRACT: Hurricanes will play a major role in the long-term management of the red-cockaded woodpecker (*Picoides borealis*) [RCW]. Historical evidence indicates that hurricanes can destroy extensive areas of forests. Forests are helpless in the face of catastrophic winds, but such winds are limited in extent in hurricanes. Most damage is caused by lesser winds interacting with tree and site factors. It may be possible to enhance tree survival silviculturally by modifying tree factors. The vulnerability of RCW recovery areas to hurricanes is largely determined by their distance from the coast. The 4 most inland of the 15 recovery areas are essentially immune from severe hurricane-induced winds. The mean time between category I and III force winds occurring over 1 or more of the 11 most vulnerable recovery areas is 2 and 16 years, respectively. On average in 100 years, 45 hurricanes are expected to deliver at least category I force winds to 10 of the 11 most vulnerable recovery areas. Six of these hurricanes are expected to bring very destructive winds of at least category III force to 4 of the recovery areas. Repeated hits to the same areas could extirpate their RCW populations. It seems certain that at least 1 recovery area will always be in the process of restoration from hurricane damage. Currently, at least 2 recovery areas are in that process.
KEYWORDS: catastrophic events, hurricanes, *Picoides borealis*, recovery areas, wind damage

Hurricanes have the potential to instantly destroy the habitat of red-cockaded woodpecker (RCW) over large areas. A hurricane's effects can be long lasting, and some probability exists that an area will get hit again before it has had time to recover from its last hurricane.

From 1970 when the RCW was listed as endangered, through 1991, there had been only 0.68 hurricanes per year making landfall within the range of the RCW (from Galveston, TX to the NC-VA border). This is the lowest frequency of hurricanes for any period of similar length since 1899, and is half the frequency from 1899 through 1969 (Neumann et al. 1987). Thus, the first 2 decades of recovery of the RCW were conducted during a lull in hurricane occurrence. During this lull, 5 recovery areas were impacted by hurricanes, and the populations on 2 of these areas suffered significantly. We can probably expect an increase in frequency of hurricanes in the coming decades (Gray 1990). In addition, there is some concern that global climate changes may significantly increase the frequency and intensity of future hurricanes (Walker et al. 1991). It seems evident that hurricanes will have some role in the long-term management of RCW recovery areas (**Figure 1**). Our paper explores this role.

THE HISTORICAL RECORD

Damage to the Southern Pine Forest

In assessing the risk to RCW recovery areas from future hurricanes, it is important to understand the damage these storms can cause to pine forests. One approach is to examine the damage inflicted by past hurricanes (**Table 1**). Given the human tragedy associated with hurricanes, we suspect that forest damage has frequently gone unreported. Some evidence exists for hurricane damage to forests prior to 1875 (Ludlum 1963), but it is not quantitative either in the extent or degree of damage. Nonetheless, it appears that earlier hurricanes were very destructive to the virgin

Figure 1. Location of red-cockaded woodpecker recovery sites: (1) Sam Houston National Forest [NF]; (2) Ft. Polk—Vernon District, Kisatchie NF; (3) Bienville NF; (4) DeSoto NF; (5) Oakmulge District, Talladega NF; (6) Talladega—Shoal Creek Districts, Talladega NF; (7) Eglin Air Force Base—Blackwater State Forest—Conecuh NF; (8) Apalachicola NF; (9) Osceola NF; (10) Piedmont National Wildlife Refuge—Oconee NF; (11) Ft. Stewart; (12) Francis Marion NF; (13) Carolina Sandhills National Wildlife Refuge—Sandhills State Forest; (14) Ft. Bragg; (15) Croatan NF (from Lennartz and Henry 1985).

forests of the time.

By 1900, there was increased economic interest in forest resources in the South (Walker 1991:97-101), and foresters and lumbermen were making quantitative reference to hurricane damage. But we suspect that hurricanes that hit after the old-growth forests had been cut (1885-1930 depending upon region) had relatively little to destroy until fairly recently (say after 1960). Thus, it is difficult to get a clear picture of potential hurricane damage to forests from the historical record alone.

The hurricanes of 1875 and 1900 in Texas did impressive damage to large areas of forests, as did the 1906 hurricane in Mississippi (**Table 1**). The 1906 hurricane may have done more damage than Hugo. There seems to be a pattern of increasing damage over time beginning with Hazel in 1954 and continuing with Audrey in 1957, Gracie in 1959, Camille in 1969, Federick in 1979 and Hugo in 1989. This trend may be due in part to the increasing age of the southern forest. Hugo was 5 times more destructive to forests in the South than any previous hurricane in the past 4 decades. There can be little doubt that much of the damage resulted from the extensive areas of pine sawtimber in Hugo's path (Sheffield and Thompson 1992). Bray (1901), Foster (1912), and Derr and Enghardt (1957) made special note of hurricane destruction to the old-growth pine forest (**Table 1**).

Damage to RCW Cavity Trees and Recovery Areas

Note that the references to hurricane strength in the following section are hourly interpolations of maximum sustained winds (1-minute average) near the storm center using estimates of such winds that had previously been made at 6-hour intervals. These data were obtained as part of the print out from the

Table 1. Hurricanes that occurred within the range of the red-cockaded woodpecker since 1875 for which forest damage was reported.

Year	State	Damage	Reference
1875	TX	Almost completely destroyed old-growth loblolly in 20X100 mile area.	Bray 1901
1896	FL	Enormous damage to pine in 50 mile wide strip.	Henry 1896
1900	TX	50% of oak and pine timber in 2000 square mile area blown down. Nearly 100% of trees down on thousands of acres.	Bray 1901
1906	MS	30-90% of timber in 50X150 mile area blown down.	Dunston 1910
1909	LA	10% of timber in 2 parishes lost. 65% of virgin longleaf blown down.	Foster 1912
1911	AL	Extensive areas of even-aged longleaf established following hurricane.	Boyer 1991
1928	FL, GA	10% of turpentine trees killed.	Annon. 1928
1935	FL	50-75 million board feet of pine blown over in 30 mile wide strip.	Annon. 1935
1947	GA	Many pines and hardwoods in 50X50 mile area were destroyed.	Bradwell 1947
1954	NC	Hazel: wide spread but unquantified damage.	Carney and Hardy 1962 Seamon 1954
1957	LA	Audrey: 212 million board feet of merchantable timber killed. 90 miles inland 10 million board feet of virgin loblolly killed.	Derr and Enghardt 1957
1959	SC	Gracie: 580 million board feet of timber damaged.	Flory 1960
1964	LA	Hilda: about 4 million board feet expected to be salvaged.	Hurst 1965
1965	LA	Betsy: 200 million board feet of timber killed.	Robinson 1965 Annon. 1965
1969	MS, LA	Camille: 1.2 billion board feet of sawtimber and 1.3 million cords of pulpwood on 115 mile strip 20-60 miles wide damaged.	Van Sickle and Hedlund 1969; Touliatos and Roth 1971
1975	FL	Eloise: 400,000 cords of pine and 270,000 cords of hardwood damaged.	Wilkinson et al. 1978
1979	MS, AL	Frederick: 3.4 billion board feet of sawtimber (68% pine), plus 881,000 cords of pulpwood damaged.	MFC 1979
1983	TX	Alicia: 56 million board feet damaged in Sam Houston National Forest.	L.J. Carmical, pers. commun.
1985	FL	Kate: 80,000 mostly older longleaf pines destroyed on Apalachicola NF. 100 miles inland, 10% of old-growth longleaf killed.	Bodie 1985; Platt et al. 1988
1989	SC, NC	Hugo: In SC 10.8 billion board feet killed, plus 9.6 billion in critical condition. In NC, 1 billion board feet destroyed.	Sheffield and Thompson 1992; Doggett 1989

HURISK model (Neumann 1987) we used to estimate average time between hurricane occurrences. The threshold maximum sustained winds for Saffir-Simpson category I, II, III, IV, and V hurricanes are 74, 96, 111, 131, 155 mph, respectively (Saffir 1977).

Hurricane Hugo. The Francis Marion National Forest (NF) **(Figure 1)** suffered catastrophic loss to its RCW population, cavity trees and foraging habitat from Hugo. The storm destroyed 87% of the cavity trees, killed 63% of the birds, and destroyed 70% of the foraging habitat (Hooper et al. 1990). Overnight, the population was reduced from 477 groups down to 239 groups. The 239 groups reflected a response to installation of 537 artificial cavities prior to the first nesting season following Hugo. Without the artificial cavities the number of groups may have been as low as 100 (Watson et al. 1995).

Hugo was a category IV hurricane at

landfall, and a category III and then II as it passed over the Francis Marion. The Carolina Sandhills National Wildlife Refuge **(Figure 1)** lost 10% of its cavity trees from Hugo (David H. Robinson, pers. commun.). Had Hugo turned more northerly, loss of cavity trees on the refuge could have been substantial. Severe loss of cavity trees occurred over an area > 50 miles wide 15 miles from the coast, to at least 10 miles wide 110 miles inland **(Figure 2)**.

Hooper et al. (1990) thought RCW cavity trees were 2-4 times more vulnerable to wind than trees without cavities. Lipscomb and Williams (1995a) found that cavity trees 45 miles from the center of Hugo's storm track were 4 times more vulnerable to wind damage than trees without cavities. Eighty-six percent of the heavily damaged cavity trees broke at the cavity (Lipscomb and Williams 1995a).

The hypothesis that RCW cavities weaken trees is supported by Watson et al. (1995), who found 53% of the cavity trees lost in Francis Marion NF broke at the cavity; 10% broke above the cavity and 19% below the cavity, and the remaining 18% were uprooted. Engstrom and Evans (1990) reported 6 of 8 wind-killed cavity trees broke at the cavity and 2 snapped below the cavity.

Other hurricanes. Kate passed over the western edge of the Apalachicola NF **(Figure 1)** in 1985 with category I force winds. About 80,000 trees, mostly older longleaf pines, were blown down or broken (Bodie 1985). Balboni (1985) estimated more than 80 cavity trees were killed. This number was only about 4% of the cavity trees on the western half of the Apalachicola (Susan M. Fitzgerald, pers. commun.).

Alicia, damaged about 56 million board feet of timber in the Sam Houston NF **(Figure 1)** in 1983 (L.J. Carmical, pers. commun.). The storm did not have maximum sustained winds of hurricane force by the time it was abreast of the center of the Sam Houston. Wind damage exacerbated a southern pine beetle (*Dendroctonus frontalis*) epidemic which eventually destroyed more timber than Alicia. Conner et al. (1991a) reported about 3% of the RCW cavity trees were killed outright by Alicia, but that during the 5 years after Alicia 26% of the cavity trees were lost.

Eloise passed over Eglin Air Force Base **(Figure 1)** in 1975 as a category III and II hurricane. Significant forest damage occurred on 55-60 % of the 400,000 acre area (Wilkinson et al. 1978). Most of this damage was to sand pine (*Pinus clausa*) (Stephen M. Seiber, pers. commun.), a species of little or no value to RCW. Seiber stated that little damage was done to RCW cavity trees because most were west of the storm track.

The DeSoto NF **(Figure 1)** suffered significant damage from 3 hurricanes starting with Camille in 1969 (Wayne H. Stone, pers. commun.). Camille was a category V hurricane at landfall but had decayed to a category II when abreast of the west side of the DeSoto. Ninety million board feet of timber were salvaged from the DeSoto and it was estimated that at least 12 clusters of RCW cavity trees were destroyed by Camille. In 1975, Frederick passed on the east side of the DeSoto as a category II hurricane when abreast of the DeSoto. After Frederick, 110 million board feet were salvaged from the DeSoto and 9 clusters of RCW cavity trees were destroyed or heavily damaged.

In 1985, Elena passed to the southwest of the DeSoto as a category II hurricane. After Elena, 8 million board feet of timber were salvaged from the DeSoto, and 4 RCW clusters were destroyed or heavily impacted. The DeSoto had only 18 active clusters in 1989 (Dennis L. Krusac, pers. commun.).

FACTORS AFFECTING FOREST DAMAGE

Conceptual Forest Damage Model

Damage to forests varies greatly within the path of a hurricane. A better understanding of this variation might help in derivation of a strategy for dealing with future hurricanes. We can think of the total damage from a hurricane as simply the sum of the damage to individual stands of trees. Stand damage is a function of the hurricane wind field parameters, the position of the stand in the wind field, rainfall, tree factors, and site factors **(Figure 3)**.

Hurricane wind field parameters affecting the potential for forest damage are maximum sustained winds, horizontal wind profile, forward speed of the storm, decay rate as the hurricane progresses inland, associated rainfall, and the frequency, intensity and aerial extent of down bursts. The damage potential of a hurricane is filtered through site and tree factors, which either exacerbate or ameliorate the actual damage that occurs. With

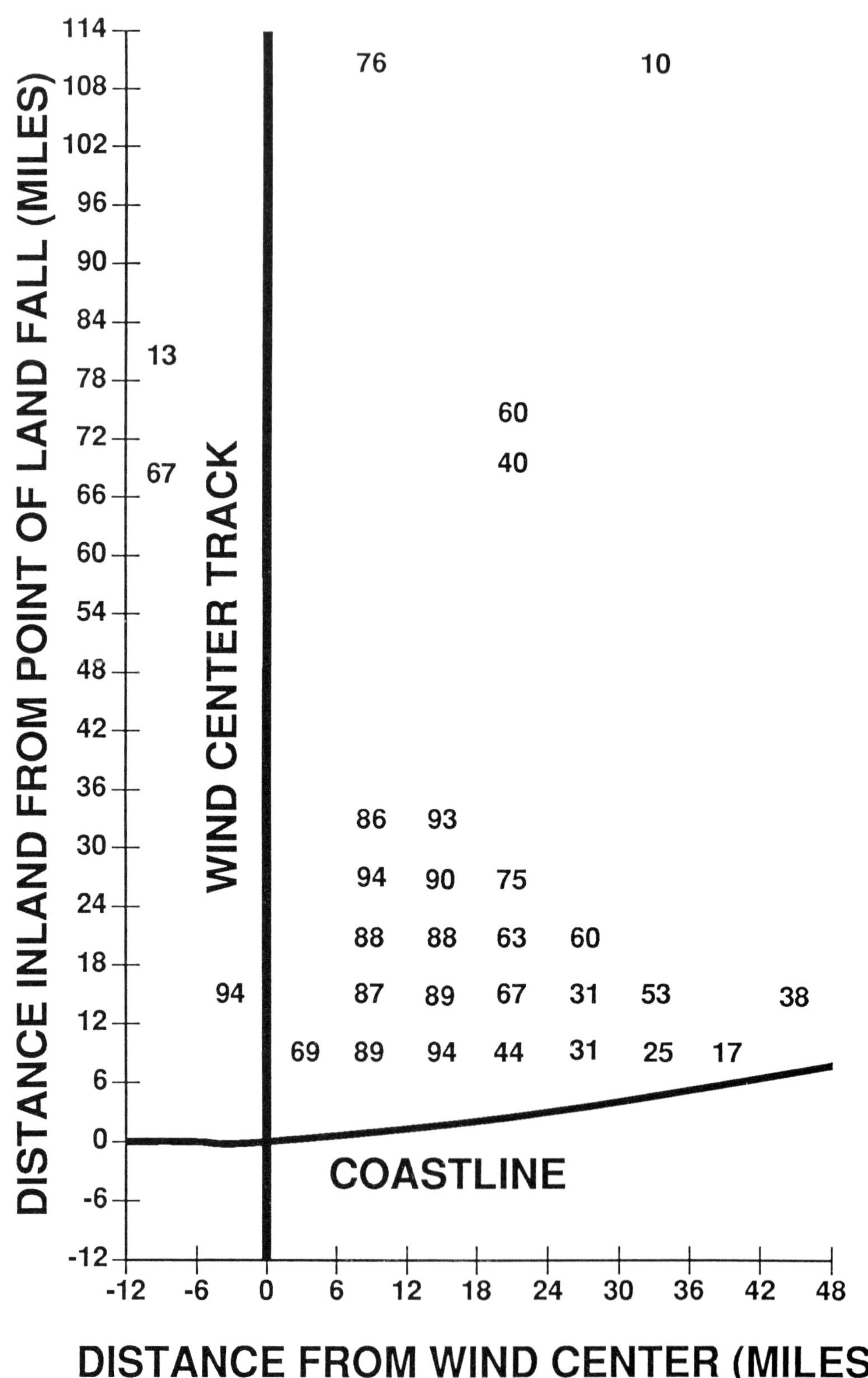

Figure 2. Percentages of red-cockaded woodpecker cavity trees in 6 by 6 mile areas that were destroyed by Hurricane Hugo, relative to the distances from the wind center tract and the coast. Data are from Cely (1991); Lipscomb and Williams (1995a); D.L. Carlson, J.E. Cely, C.J. Townsend, D.H. Robinson, J.C. Watson (pers. communications).

catastrophic winds (probably category V force winds), site and tree factors are overridden; any stand subjected to such winds will be destroyed (Hook et al. 1991). However, such winds over land usually occur over a relatively small portion of the area affected by most hurricanes. Thus, most forest damage occurs from an interaction of wind field parameters, tree factors and site factors (**Figure 3**).

Wind Field Parameters

Horizontal wind profiles. In the northern hemisphere the highest maximum sustained winds usually occur on the right side (facing the direction of forward motion) of hurricanes. The high velocity of wind near the eyewall is supplemented by the hurricane's forward speed (Simpson and Riehl 1981). Thus, forest damage tends to be more severe on the right than on the left side of hurricane tracks. This skewed effect was prominent in Hugo (Sheffield and Thompson 1992, Doggett 1989) and was seen to a lesser extent in Camille (Touliatos and Roth 1971). An exception to this rule occurred with Hazel. More forest damage occurred on the left side of the track because the heaviest rainfall occurred there (Trousdell 1955).

The radius of maximum winds (RMW) is the "radial distance outward from the center of the eye where the winds reach a maximum value. This typically occurs a few miles radially outward from the eyewall" (Neumann 1987). On average, the RMW varies directly with latitude and inversely with storm intensity, with latitude having the greatest effect. The difference in latitude between the southernmost and northernmost recovery areas is about 5 degrees, resulting in an average increase in RMW of 5 miles (Neumann 1987). Beyond the RMW, wind velocity decreases with distance from the center of circulation. Wind damage follows this same pattern: damage is greatest near the RMW and decreases as one moves away from the hurricane center.

Gusts. Much of the damage to forests may be caused by short-duration gusts (2 - 4 seconds) resulting from turbulence (Savill 1983); and from extreme gusts associated with extremely convective rainbands, both within and external to the eyewall (Hook et al. 1991, Powell et al. 1991). These gusts may partially explain the patchy occurrence of damage to forests from hurricanes (Bray 1901, Curtis 1943, Derr and Enghardt 1957, Flory 1960, Robinson 1965, Hurst 1965, Wilkinson et al. 1978, Doggett 1989, Hook et al. 1991, Sheffield and Thompson 1992). Damage caused by extreme gusts is probably mistaken for tornadic damage. Tornadoes are associated with about 25% of the hurricanes making landfall in the United States. Typically, when they do occur, tornadoes are found in the right front quadrant of large hurricanes (Anthes 1982:62).

Decay rate. Barometric pressure increases and wind velocity decreases after hurricanes make landfall. The decay rate strongly affects potential damage from a hurricane. Three major physical effects are responsible for the decay of hurricanes over land (Anthes 1982:62). The most important factor may be the sudden reduction in evaporation as the storm leaves the ocean. A second factor is that during the hurricane season, the land is cooler than the ocean and the low-level air is cooled rather than heated by the underlying surface. Lastly, the rougher surface of the land causes more friction than the ocean.

Rainfall

The amount of rain prior to and during hurricanes interact with soils to greatly increase damage to forests, especially damage from less intense storms. Saturated soils can behave as a liquid, greatly reducing the holding ability of tree roots (Day 1950, Derr and Enghardt 1957, Trousdell et al. 1965).

Damage from Hurricane Hazel was probably lessened considerably by a dry summer and by the rainfall distribution resulting from the hurricane (Trousdell 1955). Heavy rainfall for the 9 months preceding the 1900 Galveston Hurricane was credited with contributing to the enormous forest damage from that storm (Bray 1901). Others have also implicated rainfall as a major factor in wind-related forest damage (Curtis 1943, Grano 1953, Croker 1958, Hurst 1965, Trousdell et al. 1965).

Influence of Tree Factors on Stand Damage

Species. Derr and Enghard (1957) thought slash pine (*Pinus elliottii*) plantations on wet, shallow soils, were more susceptible to windthrow during Hurricane Audrey than loblolly (*P. taeda*) plantations on similar soils. Grano (1953) found loblolly to be many times more wind resistant than shortleaf pine (*P.*

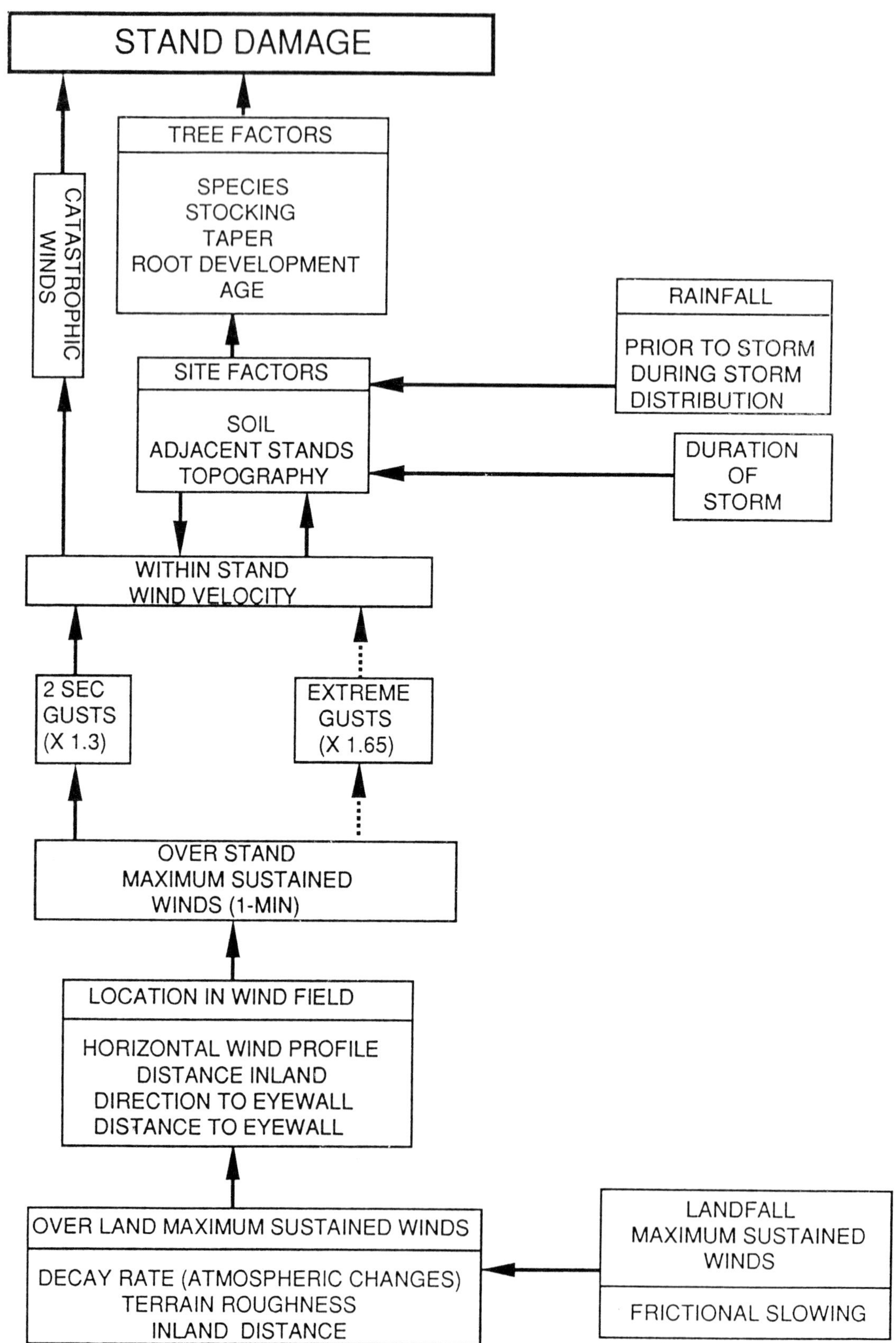

Figure 3. Major factors that affect hurricane damage to an individual forest stand. The total damage to a region from a hurricane is the sum of the damages to individual stands. Because of the relationships shown here, forest damage tends to be patchy.

echinata) on soils with poor internal drainage. From their experience with Hurricane Camille, Touliatos and Roth (1971) ranked longleaf(*P. palustris*), slash, and loblolly in that decreasing order of resistance to breakage, uprooting, insects and disease.

Hook et al. (1991) found no difference in the loss of longleaf (89%) and loblolly (91%) pines that were subjected to eyewall winds during Hugo. However, in an area 45 miles to the right of the storm track, only 27% of the longleaf was damaged, compared to 52% of the loblolly. In the same area, there was little difference in the percentages of longleaf and loblolly that suffered fatal damage (7.8 vs 9.0%, respectively)(Gresham et al. 1991). Loblolly, however, was much more susceptible to defoliation and top breakage than longleaf (17.5 vs 4.5 %, respectively). Some 26.7 percent of the pond pine (*P. serotina*) suffered fatal damage.

Sheffield and Thompson (1992) saw no difference in resistance to wind damage among loblolly, longleaf, slash and pond pine throughout South Carolina from Hugo. But in the Francis Marion NF, where some of the strongest winds occurred, 56% of the longleaf ≥ 10 inches DBH survived, compared to 40% of the loblolly (comparison based on inventories made in 1986 and 1990 by the Forest Inventory and Analysis Work Unit, Southeastern Forest Experiment Station, Asheville, NC).

Pond pine seems highly vulnerable to wind damage. Pond pine often have heartrot and grow on unstable soils (Bramlett 1990). These factors probably account for its vulnerability. Among the other pine species, none seems to have a major advantage over the others in the ability to survive hurricanes. Loblolly may be more windfirm than slash on wetter sites. Longleaf seems to be more resistant to fatal wind damage than loblolly. But, because site and tree factors have not been accounted for in the assessment of vulnerability among species, we are not certain these apparent advantages are real. Comparisons among species are also confounded by the patchy distribution of hurricane damage caused by extreme gusts and downbursts.

Tree age and size. Age and size affect tree survivability in strong winds. Typically, the older and larger trees are more susceptible to fatal wind damage than younger and smaller trees. In what is now the Francis Marion NF, Chapman (1905) reported that windfalls were common in over-mature loblolly (some of the windfall was due to earlier fire damage). In Louisiana, the major causes of tree loss over a 5-year period in a virgin longleaf stand with trees 175-240 years old were windthrow (not of hurricane origin) and bark beetles (Chapman 1923). Windthrow usually occurred after heavy rains (implicating soils) and usually involved the larger, older trees. The major sources of mortality among longleaf > 11 inches DBH in a virgin stand in Georgia over a 4-year period were lightning and wind (54 and 31% of the mortality, respectively) (Platt et al. 1988). Later, Hurricane Kate killed 10% of the larger trees in the stand, about 2.5 times the mortality observed during the 4 years of study. Chapman (1923) stated that hurricanes (as evidenced by the orientation of windthrow mounds) had created even-aged stands in former old-growth longleaf. This statement suggests that the vast majority of canopy trees had been destroyed.

Trousdell (1955) reported that young stands suffered minor damage while older stands, especially those that were recently thinned or had low stocking, sustained significant damage during Hazel. In a 1,365 acre area 50 miles east of the storm track and 170 miles after landfall, Hazel destroyed 6.1% of the pines > 10 inches DBH. Derr and Enghardt (1957) found that Audrey damaged older stands more than younger ones. Stands below merchantable size and 30- to 40- year-old longleaf were not seriously damaged. Managed stands of intermediate age were damaged to varying degrees, and virgin stands of loblolly were heavily damaged. Much of the bole breakage in large loblolly pines was thought to be associated with heartrot. Pulpwood-sized stands suffered less damage from Betsy than sawtimber stands (Robinson 1965). Touliatos and Roth (1971) reported that older stands suffered the most damage from Camille. Bodie (1985) stated that the majority of longleaf pines uprooted and broken during Kate were older trees.

In an area 45 miles to the right of Hugo's track, mean DBH's of undamaged longleaf, loblolly and pond pine were 11, 7, and 6 inches, respectively; means for trees with fatal damage were 15, 11, and 13 inches, respectively (Gresham et al. 1991). In the Francis Marion NF, loss of pines from Hugo increased as DBH increased **(Figure 4)**.

Hooper et al.(1990) reported that sapling and pole stands were frequently heavily damaged but that enough trees survived to make a manageable stand. Many 30-40 year old pine stands in the Francis Marion were moderately damaged and many mature stands suffered very heavy damage. In Congaree Swamp, 90 miles inland, Hugo killed 71% of the loblolly pines 24-30 inches DBH, 45% of those 32-38 inches DBH, and 24% of those more than 40 inches DBH (Putz and Sharitz 1991). It is not clear why those results are contrary to the other studies cited in this section.

Stand density. Stand damage is inversely related to stand density (stocking), with denser stands surviving better than more sparsely stocked ones. Closed canopies tend to keep wind out of stands (Curtis 1943, Alexander 1964, Gorden 1973, Wilson 1984, Foster 1988). Trees also gain mutual support from adjacent crowns. Recently thinned stands are especially vulnerable to winds if the individual trees have not had sufficient time to develop strong roots and boles, but such trees gradually become less vulnerable as they respond to the increased exposure to wind and the temporary reduction in competition (Mergen 1954, Savill 1983, Kozlowski et al. 1991:491). Hooper et al. (1990) observed that 100-year-old longleaf stands with 110 square feet of basal area were reduced to about 70 square feet of basal area by Hugo. Similar stands thinned to 70 square feet 3 years before Hugo were reduced to less than 10 square feet of basal area. Seed tree and shelterwood stands are vulnerable to windthrow for a long time (Trousdell 1955, Robinson 1965, Savill 1983).

The effect of stand density may be misleading in that trees grown in stands where competition is less intense throughout their life, tend to develop superior wind resistance. Thus, we do not mean to imply here that growing trees in dense stands would necessarily be the best long-term approach to reducing hurricane damage.

Taper. Open-grown conifers with pronounced stem taper are much more resistant to wind breakage from wind than trees grown in dense stands (Curtis 1943, Mergen 1954, Gratkowski 1956, Larson 1963, Assmann 1970:60, Savill 1983, Smith 1986:398). Increased tree taper may be partially a response to swaying caused by the wind; this effect is inhibited in stands with closed canopies (Mergen 1954, Savill 1983, Burton and Smith 1972, Wilson and Archer 1979).

Curtis (1943) found a direct relationship between crown development and resistance to windthrow and breakage. He thought the longer crowns had their geometric centers lower down the bole, which reduced the leverage affect of the wind on the bole. More likely, the trees with longer crowns had more taper and stronger boles (Assmann 1970:60, Smith 1986:100, Farrar and Murphy 1987, Oliver and Larson 1990:83, Nix and Ruckelshaus 1990).

Root development. Mergen (1954) thought that windthrow resulted from compression failure of roots on the leeward side of trees and that roots could be strengthened by exposure to wind during their development. Resistance to uprooting increases as the overall size of the root system increases (Frazer 1962, Coutts 1983). It is generally thought that open-grown trees or at least trees with well developed crowns, have stronger root systems than trees grown in dense stands, and that thinning of stands can improve root development, especially if done when the stands are young, and continued into older age (Mergen 1954, Zahner and Whitmore 1960, Larson 1963, Smith 1986: 399, Oliver and Larson 1990, Kuiper and Coutts 1992).

Influence of Site Factors on Stand Damage

Soil. Hardpans and other restrictive layers can halt the development of a taproot in the pines important to the RCW (Boyer 1990, Baker and Langdon 1990, Lohrey and Kossuth 1990, Lawson 1990). Pulling studies have shown that relatively small reductions in root development result in significant loss of wind firmness (Frazer 1962). Thus, it is not surprising that soils have frequently been cited as a primary factor in wind damage to stands, especially when the storms were accompanied by heavy rainfall (Bray 1901, Foster 1912, Pessin 1933, Grano 1953, Croker 1958, Trousdell et al. 1965, Foster 1988). The same restrictive layers that retard root development can also inhibit internal soil drainage. Waterlogged soils behave essentially as a fluid with greatly reduced root holding properties (Day 1950, Derr and Enghardt 1957, Trousdell et al. 1965).

Soil texture can also affect wind firmness (Mergen 1954). Trousdell et al. (1965)

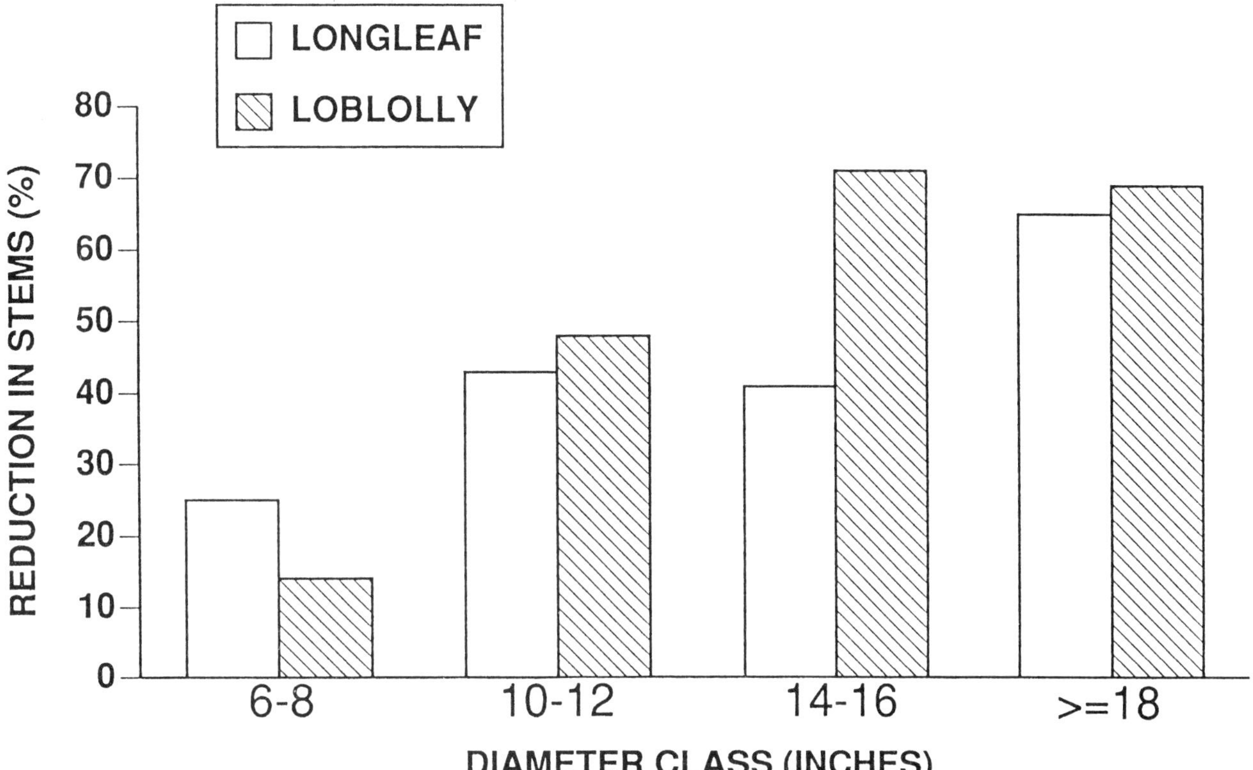

Figure 4. Reduction in the number of longleaf and loblolly pines by diameter classes in Francis Marion National Forest as a result of Hurricane Hugo. Under normal conditions, the percentage change in any diameter class would have been < 5%. Data are from inventories conducted in 1986 and 1990 by the Forest Inventory and Analysis Research Unit, Southeastern Forest Experiment Station, Asheville, NC.

observed that stands on moderately coarse textured soils lost 30% of their trees, compared to 5% loss in stands on medium and fine textured soils, and 51% loss on soils with restrictive layers.

The effects of soils on root development and wind firmness are important because relatively low velocity winds can cause major forest damage to stands on shallow, coarse or restrictive soils (Bray 1901, Foster 1912, Trousdell 1955, Foster 1988). Thus, if a recovery area has such soils it will be at great risk from even a category I hurricane and category III force winds would cause catastrophic damage.

Topography. Topography strongly influences wind damage to trees (Curtis 1943, Gratkowski 1956, Alexander 1964, DeWalle 1983, Foster 1988, Bellingham 1991). Ridges and valleys accelerate winds in some areas and create protected zones in others. On the lower coastal plain, where hurricane-induced winds are strongest, however, topography is not very pronounced. Topographic influences on hurricane damage to forests, are probably greater in the more inland recovery areas.

Adjacent stands. The more uniform a canopy, the less the damage from wind (Curtis 1943). Gaps in the canopy allow wind to enter the stand and cause damage (Curtis 1943, Gratkowski 1956). In addition, an uneven canopy causes turbulence, a principal factor in gusting. The rougher the canopy and the faster the wind velocity, the greater the turbulence (Savill 1983). In some areas, partial cuts that created small gaps, have led to more wind damage than clearcuts that created large gaps (Alexander 1964). Several studies outside the range of the RCW suggested that overall wind loss decreases as gap size increases (Alexander 1964, Neustein 1965, Gorden 1973).

Two additional studies found no relationship between gap size and damage to adjacent stands (Ruth and Yoder 1953, Gratkowski 1956). The shape of adjacent stand boundaries can funnel and accelerate winds, multiplying damage (Gratkowski 1956, Alexander 1964, Gorden 1973, DeWalle 1983,

Conner et al. 1991a). The effect of gaps on wind damage decreases if residual trees have time to respond to increased growing space and become more windfirm.

EXPECTED FREQUENCY OF FUTURE HURRICANES

The HURISK model

No one can predict when a hurricane will hit a recovery area, however, it is possible to estimate the mean elapsed time between hurricanes (return period). Neumann (1987) developed a model, HURISK, for estimating return periods for tropical cyclones with sustained winds of at least some threshold strength at specified locations within the Atlantic tropical cyclone basin. HURISK fits site-specific historical information for tropical cyclones to probability distributions or functions. The primary parameters in the model are storm distance from the site, maximum winds, radius of maximum winds, effects of latitude, horizontal wind profile, and frictional effects. The model uses a Monte Carlo simulation to select sequentially from the distributions 10,000 tropical cyclones within a 150-nautical mile radius of a site. The number of years that it would take for the 10,000 storms to occur is based on the observed frequency of tropical cyclones within the 150-mile radius. For example, with 2.00 storms per year, it would take 5,000 years to have 10,000 cyclones.

Return periods (mean time between events) are then calculated from the simulated frequencies for specified winds occurring at the site (e.g. the center of each recovery area), based on the number of years it would take to have 10,000 storms. For example, out of the 10,000 simulated storms, 25 might have caused category III force winds to occur over the site. Thus, 5,000 years divided by 25 equals a return period of 200 years for category III force winds over the site.

HURISK uses a Weibull distribution of maximum wind velocities associated with tropical cyclones within 75 nautical miles of a site and a storm count-distance function, to create a family of lines representing return periods for specified winds at different distances from the site. The combination of the site and areal return periods allows one to determine the return periods for winds having at least the specified strength occurring anywhere over an area of specified size and stated location within the Atlantic tropical cyclone basin. The data in **Tables 2 and 3** were derived with these procedures.

Based on comparisons to actual events, the estimated return periods generated by HURISK are considered to be reasonably valid for periods up to 100 years. While the validity of the estimates beyond 100 years are less certain, they are still the best available (Neumann 1987).

Relative vulnerability of recovery areas

We used HURISK to estimate return periods for hurricane-induced winds of a specified or greater force (category I—IV, Saffir 1977) occurring within the boundaries of a specified RCW recovery area **(Figure 1, Table 2)**. For application in HURISK, we defined the recovery area boundary as the area of a circle equal to the area of the irregular convex polygon that enclosed the recovery area. The area of any private lands enclosed by the polygon was included as part of the circle.

The threshold maximum sustained wind (1-minute average) expected to occur within the boundary of each recovery area on average in 100 years as a result of tropical cyclones **(Table 3)**, was used as an index of the relative vulnerability of each recovery area to hurricanes. The 100-year expected winds were regressed against distance from the center of the recovery areas to the coast. Distance from the coast accounted for most of the variation in the relative vulnerability among the 15 recovery areas **(Figure 5).** The closer a recovery area is to the coast, the more frequently it will experience hurricane-induced winds and the more intense those winds are likely to be.

The least vulnerable. Least vulnerable to hurricanes are the Bienville NF, Piedmont National Wildlife Refuge—Oconee NF, and the Oakmulgee and Talladega Ranger Districts of the Talladega NF. For these areas, estimated return periods for hurricane-induced winds of at least category I strength were $\geq$ 170 years **(Table 2)**. HURISK output indicates the return periods for hurricane-induced maximum sustained winds of at least category III force are more than 5,000 years for these 4 recovery areas. These values greatly exceed the model in several respects, but do suggest that the 4 most inland recovery areas may be essentially immune to the more destructive category III force winds.

The 2 least likely recovery areas to be subjected to hurricane force winds are the Oakmulgee and Talladega Ranger Districts with return periods of 350 and more than 500 years, respectively, for winds of category I force or greater (**Table 2**). Because of the paucity of observed hurricanes near these two areas, and the fact that HURISK does not take into account the physical constraints imposed by sites located well inland, our estimated return periods for the 2 areas are questionable. Nonetheless, tropical cyclones can still have hurricane force winds that far inland in the southern United States, and given the uncertainties involved, the Oakmulgee and Talladega Ranger Districts should not be thought of as being totally immune to minimum hurricane force winds.

The most vulnerable. The 11 most vulnerable recovery areas represent a wide range of return periods: 14 to 90 years for category I force winds, and 90 to more than 500 for category III force winds of hurricane origin (**Table 2**). Compared to say the Francis Marion NF, Ft. Polk—Vernon Ranger District is not very vulnerable to hurricanes. HURISK indicates the return period for category III force winds occurring over Ft. Polk - Vernon RD and Sam Houston NF may be more than 1,000 years. Nonetheless, those 2 recovery areas are relatively vulnerable to category I force winds. The 5 most vulnerable recovery areas are Francis Marion, Croatan, Apalachicola NF, DeSoto NF, and Eglin AFB —Blackwater SF—Conecuh NF (**Table 2**). These areas have return periods of ≤ 130 years for category III force winds of hurricane origin, and ≤ 55 years for category II force winds. Osceola NF, Ft. Stewart and Ft. Bragg have relatively short return periods for category I hurricanes (≤ 38 years), but relatively long ones (≥ 240 years) for category III force winds.

The shorter the return period, the more likely a recovery area will be hit by another hurricane before its forests can recover from previous hurricane damage. For example, it may take the Francis Marion NF more than 50 years to recovery to its pre-Hugo condition (Hooper et al. 1990, Watson et al. 1995).

The return time for hurricane-induced winds of at least category I force for the Francis Marion is only 14 years (**Table 2**). Such a storm would be capable of destroying most of the trees with artificial cavities and many of the trees without cavities that were injured by Hugo. Thus, RCW populations on the 5 most vulnerable recovery areas are at considerably greater risk of extirpation from hurricanes than any of the other recovery areas.

N-year events

A useful question to ask for a given recovery area is what intensity of wind on average, can be expected in a specified number of years? N-year events for each recovery area were determined from the same HURISK output as return periods (**Table 3**). The threshold maximum sustained wind that the Francis Marion NF, Croatan NF, Apalachicola NF, Eglin AFB - Blackwater SF - Conecuh NF, and DeSoto NF will experience on average in 25 years as the result of tropical cyclones is ≥ 82 mph, sufficient to destroy a significant number of cavity trees. The seven most vulnerable recovery areas can expect at least 95 mph maximum sustained winds every 100 years on average. Winds of that velocity can be expected to cause serious loss of both cavity trees and foraging habitat. In comparison, the four least vulnerable recovery areas would expect to experience winds of ≤ 76 mph in 200 years on average.

N-year events and return periods are averages from a distribution of events. Therefore, a 100-year event will not happen every 100-years, nor will a 25-year event happen every 25 years, etc. The frequency of N-year events occurring through time can be described by the binomial distribution, from which one can determine the probability of an event of any N-years occurring one or more times in any number of years (**Formula 4, Appendix I**). The relationship of the probabilities of several N-year events to elapsed time is shown in **Figure 6**. This figure essentially describes the recurrence of hurricanes over a recovery area. For example, to be nearly certain that a 100-year event will occur at least once, one would have to wait 500 years. Stated another way, every time a 100-year event occurs, it will almost certainly have been less than 500 years since the last such event occurred and it will almost certainly be less than 500 years before such an event occurs again. But six out of ten times a 100-year event will occur, on average, within 100 years of the last such event. And two times out of ten, a

Table 2. Return periods for hurricane-induced winds of at least the specified force (Saffir-Simpson Category I-IV[1]) occurring within the boundaries of red-cockaded woodpecker recovery areas. Return periods were estimated by HURISK

Recovery Area	Miles From Coast[2]	Category of Wind[1]			
		I	II	III	IV
		Return Periods (years)			
Francis Marion NF	15	14	43	90	260
Croatan NF	15	15	55	130	400
Apalachicola NF	20	16	50	100	325
Eglin-Blackwater-Conecuh[3]	25	20	52	99	250
DeSoto NF	50	21	55	105	260
Ft. Stewart	35	25	110	280	>500
Osceola NF	60	27	140	240	>500
Ft. Bragg	105	38	120	270	>500
Sam Houston NF	90	48	290	>500	>500
Carolina Sandhills NWR-SF	105	50	210	>500	>500
Ft. Polk-Vernon RD[4]	90	90	420	>500	>500
Bienville NF	140	170	>500	>500	>500
Piedmont NWR-Oconee NF	180	200	>500	>500	>500
Oakmulgee RD[5]	175	350	>500	>500	>500
Talladega-Shoal Cr. RD[5]	200	>500	>500	>500	>500

[1]Maximum sustained winds (1-minute average) Saffir-Simpson Category I: 74-95 mph. Category II: 96-110 mph. Category III: 111-130 mph. Category IV: 131-155 mph.

[2]Miles from center of recovery areas to the coast.

[3]Eglin Air Force Base - Blackwater State Forest -Conecuh NF.

[4]Ft. Polk and Vernon Ranger District of the Kisatchie NF.

[5]Talladega NF.

100-year event will occur within 22 years of the last event. There are numerous examples of several hurricanes making landfall near the same location in a relatively few years, but then many years elapsing before the same location is again visited (Neumann et al. 1987).

Likelihood of Hurricanes Among Recovery Areas

There are 15 recovery areas (**Figure 1**) with varying degrees of risk from hurricanes (**Figure 5**). The 4 most inland recovery areas have such a low probability of hurricane force winds occurring over them, we excluded them from the following analyses. The chance of any one or more of the 11 most vulnerable recovery areas experiencing hurricane force winds is much greater than for a specified recovery area. For example, the probability of the Francis Marion NF experiencing hurricane-induced winds of category I and III or greater, in a year is 0.069 and 0.011, respectively (**Formula 1, Appendix I**). In contrast, the probability of at least 1 of the 11 most vulnerable recovery areas experiencing hurricane-induce winds of category I and III force in a year, is 0.319 and 0.060, respectively (**Formula 2, Appendix I**).

Similarly, the return period for any 1 or more of the 11 most vulnerable recovery areas experiencing hurricane-induced winds of at least a specified intensity, is much less than for a given recovery area. The return period for category III force winds occurring over any 1 or more of the 11 recovery areas is only 16 years (**Formula 3, Appendix I**), compared to a return period of 90 years for the Francis Marion NF (**Table 2**). The return period for category I force winds occurring over any 1 or more of the 11 most vulnerable recovery populations is only 2 years. Again this is much shorter than the 14 year return period for category I force winds occurring over the Francis Marion.

Expected Number of Hurricanes

We used the Poisson distribution (**Formula 5, Appendix I**) to estimate the total number of

Table 3. Threshold maximum sustained winds[1] expected to occur within the borders of red-cockaded woodpecker recovery areas in N-years, on average, as a result of tropical cyclones. Expected values were estimated by HURISK.

Recovery Area	N - Year Event				
	10	25	50	100	200
	Maximum Sustained Winds (mph)[1]				
Francis Marion NF	68	86	101	116	131
Croatan NF	66	83	95	109	115
Apalachicola NF	66	84	98	112	125
Eglin-Blackwater-Conecuh[2]	60	83	99	115	132
DeSota NF	59	82	95	112	135
Ft. Stewart	59	74	84	95	106
Osceola NF	59	71	86	97	110
Ft. Bragg	48	65	77	90	106
Sam Houston NF	53	66	76	84	94
Carolina Sandhills NWR-SF	49	62	77	84	95
Ft. Polk-Vernon RD[3]	44	55	66	76	84
Bienville NF	41	52	61	69	76
Piedmont NWR-Oconee NF	<40	49	58	66	73
Oakmulgee RD[4]	<40	46	53	60	68
Talladega-Shoal Cr. RD[4]	<40	43	54	56	64

[1]Maximum sustained winds averaged over 1-minute.

[2]Eglin Air Force Base - Blackwater State Forest - Conecuh NF.

[3]Ft. Polk and Vernon Ranger District of the Kisatchie NF.

[4]Talladega NF.

recovery areas that can be expected to be affected by at least 1 hurricane-induced wind event of designated intensity over t-years (**Figure 7**). For example, it will on average take about 500 years for all 11 of the most vulnerable recovery areas to experience at least category I force winds within their boundaries. However, 8 rccovery areas will on average experience at least 1 such event in only 50 years, and 10 of the 11 areas will do so in 80 years (**Figure 7**). Similarly, 4 of the recovery areas are expected to have at least category III force winds occur over them in 100 years on average (**Figure 7**).

While we wait for all 11 of the most vulnerable recovery areas to experience hurricane-induced winds of a given intensity, some of the 11 areas will have numerous such events. In other words, more than 11 hurricanes will occur among the 11 recovery areas before all 11 areas experience winds of specified intensity at least once. This relationship (**Figure 8**) was estimated by **Formula 6, Appendix I**. For example, in the 500 year period that it takes for all 11 recovery areas to experience at least 1 event with category I force winds (**Figure 7**), 226 such events are expected among the 11 recovery areas (**Figure 8**). Obviously, some of the 11 areas will experience many hurricanes before all the areas have experienced at least 1. Similarly, while category III or greater winds are expected to occur over 4 recovery areas in 100 years on average (**Figure 7**), 2 of those 4 areas are expected to have 2 such events in that time (**Figure 8**).

DISCUSSION

Uncertainty

The stochastic nature of tropical cyclones adds considerable uncertainty to the long-term recovery and management of the RCW. We presented estimated average return times (**Tables 2, 3**) and average numbers of recovery areas expected to be hit by hurricanes (**Figures. 7, 8**). Because our estimates are averages, recovery areas could be impacted fewer times by hurricanes in say the next 100 years than our data suggest, or just as likely, they could be impacted more times than we estimate. However, the longer the period of time that we define as "long-term", the less affect stochasticity has and in that sense, the more closely our estimates may predict the future.

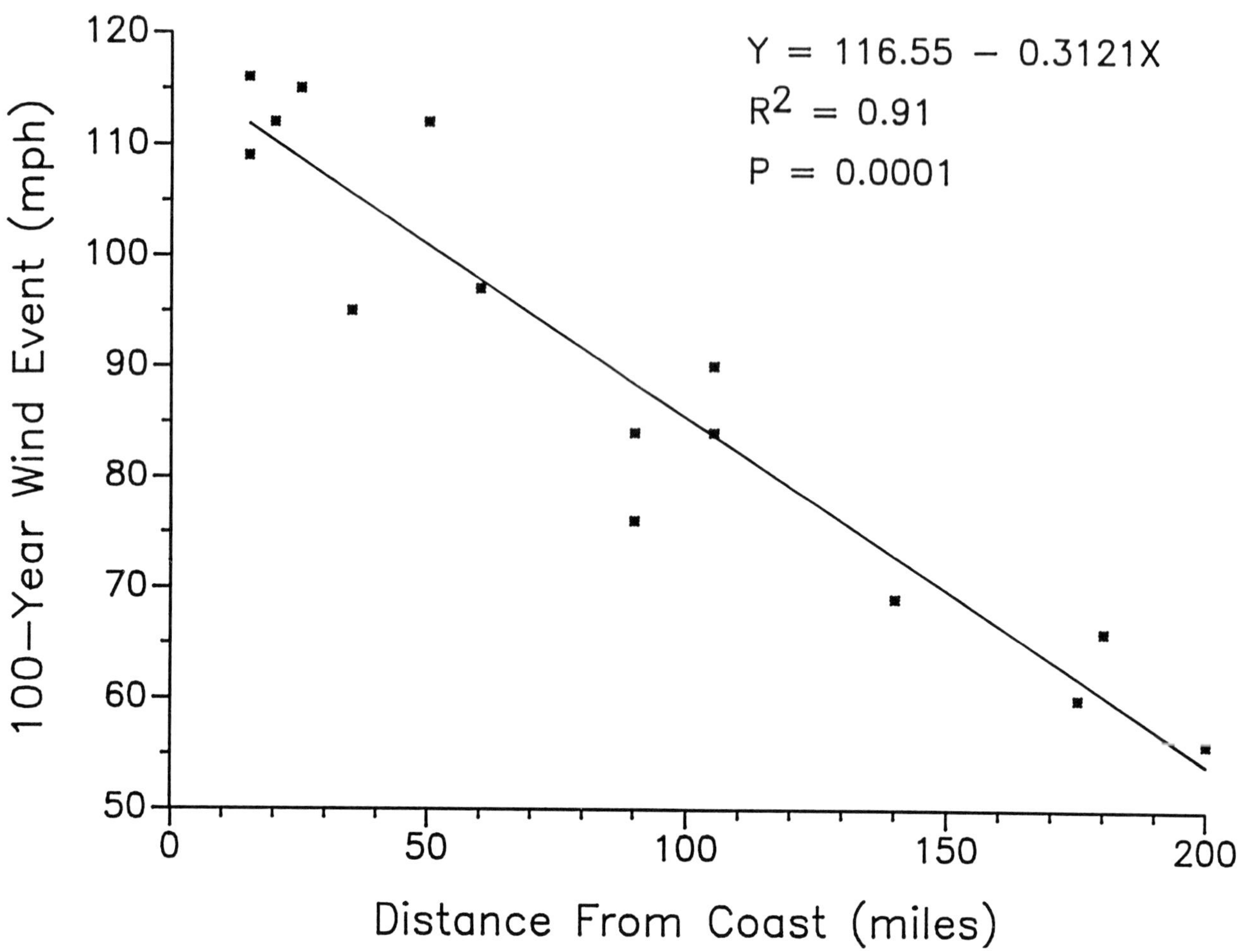

Figure 5. Relationship between tropical cyclone-induced 100-year wind events that are expected to occur within the borders of red-cockaded woodpecker recovery areas, and the distance from the center of recovery areas to the coast. The 100-year wind events are the average threshold maximum sustained winds (1-minute average) expected in 100 years. Expected values were estimated by HURISK.

The average worst case scenario

Hurricane-induced maximum sustained winds of category I force can destroy a significant number of cavity trees. Most recovery populations have demographic problems and suffer from a paucity of trees suitable for cavities (Costa and Escano 1989). A significant loss of cavity trees can have major impact on these floundering populations. Category I force winds can have a devastating effect on a recovery area that has recently been hit by a major hurricane. Trees with artificial cavities are probably more susceptible to breakage from winds than trees with natural RCW cavities (Allen 1991, Taylor and Hooper 1991). Also, trees that have survived major winds frequently have compression and root damage that render them more susceptible to breakage in future winds (Mergen 1954, Trousdell 1955, Watson et al. 1995). Recovery areas with soil limitations can suffer major damage from category I force winds. Finally, catastrophic loss of trees can occur when soils are waterlogged.

We expect that hurricane-induced maximum sustained winds of category III or greater force occurring over a recovery area will cause tremendous damage to cavity trees, potential cavity trees and foraging habitat. Such an event in 1 of the floundering populations can cause extirpation. Major hurricanes passing over large and relatively healthy populations can greatly reduce the population and create isolated sub-populations.

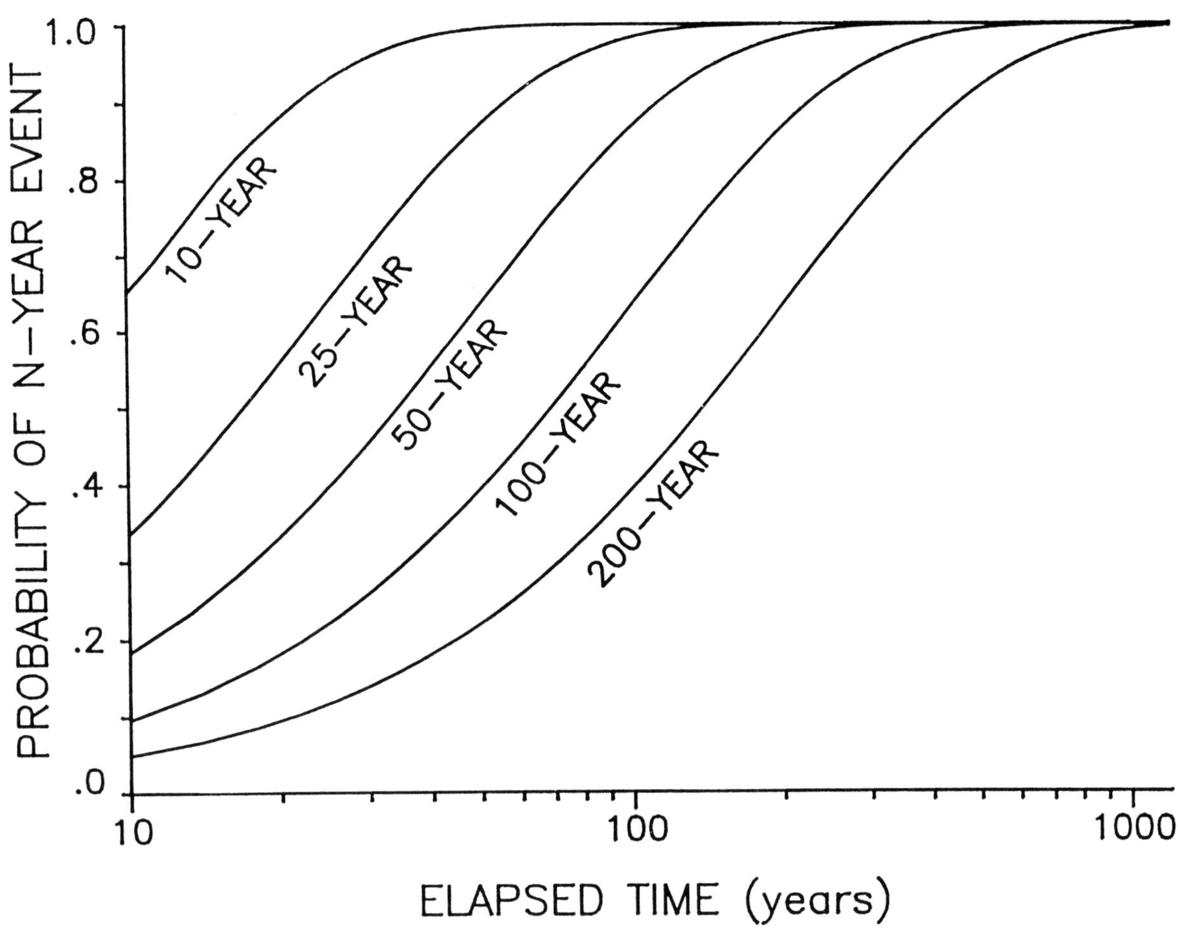

Figure 6. Probabilities of N-year events occurring over time. Values are calculated from the binomial distribution using Formula 4, Appendix I.

In 100 years, on average, we expect 4 recovery areas to suffer the effects of maximum sustained winds of category III or greater force (**Figure 7**). Two of those areas will probably be affected twice (**Figure 8**). In addition, 39 instances of category I or II force winds are expected to occur among the 11 most vulnerable recovery areas. Some of those could as likely hit a recovery area following a category III hurricane as anywhere else. Thus, even in 100 years we can expect major hurricane-induced problems in recovery populations. Clearly, future managers of 1 or more RCW recovery areas will be dealing with the effects of hurricanes on a continuing basis. Currently, at least 2 recovery areas are suffering from past hurricanes. Hugo devastated the Francis Marion NF. The RCW population on the DeSoto NF is small and has suffered a decline, partially at least, as a result of Camille, Frederick and Elena.

What Can We Do?

The inclusion of inland populations as part of the recovery of the species was wise (Lennartz and Henry 1985). The 4 most inland recovery areas are essentially immune from major hurricanes and are at low risk from minor ones. These inland areas should receive renewed emphasis. Establishment of additional inland recovery areas has considerable merit; several opportunities exist for doing that on public lands.

Catastrophic hurricane damage is fairly limited geographically and overall damage is typically patchy. Thus, anything that can be done to increase the geographic extent of a population associated with a recovery area will enhance its potential for surviving hurricanes and its ability to recover from them. Probably

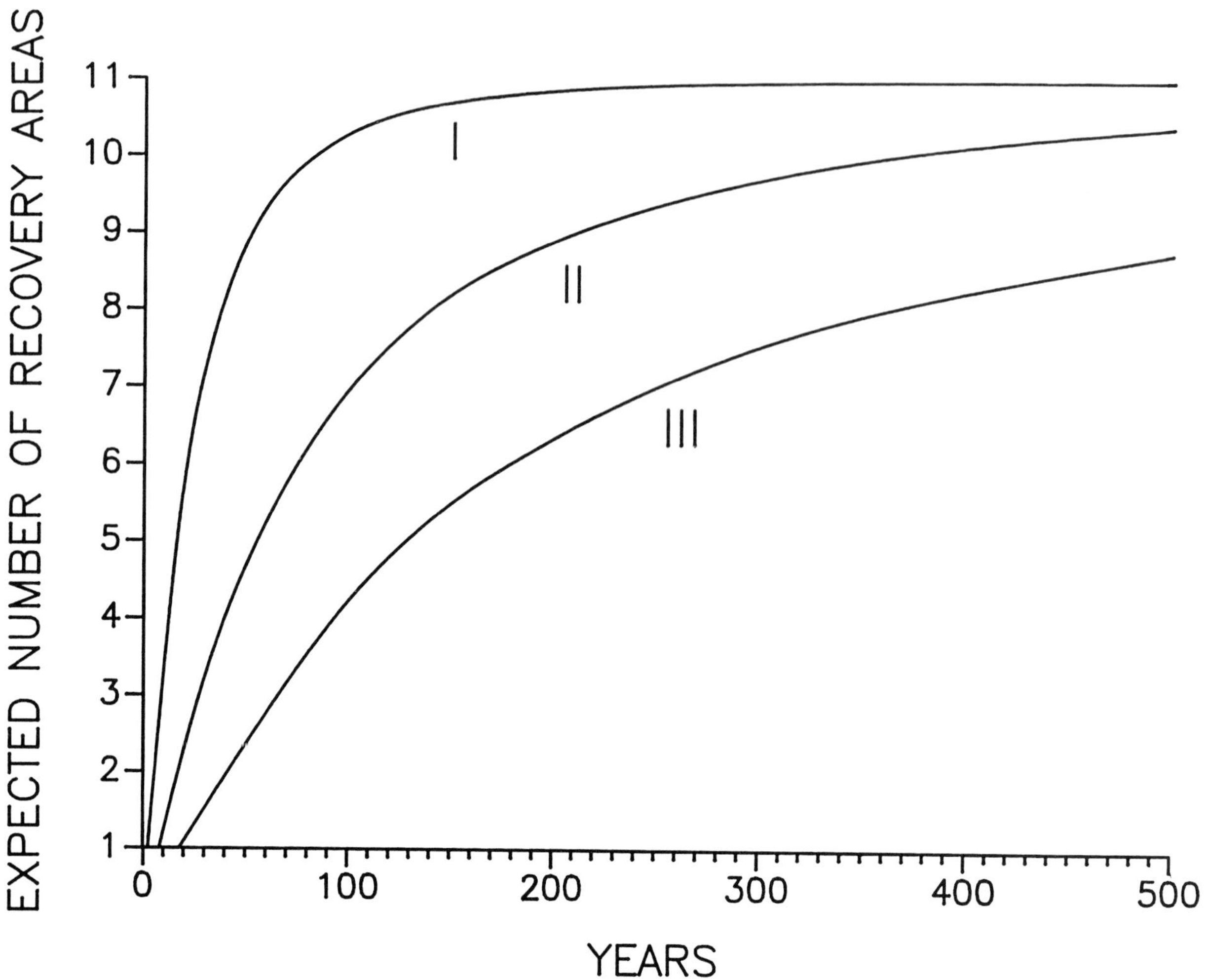

Figure 7. Expected number of red-cockaded woodpecker recovery areas from among the 11 most vulnerable, that will experience at least once over time, a hurricane-induced wind event within their respective boundaries of at least the specified Saffir-Simpson category (I, II, or III). Expected numbers of areas were calculated by Formula 5, Appendix I, using return periods estimated by HURISK (Table 1).

the worst management decision that could be made would be to concentrate a population into a portion of a recovery area. In our opinion, it is better to have a given number of groups at a lower density on a larger area, than to have the same number of groups concentrated at a higher density in a smaller area.

Recovery areas under even-aged management have an advantage of having stands of different age classes spatially distributed. In general, the younger age classes survive better than the older ones. Recovery from hurricanes is enhanced by having stands surviving that are 10-50 years old (Watson et al. 1995). Some reasons to think that net tree survival may not be as great under uneven-aged management are the extended periods of growth suppression for younger trees (Farrar and Boyer 1990) and the increased turbulence from an uneven canopy (Curtis 1943, Savill 1983).

There is general agreement that nothing can be done to protect trees from catastrophic winds. However, much of the damage from hurricanes comes from lesser winds interacting with tree characteristics **(Figure 3)**. Considerable interest has been shown in wind-silvicultural relationships in other regions, but almost none has been shown within the range of the RCW. One reason for this lack of interest has been the typically short rotations (ca. 30 years) used to grow the southern pines. Given that recovery of the RCW will require extended rotations (ca. 80-120 years) in order to provide

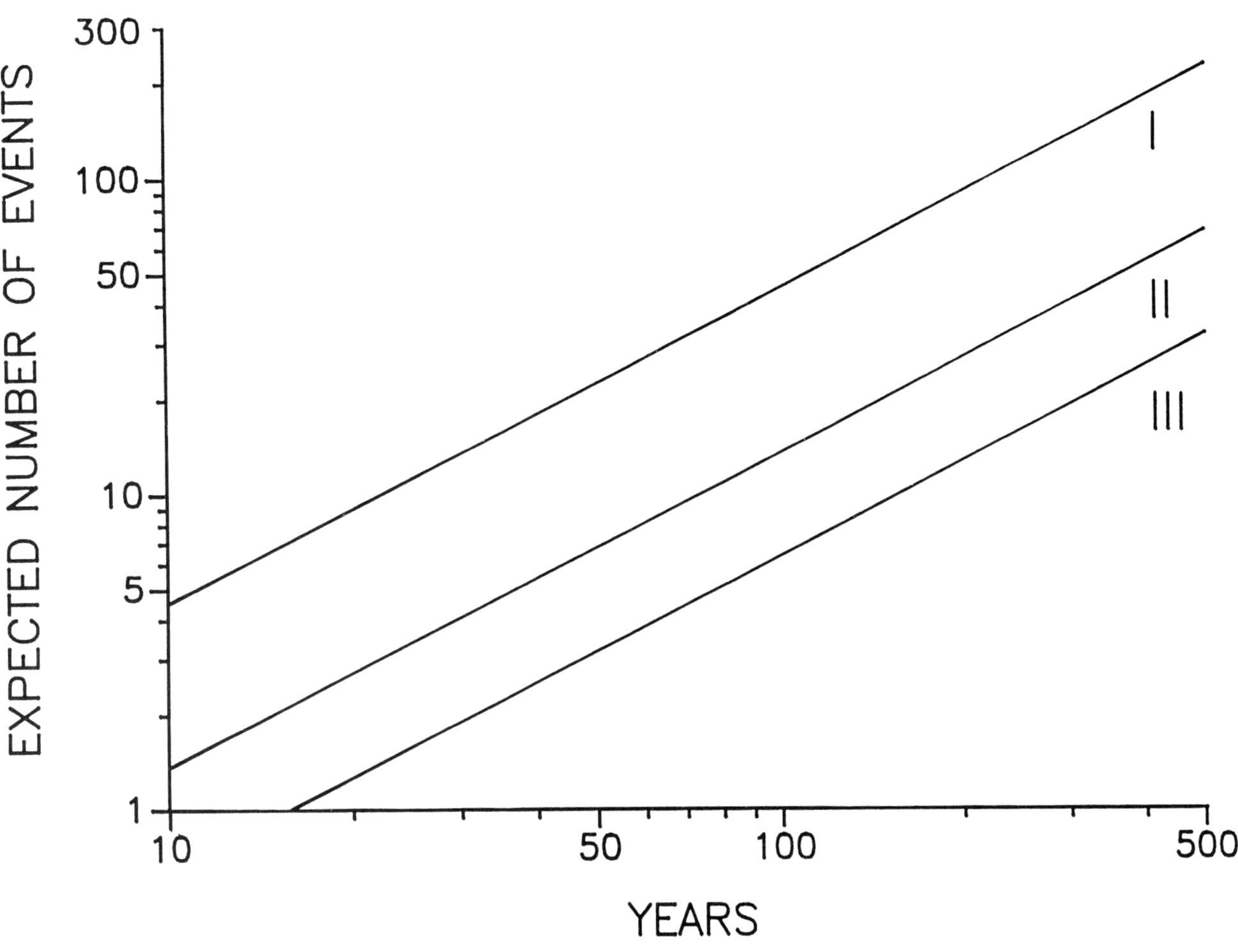

Figure 8. Expected number of events in which hurricane-induced winds of at least the specified Saffir-Simpson category (I, II or III), will occur within the boundaries of any of the 11 most vulnerable red-cockaded woodpecker recovery areas. Expected numbers of events were calculated by Formula 6, Appendix I, using return periods that were estimated by HURISK (Table 1).

old trees for cavities, we think an analysis of the best way to grow stronger and more windfirm trees would be worthwhile. There is much literature that suggests it is possible to grow more wind resistant trees. Basically trees need to be grown throughout their lives in stands that are not overcrowded in order to enhance crown development, root development, taper, and bole strength. This approach to growing trees in recovery areas would also reduce the threat of southern pine beetle epidemics (Thatcher et al. 1986, Belanger et al. 1988), another very serious threat to long-term management of the RCW (Rudolph and Conner 1995).

The importance of recent technological advances that can now be used to counter the effects of hurricanes can not be overstated. The artificial cavities developed by Copeyon (1990) and Allen (1991) have provided an effective response to the loss of RCW cavity trees. The use of artificial cavities, along with augmentation of populations (Allen et al. 1993) when necessary, can reduce the effects of hurricanes on healthy RCW populations.

This prediction assumes:

(1) the responsible agency can respond to the crisis,

(2) enough trees for foraging and artificial cavities survive to support a population, and

(3) that the recovery area is not soon hit by other hurricanes. We may very well lose some populations from repeated hits by hurricanes. Also at great risk are smaller populations that are already struggling.

Appendix I. Formulae used to calculate the probabilities and expected values for hurricanes that were used in the text and figures.

Our basic data were the return periods for hurricane-induced winds of at least specified force occurring over each of the 15 red-cockaded woodpecker recovery areas **(Table 2)**. Return periods were generated by HURISK (Neumann 1987). A brief description of HURISK is provided in the text.

The mean number of hurricane-induced wind events per year, m_i, is the reciprocal of the return periods in **Table 2**.

The yearly probability of 1 or more hurricane-induced wind event of at least a specified intensity occurring over a specified recovery area is equal to,

$$1 - e^{-m_i} \quad (1)$$

The yearly probability of 1 or more hurricane-induced wind events of a specified intensity occurring over any 1 or more of the 15 recovery areas is equal to,

$$1 - \prod_{i=1}^{15} e^{-m_i} \quad (2)$$

The return period for 1 or more hurricane-induced wind events of specified intensity occurring over any 1 or more of the 15 recovery areas is equal to,

$$1 \Big/ \sum_{i-1}^{15} e^{-m_i} \quad (3)$$

The probability of having 1 or more **N**-year events in **t** years is equal to,

$$1 - [1 - (1/N)]^t \quad (4)$$

The expected number of recovery areas, from among the 11 most vulnerable, that will experience 1 or more hurricane-induced wind events of at least a specified intensity over **t** years is equal to,

$$\sum_{i=1}^{11} \left(e^{-m_i t} \right) \quad (5)$$

The expected number of hurricane-induced wind events of at least a specified intensity occurring over any of the 11 most vulnerable recovery areas over **t** years is equal to,

$$\sum_{i=1}^{11} \left(m_i t \right) \quad (6)$$

Impact of Hurricane Hugo on Cavity Trees of a Red-cockaded Woodpecker Population and Natural Recovery After Two and a Half Years[1]

Donald J. Lipscomb and Thomas M. Williams, The Baruch Forest Science Institute, Clemson University, Georgetown, SC 29442

ABSTRACT: The impact of hurricane Hugo on a relatively isolated population of red-cockaded woodpeckers (*Picoides borealis*) was evaluated by the loss of cavity trees to wind, salt water, and bark beetles. The population is located 90 km northeast of Charleston, South Carolina on a peninsula between Winyah Bay and the Atlantic ocean. Damage to red-cockaded woodpecker cavity trees was significantly more severe than damage to the surrounding forest and 45 of 127 active cavity trees were destroyed. New cavities were begun in 47 trees and 38 cavities completed two and a half years after the storm. However, additional cavity tree mortality from salt intrusion and bark beetles resulted in an eighteen month delay before a net gain in active cavity trees occurred. Dynamics of cavity activation, deactivation, and tree mortality show the complexity of cavity tree replacement in a red-cockaded woodpecker population after a catastrophic event.
KEYWORDS: red-cockaded woodpecker (*Picoides borealis*), cavity tree, geographic information system, wind damage, storm surge, bark beetles, recovery.

Hobcaw Barony is located on the South Carolina coast approximately 90 kilometers northeast of Charleston and 4km east of Georgetown. Hobcaw Forest is approximately 3077 hectares surrounded by 3800 hectares of non-forested marshes and islands between Winyah Bay and the Atlantic ocean. The forest is 80 percent pine and pine-hardwood types. Age class distribution of the pine types was skewed heavily toward the over 100 year class before hurricane Hugo (**Figure 1**). Less than four percent of the longleaf area was under 80 years old. Hobcaw Forest was prescribe burned regularly since the mid 1970's. Thus there was an abundance of open old pine stands on the forest.

Red-cockaded woodpeckers (RCW) have been reported on Hobcaw for more than 20 years. Dennis (1971) reported 138 cavity trees (active and inactive combined) on Hobcaw in the early 1970's. Grimes (1977) studied 77 active cavity trees in the late 1970's. A geographic information system (GIS) red-cockaded woodpecker data base, initiated in 1988 and completed before August 1989, recorded 194 active and inactive cavity trees. The cutting on Hobcaw between 1955 and 1965 (Williams and Lipscomb 1983) combined with the burning program of the 1970's and 1980's resulted in a density of approximately 1 cavity tree per 13 hectares of pine type.

On September 22, 1989, hurricane Hugo crossed the South Carolina coast 20 kilometers east of Charleston with maximum sustained winds of 222 km/hr (Purvis et al. 1990). However, winds at Georgetown were lower at 87 km/h and gusts to 138km/h. Hobcaw Forest was still significantly impacted by the wind with 27 percent of the trees receiving moderate to heavy damage (Gresham et al. 1991). In addition, the storm surge inundated roughly 600 hectares on the east and south portions of the forest. The combination of salt water in the soil and bark beetles killed more trees than were killed by wind. Latter mortality occurred primarily during 1990 but has continued until 1993. These combined forces killed trees accounting for 40 percent of the merchantable volume on the forest.

We observed the effect of Hurricane Hugo

[1]Belle W. Baruch Forest Science Institute Technical Contribution: 93-01.

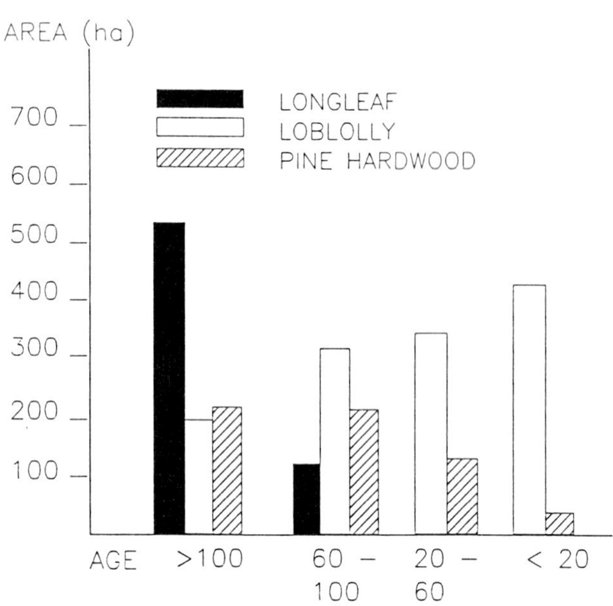

Figure 1. Age class distribution of pine types on Hobcaw Forest, South Carolina, before Hurricane Hugo.

on RCW cavity trees and the dynamics of cavity tree replacement by the surviving RCW population on Hobcaw. Our objective was to observe the interaction of forest condition, storm impact, and nesting recovery with a goal of defining forest conditions that allow recovery of RCW cavity trees without resort to artificial cavities. We chose this goal rather than artificial cavity construction because the area studied had a surviving woodpecker population and extensive nesting habitat remaining after the storm.

METHODS

Maps of RCW cavity trees were created by RCW research in the early 1970's. Annual inventories of Hobcaw Forest were conducted in 1976 and 1986. In each inventory all stands in the forest were surveyed. RCW cavities found during these surveys were also added to the maps. In 1988 and the early part of 1989 all known red-cockaded woodpecker cavity trees were visited and evaluated. Information recorded included the height and activity of the cavity or cavities; the species of the cavity tree and its dbh, total height, and crown class. Information about the surrounding vegetation included basal area of pines 10 inches dbh and larger and basal area of midstory pine and hardwoods.

This data and mapped locations were put in the Hobcaw GIS (Lipscomb and Williams 1988) which was established in 1986. The RCW data layer could then be overlain on existing stand, soil, and other information already in the data base (Lipscomb and Williams 1990). All known RCW cavity trees were entered into the data base before Hurricane Hugo.

A 1.08 percent sample of the forest was taken during the winter of 1989-90 to evaluate damage to all species on the forest. Over 16,000 trees were measured and placed into one of eight wind damage categories (Gresham et al. 1991). All 194 cavity trees were visited, examined for cavity activity, and placed into the one of the same damage categories immediately after Hugo in November 1989.

We re-evaluated the surviving trees in April 1991 and again in April 1992. Areas likely to have new cavities (based on either distance from a known active or inactive colony or stand characteristics) were searched for new cavities from December through March prior to each evaluation. The same parameters were measured on new trees as on previously known cavity trees and the new trees were added to the permanent record.

RESULTS

RCW cavities occurred in three pine species on Hobcaw: loblolly pine (*Pinus taeda* L.), pond pine (*Pinus serotina* Michx.) and longleaf pine (*Pinus palustris* Mill.). For each species the three data sets were prepared; one containing the cavity trees, one containing all trees in the damage survey, and one from the damage survey with tree diameter distributions matched to the cavity trees. The eight damage classes of Gresham et al. (1991) were grouped into light (undamaged, bent, branches broken), medium (defoliated, top broken), and heavy (bole broken, uprooted, downed) and the three data sets compared. Heavy damage in cavity trees was four times the damage of all trees of the same species, and at least twice as high as in trees matched by diameter (**Figure 2**). RCW excavate cavities in old, large pines (Conner and O'Halloran 1987, Jackson et al. 1979). Larger trees were subject to greater wind stress and were more heavily damaged as shown by the matched data set in **Figure 2**. Note that there is little difference in longleaf pine since 96% of the longleaf on the forest was over 80 years old. In addition, 69 of the 80 cavity trees

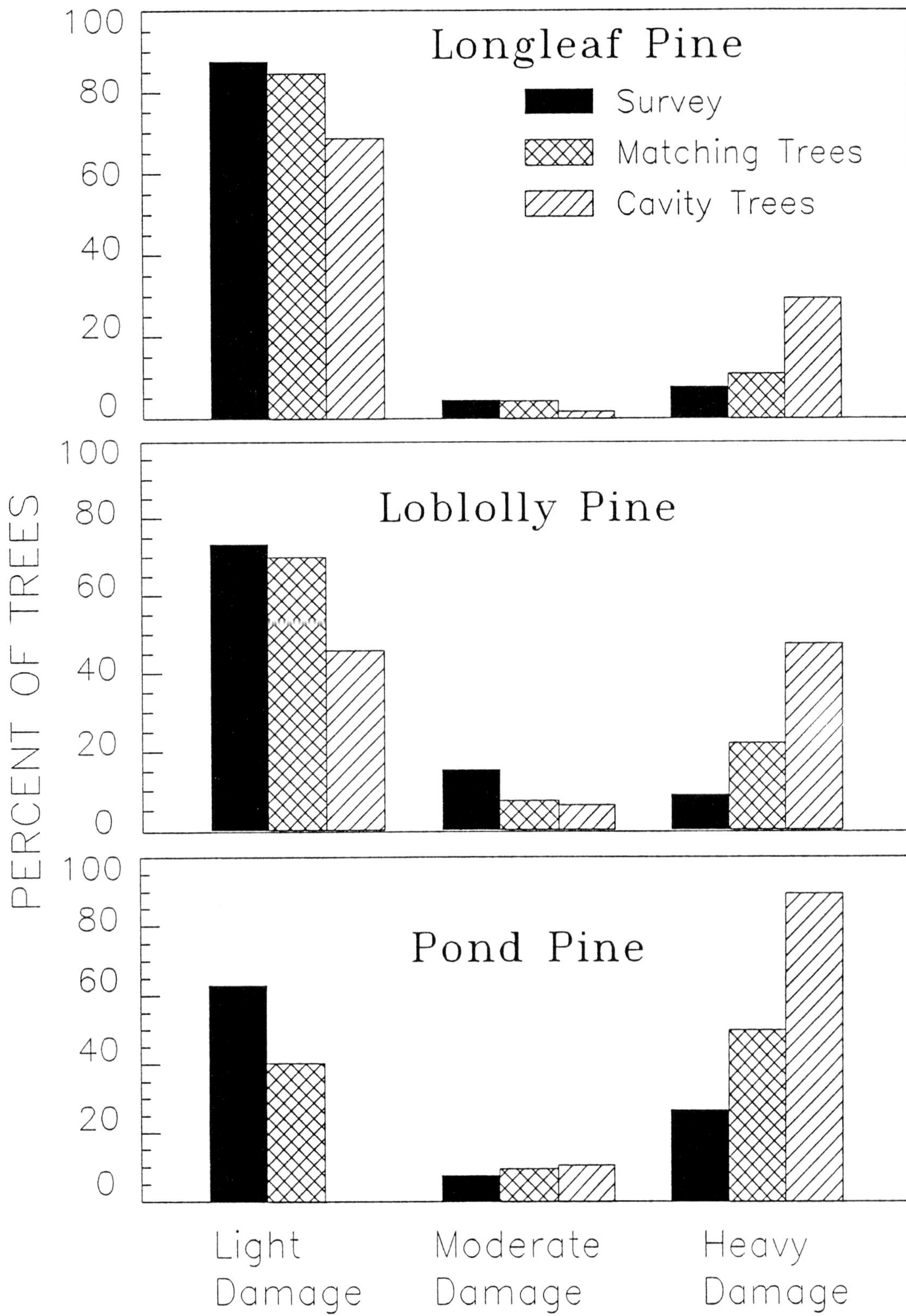

Figure 2. Wind damage to pines on Hobcaw Forest following Hurricane Hugo. Survey was a 1.08 percent sample of the entire forest. Matching was a subset of the survey with the same diameter distribution as the cavity trees. Light damage: none, bent, or broken branches; Moderate damage: defoliated or top broken; Heavy damage: bole broken, or uprooted and leaning, or uprooted and down.

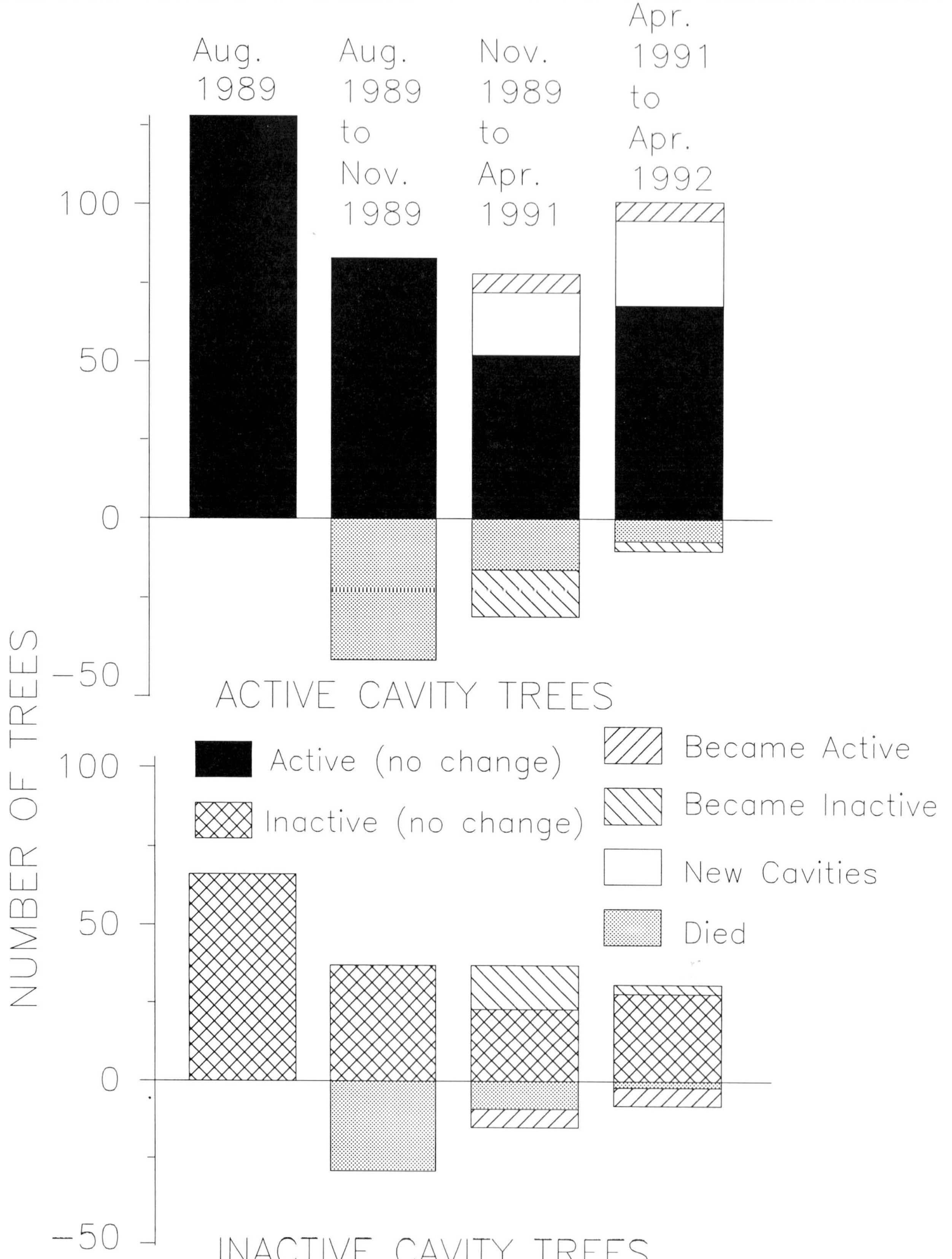

Figure 3. Net changes in red-cockaded woodpecker cavity trees on Hobcaw Forest from August 1989 to April 1992. Active cavities includes 6 starts in April 1991 and 9 starts in April 1992.

heavily damaged by the wind, broke at the cavity. For the general forest trees matched by diameter the greatest damage was by uprooting. Not only were the cavities in large trees which were likely to be uprooted but apparently the cavity weakened the tree to the point that it broke at the cavity.

While 80 of the 194 cavity trees were severely damaged by wind, six were still alive in November 1989. Of the 120 surviving cavity trees 83 were active. Since wind mortality in the forest in general was less than 20 percent and more than 87 percent of the longleaf pine survived (Gresham et al. 1991), there was an abundant supply of potential cavity trees available.

Forty-five active cavity trees were lost during hurricane Hugo, 20 new cavity trees (including starts) were found by April 1991, and 27 more during the following year for a total of 47 by April 1992. Of these 47 new cavity trees 38 had completed cavities 30 months after the hurricane. However, the number of active cavity trees did not reach pre-hurricane level since normal mortality, saltwater intrusion, and bark beetles continued to take a toll (**Figure 3**).

Between November 1989 and April 1991, 20 new cavity trees were found (14 completed cavities) and 16 active cavity trees died. Fifteen active cavity trees became inactive while 6 previously inactive cavity trees (some having been inactive for more than three years) were reactivated. These changes in activity affected more than one third of the total remaining active cavity trees. However, the number of active cavity trees declined by five.

Between April 1991 and April 1992, 27 new cavities (for a total of 47 new cavities with 38 completed) were found and seven active cavity trees died. Again six previously inactive cavities became active, but only three previously active trees became inactive. Less than thirteen percent of active cavities was subject to change during 1991 - 1992. One might surmise that the large number trees becoming inactive in the first time period compared with the second, were a result of bird mortality during the storm. The combination of lower mortality and lower de-activation allowed a twenty-nine percent net increase of active cavity trees during this time period. By April 1991 the woodpeckers had completed 38 new cavities and had nine starts. If this rate were to continue the number of active cavity trees would reach pre-hurricane levels in a few years. However, by 1991 the hurricane effects killed 40 percent of the pine foraging habitat as well as 97 active cavity trees.

CONCLUSIONS

There was significant damage to RCW cavity trees on Hobcaw Forest located 70 kilometers from the center of the path of Hurricane Hugo's eye. Damage to the RCW cavity trees was significantly greater than the damage to similar trees in the surrounding forest. Within two and a half years the birds had constructed more new active cavity trees (and had 38 of 47 new cavities completed) than were initially lost to hurricane winds. However, additional losses of cavity trees occurred after Hugo from salt intrusion and bark beetles. Thus, thirty months after the hurricane there were still 21 percent fewer active cavity trees than before the hurricane.

Hurricanes are a frequent episodic event throughout the red-cockaded range (Hooper and McAdie 1995). We found RCW cavity trees are especially vulnerable to wind damage. Over forty percent of the cavities were lost 70 km from the eye with wind gusts under 140 km/h. Old, large pines are both preferred for cavities and subject to strong wind stress. Cavity trees are more likely to break at the cavity probably due to weakness caused by either the cavity itself or red heart fungus often associated with the cavity.

The RCW population on Hobcaw Forest was able to establish new active cavity trees in a relatively short time period because there were abundant old pines that survived. With sufficient nesting habitat the birds were able to complete a large number of cavities in a short period of time. Artificial cavity placement (Copeyon 1990, Allen 1991) is an important strategy but it was unnecessary for this population. The criteria we used to choose natural recovery was presence of a surviving cavity tree in each colony and survival of 87 percent of the longleaf which was over 80 years old.

Restoration of the Red-cockaded Woodpecker Population on the Francis Marion National Forest: Three Years Post Hugo

J. Craig Watson USDA Forest Service, Francis Marion National Forest
Moncks Corner, SC, 29461
Robert G. Hooper USDA Forest, Service Charleston,
Southern Research Station, 29414
Danny L. Carlson USDA Forest Service, Francis Marion National Forest
McClellanville, SC 29458
William E. Taylor USDA Forest Service, Francis Marion National Forest
Moncks Corner, SC, 29461
Timothy E. Milling USDA Forest Service, Southern Research Station
Charleston, SC 29414

ABSTRACT: The Francis Marion National Forest (FMNF) once supported the second largest and only documented naturally increasing population of red-cockaded woodpeckers (RCW). Prior to Hurricane Hugo, an estimated 477 RCW groups existed on the FMNF. Hugo destroyed 87% of the active red-cockaded woodpecker cavity trees on the Forest and killed 63% of the woodpeckers. In addition, 59% of the birds' foraging habitat, nesting habitat, and potential cavity trees were destroyed. By the nesting season of 1990, 537 artificial cavities had been installed and there were 249 groups. Forty five percent of RCW nests were in artificial cavities and 302 young were fledged. In the 1991 breeding season, there were 306 groups, 483 young were fledged and 60% of the nests were in artificial cavities. In 1992, there were 332 groups, 500 young were fledged, and 61% of the nests were in artificial cavities. Following the catastrophic loss of woodpeckers and their habitat in 1989 from Hugo the number of adult RCWs and the number of groups had increased 33% by 1992. This remarkable increase, despite the severe losses of nesting and foraging habitat, cavity trees, and individual birds, resulted primarily from installation of artificial cavities.
KEYWORDS: *Picoides borealis*, Hurricane Hugo, Francis Marion National Forest, artificial cavities, recovery, restoration.

The Francis Marion National Forest (FMNF) is unique in several respects in regard to the red-cockaded woodpecker (*Picoides borealis*, {RCW}). The RCW population on the FMNF was the second largest, one of the densest, and the only population known to be naturally increasing (Hooper et al. 1990); and it was one of the few populations near a recovered level. It was also the first population, following classification of the bird as endangered, to have been impacted by a major hurricane (Hooper et al. in press).

Hurricane Hugo, a Category IV hurricane at landfall, hit the FMNF on 21-22 September, 1989 (Townsend 1990). The center of the eye passed within 5 miles of the FMNF and the track of the hurricane was such that the FMNF received the strongest winds generated by the hurricane. Hurricane force winds occurred in a band greater than 50 miles wide and affected the entire FMNF (Hooper and McAdie 1995). As a result, catastrophic loss to RCW cavity trees, RCWs, and RCW foraging habitat occurred.

Hurricanes will be a continual threat to most recovery areas. Hooper and McAdie (1995) estimated there will be 4 recovery areas hit by major hurricanes every 100 years on average with 39 less intense hurricanes hitting the 11 most vulnerable recovery areas. The FMNF, recently impacted by a major hurricane, is one of these recovery areas. Clearly managers will be dealing with restoration efforts in the future because of the periodicity of hurricanes impacting RCW populations. Our experience following Hugo should be of value to managers because our efforts have led to a remarkable and unprecedented increase in a RCW population following a catastrophic event.

IMPACTS OF HURRICANE HUGO

Impacts to the Forest

Extensive damage to timber resources from hurricanes is not an unusual occurrence (Van Hooser and Hedlund 1969, Hook et al. 1991, Hooper and McAdie 1994). In South Carolina, approximately 10.8 billion board feet (bf) of sawtimber on 4.5 million acres of forested lands were destroyed by Hurricane Hugo (Sheffield and Thompson 1992).

Destruction of pine trees on the FMNF that represented RCW habitat was catastrophic. Initial estimates of destroyed pine sawtimber on the FMNF was between 700 million to 1 billion bf or about 50-60% of pine sawtimber (Hooper et al. 1990). Data from the Forest Inventory and Analysis (FIA) Research Work Unit of the Forest Service's Southeastern Forest Experiment Station indicated a 56% loss of pine trees greater than or equal to 10" dbh with about 2.44 million stems surviving **(Table 1)**. CISC (Continuous Inventory of Stand Condition) database information indicated a slightly higher loss of pine sawtimber with approximately 2.28 million pine trees greater than or equal to 10" dbh surviving Hugo **(Table 1)**. This represents a 59% loss of pine sawtimber on the Forest.

Not all pine stands were equally damaged by winds of the same force (Hooper et al. 1990). Damage appeared to be a function of both tree age and basal area (ba). Sapling and young pole stands survived well with varying degrees of damage. Stands in the 30-40 year age class generally escaped with moderate damage **(Figure 1)**. Most of these younger stands can be managed without regeneration. Mature pine stands with low to moderate basal areas (70 sq/ft ba) typically had less than 10 sq/ft ba after the hurricane. Most of the killed pine trees broke off at about 1/3 the height of the tree and the remaining were uprooted. Mature pine stands with high basal areas (90 sq/ft ba) escaped some of the strongest winds with moderate damage.

Table 1. Number of pine stems in Francis Marion National Forest (x 1000)

Size	1936[1]	Pre-Hugo 1986[2]	Post-Hugo 1990[3]	Survival (%)
5-9" dbh	3,102	--	6,681	--
≥10" dbh	2,960	5,539	2,280 (2,442)[2]	44 41[2]

[1]Grumbine, A.A. 1936.

[2]Data are from inventories conducted in 1986 and 1990 by the Forest Inventory and Analysis Research Work Unit, Southeastern Forest Experiment Station, Asheville, NC.

[3]Data are from updated Continuous Inventory of Stand Condition (CISC) database from the Wambaw and Witherbee Ranger Districts of the FMNF after Hurricane Hugo.

Hugo not only inflicted massive destruction to RCW nesting and foraging habitat in the form of broken and wind thrown trees, but also created conditions for a continuing process of short term habitat loss. Many pine trees that initially survived suffered root, bole, and crown damage that contributed to additional mortality. Remaining trees were subject to attack from engraver beetles (*Ips* spp.), turpentine beetles (*Dendroctonus terebrans*), and southern pine beetles (*Dendroctonus frontalis* [SPB]).

Insect populations, suspected to be SPB, reached outbreak status within 9 months of Hugo's landfall and lasted through the summer of 1990. Through October 1990 there were 343 distinct infestations, mostly from Ips beetles in pine sapling stands on the FMNF. Long periods of hot dry weather during the summer of 1990 are thought to have considerably slowed down further spread of Ips infestations by making host material unsuitable to attack (P. Barry pers. comm.). Basal area in sawtimber stands was so reduced from the storm that SPB outbreaks were limited and confined. Not all acres infested required control, but a total of 526 acres of potential and current RCW habitat was lost to insect infestations. Although not as dramatic as the loss of habitat from Hugo, the attrition of pine stems from these cumulative impacts is continuing to contribute to loss of RCW habitat.

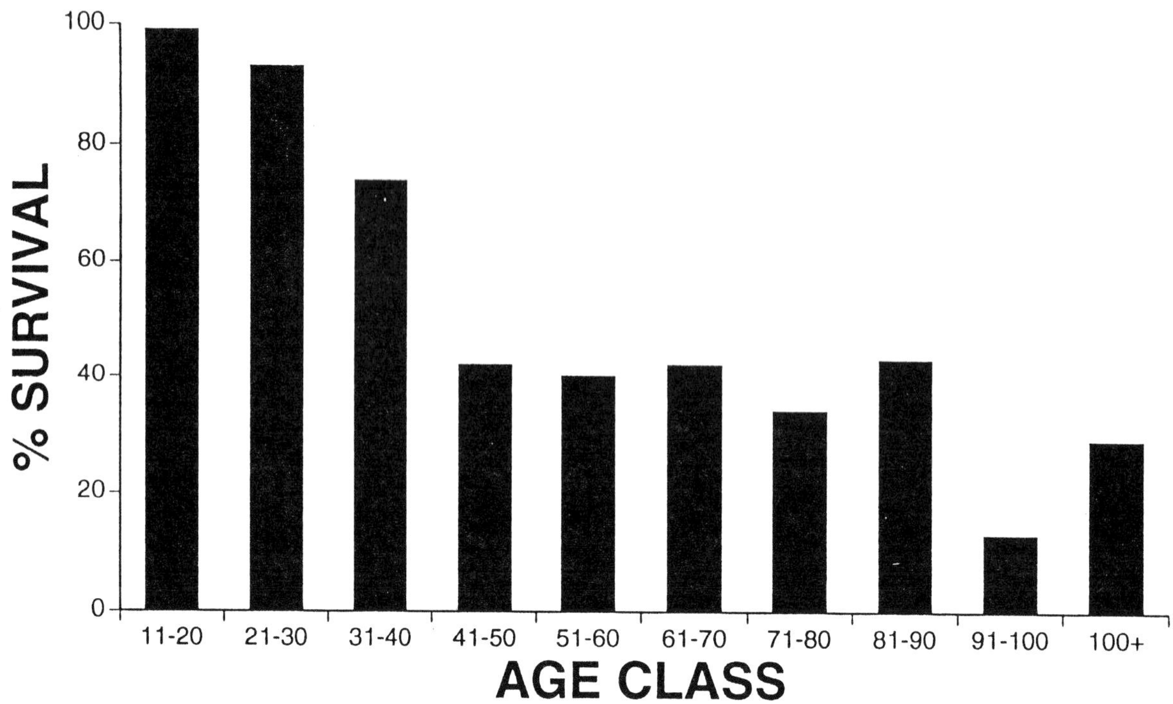

Figure 1. Percent survival of pine on the Francis Marion National Forest after Hurricane Hugo.

Impacts to the Red-cockaded Woodpecker Cavity Trees

Impacts to the RCW from Hurricane Hugo were also catastrophic. Ligon (1971) suggests that the cavity tree is the single most important feature in the life of the RCW, and the lack of potential cavity trees for excavation was one reason cited for the species decline (USFWS 1985).

Hugo destroyed 87% of the estimated 1765 active cavity trees on the FMNF; only 229 RCW cavity trees remained standing after the storm. Most of the destroyed RCW cavity trees were broken at the cavity (52.8%) while loss of cavity trees from the tree being uprooted or being broken between the cavity entrance and stump was about equal **(Table 2)**. A small percentage (9.9%) of RCW cavity trees died as a result of varying degrees of crown damage. Fifty percent (269/538) of all RCW clusters on the FMNF lost all cavity trees. Forty six percent (218/477) of the active clusters lost all cavity trees (this included active, inactive, and incomplete cavities), and 57% (272/477) lost all active cavity trees. Only in 2% (12/538) of the clusters did all cavity trees survive. The average number of active cavities per cluster in a sample of 205 clusters surviving Hugo was 1.9 compared to 3.7 active cavities in 100 randomly selected clusters in 1988. Thus, the quality of the surviving clusters was greatly reduced.

Table 2. Type of damage to RCW cavity trees on the Francis Marion National Forest from Hurricane Hugo (n=1028).

Type of damage	# cavity trees damaged	Percent damage by category
Crown gone/cavity present	102	9.9
Broken at cavity	543	52.8
Broken below cavity	194	18.9
Cavity tree uprooted	189	18.4
Total	**1028**	**100.0**

Initially, non-lethal damage from Hugo was expected to cause continuing tree mortality (Sheffield and Thompson 1992). A stratified random sample of clusters indicated a net post-Hugo loss of cavity trees prior to the 1991 nesting season of 4.9% and 10.7% prior to the 1992 nesting season **(Table 3)**. These losses included cavity trees that had natural or artificial cavities. Sixteen percent of the loss

Table 3. Percent mortality of RCW cavity trees by type of cavity on the FMNF after Hurricane Hugo.

	% natural cavities	% inserts	% drilled cavities	% drilled starts	% overall mortality
1991[1]	5.7	4.7	3.0	7.1	4.9[1]
1992[2]	16.2	4.9	7.7	7.7	10.7[2]

[1]20 of 409 cavity trees sampled died.

[2]43 of 406 cavity trees sampled died.

was attributable to lightning strikes and 14% percent was the result of Hugo induced wind damage. Causes for the remainder of the losses were difficult to determine although most such trees showed signs of Ips beetle attack. Mortality was most likely a combination of crown and bole damage from Hugo and secondary attack by insects. Estimated mortality of cavity trees on the FMNF in 1991 was 65 trees and 137 in 1992.

Mortality of RCW cavity trees in 1991 was highest for natural cavity trees and cavity trees with drilled starts, while cavity trees with inserts and drilled cavities experienced the least mortality (**Table 3**). Mortality of cavity trees with drilled starts and drilled cavities was about equal in 1992, while mortality of natural cavity trees was highest. Mortality of trees with inserts increased slightly (**Table 3**). Mortality for all types of cavity trees increased from 1991 to 1992, but it was apparent that natural cavity tree mortality exceeded that of trees with artificial cavities.

RCWs

When Hugo hit there were an estimated 1908 RCWs on the FMNF (Hooper et al. 1991). The mean number of RCWs per cluster on the FMNF in September prior to Hugo making landfall was 4.0 birds. Although only twelve dead red-cockaded woodpeckers were found in post-Hugo inventory efforts, the mean number of birds per cluster surviving Hurricane Hugo was 1.5 birds. Seven hundred six (706) RCWs were accounted for after Hugo which indicated a loss of 63% of individual RCWs on the FMNF (Hooper et al. 1990).

Seventy two percent (344/477) of all formerly active clusters, 91% (227/248) of clusters with surviving cavity trees, and 47% (117/248) of those with no surviving cavity trees still supported one or more RCWs after the hurricane. Thus, there were 344 sites that still had one or more RCWs; 117 of these had no cavity trees and 227 had severely reduced numbers of cavity trees. Only 68% (233/344) of these had two or more RCWs, and thus a potential breeding group.

Further significant losses of RCWs most likely occurred from severe winter weather. Most RCWs that survived Hugo were roosting in the open. Temperature dropped to an average of 15.3°F during a 4 day period from 22-24 December 1991 with a low of 11°F on 24 December (NOAA 1989, temperature data obtained from the Santee Experimental Forest headquarters on the Witherbee Ranger District). A stratified random sample indicated that 579 individual RCWs existed prior to the 1990 nesting season, representing a further loss of 18% of individual RCWs after Hugo hit. Thus, a 70% loss of RCWs occurred on the FMNF after Hurricane Hugo.

RECOVERY EFFORTS

Colony Damage Assessment and Prescriptions

Damage assessment was evaluated and management prescriptions were developed for each cluster. Biologists visited all known clusters within 4 months after Hugo. The number of surviving RCWs and cavity trees were determined and an account was made of the trees suitable for natural and artificial cavities. The extent of foraging habitat for each cluster was also determined. This data provided a basis for our prescription of each clusters' potential for restoration. Prescriptions provided information for the priority of artificial cavity installation, placement of restrictors, removal of slash and debris from around cavity trees, needed midstory treatments, and silvicultural and timber management needs.

Priority for rehabilitation of RCW clusters with artificial cavities (given adequate foraging habitat being available) was where:

1) 2 or more birds survived with no surviving cavity trees,

2) 1 bird survived with no surviving cavity

Table 4. Number of artificial RCW cavities installed on the Francis Marion National Forest since Hurricane Hugo prior to each nesting season (nesting season year here is defined as occurring from April 1 of each year to March 31 of the following year).

	1990	1991	1992	1993[1]	Total
Inserts	277	52	27	72	428
Drilled cavities	126	65	89	39	319
Start cavities	134	34	24	50	242
Total	**537**	**151**	**140**	**161**	**989[2]**

[1]Number of artificial cavities installed from April 1, 1992 to September 30, 1992.

[2]An additional 70 artificial cavities are projected to be installed prior to the 1993 nesting season for a total of 1,059.

trees,

3) 2 or more birds survived with one functional cavity present, and

4) 1 bird survived with one functional cavity. Care was taken not to force the process of recovery by rehabilitating colonies where suitable habitat did not exist.

Artificial Cavity Program

Prior to Hugo, three experimental techniques for constructing artificial RCW cavities had been introduced (Copeyon 1990, Allen 1991). These techniques were a major technological breakthrough in RCW management. Copeyon's techniques (drilled cavities and drilled starts) had been tested in North Carolina and were successful in providing cavities that were used for nesting by RCWs (Copeyon 1991). Allen's technique (inserts) had not been tested during a RCW nesting season prior to Hugo. Copeyon and Allen came to the FMNF after Hugo to teach us how to construct artificial cavities for RCWs.

Our immediate goal prior to the 1990 nesting season was to ensure at least 2 functional cavities per cluster where at least 1 RCW survived and where minimal or adequate foraging habitat existed. The long term goal is to have at least 4 functional cavities (natural or artificial) per cluster. These actions should contribute to the overall survival of RCWs, provide functional cavities necessary for nesting, and provide sufficient cavities for dispersing RCWs.

As of this writing, 989 artificial cavities of the three types have been constructed in 328 clusters. A major effort prior to the 1990 nesting season resulted in the installation of 537 artificial cavities in 246 clusters, of which 117 clusters did not have any surviving cavity trees. A total of 428 inserts, 319 drilled cavities, and 242 drilled starts have been installed since Hugo, with the number of artificial cavities installed being about equal each of the following nesting years after 1990 (**Table 4**). The average number of potential roosting cavities per cluster, as of the 1992 nesting season (including natural and artificial cavities), is now 3.6 and essentially the same as the number of active cavities per cluster (3.7) before Hugo hit.

Initial monitoring of artificial cavities to determine use was intensive. At this writing, greater than 90% of all artificial cavities were estimated to be used to some extent by RCWs (maintenance of resin wells, chipping of the entrance tunnel or cavity, roosting, or nesting). As the utilization rate of artificial cavities and the average number of cavities per cluster increased, monitoring activities were reduced. The elapsed time for cavities to be used varied, dependent upon the number of available cavities in the cluster and the number of birds in the group. Some cavities were utilized the same day of construction and approximately 30 are not being used three years after installation. Those not utilized may not have sufficient habitat in those sites to support RCWs. We expect nearly all artificial cavities installed in suitable habitat to be utilized eventually. Time till first use of an artificial cavity also appeared to be related to the degree of simulation of a natural cavity tree. Trees with artificial cavities that simulated the copious resin flow of natural cavity trees were utilized quicker than trees with artificial cavities that were not treated with artificial resin wells (Taylor and Hooper 1991, Copeyon 1990, Allen 1991). Some artificial cavity trees that are now active have enough resin flow developed to be indistinguishable from natural cavity trees. Many of the drilled starts were excavated to complete cavities within 2 months.

Approximately 540 artificial cavities were constructed prior to the 1990 nesting season by Copeyon's and Allen's technique. A modification of Copeyon's technique by Taylor and Hooper (1991) has reduced training time, safety hazards, and improved mobility of work crews. Allen's technique was modified by increasing the depth of the cavity, primarily to reduce competition from secondary cavity nesters. Artificial cavities constructed after the 1990 RCW nesting season utilized the modified techniques. These changes improved already successful techniques, allowed increased efficiency of construction, and resulted in higher utilization of artificial cavities.

Restrictors

Since Hugo, 306 cavity restrictors have been installed. Traditional restrictors recommended by Carter et al. (1989) were installed where cavity enlargement was observed. Full plate restrictors similar to Type A restrictors described by Carter et al. (1989) were placed on inserts when enlargement occurred. Approximately half of the inserts installed have been outfitted with restrictors. The width of the restrictor entrance was increased to 1.75" to allow easy entry and exit by RCWs, and to reduce monitoring efforts. Although no quantitative data are available, we feel that restrictors have most likely contributed to the successful nesting efforts of RCWs since Hugo. Data from the stratified random sample indicates that 25%, 21%, and 20% of RCW nests were in cavities with restrictors in 1990, 1991, and 1992, respectively.

Reproductive Attainment and Population Trend

For the first three nesting seasons following Hugo a stratified random sample of RCW clusters (n=90) was used to estimate reproductive attainment and population trend. The sample was stratified by the level of damage to the clusters: high, medium, and low (this is the same stratified random sample mentioned throughout the text to estimate other parameters of population condition). Clutch size, number of eggs hatching, and the number of nestlings reaching fledgling age was determined. An additional 30 clusters were examined throughout the nesting season for the same nesting activity. Data on competition was also collected in these 120 clusters, 60 of which had 3 bluebird boxes erected in the cluster. The purpose of the bluebird boxes was in part to reduce competition from secondary cavity nesters. Although data on competition has yet to be analyzed, juvenile RCW's were observed using bluebird boxes for roosting in clusters where the number of RCWs in a group was greater than the number of available cavities.

Three years after Hurricane Hugo, all evidence supports the conclusion of a continued improvement in the FMNF RCW population size. An estimated 250 groups with two or more adults were present in the 1990 nesting season (**Table 5**). The number of groups with 2 or more adults increased each of the first three years since Hugo. Between 1990-91, the number of 2 bird groups increased 23.0%, from 249 to 306. The rate of increase between 1991-92 slowed, but was still 8.5%. Overall, increase in 2 bird groups from 1990 to 1992 was 33.0%, from 249 to 333.

The increase came mostly from either single birds that successfully attracted a mate or from colonization of vacant clusters (**Table 5**). The number of clusters with 1 bird decreased by 67% from 1990-92, while the number of clusters with 0 birds decreased by 40% for the same period (**Table 5**). Relatively large numbers of dispersing RCWs (302, 483, and 500 RCWs fledged in 1990, 1991, and 1992 respectively, **Table 6**) and the presence of suitable cavities (mostly artificial cavities) has resulted in the occupation of clusters that formerly had 0-1 birds. Additional population increases resulted from an increase in the mean number of birds per group for the three year period since Hugo. The mean number of adult birds per group during the nesting season has increased each year from 1.5 in 1990, and to 1.9 and 2.1 birds in 1991 and 1992, respectively.

In 1990, 1991, and 1992, respectively, 45, 60, and 61% of RCW nests were in artificial cavities (**Table 6**). Without the use of artificial cavities by RCWs, the current population would most likely be 100-125 groups.

Recruitment Stands

Further evidence of the excellent nesting attainment, dispersal of RCWs, and the utility of artificial cavities, has been the colonization of recruitment stands provisioned with artificial cavities. Two kinds of recruitment stands were designated. The first was the traditional recruitment stand with no history of occupation by RCWs (USFS 1985). The second type of recruitment stand designated were clusters totally destroyed by Hugo having suitable habitat available to support a group and not having birds two years after Hugo. Utilizing CISC database information and Geographical Information System (GIS) maps displaying RCW habitat elements, the most likely

Table 5. Estimated values for RCW population in the Francis Marion National Forest after Hurricane Hugo, based on a stratified random sample of 90 clusters.

	1990	1991	1992
# of clusters	384	384	391[1]
# of adults	607	729	806
Clusters with 2+ birds	249	306	332
Clusters with 1 bird	83	30	27
Clusters with no birds	52	48	31

[1]The increase in the number of clusters from 1991 to 1992 is due to artificial cavities being installed in seven recruitment stands and being managed as clusters.

Table 6. Estimated nesting attainment of RCW population on the Francis Marion National Forest after Hurricane Hugo, based on a stratified random sample of 90 clusters.

	1990	1991	1992
# of young fledged	302	483	500
Mean young fledged/group with 2+ birds	1.2	1.58	1.50
% of groups with 2+ birds nesting successfully	72	81	81
% of nest in artificial cavities	45	60	61

locations for recruitment stands were determined. Criteria used to delineate recruitment stands were:

1) stands with greater than 40 sq/ft basal area (preferably higher),

2) pine trees ≥ 14" dbh and,

3) the presence of suitable foraging habitat within .5 mile of the proposed site. The value of this criteria was verified as one recruitment stand, identified through use of GIS, was found to have an active cluster (not previously known) when the stand was visited. Since the 1991 nesting season artificial cavities were installed in 9 recruitment stands; six were in previously unoccupied habitat and two were formerly active, but now destroyed clusters sites. The remaining recruitment stand was a shift of a formerly inactive cluster where all cavity trees had died. The new cluster, approximately .2 mile away from the original site, contained larger trees suitable for artificial cavity excavation. Seven of the 9 recruitment stands were created prior to the 1992 nesting season. Of these 7, five successfully fledged young in 1992 (including the inactive cluster we shifted). Of the remaining 2, one is inactive and the other had 1 RCW in 1992. The cluster with 1 bird attracted a second bird after the 1992 nesting season. Two recruitment stands were created after the 1992 nesting season; both are active, with 1 bird at one cluster and two birds at the other. The mean period for colonization was 49 days, and all recruitment stands outfitted with artificial cavities are expected to be occupied by functional groups in time. All recruitment stands were located within .75 mile of an active cluster except one that was located 1.25 miles from an active cluster; it became active in 30 days.

Augmentation

Due to the excellent dispersal of RCWs into unoccupied and single bird clusters since Hugo, augmentation efforts were minimal. Three female RCWs were moved from the FMNF to the Savannah River Site after Hugo because these solitary birds had little chance of surviving. Four translocations were conducted within the FMNF similar to those described by DeFazio et al. (1987); two resulted in group formation and successful nests. Because natural RCW dispersal appears to be contributing to the overall growth and health of the population, intensive augmentation efforts within the FMNF are not anticipated.

HABITAT MANAGEMENT

As critical as the artificial cavity program has been in the stabilization and recovery of the RCW on the FMNF, other successful management efforts have played an important role. Salvage of downed timber, fire suppression, and prescribed burning, have been an integral part of the short term recovery of the RCW. Habitat restoration and timber management efforts will no doubt play an integral role in the long term recovery of the RCW by establishing much needed RCW habitat.

Salvage Efforts

Post-Hugo, the greatest risks to the remaining pine resource on the FMNF were fire and insect attack. Normal fuel loads in coastal

Table 7. Timber salvage and thinning volumes in MBF on the Francis Marion National Forest since Hurricane Hugo (MBF=thousand board feet).

	1990	1991	1992	1993
Net salvage volume	249,235	37,805	11,850	298,980
Green volume cut	13,149	-----	-----	13,149
Thinning volume	2,786[1]	793	1,663	5,242
Total volume cut	265,170	38,598	13,513	317,281

[1]Volume cut before Hurricane Hugo struck the FMNF.

plain forests prior to Hugo were between 1.7-7.1 tons/acre (Hook et al. 1991). Fuel loads after Hugo were in excess of 60 tons/acre, increasing the chances of catastrophic wildfires (and difficulty of suppression) and insect attacks. Salvage guidelines were designed to protect the RCW and facilitate removal of downed timber. Guidelines included surveying proposed salvage areas for RCW clusters before salvage, leaving trees that were leaning less than 45° for RCW foraging, leaving 5 snags/acre for secondary cavity nesters, restrictions on salvage activities during the RCW nesting season, and penalties for damaging cavity trees and potential cavity trees during salvage operations. Following these guidelines resulted in the protection of RCW habitat, reduced fuel loads, and reduced dead and dying wood that attracted insects.

Efforts to salvage the estimated 1 billion bf of downed timber began immediately, and by the end of 1992 approximately 290 million bf of timber had been salvaged, representing about 30% of the downed volume (**Table 7**). The largest effort was in 1990 with 250 million bf salvaged. In addition, debris was removed and scattered from within 50 feet of existing cavity trees to further reduce the threat of insect attack and damaging wildfires.

Fire Suppression

Efforts for fire suppression after Hugo included an intensive fire prevention program by the South Carolina Forestry Commission and the Forest Service, and the rapid detection and suppression of wildfires. One hundred ninety miles of fuel breaks were constructed between private lands and Forest Service lands to minimize wildfire hazards to all landholders. Prior to January 1993, 242 wildfires had started on the FMNF. They were rapidly suppressed and only 3,308 acres were burned. No wildfires have resulted in the mortality of cavity trees or RCWs. Many of the wildfires suppressed were determined to be roadside arson, thus salvage efforts played a major role in prevention of catastrophic wildfires by reducing fuel loads near roadsides.

Prescribed Burning

The importance of prescribed fire to the RCW can not be over stated. The RCW does not tolerate hardwood encroachment into the colony site (Val Balen and Doerr 1978, Locke et al. 1983, Loeb et al. 1991, Hooper et al. 1991). Prescribed fire is the most cost effective and ecologically sound method of controlling unwanted hardwoods. No prescribed burning was conducted on the FMNF the first year after Hugo. However, prescribed burning resumed the second year after Hugo with 21,000 acres burned in 1991 and 31,000 acres burned in 1992. In a five year period (1985-1989) prior to Hugo, an average of 37,600 acres were burned annually on the FMNF, and this level of burning is expected to resume in the near future. Growing season burns, thought to be more efficient in controlling encroachment of hardwoods into clusters, were conducted on approximately 4000 acres in 1992. To further control encroaching hardwoods in clusters after Hugo, 1900 acres were treated with a combination of chainsaws and herbicide.

Reforestation Efforts

Prescribed burning is also playing a key role in reforestation efforts necessary to reestablish suitable RCW habitat, particularly in the recovery of longleaf pine (Pinus palustris) stands. Approximately 9,000 acres of longleaf stands have been burned for seedbed preparation and/or brown spot needle blight (*Scirrhia acicola*) control. An additional 6,000 acres have been site prepared and planted to longleaf pine since Hugo.

Efforts to manage loblolly pine (Pinus taeda) have also been intensive with 13,700 acres of loblolly pine seedlings fertilized. An additional 12,500 acres of naturally regenerating loblolly pine stands were site prepared by injecting hardwoods with herbicide. Approximately 32,000 acres on the FMNF have regenerated naturally to loblolly pine without any post-Hugo silvicultural treatments (J. Dupre, A. Kiser, pers. comm.).

Timber Management

Timber management practices that promote the growth of pine stands will also have a profound positive effect on future RCW habitat. Hooper et al. (1991) reported that 36% of the pine stands on the FMNF were regenerated from 1957-1987. These stands are now 6-36 years old. Although only 2,224 acres have been pre-commercially thinned since Hugo, the potential for these thinnings is enormous, as is the potential to thin loblolly pine stands naturally regenerating after Hugo. Thinned stands should be suitable as foraging habitat sooner than unthinned stands (Hooper and Harlow 1986). Except for salvage, all timber harvested since Hugo has been from first time thinning operations (**Table 7**). Projected thinning volume for the next three years is 38.3 million bf (**Table 8**), all in pine poletimber 20-30 years old. These young stands suffered little damage and because of the amount of pine stands in younger age classes present when Hugo hit a relatively large number of young pine stands are available for management.

DISCUSSION

Three years after Hurricane Hugo destroyed 87% of the active cavity trees and 63% of the RCWs on the FMNF, the RCW population has recovered remarkably. Popular opinion and current literature would have suggested that a continued and severe decline of the RCW population could be expected. Losses from Hugo were thought to be insurmountable: loss of birds, disruption of group composition, loss of cavity trees, the tremendous loss of foraging habitat, and fragmentation of the forest. Instead, the population is at 333 groups, instead of perhaps 100-125 groups or less had there been no restoration efforts. This recovery is primarily due to the implementation of an artificial cavity program and the efficacy of artificial cavities. Forty five, 60, and 61% of RCW nests were in artificial cavities in 1990, 1991, and 1992, respectively.

Excellent reproductive attainment (**Table 6**) and dispersal of RCWs produced during the past three nesting seasons has led to this increase. New groups are being formed by RCWs occupying recruitment clusters or by single birds attracting mates. We estimate 1285 RCWs successfully fledged in the three years since Hugo, and these birds have taken advantage of the opportunities provided by single bird clusters and vacant recruitment clusters. Although no quantitative data on the number of floaters in the population exists, they are present and likely responsible for the quick occupation of recruitment stands with artificial cavities. The number of 2 bird groups has increased annually, but has slowed (8.5% increase in 1992 as compared to 23.0% in 1991). It is unlikely that future increases will be this high because not enough suitable habitat exists at this time to sustain this continued rate of growth.

Currently we see the biggest threats to the population to be:

1) the desire by some individuals, agencies, and organizations for more hardwoods throughout the FMNF;

2) encroachment of both pine and hardwood midstory into RCW clusters;

3) the general lack of potential cavity trees and increased cavity tree mortality;

4) another hurricane.

If successfully implemented, management programs designed to grow more hardwoods throughout the FMNF would have a negative affect on the RCW. Significant increases of hardwood in RCW habitat and potentially suitable habitat would not only most likely cause a loss of groups, but also decrease the potential for growth and expansion of the existing population.

Midstory problems on the FMNF are compounded by the flush of new growth in clusters, in the wake of Hugo opening the forest canopy and stimulating growth of shade intolerant species. It is hoped that an aggressive prescribed burning program combined with mechanical treatments and the application of herbicide can control most of this growth.

Another management challenge includes controlling loblolly pine regeneration in clusters with low basal areas. Our goal is to reestablish foraging and nesting habitat as quickly as possible for long term recovery of the RCW. However, with loblolly pine regenerating in clusters, fire may be excluded to protect the seedlings and saplings, creating a pine midstory in clusters. Pine encroachment will also cause abandonment of clusters if not

Table 8. Projected salvage and thinning volumes for the Francis Marion National Forest in MBF for 1993-1995 (MBF=thousand board feet)

	1993	1994	1995	Total
Salvage volume	3,211	1	1	3,211
Thinning volume	18,939	16,022	3,408	38,369[2]
Total	22,150	16,022	3,408	41,580

[1]No projected salvage volume although salvage volume will most likely occur.

[2]Additional volume is being proposed on the Wambaw Ranger District in this three year period.

controlled (Loeb et al. 1991). No doubt we will have to develop a strategy that both controls pine midstory and encroaching hardwoods, and facilitates pine regeneration for improvement of RCW habitat.

An important factor that became apparent during the last two years was the increasing mortality of cavity trees. RCWs are making cavities on their own, but the general lack of potential cavity trees and increased mortality of cavity trees may affect some groups. It is likely that some groups can not locate or construct enough cavity trees to offset the mortality of existing cavity trees. A shortage of trees suitable for artificial cavities is anticipated, but the extent of this shortage is unknown. Hopefully, in most cases it will be possible to shift the cluster to another stand. There is no doubt that the current RCW population would be a fraction of what it is now without the use of artificial cavities, and we doubt that the current population can maintain its current level without the continued construction of artificial cavities. However, the rapid development of nesting habitat and the potential for provisioning recruitment stands with artificial cavities will hopefully provide suitable habitat and clusters necessary to recover the population.

Ironically, the amount of sawtimber surviving Hugo is similar to the amount of pine sawtimber that existed when the FMNF was acquired, and the number of pine stems 5-9" dbh on the FMNF now is twice what is was in 1936 **(Table 1)** (Grumbine 1936). Pine stands 40 years old and younger survived well **(Figure 1)**. Stands that survived Hugo with less than 30 sq. ft. ba were considered sparse or destroyed and were placed into the 0-10 year age class, although some stands categorized as such support RCW clusters and residual trees that represent potential RCW cavity trees. In addition, the number of pine stems ≥10" dbh is only slightly less than what was present in 1936 **(Table 1)** (Grumbine 1936). The FMNFs' RCW population recovered largely on its own from the conditions that existed in 1936, despite the fact that the first 30 years of management included cutting of cavity trees and entire clusters in thinnings and regeneration harvests (Hooper et al. in press). By 1989 the population was essentially recovered. Presently, conditions are similar to those in 1936. The number of potential cavity trees surviving Hugo, the potential for developing foraging and nesting habitat quickly, and the amount of pine in regeneration will contribute greatly to the long term recovery of the RCW on the FMNF. The more foraging habitat and potential cavity trees, the more probable the recovery of the population.

Considering the potential for managing future RCW habitat, it is likely that the RCW population on the FMNF can reach a recovered level (500 active groups or 250 effective groups) in time, assuming another hurricane causing catastrophic losses does not impact the FMNF. Hurricanes hitting the FMNF within the next 50 years would be capable of destroying most of the trees with artificial cavities and many of the trees without cavities that were injured by Hugo. In the next 100 years, 45 hurricanes are expected to make direct hits on the 11 most vulnerable recovery areas (Hooper and McAdie 1995). Four of these hurricanes will bring category III winds to 4 of the recovery areas, 2 of the 4 getting hit twice by such winds. Perhaps the most vulnerable recovery area is the FMNF (Hooper and McAdie 1995). Fortunately, not all of the RCW population and habitat was destroyed, although 3 of 4 major concentrations of RCWs on the FMNF were severely impacted by Hugo. A more uniform spatial distribution of clusters on the FMNF would have resulted in less impacts to the entire population and would be desirable so that concentrations of RCWs would not be severely impacted.

The big lesson from Hurricane Hugo is that a RCW population suffering catastrophic losses can be stabilized and increased, at least over the short term, given that integrated management efforts strive for common goals. This is important considering the expected number of hurricanes to hit recovery areas in the next 100 years (Hooper and McAdie 1995). A myriad of management efforts led to an increase of the RCW population on the FMNF before Hugo, and the continuation of these efforts will no doubt contribute to the eventual recovery of the RCW on the FMNF after Hurricane Hugo. In only three years, the population of RCWs on the FMNF has increased greatly; breeding groups fledged 1285 young, much of this recruitment coming from artificial cavities. Given the enormous impacts to the RCW population on the FMNF in 1989 and its' apparent subsequent recovery, it is highly probable and technologically possible that similar results could be achieved on other recovery areas not impacted by hurricanes.

ACKNOWLEDGEMENTS

We would like to express thanks to the many persons that have contributed to the success of RCW recovery on the FMNF. District Rangers Jim Brotherton and Glen Stapleton, their immediate staff, support personnel, and District employees, all gave full support and cooperation for this effort. The Southeastern Forest Experiment Station (SEFES) was primarily responsible for the initiation of the artificial cavity program, providing initial and continued support for the project. The Supervisor's Office of the FMNF, the Southern Region Office, and the Washington Office of the Forest Service are to be commended for the quick response for funding and support. Jerry Henderson, Oscar M. Stewart, Ronald E.F. Escano, Michael Lennartz, Charles C. VanSickle, and Eldon W. Ross were instrumental in these efforts. Wildlife biologists and technicians from Region 8 detailed to the FMNF after Hurricane Hugo spent many months collecting the data necessary to plan and build a program of recovery for the RCW. Carol Copeyon, Jay Carter, John Hammond, Dave Allen, and Sheryl Sanders taught us the original techniques for making artificial cavities. Fort Bragg and Camp Lejeune provided personnel and vehicles necessary to initiate the recovery work. This remarkable recovery could not have been accomplished without the personnel that have to date installed approximately 1000 artificial cavities for the RCW on the FMNF. These people are Carolyn Short, Doug Short, Anne Ripley, Joseph Zisa, John Young, Richard Lamma, Michael Weeks, Bradley Morris, Roger Dennis, Michael Leslie, Michael Ayers, Brian Grant, Chuck Kinsey, Susanne Shipper, Will Howard, and Hal White. Thanks to the Pine Knot Job Corps Center, KY, for supplying inserts. Finally, thanks to Jean Reiher of the SEFES for continuing to monitor, collect, and compile data necessary to our successful program.

Wind Damage to Red-cockaded Woodpecker Cavity Trees on Eastern Texas National Forests

Richard N. Conner and D. Craig Rudolph
Wildlife Habitat and Silviculture Laboratory[1], Southern Forest Experiment Station, USDA Forest Service, Nacogdoches, Texas 75962

ABSTRACT: Wind damage is the second major cause of red-cockaded woodpecker (*Picoides borealis*) cavity tree loss in eastern Texas. The distance between cavity trees and openings in the forest canopy (primarily clearcuts) affects the rate at which wind damage occurs. This distance is largely affected by the 61-m cluster protection buffer established around each cluster of cavity trees by forest management. However, after buffers are established, woodpeckers often excavate cavity trees closer than 61 m to the edge of the cluster, especially if excessive hardwood midstory foliage is present within the cavity tree cluster area. In Louisiana and Texas we examined wind damage to 60 cavity trees and to 44 wind-thrown longleaf pines in open forest habitat adjacent to clearcuts to evaluate the effectiveness of the 61-m buffer distance. Wind-damaged cavity trees in Texas averaged 110 m from forest openings whereas randomly selected undamaged cavity trees averaged 311 m. Most damaged cavity trees were wind-snapped, and 35 % were within 25 m and 48 % were within 50 m of openings in the forest canopy. However, there may be secondary zones of damage between 75 and 200 m of forest edges perhaps caused by turbulence of wind swirling above trees after passing over the forest edge. Wind throw and wind-snap at some rate should be expected to occur to cavity trees as a natural process. Because woodpeckers often excavate cavities within existing buffer zones, we are uncertain as to whether zones wider than 61 m would provide additional protection for cavity trees.
KEYWORDS: Cavity trees, wind-snap, clearcuts, *Picoides borealis,* wind damage.

[1]Maintained in cooperation with the College of Forestry, Stephen F. Austin State University, Nacogdoches, Texas, 75962.

On 21 September 1989 hurricane Hugo made land fall on the coast of South Carolina with the main force of its winds directly hitting red-cockaded woodpecker habitat on the Frances Marion National Forest (Hooper et al. 1990). Earlier in the decade tropical storm Juan and hurricanes Elena and Kate had destroyed red-cockaded woodpecker cavity trees on the Wade Tract managed by Tall Timbers Research Station in Georgia (Engstrom and Evans 1990). Major storms such as Hurricanes cannot be predicted far in advance and are difficult to deal with using anticipatory management. Buffer zones of uncut trees left to protect cavity trees from wind damage have little or no effect during storms of such magnitude.

Isolated thunderstorms, however, are a regular occurrence during most seasons of the year in the South (Anon. 1952, Jackson et al. 1986). Although winds of these small localized storms do not reach the velocities of the major hurricanes, wind gusts exceeding 80 km/h (50 mph) are common. Many severe thunderstorms precipitate tornadic wind activity. Although such conditions do not always produce a tornado, they can cause significant isolated gusts of wind. Conner et al. (1991a) examined the effects of wind damage from isolated severe thunderstorms and reported that wind damage to red-cockaded woodpecker cavity trees was the second greatest cause of cavity tree mortality in a 12-year study in eastern Texas. Wind damages and kills cavity trees by either snapping the tree at a nest cavity (wind-snap) or blowing trees over and exposing roots (wind throw). Of 85 cavity trees destroyed by wind in the Texas study, 75 were wind-snapped and 10 were wind-thrown (Conner et al. 1991a).

Forest management activities can affect the rate of cavity tree loss from wind damage associated with isolated severe thunderstorms. Jackson et al. (1978) observed wind damage to cavity trees on the Angelina, Davy Crockett, and Sabine National Forests and suspected the close proximity of clearcuts as the possible cause. The current U.S. Forest Service Wildlife Management Handbook Guidelines for Red-cockaded Woodpeckers require that a 61-m (200 ft.) buffer of uncut trees should be left between cavity trees and regeneration cuts (U.S. Forest Service 1985). The 61-m distance was a rough estimate made during the early days of the first recovery team as to the distance needed to protect cavity trees from wind damage, construction, or whatever else might damage cavity trees (Jerome A. Jackson, Pers. Commun.).

This study examines wind damage to red-cockaded woodpecker cavity trees as a function of distance from openings in the forest canopy. The primary objective of the research was to evaluate the effectiveness of the 61-m buffer distance for the protection of red-cockaded woodpecker cavity trees. Secondarily, we evaluated the relationship of cavity tree diameter with probability of cavity tree wind-snap. We also examined wind-thrown longleaf pines following a spring storm on the Vernon Ranger District of the Kisatchie National Forest to gain additional insight into relationships of wind damage and distance from forest edge.

METHODS

Red-cockaded woodpecker cavity tree mortality was measured during annual visits to the Angelina National Forest from 1978 to 1991 and the Davy Crockett and Sabine National Forests from 1987 to 1991. Causes of cavity tree mortality were recorded in field notes. Distances of 60 wind-damaged cavity trees to clearcuts and other forest openings were measured. We defined a forest opening as any area > 2 ha with dominant and codominant trees removed. Fifty-eight other live, undamaged cavity trees were randomly selected using a random numbers table to select cavity tree numbers (Rohlf and Sokal 1969). Distances to openings in the canopy were determined from computer generated, digitized maps of cavity trees and aerial photographs. Diameters at breast height (DBH) of cavity trees were also recorded for random and wind-snapped cavity trees when possible. A *t*-test was used to compare DBHs of wind-snapped and undamaged cavity trees. A Kolmogorov-Smirnov two-sample test was used to compare distributions of wind-damaged and undamaged cavity trees.

The distances (to the nearest 25 m) from openings in the canopy of an additional 644 cavity trees were measured using computer generated maps of cavity trees and aerial photographs to obtain a relative estimate of the number of cavity trees in each distance zone. Wind-damaged cavity trees were assigned to distance zones from forest openings (25 m increments) in order to calculate the relative frequency of wind-damaged cavity trees (n = 60) in each zone as a percent of the total available cavity tree population (n = 644) on the Angelina and Davy Crockett National Forests that were also assigned to a 25 m increment distance zone. This procedure was done to yield a probability of cavity tree wind damage in each distance zone.

On 3 June 1992 at 1528 CDT a severe thunderstorm passed over the Vernon Ranger District of the Kisatchie National Forest, Louisiana. The Fort Polk Army Airfield Weather Service adjacent to the forest measured wind velocities up to 72.3 km/h (39 knots) and a total precipitation of 2.3 cm. The weather service also indicated that tornadic winds were not present during the storm. Wind throw of longleaf pines occurred near the edges of two clearcuts on the Vernon Ranger District. We measured the number and diameters of wind-thrown and surviving longleaf pines in 25 m increment distance zones from each clearcut within approximately one month after the storm. Six distance zones (25 m deep x 100 m wide) were measured at one site (150 m from clearcut edge) and four at the second site (100 m from clearcut edge). Successive zones were measured until the wind damage stopped. The proportion of wind-thrown pines in each distance zone was calculated. The distance from each wind-thrown longleaf pine to the clearcut edge was also measured. Diameters of wind-thrown, wind-snapped, and surviving pines were compared with a *t*-test. A Kolmogorov-Smirnov two-sample test was used to compare distributions of wind-thrown and undamaged longleaf pines.

RESULTS

A total of 45 longleaf pines were wind-damaged on the two Vernon Ranger District study sites (**Table 1**). Forty-four of these pines were wind-thrown and only one was wind-snapped. Wind-thrown pines (n = 44, = 42.3 cm DBH) were not significantly larger than undamaged pines (n = 93, = 39.6 cm DBH) (t = 2.62, $P < 0.01$). Diameter of wind-thrown pines was negatively correlated to distance from clearcut edge (r = -0.40, P = 0.006) suggesting that larger diameter pines had a lower probability of wind throw than smaller diameter pines as distance from edge increased.

Raw numbers of wind-thrown pines (**Figure 1**) suggest that damage in the 0-25 m and 50-75 m zones is somewhat similar, with most wind throw occurring 25-50 m from the edge of the clearcuts. The percentage of wind-thrown pines based on the number of pines originally subjected to the storm indicates that damage between the 0-25 and 25-50 m zones is actually very similar, while damage in the 50-75 m zone is lower (**Figure 2**). Wind throw was lowest in the 75-100 m zone with virtually no wind damage beyond 100 m (only 2 pines were wind-thrown in the 125-150 m zone - not shown in **Figures 1 & 2**). The distribution of wind-thrown longleaf pines was significantly different from undamaged pines (KS_a = 1.49, P = 0.024).

On National Forests in Texas, wind-damaged red-cockaded woodpecker cavity trees averaged 110 m from canopy openings whereas undamaged cavity trees averaged 311 m (t = 3.50, $P < 0.001$). Of the 60 wind-damaged cavity trees 53 were wind-snapped and 7 were wind-thrown. Thirty-five percent (21/60) of the wind-damaged cavity trees were within 25 m of an opening in the forest canopy (primarily clearcuts or young plantations), and 48 % (29/60) were within 50 m (**Figure 3**). Twenty-two percent (142/644) of available cavity trees were found within 50 m of forest openings (**Figure 4**). In general, damaged and undamaged cavity trees were found in highest frequencies close to forest openings and

Table 1. Distances of Wind Thrown and Surviving Longleaf Pines from Two Clearcuts on the Vernon Ranger District of the Kisatchie National Forest, Louisiana, Following a Wind Storm During Spring, 1992

	Distance Zones from Clearcuts (m)			
Tree Status	0-25	25-50	50-75	75-100
Live standing	30	38	41	40
Wind thrown	13	18	11	2
% Wind thrown	43.3	47.4	26.8	5.0

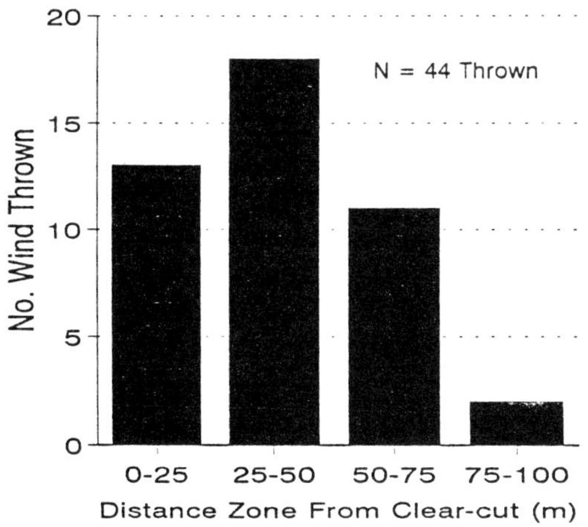

Figure 1. Numbers of wind-thrown longleaf pines in 25 m distance zones from the edges of two clearcuts on the Vernon Ranger District of the Kisatchie National Forest in Louisiana during 1992.

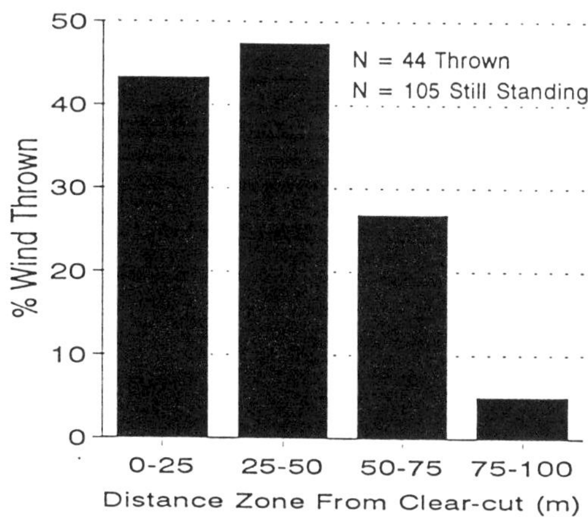

Figure 2. Percentages of wind-thrown longleaf pines in 25 m increment distance zones from the edges of two clearcuts on the Vernon Ranger District of the Kisatchie National Forest in Louisiana during 1992.

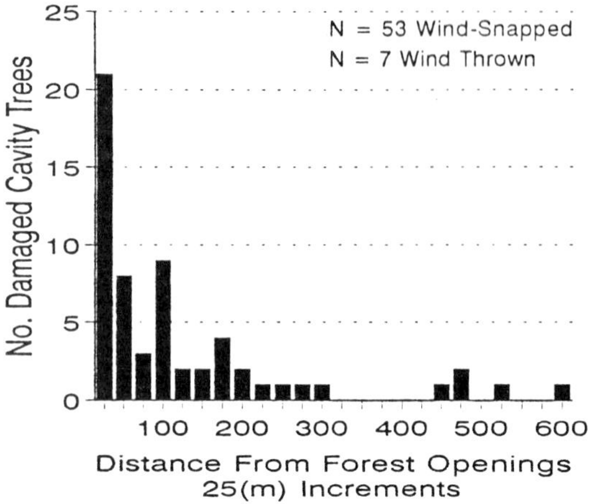

Figure 3. Numbers of wind-damaged red-cockaded woodpecker cavity trees in 25 m increment distance zones from forest openings (> 2 ha) on the Angelina and Davy Crockett National Forests in eastern Texas from 1978 to 1991.

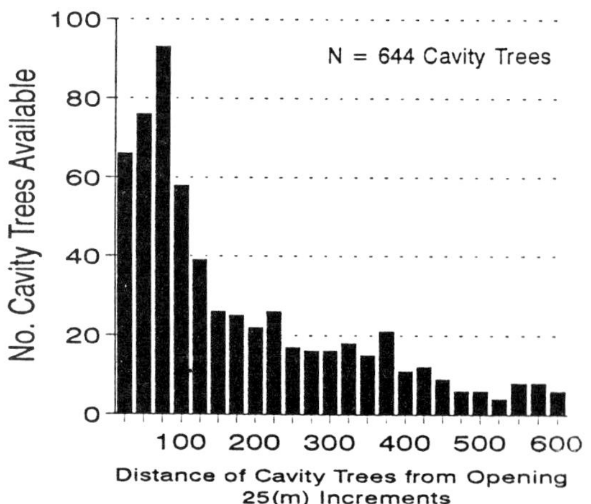

Figure 4. Distribution of 644 red-cockaded woodpecker cavity trees in 25 m increment distance zones from forest openings (> 2 ha) on the Angelina and Davy Crockett National Forests in eastern Texas from 1983 to 1991.

decreased in frequency as distance from forest openings increased **(Figures 3, 4).** However, the distributions of wind-damaged and undamaged cavity trees were significantly different ($KS_a = 1.99$, $\underline{P} < 0.001$).

The raw numbers of cavity trees damaged by wind in each distance zone can be misleading. The frequency of wind damage is directly influenced by the availability of cavity trees in each distance zone. The probability of wind damage to cavity trees was calculated for 25 m increment distance zones from the edge of forest openings by dividing the number of cavity trees damaged in each zone by the number of cavity trees available to be damaged in each zone **(Table 2)**. The baseline, or "background" level of damage to cavity trees resulting from windaction within a continuous forest canopy is approximately 4 % based on data from 250 to 600 m **(Table 2)**. Probability of wind damage in the 25 m (31.8 %), 50 m (10.5 %), 100 m (15.5 %), and 175 m (16.0 %) zones all appear to be somewhat higher than baseline damage rates **(Figure 5, Table 2)**. Data beyond the 300 m zone are probably less meaningful than zones closer to cavity trees because of small sample sizes **(Table 2)**.

We suspected that wind-snapped cavity trees may have been smaller in diameter (DBH) than cavity trees that were not damaged by wind. However, cavity trees that had been wind-snapped (48.4 cm) were not significantly smaller in diameter than randomly selected undamaged cavity trees (49.6 cm)**(Table 3).**

DISCUSSION

Wind-thrown trees within a forest are a natural phenomenon and provide habitat features that are important for some species. In southern pine forests up-turned roots associated with wind-thrown pines provide hollows (surface depressions) for Louisiana pine snakes (*Pituophis melanoleucus*) and other herpetofauna. This is particularly important in longleaf pine communities where such hollows provide refugia from predators and frequent natural fires (Todd Engstrom, Pers. Commun.). Wind throw is also important for the turnover of forest soils, soil development, and soil spatial heterogeneity (Norton 1989).

Factors affecting wind throw are numerous and include tree diameter, bole shape, crown size and shape, strength of wood (Curtis 1943, Swain 1979), root rot (Conner et al. 1991a),

Table 2. Distances of red-cockaded woodpecker cavity trees from forest openings on the Angelina, Davy Crockett, and Sabine National Forests in eastern Texas 1978-1991.

	Distance zones from forest openings (m)									
	0-25	25-50	50-75	75-100	100-125	125-150	150-175	175-200	200-225	225-250
Wind-damaged (#)	21	8	3	9	2	2	4	2	1	1
Available undamaged cavity trees (#)	66	76	93	58	39	26	25	22	26	17
% wind damaged	31.8	10.5	3.2	15.5	5.1	7.7	16.0	9.1	3.8	5.9
	250-275	275-300	300-325	325-350	350-375	375-400	400-425	425-450	450-475	475+
Wind-damaged (#)	1	1	0	0	0	0	0	1	2	2
Available undamaged cavity trees (#)	16	16	18	15	21	11	12	9	6	72
%wind damaged	6.3	6.3	0	0	0	0	0	11.1	33.3	2.8

Table 3. Average diameters of randomly selected and wind-snapped red-cockaded woodpecker cavity trees on the Angelina, Davy Crockett, and Sabine National Forests in eastern Texas.

	Mean	S.D.	N	t	P
Wind-snapped cavity trees (cm)	48.4	6.6	35	0.70	0.48
Random cavity trees (cm)	49.6	9.4	53		

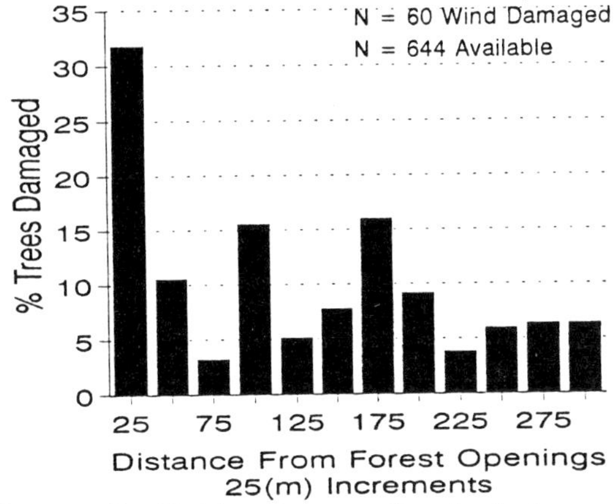

Figure 5. Percentage of wind-damaged woodpecker cavity trees per cavity trees available within 25 m increment distance zones from forest openings (> 2 ha) on the Angelina and Davy Crockett National Forests in eastern Texas from 1978 to 1991.

root size, configuration and condition, soil composition and moisture (Irvine 1970, Cutler et al. 1990), and land contour (Weir 1973). Sandy and gravel soils afford less protection from wind throw than clay soils (Cutler et al. 1990). Also, the relative flexibility of loblolly pine (*Pinus taeda*) and Fraser fir (*Abies fraseri*) has been shown to be related to ethylene production within the trees which results from mechanical perturbation (Telewski and Jaffe 1986). Recency of cut also affects the risk of wind damage. Trees on the edges of cuts are more vulnerable to wind damage immediately following a cut than several years later. Expanded root development that occurs post-harvest gives the trees on the edge of cuts additional resistance (Weidman 1920).

Wind throw of longleaf pines on the Vernon Ranger District of the Kisatchie National Forest in Louisiana occurred up to 100 m from the edge of the two clearcuts examined (**Figures 1, 2**). Although wind throw occurred in the 75 - 100 m zone, damage occurred in only 5 % of the pines in that zone, suggesting that a buffer distance of 75 m may be adequate to protect pines from severe wind throw damage during isolated severe thunderstorms. Wind throw in the three zones closest to the clearcut edge (0 - 75 m) was substantial ranging from 26.8 to nearly 50 % of the pines present (**Table 1**). The current buffer distance to protect red-cockaded woodpecker cavity trees is 61 m, which falls nearly mid-distance within our 50 - 75 m zone. A closer examination of the wind throw data within this zone indicates that 5 pines were wind-thrown between 50 and 61 m and 6 pines were wind-thrown between 61 and 75 m.

Does this suggest that a buffer distance of 75 m may be required for adequate protection of pines from wind throw? It is possible that the wind damage we observed could have occurred at the locations on the Vernon even if the cuts had not been present. It is also possible that ground contour could have funneled the wind into the edge created by the clearcuts, and the wind throw we observed was the result of the combined effects of the clearcut and land contour. A third possibility is that tornadic wind activity occurred at the specific sites of the wind throw we observed.

Tornadic winds were probably not the cause of the wind throw we observed. Weather reports from the Fort Polk Army Airfield Weather Service did not indicate the presence of such winds. Also, the wind-thrown pines all lay in the same general compass direction, and there was no indication of twisted tree tops and trunks as is typically observed with tornados.

Although land contour may have contributed to the damage, wind damage is known to be associated with trees adjacent to clearcuts (Curtis 1943, Gordon 1973, DeWalle 1983). Gilmour (1926) noted that the greatest damage from wind occurred adjacent to clear-cuts. Gordon (1973) noticed substantial wind damage to fir (*Abies* spp.) stands in California in both an inner zone (0 - 34 m) and an outer zone (34 - 68 m) of uncut timber around clearcuts. Our personal observations of wind damage in forests also suggests that damage is most frequently associated with forest edges. The most severe wind damage is typically found at the ends of valleys or hollows, and as a result of V-shaped or egg-shaped cuts where "funneling" occurs (Curtis 1943).

Wind damage to cavity trees was typically wind-snap rather than wind throw. Cavity trees that were wind-thrown during our study often had root decay which increased their susceptibility to wind throw. Wind-snap nearly always (only several exceptions) occurred at a cavity site, most likely because the pines were weaker at those sites as a result of wood loss during cavity excavation and heart wood decay (Conner et al. 1975, 1976, Griffin et al. 1975, Conner and Locke 1982).

More than 180 cavity trees were less than 61 m (cavity tree cluster protection buffer zone) from forest openings, which were primarily regeneration cuts. Many clearcuts that were closely associated with cavity trees were on private lands primarily because of the fragmented ownership pattern of the forests. Relatively few openings near cavity trees were created by treatment to control southern pine beetle (*Dendroctonus frontalis* Zimmermann) infestations. Red-cockaded woodpecker cavity trees were primarily close to clearcuts because the woodpeckers selected pines for cavity excavation that were close to cuts. Our past observations have suggested that red-cockaded woodpeckers select pines close to cuts because of the relative openness of the habitat (Conner and Rudolph 1989, Conner et al. 1991a). Selection of such pines appears to be the result of excessive hardwood midstory in the areas of continuous forest where the rest of the cavity trees are located. The baseline frequency of wind-snap damage to cavity trees appears to be about 4 % (**Table 2, Figure 5**). With this in mind the pattern of wind-snap to cavity trees appears to be somewhat different than what was observed in wind-thrown areas on the Vernon Ranger District. Frequency of wind damage to cavity trees was highest in the 0 - 25 m zone, high in the 25 - 50 m zone, but unexpectedly high in the 75 - 100 m and 150 - 175 m zones. This pattern suggests an initial damage zone on the edge of the forest and a secondary damage zone(s) somewhat distant from the edge (**Figure 5**).

The causes of these secondary damage zones may be several fold. Small sample size might be the cause for the elevated damage that was observed in the 150 -175 m zone (4 of 25 cavity trees were snapped). However, such is not the case for the 75 - 100 m zone as 9 of 58 cavity trees were wind-snapped (**Table 2**). A second alternative is that wind damage might occur to cavity trees in secondary zones because they were taller than the forest that surrounded them as a result of past harvesting practices. However, on the forests we examined, forest canopies surrounding the cavity trees and cluster areas in general were almost always the same height or taller.

We suggest that wind turbulence resulting from moving air striking the forest edge and swirling above the forest canopy may be the cause of the secondary damage zones (**Figure 6**). Differential wind velocities within and above forests are known to occur (Fons 1940). Turbulence and vortices are formed when moving air encounters an object or boundary zone. The greater the distance between the ground and the top of the forest canopy, the greater the turbulence encountered over the canopy (Fons 1940). A mature forest edge would be expected to create more turbulence than edges associated with shorter, younger stages of regeneration. Experiments with model forests using trees constructed with metal and plastic have demonstrated highest drag coefficients (turbulence) on the edge of the forest, a sharp decrease in drag immediately in from the model forest edge, and a secondary zone of increased drag further away from the edge (Meroney 1968, Hsi and Nath 1970:600). This is consistent with our hypothesis of secondary zones of wind damage as observed with wind-snapped cavity trees.

Observations of wind damage to grain fields also supports our hypothesis. The phenomenon termed "lodging" refers to permanent bending of the stem of grain crops (Neenan and Spencer-Smith 1975, Carter and Hudelson 1988, Baker et al. 1990). The structural breakage of grain stems during lodging is quite similar to what occurs during wind-snapping of trees. Lodging frequently occurs on the edge of grain fields, but is also observed in areas offset from the edge as a result of wind turbulence.

Our results also suggest that pine diameter may have an effect on wind throw but not on wind-snapping of cavity trees. Larger diameter pines appeared to have a lower probability of wind throw than smaller diameter pines of equivalent height. A New England study indicated that white pines (*Pinus strobus*) that were 46 cm in diameter could withstand nearly twice the wind force of 31 cm diameter white pines before wind throw occurred (Curtis 1943).

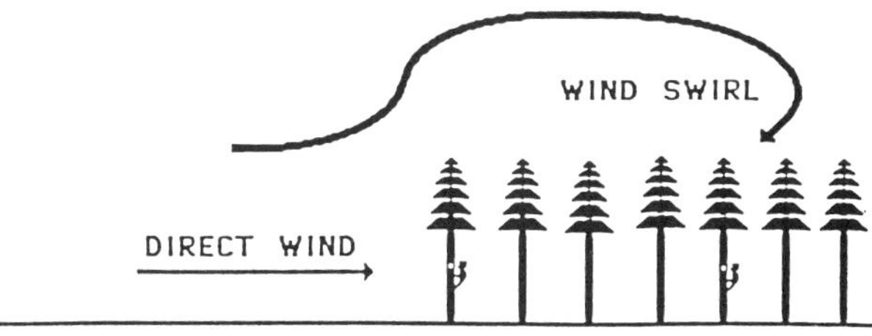

Figure 6. Schematic representation of wind action as wind strikes the edge of forest stand next to an opening, swirls over the forest causing turbulence, and damages trees both on the edge and in secondary zones distant from the edge

We are uncertain whether any further protection of cavity trees from wind damage would be achieved by a buffer increase. Simply increasing the buffer width may not increase the eventual average distance between cavity trees and forest edges. Red-cockaded woodpeckers regularly excavated new cavity trees quite close to the forest edge, and in a few cases even after the midstory had been reduced to within acceptable ranges for the woodpecker.

We are also uncertain how woodpeckers will select cavity tree locations relative to forest openings when hardwood midstory within all cluster areas is appropriately reduced. Criteria for selection of cavity trees by red-cockaded woodpeckers also include special characteristics of pines, e.g., old age, heartwood decay, crown volume, etc. (Conner and O'Halloran 1987, DeLotelle and Epting 1988). Suitable pines must be available in an area before a woodpecker can excavate a cavity. There are insufficient data to evaluate red-cockaded woodpecker selection of cavity trees in areas where midstory is absent, and extensive use of artificially constructed cavities currently negates the ability to collect such data.

The number of cavity trees lost to wind damage in these secondary damage zones was not great. Only 9 cavity trees were wind-damaged in the 75-100 m zone and 4 cavity trees in the 150-175 m zone (**Figure 3**). This yields only 13 cavity trees lost over an average of approximately seven years. Several of these cavity trees would have been lost anyway due to background wind damage rates. An average of about 50 clans of woodpeckers (Angelina and Davy Crockett National Forest populations combined) were excavating cavity trees over the course of the study, indicating that only 0.037 cavity trees were lost per year, per clan, in the secondary damage zones. Clearly, cavity tree losses to wind damage in these secondary zones would not significantly impact woodpecker populations.

Negative impacts resulting from the losses of cavity trees to wind damage can be mitigated by the installation of artificial cavities (Odom et al. 1982, Copeyon 1990, Allen 1991, Taylor and Hooper 1991). Both the drilled and insert cavity techniques were extensively used to help offset the negative impacts of Hurricane Hugo. Artificial cavities could be of major benefit for small red-cockaded woodpecker populations that have insufficient supplies of cavity trees and potential cavity trees and following catastrophic disturbances such as hurricanes (Hooper et al. 1990). However, we do not have sufficient data to evaluate the vulnerability of pines containing cavity inserts to wind damage or bark beetle infestation. Recovered populations should not need artificial cavities to compensate for cavity tree losses to wind damage. The long term goal should be a self-sustaining system of cavity tree loss and natural replacement that is independent of artificial management techniques.

ACKNOWLEDGMENTS

We thank D. Saenz and R.R. Schaefer for their assistance in the field, and R.S. DeLotelle, R.T. Engstrom, and J.A. Jackson for constructive comments leading to the improvement of the manuscript.

Red-cockaded Woodpecker Habitat Management and Southern Pine Beetle Infestations

Robert N. Coulson and Jeffrey W. Fitzgerald, Knowledge Engineering Laboratory Department of Entomology, Texas A&M University, College Station TX 77843
Forrest L. Oliveria, USDA Forest Service, Forest Pest Management, Pineville, LA 71360
Richard N. Conner and D. Craig Rudolph, USDA Forest Service, Southern Research Station, Nacogdoches, TX 75962

ABSTRACT: During the past decade, a large number of red-cockaded woodpecker (RCW) cavity trees (ca. 530) occurring on National Forests in Texas have been infested and killed by the southern pine beetle (SPB). Many of the infested trees occurred on sites prepared in accordance with the RCW management guidelines employed by the USDA Forest Service. Most of the infestations on these sites originated in single active cavity trees. In this paper we examine the association of SPB in RCW cavity trees and foraging areas. We found evidence from the published literature to suggest that initiation of infestations in cavity trees was similar to the process as it occurs in lightning-struck hosts. Growth of infestations in cluster sites, maintained in accordance with management guidelines, was not likely because of the low basal area and wide spacing of trees. However, the proximity of infested cavity trees to high hazard stands could result in insect outbreaks in adjacent foraging habitat.
KEYWORDS: southern pine beetle, red-cockaded woodpecker, IPM, disturbance, endangered species, landscape ecology

The red-cockaded woodpecker (RCW), *Picoides borealis,* is indigenous to pine forests of the southeastern United States. It is an endangered species protected under mandates of the Endangered Species Act of 1973, as amended. The southern pine beetle (SPB), *Dendroctonus frontalis*, is the most important pest species that infests pines throughout the range of RCW. Because infestations by SPB result in mortality to the host trees, this insect represents a direct threat to RCW cluster sites as well as foraging habitat. Forest management guidelines for RCW have been developed by the USDA Forest Service and implemented on Federal lands in the South. The guidelines specifically consider issues associated with cluster[1] protection and management, habitat management, and clearing for non-timber management purposes. Intensive management includes removal of all hardwood vegetation and thinning of pines to 60 ft^2/ac within the designated cluster area. This management practice, when hardwood removal is accomplished only in cluster areas, results in introduced open patches (the clusters) in a forest matrix that contains extensive hardwood midstory.

The seasonal behavior of SPB results in two different types of impact of the insect on RCW. First, behavior associated with the establishment of populations of SPB in refuge hosts (Coulson et al. 1985) can result in mortality to cavity trees. (Conner et al. 1991a, Rudolph and Conner 1995). Second, behavior associated with growth of SPB infestations can result in destruction of RCW foraging habitat. These impacts create difficult management decisions. As a consequence of the different mandates associated with the Southern Pine Beetle Environmental Impact Statement (SPB/EIS), Endangered Species Act, Wilderness Act, and a 1988 Federal Court order, recommendations for suppression of

[1]Cluster: Aggregations of cavity trees used by groups of woodpeckers (Walters et al. 1988), the sum of all cavity trees owned by the clan (Ligon et al. 1986).

SPB on Federal lands are often contradictory. For example, whereas the management guidelines for RCW (SPB/EIS) permit suppression of SPB infestation growth that threatens foraging and nesting habitat, there is no provision (operational procedure) for protection of specific cavity trees within a cluster site.

The goal of this paper is to consider the ecological association of SPB and RCW in the context of forest landscape management. In particular we are interested in how management guidelines for protection of RCW may affect infestation of active cavity trees and surrounding foraging habitat.

THE ASSOCIATION OF SOUTHERN PINE BEETLE AND RED-COCKADED WOODPECKER CAVITY TREES

Southern pine beetle infestation of single RCW cavity trees, cluster sites, and foraging habitat is a problem of paramount significance to conservation and recovery of RCWs on federal lands (Conner et al. 1991a, Rudolph and Conner 1995). The seasonal behavior of SPB creates two separate problems for RCW management: one associated with colonization of single active cavity trees, primarily during the fall, and one associated with infestation growth in foraging habitat or trees within the general cluster area.

SOUTHERN PINE BEETLE INFESTATION OF SINGLE ACTIVE CAVITY TREES

Infestation of active cavity trees appears to be a direct consequence of SPB behavior associated with establishment of populations in refuge hosts. Although lightning-struck trees are commonly colonized, hosts disturbed by other means (e.g., cultural damage, wind-throw etc.) or in a declining state resulting from root disease or poor site conditions are also suitable (Lorio 1966, 1968). An integral component of the natural history of SPB centers on the beetles' use of lightning-struck hosts (Hodges and Pickard 1971, Lorio and Bennet 1974, Lorio and Yandel 1978, Coulson et al. 1986, Lovelady et al. 1991) as habitat.

Response of SPB to Host Disturbance

There are several aspects of the relation of SPB and lightning. Lightning-struck pines are readily identified by SPB. It is likely that the insects aggregate in response to the host terpenes released as a consequence of the disturbance (Payne and Coulson 1985). Lightning-struck pines also have much reduced defensive capability and, therefore, are easily colonized by small numbers of SPB as well as the less aggressive *Ips* spp. Lightning is a reliable disturbance and struck trees are available for colonization throughout the year. Trees struck by lightning persist as attractive centers for an extended period of time (as long as 130 days) (Coulson et al. 1986). Lightning-struck trees are numerous and occur at densities that place vulnerable hosts within the dispersal range of SPB, i.e., within one km. Although lightning-struck pines are available to and can be colonized by SPB throughout the year, these hosts appear to be of critical importance to the insect during periods when populations are dispersed and small in size, and when environmental conditions are not favorable for colonization of undisturbed trees.

Characteristics of RCW Cavity Trees

Research on RCWs has identified the characteristics of pines and habitat necessary to provide nesting and roosting sites (Hovis and Labisky 1985, Conner and O'Halloran 1987, DeLotelle and Epting 1988, Hooper 1988, Hooper et al. 1991b, Rudolph and Conner 1991, Loeb et al. 1992). Among the attributes identified the following are most important. The predominant species used for cavity trees are longleaf, shortleaf, loblolly, and slash pines. Mature, living pines 60 to 80 years of age can be suitable for cavity establishment, but trees 90 to 130+ are preferred (Conner and O'Halloran 1987, Rudolph and Conner 1991). Living trees are essential, as the woodpeckers excavate "resin wells" around the entrance to the cavity. The resin exudate from the tree serves to deter predatory rat snakes (*Elaphe* spp.), i.e., the various subspecies of *Elaphe obsoleta* (Jackson 1974, Rudolph et al. 1990b). Cavity trees are often infected with red heart fungus, *Phellinus pini* (Steirly 1957a, Jackson 1977c, Conner and Locke 1982) and are known to be actively selected by the woodpeckers (Hooper et al. 1991b, Rudolph et al. 1995). This disease, which rarely occurs in pines less than 60 years of age, decays heartwood of living pines and significantly facilitates the excavation of cavities by RCWs (Conner and Rudolph 1995a). Encroachment of hardwood under- and midstory vegetation creates a forest structure that leads to cluster abandonment by RCWs (Conner and Rudolph 1989, Loeb et al. 1992). Excessive hardwood vegetation can also lead to increased densities of other woodpecker species such as pileated woodpeckers (*Dryocopus pileatus*) that enlarge cavities making them unsuitable for the RCW. Preferred cluster sites are typically open and "park-like" in appearance, a condition that existed naturally with longleaf pine as a consequence of fire in pre-settlement forests in the South. Where SPB infestation is a concern on National Forests in Texas, management guidelines ordered by Federal Judge Robert M. Parker, following a 1988 court case, direct the maintenance of an overstory pine basal of ca. 60 sq. ft/acre with 20 to 25 ft spacing between trees.

RCW Cavity Trees as Hosts for SPB.

The process of initiation of new infestations by SPB has been investigated in considerable detail. In the past much of the research emphasis has focused on the behavior of SPB in response to natural and cultural disturbance. Lightning struck hosts were identified to be of particular significance in the process, as this disturbance is reliable in space and time. Active RCW cavity trees and managed cluster areas may also act as disturbance centers and be recognizable by SPB. There are several similarities with other types of disturbances associated with the initiation of SPB infestations.

(i) SPB may be able to readily identify cavity trees because the open wounds (resin wells), created by the woodpeckers, serve to wick host volatiles that are attractive to the insect (Payne and Coulson 1985). It is likely that active cavity trees, like lightning-struck hosts, are identified and colonized most often during periods when refuges play the dominant role in the epidemiology of the insect. Further support for the contention that host volatiles are important in locating RCW cavity trees can be found by examining response of SPB to artificial cavity inserts. Cavity inserts (Allen 1991) have been widely implemented throughout the southern US as a vital management tool to slow or reverse RCW population declines. Existing data indicate that pines with cavity inserts are not infested by SPB following installation. Only when RCWs begin to use the inserts and excavate resin wells do the trees suffer the same mortality rates as naturally excavated cavity trees (Rudolph and Conner 1995).

(ii) The cavity trees may be easily colonized and killed by SPB, because they normally occur in mature and diseased pines that have diminished defense capability. Lightning struck hosts are also thought to have weakened defenses against infestations by the insect.

(iii) The open conditions of managed woodpecker cluster areas (due to hardwood midstory removal and pine thinning) may facilitate location and infestation of cavity trees. Although the mechanism of response of SPB is not fully understood, there is evidence to suggest that forest stand conditions greatly influence dispersion characteristics of host- and insect-produced volatiles (Fares et al. 1980). Generally, the forest surrounding the cluster site has a dense hardwood midstory component that could serve as a barrier to dispersing insects.

(iv) It is also possible that the practices used for the cluster area preparation (thinning of pines and removal of hardwoods) represents a disturbance severe enough to increase vulnerability of the hosts to SPB (Jackson 1990, Mitchell et al. 1991).

(v) Cavity trees with heartwood decay and years of trauma associated with resin well excavation may resemble trees suffering from feeder-root pathogens and prolonged water logging. These conditions profoundly affect the root system (P. L. Lorio, personal communication). Also, lightning typically damages the roots of struck trees. Such trees represent refuges that SPB could easily colonize.

The Occurrence of SPB in Red-cockaded Woodpecker Cavity Trees

During the past decade, a large number of RCW cavity trees (ca. 530), occurring on National Forests in Texas, have been infested and killed by SPB. Conner et al. (1991a) reported that SPB was responsible for more than 50% of cavity tree mortality (single tree infestations) on National Forests investigated in Texas. This mortality occurred during periods when insect populations were at non-epidemic levels. Most of the cavity trees infested by SPB were active. However, some cavity trees were inactive.

The SPB activity that occurred on the National Forests in Texas between 1991 and

1993 resulted in the infestation and subsequent death of numerous cavity trees on sites where RCW management guidelines had been implemented within the woodpecker cluster area. On such sites, one fourth to nearly one half of all active RCW cavity trees (7 of 30 in 1992 and 13 of 30 in 1993) on the north side of the Angelina National Forest were infested and killed by SPB. A similar phenomenon was observed on the Neches Ranger District of the Davy Crockett National Forest. The number of active cavity trees destroyed by SPB during these two years was so large that simple chance appears to be an unlikely explanation. Although cavity loss may be associated with increases in local SPB populations, intensified RCW management may be contributing to conditions conducive to colonization of active cavity trees.

SOUTHERN PINE BEETLE IN RED-COCKADED WOODPECKER CLUSTER AREAS

In recent decades most RCW populations have been declining. In some populations the declines have been severe (Baker 1983b, Conner and Rudolph 1989, Costa and Escano 1989). Several causes for these declines have been identified (Hooper 1988, Conner and Rudolph 1989, Walters et al. 1992a) and intensive management guidelines for cluster areas are being developed to address the problem. Encroachment of midstory vegetation (due to changes in the historic fire regime), lack of high quality potential cavity trees (due to shortened rotations), and high rates of cavity turnover (due in part to SPB caused mortality) necessitate management intervention. Midstory control (using mechanical methods), installation of artificial cavities, and thinning of pines to reduce SPB hazard are the methods used to reverse RCW population declines. The efficacy of these management approaches has been demonstrated in National Forests in Texas. Implementation of the approaches has reversed severe population declines. Continued human intervention through implementation of intensive forest management is crucial to the survival of RCW, given the current condition of habitat and demographic state of woodpecker populations. Research directed to understanding the process of SPB infestation of managed RCW cluster sites is of paramount importance to the goal of protecting this endangered species.

SOUTHERN PINE BEETLE INFESTATION IN RED-COCKADED WOODPECKER FORAGING HABITAT

In addition to cluster areas where cavity trees are located, red-cockaded woodpecker management guidelines also consider woodpecker foraging habitat. Suitable RCW foraging habitat is defined to include pine and pine-hardwood stands 30 years or older. This habitat is judged to be available to the red-cockaded woodpecker if it is within 0.5 miles and continuous with the cluster area (active or inactive) or a recruitment stand. Cluster areas are also used as foraging habitat by RCWs. Shelterwood cuts with a basal area of 30 sq. ft/acre of pine or greater can be considered suitable foraging habitat as well (USFWS 1985).

Considerable research has been directed toward identification of site and stand conditions associated with pine infestation by SPB (Lorio 1980, Branham and Thatcher 1985, Mason et al. 1985). Several different methods have been devised to classify stands according to SPB hazard and risk. Each rating system has novel features but all involve integration of a subset of variables: tree species, radial growth, height, and DBH; stand basal area (pine, hardwood, total), species composition, site index, and degree of crown closure; and landform classification. In general, stands containing mature loblolly or shortleaf pines with high basal area and stagnant radial growth are considered to be high hazard. These stands are suitable for SPB infestation growth. Most of the annual timber volume loss attributed to the insect occurs in high hazard stands during Spring, Summer and Fall seasons.

A significant portion of RCW foraging habitat for a group of woodpeckers consists of mature older age-class loblolly and shortleaf pine forest. If these stands are not maintained in low or medium SPB hazard, infestation growth can occur. It is noteworthy that because of the low basal area and wide spacing of trees, cluster area habitat is generally not well-suited for SPB infestation growth. However, for the reasons outlined above, the cavity trees can function as refuges for dispersing SPB in the same manner as lightning-struck hosts. As such, the cavity trees, as well as lightning-struck pines, may function as epicenters for the initiation of infestations. The proximity of the cavity trees and potentially high hazard foraging habitat creates a forest environment that can be conducive to SPB outbreaks.

CONCLUSIONS

A large number of RCW cavity trees on sites prepared in accordance with management guidelines have been infested and killed by the SPB (Conner et al. 1991a, Conner and Rudolph 1995a). It is possible that aspects of cluster area management may create forest conditions that enhance the likelihood for colonization of active cavity trees by the insect. RCW management guidelines have not been considered in the context of the seasonal behavior of SPB. We suggest that active cavity trees can be attractive centers for SPB and that these trees may serve as epicenters for colonization by the insect. Our suggestions are based on observations of the behavior of SPB in relation to lightning struck trees and information on the response of the insect to other types of disturbance. We further suggest that subsequent growth of infestations in RCW cluster areas is unlikely because of the low basal area and wide spacing of trees. However, the proximity of infested cavity trees to high hazard stands could result in insect outbreaks in adjacent foraging habitat.

Our interpretation of the relation of SPB and RCW and the potential influence that forest management guidelines have on this association leads to several important questions that can be answered only through further research. Specific questions include the following:

(i) do RCW cavity trees serve as resin wicks that attract SPB,

(ii) does mid-story hardwood removal and thinning of pines in colonies increase the likelihood of SPB infestation by serving as a trap/sink for SPB that increases local beetle abundance,

(iii) does recent site disturbance associated with RCW management increase SPB abundance and the incidence of RCW cavity tree infestation, and

(iv) are multiple-tree infestations that develop in the adjacent RCW foraging area associated with SPB activity within woodpecker colonies.

ACKNOWLEDGMENTS

We acknowledge and thank A. M. Bunting for assistance in preparation of the manuscript of this paper. This paper is TA No. 31258 of the Texas Agricultural Experiment Station.

Potential Impact of Southern Pine Beetle Infestations on Red-cockaded Woodpecker Colonies on the Noxubee National Wildlife Refuge: Hazard Rating Comparisons and Spot-Growth Potentials

T. Evan Nebeker[1], Matthew Pelligrine[2], Robert A. Tisdale[1], and John D. Hodges[2]
[1]Department of Entomology and Plant Pathology, [2]Department of Forestry
Mississippi State University, Mississippi State, MS 39762

ABSTRACT: Sixteen active red-cockaded woodpecker (RCW), *Picoides borealis*, colonies on the Noxubee National Wildlife Refuge were evaluated for potential impacts of southern pine beetle (SPB), *Dendroctonus frontalis*, using hazard/risk rating models, and infestation-growth models. Predictive models suggested that the colonies were of low risk to SPB attack. Hazard rating, however, was uncertain and varied with the choice of model. Infestation-growth models indicated little danger of infestations spreading rapidly within the colonies (assuming low pine basal area and endemic SPB population levels). However, even a small scale infestation could extirpate the RCW from a colony site. In light of these results, the best course of action for the refuge is to continue its active SPB suppression program and its program of maintaining low pine basal area within the colony sites.
KEYWORDS: red-cockaded woodpecker, southern pine beetle, hazard rating, risk rating

The Problem

The southern pine beetle (SPB) is the most serious pest of southern pine forests (Payne 1980; Blanche et al. 1983; Flamm et al. 1988). The worst recorded outbreak occurred from 1982 through 1985, when beetles destroyed approximately 2,910,200 m^3 of timber valued at $92 million in the national forests alone (Kelly et al. 1986). In addition, SPB has been identified as a major threat to RCW cavity trees (Conner et al. 1991a; Rudolph and Conner 1995).

The SPB has been studied on the Noxubee National Wildlife Refuge (NWR) since the late 1970's. Epidemic populations were reported on Noxubee in 1982 and in 1986 (Connor and Wallace 1982; Mangini et al. 1988). Mangini et al. (1988) reported epidemic population levels of SPB on Noxubee and expressed concern over the potential loss of red-cockaded woodpecker (RCW) cavity trees. Forest Pest Management (Forest Health) evaluated 10,117 ha of pine and mixed pine forest on the Noxubee and determined that on average, 2.1 SPB spots occurred per 40.5 ha of host type indicating that beetle problems were likely to continue in the future (Mangini et al. 1988). In 1989, when this study was initiated, only 8 spots warranted control on Noxubee. No SPB spots were reported in 1990, although *Ips* spp. infestations were common. Black turpentine (*D. terebrans* Olivier) and *Ips* beetles attacked some of the colony trees (sites 2, 5, 10) during the summer of 1990. One tree on site 5 was heavily infested with turpentine beetles and later died although it is unlikely that beetles were the actual cause of death.

SPB activity increased in 1991 on the Noxubee and over 120 spots were controlled. Two cavity trees in colony 4 were infested late in the fall and later died, but were not removed. Lightning strikes were a common occurrence within the colonies. Two trees in colony 9 and 1 tree in colony 3 were struck by lightning, became infested, and were later removed (D. M. Richardson pers. commun.).

Successful SPB control has been hampered by a lack of understanding of the basic interactions that allow the SPB, and its associated fungi, to overcome the tree's natural defenses, to successfully colonize and reproduce, and to finally kill the tree (Nebeker et al. 1988). When present in epidemic populations, SPB can successfully attack and kill healthy, vigorous trees. Small (endemic) SPB populations concentrate their attacks on weakened trees. Overstocked stands, and stands showing reduced radial growth, are the most susceptible to attack in the Coastal Plain (Payne 1980; Hicks 1980; Blanche et al. 1983). SPB will attack all species of southern pine, but prefer loblolly (*Pinus taeda* L.) and shortleaf (*P. echinata* Mill.). Only living trees are attacked. Mature stands of high basal area (over 27 m^2/ha) are the most prone to infestation and potential spot growth.

Hazard/Risk Rating Systems

Hazard and risk rating systems are a preventative approach to insect outbreaks, based on the ability of the observer to recognize stand and environmental factors that may lead to infestation. The need for such systems was colorfully stated by Lorio and Sommers (1981) who pointed out that outbreaks of SPB result in a "flurry of direct control activities with at best doubtful benefit... preventative medicine would be much preferable to the rough and ready first aid applied during outbreaks." Hazard and risk rating systems allow the manager to rate a stand's susceptibility to attack, as well as its future hazard condition. Such knowledge allows the manager to determine which stands require silvicultural treatment to reduce the probability of infestation, to set priorities among the stands requiring treatment, and to assess outbreak and loss potential (Mason et al. 1985).

The terms "hazard" and "risk", often used interchangeably, have distinct meanings regarding hazard rating systems. "Hazard" refers to the relative susceptibility of an area to insect infestations based on site, stand and host factors, while "risk" refers to the probability the infestation will occur within a given time span (Billings et al. 1985) or the probability the infestation will grow (Hedden, 1985).

Researchers develop hazard rating systems by measuring site and stand variables and then attempting to establish associations between these variables and SPB attack. Some relationships, such as the influence of soil moisture on the probability of SPB infestation, have proven difficult to model (Hicks et al. 1987).

The data inputs for hazard rating systems can be generated from ground cruises or aerial imagery. Generally, if the timber has been cruised for stand inventory purposes, no further cruising is required since these records usually include all pertinent information. The use of appropriate computer software can simplify analysis (Mason et al. 1985).

Aerial imagery has the advantage of providing nearly a 100 percent sample of stand conditions (Mason and Bryant 1984), and is faster than ground cruising. Species composition, height, diameter, basal area and land form type can be detected from large scale, low altitude, color infrared aerial imagery and input into hazard-rating systems (Mason et al. 1981).

Dozens of hazard rating systems have been developed for the SPB and other pests. The choice of a hazard rating system depends largely on the specific geographic or physiographic region, the availability of the model, the landowner's management objectives, and the input requirements of the model (Mason et al. 1985). The choice of model may have economic implications as well. To prevent timber loss, managers may decide, based on hazard rating systems, to harvest or thin stands before they have reached their economically optimum rotation age, and thus lose potential income (Honea et al. 1987).

Mason et al. (1985) discussed many of the considerations in evaluating a hazard rating system. To be effective, hazard rating systems should allow the manager to "identify the smallest land unit on which the greatest loss due to SPB might be expected." Hazard rating systems should not classify all stands, where SPB spots occur, as high hazard and thus overestimate the number of spots requiring immediate management. Judgment of a model's performance should be based on its overall accuracy for a large area of timber and beetle population conditions, rather than for a few stands containing a few SPB spots. The authors contended that "stand ratings should be based on a comparison of spot occurrence, tree mortality, and resource value." Hazard rating results may be overridden by factors beyond the model's power of prediction such as lightning strikes or logging damage. High risk "pockets" may occur in low risk stands.

Our study is based on a similar study performed in the Piedmont NWR and the Hitchiti Experimental Forest in the Georgia Piedmont by Belanger et al. (1988). The Georgia Piedmont sites consisted of approximately 15,054 hectares and supported 34 active colonies. Epidemic SPB outbreaks had occurred as recently as 1986. To hazard and risk-rate the colonies, the authors used various systems which had been developed for the Piedmont region. Susceptibility to attack (hazard) was determined using BEST (Belanger et al. 1981) and PIEDMONT RISK (Hedden 1984). Probability of attack (risk) and potential tree mortality were estimated using CLEMBEETLE (Hedden 1985), which predicts beetle-caused damage over time. The authors projected over a 30-year period based on typical SPB populations in central Georgia for endemic levels of 2 spots per 405 hectares of pine and for epidemic levels of 5 spots per 405 hectares.

Mitchell et al. (1991) performed a similar study in the Bannister Wildlife Management Area of the Angelina National Forest in the lower Coastal Plain of Texas. The site supported 7 colonies, 6 of which were located in high risk blocks as determined using U.S. Forest Service Continuous Forest Inventory data. Hazard rating was accomplished for a 400 m radius around each colony using TX HAZARD (Mason et al. 1981) and NF RISK. NF RISK is based on data from the Continuous Inventory of Stand Conditions (CISC) data base for national forests (Lorio and Sommers 1981). The stands were also hazard-rated within a 7,287 hectare area using the Texas Forest Service Grid Hazard System (Billings et al. 1985). Risk-rating was accomplished using the Coastal Plain version of CLEMBEETLE (Hedden and Belanger 1985).

The basic objective of our study was to determine the likelihood of SPB attack, spot spread and losses of trees in RCW colonies on the Noxubee National Wildlife Refuge. To do this, we compared various hazard rating models and models, described above, that predicted beetle-caused damage over time.

STUDY AREA

This study was conducted on the Noxubee National Wildlife Refuge in eastern Mississippi in Noxubee, Winston and Oktibbeha Counties. Noxubee is 19,020 ha, with approximately 10,117 ha of pine and mixed pine/hardwood forest. Management on Noxubee was designed to provide habitat for native wildlife. When the study was initiated in 1989, 16 active and approximately 50 abandoned RCW colonies existed on Noxubee. By 1990, 2 more active colonies were located, although these were not included in the study. Today (1993) there are 32 active colonies (Richardson and Stockie 1995). Active colonies are intensively managed for the RCW, while efforts are being made to revitalize abandoned colonies through vigorous mid story hardwood control.

MODELS TESTED

The data utilized and methods of collection for this study were presented in Nebeker et al. (1995). For comparison, hazard/risk rating systems chosen for this study were similar to those used in previous studies in Georgia (Belanger et al. 1988) and Texas (Mitchell et al 1991). Piedmont models have not been applied historically to the Coastal Plain due to differences in site conditions. However, it was possible that these differences were irrelevant, partly because forest management favorable to the RCW resulted in little variation among stands, and partly because the biology of the SPB may not be effected by regional differences. Rating systems initially designed for Mississippi were also used. Comparisons with the Piedmont were made using BEST and LAND MANAGER'S models, and PIEDMONT RISK to determine susceptibility to attack. Probability of attack and tree mortality were estimated using CLEMBEETLE to predict beetle-caused damage over time.

BEST model was developed specifically for the Georgia Piedmont by Belanger et al. (1981). The model rates susceptibility to attack through a scoring system based on soil and stand characteristics, and is based on the following suppositions about the Piedmont: high hazard stands are characterized by a high proportion of shortleaf pine; are over-mature, over-stocked and slow growing (poor radial growth); and are on poorly drained sites with high clay content in the surface and subsurface soil horizons. Thus, the driving variables of the model include soil texture, depth of the A horizon, and stand parameters, such as species composition (percent loblolly in the pine component), average radial growth and percent live crown. The model was originally tested in both the Upper and Lower portions of the Georgia Piedmont but was found effective only in the Upper Piedmont. LAND MANAGER requires less soils input and is not quite as accurate as BEST but it is easier to apply.

PIEDMONT RISK (Hedden 1984) was developed specifically for shortleaf pine growing on the Piedmont of South Carolina and Georgia. The model provides an estimate of spot occurrence (hazard), spot spread (risk), and expected economic loss[1]. Model inputs for risk include slope, clay content of the surface soil, percent shortleaf pine in the stand, and radial growth (optional). Inputs for hazard include pine basal area/acre. Expected loss is based on the products to be harvested (sawtimber or pulpwood) and local stumpage values.

CLEMBEETLE (Hedden 1985) is a simulation model, applicable to both the Piedmont and Gulf Coastal Plain, which predicts beetle-caused damage over time. The current version ranks spots per 40.5 ha of host type. Inputs for the Piedmont include total basal area, density, site index, stand origin (natural or artificial), predominant pine species (loblolly or shortleaf), percent slope, clay content of the surface soil, and disturbance (logging, hail, fire, etc.). Variables for the Coastal Plain are similar, but land form (bottom, ridge or sideslope), age and percent pine in the stand substitute for stand origin, pine species, percent slope and clay content of the surface soil. Hedden compared simulated results of volume lost in the Piedmont with those published by Price and Doggett (1982). CLEMBEETLE tended to over predict volume killed by SPB. The poorest predictions were for Arkansas and Tennessee, while the most reliable predictions were for Georgia, North and South Carolina and Virginia.

The coastal plain version of CLEMBEETLE was tested throughout the Coastal Plain and has proved highly effective (Hedden and Belanger 1985). On the Coastal Plain, probability of infestation increased with total basal area, age and percent pine. Probability of infestation was higher for disturbed stands and higher on ridges and bottoms, than sideslopes. Overstocked stands with basal areas of 46 m^2/ha were viewed by the model as having the same probability of infestation as disturbed stands.

The hazard rating system developed in Texas (Mason et al. 1981) and modified by Billings was used to evaluate stands as high, moderate or low hazard. This method was used by Mitchell (Mitchell 1987, Mitchell et al. 1991) for hazard rating of RCW colonies in Texas.

The MISSISSIPPI A & B systems, developed by Nebeker and Honea (cited in Mason et al. 1985), are particularly applicable to Mississippi and Alabama. Much of the infestation data used in developing the model was based on work done in Mississippi, Louisiana and Arkansas. MS A provides the manager with a hazard classification based on pine basal area and radial growth. When tested, the model classified 68 out of 111 SPB spots as being high or very high hazard (Mason et al. 1985). The MS B model is similar, but takes more data into consideration. Inputs include pine and total basal area per acre, stand age, stand density, and site index.

Infestation Growth Models.

Infestation growth models predict the probability of spot spread once an infestation has occurred, based on the numbers of trees infested and stand parameters. Since no infestations were occurring in the colony sites in 1990, the models were run by inputting infestations of various sizes. For example, HOG Model (Taha and Stephen 1984, Lih and Stephen 1986) was run by inputting the stand parameters for each colony and for the average colony, and then varying the infestation size from 1 to 40 trees. This provides short-term (1-3 months) predictions of the growth of SPB infestations and of expected timber losses in currently infested stands. The Texas infestation growth models developed by Billings and Hynum (1980) were also used.

RESULTS AND DISCUSSION

Hazard Rating for RCW Colonies

LANDMANAGER ranked 10 of the 16 colonies (63%) as low hazard, and the rest as moderate hazard, while BEST ranked all the colonies as low hazard **(Table 1)**. Thus, the 2 models were in agreement 63% of the time. Belanger et al. (1981) found that in the Piedmont, the LANDMANAGER was 13% less accurate than the BEST in classifying

[1]Hedden in both the PIEDMONT RISK (1984) and CLEMBEETLE (1985) models used the terms "risk" for spot occurrence and "hazard" for spot spread. The terms are presented here to be consistent with the common usage as defined by Billings et al. (1985).

Table 1. Comparison of different southern pine beetle hazard rating models applied to red-cockaded woodpecker colonies at the Noxubee National Wildlife Refuge, Mississippi.

Colony No.	Piedmont Models				Coastal Plain Models			
	Best	Land Manager	Piedmont Risk	Clembeetle	Clembeetle	Texas	MS A	MS B
1	L	L	L	L	E	L	VL	VL
2	L	L	L	M - H	E	M	L	M
3	L	L	L	M - H	E	L	VL	L
4	L	M	L	M - H	E	M	M	M
5	L	M	H	M - H	E	L	VL	L
6	L	L	M	E	E	M	L	M
7	L	L	L	E	E	L	L	M
8	L	M	L	E	E	L	L	M
9	L	L	M	M - H	E	M - H	M	M
10	L	L	L	E	E	L	L	M
11	L	M	M	M - H	E	M - H	M	M
12	L	M	H	E	E	M - H	L	M
13	L	L	L	M - H	E	H	M	M
14	L	L	L	M - H	E	M	L	M
15	L	M	L	M - H	E	M	M	M
16	L	L	M - H	E	E	L	L	M
Mean	L	L - M	L - M	M - H	E	M	L	M

*L=Low, M=Moderate, H=High, E=Extreme

study plots. However, LANDMANAGER may be more useful and appropriate for the Coastal Plain because it has fewer soil inputs specific to Piedmont conditions.

Using the Noxubee data in LANDMANAGER, it was found that average radial growth was the most important variable. Colonies with low hazard rankings tended to demonstrate better radial growth (13 mm) than colonies with moderate rankings (7.8 mm). Percent live crown was the next most important input. As previously mentioned, in the Piedmont a large live crown can be a liability. Trees in low hazard stands tended to have smaller crowns (48%) than trees in moderate hazard stands (51%). Percent loblolly (in the pine component) was the next most important input. In the Piedmont, it is desirable to have a greater proportion of loblolly than shortleaf pine due to the incidence of littleleaf disease. Low hazard stands tended to have more loblolly pine (98%) than moderate hazard stands (85%). Surface depth was the least important input. In the Piedmont, deep soils with high clay content encourage the development of littleleaf disease. Low hazard stands tended to have shallower soils (0.75 cm) than moderate hazard stands (1 cm).

Belanger et al. (1988) applied BEST to the RCW colonies in the Piedmont, and found that 71% of the colonies were of low hazard. The rest were of moderate hazard. Differences in the results of the Belanger et al. (1988) study and the current Noxubee study may be due to differences in live crown ratio (Nebeker et al. 1995). Noxubee sites had 4% more live crown than Belanger's piedmont sites. This would cause Noxubee sites to rank as higher hazard than piedmont sites, because, in the Piedmont, where the model was developed, a large live crown can be a liability if the root system is poorly developed due to littleleaf disease (R. P. Belanger pers. commun.).

PIEDMONT RISK rated 11 of the 16 stands (69%) as being of low hazard, 3 as moderate (19%), 2 as high (12%), and 1 as moderate to high (6%). Belanger et al. (1988) applied PIEDMONT RISK to the RCW colonies in the Piedmont and found that 71% of the colonies were of low hazard, 23% were moderate hazard, and 6% were high hazard. The difference in the two locations is difficult to infer. The 2 high hazard sites at Noxubee were the result of an exceptionally high shortleaf pine component (67%) in colony 5 and a steep slope in colony 12. PIEDMONT RISK also provided a risk rating for spot spread based on pine basal area. All colony sites, except 1, were ranked as low risk. Colony 13 had an unusually high basal area of 23.9 m^2/ha.

and was ranked as moderate.

The low hazard sites had low percentages of shortleaf, little slope, and little clay in the surface soil. However these variables are very specific to the Piedmont where high clay content, poor drainage and an abundance of shortleaf pine result in littleleaf disease, which predisposes the trees to SPB infestation. They have little relevance to Noxubee which is mostly flat, loblolly pine stands.

CLEMBEETLE is unique in having simulation programs suitable to both the Piedmont and Coastal Plain. All coastal plain colonies ranked extreme hazard, with 1.3 beetle spots per 40.4 ha of host type. Risk ranking of all colonies was low, with an average spot growth of 3.4 trees, and expected losses were low to moderate with an average expected loss of 4.4 trees over a 45 day period.

The overall results using the Piedmont version were similar to those of the Coastal Plain version. Mean hazard for the colonies was moderate to high, with an average of 0.5 beetle spots per 40.4 ha of host type. All colonies ranked as being of low risk, with an average spot growth of 3.4 trees. Expected losses were low with an expected loss of 1.5 trees.

Belanger et al. (1988) used the Piedmont version to risk rate the Piedmont colonies and found the colonies to be of low hazard. Predicted spot growth was low, at 3 trees per 40.5 ha. Projected losses were extremely low at 2 trees per ha. Although Noxubee and Piedmont sites are fairly similar in regards to the inputs for the model, the results of the risk rating differed. Noxubee sites were of moderate to high risk, whereas piedmont sites were of moderate to low risk. Again, the differences were probably caused by colony 5, with its high shortleaf pine component, and colony 12 with its steep slope.

Mitchell (1987) used the coastal plain version of CLEMBEETLE to risk rate 7 RCW colonies in the Texas Lower Coastal Plain. One colony was classified as low risk, 5 as moderate, and 1 as high. Noxubee sites had higher total basal areas (20 m^2/ha) than Texas (16 m^2/ha) sites. The other inputs were similar, except possibly for land form. Unfortunately, landforms were not discussed in Mitchell et al. (1991) or in previous references (Mitchell, 1987; Kulhavy et al. 1988).

The CLEMBEETLE model was developed using "dummy" variables, which make the model difficult to analyze. Several factors come into play in calculating hazard. For the Coastal Plain, these include total basal area, land form, age, percent pine, disturbance, and the interaction of disturbance and total basal area. The variables for natural stands in the Piedmont include pine species, percent slope, soil texture, and disturbance. The formula for planted piedmont stands also includes site index. Risk for both regions is dependent on only 2 variables, trees per hectare and total basal area, while loss is dependent on all variables used to calculate risk and hazard.

Disturbance is considered the most important variable in calculating hazard and expected loss (R.L. Hedden pers. commun.). Major causes of disturbance in the woodpecker colonies are thinning, lightning strikes, wind related effects (wind throws and snaps) and SPB. Logging might be considered a minor disturbance if done properly. Lightning is a major disturbance, but if the trees are salvaged, the effects (especially their attractiveness to SPB) are drastically reduced. With this in mind, CLEMBEETLE was run a second time, without disturbance. This trial showed how changing 1 important variable in a hazard/risk model can drastically alter the results (**Table 2**). For the Coastal Plain, mean hazard dropped from extreme to high, and mean loss dropped from moderate to low. The results for the Piedmont were even more drastic. Mean hazard dropped from extreme (1.8) to low (1.4) and mean loss dropped from moderately high (10.3) to low (0.5).

Land form was the next most important variable affecting hazard and loss for the Coastal Plain and was also altered (**Table 3**). Using mean values for a disturbed colony site in the Coastal Plain, it was found that all landforms were of extreme risk. However a bottom was of highest hazard (1.66 spots per 40.4 ha), a ridge was of moderate hazard (1.3 spots), and a slope was of the lowest hazard (0.57). Expected losses were from low to moderate and again ranked highest for the bottom and lowest for the slope. Land form had no apparent affect on spot growth.

The Piedmont version was tested utilizing a "dummy" set of variables. Different results were found by trying different combinations of the major inputs in the equation (**Table 4**). The results showed that an extreme hazard Piedmont site would have an abundance of shortleaf, over 10% slope, over 28% clay

Table 2. Results of running southern pine beetle hazard rating model, CLEMBEETLE, coastal plain version with and without disturbance for the 16 red-cockaded colonies on the Noxubee National Wildlife Refuge, Mississippi.

	Disturbance				No Disturbance			
Colony No.	Number Spots/ 40.4 ha	Hazard* Rating	Number Trees Killed	Potential Loss*	Number Spots/ 40.4 ha	Hazard Rating	Number Trees Killed	Potential Loss*
1	0.6	E	0.4	L	0.06	M	0.04	L
2	1.6	E	1.8	L	0.14	M - H	0.22	L
3	0.5	E	1.3	L	0.06	M	0.14	L
4	1.3	E	10.8	M - H	0.17	H	1.51	L
5	0.5	E	1.0	L	0.06	M	0.12	L
6	1.0	E	3.7	L - M	0.15	M - H	0.52	L
7	1.0	E	5.6	M	0.13	M - H	0.68	L
8	0.8	E	6.5	M	0.11	M - H	0.81	L
9	1.2	E	1.8	L	0.16	H	0.25	L
10	1.0	E	0.4	L	0.11	M - H	0.04	L
11	1.2	E	10.2	H	0.18	H	1.48	L
12	0.5	E	1.4	L	0.07	M - H	0.19	L
13	1.9	E	13.6	M - H	0.29	H	2.04	L
14	0.9	E	4.9	M	0.11	M - H	0.63	L
15	1.4	E	0.9	L	0.18	H	0.11	L
16	1.2	E	7.0	M	0.15	H	0.87	L
Mean	1.3	E	4.4	L - M	0.17	H	0.56	L

*Hazard and potential loss of additional trees within the colony areas are L = Low M = Moderate H = High or E = Extreme

Table 3. Results of running southern pine beetle hazard rating model, CLEMBEETLE, coastal plain version, using different landforms for the 16 red-cockaded woodpecker colonies on the Noxubee National Wildlife Refuge, Mississippi.

LAND FORM	NUMBER SPOTS/40.4 ha	HAZARD RATING	NUMBER TREES KILLED	POTENTIAL LOSS
RIDGE	1.30	E	4.4	L - M
SLOPE	0.57	E	1.9	L
BOTTOM	1.66	E	5.6	M

content in the soil and would be disturbed. A low hazard site would demonstrate the opposite characteristics: loblolly pine would be the primary species, slope would be under 10%, clay content would be under 28%, and the site would be undisturbed.

Although the MS A and MS B models were both developed specifically for the Mississippi Coastal Plain, the results from the 2 models were quite different. MS A classified 8 of the colonies as low hazard, 3 as very low hazard and the remainder as moderate hazard. MS B classified only 3 colonies as low or very low hazard. The rest were classified as moderate hazard. The 2 models agreed on only 6 of the 16 colonies as having the same hazard rating.

For MS A, pine basal area was the most important of the 2 inputs. Colonies which were rated as low hazard tended to have low basal areas (13 m^2/ha), while moderate hazard colonies had higher basal areas (20.3 m^2/ha). Low hazard colonies also had higher radial growth over the past 10 years (28 mm) as opposed to moderate hazard colonies (17 mm).

The MS B equation, applied to Noxubee data, ranked pine basal area as the most important variable, followed by site index and density, age, and total basal area. Low hazard stands tended to have lower basal areas and densities and were younger than moderate hazard stands. Low hazard stands also tended to have lower site indices (88) than moderate hazard stands (92). The relationship between site index and SPB infestation is poorly

Table 4. An analysis of the southern pine beetle hazard rating model, CLEMBEETLE, piedmont version, by varying major inputs in the model for the 16 red-cockaded woodpecker colonies on the Noxubee National Wildlife Refuge, Mississippi.

Species	Slope	Clay	Disturbed	Spots/ 40.4 Ha	Hazard	Trees Killed	Potential Loss
Shortleaf	>10	>28	YES	9.25	E	31.13	H
Shortleaf	<10	>28	YES	4.21	E	14.17	M-H
Shortleaf	>10	>28	NO	3.04	E	10.22	M-H
Shortleaf	>10	<28	YES	2.47	E	8.32	M-L
Loblolly	<10	>28	YES	1.79	E	6.02	M
Shortleaf	<10	>28	NO	1.33	H	4.49	M
Loblolly	>10	<28	NO	1.28	H	4.31	L-M
Shortleaf	<10	<28	YES	1.08	H	3.64	L-M
Loblolly	>10	<28	YES	1.04	H	3.49	L-M
Shortleaf	>10	<28	NO	0.77	H	2.60	L-M
Loblolly	<10	>28	NO	0.56	M-H	1.87	L
Loblolly	<10	<28	YES	0.45	M-H	1.52	L
Shortleaf	<10	>28	NO	0.33	M	1.13	L
Loblolly	>10	>28	NO	0.32	M	1.08	L
Loblolly	>10	>28	YES	0.32	M	1.08	L
Loblolly	<10	<28	NO	0.14	L	0.47	L

understood. Lorio and Sommers (1985) suggested that prolonged rapid growth increases the demand for photosynthates, limiting the production of oleoresin and hence increasing susceptibility to SPB.

TX HAZARD evaluated 7 of the 16 colonies (44%) as low hazard, 6 (38%) as moderate, and 1 (6%) as high hazard. The remaining 2 sites were of varied topography and were evaluated as moderate-high (12%) depending on the type of topography selected (these sites will be considered as moderate for purposes of comparison with previous work in Texas). The model depends heavily on topography and rates upland sites as being of low hazard, and bottomland sites as being of high hazard.

Mitchell et al. (1991) used TX HAZARD to hazard rate 7 colonies in the Texas Lower Coastal Plain. They found 4 colonies (57%) to be in low hazard stands, 2 (29%) in moderate and 1 (14%) in extreme hazard stands. Differences in the results at the Noxubee and Texas sites are difficult to infer because both locations have similar pine basal areas and tree height. Land form in Texas was taken using U. S. Geological Survey maps; the landform was averaged over the sites resulting in a hazard rating for SPB. Thus, the difference must be in land form.

TX HAZARD depends heavily on topography and rates upland sites as lower hazard than bottomland sites. Sites with low pine basal area and small trees (height or diameter) are ranked as least hazardous. Low hazard stands had less pine basal (12.5 m^2/ha) area than moderate hazard stands (18 m^2/ha). Low hazard stands also had shorter trees (29 m) than moderate hazard stands (32 m).

Infestation growth models were applied to Noxubee by taking the data for each colony site and running simulations with varied infestation sizes. The model developed by Billings and Hynum (1980) depended largely on basal area. The higher the basal area, the greater the number of trees killed and infested **(Table 5)**. Thus, the colonies with the highest basal areas were in greatest danger of losses to SPB.

The HOG model was run by inputting the stand parameters for each colony and for the average colony, and then varying the infestation size from 1 to 40 trees. The maximum number of trees killed grew in a fairly linear fashion as infestation size increased indicating that infestation size was the most important consideration. The importance of the other inputs was difficult to infer. At the maximum infestation size of 40 trees, the maximum number of trees killed varied from 82 to 146. The average number of trees killed was 93. The colony with the greatest number of trees killed had an unusually high basal area of 24 m^2/ha. Generally, infestations of less than 20 trees failed to grow

Table 5. Expected spread of southern pine beetle induced tree mortality in 16 red-cockaded woodpecker colonies on Noxubee National Wildlife Refuge, Mississippi, based on Billings & Hynum infestation growth model.

Colony	Stand BA (M^2/Ha)	Initial Infestation Size (# of trees) 5	10	20	30	50	75	100
		Expected spread (Additional trees killed)						
1	8.7	0	0	0	2	9	16	24
2	18.0	0	0	5	12	24	39	54
3	14.0	0	0	0	2	9	16	24
4	24.9	0	0	5	12	24	39	54
5	16.6	0	0	5	12	24	39	54
6	25.8	0	0	5	12	24	39	54
7	18.2	0	0	5	12	24	39	54
8	19.3	0	0	5	12	24	39	54
9	23.7	0	0	5	12	24	39	54
10	15.0	0	0	0	12	24	39	54
11	26.5	0	2	12	21	39	62	84
12	22.8	0	0	5	12	24	39	54
13	28.1	0	2	12	21	39	62	84
14	20.0	0	0	5	12	24	39	54
15	19.3	0	0	5	12	24	39	54
16	18.6	0	0	5	12	24	39	54
Mcan	20.0	0	0	5	12	24	39	54

after 40 days, while infestations of 25 or more continued to grow. HOG is believed to be a poor infestation growth model for RCW colonies due to the low total basal areas in the sites. The authors of the model advise caution when interpreting the results for sites with low basal areas (M. Lih, pers. commun.).

Comparison Of Hazard/Risk Rating Models

Hazard ratings provided by the various models for the Noxubee colonies ranged from low to extreme (**Table 1**). Although stand conditions were similar due to RCW management, other environmental factors such as soil characteristics were significantly different among colony sites.

Unlike the hazard rating models, the risk rating models were in almost perfect agreement. PIEDMONT RISK and both versions of CLEMBEETLE rated all colonies as low risk except for colony 13, which PIEDMONT RISK rated as moderate.

Appropriate hazard rating models for use in the RCW colonies at Noxubee should incorporate factors typical of the mature, low density pine stands on Noxubee. Disagreement (**Table 6**) in the results of the hazard rating models indicates that models used in this study may not incorporate these factors or may not weigh them properly. Potential factors include total and pine basal area, site index, age and radial growth. Coastal Plain models, used in this study, all incorporate at least 2 of these factors. Field testing of the models on actual colony sites or on stands manipulated to simulate colony sites would be necessary to determine which model is best for use on Noxubee.

CLEMBEETLE was the only risk rating model used which was appropriate for the Coastal Plain. Thus, it cannot be determined whether it is the best choice for RCW colonies on Noxubee. As for infestation growth models, the Billings and Hynum model is preferable over HOG. The latter was not developed to deal with the low basal area conditions found in the RCW colonies.

Comparison of Regional Hazard/Risk Rating Studies

The results of the Noxubee and Belanger et al. (1988) studies were fairly similar. LANDMANAGER ranked Noxubee colonies as low to moderate hazard, while BEST ranked all the colonies as low hazard. Belanger applied BEST to the RCW colonies in the Piedmont, and also ranked the colonies as low to moderate

Table 6. Agreement[1] among results of piedmont and coastal plain southern pine beetle hazard models applied to the red-cockaded woodpecker colonies at the Noxubee National Wildlife Refuge, Mississippi.

COMPARISON	AGREEMENT[1] (%)	SIMILARITY
BEST vs. CLEMBEETLE CP[2]	0	VERY DISSIMILAR
BEST vs. TEXAS	44	DISSIMILAR
BEST vs. MS A	69	SIMILAR
BEST vs. MS B	19	VERY DISSIMILAR
LANDMANAGER vs. CLEMBEETLE CP[2]	63	SIMILAR
LANDMANAGER vs. TEXAS	57	SIMILAR
LANDMANAGER vs. MS A	63	SIMILAR
LANDMANAGER vs. MS B	44	DISSIMILAR
PIEDMONT RISK vs. CLEMBEETLE CP[2]	0	VERY DISSIMILAR
PIEDMONT RISK vs. TEXAS	50	DISSIMILAR
PIEDMONT RISK vs. MS A	57	SIMILAR
PIEDMONT RISK vs. MS B	38	DISSIMILAR
CLEMBEETLE P vs. CLEMBEETLE CP[2]	38	DISSIMILAR
CLEMBEETLE P vs. TEXAS	44	DISSIMILAR
CLEMBEETLE P vs. MS A	38	DISSIMILAR
CLEMBEETLE P vs. MS B	50	DISSIMILAR

[1]Agreement: Percentage of times 2 different models rated red-cockaded woodpecker models the same.

[2]CP = Coastal Plain

hazard.

PIEDMONT RISK rated the majority of Noxubee colonies as low hazard for spot occurrence. The remaining colonies were of moderate to high hazard. Belanger et al. (1988) achieved similar results in the Piedmont.

The Piedmont version of CLEMBEETLE rated Noxubee colonies as moderate to high mean hazard, but the Coastal Plain version rated the mean hazard as extreme. For both versions, all colonies ranked as having low risk and expected losses were low to moderate. Belanger et al. (1988) used the Piedmont version to hazard/risk rate the Piedmont colonies and found the colonies to be of moderate to low hazard. Predicted spot growth was low, as were projected losses.

The results of the Noxubee and Texas studies were also similar. TX HAZARD evaluated Noxubee sites as moderate to low hazard. Mitchell et al. (1991) used TX HAZARD on the Texas sites and found the majority of the colonies to be low hazard, the rest were moderate to extreme hazard. Using CLEMBEETLE all Noxubee colonies ranked as low risk. Mitchell (1987) however found the majority of his colonies to be moderate risk, the rest were moderate to high.

It is important to remember that the results of the Noxubee study and of previous studies in the Piedmont and in Texas are based on modeling. Models are developed by establishing relationships between stand conditions and SPB infestations. The resulting model provides predictions on the outcome of SPB infestations, the validity of the predictions depends on the selection of inputs. Thus, models cannot replace controlled experiments with actual stands and insects.

In closing, it seems that RCW colonies in the 3 regions are of low to moderate susceptibility to attack by SPB and of low to moderate probability of spot spread. Colonies should be closely monitored since even a small infestation could destroy several cavity trees. Further study in other regions are necessary to provide data throughout the range of the RCW.

CONCLUSIONS

We found that the susceptibility of colony sites, the probability of infestation in colony stands, and the probability of spot growth were generally low. Projected losses of trees were low overall, but such losses could eliminate the RCW from some colony sites. We attributed the overall resistance of the colony sites to SPB attack and spread to intensive silvicultural practices which promoted low stand density and healthy, vigorous trees. None the less, it is important to closely monitor colony sites so that SPB spots can be controlled should they occur.

Although colony sites had low risk and low to moderate hazard, high hazard/risk areas were found within 400 meters of each colony. Six of the 7 colony sites that were rated as having high risk had recently experienced logging activity and lightning strikes, which increased their chances of infestation. Within the colonies, 4 of the 17 trees killed by bark beetles contained active cavities.

Management Of SPB In RCW Colonies

National forests, national wildlife refuges and military bases, are required by law to protect and manage RCW colonies (Lennartz and Henry 1985). Loss of cavity trees to SPB is the most serious cause of cavity tree mortality in some areas (Conner et al 1991a). RCW cavity trees are generally considered over-mature by foresters and thus, would seem particularly susceptible to SPB. Since regeneration of over-mature pine stands is not a viable option for RCW colonies, appropriate silvicultural practices must be used to protect the site. RCW's tolerate a high degree of disturbance within their colonies (J. A. Jackson 1989 pers. commun.), allowing for a wide range of management procedures. Much has been written on silvicultural prescriptions to reduce the likelihood of SPB attack and spread (Belanger, 1980, Nebeker et al. 1985). These recommendations can easily be modified to reduce SPB hazard/risk ratings of RCW colonies:

(1) when regenerating areas surrounding the colonies, select species most suitable to the site and most resistant to attack;

(2) perform sanitation cuts within and around the colony to remove injured trees to reduce the probability of SPB attack and infested trees (to reduce the probability of spot growth and spread);

(3) promote proper stand density by thinning; and

(4) minimize logging damage (wounded trees will attract beetles to the site).

Thinning, the principle silvicultural tool for managing woodpecker colony sites, is an essential practice for discouraging bark beetles. Low density sites are less likely to become infested (Belanger et al 1988). Nebeker and Hodges (1983) suggested that proper stand management can increase the vigor of trees on the site, decreasing the likelihood of successful attack. Spot spread of pine beetle infestations is also related to density (Hedden and Billings 1979). The further apart the trees, the less likely infestations are to spread from one tree to another (Johnson and Coster 1978). Maintaining a spacing of 6.1 to 7.6 m will minimize the chances of infestations dispersing to adjacent trees (Gara and Coster 1968) and the chances of losing the entire colony (Lennartz and Henry 1985).

Mason (1971) reported that the average resin flow was 40% higher in loblolly pines growing in thinned plots than in unthinned plots. However, this was not a direct effect of thinning but was a result of removing suppressed trees with poor resin flow from the mean. In spite of these positive effects, improperly performed thinning practices can negatively affect resin flow rates, increasing the susceptibility of the pines to attack. Blanche et al. (1985) found that root pruning, in simulated thinning experiments, temporarily lowered resin flow. Damage to trees, during thinning operations, may attract beetles, as well as other pests and diseases (Nebeker et al. 1985).

There are other negative effects of thinning colony sites. Thinning too heavily can open the stand up to wind and ice damage, which can attract beetles (Belanger and Brender 1968). Reduced stand densities increases the probability that trees within the colonies would be struck by lightning and thus become attractive to SPB attack. These trees are the tallest objects in a relatively open area and thus may act like lightning rods.

The use of herbicides for understory control can enhance conditions for SPB infestations. Smith (1981) tried using herbicides to control small pines in RCW colonies in the Bienville National Forest in Mississippi and found that the dying trees attracted SPB to the site. As a result, 22 percent of the cavity trees became infested with SPBs.

Successful management of our natural resources is dependent on our understanding of the interactions which take place between the components of the system. Management can not have a single focus. It is critical that we consider the consequences of our silvicultural (management) recommendations and especially our efforts to enhance habitat for species such as the RCW. Disturbances play a key role in attracting SPB populations.

ACKNOWLEDGMENT

We would like to thank the personnel of the Noxubee National Wildlife Refuge for their assistance in completing this study. We also wish to acknowledge all that had verbal input such as Roy Hedden (Clemson University), Dick Conner (U.S. Forest Service, Nacogdoches, TX), Dave Richardson (Noxubee NWR, MS), Marita Lih (University of Arkansas), Jerry Jackson (Mississippi State University), Roger Belanger (U.S. Forest Service, Athens Georgia), and Joe Pase (Texas Forest Service, Lufkin, TX). This study was supported by the Mississippi Agricultural and Forestry Experiment Station, Mississippi State, Mississippi.

The Impact of Southern Pine Beetle Induced Mortality on Red-cockaded Woodpecker Cavity Trees

D. Craig Rudolph and Richard N. Conner
Wildlife Habitat and Silviculture Laboratory[1]
Southern Research Station, Nacogdoches, Texas 75962

ABSTRACT: Southern pine beetle (*Dendroctonus frontalis*) infestation was the major cause of red-cockaded woodpecker (*Picoides borealis*) cavity tree loss in eastern Texas. Infestation of individual cavity trees, primarily during the fall, resulted in annual mortality rates of active cavity trees exceeding 30%. Active cavity trees experienced higher mortality rates than inactive cavity trees. This difference was highly significant for loblolly pine (*Pinus taeda*) and shortleaf pine (*P. echinata*). Longleaf pine (*P. palustris*) was significantly more resistant to southern pine beetle mortality than the other 2 species. The specific causes of the increased mortality of active cavity trees was not determined. Cavity tree mortality from beetles in eastern Texas can seriously threaten the continued existence of those red-cockaded woodpecker populations.
KEYWORDS: *Picoides borealis*, *Dendroctonus frontalis*, cavity trees, red-cockaded woodpecker, southern pine beetle

[1]Maintained in cooperation with the College of Forestry, Stephen F. Austin State University, Nacogdoches, Texas, 75962.

Red-cockaded woodpeckers (*Picoides borealis*) are unique among woodpeckers in that they excavate their roost and nest cavities nearly exclusively in living pines (Steirly 1957a, Ligon 1970, Rudolph and Conner 1991). This behavior is presumably an adaptation to the fire climax pine ecosystem of the southeastern United States in which frequent fires limited the abundance of snags for cavity excavation (Jackson et al. 1986). The excavation of cavities in living pines presents additional constraints that the birds must contend with during cavity excavation. Consequently, the average time required for red-cockaded woodpeckers to complete cavity excavation averages 2-3 years in loblolly and shortleaf and 6.3 yrs in longleaf pines (Conner and Rudolph 1995a). Selection of older pines, often with red heart (*Phellinus pini*) infection, is an adaptation presumably allowing more efficient excavation in living pines (Conner and Locke 1982, Jackson and Jackson 1986, Rudolph and Conner 1991). Walters and coworkers (Walters et al. 1988, Walters 1990, Walters et al. 1992a) presented convincing evidence that the time and energy required to excavate cavities has resulted in cavities constituting a critical resource. Available cavities provide the variation in habitat quality that drives the evolution of cooperative breeding as hypothesized in the ecological constraints model (Emlen 1982a) and its variations (Stacey and Ligon 1991, Walters et al. 1992a).

Red-cockaded woodpeckers also continually excavate resin wells, small excavations in the phloem, that produce a continuing flow of oleoresins onto the bole of the pine in the vicinity of the cavity. This behavior produces a resin barrier that provides substantial protection from predatory snakes (Jackson 1974, Rudolph et al. 1990b).

Southern pine beetles have presumably been an important factor in the dynamics of pine ecosystems in the southeastern United States for thousands of years. Accounts from the 1700's describe pine mortality that is consistent with bark beetle infestations (Price and Doggett 1978). Population levels are extremely variable in time and space and epidemic levels tend to occur every 7-10 years in a given geographic area. Contemporary data

(Hedden 1978, Price and Doggett 1978, Texas Forest Service 1978) on number and size of infestations suggest that general population levels have increased in recent decades.

The southern pine beetle occurs throughout the range of the red-cockaded woodpecker (Payne 1980). All native pine species in the southeastern United States are potential hosts of southern pine beetles although susceptibility varies. Southern pine beetles use an elaborate pheromone communication system to coordinate a mass attack of many hundreds of adult beetles on individual trees (Coster et al. 1977). The mass attack allows the beetles to overcome the defense mechanisms of the pines, primarily the release of copious amounts of oleoresin into the entrance tube the female beetle must bore into the phloem to reach the cambial layer (Coster et al. 1977, Fargo et al. 1979). If sufficient numbers of adult beetles colonize a pine, the tree dies and in the process provides substrate for brood habitat (Goldman and Franklin 1977).

Southern pine beetle behavior varies by season (Thatcher and Pickard 1964, 1967). During the warmer months brood emerging from an infested tree may attack adjacent trees resulting in a multi-tree infestation (Coulson 1980). Infestations can ultimately involve many hundreds of hectares and result in the death of essentially all pines in the infestation area (Hedden and Billings 1979, Billings and Varner 1986). During the cooler months multi-tree infestations are less likely to be initiated. Most of the beetle population is dispersed in single trees scattered throughout the forested landscape (Hodges and Pickard 1971, Payne 1980). Adults emerging from these trees in the spring typically disperse widely and initiate infestations at other sites (Payne 1980). Lorio (1986) has proposed a hypothesis incorporating a balance between tree growth and differentiation as a basis for beetle-host tree interactions.

Southern pine beetles have the potential to impact red-cockaded woodpeckers in several ways. Pines infested with southern pine beetles, and typically a diverse array of other bark beetles and secondary insects, provide a significant prey source for red-cockaded and other woodpecker species (Kroll and Fleet 1979, pers. obs.). Infested trees, especially during the period when later instar larvae and pupae are present, are heavily used as foraging substrates. Red-cockaded and other woodpeckers often completely remove the outer layers of bark from the boles of pines while foraging on infested trees. The interval during which an individual tree is highly attractive to red-cockaded woodpeckers is relatively brief; however, beetle populations are often abundant enough to provide a significant portion of red-cockaded woodpecker prey. During southern pine beetle epidemics prey associated with infested trees is superabundant.

Multi-tree infestations can kill all, or substantial portions of the pines within the foraging range of individual red-cockaded woodpecker groups altering foraging patterns. Infestations can also result in the death of cavity trees, killing most or all of a group's cavity trees during a brief period. During beetle epidemics losses can be severe. We estimated that approximately 50 cavity tree clusters were completely eliminated by southern pine beetles on the Sam Houston National Forest in Texas during the epidemic of 1985-86 (Conner et al. 1991a). In order to survive as breeding units, red-cockaded woodpecker groups must excavate new cavities in surviving pines in the immediate area, or colonize new sites beyond the limits of the beetle infestation.

A second pattern of cavity tree mortality due to beetle infestation also impacts red-cockaded woodpeckers. During the fall and winter months southern pine beetles attack and overwinter in susceptible pines (Payne 1980, Lorio 1986). Lightning damaged trees (Hodges and Pickard 1971, Coulson et al. 1983) and woodpecker cavity trees (Conner et al. 1991a) are often infested. Typically single trees are infested and adult beetles emerging from these trees disperse widely (Gara 1967, Hedden and Billings 1977). Multi-tree infestations rarely develop in the immediate vicinity (pers. obs.). Losses of individual cavity trees increase the rate at which red-cockaded woodpeckers must excavate new cavities. In Texas, red-cockaded woodpeckers are unable to maintain sufficient cavities due to the high loss rates resulting from beetle mortality and other causes (Conner et al. 1991a).

We have been monitoring causes of red-cockaded woodpecker cavity tree mortality on 3 National Forests in Texas for several years. Cavity tree mortality due to southern pine beetle infestation has been the primary cause of cavity losses during this period (Conner et al. 1991a). In this paper we focus on cavity tree mortality due to single tree infestations during the fall and winter months.

STUDY AREA AND METHODS

We conducted annual surveys of red-cockaded woodpecker cavity tree clusters on the Angelina (ANF), Davy Crockett (DCNF), and Sabine (SNF) National Forests in Texas to evaluate status and condition of all cavity trees (Conner et al. 1991a). Red-cockaded woodpecker habitat on the DCNF consisted of loblolly and shortleaf pine stands. Mean cavity tree ages were 97 and 105 years, respectively (Rudolph and Conner 1991). Habitat on the ANF and SNF consisted of loblolly and shortleaf pine stands in the northern portions and longleaf pine stands in the southern portions of the forests. Mean cavity tree ages were 87, 101, and 130 years for loblolly, shortleaf, and longleaf pine respectively on the ANF (Rudolph and Conner 1991).

Red-cockaded woodpecker populations consisted of fewer than 30 groups on each forest during the period covered by this study. Populations had been declining on all 3 forests until recent intensification of management efforts (Conner and Rudolph 1989, unpub. data).

Cavity tree surveys were conducted between February and May from 1978-92 on the ANF, and from 1987-92 on the DCNF and SNF. We recorded tree species and tree condition (live or dead). If cavity trees were dead or dying the apparent cause was recorded. Since most trees damaged by lightning or other severe types of damage were usually rapidly attacked by bark beetles, including southern pine beetles, details were recorded and cause of death was assigned to the initial damage agent. If the sequence of events was unclear, cause was assigned to the non-beetle injury. Cavity trees originating from the construction of artificial cavities (Copeyon 1990, Allen 1991) were so designated, and included in the analysis.

Presence or absence of red-cockaded woodpeckers in each cavity tree cluster was generally known from ongoing survey activities. In addition, the status (active or inactive) of each cavity tree was determined annually using established criteria (Jackson 1978d, Conner and Rudolph 1989). Active cavity trees exhibited signs of recent red-cockaded woodpecker activity, including recently worked resin wells and fresh resin flow. Trees lacking evidence of recent woodpecker activity were considered inactive. Trees with cavity starts that lacked resin wells were recorded and included in the analysis as inactive cavity trees. They comprised a small minority of the total cavity tree population. Mortality comparisons among tree species and between active and inactive cavity trees were based on the log likelihood ratio test, called the G-test (Sokal and Rohlf 1969, p. 591).

RESULTS

Our primary survey period was in the early spring prior to the woodpecker breeding season. During this period numerous trees infested with southern pine beetles were present, but new infestations were rare. Opportunistic observations suggested that a large majority of the single tree infestations were initiated during October and November.

On the ANF and SNF prior to 1991-92, annual loss of active cavity trees as a result of southern pine beetles was 11.8% for loblolly pine, 2.6% for shortleaf pine, and 0.5% for longleaf pine. Mortality rates for inactive cavity trees were 1.1% for loblolly pine, 0% for shortleaf pine, and 0.4% for longleaf pine **(Table 1)**.

Annual mortality rates of active cavity trees on the DCNF prior to 1991-92 were 4.4% for loblolly pine and 5.3% for shortleaf pine. Mortality rates for inactive cavity trees were 0.3% for loblolly pine and 1.5% for shortleaf pine **(Table 1)**.

Southern pine beetle populations were substantially higher during 1991-92 and cavity tree mortality rates were also higher. On the ANF annual mortality rates for active and inactive cavity trees were 36.0 % and 0% for loblolly and shortleaf pines combined **(Table 2)**. Likewise, on the DCNF annual mortality rates for active and inactive cavity trees were 27.5% and 3.4% for loblolly and shortleaf pines combined **(Table 2)**.

Preliminary data provided by the National Forests for fall 1992 suggested that SPB mortality rates for 1992-93 would exceed the rates observed in 1991-92. As of 2 December 1992 40 cavity trees had been infested by southern pine beetles on the ANF and DCNF, nearly all of which were expected to die. This compares with a total of 36 cavity tree deaths in 1991-92 due to southern pine beetles on these 2 forests.

Mortality rates varied among the 3 pine species. For all forests combined the annual

Table 1. Red-cockaded woodpecker cavity tree mortality due to southern pine beetles on three national forests in Texas.

Tree Species	Cavity Tree Deaths (annual %) Active	Inactive
Angelina[1] and Sabine[2] NF		
Loblolly pine	9 (11.8%) n=76	5 (1.1%) n=459
Shortleaf pine	2 (2.6%) n=76	0 n=150
Longleaf pine	3 (0.5%) n=563	4 (0.4%) n=1134
Davy Crockett[2] NF		
Loblolly pine	4 (4.4%) n=90	2 (0.3%) n=607
Shortleaf pine	9 (5.3%) n=169	11(1.5%) n=711

[1]Data for 1978-91
[2]Data for 1987-91

Table 2. Annual red-cockaded woodpecker cavity tree mortality due to southern pine beetles during 1991-92 (1 yr) on the Angelina and Davy Crockett National Forests in Texas.[1]

National Forest	Cavity Tree Deaths (annual %) Active	Inactive
Angelina	9 (36.0%) n=25	0 n=47
Davy Crockett	19 (27.5%) n=69	8 (3.4%) n=237

[1]Loblolly and shortleaf pine only.

mortality rates for active cavity trees were 6.2%, 7.6%, and 0.5% for loblolly, shortleaf, and longleaf pine, respectively. The corresponding rates were 0.7%, 1.8%, 0.3% for inactive cavity trees (**Table 3**). Mortality rates of active and inactive cavity trees were not significantly different between loblolly and shortleaf pines (**Table 3**). However, mortality rates of active longleaf pine cavity trees were significantly less than corresponding rates for loblolly ($G = 27.52$, $P < 0.001$) and shortleaf ($G = 37.86$, $P < 0.001$) pines. For inactive cavity trees the trends were the same but the differences were not statistically significant.

Comparisons between active and inactive cavity trees (**Table 3**) indicated highly significant differences in southern pine beetle

Table 3. Annual southern pine beetle mortality rates of active and inactive red-cockaded woodpecker cavity trees on the Angelina, Davy Crockett, and Sabine National Forests in Texas.[1]

Tree Species	Annual Mortality Rate[2] Active	Inactive
Loblolly pine	6.2%[a,a] n=321	0.7%[a,b] n=1169
Shortleaf pine	7.6%[a,a] n=354	1.8%[a,b] n=994
Longleaf pine	0.5%[b,a] n=597	0.3%[a,a] n=1287

[1]Data for 1987-92 for DCNF and SNF: 1978-92 for ANF.
[2]Letters indicate significant (different letters) and non-significant (same letters) differences at the P=0.05 level by the G-test. First letter indicates comparisons among tree species (compare vertically); second letter indicates comparisons between active and inactive cavity trees (compare horizontally).

related mortality rates for loblolly pine ($G = 32.60$, $P < 0.001$) and shortleaf pine ($G = 23.54$, $P < 0.001$). Active cavity trees were infested by southern pine beetles and experienced greater mortality due to southern pine beetle infestation than inactive cavity trees. There was no significant difference between the mortality rates of active and inactive longleaf pine cavity trees, although the trend was similar.

DISCUSSION

Red-cockaded woodpecker cavity tree mortality due to southern pine beetle infestation was the primary cause of cavity loss in Texas (Conner et al. 1991a). The majority of the beetle mortality was initiated during the period October-November and annual mortality sometimes exceeded 30% of active cavity trees, even during non-epidemic years. In addition, a portion of the inactive trees that suffered beetle mortality were still suitable for use by red-cockaded woodpeckers prior to infestation. Combined with other causes of cavity loss the rate of loss of usable cavities exceeded the rate of excavation of new cavities in the study populations (Conner and Rudolph 1995a).

Lack of sufficient cavities is a limiting factor in many red-cockaded woodpecker populations, including those in Texas (Costa and Escano 1989, Rudolph et al. 1990b, Conner et al. 1991a, Walters et al. 1992a). Loss of cavity trees directly contributed to loss of groups in numerous instances in our study populations (Conner and Rudolph 1989, Pers. obs.). Combined with the extended time required to excavate cavities (Conner and Rudolph 1995a) and the limited availability of suitable potential cavity trees in most current habitats (Rudolph and Conner 1991), losses of the magnitude that we observed can have major population consequences. The consequences apply both to the availability of sufficient cavities to maintain existing red-cockaded woodpecker groups, and the availability of sufficient cavities to support a population increase.

The susceptibility of pines to southern pine beetle mortality varies with species (Coster et al. 1977, Fargo et al. 1979). Of the 3 pine species available for cavity excavation in Texas, longleaf pine was significantly more resistant to beetle mortality than loblolly or shortleaf pine. The greater resistance of longleaf pine was primarily a result of its ability to produce and transport more oleoresins to injury sites (Hicks 1980) and the physical and chemical properties of the oleoresin produced (Coyne and Lott 1976, Hodges et al. 1977). In the present study annual longleaf pine mortality due to southern pine beetles was consistently less than 1%, and did not vary significantly between active and inactive cavity trees. Loblolly and shortleaf pines were significantly more susceptible to beetle mortality and there was a significant (4-8 fold) increase in mortality of active compared to inactive trees.

The increased mortality rates for active cavity trees were greater than for inactive cavity trees. The impact on red-cockaded woodpecker populations is therefore more severe than the overall mortality rates would suggest. Impact is further increased because the cavity trees affected tend to be some of the more recently excavated trees, often the nest trees of the previous season. These are the trees that have the highest future value to the woodpeckers.

The cause of the increased susceptibility of the active cavity trees was not determined, but the pattern strongly suggested that some wound response exhibited by the trees was involved. Obvious possibilities included potential attractive properties of the oleoresins released at active resin wells, other chemical changes in active trees that beetles could detect, or weakening of the trees due to the continual working of the resin wells by the woodpeckers. A preliminary study in Texas suggested that tree physiology, specifically oleoresin flow, varies in complex patterns between active and inactive cavity trees (Ross et al. 1991). Alpha-pinene, a resin component, and other host odors apparently do not function as attractants for southern pine beetles (Renwick and Vite 1969). However, in synergism with the pheromone frontalin, alpha-pinene is attractive to beetles (Kinzer et al. 1969). Further research in this area is warranted. Further evidence that implicated activity at the resin wells was the observation that inactive trees with artificial cavities had low susceptibility. However, once the trees were activated by the woodpeckers and resin wells were being worked, tree susceptibility increased.

During most of the present study annual mortality ranged from 2-5% for active loblolly and shortleaf pine cavity trees, and 1.5% or less for inactive cavity trees. During the 1991-92 period there was a fairly uniform 5-fold increase in mortality rates for both active and inactive cavity trees. Recent reports indicate that the 1992-93 period may remain at a similar level. Mortality rates for longleaf pine cavity trees were uniformly low during all years.

During the study, management of red-cockaded woodpecker habitat changed substantially, due in part to a 1988 Federal Court ruling. The possibility exists that intensified management required to halt woodpecker population declines (Conner and Rudolph 1989) may be increasing the likelihood of southern pine beetle infestation of active cavity trees. The removal of midstory vegetation, primarily by mechanical means with heavy equipment, resulted in disturbance of the soil, a change in forest structure, and occasional damage to the remaining pines. However, these changes occurred primarily during the 1988-89 period, substantially prior to the increase in beetle mortality in the fall of 1991. A more consistent hypothesis is that the mortality rates of cavity trees was determined primarily by the population characteristics of beetles in the general forest area.

Major changes occurred in the pine ecosystem of the region since the early 1900s. Most important was a major reduction in the area dominated by the relatively beetle resistant longleaf pine, and a corresponding increase in the area dominated by the more susceptible loblolly pine (Wahlenberg 1960). Forestry practices have also resulted in economically preferred species occurring beyond their original geographic and ecological ranges (Bridges and Orzell 1989). Changes in the fire regime, specifically a reduction in frequency of fire, and a temporal shift from primarily growing season fires to non-growing season fires (Komarek 1968, Christensen 1981a), and the more recent active suppression of hardwoods have also resulted in major changes in the forest ecosystem. Even-aged silvicultural practices have increased stand homogeneity and pine density of many existing stands. These changes have undoubtedly resulted in major changes in the ecological interaction of pines and southern pine beetles.

A variety of factors have been correlated with the susceptibility of individual pines and stands to infestation by southern pine beetles (Hicks 1980, Lorio 1986, Turchin et al. 1991). In general, stress resulting in reduced vigor and decreased radial growth are positively associated with beetle infestation (Bennett 1968, 1971; Hicks 1980). A variety of climatic (King 1972, Kalkstein 1976) and edaphic factors (Belanger et al. 1977, Hicks et al. 1978) that reduce vigor and growth are associated with increased susceptibility. Injuries increase susceptibility of pines to beetle attack including lightning (Hodges and Pickard 1971, Coulson et al. 1983), logging damage (Porterfield and Rowell 1980, Hicks et al. 1987), wind damage (Hicks 1980), and disease (Skelly 1976, Belanger et al. 1977). High pine density, which results in decreased pine vigor, can also increase susceptibility of stands to infestation (Leuschner et al. 1976, Hicks et al. 1978). Tree age is often cited as a factor increasing susceptibility to southern pine beetles (Hicks 1980). Specific studies, however, have produced variable results (Hicks 1980, Leuschner et al. 1976, Ku et al. 1980), due in part to correlation of age with other factors.

A management solution to these losses of cavity trees to southern pine beetles is urgently needed. Southern pine beetle infestations involving multiple trees are amenable to a variety of suppression techniques capable of limiting further spread (Billings 1980). However, the single tree losses currently impacting red-cockaded woodpeckers cannot be reduced by after-the-event methods. Once the tree is infested the damage is done. One solution would be to reduce the beetle hazard of the general forest area, thereby reducing the beetle population. Basal area thinnings are currently being implemented to achieve this effect. A more direct solution would be to use beetle pheromones to disrupt initial beetle attacks. Research is currently underway to test the use of verbenone to prevent beetle infestation (Ronald F. Billings pers. commun.).

In the absence of an effective solution to this problem, mitigation using artificial cavities is an option. In the populations we studied, National Forest System personnel have done an excellent job of maintaining adequate cavities during the recent increases in southern pine beetle mortality. This option is critical in the short term, but a long-term solution based on the rehabilitation of the ecosystem, ensuring that red-cockaded woodpeckers can maintain viable populations without intensive intervention is required.

ACKNOWLEDGMENTS

We thank P.L. Lorio Jr., R.T. Engstrom, and R.G. Hooper for comments on an earlier draft of the manuscript that resulted in substantial improvements.

Forest Canopy Gap Size in Red-cockaded Woodpecker Cavity Tree Clusters: Implications for Management

G. M. Chrismer, W. G. Ross and D. L. Kulhavy
College of Forestry, Stephen F. Austin State University, Nacogdoches, TX, 75962

ABSTRACT: We examined forest canopy gaps resulting from disturbances from a variety of causes in 11 red-cockaded woodpecker (*Picoides borealis*) (RCW) cavity tree clusters, 5 RCW replacement stands, and 6 control (non-RCW) areas in eastern Texas. The stands, dominated by loblolly (*Pinus taeda*) and shortleaf (*Pinus echinata*) pines, were located in the Angelina and Davy Crockett National Forests. The ratio of disturbed area to total area surveyed did not vary among stand types, but size and distribution of disturbed areas varied substantially. Canopy gaps in the control stands were more numerous per hectare, but significantly smaller than those in the RCW cavity tree clusters and replacement stands. Bark beetles and lightning were the most important disturbance factors in all stand types, both in terms of number of canopy gaps created and total area influenced.
KEYWORDS: disturbance, lightning, bark beetles, canopy gaps, hardwoods

All forest landscapes, including those occupied by the red-cockaded woodpecker, *Picoides borealis* (Vieillot) (RCW), are continually changing over time as a result of natural occurrences including lightning, wind, storms, flooding, insect outbreaks, and fire. Forest management with thinning, harvest/regeneration of varying intensity, prescribed fire, and other control of unwanted woody and herbaceous species also modifies forest composition, structure and function. Such events, natural or anthropogenic, that open the forest canopy produce an ever-changing, heterogeneous mosaic of forest matrix, patches, and corridors (Forman and Godron 1986).

Disturbances cause forest stands to deviate from theoretical successional patterns by creating canopy gaps that favor early to mid successional species (Spurr and Barnes 1980). For example, in much of the RCW's pre-European settlement range, frequent fires from lightning and Native American land management produced fire climax longleaf pine forests characterized by an open stand habit and diverse herbaceous communities with few hardwoods except on moist bottomland sites where fire was naturally limited. Fire suppression in these ecosystems results in continued succession, with the shade-intolerant longleaf being replaced by more shade-tolerant species formerly limited by fire.

In much of the RCW's modern range, cavity tree clusters are in relatively isolated patches in forests fragmented by cities, agriculture, and short-rotation industrial forests. Natural and anthropogenic disturbances have a profound effect on RCW habitat and local populations under these conditions (Conner and Rudolph 1991b). In the loblolly and shortleaf pine forests of eastern Texas, disturbances such as lightning in conjunction with high southern pine beetle populations can lead to massive losses of mature pines, destroying both RCW cavity trees and future habitat (Conner et al. 1991a).

Investigating disturbance patterns and effects in RCW stands is necessary if negative disturbance impacts are to be minimized and positive effects encouraged. This paper is part of a larger, continuing study of disturbances in RCW habitat.

Objectives of this paper are to:

1) describe disturbance patch occurrence and cause in active RCW cavity tree clusters, replacement stands, and non-RCW stands in the loblolly and shortleaf forests of eastern Texas;

2) quantify disturbance patches by size,

number and cause over time; and

3) develop management recommendations to minimize negative disturbance impacts in these and similar stands.

METHODS

Disturbance canopy gaps were measured in 11 red-cockaded woodpecker cavity tree clusters, 5 replacement stands, and 6 non-RCW (Control) areas periodically from the summer of 1987 through the spring of 1992 in the Angelina Ranger District of the Angelina National Forest (ANF) and the Neches Ranger District of the Davy Crockett National Forest (DCNF). RCW cluster and replacement stand boundaries as designated by the U.S.D.A. Forest Service were used to delineate those areas of study. Control areas were mature loblolly and shortleaf pine stands as similar as possible in structure and composition to nearby RCW stands, but were unaltered (at the time) by any management designed to favor the RCW, and were without active or inactive cavity trees.

Each area was examined for all canopy-opening disturbances. A disturbance was defined as an event that caused the loss of a canopy tree (or trees) and thereby resulted in a gap in the upper tree canopy and the resultant patch in the landscape matrix (Runkle 1992).

Disturbance length and width were measured to the nearest 0.1 m directly beneath the canopy opening caused by each disturbance. The height (HT) and diameter at breast height (DBH) of the tree or trees involved in each disturbance was also recorded.

Probable cause and year of each disturbance event were recorded. Time of occurrence was estimated for each disturbance by observing the presence or absence of needles, twigs and cones, bark condition, and decay of the involved trees. Occurrences found to be over 8 to 10 years old were considered as part of the background matrix. The dates of forest harvest, thinnings, burns, and cavity inserts were obtained from USDA Forest Service records for the ANF and DCNF.

ANALYSIS

Data analyses were completed on the CCS-VAX mainframe computer at Stephen F. Austin State University using the SPSS statistical software package (Norusis 1985). One-way analysis of variance or the k-samples median test was used to evaluate disturbance areas by stand type. The Kolmogorov-Smirnoff two-sample test was used to compare the distribution of disturbance events by area affected for the different stand types.

RESULTS

Disturbances in RCW clusters and RCW replacement stands resulted in larger canopy openings than disturbances in the control stands **(Tables 1 and 2)**, but control stands had more disturbances per hectare than RCW clusters and replacement stands **(Table 3)**. Disturbance area as a percentage of total area surveyed did not vary among stand types **(Table 3)**. Diameter at breast height (DBH) of trees killed in RCW clusters was greater than DBH for trees killed in both replacement stands and control stands **(Table 4)**. A comparison of frequency distributions of disturbance events by area affected revealed statistically significant differences, however, with control stands differing substantially ($p < 0.001$, Kolmogorov-Smirnoff two-sample test, Norusis 1985) from both RCW clusters and RCW replacement stands.

Bark beetles (mostly southern pine beetle) and lightning together constituted the majority of disturbance events in all stand types: 72.9 percent for RCW clusters, 74.2 percent for replacement stands, and 65.8 percent for control stands **(Table 5)**. In addition to being the most common types of disturbance, bark beetles and lightning accounted for 73.9 percent of total disturbed area in RCW clusters, 77.0 percent in RCW replacement stands, and 68.7 percent in control areas.

Number of disturbances per year for 1986 through 1992 peaked in 1989 in all stand types **(Table 6)**. This occurred because of a significant increase in tree mortality from bark beetles from 1988. Generally, number and area of canopy-opening disturbances varied widely from year to year as well as among stand types. Data will need to be collected for several more years before tests for trends and patterns can be meaningful.

DISCUSSION

Differences in disturbance canopy gap size between RCW management areas (both cavity tree clusters and replacement stands) and control stands are probably the result of differences in stand density, structure, and

Table 1. Canopy gap size (m^2) by stand type, Angelina and Davy Crockett National Forests, Texas, 1985 - 1992.

Stand Type	N	Mean (SD)	Median[1]	Minimum	Maximum
RCW	344	177.8 (630.5)	85.9a	5.9	9212.7
Replacement	99	140.3 (251.6)	84.3a	14.3	2073.8
Control	164	94.8 (325.3)	43.6b	7.3	4074.1

[1]Among stand types medians followed by same letter do not differ significantly ($p > 0.05$, k-Samples Median Test [Norusis 1985])

Table 2. Canopy gap size (m^2) by percentiles and stand type, Angelina and Davy Crockett National Forests, Texas, 1985 - 1992.

		PERCENTILES			
Stand Type	N	25	50	75	95
RCW	344	42.0	85.9	149.7	402.8
Replacement	99	42.0	84.3	146.6	490.1
Control	164	24.5	43.6	76.6	286.3

Table 3. Total surveyed area by stand type, with total disturbed area, percent of area disturbed and average number of disturbances per hectare, Angelina and Davy Crockett National Forests, Texas.

Stand Type	Total Area Surveyed (ha)	N	Total Disturbed Area (ha)	% of Area Disturbed[1] Mean (SD)	Disturbances per ha Mean (SD)	Disturbances per ha Median[2]
RCW	68.6	344	6.1	8.7 (5.3)	5.0 (3.5)	4.4b
Replacement	28.6	99	1.4	5.0 (2.7)	3.5 (1.3)	3.6b
Control	23.4	164	1.6	6.5 (3.4)	6.8 (1.2)	7.0a
Totals	120.6	607	9.1	7.2 (4.5)	5.0 (2.8)	4.8

[1]Does not vary significantly among stand types ($p > 0.05$, One-way Analysis of Variance [Norusis 1985]).

[2]Among stand types medians followed by same letter do not differ significantly ($p > 0.05$, k-Samples Median Test [Norusis 1985]).

Table 4. Height (HT) and diameter at breast height (DBH) of trees killed, and general range of basal area by stand type, Angelina and Davy Crockett National Forests, Texas.

			Basal Area ($m^2 ha^{-1}$)	
Stand Type	Median HT (m)[1]	Median DBH (cm)[1]	Pine	Hardwood
RCW	24.7a	41.9a	14-21	0-3
Replacement	24.4ab	37.8b	14-21	0-3
Control	22.9b	35.6b	17-28	5-12

[1]Among stand types medians followed by same letter do not differ significantly ($p > 0.05$, k-Samples Median Test [Norusis 1985])

Table 5. Number and area of disturbances by cause and stand type, Angelina and Davy Crockett National Forests, Texas, 1985-1992.

	STAND TYPE					
	RCW		Replacement		Control	
Cause of Disturbance	N	Mean Area (SD)	N	Mean Area (SD)	N	Mean Area (SD)
Bark Beetles	143	182.2 (776.5)	37	185.6 (384.8)	66	119.3 (502.8)
Lightning	104	112.0 (67.6)	35	107.3 (97.9)	42	67.0 (58.2)
Wind	42	129.6 (152.5)	11	104.9 (70.3)	12	104.8 (115.8)
Management	12	1049.8 (1836.7)	6	194.2 (241.5)	13	120.0 (134.9)
Other	38	121.7 (249.6)	8	107.4 (71.4)	31	65.9 (81.2)

Table 6. Canopy gap size by stand type and year, Angelina and Davy Crockett National Forests, Texas.

	STAND TYPE					
	RCW		Replacement		Control	
Year	N	Mean (SD)	N	Mean (SD)	N	Mean (SD)
≤ 1985[1]	35	380.1 (1539.9)	16	235.0 (494.4)	52	67.6 (81.7)
1986	24	87.2 (58.1)	7	51.1 (27.7)	15	363.0 (1038.1)
1987	34	308.6 (1095.5)	14	141.9 (183.3)	14	54.1 (58.6)
1988	44	110.9 (121.0)	16	94.7 (71.2)	19	60.9 (64.6)
1989	73	165.7 (355.8)	19	86.0 (70.2)	32	63.0 (62.7)
1990	61	106.2 (132.8)	11	202.6 (354.1)	22	53.8 (49.4)
1991	47	134.2 (91.7)	9	188.5 (194.0)	6	77.5 (36.9)
1992	20	158.2 (106.6)	7	102.4 (51.1)	0	0

[1]Includes all disturbances happening before 1986.

composition. Stands managed for RCW had lower pine basal area (14 to 21 m^2ha^{-1}). Most pines in suppressed and intermediate crown classes and practically all hardwoods were removed in basal area reduction thinning and midstory control (Cahal et al. 1995).

Control areas had generally higher pine basal area (17 to 28 m^2ha^{-1}) with hardwood basal area ranging from 5 to 12 m^2ha^{-1}. Disturbances in red-cockaded woodpecker management areas resulted in larger canopy gaps because of lower density. An absence of vertical layers in the forest structure of intensely managed stands, combined with larger, more widely spaced crowns, would encourage increasingly larger areas to be opened up with each disturbance. This is shown in the significantly larger disturbance area for the RCW clusters and replacement stands in comparison to smaller disturbance area for controls **(Tables 1 and 2)**.

RCW clusters and replacement stands were similar to non-RCW control stands before midstory control and basal area reduction gave them a more open, even-age character. Lower basal areas allow more sunlight to reach the forest floor and, as a result, shade intolerant pines are regenerating in some of the larger openings. Hardwoods are sprouting profusely in the understory in competition with the pines in cavity tree cluster stands and replacement stands.

Control stands or non-red-cockaded woodpecker areas were structurally and compositionally representative of typical southern "old field" succession at a mid-successional stage (Spurr and Barnes 1980). These stands have mature pine overstory, approximately 60-80 years old, some overstory hardwoods, and emerging hardwood midstory. The shaded understory consisted of leaf litter and some sparse grasses. Pine regeneration was sparse. In normal succession without fire, the hardwoods in the midstory would eventually dominate these stands (Spurr and Barnes 1980). As pines are lost to disturbance or die out from old age, the hardwood midstory would expand and fill the canopy gaps (Schowalter et al. 1981, Coulson et al. 1985).

An increase in tree mortality in some areas managed for RCWs may be noticed following thinning and midstory control. These stands developed under dense conditions with lower bole strength and root development than trees developing in the open. Although basal area in the stands surveyed was not reduced by more than the generally recognized thinning limit of one-third the original basal area (Smith 1986), this guideline was devised for vigorous trees being grown for harvest. Senescent trees may prove less tolerant of drastic changes in stand structure. Most of the midstory removal and thinning in the stands surveyed was accomplished in 1989 and 1990. Increased mortality and larger total disturbance area (mostly from bark beetles) are seen in these two years (**Table 6**). Control stands generally also showed a larger number of bark beetle events and total disturbance area for this period, however, and data will need to be collected for several more years before tests for trends in mortality and disturbance size after thinning and midstory control will be meaningful.

Forest management problems and opportunities are presented by disturbance dynamics in the loblolly/shortleaf stands surveyed. Although hardwood midstory removal has been completed in red-cockaded woodpecker clusters and replacement stands, hardwood understory and sprouting are prolific. Reduced basal area and large canopy gaps may serve to release hardwoods rather than enhance pine regeneration. Reversing this trend will require more than bush-hogging and periodic winter prescribed fire. Many of the stands surveyed would benefit from application of hardwood selective herbicides followed by hot growing season prescribed fire. Pine regeneration, including longleaf pine reintroduction on appropriate sites, would be greatly enhanced by drastically reduced hardwood competition and large canopy gaps. Further reduction of pine basal area would reduce wildfire and bark beetle hazard, as well as create superior red-cockaded woodpecker habitat.

Economic Valuation of the Red-cockaded Woodpecker and Its Habitat

Dixie Watts Reaves, Department of Economics, Duke University, Durham 27707
Randall A. Kramer, School of the Environment, Duke University, Durham 27707
Thomas P. Holmes, USDA Forest Service, Southern Research Station
Research Triangle Park, NC 27709

ABSTRACT: The issue of how to assign economic values to goods which are not traded in the marketplace is a controversial one. When goods have no proxy in an established market, a non-market valuation technique can be used. Natural and environmental resources generally fall into this category, and the endangered red-cockaded woodpecker (*Picoides borealis*) is one such example. The red-cockaded woodpecker population and its habitat in the Francis Marion National Forest (near Charleston, South Carolina) suffered damages from Hurricane Hugo. The economic value of these losses was assessed using the contingent valuation method. A hypothetical market was developed for the good to be valued, and individuals were asked how much they would be willing to pay for the good, contingent on the described setting. Individuals were asked to value different population restoration levels of the red-cockaded woodpecker in Francis Marion National Forest, to value restoration of the Francis Marion National Forest if the endangered RCW did not exist there, and to value more intensive management of all red-cockaded woodpecker populations in the United States. Data were collected by mail survey from a randomly drawn national sample. Results indicated that individuals were willing to pay between 8 and 14 dollars, on average for restoration activities.
KEYWORDS: valuation, population restoration, preservation, contingent valuation method

BACKGROUND AND JUSTIFICATION

With the renewed interest in environmental quality, recent research has focused on improving methods to value environmental amenities. In order to value the commercial components of an environmental good, one merely uses market prices. However, there is often a need to value non-market components, and that was the focus of this research: to use the contingent valuation method (CVM) to determine the social value of protecting and restoring the endangered red-cockaded woodpecker (RCW) and its habitat, the old-growth longleaf pine (*Pinus palustris*) ecosystem.

Since the United States Forest Service is mandated to manage for multiple uses, there is a specific need for the valuation of these non-market goods:

> The Forest Service, U.S. Department of Agriculture, is dedicated to the principle of multiple use management of the Nation's forest resources for sustained yields of wood, water, forage, wildlife, and recreation (USDA Forest Service).

The Forest Service, given this multiple use mandate, needs to know how to allocate limited budgets among competing needs. In an economic sense, the Forest Service needs to know the appropriate weights to use in a multiple use objective function. Market prices can be used as the weights for commercial outputs of a National Forest, but non-market techniques must be used to estimate weights for the non-commercial outputs.

The Longleaf Pine Ecosystem

The longleaf pine ecosystem, ranging naturally from Virginia to Florida and west to Texas, is one of the most unique and diverse ecosystems in the temperate zone. It serves as habitat for numerous species including deer, quail, turkey, and squirrel, as well as providing forage for cattle. It is also home to the RCW, a federally-listed endangered species. The longleaf pine forest, when regularly burned, has an open, park-like appearance, and provides numerous recreation opportunities. The longleaf pine is an excellent timber tree, offering a wide range of forest products. Given its characteristics, the longleaf timber type is well adapted to multiple-use management because of the many forest products it supplies, the forage it produces, the wildlife it supports, and the recreation it affords (Boyer and Peterson, 1983).

Only the forest product component of the longleaf pine ecosystem is valued in the marketplace. One objective of this research was to provide the U.S. Forest Service with an estimate of value for some of the non-commercial components. Valuation focused on one particular geographic component of the overall ecosystem (the Francis Marion National Forest), as well as one particular species (the RCW) which uses the old-growth longleaf pine forest as habitat.

Hurricane Hugo

On the evening of September 21, 1989, Hurricane Hugo scored a direct hit on the Francis Marion National Forest (FMNF), located in coastal South Carolina. The FMNF contained a significant remnant of old-growth longleaf pine forest, and served as the home to the RCW. The RCW population in the FMNF was the densest, the second largest, and the only known naturally increasing population. Most of the RCW cavity trees for nesting and roosting were destroyed (87%), much of its foraging habitat was damaged (50-60%), and many of the birds were killed (63%) (Hooper, et al. 1990). The Forest Service is mandated by the Endangered Species Act to take steps to recover the species, and much restoration work has already been carried out, including the drilling of artificial cavities (Watson et al 1995). For future management decisions, the Forest Service needs to assess the benefits and costs of further restoration, and of potential prevention of similar destruction. Thus, they need estimates of the value of the endangered species, its habitat, and the unique ecosystem in which it lives. This valuation can be extended beyond the damages to FMNF to include the overall valuation of the unique and diverse old-growth longleaf pine ecosystem as habitat for the RCW.

STUDY OBJECTIVES

Since these environmental goods are not typically valued in the marketplace, non-market valuation can be employed to estimate values for the goods. The contingent valuation method (CVM) was used to determine dollar values for the following:

(i) restoration of the RCW population to two different population levels in the FMNF in South Carolina,

(ii) preservation of all RCW populations in the United States, and

(iii) restoration of the FMNF if the RCW did not exist there.

A second objective of this research was to make contributions to the non-market valuation literature by testing hypotheses about the values obtained when using the contingent valuation method.

THE VALUATION OF NON-MARKET GOODS

Non-market goods are goods for which markets are nonexistent, incomplete, or institutionally restrained from reflecting the free interplay of supply and demand (Peterson and Randall, 1984). Two common characteristics of non-market goods are nonrivalry, where consumption by one individual does not decrease the amount of the good available to other individuals (up to the point of congestion), and nonexclusiveness, which results from incomplete property rights. One public good that fits this category is the nation's highway and bridge system. Less common examples of this type of good are ecological diversity and stability, and the continued existence of endangered species. These environmental amenities can be enjoyed by many people simultaneously, either through site-seeing, reading about them, looking at photographs, or just appreciating the fact that they exist in an undisturbed state. This represents the nonrivalry characteristic of the goods. The fact that these amenities can be enjoyed by people who do not pay directly for the preservation of the goods represents the nonexclusiveness of the goods.

When prices do not exist for goods in the market place, or when prices are institutionally set and do not truly reflect consumer preferences, such as in park entrance fees, some method of valuation is needed. Economists often turn to Hicksian measures (Mitchell and

Carlson 1989). An equivalent measure is the amount of compensation, paid or received, which would bring the consumer to his subsequent welfare level if the change did not take place. For example, if environmental quality increases from Q to Q', an individual consumer's utility increases from U to U'. If quality did not increase, willingness to accept (WTA) indicates the amount that the individual would have to be paid to attain utility level U'. Another Hicksian measure is the compensating measure and is the amount of compensation, paid or received, which would keep the consumer at his initial welfare level if change did occur. As an example, consider the case described above where an increase in quality from Q to Q' leads to an increase in utility from U to U'. Now, assume the change does take place. Willingness to pay (WTP) indicates the amount of money that one could take from the consumer and leave him at his initial utility level, U.

There are 4 Hicksian measures. In the compensated case, WTP^c refers to the willingness to pay for an increase in Q and is a measure of individual benefit. For example, an individual might be willing to pay to ensure a higher population level of the red-cockaded woodpecker. WTA^c refers to the willingness to accept a decrease in Q and measures individual costs or negative benefits. On the other hand, the equivalent measure, WTP^e, is the willingness to pay to avoid a threatened decrease in Q (for example, a decrease in red-cockaded woodpecker population size), and WTA^e is the willingness to accept compensation in lieu of a promised increase in Q. In this study, both types of willingness to pay measures were estimated.

Benefit-Cost Analysis (BCA) is one empirical test for potential Pareto-improvements (PPI's). A PPI is one where those who would gain from a proposed change could compensate those who would lose, to the full extent of their perceived losses. If this were the case, i.e. if the sum of the money-valued gains as judged by the gainers exceeded the sum of the money-valued losses as judged by the losers, then the change would be acceptable. In order to make use of the benefit-cost criterion in public policy decisions concerning environmental goods, one must have monetary measures of benefits and costs. Shaw (1984) raises 3 key questions that are relevant for the valuation of environmental resources:

(i) What are the environmental products that are being valued (recreational, aesthetic, educational, biological, social, and/or commercial uses)?

(ii) Who is benefiting from the resource (direct consumptive users, direct non consumptive users, and/or indirect or vicarious users)?

(iii) How well can the resource value be assessed in quantitative terms (i.e., how precise and valid are the valuation approaches, and how much of the total resource is being valued)?

This research addressed these 3 questions as they related to the valuation of restoration of Hurricane Hugo damages to the FMNF and the valuation of the unique ecosystem, the old-growth longleaf pine forest.

EMPIRICAL APPLICATION OF THE CONTINGENT VALUATION METHOD

The CVM is based on utility theory where an individual consumer derives utility from market goods, X, and non-market environmental goods, Q, and maximizes that utility subject to a budget constraint. Borrowing from Bergstrom and Stoll (1987), let a multidimensional non-market good be denoted as

$$Q=Q(a_1, a_2, ..., a_n),$$

where, in this case, Q is the forest ecosystem and each a_i is a different component of the ecosystem, eg. the RCW. The components of Q are assumed to be provided in quantities exogenously determined by policy. An individual consumer's utility function can be defined as

$$U=U[X,Q(a_1, a_2, ..., a_n);S],$$

where S is a vector of attributes of the individual that may affect preferences (eg., age, sex, and participation in outdoor activities). Each component of Q is assumed to have positive marginal utility:

$$\frac{\partial U}{\partial Q} \times \frac{\partial Q}{\partial a_i} > 0$$

Simply stated, an increase in the RCW

population increases the individual's utility.

Suppose a policy change will increase all components of Q from the original level, $Q^0 = Q(a_{10}, a_{20}, \ldots a_{n0})$, to some higher level, $Q^1 = Q(a_{11}, a_{21}, \ldots a_{n1})$. To determine willingness to pay for this increase in Q, one needs to specify the indirect utility function, which is derived by maximizing utility with respect to the budget constraint, and substituting the resulting demand equations into the utility function. This yields the indirect utility function,

$$V = V[P, Q(a_1, a_2, \ldots, a_n), U^*; S],$$

where P is the price vector for market goods and M is money income. The expenditure function is the inverse of the indirect utility function,

$$M = E[P, Q(a_1, a_2, \ldots, a_n), U^*; S],$$

which is derived by minimizing expenditures subject to attaining a given level of utility, U^*.

The derivative of the expenditure function, with respect to Q, yields the inverse Hicksian demand function for changes in all components of the non-market good, Q:

$$g = \frac{\partial E}{\partial Q} = E_Q\left[P, Q, U^*\right],$$

where g is the marginal willingness to pay (WTP) for a change in Q, the vector of ecosystem outputs. Total WTP for a non-marginal change in Q from Q^0 to Q^1 is calculated as

$$WTP_Q = \int_{Q^0}^{Q^1} \left[E_Q\left(P, Q, U^*\right) \right] dQ$$

One may also be interested in calculating the willingness to pay for an increase in one component of Q, say a_1, the RCW population. This can be done by differentiating the expenditure function with respect to a_1 while holding all other components constant. The resulting Hicksian demand function for a_1 is denoted by

$$Z = \frac{\partial E}{\partial a_1} = \frac{\partial E\left[P, Q(a_1, a_2, \ldots, a_n), U^*\right]}{\partial Q} \times \frac{\partial Q}{\partial a_1}$$

Z represents the marginal WTP for a marginal change in a_1. Total WTP for a change in a_1 from a_{10} (the current RCW population size) to a_{11} (the desired RCW population size) can be calculated by integrating the Z function from a_{10} to a_{11}. This is one value that was estimated in this research project using the contingent valuation method. WTP for other components can be derived theoretically in the same manner.

For the environmental commodities of interest in this study (an endangered species, its habitat, and associated ecosystem), the primary sources of value are off-site use (eg., viewing photographs and watching television specials) and the knowledge of their existence. Randall, Hoehn, and Swanson (1990) state that current methodology suggests the measurement of off-site values concurrently with existence values in a contingent valuation framework. Willingness to pay for an improvement in the commodity can be denoted EV (as in existence value), where EV is calculated as the decrease in income that the consumer is willing to accept in order to achieve an increase in Q. Initial utility is equal for initial levels, Q^0 and M, and subsequent levels, Q^1 and M-EV:

$$U^0 = V(P, Q^0, M; S) = V(P, Q^1, M-EV; S)$$

The preference shifter, S, will be suppressed in future discussion. For the purposes of this study, the vector Q can be thought of as representing the unique and diverse old growth longleaf pine ecosystem. Components of the ecosystem include numerous plant and animal species, including the endangered red-cockaded woodpecker. Individuals were asked to value the specific species, the species' habitat, and the ecosystem in and of itself.

A NATIONAL SURVEY: DISCUSSION AND RESULTS

To elicit individual consumers' values for the goods in question, Dillman's (1978) Total Design Method (TDM) was used in a mail survey. In a pretest, the survey was sent to 200 individuals: 50 in South Carolina, 50 in Mississippi, 50 in Indiana, and 50 in Montana. *A priori*, it was expected that those people in the states where the RCW exists (in the pretest, South Carolina and Mississippi) were more likely to be aware of the good, to have a positive value for the good, and to respond to the survey to indicate that value.

The goals of the pretest were to test the survey instrument, to determine appropriate dollar value ranges to use in the full survey, and to obtain some preliminary estimates of value for RCW population restoration. The overall response rate for the pretest was 21%. This level of response is to be expected for a mail survey which included only a reminder postcard follow-up. As expected, individuals in South Carolina seemed more concerned for the good in question, as proxied by their higher response rate to the survey (26%). Since the longleaf pine ecosystem and the RCW are also found in Mississippi, but not in Indiana or Montana, it was expected that Mississippi, too, would have a higher interest in the good and therefore a higher response rate. This was not true, however, as both Mississippi and Indiana had a response rate of 20%. Seventeen percent of the individuals surveyed in Montana responded to the survey.

In terms of the valuation questions, the pretest was not meant to reveal final valuations, but rather to give an idea of the range of expected values. The valuation questions were prefaced with a discussion of the contingent setting: first, individuals were asked to assume that the Endangered Species Act would not be reauthorized by Congress this year. Therefore, there would no longer be a federal mandate for the protection of endangered species, and there would be no provision for further federal funding for their protection. They were then informed of the hurricane damage to the RCW population in the FMNF. The probability of population survival was stated for different population sizes: at the post-hurricane population size, the population had a 50% chance of survival if minimal restoration activities were undertaken; if no restoration activities were undertaken, the population had a 0% chance of survival; and if intensive restoration activities were undertaken and the population was returned to its pre-hurricane size, it had a 99% chance of survival. These probabilities of survival were chosen so as to be able to estimate how individuals value the reduced risk of losing the population.

Given the contingent setting, individuals were asked how much they would be willing to pay into an independent foundation whose sole purpose would be the protection and restoration of the RCW. Four separate valuation questions were asked:

(i) How much would you be willing to pay into the preservation fund each year to keep the population chance of survival at 50% and to prevent it from dropping to 0%?

(ii) How much would you be willing to pay every year, over and above the amount already stated, to improve the population's chance of survival from 50% to 99%?

(iii) Rather than paying into a fund to restore and protect one single population, how much would you be willing to pay each year into a preservation fund to give all remaining RCW populations in the United States a 99% chance of survival?

(iv) Suppose the RCW did not exist in FMNF. How much would you be willing to pay to restore that ecosystem to its pre-hurricane condition?

The average willingness to pay for each of the 4 valuation questions fell in the 8 to 13 dollar range. Individuals in South Carolina had the highest valuations as expected, followed by Indiana, Montana, and Mississippi. Individuals were willing to pay twice as much, or more, for the higher probability of survival, indicating the higher value people place on reduced risk. However, they were not willing to pay substantially more for the restoration and protection of all U.S. populations versus a single population. This may indicate that a large component of the value that people place on endangered species is existence value: they are concerned primarily with the knowledge that the species exists and not so much with the number of remaining populations. However, this result may also be due to the small sample size used in the pretest. Recall that the primary goal of the pretest was to refine the survey in preparation for the full survey.

After making modifications to the survey instrument based on results and comments in the pretest, the survey was sent to 1,500 individuals with 750 people randomly chosen in South Carolina, where the primary good being valued is located, and with 750 people randomly chosen throughout the other states of the U.S. This stratification was undertaken in keeping with suggestions in the literature to sample more heavily in areas where a rare trait (here, a positive valuation for the good in question) is most likely to occur (Kish 1965).

The response rate[1] for the South Carolina subsample (58.6%) was higher than for the remainder of the United States (50.70%), although not significantly so. Average willingness to pay values for the 4 restoration activities for the two subsamples were as follows. For the 50% chance of survival of the RCW in the FMNF, South Carolina had an average willingness to pay $9.52 on average. For the 99% chance of survival of the RCW in the FMNF, the values were $13.18 and $8.74 for South Carolina and the U. S., respectively. To give all RCW populations in the U. S. a 99% chance of survival, South Carolina residents were willing to pay $8.56. Finally, willingness to pay to restore the FMNF if the RCW did not live there was $13.81 for South Carolina residents and $7.89 for the remainder of the U. S. These average willingness to pay values are significantly different from zero. However, it should be noted that 56% of all citizens surveyed indicated that they would be willing to pay $0[2] for restoration activities for the RCW.

As was expected, average values were higher for South Carolina than for the remainder of the U.S. However, since the U.S. subsample (many of whom may never expect to see a red-cockaded woodpecker or visit the Francis Marion National Forest) reported a positive average value as well, it can be concluded that a primary portion of the value that individuals place on the endangered RCW is existence value. Individuals all across the United States valued the restoration of the endangered RCW. Not only were U.S. citizens concerned about endangered species, but they expressed concern about ecosystems as well, as reflected in the positive valuation for restoration of the FMNF even if the endangered RCW did not live there. This survey has indicated support for continued efforts for the RCW and its habitat.

SUMMARY

This paper has provided a discussion of the theory underlying the valuation of non-market goods, an outline of an empirical model for determining the value of a particular environmental good, and the results from a contingent valuation survey . These preliminary results suggested that individuals did hold some positive value for the endangered RCW and that they would be willing to forego part of their income to improve its chance of survival. Full survey results and a more detailed discussion of policy implications will be presented following the final analysis of the survey data.

[1]In the full survey, the contingent valuation question was asked using three different elicitation techniques: open-ended, payment card, and dichotomous choice. Results presented here are for the open-ended elicitation technique only.

[2]These zero valuations are included in the calculation of average willingness to pay.

SECTION 4

New Insights on the Biology

New Insights on the Biology of the Red-cockaded Woodpecker

Ernest E. Stevens

The red-cockaded woodpecker is arguably one of the best studied species on the Federal list of Threatened and Endangered species. With a cooperative breeding system, year-round territories centered on clusters of cavities it constructs in living pines, and strong ecological ties to fire-maintained habitat, the red-cockaded woodpecker is an intrinsically fascinating subject. But no matter how fascinating or accessible a subject it is, the red-cockaded woodpecker's listing as endangered greatly accelerated its study. As one of the first species listed (1970), pressure to conserve the red-cockaded woodpecker has increased for two and a half decades. And when, as with the RCW, legal and financial concerns collide, the pressure for a knowledge-based resolution is intense.

The result is a body of red-cockaded woodpecker literature that now exceeds 750 publications and 3 symposia, covering a diverse array of topics. So what new insights can the papers of this section add to this body of knowledge? Perhaps the most difficult perspective to develop for any species is the historical: How and why populations change over time. With the red-cockaded woodpecker we are limited to quantitative studies extending barely a couple of decades. Results from 2 of the longest, continuous studies of red-cockaded woodpecker populations are presented in this section. Both cover virtually the same 12 year time period. Carter et al. compare changes in abundance and demographic parameters for 3 large subpopulations in the North Carolina Sandhills that are subject to different habitat "management" regimes, one of which, the residential/commercial setting, is virtually unique to this study site. DeLotelle et al. report on the demography, ecology, and social behavior for a small study population in the sparse pine forests of central Florida. Both studies report declines in red-cockaded woodpecker abundance and active clusters, with substantial year-to-year variation, and both point to habitat constraints as the reason for the declines.

Masters et al. take a different tack to develop a historical perspective. Working from recorded observations and evidence from hardwood tree rings, they reconstruct the fire history, stretching back to the early 1700s, for the McCurtain County Wilderness Area, Oklahoma, containing the largest remaining tract of old-growth shortleaf pine (*Pinus echinata*). Against this backdrop they track changing patterns of human use of fire for habitat management and its implications for red-cockaded woodpecker populations. The relationship they develop between fire suppression and red-cockaded woodpecker populations is generally a familiar one, but has seldom been documented so thoroughly.

Stevens carries the long-term perspective in the other direction projecting modeled populations centuries into the future. Management for viable populations requires large, genetically diverse populations. Using both deterministic and stochastic models, Stevens investigates the size and genetic diversity needed to meet current recovery objectives, and explores how best to reach these objectives.

Most remaining RCW populations are small and isolated, a situation that could reduce genetic fitness and hasten population extirpation. Stangel and Dixon use asymmetry in normally symmetrical traits (e.g., wing length) as an indicator of reduced fitness in red-cockaded woodpecker populations from North Carolina to Texas. Fortunately, by this measure small population size is not an immediate threat to red-cockaded woodpecker populations. Stangel and Dixon report that small populations with reduced heterozygosity showed no signs of increased morphological asymmetry. As an aside, it is also worth noting that this paper contains the largest array of morphometric data from a wide geographic area since Mengel and Jackson (1977). While it was not the purpose of this paper to examine geographic variation in morphology, such an extension with these data would be natural

A critical concern for red-cockaded woodpecker recovery is providing suitable foraging habitat, yet much remains to be learned about foraging requirements and determinants of habitat use for red-cockaded woodpeckers. Hooper and Lennartz illustrate why this task has proven formidable. They describe an experimental removal of nearly half the foraging habitat in a dense population of woodpeckers. Contrary to expectation, the mean number of red-cockaded woodpecker groups actually increased following habitat removal. Unfortunately Hurricane Hugo forced an untimely end to the study, precluding any long-term follow-up that would have helped make sense of these perplexing results. For now, this and other cited studies suggest that, except at low population densities, red-cockaded woodpeckers are not very sensitive to removal of foraging habitat.

An opportunistic study by Jackson and Parris fits better with expectations for RCW response to loss of foraging habitat. A newly created military training facility on Fort Polk, Louisiana was "shoehorned" in among several red-cockaded woodpecker groups. Jackson and Parris studied 8 groups in the area, starting two years before construction and continuing several years into its operation. Comparative data on habitat use before and after site construction were collected for 3 of these groups. Home range size increased dramatically for 2 groups with reduced foraging habitat, while remaining about the same for the 1 group that was outside the construction site. Similar studies of an experimental nature are few in number, and further work is certainly needed.

With the number of studies from throughout the red-cockaded woodpeckers range mushrooming, differences in field methodology and data analysis have become glaringly apparent. Such discrepancies pose a formidable hurdle for comparative studies among populations. Epting et al. tackle this problem head-on, adopting a common set of procedures to compare determinates of habitat use for red-cockaded woodpecker populations in Florida and Georgia. Thus they are able to distinguish ecological factors contributing to observed patterns of habitat use. Interestingly, rankings of determinants of use are similar in the 2 populations, with (in descending order) stand size, distance to cluster and interaction sites, and density and age accounting for most of the variation in both Georgia (55%) and Florida (63%). Only in the third principal component is there a difference in the relative rankings of determinates with hardwoods a more significant factor in Georgia.

The focus on declining red-cockaded woodpecker populations underscores a much more pervasive problem, decline of the great pine forests of the southeast. Brennan et al. examine the consequences of what is an increasingly prominent agent of change in the region's publicly-owned forests, management for red-cockaded woodpeckers. Their focus is the forest vertebrate community, and their objective is to assess changes in its abundance and distribution in response to burning rotation and midstory control. Although preliminary, their results suggest that remarkably divergent communities can develop under different red-cockaded woodpecker habitat management regimes. Perhaps of greater value, however, the study underscores the need to expand this type of study. Determining patterns of change in forest communities throughout the red-cockaded woodpeckers' range will require long-term studies based on consistent experimental regimes.

Several studies described in this section offer promise for furthering our knowledge of red-cockaded woodpecker biology and management. Ortego et al. expand demographic coverage of the red-cockaded woodpecker to the westernmost extent of the species' range. Since population persistence is often most tenuous on the edge of a species' distribution, studies such as theirs may, with time, prove especially revealing about the conditions necessary for population persistence. Sometimes the introduction of new tools and techniques punctuates rapid advances in knowledge. Franzreb and Barnhill introduce the use of Global Positioning System technology as a preferable alternative to more traditional methods for measuring animal movement patterns, especially when habitat associations are to be derived. And sometimes it is a new perspective that proves insightful. While concern for genetic differentiation among populations has been a cornerstone for red-cockaded woodpecker translocations, Luttrel and Stangel caution that transmission of disease and parasites is an associated, but largely overlooked problem. They bolster their stance by documenting different blood parasites among some red-cockaded woodpecker populations. The consequences of introducing new parasites into populations through translocations of their woodpecker hosts are unknown and should be examined. McFarlane also takes a novel perspective to examine the apparently large territories occupied by red-cockaded woodpeckers. He uses a comparative approach to correlate allometric and trophic differences with territory size among North American woodpeckers.

The papers in this section advance our understanding of the red-cockaded woodpecker and its role in southern pine ecosystems. Readers are encouraged to use this resource to make their own connections and to draw their own insights concerning this most interesting species.

Population Viability Considerations for Red-cockaded Woodpecker Recovery

Ernest E. Stevens, USDA Forest Service,
Southern Research Station, Department of Forest Resources,
Clemson University, Clemson, SC 29634-1003

ABSTRACT: Conservation strategies for the red-cockaded woodpecker (*Picoides borealis*; RCW) must provide for the long-term viability of populations, and develop the means for achieving it. Recovery guidelines for RCWs stipulate an effective population size (N_e) of 500 breeding adults for population viability. I calculated the population sizes necessary to give N_e = 500 using two models: **1)** a deterministic, analytical model for effective population size (Johnson-Emigh-Pollak) and **2)** a stochastic simulation model incorporating chance demographic, environmental and genetic events (Vortex). The former predicted that 556 adults, with a male:female sex ratio of 1.565, were needed to have an effective population of 500. The latter predicted the need for more than 1000 adults using demographic variability alone, and more than 1200 with the inclusion of environmental variability. Reasons for the disparate results and their applicability to field populations are discussed. Analytic and numerical sensitivity analyses were used to evaluate potential recovery tactics. Increasing the survival of female fledglings had the greatest effect of any single life-history component on population growth rate. Overall, adult survival produced greater increases in population growth and N_e, but was judged more difficult to manage in the field. Augmentation programs extended population persistence and restrained the loss of genetic variation.
KEYWORDS: population viability analysis, Vortex, Leslie matrix, effective population size, extinction, management, red-cockaded woodpecker.

Conservation strategies for the endangered RCW must confront 3 critical questions regarding the establishment of viable populations:

(1) How large a population is needed for viability?;

(2) How can small populations be recovered? and

(3) Can sufficient land be provided for viable populations? By definition, a viable population has a reasonable probability of long-term persistence, and sufficient genetic variation to be able to adapt genetically to changing environments. Population size and rate of growth are fundamental determinants of both. Because viability is typically expressed as the probability of population persistence over centuries or millennia, populations should be large enough to balance losses in quantitative genetic variation resulting from random genetic drift with gains through mutation (Lande and Barrowclough 1987). In estimating viability, the critical parameter is not the total number of individuals, but rather the effective population size, which is determined by the number of parents producing the next generation (Wright 1931). Effective population sizes in the order of 500 individuals may be required for viability (Franklin 1980, Lande and Barrowclough 1987). Recovery guidelines for RCWs in fact specify an N_e = 500 for population viability (USFWS 1985).

The problem for RCWs is that most remaining populations are small and geographically isolated. Only 6 populations contain more than 175 groups (1 or more birds per territory); 70% of populations on Federal Land contain fewer than 50 (Costa 1995). Small size and genetic isolation threaten these populations with imminent extinction through habitat loss, chance demographic events, and reduced fitness resulting from inbreeding (Shaffer 1981, 1987; Ralls et al. 1988). A number of management tools, including midstory control, artificial cavities, and augmentation, have already proven successful in securing the short-term persistence of populations. Their integration into concerted programs for recovery and genetic management of small RCW populations is only beginning to be implemented (Allen et al. 1994, Gaines et al. 1995, Haig et al. 1993a, Reinman 1995, Richardson and Stockie 1995, Walters et al. 1995b).

Thus, the immediate challenges for RCW management are to:

(1) set a target population size that will produce a rate of loss in genetic variation no greater than expected with $N_e = 500$, and

(2) prioritize methods for recovering populations.

Both can be addressed empirically with population projection models; the problem is that the RCW, because of its complex social structure and site-limited breeding, does not conform well to existing models. Nevertheless, application of these models to RCW populations is heuristically valuable and will help in developing interim guidelines. I compared 2 methods for predicting the number of adults necessary to obtain $N_e = 500$, and evaluated alternative tactics for the expansion and genetic management of RCW populations. I did not address the subsequent challenge of providing sufficient habitat for sustaining viable populations.

METHODS

Overview. I first constructed a conceptual RCW population that was demographically stationary and closed to migration. I then analyzed population viability using 2 models:

(1) Johnson-Emigh-Pollak (JEP) model (Heckel and Lennartz in review), an analytic formulation for effective population size and

(2) Vortex (V. 4.1; Grier 1980a, 1980b; Lacy 1991), a simulation model of stochastic population change. My comparison emphasized the effects of stochastic variation (i.e., chance demographic, environmental, and genetic events) in determining population viability. Next, the sensitivity of population growth rate to changes in fecundity and survival probability was examined using Leslie matrix population models (Caswell 1989). The analyses were then extended to consider consequences of these changes for population viability. Finally, Vortex was used to test the efficacy of retaining dispersing female fledglings and population augmentation as management tools.

Parameter Estimates.

Parameter estimates were derived for the RCW population on the Piedmont National Wildlife Refuge and Hitchiti Experimental Forest in central Georgia (described in Lennartz and Heckel 1987). This small population of some 34 active groups is the subject of on-going, long-term demographic studies. The number of groups and breeders in the population remained relatively stationary between 1983 and 1992, the period for which the demographic estimates below were derived (Stevens in prep.). All parameter estimates were based on a post-birth-pulse census. Each model required a different set of parameter estimates. Because Vortex required the most restrictive set, I determined the appropriate estimators for it first. Then, I used these estimates to construct corresponding sets for the other two models. When applied to the appropriate model, each set of parameter estimates produced the same population characteristics (i.e., generation time, growth rate, stable age distribution, reproductive value).

Several estimates used in Vortex **(Table 1)** require clarification:

(1) I assumed no physiological constraints on breeding at age 1 for either males or females. In natural populations, breeding among females can be constrained by the availability of clan sites, and among males by social hierarchy; both were ignored in the models.

(2) I used the proportion of potentially breeding females—those with mates and associated cavity trees—to calculate the proportion of females producing broods of various sizes. This approach indicated that on average 26.08% of females failed to produce young each year. For heuristic purposes, however, I decreased this value to 24.94% for these analyses to create a combination of demographic characteristics that produced a stationary population. In demographic terms, a stationary population is a special case of the stable population that also has an unchanging age distribution over time, due to the repeated application of constant age-specific birth and death rates. The unique feature of the stationary population is that it remains constant in size (intrinsic rate of natural increase [r] = 0.0000, or finite rate of increase [λ] = e^r = 1.0000).

(3) I set high carrying capacities for the model runs, making carrying capacity largely irrelevant as a constraint to population growth.

The JEP and Leslie matrix models required estimators for standard life table parameters. **Table 2** presents the fecundity and survivorship schedules corresponding to **Table 1**. The corresponding open-aged Leslie matrix model for females was:

$$N_{t+1} = \begin{bmatrix} 0.309 & 0.552 & 0.552 & 0.552 & 0.552 & 0.552 \\ 0.387 & 0.000 & 0.000 & 0.000 & 0.000 & 0.000 \\ 0.000 & 0.691 & 0.000 & 0.000 & 0.000 & 0.000 \\ 0.000 & 0.000 & 0.691 & 0.000 & 0.000 & 0.000 \\ 0.000 & 0.000 & 0.000 & 0.691 & 0.000 & 0.000 \\ 0.000 & 0.000 & 0.000 & 0.000 & 0.691 & 0.691 \end{bmatrix} N_t$$

where N is a vector representing the size of each age class at time t and t+1.

Table 1. Initial parameter estimates for the Vortex model based on the Piedmont-Hitchiti population.

REPRODUCTION
Age at first reproduction: M=F=1
Maximum brood size 4
Proportion of females producing:

0 fledglings	0.2494[a]
1 fledgling	0.1523
2 fledglings	0.2683
3 fledglings	0.2973
4 fledglings	0.0327

Sex ratio at fledging (proportion male): 0.533

MORTALITY
Maximum age: M=F=18
Age-specific mortality

Age	Females	Males
0-1	0.613	0.582
≥ 1	0.309	0.242

POPULATION REGULATION
Carrying capacity 1500[b]

[a]**adjusted to produce a growth rate of r=0.0000**
[b]**5000 for populations with initial size ≥1000**

Model Assumptions.

All the models incorporated 2 critical assumptions about population structure that were inappropriate for RCWs. First, they assumed that all adult females bred each year. Theory (Brown 1974, 1978; Emlen 1982a, 1991) and empirical evidence (Walters et al. 1992a, Richardson and Stockie 1995, Stevens pers. obs.) suggest rather that breeding opportunities for RCWs were site limited. This had no relevance for the JEP or Leslie matrix models, which incorporated stationary populations, but because Vortex populations were not constrained by the availability of breeding sites, growth under this model could exceed rates observed in field populations. Second, all 3 models assumed random mating to some extent. The Vortex and Leslie Matrix models assumed random mating for the whole population; JEP assumed random mating within the subgroup of males aged i and females aged j. Mating is probably never random in the socially structured populations of RCWs. Most well studied populations reported relatively high degree of mate/site faithfulness by the putative breeders (Lennartz et al. 1987, Lennartz and Heckel 1987, Walters et al. 1988) and auxiliary males are apparently behaviorally precluded from breeding (Lennartz et al. 1987, Haig et al. 1993b). Thus, the models maximized the mixing that occurred in the transmission of alleles from one generation to the next. No direct corrections for these assumptions were applied to these models, although specifying a stationary population precluded additional breeders in the two analytic models.

JEP Model.

The JEP model (Johnson 1977, Emigh and Pollak 1979) calculates effective population size for two-sex, diploid, populations with overlapping generations. The model underlying its derivation constrains the demography of the population to be deterministic, but allows stochastic variation in allele frequencies due to genetic drift. Here I used the Heckel and Lennartz (in review) version of the formula, which casts Ne in terms of standard life table entries;

$$N_e = N1T(4P_mP_f)\left\{1 + P_f\sum_{i=1}^{k_m^{-1}}(q_{i+1}^{m})^2\left(\frac{1}{l^m_{(i)}} - \frac{1}{l^m_{(i-1)}}\right) + P_m\sum_{j=1}^{k_f^{-1}}(q_{j+1}^{f})^2\left(\frac{1}{l_{f(i)}} - \frac{1}{l_{f(j-1)}}\right)\right\}^{-1}$$

Table 2. Survivorship [l(x)] and fertility [m(x)] schedules corresponding to age-specific survival rates [p(x)] used in Vortex.

	Females				Males			
x	$p^f(x)$	$l^f(x)$	$m^f(x)$	$l^f(x)m^f(x)$	$p^m(x)$	$l^m(x)$	$m^m(x)$	$l^m(x)m^m(x)$
0	0.3870	1.0000	0.0000	0.0000	0.4180	1.0000	0.0000	0.0000
1	0.6910	0.3870	0.7993	0.3093	0.7580	0.4180	0.5829	0.2437
2	0.6910	0.2674	0.7993	0.2137	0.7580	0.3168	0.5829	0.1847
3	0.6910	0.1848	0.7993	0.1477	0.7580	0.2401	0.5829	0.1400
4	0.6910	0.1277	0.7993	0.1021	0.7580	0.1820	0.5829	0.1061
5	0.6910	0.0882	0.7993	0.0705	0.7580	0.1380	0.5829	0.0804
6	0.6910	0.0609	0.7993	0.0487	0.7580	0.1046	0.5829	0.0610
7	0.6910	0.0421	0.7993	0.0337	0.7580	0.0793	0.5829	0.0462
8	0.6910	0.0291	0.7993	0.0233	0.7580	0.0601	0.5829	0.0350
9	0.6910	0.0201	0.7993	0.0161	0.7580	0.0456	0.5829	0.0266
10	0.6910	0.0139	0.7993	0.0111	0.7580	0.0346	0.5829	0.0202
11	0.6910	0.0096	0.7993	0.0077	0.7580	0.0262	0.5829	0.0153
12	0.6910	0.0066	0.7993	0.0053	0.7580	0.0199	0.5829	0.0116
13	0.6910	0.0046	0.7993	0.0037	0.7580	0.0151	0.5829	0.0088
14	0.6910	0.0032	0.7993	0.0026	0.7580	0.0114	0.5829	0.0066
15	0.6910	0.0022	0.7993	0.0018	0.7580	0.0086	0.5829	0.0050
16	0.6910	0.0015	0.7993	0.0012	0.7580	0.0065	0.5829	0.0038
17	0.6910	0.0010	0.7993	0.0008	0.7580	0.0049	0.5829	0.0029
18	0.6910	0.0007	0.7993	0.0006	0.7580	0.0037	0.5829	0.0022
Net reproductive rate				$R^f_0 = 0.9999$		$R^m_0 = 1.0001$		
Generation time				$T^f = 3.21y$		$T^m = 4.01y$		

where N_e = effective population size; N1 = number of fledglings; T = average generation length; P_m = percentage of male fledglings; l^m_i = the survivorship probability for age class i, which equals $l^m(i-1)$, survival to age i-1; $m^m(i)$ the fecundity of males of age i; k_m = the number of age classes in males; and

$$q^M_j = \sum_{i=j}^{km} p_{Mi} \quad \text{with} \quad p_{Mi} = \frac{l^m_{(i)} m^m_{(i)}}{\sum_{z=0}^{km} l^m_{(z)} m^m_{(z)}}$$

Comparable parameters for females are noted with "f".

The model was more conveniently rearranged to produce a left-hand side of N_e/N1, the ratio of effective population size to number of fledglings. The resulting right-hand side of the equation was then totally defined by life-table parameters.

Vortex Model. Vortex simulates stochastic population change, revealing variations in population size and genetic composition that result from variability in demographic, environmental, and genetic processes. Population processes are simulated as discrete, sequential events, with probabilistic outcomes determined for each individual in Monte Carlo fashion. Vortex generates pseudo-random numbers to determine whether each individual lives or dies, whether each adult female produces broods of 0, 1, 2, 3, or 4 fledglings, and which of 2 alleles are transmitted to young from each of their parents. Fecundity and mortality are sex-specific. Fecundity is independent of age for adult (reproductive) animals, while mortality is age-specific for sub-adults and age-independent for adults. The mating system was specified as monogamous.

Each run began with user-specified distributions of males and females among the various age classes from juveniles at age zero to a maximum age set at 18 (**Table 2**). Vortex assigned 2 unique alleles to each of these individuals. The program tracked the fate of these individuals, their offspring, and the alleles they carried over 200 years, with the population subject to the specified demographic characteristics and to chance demographic, environmental and genetic events. One hundred such runs were completed for each set of conditions.

Summary statistics were presented for:

(1) probability of population extinction over time;

(2) mean time to extinction for populations going extinct; and, for extant populations,

(3) mean population size,

(4) remaining level of heterozygosity, and

(5) number of remaining alleles.

Analysis.

I first calculated the ratio of effective population size to number of fledglings ($N_e/N1$) from the JEP model and used it to determine the number of fledglings produced by a population with an $N_e = 500$. I next calculated the number of successfully breeding groups that an actual population would need to produce this number of fledglings.

Based on this population size, I then used Vortex to project the rate of loss in heterozygosity under more natural conditions, focusing specifically on the effects of demographic and environmental stochasticity. Demographic variation is intrinsic to Vortex; I followed Haig et al. (1993a) in assigning arbitrary levels of low (0.01%), medium (0.05%), and high (0.10%) environmental variation. I started each run using the total number of adults calculated from the JEP model with $N_e = 500$, allowing the program to distribute the birds according to the stable age distribution for each sex. I then compared the rate of loss of heterozygosity among several larger populations. Following Harris and Allendorf (1989), I directly measured the rate of loss in heterozygosity as the slope of the natural log of observed heterozygosity regressed on time ($m = \partial(\ln Ht)/\partial t$). I used $m \approx -1/(2Ne)$ for backcalculating N_e, because it provides a good approximation when $N_e > 20$.

I used a sequential approach to examine strategies for recovering small populations. First I conducted a sensitivity analysis to determine which demographic parameters had the greatest effect on the rate of population growth (λ), as estimated by the principal eigenvalue in the Leslie matrix model, and on the $N_e/N1$ ratio from the JEP model. I initially focused only on females, since they were the limiting sex in these models, and used elasticity (Caswell 1989) to measure the proportional change in λ resulting from small, proportional changes in their reproduction and survival. The elasticity approach to sensitivity analyses provided comparable estimates of eigenvalue sensitivity for demographic parameters measured on different scales.

I next tested the sensitivity of λ and $N_e/N1$ to larger changes in demography by direct numerical substitution into the Leslie matrix and JEP models. I used the Leslie matrix to test the sensitivity of λ to changes in the survival of adult and fledgling females, and in the proportion of nests failing to produce fledglings. For the last analysis I used the JEP model to compare the sensitivity of the $N_e/N1$ ratio to changes in the survival of fledgling females and to the survival of all adults.

I then tested how best to exploit these sensitivities to increase the size of a small population. I used Vortex to assess population persistence and loss of heterozygosity in a population begun with 90 adults (approximately the present size of the Piedmont-Hitchiti population), distributed according to the stable age distribution for each sex. Because inbreeding can prove deleterious in small populations (Ralls et al. 1988), I arbitrarily assigned 1 lethal equivalent to each bird in the starting population to illustrate its potential ramifications for management strategies. In this context, I then tested two alternatives for increasing the number of breeding females in the population. Because the loss (including mortality and emigration) of fledgling females in the Piedmont-Hitchiti population is approximately 15% greater than it is for males (pers. obs.), I first considered retaining some females who would otherwise have migrated out of the population. Assuming arbitrarily that 25% of these females that either disperse or die could be retained by creating suitable cavities in suitable sites, I decreased the mortality rate for fledgling females in Vortex to 60%, which increased the rate of deterministic population growth for females to 1% per generation. I then tested the effectiveness of augmenting populations that were either stationary or growing slightly. I focused only on the addition of females, since the sex ratio for adults and fledglings is biased toward males (**Table 1**), and only on a scheme adding 5 first-year females for each of the first 10 years of the model runs.

RESULTS AND DISCUSSION

Population viability. The JEP model gave an $N_e/N1$ ratio of 1.35. For $N_e = 500$ and the

Table 3. Calculation of effective population size (N_e) for the population with demographic estimates described in Table 2.

$N_e/N1 = 1.35$ from the JEP model, $N1 = 370$ for $N_e = 500$		
Stable age distribution for population		
Age	**Males**	**Females**
Fledgling (pf)	0.213	0.187
Adult (pa)	0.366	0.234
Population (pt)	0.579	0.421
Nt = 370*(pt/pf) = 926		
Na females = 926*0.234 = 217		

specified demographic characteristics this ratio meant an annual production of 370 fledglings (N1). Using the stable age distribution generated by Vortex for the population defined by the parameter estimates in **Table 1**, and confirmed by life-table analysis for the estimates in **Table 2**, 370 fledglings corresponded to a stationary population of 556 adults, 217 females and 339 males **(Table 3)**. Because the JEP model did not address the formation of groups or the status of birds as breeders, helpers, or solitary individuals, attempts to fit these results to socially structured field populations are questionable. Probably the most reasonable configuration for such a field population would be 217 groups, each containing a breeding male and female, with most of the remaining males distributed as helpers, and a few in male-only groups. However, such a field population would need to be stationary, with random mating within age and sex classes to meet the model assumptions. No known RCW population meets these criteria.

An "ideally" behaving population loses heterozygosity at a rate equal to the inverse of twice its effective population size ($1/2N_e$) each generation (Wright 1931). A population of 556 adult woodpeckers would thus lose heterozygosity at the rate m = 0.1% per generation. In the more realistic situations modeled by Vortex, heterozygosity always declined much more rapidly. Even when considering only demographic variability, an intrinsic component of all natural systems, heterozygosity decreased more than twice as fast (m=0.23%) as predicted for the deterministic case **(Figure 1)**. Rearranging the equation for rate of loss of heterozygosity ($m \approx -1/2N_e$) and calculating for N_e, shows that this rate of loss in heterozygosity would actually correspond to an effective population of just 218 adults.

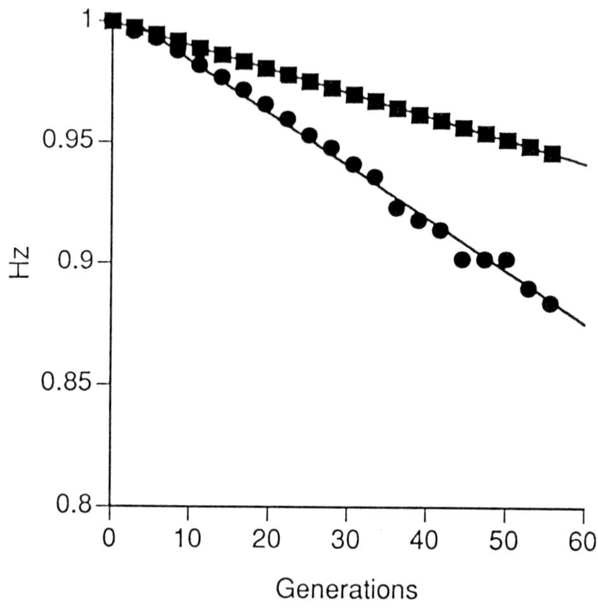

Figure 1. Decline in mean heterozygosity (Hz) over time for RCW population started with 556 adults (217 females and 339 males). Squares indicate projected declines for the Johnson-Emigh-Pollak model, and circles the projected declines for the Vortex model with demographic stochasticity. Generation time = 3.61 years.

The principal reason for the more rapid loss of heterozygosity with Vortex was variation in population size. Unlike the JEP model which held population size constant, Vortex allowed population size to vary stochastically. Although mean size remained fairly constant in the Vortex runs **(Figure 2)**, demographic stochasticity caused substantial variation in population size, and even resulted in the extinction of two populations by year 170. While Vortex does not provide a convenient summary of final population sizes, even a

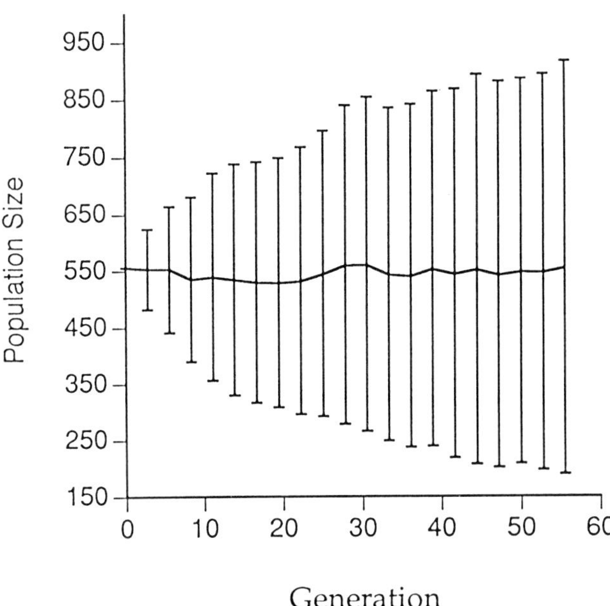

Figure 2. Mean population size (±1sd) for 100 RCW populations started with 556 adults (217 females and 339 males) in the Vortex model with demographic stochasticity.

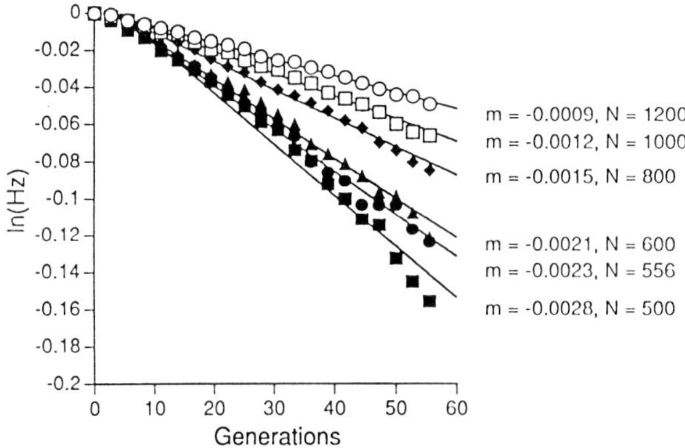

Figure 3. Rate of decline in heterozygosity (Hz) corresponding to RCW populations of various sizes (N). m = slope of natural log of mean heterozygosity regressed on time.

rough estimation based on a normal distribution with mean and standard deviation as observed at year 200 suggests that 15.73% (~15) of the remaining 98 populations contained fewer than 190 birds (mean - 1 sd), and an equal number contained more than 918.

Vortex required an initial population of between 1000 and 1200 adults to produce an average rate of decline in heterozygosity approximating 0.1% (**Figure 3**). However, even these estimates proved too low for more realistic populations. They pertained to populations experiencing demographic stochasticity only. More realistic simulations should incorporate environmental and genetic variation, and catastrophes as well. For example, the mean rate of decline in heterozygosity for populations started with 800 adults and subject only to demographic variation was 0.15% per generation. Under regimes of low (1%), medium (5%), and high (10%) environmental variability, the decline was 0.15%, 0.35%, and 0.55% per generation, corresponding to effective population sizes of 334 , 143 and 91. In addition, population persistence declined precipitously with increasing environmental variability. While all populations started with 800 birds survived for the full 200 years under low environmental variability, population survival was 89% with moderate variability, and only 40% with high variability. Greater system variability necessitated larger populations for viability.

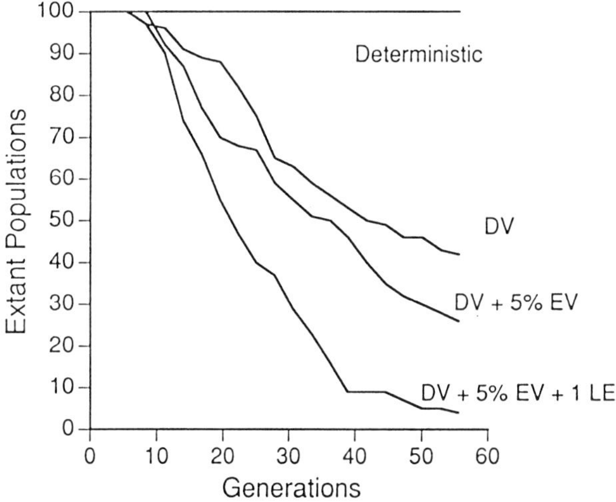

Figure 4. Population persistence for populations started with 90 adults and cumulative exposure to demographic (DV) and environmental variation (EV) and to 1 lethal equivalent (LE) per initial individual.

Population Recovery.

The immediate threat to small populations is extinction rather than loss of genetic variation. While a stationary population will persist indefinitely under deterministic conditions, the likelihood of extinction increases with the successive addition of stochasticity. **Figure 4** illustrates the persistence probabilities for populations started with 90 RCWs in the Vortex model. This example incorporates the same demographic estimates as above, a moderate rate of environmental stochasticity (5%), and 1 lethal equivalent per individual. With only demographic stochasticity (DV), 65% of these populations persisted for 100 years, and 42% for 200 years. With the addition of environmental stochasticity and inbreeding depression (DV + 5%EV + 1LE), persistence over the same periods dropped to 37% and 4%, respectively. Although short-term population persistence for up to 40 years (~11 generations) was still greater than 90%, even for the worst-case scenario (DV + 5%EV + 1LE), the decline thereafter was rapid for all scenarios.

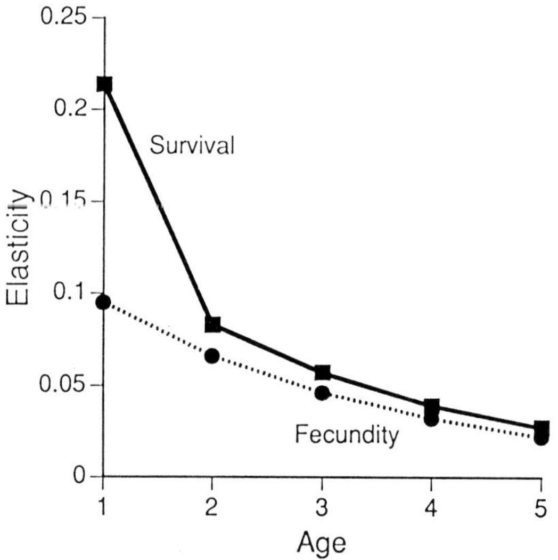

Figure 5. Elasticities of population growth rate (λ) in response to small changes in age-specific survivorship and fecundity.

The probability of a population's persistence is closely tied to its size and rate of growth: the larger a population and the greater its rate of growth, the less likely is its extinction (Goodman 1987). Therefore, recovery strategies should focus on methods for expediting population growth. Sensitivity analyses can help select the most appropriate strategies by testing the effects of alternative recovery strategies on the rate of population growth (λ) and on effective population size. An initial approach simply examined the sensitivity (elasticity; Caswell 1989) of λ to proportional changes in age-specific fecundity and survival. For females, the limiting sex in these RCW simulations, λ was consistently more sensitive to survival than fecundity (**Figure 5**). Fledgling survival had the greatest impact on λ, more than double the effect of first-year fecundity, the next most influential. For both parameters, sensitivity declined progressively with age. Management strategies, however, can seldom target either survival or fecundity on an age-specific basis. A more meaningful analysis compared the sensitivity of λ to fledgling survival, a combined adult survival parameter, and mean female fecundity. Of these three, λ was most sensitive to adult survival, not unexpected since this compounded survival over all adult age classes (**Figure 6**). Here I actually measured the population growth rate resulting from direct numerical substitution into the Leslie matrix model for females. The change in population growth rate for adult survival (slope = 0.708) was nearly twice that for a comparable change in fledgling survival (slope = 0.313), and 7 times the magnitude for mean female fecundity (slope = 0.103).

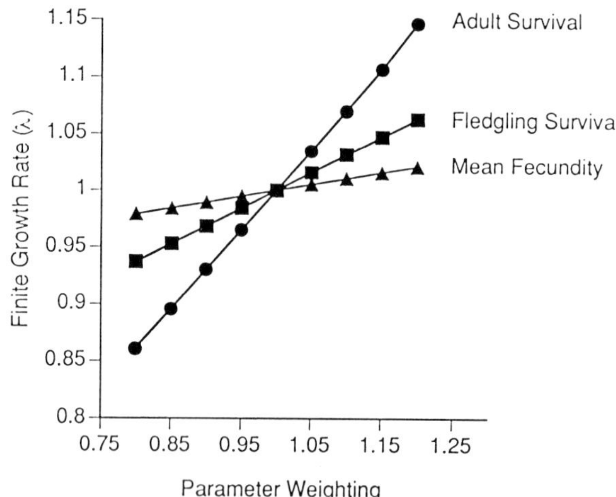

Figure 6. Sensitivity of population growth rate (λ) for females to changes in survivorship and fecundity.

The $N_e/N1$ ratio responded differently to changes in demographic parameters than did λ Changing the survival of female fledglings in the JEP model produced nearly linear change in $N_e/N1$ (slope 0.566; R2 > 0.99), but exponential change in the combined survival of male and females adults ($f(x) = .041^* e^{1.279x}$; $R^2 = 0.88$; **Figure 7**). The different shapes of these curves resulted from the way in which changes in survival affected the second-order terms, q^M and q^F, in the JEP model. Since no breeding occurred in the fledgling class ($m^m(0)$ & $m^f(0)$; **Table 2**), the product $l^m(0)m^m(0)$ in the q^M term (and for the corresponding female term) was always 0, regardless of the value for fledgling survival. Adult survival, however, was always compounded in the q^M and q^F terms. As a consequence, the N_e/N1 ratio was always more sensitive to changes in adult survival.

The magnitude of any increase in λ and N_e/N1 resulting from enhancements in nesting success and in adult and fledgling survival is only one consideration in the development of management strategies. Equally important is the likelihood of affecting a change in each of these parameters. Given our present knowledge of RCWs, nesting success and fledgling survival seem much more manageable than adult survival. Methods for minimizing cavity competition and nest predation are already in use and their efficacy is being evaluated. Artificial cavities offer the greatest potential for improving fledgling survival, because unlike most resident adults, they often lack roosting cavities. Fledglings unable to find vacant cavities in breeding sites must either immigrate or remain in the population to become nonbreeding adults. With the creation of artificial cavities in new breeding sites, these birds could increase the size of the breeding population instead. Consequently, I used Vortex to test two methods for increasing the number of fledgling females in the population:

(1) retaining female fledglings that would otherwise depart the population, and
(2) augmenting the population by translocating fledgling females from outside the population.

This focus on females assumes a surplus of unmated adult males in the Vortex populations. A male-biased sex ratio for fledglings was specified in the models **(Table 1)**, and this difference was further exaggerated by the greater survival of fledgling and adult males **(Table 2)**. The resulting sex ratio for adults in the stable age distribution was 1.565 males per female.

Retaining some of the fledgling females that presumably migrate out from populations in search of potential breeding sites is essentially equivalent to increasing their survival with respect to their natal population. When, as in the Vortex examples, there are no limitations on the number of breeding females, the resulting increase in growth rate leads to greater population persistence. For example, taking the most extreme example depicted in Figure 4, 37% of the initial populations remained after 100 years and only 4% after 200 years (**Figure 8**). The comparable persistence probabilities for populations retaining 25% of the female fledglings that would otherwise have dispersed were 42% and 12%, respectively. The extra fledglings resulted in longer population persistence times, but the difference was only of consequence after a period of some 50 years (~14 generations). These additional female fledglings prolonged the persistence of populations started with 90 adults, but could not preclude extinction from purely random events.

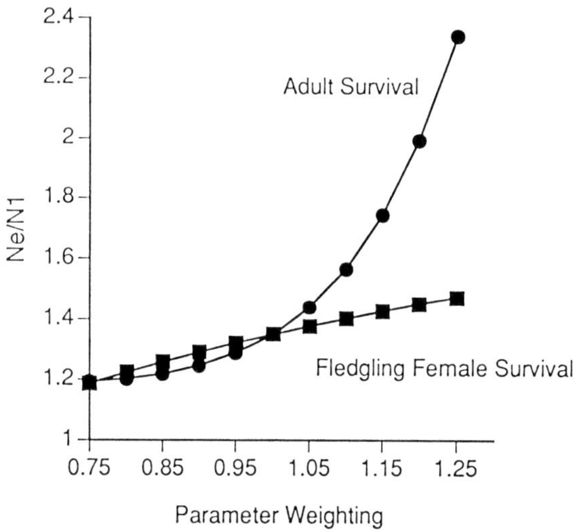

Figure 7. Sensitivity of the Ne/N1 ratio to changes in fledgling female survival and to adult survival.

Retention of female fledglings changed the underlying demographic characteristics of the modeled population, and therefore its effects pertained throughout the population's existence. The effects of augmentation were fundamentally different. The addition of 5 translocated birds in each of the first 10 years increased the size and genetic diversity of the population, but left its demography unchanged. Once past the period of augmentation, both the "stationary" and "growing" populations assumed mean trajectories that reflected their larger mean sizes **(Figure 8)**. Without augmentation the "stationary" populations averaged 101.31 (47.24 sd) birds after 10 years. Augmented populations averaged significantly larger, 185.13 (53.00 sd; $P < 0.001$, 1-tailed t-test). Larger size meant greater persistence. The mean number of years to the first extinction for the unaugmented populations was 82.97 (39.16

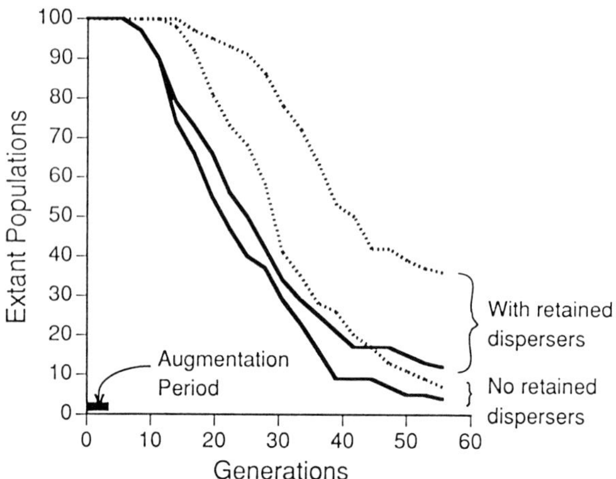

Figure 8. Effects of augmentation and the retention of fledgling females on population persistence for populations of 90 adults. Dashed lines indicate augmented population.

sd); it took significantly longer in the augmented ones 106.91 (37.18 sd; $P < 0.001$, 1 way t-test). Nevertheless, both sets of populations declined at similar rates, as indicated by their roughly parallel curves. The augmented populations just did so at a later time.

The increase in population persistence due to augmentation was greater for populations with retained female fledglings, since their higher growth rate was compounded annually. In these populations, the mean population size after 10 years without augmentation was 101.79 (53.10 sd), whereas following 10 years of augmentation with 5 females per year it was 196.54 (73.21 sd). Comparable figures for populations in which fledgling females were not retained were 101.31 (47.24 sd) without augmentation, and 185.13 (53.00 sd) with it. Thus, among augmented populations, those that retained 25% of the fledgling females that would have dispersed were about 6% larger at the end of the augmentation period. The net effect of augmentation in both these sets of populations was to delay extinction, without substantially altering the extinction rate. Augmentation just had a greater effect in the population with the greater rate of growth.

CONCLUSIONS

All models are necessarily abstractions of the natural world. How well they succeed in capturing the salient biological features of a system largely determines the scope of their usefulness. The questions that must be addressed for this examination of population viability for RCWs are therefore:

(1) How applicable are these models for extant RCW populations?; and, despite their limitations,

(2) What can these models contribute to the development of conservation strategies for RCWs?

The models examined here were generic: they captured the fundamental demographic features of RCWs, but ignored the significant social and ecological factors unique to this species. They were not designed to accommodate the group-based social system of this species, the possibility of site-limited breeding, or restricted dispersal, all features of actual or potential significance for the viability of these woodpeckers. The models instead imposed a random mating regime in which all females were able to breed. Clearly these assumptions are inappropriate for most, if not all, RCW populations. Furthermore, the demographic parameter estimates used in the models were specific to the Piedmont-Hitchiti population, and even these were adjusted to produce a stationary population. Generalization to other populations is therefore limited by interpopulation variation. Most well-studied populations vary substantially in demographic characteristics (Walters and Stevens in prep.), and few if any can be considered stationary.

Despite these limitations, the models still prove useful for shaping the development of conservation strategies for RCWs. The existing recovery plan adopted $N_e \doteq 500$ as the criterion for population viability and, for lack of better information, equated this with a breeding population of 500 adults, or 250 breeding pairs (USFS 1985). The JEP model indicated that a slightly larger population (556 adults) would be needed to meet this criterion. However, this result pertained to the conceptual model. Whether it is meaningful for actual field populations needs to be established. As it turns out, $N_e/N1$ for the Piedmont-Hitchiti population is 1.19, meaning a population of 668 adults (246 females and 422 males) would be needed for a viable population with $N_e = 500$. Long-term demographic data are available for 1 other population, the North Carolina Sandhills. Reed et al. (1988b) summarize 6 years of demographic data for this population in their Table 1, which after reorganization can be cast in terms of conventional reproductive and survivorship schedules. $N_e/N1$ calculated for these data is 1.24, remarkably similar to the value calculated for the Piedmont-Hitchiti population. For the stationary equivalent population required by the JEP model (Heckel and Lennartz in review), $N_e = 500$ translates to a population of 664 adults, 226 females and 438 males. Using a different model, Reed et al. (1988b) calculated the needed number of breeders to be 1018. However, their estimate is unquestionably too high, because the model they used was inappropriate for RCWs (Heckel and Lennartz in review). Thus, according to the JEP model, viable populations in both the Piedmont-Hitchiti and North Carolina Sandhills will be virtually the same size. The similarity is especially significant considering that the two populations differ in dominant tree species, vegetation structure, and possibly RCW prey species.

Viable populations of RCWs almost certainly will have to be larger than those indicated by the JEP model. The major argument against the unqualified application of the JEP estimate is that it assumes a stationary population in a constant environment. Vortex indicated that a more realistic environment, one incorporating chance demographic and environmental events, as well as genetic effects on fitness, would require a much larger population. For example, even when starting with 1200 adults (468 females and 732 males) and a modest level of environmental stochasticity (5%), 11 of 100 populations went extinct within 200 years, and their mean rate of loss of heterozygosity was 0.17%, still 70% greater than the stationary population. Even larger populations would be needed to survive the effects of catastrophes, such as the hurricane that devastated the RCW population on the Francis Marion National Forest (Watson et al. 1995). Such concerns are not trivial given the high probability of hurricanes striking recovery populations (Hooper and McAdie 1995). However, dealing with these expected catastrophes extends viability considerations from the single population perspective to the development of regional metapopulation plans.

The principal reason Vortex predicted such rapid rates of decline in heterozygosity for even large populations was fluctuating population size. Even though I used a stationary population as an example, chance demographic and environmental events could, and did, produce declines and even extinctions in populations that were initially large. Since the rate of loss in heterozygosity is inversely related to population size, these declining populations greatly increased the mean rate of loss in heterozygosity measured over the 100 populations in the Vortex model. Growing populations would have retained a larger fraction of their original allelic diversity, and would in this random mating regime also retain high levels of heterozygosity. In contrast, genetic drift would eliminate alleles quickly in small populations, decreasing their heterozygosity. Because the trajectories of individual populations were not available with Vortex, it was impossible to directly assess the effects of population fluctuations on heterozygosity. Had we, like Harris and Allendorf (1989) in their simulations of grizzly bear (*Ursus horribilis*), introduced some density-dependent feedback mechanism to maintain populations near or at their original sizes, the observed rate of loss would undoubtedly have better fit the idealized rate of decline. Whether such a compensatory mechanism exists in RCW populations is unknown; even if it does, populations in apparently "good" habitat have still declined or disappeared (Baker 1983b).

If managers could, in effect, take on such a compensatory role, however, population size for RCW viability would actually shift closer to the value predicted by the JEP model. This is not to say that the JEP estimate is "correct" - it still leaves out critical aspects of RCW biology. But the consequences of demographic and environmental variability and of deleterious genetic combinations that necessitate larger populations can to some extent be offset through intervention. As a first step toward intervention, the current sizes and growth rates of populations must be documented. Beyond that a strategy is needed for maintaining or recovering populations. Artificial cavities, exclusion of cavity competitors and predators, and augmentation can all be used to increase or maintain the size of populations, and translocations among and within populations can reduce matings between close relatives and retard the loss of genetic variation.

Haig et al. (1993a) provides an exhaustive example of the development of such a recovery strategy for a small population. Their strategy, which relied on a combination of augmentation and the retention of potentially dispersing females, is currently being implemented by the USDA Forest Service on the Savannah River Site and evaluated by the Southeastern Forest Experiment Station (SEFES). As discussed by Gaines et al. (1995), augmentation, translocations, exclusions of cavity competitors, and construction of artificial cavities have all been instrumental in arresting the decline of this population and stimulating its growth. Despite an abundance of artificial cavities in seemingly suitable habitat, retention of fledgling females is still very low in this population. It seems likely that female fledglings will not remain in this population until the number of unmated males increases (K. Laves, pers. com.).

SEFES is also studying the induced expansion of the Piedmont-Hitchiti population to identify the demographic and social consequences of managing for population growth. Since the entire population is banded and have established demographic and social histories have been established for this population over the past 10 years, it should be possible to directly evaluate the efficacy of various management tactics.

The results of the sensitivity analyses conducted here should provide guidance in the development of recovery strategies. While it may prove fairly simple to exclude some cavity competitors and reptilian nest predators (Richardson and Stockie 1995, Withgott et al. 1995), increased fledgling productivity contributed the least toward increasing population growth rate in this study. In contrast, a strategy to construct artificial cavities and provide suitable habitat to intercept dispersing female fledglings requires considerably greater effort and expense. However, since this strategy provides the best way to realistically increase population size and stimulate growth, the costs of its implementation will be justified in most cases. Augmentation and translocations will always prove advantageous, but will usually require habitat improvement and possibly the construction of artificial cavities. Again, expenditures of time and money will be sizeable, but both methods provide means to manage the genetic structure of populations that others do not. Augmentation mimics migration between populations, and even low rates can increase the genetic variation of populations. Translocations within populations can be used to reduce matings between close relatives, further retarding declines in genetic variation. Short-term successes with the restoration of a post-catastrophe population (Francis Marion National Forest, Watson 1995) and with the recovery efforts for smaller populations (Savannah River Site, Gaines 1995; St. Marks National Wildlife Refuge, Reinman 1995; Noxubee Wildlife Refuge, Richardson and Stockie 1995) are extremely encouraging, and testify to the role that recovery strategies built around these tools can play in creating viable populations of RCWs.

The results of this modeling exercise emphasize that genetic management should be a principal concern for the long-term maintenance of red-cockaded woodpecker populations. Retention of genetic variation is a formidable challenge, requiring much larger populations than are necessary to ensure short-term population persistence. A viable population of red-cockaded woodpeckers will likely have to be larger than the 250 breeding groups stipulated by the recovery guidelines to produce a rate of loss of heterozygosity consistent with an effective population size of 500. How much larger could not be determined from these models satisfactorily, although Vortex suggests that populations of more than double this size might be needed. More realistic, spatially-explicit models incorporating the key social and ecological characteristics for this species are under development and should provide more informed estimates. Until that time, forest managers should heed the recommendations of the recovery guidelines and "establish population goals based on the potential carrying capacity of their properties," whenever possible managing for populations larger than 250 breeding pairs. Further stipulating that populations be stable or growing in size will reduce the rate at which genetic variation is lost, and enhance their capacity to cope with future environmental changes.

Associations Between Fluctuating Asymmetry and Heterozygosity in the Red-cockaded Woodpecker

Peter W. Stangel[1] and Philip M. Dixon,
Savannah River Ecology Laboratory, Drawer E, Aiken, SC 29802

ABSTRACT: Some small, isolated populations of red-cockaded woodpeckers (RCW) exhibit reduced heterozygosity. We examined the relationship between small population size, heterozygosity, and fluctuating asymmetry, a measure of developmental stability and indirectly of fitness. We found no clear association between either population size or heterozygosity and fluctuating asymmetry. Although the techniques we employed were relatively crude, our results indicate that even small, low heterozygosity populations of RCWs are not exhibiting reduced developmental stability. Failure to associate small population size and low heterozygosity with fluctuating asymmetry may also be the result of procedural and statistical techniques employed, or natural history traits. Despite our findings, the weight of management evidence suggests that allowing populations to reach small size and low heterozygosity places populations at risk.
KEYWORDS: genetics, variability, heterozygosity, asymmetry, populations
[1]Current Address: National Fish & Wildlife Foundation, 1120 Connecticut Ave., NW, Suite 900, Washington, DC 20036.

Small populations face increased susceptibility to extinction from a variety of biological and environmental factors (Terborgh and Winter 1980, Wilcox 1982). One potential threat is the loss of genetic variability that may accompany bottlenecks in population numbers (Nei et al. 1975). Over the long term, reduced genetic variability may affect a species' survival by limiting its adaptive response to changing environmental conditions (Soulé 1980). Over the short term, lost genetic variability may reduce fitness (Mitton and Grant 1984, Allendorf and Leary 1986).

The widespread application of electrophoretic techniques has made assessment of genetic variability in natural populations routine (M.W. Smith et al. 1982). Electrophoretic surveys indicate that population fragmentation may in some cases lead to reduced heterozygosity (Vrijenhoek et al. 1985, Stangel et al. 1992). Determining the effect of reduced heterozygosity on the fitness of individuals within these small populations has proved more challenging. Fitness itself is very difficult to measure (Lewontin 1974). Most studies have therefore examined phenotypic effects that are likely to affect or be correlated with fitness. One common approach in electrophoretic studies is to test for associations between heterozygosity and developmental stability. Waddington (1942, 1948) proposed that an individual's development was stabilized by natural selection to resist variation due to genetic or environmental perturbation. Stabilization would increase the likelihood that development would result in a phenotypically normal individual that would in turn have higher fitness. Lerner (1954) pointed out that heterozygous individuals or strains generally exhibited reduced morphological variance, implying greater developmental stability. Although the causal mechanisms responsible for the relationship between heterozygosity and morphological variance are debated (Mitton and Grant 1984, Allendorf and Leary 1986), this association has been used to examine the effect of heterozygosity on fitness. Associations between heterozygosity and developmental stability are often tested by comparing the amount of variation in morphological traits with heterozygosity levels. This comparison has been made at both the individual and population level. Unpaired traits are sometimes used (e.g., Handford 1980, Fleischer et al. 1983), but more commonly fluctuating asymmetry (FA) of paired characters is used as an index of developmental stability (Leary and Allendorf 1989). FA is the normally distributed deviation about zero of the difference between sides in a typically symmetric trait (Van Valen 1962). The subtle departures from symmetry provide a measure of the sensitivity of development to perturbations. The purpose of our research was to explore the relationship between heterozygosity and FA in the RCW. This species is endemic to old-growth pine forests of the southern United States (USFWS 1985), where land management practices have contracted and fragmented RCW populations (Jackson 1971, Jackson 1978a, Lennartz et al. 1983a). RCWs are nonmigratory, with mean juvenile dispersal distances in a large North Carolina population estimated at less than 5 km (Walters et al. 1988). Species wide population estimates for the RCW are relatively high (4,500-10,500 individuals; Jackson 1978a, Lennartz et al. 1983a), but many populations are small and declining (Costa and Escano 1989, James 1995). These small populations are found at both the center and periphery of the bird's historical distribution. Estimates of heterozygosity range from 2.5-9.4% among populations with some indication that very small, isolated populations have reduced heterozygosity (Stangel et al. 1992). With no historical estimates against which to compare current population sizes and heterozygosity levels, it is difficult to determine potential negative effects in populations with small population sizes or low heterozygosity. The objective here was to estimate FA in the RCW, and to look at the associations between FA and population size and heterozygosity. The null hypothesis tested was that FA was not correlated with either heterozygosity or population size.

METHODS

Measurements of 7 paired morphological characters were taken on 276 live, adult RCWs from 25 sites throughout the species' distribution. Birds were captured during July-September, 1985-1988. All measurements were taken by PWS. The following paired characters were measured with digital calipers to the nearest 0.01 mm: tarsus length (TARSUS) from the junction of the tibiotarsus and tarso metatarsus to the tip of the heel scale; and lengths of the proximate phalange on toes III (TOE3) and IV (TOE4). Foot length (FOOT) was measured to the nearest 0.5 mm as the flattened length between the distal end of the pads of toe I to toe III. Wing length (WING) was measured to the nearest 1.0 mm from the bend of the unflattened wing to the tip of the distal primary. The number of barbs (BARBS) projecting from the lateral edges of the tongue was counted with the aid of a magnifying glass. On males the number of red feathers (FEATHERS) comprising the cockade on either side of the head was counted. Feather pulp was collected for electrophoresis. Sixteen presumptive gene loci were resolved; 11 were polymorphic (Stangel et al. 1992). The statistical program BIOSYS-1 (Swofford and Selander 1981) was used to calculate mean population heterozygosity (H). Individual heterozygosity (h) was calculated by dividing the number of polymorphic loci per individual by 16, the number of loci examined. In this analysis, heterozygosity estimates are derived from only those individuals for which morphological data are available. Details of the electrophoretic analysis are in Stangel et al. (1992).

For metrical (continuously distributed) characters, plots of left minus right differences vs size indicated size effects; analyses of WING, TARSUS, TOE3, TOE4 and FOOT were therefore conducted as size-scaled measurements. Observations were scaled by dividing the left and right side measurements by the average size [(R+L)/2]. Comparison of size-scaled WING and TOE3 measurements between males and females indicated a slight but statistically significant sexual dimorphism in size-scaled asymmetry. Hence, each trait was adjusted for sex dimorphism by estimating the average size-scaled difference between males and females and using that estimate to adjust all male values. FA was estimated by calculating the variance (among individuals) of the scaled difference between right and left sizes (Index 6 of Palmer & Strobeck 1986). For comparison Index 6 and Index 8 were also calculated on data that were not adjusted for sex differences.

Meristic (count) characters (BARBS and FEATHERS) showed no size effect. FA was estimated by the variance among individuals of the difference between sides (Index 4 of Palmer and Stobeck 1986). When one of the pair of bilateral measurements was missing, that individual was eliminated from analysis of that trait. Calculations were done using SAS Version 6.07 (SAS Institute 1989) and S-plus (Statistical Sciences 1991). Normality of the difference between sides was tested using normal quantile-quantile plots (Madansky 1988).

Differences in estimates of FA among populations were examined in two ways. First, the hypothesis of no differences in FA among populations was tested using a variant of Levene's (1960) test of equality of variances. The variant is a one-way ANOVA on the absolute value of the difference between sides, possibly size scaled (Madansky 1988). Second, Spearman's rank-order correlation coefficients were calculated between estimates of FA and either H or population size.

The relationship between h and FA was examined by extending Levene's test to an analysis of covariance. The absolute value of the difference between sides was modelled as a population-specific intercept plus the product of a common slope and the observed heterozygosity. The effect of h on FA was tested by estimating the common slope and testing if it was significantly different from 0. The assumption of equal slopes across populations was tested by a homogeneity of slopes test; no significant differences in slopes among populations were found.

To estimate the contribution of measurement error to FA, repeated measures were taken on a subset of 11 birds. On each of 11 mornings, one bird was randomly selected from the 5-10 birds captured. After being measured, this bird was returned to the group of unmeasured birds and remeasured 15-60 minutes later. The identity of the bird to be remeasured was not revealed to the measurer until after all birds were processed. Measurement error was estimated by the variance between replicate measurements of the difference between sides, scaled by mean size if the character was metrical. The F-ratio, was used to test if the estimate of FA was larger than would be expected from measurement error alone.

RESULTS

H in the red-cockaded woodpecker varied from 2.5-9.4% (**Table 1**); h ranged from 0 to 18.75%. Pooled estimates of FA for WING, TOES, FOOT, BARBS and FEATHERS were significantly larger than the estimated measurement error (**Table 2**). For TARSUS and TOE4, FA was less than three times the measurement error and not significantly different from it (**Table 2**). Significant differences in FA among populations were found for WING, FOOT, and BARBS but not for TARSUS, TOE3, TOE4, and FEATHERS (**Table 2**). However, there were no significant correlations between FA and H (**Table 3**). Fluctuating asymmetry in two of seven traits (TOE3, FEATHERS) was significantly positively associated with population size (**Table 3**). Similar results were found when the estimated FA was not corrected for sex-specific differences and when Index 8, the variance of log (left/right), was used to estimate FA. At the individual level, there were significant differences in FA among h levels for only one trait (TOE3; $p = 0.004$). The slope for this relationship was negative, indicating decreasing FA with increasing h.

DISCUSSION

No convincing evidence was found that FA of paired characters varied significantly with H across RCW populations. There were, however, some weak trends for increasing FA with increasing heterozygosity classes for BARBS, WING and TOE3. FA of two characters, TOE3 and BARBS, were significantly and positively associated with population size. Because most of these analyses do not reject the null hypothesis of no correlation between population FA and either heterozygosity or population size, it is useful to consider the statistical power of these analyses. Estimates of FA from small samples are quite variable (B.H. Smith et al. 1982) and there is a possibility that relationships are hidden by this variability. It is difficult to calculate the power of the analyses used here because of the two levels of sampling: populations and individuals. An approximation is to calculate the power of the Spearman rank correlation; this includes individual-individual variability with the population-population variation. By standard calculations (Lenth 1987) a Spearman rank correlation with 23 populations has a power of 80% if the true correlation is 0.5 and a 95% power if the true correlation is 0.6. We used a sufficiently large number of populations to detect a strong correlation, if one existed.

Table 1. Estimated population size, sample size, and mean heterozygosity in 25 populations of the red-cockaded woodpecker.

Population	Estimated Population Size[1]	Sample Size (# males)	Mean Percent Heterozygosity (H)
Manchester, SC	5	5 (2)	2.5
Pee Dee, NC	5	2 (1)	3.1
Gamelands, NC	126	22 (15)	4.0
Robeson, NC	9	5 (1)	5.0
DeSoto, MI	16	5 (4)	6.3
Talladega, AL	14	2 (2)	6.3
Savannah River, SC	7	4 (3)	6.3
Pineland, SC	9	2 (1)	6.3
Osceola, FL	97	11 (8)	6.8
Sandhills, SC	440	29 (15)	6.9
Evangeline, LA	71	7 (4)	7.1
Sabine, TX	25	6 (3)	7.3
Ocmulgee, AL	348	6 (4)	7.3
Bienville, MI	212	15 (8)	7.5
Francis Marion, SC	1034	15 (9)	7.5
Webb, SC	14	14 (3)	7.8
Ocala, FL	33	12 (9)	7.8
Boiling Springs, NC	66	12 (6)	8.3
Sam Houston, TX	264	9 (5)	8.3
Stanton, FL	47	12 (7)	8.3
Vernon, LA	390	20 (17)	8.5
Bragg, NC	322	34 (20)	8.6
Apalachicola, FL	1122	21 (14)	8.9
Kisatchie, LA	170	9 (7)	9.0
Cat Island, SC	14	8 (5)	9.4

[1]Estimated by multiplying the number of active clusters by 2.2, the average number of individuals inhabiting a cluster (Walters et al. 1988), and rounding up.

Zink et al. (1985) pointed out that at the population level, either a positive or a negative relationship between H and morphological variability might be expected, confounding interpretation of our results. A negative relationship would result from Lerner's (1954) predictions, and a positive relationship could occur if increasing H resulted in increased FA. Previous studies on other organisms have revealed many negative relationships between H and asymmetry (lizards, Soulé 1979; bivalves, Kat 1982; fish, Vrijenhoek and Lerman 1982, Quattro and Vrijenhoek 1989). In contrast, Zink et al. (1985) found that morphological variance and heterozygosity did not covary significantly across populations of fox sparrows (*Paserella iliaca*) or pocket gophers (*Thomomys bottae*). Trends for increasing morphological variability with increasing population size may also reflect heterogenous age distributions in larger populations. Age may contribute significantly to morphological variance (Straney 1978). Although age of birds sampled in this study as unknown, it might be assumed that larger populations have a more heterogeneous age distribution than do small populations. Comparison of populations with homogeneous vs heterogeneous age distributions could result in the increased variance of morphological traits in large populations documented here.

At the individual level there was a significant negative relationship between FA and h for only 1 of 7 paired characters examined. Numerous previous studies have established a strong case for increasing morphological variability with decreasing h (reviews in Mitton and Grant 1984, Allendorf and Leary 1986, Leary and Allendorf 1989). This evidence for a negative relationship between morphological variability and h has come primarily from poikilothermic organisms. This led Handford (1980) to suggest a potential dichotomy between homeotherms and

Table 2. Estimates of fluctuating asymmetry [Var (R - L), scaled by size where appropriate] and measurement error for 7 paired characters in the red-cockaded woodpecker.

Sample site[1]	Wing *[2]	Tarsus	Toe3	Toe4	Foot **[2]	Barbs ***[2]	Feather[3]
Manchester, SC	1.1	12.8	4.9	14.5	7.5	1.01	2.0
Pee Dee, NC	2.2	3.7	5.9	0.5	0.0	0.00	---
Gamelands, NC	0.6	3.4	16.9	9.9	8.7	1.10	12.7
Robeson, NC	1.7	3.1	2.7	1.3	8.9	2.29	---
DeSoto, MS	0.1	1.4	2.1	7.5	1.4	1.73	4.9
Talladega, AL	0.4	0.0	3.7	0.1	---	0.50	---
Savannah River, SC	0.2	2.3	4.0	7.5	4.7	1.62	1.3
Pineland, SC	0.0	4.9	0.1	4.3	7.0	8.13	---
Osceola, FL	0.7	2.8	3.6	4.6	3.3	1.67	11.5
Sandhills, SC	1.2	2.7	6.2	11.9	14.0	0.64	5.2
Evangeline, LA	1.9	1.5	2.2	3.1	3.4	0.28	15.6
Sabine, TX	0.2	1.1	8.2	14.8	9.2	0.27	2.3
Ocmulgee, AL	0.3	0.6	5.9	7.0	3.5	0.66	12.3
Bienville, MS	0.5	1.4	2.7	7.5	0.8	1.11	12.0
Francis Marion, SC	1.7	2.2	13.9	9.9	28.0	---	18.0
Webb, SC	3.2	1.8	5.6	9.1	0.9	1.98	6.3
Ocala, FL	1.5	4.2	5.2	7.3	3.6	1.49	6.2
Boiling Springs, NC	0.5	1.8	2.7	10.5	2.6	1.66	11.4
Sam Houston, TX	0.4	2.7	3.3	6.2	1.7	0.98	2.3
Stanton, FL	1.7	1.9	4.7	6.4	6.8	1.64	2.9
Vernon, LA	2.4	2.7	9.4	5.3	3.1	1.08	10.6
Bragg, NC	1.2	2.5	10.0	9.8	4.4	0.89	7.2
Apalachicola, FL	1.6	3.9	8.0	7.4	5.4	2.64	5.6
Kisatchie, LA	2.7	2.4	4.7	4.0	2.7	1.44	5.6
Cat Island, SC	0.6	3.8	2.0	1.1	12.1	4.04	3.8
All Sites Pooled	1.3	2.8	7.4	8.5	7.2	1.34	9.0
Estimated Measurement Error	0.2	1.8	1.2	4.2	1.3	0.02	0.3

[1]Sites listed in order of increasing mean population heterozygosity (See table 1).
[2]Differences between sites significant at *: $P<0.05$, **: $P<0.01$, or ***: $P<0.0001$, using Levene's test of equality of variances.
[3]Males only.

Table 3. Spearman rank-order correlation coefficients between mean population heterozygosity and estimated population size and estimates of fluctuating asymmetry of 7 morphological traits in the red-cockaded woodpecker.

	Fluctuating Asymmetry						
	Wing	Tarsus	Toe3	Toe4	Foot	Barbs	Feathers[1]
Mean Population Heterozygosity	0.13	-0.07	0.18	0.04	-0.03	0.25	0.08
Estimated Population Size	0.10	-0.15	0.48*	0.30	0.10	-0.17	0.55*

[1]Males only *$P<0.05$.

poikilotherms with regard to development of morphological characters. Homeotherms, characterized by a regulated internal environment, may be more resilient to environmental perturbations to developing structures. Poikilotherms, in contrast, are subject to fluctuating internal temperatures and hence may be more likely to exhibit the phenotypic effects of developmental aberrations. The relative lack of relationships between FA and h in the RCW provides partial support for this dichotomy. Wooten and Smith (1986) suggested the homeotherm-poikilotherm dichotomy may actually reflect differences in the types of characters measured. They pointed out that metric characters, the type most often measured in homeotherms, may be subject to agerelated variation. These traits might therefore have different values depending on the age of the organism measured. Meristic characters, the traits most frequently measured in poikilotherms, tend to have their values fixed early in development. A bias for one trait type in either group might provide a nonbiological explanation for differences between the groups (Wooten and Smith 1986).

Two meristic traits, cockade feather number in males and tongue barbs, were included in the present study of RCWs. There are significant, negative relationships revealed in this study involving a metric trait (TOE3). We were was unable to reject the null hypothesis that H and FA vary independently across RCW. There was a significant negative relationship between FA and individual heterozygosity for only 1 of 7 paired characters. We therefore find no indication of reduced developmental stability in small, low heterozygosity populations of the RCW.

MANAGEMENT IMPLICATIONS

We found no evidence that small populations of RCWs with reduced heterozygosity were experiencing increasing morphological variability. Keeping in mind that the tests and techniques used in this study are imperfect and rather primitive, this is good news. From the management standpoint, it is risky to allow populations to reach the small sizes now characteristic of RCWs. Small populations are highly susceptible to extinction from demographic and environmental, and eventually genetic causes.

That some small RCW populations exhibit reduced heterozygosity is indicative that these populations are at risk. That no morphological abnormalities were associated with reduced heterozygosity is a relief, but should not be considered comforting. There may well be subtle morphological or other problems occurring in these populations, and we are simply unable to detect them without more precise data. Also, most RCW populations have only recently been reduced to small size, with insufficient time elapsed for genetic developmental abnormalities to occur.

The prevailing weight of management evidence suggests that populations should not be allowed to deteriorate to the small sizes found in RCWs. At best, our data indicate that small RCW populations are not immediately susceptible to obvious morphological problems. These small populations should not be considered "lost causes" and should receive appropriate consideration within the context of the RCW recovery plan.

ACKNOWLEDGEMENTS

We are indebted to D. Stangel, J. Smith, E. Smith, J. Carter, R. DeLotelle, G. Henry and many biologists with the U.S.D.A. Forest Service for field assistance. Special thanks to M. Lennartz, R. Escano, and G. Henry for administrative support. We could not have completed this analysis without guidance from J. Novak, P. Leberg, R. Kennamer, and D. Scott. Comments from R. Hooper, D. Kulhavy, A.R. Palmer, M. Smith, J. Hamrick, and R.C. Vrijenhock improved early versions of the text. M. Reneau, P. Davis, and D. Reese patiently typed this paper many times. Research and manuscript preparation were supported by contract DE-AC09-76SROO-819 between the U.S. Department of Energy and the University of Georgia's Savannah River Ecology Laboratory.

Appendix I. Means, standard errors, and sample sizes ( ) for morphological traits in 7 populations of the red-cockaded woodpecker.

Population	Weight	Bill Depth	Bill Width	Bill Length
Sam Houston, TX	47.2±0.9 (9)	6.29±0.07 (9)	6.87±0.14 (9)	16.73±0.28 (9)
Stanton, FL	42.0±0.3 (12)	5.82±0.0 (11)	7 6.86±0.10 (11)	16.90±0.16 (11)
Vernon, LA	47.1±0.5 (20)	6.06±0.0 (20)	5 6.76±0.12 (20)	16.58±0.20 (20)
Bragg, NC	47.6±0.5 (34)	6.12±0.05 (34)	6.81±0.07 (34)	16.77±0.12 (34)
Apalachicola, FL	46.3±0.5 (22)	6.24±0.54 (22)	6.84±0.08 (22)	17.00±0.17 (22)
Kisatchie, LA	45.8±0.9 (9)	6.06±0.11 (9)	6.80±0.10 (9)	16.78±0.26 (9)
Cat Island, SC	45.5±0.5 (8)	5.97±0.12 (8)	6.78±0.07 (8)	16.86±0.16 (8)

Appendix II. Means, standard errors, and sample sizes ( ) for morphological traits in 25 populations of the red-cockaded woodpecker.

Population	Rt Wing	Lt Wing	Rt Tarsus	Lt Tarsus	Rt Toe 1	Lt Toe 1
Manchester SC	116.8±1.0 (4)	117.8±1.0 (4)	24.60±0.23 (5)	24.77±0.44 (5)	7.49±0.10 (5)	7.44±0.10 (5)
Pee Dee, NC	118.5±0.5 (2)	123.0±2.0 (2)	24.51±0.72 (2)	24.90±0.37 (2)	7.59±0.04 (2)	7.61±0.15 (2)
Gamelands, NC	114.8±0.6 (21)	115.4±0.4 (21)	24.3±10.13 (21)	24.53±0.14 (21)	7.38±0.04 (21)	7.50±0.04 (21)
Robeson, NC	118.2±1.4 (5)	118.8±2.0 (5)	24.52±0.46 (5)	24.80±0.41 (5)	7.41±0.08 (5)	7.34±0.09 (5)
DeSoto, MI	115.8±1.3 (5)	115.6±1.6 (5)	24.15±0.15 (5)	24.26±0.20 (5)	7.20±0.14 (5)	7.25±0.11 (5)
Talladega, AL	118.5±1.5 (2)	119.0±1.0 (2)	25.31±0.04 (2)	24.94±0.04 (2)	7.31±0.06 (2)	7.50±0.04 (2)
Savannah River, SC	119.0±0.7 (4)	119.5±0.5 (4)	25.24±0.53 (4)	25.10±0.41 (4)	7.39±0.18 (4)	7.47±0.18 (4)
Pineland, SC	118.5±0.5 (2)	120.5±0.5 (2)	24.93±0.73 (2)	24.15±0.11 (2)	7.93±0.14 (2)	7.78±0.14 (2)
Osceola, FL	114.2±0.4 (11)	116.0±0.5 (11)	24.41±0.24 (11)	24.39±0.30 (11)	7.25±0.13 (11)	7.40±0.19 (10)
Sandhills, SC	116.3±0.4 (28)	117.1±0.4 (28)	24.66±0.16 (28)	24.80±1.14 (28)	7.53±0.09 (28)	7.50±0.08 (28)
Evangeline, LA	116.4±0.8 (7)	116.1±0.8 (7)	24.94±0.36 (7)	24.81±0.31 (7)	7.39±0.16 (7)	7.42±0.17 (7)
Sabine, TX	113.0±1.2 (6)	113.5±1.0 (6)	24.46±0.27 (6)	24.46±0.29 (6)	7.29±0.05 (6)	7.28±0.09 (6)
Ocmulgee, AL	115.5±0.3 (6)	115.5±0.3 (6)	24.31±0.08 (6)	24.50±0.10 (6)	7.19±0.12 (6)	7.28±0.08 (6)
Bienville, MI	117.7±0.4 (15)	117.9±0.4 (15)	24.49±0.20 (15)	24.51±0.21 (15)	7.37±0.09 (15)	7.45±0.10 (15)
Francis Marion, SC	117.8±0.4 (15)	119.2±0.4 (15)	24.52±0.17 (15)	24.62±0.15 (15)	7.58±0.07 (12)	7.53±0.07 (12)
Webb, SC	118.5±0.5 (2)	118.3±0.9 (3)	25.54±0.32 (4)	26.01±0.22 (4)	7.67±0.13 (4)	7.77±0.16 (4)
Ocala, FL	114.3±0.7 (12)	115.0±0.7 (12)	24.66±0.21 (12)	24.77±0.24 (12)	7.42±0.10 (11)	7.63±0.11 (12)
Boiling Springs, NC	119.8±0.6 (5)	120.2±0.6 (5)	25.04±0.23 (12)	25.27±0.19 (12)	7.54±0.06 (12)	7.57±0.07 (12)
Sam Houston TX	115.3±0.5 (9)	115.7±0.5 (9)	24.67±0.26 (9)	25.11±0.32 (9)	7.48±0.11 (9)	7.50±0.11 (9)
Stanton, FL	110.8±0.9 (12)	111.7±0.8 (12)	24.27±0.18 (12)	24.50±0.17 (12)	7.36±0.40 (12)	7.36±0.56 (12)
Vernon, LA	115.2±0.7 (20)	115.4±0.7 (20)	24.93±0.11 (20)	24.94±0.14 (20)	7.59±0.05 (20)	7.67±0.04 (20)
Bragg, NC	118.6±0.4 (34)	118.9±0.4 (34)	24.53±0.13 (34)	24.57±0.13 (34)	7.20±0.06 (34)	7.30±0.05 (34)
Apalachicola FL	119.1±0.4 (20)	119.9±0.4 (20)	25.00±0.15 (21)	25.07±0.17 (21)	7.42±0.09 (21)	7.60±0.08 (21)
Kisatchie, LA	116.4±0.9 (9)	116.1±0.6 (9)	25.08±0.28 (8)	25.17±0.31 (8)	7.67±0.09 (8)	7.80±0.12 (8)
Cat Island, SC	118.3±0.6 (8)	119.3±0.6 (8)	25.06±0.20 (8)	25.10±0.18 (8)	7.59±0.05 (8)	7.67±0.07 (8)

Appendix II continued.

Rt Toe 2	Lt Toe 2	Rt Foot	Lt Foot	Rt Feather	Lt Feather	Rt Barbs	Lt Barbs
7.50±0.05 (5)	7.64±0.17 (5)	23.7±0.3 (5)	23.8±0.4 (5)	16.0±3.0 (2)	16.0±4.0 (2)	8.6±1.8 (5)	8.6±2.1 (5)
7.62±0.07 (2)	7.94±0.12 (2)	23.3±0.3 (2)	23.8±0.3 (2)	15.0 (1)	16.0 (1)	8.5±0.5 (2)	7.5±0.5 (2)
7.41±0.04 (21)	7.57±0.06 (21)	24.5±0.1 (21)	24.7±0.1 (21)	15.7±0.6 (13)	16.1±0.8 (13)	8.8±0.5 (21)	9.0±0.4 (21)
7.58±0.10 (5)	7.63±0.12 (5)	23.0±0.4 (5)	22.9±0.6 (5)	18.0 (1)	22.0 (1)	7.4±0.5 (5)	7.0±1.1 (5)
7.37±0.14 (5)	7.36±0.15 (5)	24.5±0.20 (4)	24.3±0.3 (4)	16.3±1.8 (9)	14.5±1.9 (9)	9.4±0.2 (15)	9.6±0.6 (15)
7.51±0.12 (2)	7.41±0.14 (2)	25.3±0.8 (2)	25.3±0.8 (2)	13.0±1.0 (2)	13.0±1.0 (2)	8.5±0.5 (2)	8.0±0.0 (2)
7.51±0.13 (4)	7.74±0.16 (4)	23.3±0.5 (4)	24.0±0.6 (4)	15.3±1.2 (3)	14.0±0.6 (3)	10.5±1.0 (4)	10.3±0.5 (4)
8.16±0.10 (2)	8.06±0.01 (2)	25.0±0.5 (2)	25.0±0.0 (2)	16.0 (1)	19.0 (1)	8.5±1.5 (2)	8.5±0.5 (2)
7.26±0.10 (11)	7.41±0.10 (11)	24.2±0.2 (11)	24.1±0.2 (11)	15.3±1.2 (8)	15.1±1.8 (8)	8.2±0.4 (10)	8.4±0.4 (10)
7.71±0.09 (28)	7.74±0.09 (28)	23.4±0.2 (28)	23.4±0.2 (28)	18.7±1.6 (15)	17.9±1.3 (15)	7.3±0.3 (15)	7.3±0.4 (15)
7.54±0.19 (7)	24.8±0.3 (7)	24.9±0.3 (7)	24.9±0.3 (7)	22.5±4.5 (4)	18.3±2.6 (4)	8.3±0.6 (7)	8.7±0.6 (7)
7.32±0.06 (6)	7.60±0.07 (6)	24.3±0.2 (6)	24.8±0.2 (6)	15.3±0.3 (3)	15.0±1.0 (3)	9.3±0.9 (6)	9.7±0.8 (6)
7.30±0.07 (6)	7.42±0.12 (6)	25.0±0.4 (6)	25.0±0.3 (6)	17.8±2.2 (5)	18.6±3.0 (5)	8.7±0.3 (6)	8.3±0.2 (6)
7.37±0.82 (15)	7.53±0.10 (15)	25.1±0.2 (15)	25.0±0.2 (15)	18.8±1.4 (9)	20.9±1.8 (9)	8.7±0.4 (15)	9.1±0.4 (15)
7.72±0.06 (15)	7.83±0.07 (15)	25.3±0.7 (12)	22.4±0.8 (12)	17.1±1.69 (9)	16.7±1.1 (9)	---	---
7.76±0.16 (4)	7.88±0.15 (4)	24.4±0.5 (4)	24.5±0.5 (4)	22.6±2.3 (3)	17.3±1.7 (3)	6.8±0.5 (4)	6.8±0.5 (4)
7.60±0.09 (12)	7.77±0.10 (12)	23.7±0.3 (12)	23.7±0.3 (12)	14.6±0.8 (9)	16.2±1.0 (9)	9.5±0.3 (12)	8.8±0.4 (12)
7.72±0.09 (12)	7.77±0.08 (12)	24.6±0.3 (11)	24.3±0.4 (11)	17.6±0.9 (7)	17.1±1.1 (7)	8.1±0.4 (12)	7.8±0.4 (12)
7.51±0.11 (9)	7.59±0.09 (9)	25.2±0.3 (9)	25.3±0.3 (9)	18.8±2.4 (5)	18.2±1.8 (5)	8.4±0.7 (9)	8.4±0.7 (9)
7.39±0.05 (12)	7.55±0.56 (12)	24.3±0.2 (12)	24.5±0.1 (12)	15.3±1.5 (7)	14.6±1.9 (7)	9.5±0.3 (12)	9.5±0.3 (12)
7.61±0.05 (19)	7.72±0.04 (20)	25.1±0.2 (20)	25.2±0.2 (20)	16.1±1.0 (16)	15.9±0.9 (16)	8.4±0.3 (20)	8.0±0.4 (20)
7.25±0.04 (34)	7.51±0.06 (34)	24.7±0.1 (34)	24.8±0.1 (34)	18.5±0.9 (20)	17.8±1.0 (20)	8.0±0.2 (34)	7.9±0.2 (34)
7.56±0.07 (21)	7.71±0.07 (21)	24.3±0.1 (21)	24.7±0.1 (21)	17.7±0.9 (15)	16.5±0.8 (15)	7.7±0.2 (20)	8.2±0.3 (20)
7.80±0.10 (8)	7.88±0.13 (8)	25.3±0.5 (8)	25.3±0.4 (8)	14.0±1.2 (7)	15.4±1.6 (7)	8.4±0.6 (9)	8.2±0.8 (9)
7.79±0.07 (8)	7.78±0.09 (8)	24.4±0.4 (8)	24.0±0.2 (8)	17.0±1.9 (6)	15.7±1.4 (6)	7.5±0.6 (8)	7.5±0.6 (8)

Red-cockaded Woodpeckers in the North Carolina Sandhills: a 12-year Population Study

J.H. Carter III, J.R. Walters and P.D. Doerr
Department of Zoology, North Carolina State University, Box 7617, Raleigh, NC 27695-7617

ABSTRACT: A large, color-banded population of red-cockaded woodpeckers (*Picoides borealis*) was studied on 3 study areas (SOPI, SGL and FB) in the Sandhills of North Carolina from 1980 through 1991. There were dramatic decreases in numbers of active colonies on all study areas, particularly SGL. There were corresponding decreases in numbers of total groups, but breeding groups actually increased on FB over 1981 levels. Nesting effort showed yearly variability, being very good in 1984 and 1991, and very poor in 1982. Overall, SOPI had a slightly higher nesting success rate, but SGL had the highest average clutch size, and FB produced the most fledglings. SOPI had the highest fledging rate, and SGL produced slightly more fledglings per successful nest. Numbers of adults fell on all study areas, but especially SOPI and SGL. Groups with helpers declined sharply on SGL. The general decline in active colonies and population size were believed to be caused by deterioration of habitat quality, specifically an increase in understory encroachment in many colonies in all study areas.
KEYWORDS: red-cockaded woodpecker, population trends, demography, Sandhills

The red-cockaded woodpecker (*Picoides borealis*) is a small, black and white woodpecker endemic to mature pine forests of the southeastern United States. Formerly common throughout its broad range, it was listed as endangered in 1968 (USDI 1968b). This species excavates nest and roost cavities in old, living pines and requires large areas of mature pine for foraging habitat. Destruction of the once extensive old-growth pine forests throughout the Southeast has led to this bird's endangered status. Red-cockaded woodpeckers live in social groups called "clans" or "groups", which typically consist of a mated pair and 0-3 male "helpers", usually related young from previous years. Each group occupies a cluster of cavity trees termed a "colony" or "cluster" (Walters et al. 1988), and has a well-defined, large, home range. Ranges abut or overlap slightly in good habitat and vary from 30 to over 213 ha. Typically, only 1 brood is raised a year, and some clans may not nest at all.

The endangered status of this species has generated much interest in the last 2 decades, including 2 symposia (Thompson 1971 and Wood 1983a), 2 recovery plans (USFWS 1979 and 1985b), and numerous papers and notes. The red-cockaded woodpecker also has been the subject of much controversy because management and conservation of this species potentially impacts large amounts of private and public timberlands.

The red-cockaded woodpecker was studied in the Sandhills of North Carolina from 1980 through 1991 in order to document population status and trends. Data on population size, nesting effort, and recruitment were gathered on 3 large study areas. Goals were to locate all cavity trees and colonies, census all adults annually, locate all nests, census all fledglings, and document trends in the population as a whole and within the individual study areas. Population trends for these study areas were summarized in Carter et al. (1983b) and Walters et al. (1988).

This study is unique in having all individuals marked in such a large area (1170 sq. km.) and for such a long time (12 years). The study population comprises a major part of one of the largest remaining red-cockaded woodpecker populations. The results of this study document the current status and trends of this population and provide a basis for making management decisions for its conservation, and a benchmark for measuring future trends and responses to management. This study also provides information on the dynamics of a red-cockaded woodpecker population living in suburban habitat, a situation with unique management and legal implications.

STUDY AREAS

The North Carolina Sandhills region, like much of the Coastal Plain of the Southeast, was historically forested with almost pure stands of longleaf pine *(Pinus palustris)*, which were relatively undisturbed until the mid to late 1800's when naval stores and lumber industries moved into the area. By the early 1900's, nearly all forests had been cut-over. Our survey area was forested with second-growth longleaf pine with scattered old-growth trees, with a typical understory of 1 or more species of scrub oaks (*Quercus* spp.), of which turkey oak (*Q. laevis*) was the most common. Several decades of fire exclusion and suppression have resulted in dense understories in many areas. Some areas have failed to regenerate to pine and were vegetated with almost pure stands of scrub oaks or mixed upland hardwoods such as southern red oak *(Q. falcata)*, post oak (*Q. stellata*), hickories (*Carya* spp.), and flowering dogwood (*Cornus florida*). Pond pines (*P. serotina*) occurred along hillside drains and small creeks and in pocosins along the major streams. Pond pines were normally associated with several species of deciduous and evergreen shrubs, including blueberries (*Vaccinium* spp.), gallberries (*Ilex* spp.), fetterbushes (*Lyonia* spp. and *Leucothoe racemosa*), titi (*Cyrilla racemiflora*), and sweet bay (*Magnolia virginiana*). Old field sites supported stands of loblolly pine (*P. taeda*), often mixed with longleaf pine.

Approximately 117,020 ha in Moore, Hoke, Montgomery, Richmond and Scotland counties were surveyed, and within that area 3 major study areas were designated: Southern Pines-Pinehurst (SOPI), Sandhills Game Land (SGL) and Ft. Bragg (FB). Surveyed habitats surrounding and between the major study areas were collectively designated MIN (**Figure 1**).

SOPI consisted of approximately 16,301 ha and included the resort towns of Southern Pines, Pinehurst and Aberdeen, and 18 golf courses. Much of the area had been developed as forested residential areas, but some tracts of undeveloped pinewoods occurred between subdivisions. One substantial area supported horse farms with large forested pastures. Most land in SOPI is privately-owned with little being managed for timber production. Prescribed burning was regularly practiced only at the 175 ha Weymouth Woods-Sandhills Nature Preserve. Elsewhere, understories were generally uncontrolled, except in some areas used for fox hunting where annual mowing, and in later years, burning were practiced. Some understory openings occurred naturally, and others were provided by roads, yards, golf courses and pastures. Understory encroachment was occurring in many red-cockaded woodpecker colonies. One colony was destroyed during the study and several cavity trees were cut.

SGL consisted of 16,831 ha, with about one-half being managed by the North Carolina Wildlife Resources Commission. The remainder was private farm and forest land. The Game Land portion was undeveloped except for roads, wildlife food plots and an extensive field trial course. Most pine stands were managed on a 100-year rotation (a recent increase from a 60-year rotation) and burned on a 3-year rotation. Understory encroachment was occurring in many colonies despite the burning program. In addition, many pine stands were poorly stocked and regeneration success has been low. The privately-owned portion of this study area contained little red-cockaded woodpecker habitat and no colonies. No colonies were destroyed on SGL during our study and only 1 cavity tree was cut, but several cavity trees were killed during prescribed burns.

The 13,855 ha FB study area was undeveloped except for roads, an extensive firebreak system, wildlife food plots and 2 large, cleared parachute drop zones. About 1870 ha were in young pine plantations. Longleaf pine stands on FB were managed on a 100-year rotation and most stands were well stocked. The forest is open in many areas, due in part to a long history of wildfires. Pine stands were prescribed burned on a 5-year rotation. More natural pine savanna occurred on FB than on the other study areas, but understory encroachment still was occurring in many colonies. No colonies and only 1 cavity tree were destroyed on FB during the study, but several cavity trees were killed by wildfires and prescribed burns.

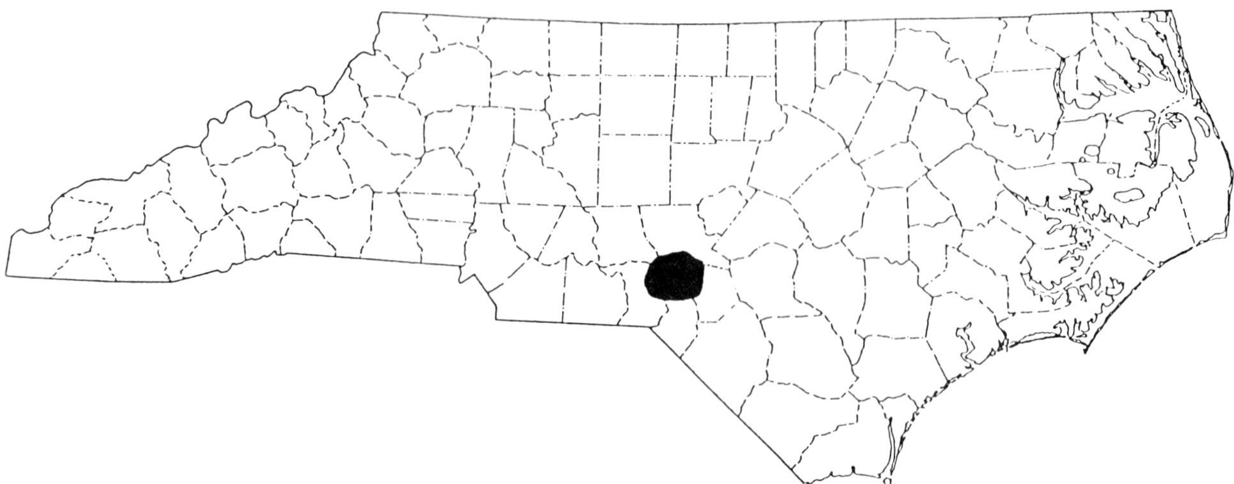

Figure 1. Red-cockaded woodpecker study areas in the Sandhills of North Carolina.

MIN contained 70,032 ha around and between the major study areas. Major habitats in MIN were longleaf pine-scrub oak forest, pine plantation, agricultural lands, and areas in varying stages of development. Also within MIN were 5 golf courses, the Towns of Pinebluff, Seven Lakes, West End, Whispering Pines, and several smaller communities. Major forested tracts were the Camp Mackall Military Reservation (2921 ha), McCain state-owned property (884 ha), and the Blue Tract (1432 ha). Two colonies were destroyed during the study, and several other cavity trees were cut (8 in one colony).

METHODS

All potential habitat within the study areas was surveyed by foot and/or vehicle beginning in 1978. Colonies were defined using criteria described in Walters et al. (1988). Beginning in 1988, artificial cavities were placed in several active and inactive colonies and in unoccupied habitat on all 3 major study areas. Data from the new colonies were not used unless the colony was occupied by red-cockaded woodpeckers, or occupied then abandoned.

Banding of adults and nestlings began in 1979. A unique combination of colored leg bands was used on 1 leg and an aluminum U. S. Fish and Wildlife Service band and a single color band was placed on the other.

Adults were captured by using a mist net sewn to a wire frame attached to a telescopic pole. Banding was not done in inclement weather or on nights when the forecast low temperature was below 0° C. Because of concern that the capturing and/or banding of the birds might compromise survival, data were analyzed for possible effect. No significant negative effect was detected (Carter et al. 1983b, Walters et al. 1988), though unhandled birds did survive slightly better than handled ones (Walters et al. 1988).

Nestlings were removed from cavities with flexible, nylon snares. Whenever possible nestlings were banded before pin feathers began to erupt. A few nestlings died during extraction and banding, and others were injured. Injured nestlings fledged at a rate comparable to uninjured nestlings (Walters et al. 1988).

Only SOPI and SGL were monitored in 1980 when intensive field study began. All colonies were checked for activity from late March to early May, and active sites were then checked for nests every 9-11 days beginning in late April. If an adult was flushed from a cavity or if woodpeckers were seen around a highly active cavity tree, the tree was climbed and the cavity's contents checked with a drop light and mirror. Even if such activity was not observed, most active cavity trees in a colony were climbed at least once during a breeding season. Whenever we failed to find a nest in an active colony by mid May to early June, the surrounding area was searched for new cavity trees. Once a nest was located and young banded, most sites were not checked again until a few days to 5 weeks (rarely) after the projected fledging date (hatching day +26). Adults were counted and identified during both nest checks and fledging checks. Nearly all adults on SOPI and SGL had been banded by the fall of 1980.

All 3 study areas were checked in 1981. Methods were the same, but the interval between checks of active colonies increased to about 2 weeks. Young from several nests on FB fledged before they could be banded, but those sites were carefully censused for adults and fledglings. About one-half of the FB adults were banded by the beginning of the breeding season and most of the remainder were banded during the fall.

The 3 study areas were again monitored from 1982 through 1991. Methods were similar, but site check intervals increased to 3 weeks at some SGL colonies in 1982, and again between 1989 and 1991. Almost all of the study population had been banded by the fall of 1982.

The number of colonies monitored on SOPI began at 99 in 1980 and ended with 107 in 1991. The number of colonies checked on SGL rose almost every year, beginning in 1980 with 92 and ending with 106 in 1991. Colonies monitored on FB increased most years, beginning with 90 in 1981, and ending with 135 in 1991. The number of colonies checked in MIN grew from 32 in 1980 to 94 in 1991.

For most comparisons, only data from the colonies that were checked in all years were used. An exception was clutch size, where data from all colonies was utilized. Generally, we did not use statistics to analyze data across years because of the non-independence of years. Tests generally used data from 1980-87, because after that, management activities potentially altered several parameters (these will be analyzed at a later time).

Efforts were made throughout the study to assure that we did not fail to detect new groups or groups that moved from known colonies. Whenever a nest was not detected in a colony that normally contained a nest, the area within one-quarter mile was searched for new cavity trees. Researchers were in the field throughout this study following birds, plotting cavity trees, and carrying out numerous research related tasks. Significant portions of all study areas were resurveyed at least once during the study, and all of FB was resurveyed in 1991-92. Our normal annual study activities detected several instances of new groups forming by territorial "budding", and we documented numerous instances of groups moving from 1 known colony to another. Resurveys detected previously unknown cavity trees, but no new colonies or groups.

RESULTS

Cavity Tree and Colony Numbers

More than 4200 cavity trees were found and grouped into 449 colonies. SOPI colonies averaged more cavity trees per colony (12.2) than SGL (8.1), FB (8.6) or MIN (6.4). Though SOPI had the most cavity trees per colony, it had the highest number and percentage of abandoned colonies (43 and 39.8%, respectively). Differences between study areas in the proportion of inactive colonies were significant both at the beginning (1981, x^2_2 = 14.4, p< 0.001) and end (1991, x^2_2 = 12.6, p< 0.01) of the study.

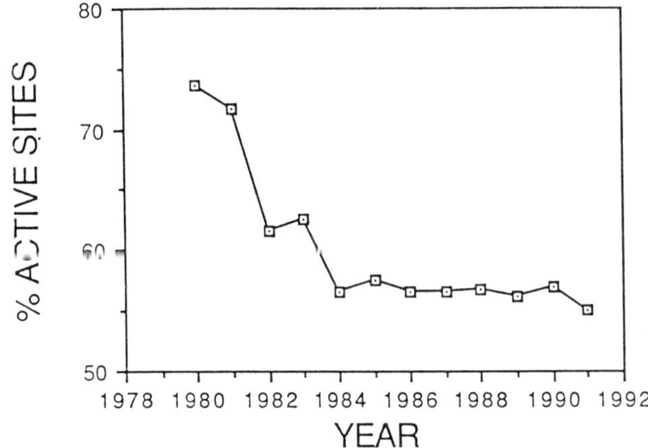

Figure 2. Percent of red-cockaded woodpecker colony sites that were active on Southern Pines-Pinehurst (SOPI) study area, 1980-1991.

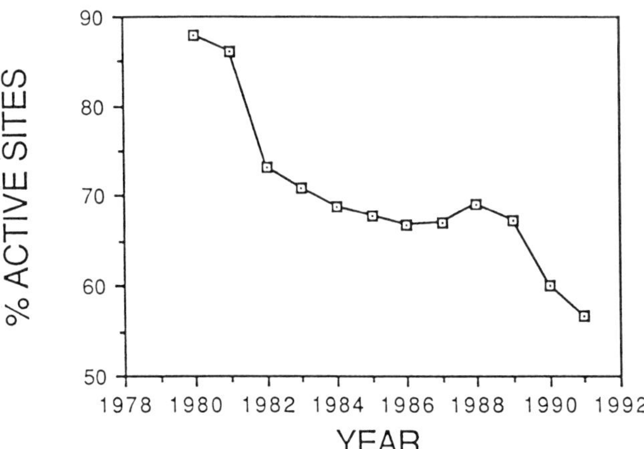

Figure 3. Percent of red-cockaded woodpecker colony sites that were active on Sandhills Game Lands (SGL) study area, 1980-1991.

Colony Activity

The numbers of active colonies on the major study areas were similar in 1980 and 1981, but decreased sharply on SOPI and SGL in 1982 **(Figures 2 and 3)**. Active colonies dropped on FB in 1982 and 1983, then gradually declined through 1991 **(Figure 4)**. Active colonies continued to decline on SOPI and SGL through 1984, but then stabilized and remained so on SOPI through 1991, and on SGL through 1989, when a further decline began.

The number of abandoned colonies increased on all 3 study areas during the study. Abandoned colonies increased on SOPI by 58%, on SGL by 291% and on FB by 200% (all percentages exclude abandoned colonies added during the study). Of all colonies monitored in 1991, 45% were abandoned on SOPI, with 43% on SGL and 24% on FB.

Groups

On SOPI the number of breeding groups, which is the same as colonies with nests, was 42 in 1980 and 1981, but fell by 21% to 33 in 1982 **(Figure 5)**. The number gradually rose to 42 in 1988, before ending at 40 in 1991. There were 47 breeding groups on SGL in 1980, but there was a 28% decline to 34 by 1982 **(Figure 6)**. SGL breeding groups recovered to 52 in 1987, then dropped to 44 in 1991. Four of the SGL breeding groups in 1990 and 1991 were in "new" colonies containing only artificial cavities. These new groups partially offset losses of breeding groups in active colonies. Breeding groups on FB began at 51 in 1981 and ended at 57 in 1991, with a high of 62 in 1984, and low of 47 in 1988 **(Figure 7)**. There was a significant study area difference in number of breeding groups for both year and site (ANOVA, F= 3.92, d.f.= 7,13, p= 0.0162; F= 37.94, d.f.= 2,13, p= 0.0001). All study areas were significantly different (p <0.05, LSD multiple-range test in GLM procedures [SAS 1985]) through 1987.

Each year there were some groups that did not breed. Such groups were of 2 primary types, pairs or pairs with helpers that either did not breed or for which nesting was not detected, and all-male groups. All-male groups also had 2 subdivisions, father-son "pairs" where a female was lacking, and solitary males (see Walters et al. 1988). Each year a substantial number of colonies were occupied by nonbreeding or all-male groups. Nonbreeding groups shared a similar pattern on all the major study areas. In years of moderate to high reproductive effort, the numbers of nonbreeding groups were low, while in years of poor reproductive effort the numbers of such groups were high **(Figures 5, 6 and 7)**. Year differences were significant (ANOVA, F= 4.73, d.f.= 7,13, p= 0.078) through 1987. The pattern for all-male groups was somewhat similar, but changes did not correspond to changes in reproductive effort as well as did variation in number of the nonbreeding groups **(Figures 5, 6 and 7)**. The number of all male groups declined or remained stable on all major study areas through most of the study, but rose toward the end on SOPI and SGL. Year

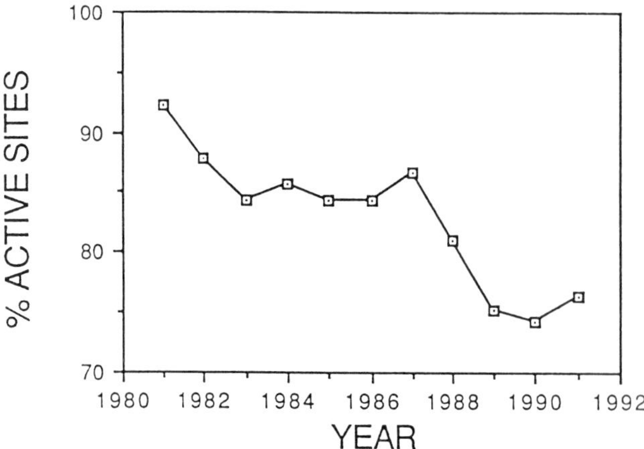

Figure 4. Percent of red-cockaded woodpecker colony sites that were active on Fort Bragg (FB) study area, 1981-1991.

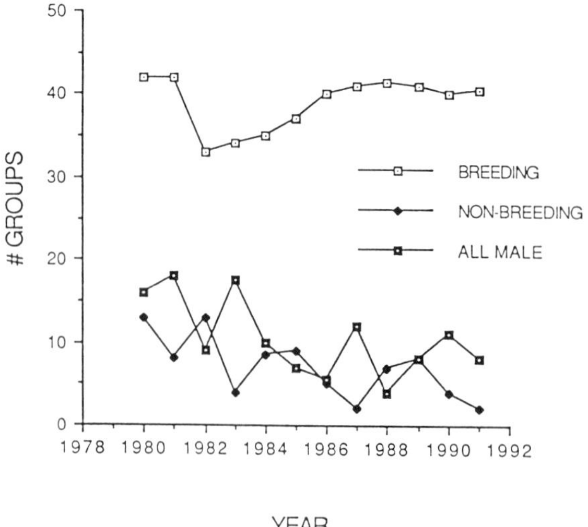

Figure 5. Number of breeding, non-breeding, and all-male red-cockaded woodpecker groups on Southern Pines-Pinehurst (SOPI) study area, 1980-1991.

differences were significant (ANOVA, F= 5.16, d.f.= 7, 13, p= 0.0054), as were differences between study areas (ANOVA, F= 4.60, d.f.= 2, 13, p= 0.0308) through 1987.

Total groups included breeding, nonbreeding, all-male groups and all-female groups (**Table 1**). Total groups fell sharply on SOPI and SGL between 1980 and 1982, recovered somewhat in the late 1980's, before drifting downward again. The decline on FB was not as dramatic, but a downward trend continued through 1983 before stabilizing. During the study the number of total groups fell 28.6% on SOPI, 22.9% on SGL and 11.7% on FB.

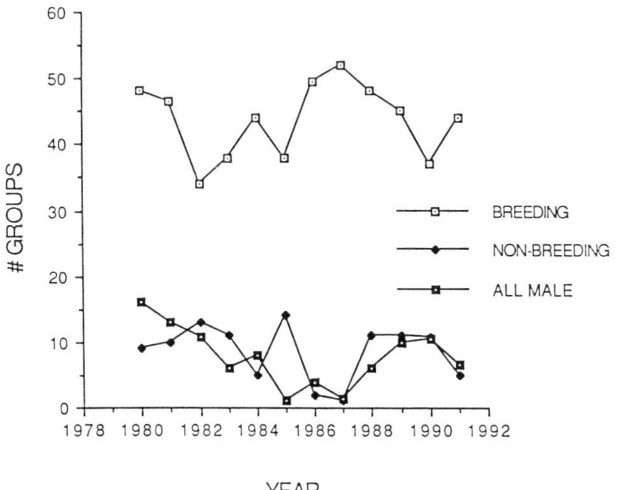

Figure 6. Number of breeding, non-breeding, and all-male red-cockaded woodpecker groups on Sandhills Game Lands (SGL) study area, 1980-1991.

Nesting Effort and Success

FB consistently had the highest number of nests, followed by SGL, and SOPI **(Table 2)**. On the 3 major study areas, 1677 nests were found during 1980-91, with 1350 (80.5%) producing fledglings. Overall success of known nests ranged from 63.3 to 94.1% **(Table 2)**. The latter figure may be misleading since monitoring techniques that year (1981 FB) were biased towards finding only successful nests. The average nest success rate across years was very similar between study areas, with SOPI highest (78.7%), followed by FB (75.7%) and SGL (73.8%). Most failures occurred at the egg or small nestling stage (LaBranche 1988). There were 117 renesting attempts from 424 failures and 64.1% were successful. Success of renesting attempts was similar between SGL (74.4%) and SOPI (72.7%), but was much lower on FB (51.9%). FB had almost half (52) of the renesting attempts on the study areas. Of the 1677 nests, only 2 were second renesting attempts. In 1991, 5 groups that had successfully raised first broods renested, and all fledged the second broods.

Clutch Size

Yearly differences in mean clutch size for each study area were more variable than expected. Clutch sizes averaged 3.14 eggs on SOPI, 3.39 on SGL and 3.22 on FB **(Table 2)**. Using data from 1980 through 1987, both site and year affected clutch size (ANOVA, F= 6.02, d.f.= 2,736, p= 0.0035; F= 2.21, d.f.= 7,736, p= 0.03, respectively; no interaction occurred between site and year, F= 1.05, d.f= 13,723, p= 0.4). SOPI and SGL differed significantly in clutch size (p <0.05, LSD multiple-range test in GLM procedures [SAS 1985]) through 1987. The highest average clutch sizes on SGL and FB were in 1984, a year of very high reproductive effort. The lower clutch size average on SOPI was caused by a high proportion of 3 egg clutches relative to 4 egg clutches. Conversely, the high SGL average was caused by relative parity between numbers of 3 and 4 egg clutches.

Table 1. Average numbers of red-cockaded woodpecker breeding, non-breeding, all male and total groups 3 North Carolina study areas, 1980-1991 (ranges are given in parentheses).

Study area	Breeding groups	Non-breeding groups	All male groups	Total groups
SOPI	38.8 (33-42)	6.6 (2-13)	10.0 (4-18)	55.4 (50-70)
SGL	43.3 (34-52)	7.8 (1-14)	7.1 (1-16)	58.4 (53-70)
FB	52.2 (47-62)	10.7 (3-19)	5.9 (1-12)	68.8 (63-77)
Mean	**44.5**	**8.4**	**7.7**	**60.7**

Table 2. Average percentage of failed nests, average clutch sizes, average percent nestlings fledged, and average number of fledglings per successful nest for red-cockaded woodpeckers on 3 North Carolina study areas, 1980-1991 (ranges are given in parentheses).

Study area	% failed nests	Clutch size	Banded nestlings fledged	# fledglings per successful nest
SOPI	21.3 (11.4-31.7)	3.14 (2.79-3.40)	86.6 (74.7-95.5)	2.08 (1.87-2.21)
SGL	26.2 (14.9-36.7)	3.39 (3.20-3.64)	81.4 (71.2-93.3)	2.11 (1.81-2.47)
FB	24.3 (5.9-32.1)	3.22 (3.00-3.48)	80.8 (69.5-90.9)	1.99 (1.46-2.39)
Mean	23.9	3.23	83.4	2.06

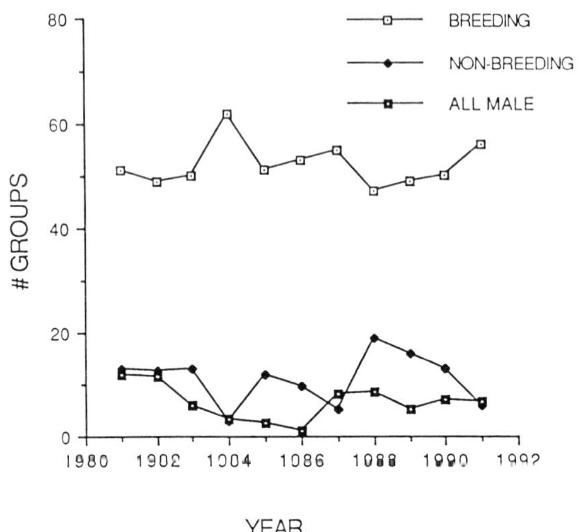

Figure 7. Number of breeding, non-breeding, and all-male red-cockaded woodpecker groups on Fort Bragg (FB) study area, 1980-1991.

Fledgling Numbers and Fledging Success

The average number of fledglings across years was lowest on SOPI and highest on FB (**Table 3**), however, the average fledging rate (% of banded nestlings seen fledged) was highest on SOPI (86.6%) and lowest on FB (80.8%) (differences not significant through 1987; X^2= 0.356, d.f.= 2, p= 0.837) (**Table 2**). Fledging success of banded nestlings varied on all study areas, with a range of 20.7% on SOPI, 22.1% on SGL and 21.4% on FB. On SGL 1983 was a particularly good year, while 1984, 1985, 1988 and 1990 were bad years. The best year on FB was 1986, with 1989 the worst. The average fledging rate across all sites and years was quite high at 83.4%. Fledging rates are conservative since some fledglings undoubtedly died or dispersed before being detected (see Walters et al. 1988). The highest numbers of fledglings on SOPI, SGL, and FB were in 1991, 1986 and 1984, respectively. In general, fledgling numbers fluctuated considerably from year to year in all study areas, but study areas shared the same good years and the same bad years.

SGL had the highest average number of fledglings per successful nest at 2.11 followed by SOPI at 2.08 and FB at 1.99 (**Table 2**). Differences between study areas were not significant through 1987 (SE's 0.07, 0.06 and 0.03, respectively). No study area was consistently higher than the others. There were similarities between years, with declines in all study areas in 1982, 1985, and 1987, and increases in 1984 and 1986. SOPI showed an

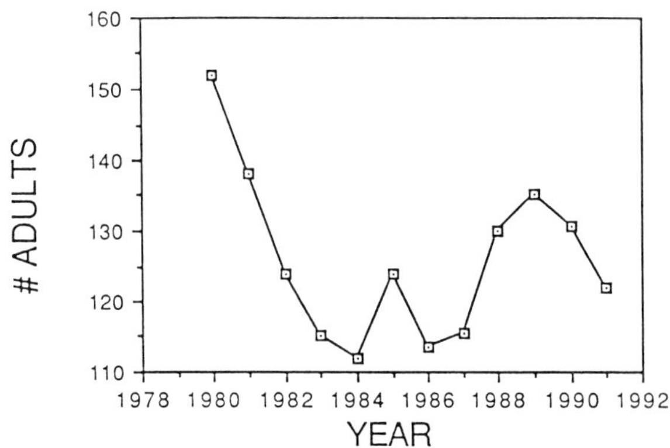

Figure 8. Number of adult red-cockaded woodpeckers on Southern Pines-Pinehurst (SOPI) study area 1980-1991.

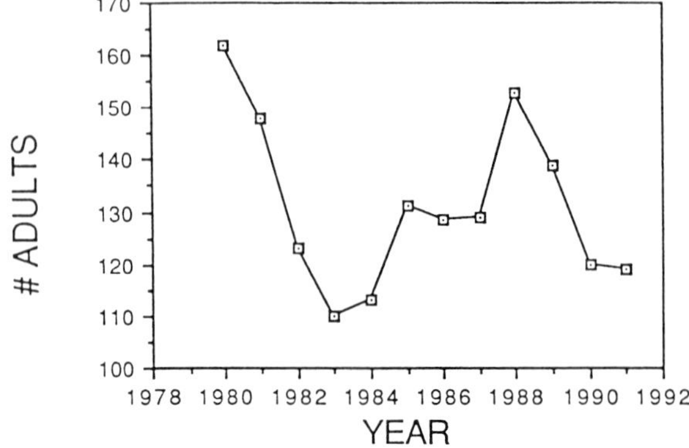

Figure 9. Number of adult red-cockaded woodpeckers on Sandhills Game Lands (SGL) study area, 1980-1991.

improvement in average numbers of fledglings per successful nest after 1987, whereas FB's average eroded after 1988.

Population Size

Numbers of adult red-cockaded woodpeckers declined steadily on SOPI and SGL from 1980 through 1983 (the sharp drop in nest numbers and young occurred in 1982) **(Table 3, Figures 8, 9 and 10)**. Adult numbers remained relatively constant on SOPI and SGL in 1984, but began climbing in 1985 following the production of a large cohort of fledglings in 1984. Adult numbers began declining again in 1986, but rose on all areas in 1988, and rose again on SOPI in 1989. Numbers declined sharply on SOPI in 1991 and on SGL in 1990, but ended on the rise in FB. Overall, adult numbers declined by 19.9% on SOPI, 27.2% on SGL, and 2.2% on FB during the study.

Groups with Helpers

The numbers of red-cockaded woodpecker groups (male-female pairs) with at least one helper varied considerably between study areas and year **(Table 3)**. SOPI had 18 groups with helpers in 1980, dropped to 15 by 1984, but rose to 23 in 1991 **(Figure 11)**. In 1980 SGL had 28 groups with helpers, fell to a low of 10 in 1983 and 1984, rose to 23 in 1987, and ended with 13 in 1991 **(Figure 12)**. Thus, SGL began with 10 more helper-assisted groups than SOPI, and ended with 10 fewer. FB showed a general decline through 1984, rebounded to a study high of 36 in 1985, declined to 20 in 1990, then jumped to 36 again in 1991 **(Figure 13)**. Across years FB averaged the most groups with helpers (26.7), followed by SOPI (18.5) and SGL (14.9) (differences between areas are significant within some years, e.g. 1991, but not others, e.g. 1981; Chi-square tests using p= 0.05). The percentage of groups (male-female pairs) with helpers showed much yearly variation between study areas, and to a lesser extent, within study areas. SGL began with the highest percentage (51.8%) and ended with the lowest (27.1%). SOPI started with the lowest percentage (32.7%), and ended second highest (56.1%). The average across all study areas and years was 37.7%.

DISCUSSION

Study Area Habitat Quality

The best red-cockaded woodpecker habitat, determined subjectively, was on FB. FB had the most open, natural forest among the study areas, and although some timber was harvested, most stands continued to be well stocked. SOPI and SGL were ranked somewhat similarly in habitat quality, but for different reasons. SGL had no development pressure, but understory encroachment was a problem almost everywhere, and timber harvesting altered a considerable amount of habitat. Development was a major influence in SOPI, as was understory encroachment in undeveloped areas. Though development usually meant loss of substantial foraging habitat, potential cavity trees and some cavity trees, it also resulted in suppression or elimination of the understory. Thus, some adverse impacts to habitat were at least partially balanced by a positive impact.

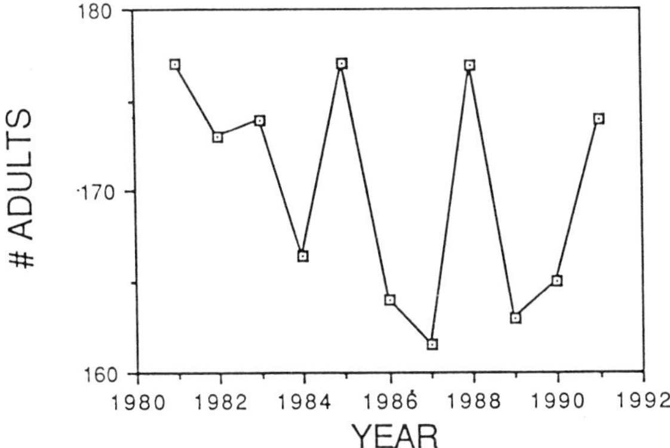

Figure 10. Number of adult red-cockaded woodpeckers on Fort Bragg (FB) study area, 1981-1991.

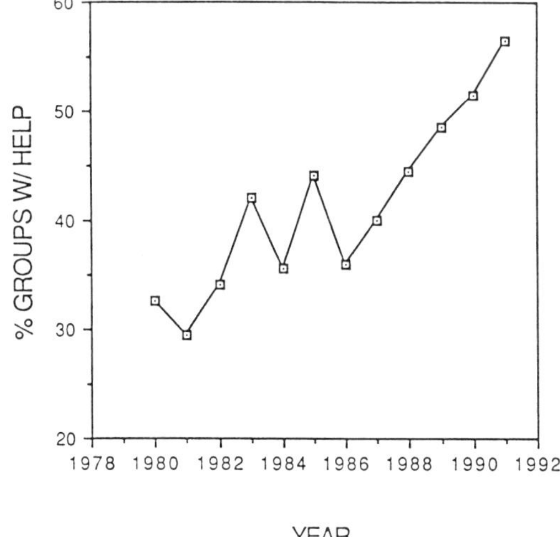

Figure 11. Percent of red-cockaded woodpecker groups with helpers on Southern Pines-Pinehurst (SOPI) study area, 1980-1991.

Table 3. Number of fledglings, number of adults and number of male-female groups with helpers and percentage of male-female groups with helpers for red-cockaded woodpeckers on 3 North Carolina study area, 1980-1991 (ranges are given in parentheses).

Study area	# of Fledglings	# of adults	# of male-female groups with helpers	% of male-female groups with helpers
SOPI	66.8 (52-85)	126.2 (113-151)	18.5 (14-24)	41.3 (29.2-56.1)
SGL	72.7 (53-99)	131.3 (111-162)	14.9 (10-28)	28.8 (20.8-51.8)
FB	85.5 (57-120)	170.3 (161-178)	26.7 (20-36)	42.5 (32.3-58.1)
Mean	72.8	141.8	19.9	37.4

Red-cockaded woodpecker habitat in a golf course-residential development was often more stable than that anywhere else in this study, because the remaining trees were protected from harvest and the understory was at least temporarily eliminated; however, long term stability was not guaranteed because of the lack of pine regeneration. Red-cockaded woodpeckers are tolerant of casual human activity, so the presence of humans in itself is not necessarily an adverse impact. However, numbers of cavity competitors such as starlings (*Sturnus vulgaris*), red-bellied woodpeckers (*Melanerpes carolinus*) and red-headed woodpeckers (*M. erythrocephalus*), appeared to be higher in developed areas. All study areas shared a negative trend in habitat quality during the study based on an increase in the severity of understory encroachment in many sites. In general, colonies with little or no understory problem at the beginning of the study were starting to show encroachment by study's end, and colonies with encroachment at the beginning got worse.

Population Trends on Major Study Areas

The number of active colonies decreased on all study areas, while the number of abandoned colonies increased. SOPI began with the most abandoned colonies, but the rate of abandonment was approximately 4 times higher on SGL and FB. Habitat quality parameters, such as degree of understory encroachment and amount of available foraging habitat, appeared to be poorest on SOPI at the beginning of the study relative to the other study areas. However, understory encroachment increased in both area affected and degree on SGL and FB during the study, while on SOPI it held steady or decreased. The prescribed burning programs on SGL and FB failed to control understory encroachment in many areas, but on SOPI land clearing for golf courses and housing construction, though destroying some habitat, often resulted in a higher quality of the remaining habitat.

There were always more active colonies than woodpecker groups and more total groups than breeding groups. A woodpecker group can keep more than one colony active by "capturing" it and using it for roosting, or by passing through the colony regularly and working on one or more cavities. The early years of the study were characterized by high numbers of nonbreeding and all-male groups on all study areas, but as colonies were abandoned, these numbers decreased temporarily and the ratio of breeding to nonbreeding groups increased (Walters et al. 1988). For example on SGL in 1980, 67.1% of all groups were breeding groups, but this increased to 81.5% in 1991. Concurrently the total number of groups decreased from 70 to 54. The number of breeding groups was variable between years on all study areas, but did not change significantly on any study area between the study's beginning and end (SOPI -2, SGL -3, FB +6). SOPI began and ended with the lowest number of breeding groups, and FB began and ended with the most. Habitat quantity and quality are assumed to be the primary factors that determine whether or not a colony is occupied (Conner and Rudolph 1989, USFWS 1985b). Numbers of groups are limited by the number of available potential nesting sites meeting certain habitat criteria. Increases in population size, as followed the large 1984 cohort, did not necessarily lead to more groups or active colonies because colonies with suitable habitat were already occupied.

As expected, SOPI, which typically had the fewest active colonies and the fewest breeding groups, had the smallest number of nests yearly. FB had the highest number of nests annually, and also had the most active colonies and breeding groups. This is consistent with a hypothesis that habitat quality and quantity govern the number of active colonies, breeding groups, and nests, since FB had the most quality habitat. However, the years of high and low nesting effort were shared by all study areas. Though nesting success varied yearly, the average nesting success rates between study areas were very similar. The high failure rate for renesting attempts on FB (49.1%) cannot be explained at this time.

SGL had the highest average clutch size, followed by FB and SOPI. It was expected that

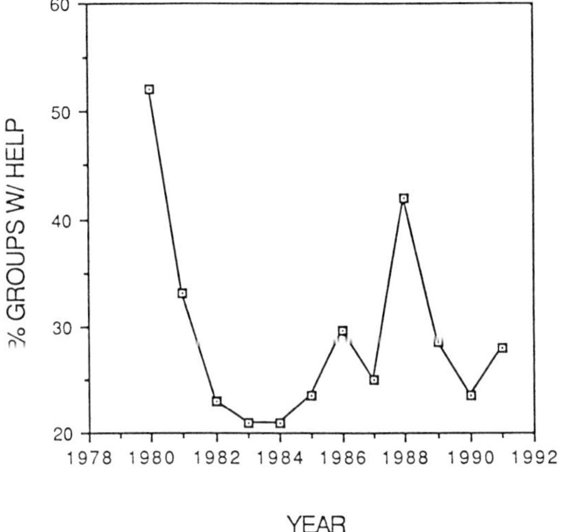

Figure 12. Percent of red-cockaded woodpecker groups with helpers on Sandhills Game Lands (SGL) study area, 1980-1991.

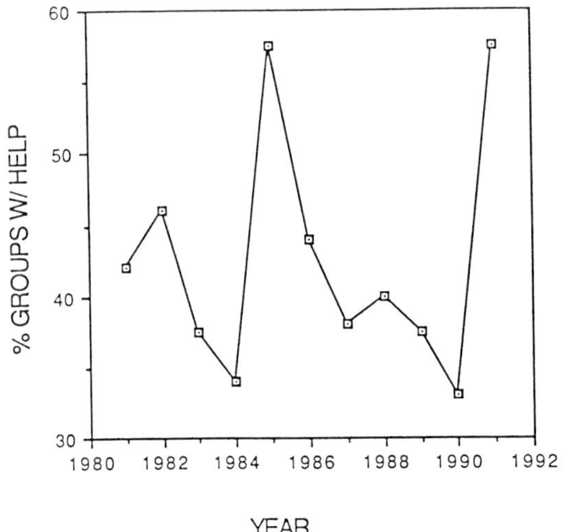

Figure 13. Percent of red-cockaded woodpecker groups with helpers on Fort Brag (FB) study area, 1981-1991.

if there was any difference in clutch size between study areas, FB would be highest, and SOPI and SGL would be similar. Therefore, these results are difficult to explain. Perhaps the remaining occupied habitat on SGL is superior to that on FB. However, the best habitat on SGL appears to be no better than the average habitat on FB. It could also be true that the woodpecker population on SGL is more stressed than those on SOPI and FB, and the response to stress is a larger clutch size; however, there is no evidence to support this view. Differing age structures of the breeding female population in each study area could be responsible, since experienced females appear to lay larger clutches on average than inexperienced ones (Clutton-Brock 1988), but no major difference in the age structure of the breeding female populations was detected. The general onset of nesting always appeared to be earlier on SOPI than on SGL and FB. In some years one-third to one-half of the nests in SOPI were detected before the end of April, while nesting was just getting started on SGL and FB during the first week of May. It is not known if these early nests could have contributed to a lower average clutch size on SOPI, but it seems doubtful. Birds with breeding experience tend to nest earlier and have larger clutches (where clutch size is variable) than inexperienced breeders (Clutton-Brock 1988). If this were true on SOPI, the earlier onset of breeding should have meant a higher average clutch size.

Total number of fledglings is directly correlated to number of active colonies, breeding groups, and nests, and potentially to average clutch size. FB consistently produced more fledglings than SOPI or SGL, except in 1989, (average difference between FB and SOPI= 19.0; between FB and SGL= 13.1), while SGL outproduced SOPI in 8 of 12 years. Variation in clutch size between study areas did not appear to significantly affect fledgling production. Average fledging rates of banded nestlings were quite similar between study areas. Thus, the higher numbers of fledglings on FB can be explained simply by the higher number of breeding groups there.

SGL had the highest numbers of fledglings per successful nest, followed by SOPI and FB. This may have resulted from the higher average clutch size on SGL and is not consistent with the hypothesis that FB contains better habitat. However, it is consistent with the hypothesis that habitat quality does not affect nesting success.

All study areas showed a major negative trend in adult population size through 1984, a sharp rebound in 1985 following the large 1984 cohort, and either a stable (SGL) or declining trend (SOPI and FB) through 1987. The adult population increased sharply on all study areas in 1988, but declines began again in 1989 on SGL and FB, and on SOPI in 1990. The decline was particularly steep on SGL in 1990 (13.1%). Adults on FB increased in 1991. FB began and ended with far more adults than SOPI or SGL, but the differences were larger at the end of the study. This was the result of more active colonies and more groups on FB, which may be explained by better habitat quality and quantity. SGL had 11 more adults than SOPI in 1980, and 14 more in 1987, but 3 fewer in 1991.

The numbers of groups with helpers varied considerably between study areas and years. All study areas showed net losses through 1984, and sharp gains in 1985, the latter due to the large 1984 cohort. SOPI and FB ended the study with more groups with helpers compared to beginning levels, but SGL had a net loss of 53.6%. Helpers are believed to act as a buffer to major population fluctuations caused by years with poor recruitment (Walters et al. 1988), so SGL is currently the most vulnerable subpopulation to decline.

GENERAL DISCUSSION

There are no comparable studies of the red-cockaded woodpecker of this size and length, and no previous studies at all in developed habitats. Many populations regionwide are known or believed to be in decline (Conner and Rudolph 1989, Costa and Escano 1989, Ligon et al. 1986), and declines have been tied directly or indirectly to decreases in habitat quality and/or quantity. Specifically, increases in understory encroachment in colonies, lack of suitable potential cavity trees, and insufficient foraging habitat have been cited as causes of colony abandonment and subsequent population declines (Costa and Escano 1989). Thus, the abandonment of colonies in this study, and subsequent reductions in groups, nests, fledglings, and overall population size are part of a regionwide trend.

In coastal South Carolina, Lennartz et al. (1987) found over a 5 year period that between .90 and .98 of all groups nested and that clutch size was 3.10 to 3.28, the 2 values representing averages with and without helpers (computed averages .94 and 3.19, respectively). Other clutch sizes in the literature include 3.33 in north-central Florida (sample size = 6; Ligon 1970), 3.25 in central Florida (DeLotelle 1983a), 3.46 in another coastal South Carolina site (Wood et al. 1985a), and 3.44 in the Georgia Piedmont (Lennartz and Heckel 1987). Overall average clutch size in this study was 3.28, and .847 of the male-female groups nested. If all-male groups are included, the nesting percentage falls to .732. Lennartz et al.'s (1987) study area supported the densest woodpecker population known, and was forested with high quality habitat.

This study produced an average of 1.60 fledglings per nest or 2.09 fledglings per successful nest. Other researchers have documented lower numbers of fledglings per successful nest, including 1.3 in Florida (DeLotelle 1983a), 1.5 in Florida (8 nests; Ligon 1970), 1.56 in coastal South Carolina (Lennartz et al. 1987), 1.9 in Texas (7 nests; Lay 1971), and 1.72 in the Georgia Piedmont (Lennartz and Heckel 1987). Grimes (1977) and Wood et al. (1985a), in coastal South Carolina, recorded 1.8 and 1.9 nestlings, respectively, surviving until the "pre-fledging period". The available literature indicates that in the Sandhills study population of red-cockaded woodpeckers, the percentage of groups nesting is lower (only 1 comparison available), clutch size intermediate, and fledglings per successful nest higher, than in other populations studied. It could be inferred from the latter 2 parameters that the Sandhills population is more productive than other populations, but comparative data on recruitment and survival are lacking.

However productive the Sandhills population may be relative to others, there has been a pronounced decrease in active colonies over the last decade, and large decreases in numbers of adults on 2 of 3 study areas. The relative stability in number of breeding groups, though subject to yearly variation, may indicate that the population has reached a plateau. Hopefully, recent increases in affirmative management will slow or reverse the abandonment of colonies, and lead to future population expansion.

A 12-Year Study of Red-cockaded Woodpeckers in Central Florida

Roy S. DeLotelle, DeLotelle and Guthrie, Inc., Gainesville, FL 32607
Robert J. Epting, Department of Resource Management, St Johns River Water Management District, Palatka, FL 32178
Greg DeMuth, Environment Division, Orlando Utilities Commission, Orlando, FL

ABSTRACT: Red-cockaded woodpeckers of the central Florida mesic longleaf pine flatwoods occurred in small scattered populations, primarily in 4 counties north of Lake Okeechobee. The ecology, demographics, and social behavior of one of these small populations were studied for 12 years. Pine density was sparse compared to more northern areas. Territory interaction site locations changed little over the 12 years. Survivability of breeders was higher and reproduction lower than for northern populations, but variable from year-to-year. During several years social conflicts were a major determinant of fledging rate. The relevance of demographic variation is discussed with respect to recovery and management for small sized populations.
KEYWORDS: demographics, reproduction, behavior, habitat, ecology, territoriality, management.

A study of the red-cockaded woodpeckers (*Picoides borealis*) (RCW) in central Florida was initiated in August 1980, as part of a site certification process for the Orlando Utilities Commission Stanton Energy Center. A management plan and long-term monitoring program were developed with the assistance of the Florida Game and Fresh Water Fish Commission and the U. S. Fish and Wildlife Service as part of the permitting process. During the initial field surveys, 9 occupied clusters occurred on site and an additional 26 clusters occurred off-site. At first we focused on territory size, foraging, and reproduction (DeLotelle et al. 1983b, DeLotelle et al. 1987). Further studies included social behavior and demographics (DeLotelle and Epting 1988, 1992).

The 12-year study provided an opportunity to observe long-term changes in a small RCW population in the southern part of the species range. Other long-term data were available for larger populations in North Carolina and South Carolina (Walters et al. 1988, Lennartz et al. 1987). These programs played a key role in developing recent management strategies by the USFWS (Lennartz and Henry 1985, Costa 1992).

Temporal variation within populations occur naturally and through disturbance, but may be more pronounced in smaller populations. Such variation may be influenced by social structure, habitat condition, and other factors. These intra-population variations and other inter-population differences may be important in long-term RCW management and recovery (Lennartz and Henry 1985).

METHODS

The 1328 ha study site was located 21 km southeast of Orlando Florida. This Floridian coastal plain habitat consisted of mesic longleaf pine (*Pinus palustris*) flatwoods, scattered cypress domes, and xeric oak scrub. The shrub layer includes saw palmetto (*Serenoa repens*), stagger bush (*Lyonia fruticosa*), fetterbush (*L.lucida*) and gallberry (*Ilex glabra*). The open pine canopy allowed the development of a rich savannah-like ground cover of herbaceous species (Myers and Ewel, 1990) including wiregrass species (*Aristida stricta* and *A. spiciformis*), beakrush (*Rhynchhospora fascicularis*), and bluestem (*Andropogon virginicus*). Range management before and after power plant construction consisted of semiannual burning of ground-cover vegetation.

Techniques used initially to find potential woodpecker habitat and cluster locations for the site and surrounding areas included aerial photographic interpretation, helicopter surveys, and ground-truthing. Subsequently, the study area was searched annually for all cavity trees. Twenty-five plots in pine stands were measured for tree density, age, dbh (diameter at breast height), and basal area (DeLotelle et al. 1987). Territories were measured in multiple years by both recording group movements at 5-min intervals and territorial interactions on copies of aerial photographs. A total of 2,005 hrs of observations were collected.

Thirteen groups were studied from the summer of 1980 to the winter of 1992. Birds were captured and banded with color coded plastic leg bands and USFWS identification bands (DeLotelle and Epting 1992). Groups were monitored for competition, breeder replacement, and social structure. Social status was based on the known history of individuals and the context of interactions with other birds. Survival, turnover rates, and breeding status were calculated by comparing the censuses of banded birds in successive years and by their involvement in nesting activities. Reproductive success was based on examining nest cavities with the aid of a mirror and flash light weekly from April-July for 9 years (81, 82, 84-85, 87, 89, 90-92) or by fledgling censuses for 3 years (83, 86, and 89). Total population estimates included mid-winter surveys of fledglings, floaters, helpers and breeding adults.

Foraging and flight behavior were obtained from 11 breeding males, 12 breeding females, 3 male helpers, and 10 fledglings. Three of the groups included pairs and fledglings, 5 of the groups consisted of pairs only, and 3 groups contained a pair, male helpers, and fledglings. Each tree used by a foraging RWC was visually divided into 4 equal parts to classify partitioning through spatial use of foraging trees. Besides foraging parameters, social interactions for each individual/group member were recorded at 5-min intervals with respect to location on the tree. Flight distance between trees was measured for individuals and groups.

RESULTS

Population

Approximately 150 clusters of RCW cavity trees occurred in the lower Florida peninsula at the southern end of the species range. The central Florida area included small RCW populations of about 18 groups each in Orange, Polk, Osceola, and Highlands counties. Eleven of these groups occurred in Orange County from 8.0 to 12 km from the study area. The remainder of the Orange County population including the study area consisted of 24 occupied territories in 1981 and 13 in 1992. Nine occupied territories occurred in the study area in 1981 and 7 in 1992. Of the 11 clusters not occupied in 1992, 7 were either clearcut or had most of the foraging habitat removed. The other 4 vacant clusters were still present in 1993. Two breeders from 2 formerly separate and adjacent groups combined in 1986 to form a single new group. The other 2 breeders disappeared. Another territory was vacated 2 years after the study began, but RCW reoccurred there 4 years later. The average density in 1981 was 1 cluster per 137 ha. The percent of occupied clusters declined from 92% in 1981 to 76% in 1992.

Habitat

Mesic longleaf pine flatwoods occurred throughout north and central Florida southward to near Lake Okeechobee. The central Florida pine flatwoods was more savannah-like (sparsely tree covered) than the north Florida pine flatwoods (Myers and Ewel 1990). Longleaf pine flatwoods in central Florida inhabited by RCW were similar to each other (Shaprio 1983, DeLotelle et al. 1983b).

Pine density was low within the study area **(Table 1)**. Nesting home ranges averaged 4,237 pine stems (>6 cm dbh), with a mean dbh of 20.3 cm. Pine stands in year-round territories averaged 121 stems/ha with a weighted mean age of 55 years. The average pine basal area per territory was 275 m^2, which included about 3401 stems ≥23 cm dbh. Tree heights extended to 15.2 m, with those over 90 years-of-age averaging 11.7 m. The most abundant age class (30-40 years) averaged 9.6 m in height.

Cypress domes occupied about 7% of the habitat. Foraging on cypress was highly variable accounting for 25% of foraging time during some years and as little as 7% in other years.

The distribution of potential cavity trees (90 years old) within the population ranged from 2 to 72 trees/ha (DeLotelle and Newman 1983). There were, however, 4 Sections that averaged

Table 1. Population and habitat characteristics for 13 groups of red-cockaded woodpeckers from central Florida.

			Reproduction					
ID	Sample Yrs	Group Size[a]	All Yrs	Successful Yrs	Pine Basal Area (m^2)	Total Pine Stems	Pine Stems >23 cm	Territory Size (ha)
A	3.0	2.0	1.3	1.3	330	14,709	3,410	155
B	2.0	2.0	1.0	1.0	291	13,547	2,497	147
C	6.0	3.0	0.8	1.3	204	8,818	2,429	102
D	5.0	3.0	1.0	1.0	241	8,733	3,804	106
E	12.0	3.0	0.9	1.8	403	13,490	4,970	121
F	12.0	2.0	1.0	1.1	295	10,418	3,500	103
G	12.0	2.0	1.1	1.6	248	6,452	3,956	87
H	12.0	3.0	0.8	1.3	229	6,320	4,263	95
I	10.0	3.0	1.3	1.6	351	10,961	5,007	97
C/D	6.0	2.0	0.6	1.3	297	8,636	3,129	96
J	2.0	3.0	2.0	2.0	221	8,210	2,405	86
K	2.0	2.0	1.0	1.0	268	8,111	2,772	140
X	6.0	1.0	0.2	1.0	201	10,237	2,068	161
Mean	6.9	2.4	1.0	1.3	275	9,896	3,401	115
S.D.	4.2	0.7	0.4	0.3	60	2,673	969	27

[a]Typical group size during nesting.

from 22 to 42 old trees/ha. Most of the unoccupied habitat surrounding the study population had a sparse stocking level of potential cavity trees, but foraging habitat was abundant with little hardwood understory.

Newly completed cavities (n=46) were occupied at a rate of 0.53 per cluster-year (n = 87). This resulted in no appreciable net change in the average number of occupied cavity trees (2.8) per cluster. The total number of completed cavities ranged from 24 to 37 trees per year. The death of completed cavity trees ranged from 1 to 22% per year. Total cavity tree mortality (32 of 144 trees) was due to unidentified causes (59%), lightening strikes (19%), fire (6%), and wind snap (6%). Three cavity trees (9%) on-site were cut down.

Predators

Nest competitors and predators occurred at low densities in the central Florida pine flatwoods. Occurrence of accipiters was infrequent and averaged 1/58 h of observation during the winter. No accipiters occurred during the summer. Sharp-shinned hawks (*Accipiter striatus*) were observed to make only two unsuccessful passes at RCW. Three times in 4,055 hrs of RCW observations, sharp-shinned hawks captured passerine species. No typical tree-climbers, such as yellow rat snakes (*Elaphe obsoleta quadrivittata*), were observed on or in cavity trees, although 1 red rat snake (*E. guttata*) was seen in the area of a cluster. No flying squirrels (*Glaucomys volans*) or red-headed woodpeckers (*Melanerpes erythrocephalus*) were observed in the study area. Visual inspection (n = 184) of RCW cavities during the nesting season revealed no predators and very few nest competitors (DeLotelle and Epting 1992).

Territoriality

On foraging trips and other territorial movements, groups traveled about 4.7 km per day at 0.5 km per h. Groups spent 21±15 (sd) min foraging in habitat patches and traveled 112±36 m between habitat patches. Within habitat patches, individual flight distance was 28±13 m. Groups attempted to forage on their neighbors' habitat and responded to intruding groups. These territory boundary flights averaged 371±182 m and occasionally included several short stops.

Territorial interactions (N=118) were highly seasonal, being significantly more frequent during the summer (1/9 h) and least frequent during the nesting (1/76 h) season. Fall (1/35 h), winter (1/26 h), and spring (1/17 h) interaction rates were intermediate (x^2= 77.0, p>0.001, df=4). Most interactions (50%) and extra-territorial movements (55%, n=101) took place within 2.5 h of sun-up. Interactions declined during mid-day and increased moderately during the afternoon. The frequency of extra-territorial movements decreased steadily throughout the day. Groups spent about 9% of foraging time outside their resident territory in the winter and about 2% in other seasons.

Territory boundaries were stable over time despite breeder turnover and changes in population size as demonstrated by the location of interactions. In 118 interactions, at least 1 breeding member of the interacting groups had previous interaction experience on the territory. Over the 12 years, 31% of the territorial interactions occurred at 5 commonly used sites. Only 3% of the interaction sites varied by an average of 265 m from these primary interaction sites, suggesting that interaction site locations changed minimally. The remaining interactions (54%) occurred near the fixed interaction sites, at a similar distance from the cluster. Other interaction sites occurred along different segments of the territory boundary. In one case, 2 adjoining groups combined territories into 1 larger territory. The external territorial boundary line was the same as it had been prior to the merger.

Home ranges averaged 129 ha and territories averaged 115 ha. Nesting home ranges average about 46.5 ha and accounted for about 46% of year-round territories. Available habitat accounted for 79% of the variation in territory size. Territories occupied about 96% of available habitat within subpopulations. On average, 86% of the 14 ha difference between the mean home range size and the territory size was part of an adjacent territory. Thus, extra-territorial habitat within subpopulation centers was reduced. Within the larger population, however, there was substantial extra-territorial habitat available.

Demographics

Thirteen groups of woodpeckers were observed over the past 12 years for a total of 90 group-years. These included 52 group-years (58%) for pairs without helpers, 33 group-years (37%) for pairs with a helper, and 5 group-years (6%) for single males. Female helpers comprised 33% of helper group-years for the first 4 years. Females accounted for 18% of all helper group-years. The percent of groups with helpers ranged from zero (1985, 1991) to a high of 63% (1982). Mean nesting season group size declined 32% from a high of 2.8 birds/group in 1982 to a low of 1.9 birds/group in 1990. The mid-winter population including fledglings, floaters, helpers, and breeding adults, showed a similar (43%) decline in size from 1982 to 1991. Substantial reductions in the population occurred in 1985 and 1991. In 1982 and 1989 population numbers increased and the population probably had surplus birds (**Table 2**).

Annual breeder survival on resident territories was 86% (n = 90) for breeding males and 80% (n = 90) for breeding females. Breeder survival varied annually from 50 to 100% for males and 38 to 100% for females (**Table 2**). Survivorship for breeding females was 69% (9/29 breeder-years) in groups with a helper male, 100% (0/6 breeder-years) in groups with a helper female, and 84% (9/55 breeder years) in groups without a helper ($x^2 = 2.70$, $p = 0.29$). There was no significant difference in survivorship of breeding males with respect to the presence (5/29) or absence (7/55 breeder-years) of a male helper.

The highest yearly average for male breeding experience (6.3 years) occurred in 1989, and the lowest levels of experience (1.9 and 2.9 years) in 1984 and 1990, respectively (**Table 2**). Helpers remained on the territory a mean duration of 2.5 years.

Behavior and Social Structure

Conflicts between group members and outside intruders occurred because of the inter-territorial movements of helpers, floaters, fledglings, and displaced breeders (DeLotelle and Epting 1992). These movements and conflicts included woodpeckers crossing several territories and engaging the resident breeders at high rates (1.8 intrusions/day) during years of high population levels, such as 1983. The rate was intermediate during 1981 and 1982 (0.60 intrusions/day) and dropped as the population size declined during 1985 and 1986 (0.29 intrusions/day).

Breeding males and females and helper males often foraged in proximity to each other without overt aggression. When foraging on the same tree, helpers occurred above the others 51% of the time, breeding males above the others 35% of the time, and breeding females above the others 14% of the time (N = 115). When pairs without a helper foraged on the same tree, the breeding male foraged above the breeding female 76% of the time. Social

Table 2. Yearly mean demographic parameters for red-cockaded woodpeckers in central Florida.

	Study Year											
	81	82	83	84	85	86	87	88	89	90	91	92
Group size[a]	2.7	2.8	2.6	2.1	2.1	2.0	2.3	2.6	2.4	1.9	2.0	2.0
Winter pop'n size[b]	60	70	65	58	48	56	66	63	66	44	40	48
Survival[c]												
Male	100	78	88	50	86	83	100	86	86	50	88	100
Female	86	86	88	38	86	83	83	86	43	86	75	100
Helpers	--	80	83	20	50	0	0	100	100	20	0	0
Nest Experience[d]												
Male	1.0	1.8	2.4	1.9	2.9	3.8	4.3	5.3	6.3	2.9	4.1	5.1
Female	1.0	1.7	2.4	1.7	3.0	3.7	3.0	5.5	5.7	2.9	2.8	4.0
Helpers	1.0	1.6	2.3	1.5	2.0	0.0	1.0	1.5	2.0	4.0	0.0	1.0
Fledging rate	1.3	1.1	1.2	0.3	1.0	1.3	1.0	1.2	0.7	0.4	1.0	1.4
Occupied clusters	24	21	21	18	18	14	14	14	14	14	13	13
Sex ratio[e]	33	25	33	0	71	50	71	40	25	0	57	55

[a]Nesting group size.

[b]January population estimate based on the average of 20 groups, including fledglings, floaters, helpers, and adults.

[c]Percent of adult breeder survival and percent of helpers remaining on natal territory.

[d]Includes the population average for years residing on a territory.

[e]Percent fledgling males.

Table 3. Comparison of demographic and ecological factors for red-cockaded woodpecker populations.

Region	Fledgling Rate	Fledgling Survivorship[a] M	Fledgling Survivorship[a] F	Helpers	Breeder Survivorship[b] M	Breeder Survivorship[b] F	Territory Size (ha)	Pine Basal Area (m^2)
Central Florida	1.00	0.85	0.63	moderate	0.86	0.80	115	275
Southern Georgia[c]	1.13	moderate		low	0.79	0.64	67	771
South Carolina[d]	1.70	0.88	0.36	high	0.78	0.68	70	832
Northern Georgia[e]	1.70	--	--	--	0.72	0.51	--	high
North Carolina[f]	1.47	0.62	--	low	0.73	0.56	59	994

[a]Survivorship after 6 months on natal territory. [b]Breeder survivorship on resident territory. [c]USFWS (1990b) and Epting et al. (1995). [d]Lennartz et al. (1987). [e]Lennartz and Heckel (1989). [f]Blue (1985); Repasky (1984); Walter et al. (1989).

interactions, particularly aggressive behavior such as chasing and displacing breeding females took place infrequently when foraging. Only one conflict between a breeding male and a male helper was observed. Experienced breeding males pecked or displaced their mate in 0.52% of the observations (N = 4,040 total observations). Conflicts between inexperienced breeders (i.e., first year mates) occurred more frequently (3.9% of the observations, n=412). All displacements and chases occurred in the tree crown, most often resulting in the female flying to the lower portion of the bole of a nearby tree.

The turnover rate for breeding females was higher in groups with male helpers than for groups with no helpers or female helpers. Several behavioral observations support the above differences. A 4 year-old male helper attacked the breeding female (his parent). During 16 observations, the helper male flew from the canopy branches down to the lower one-third of the bole of the pine tree to chase the female off the tree. The female then flew to another tree and the helper male returned to the canopy. Once this behavior continued over a period of 3.5 h of observation. On the same morning, the female breeder was chased into an adjoining occupied territory and remained there for the rest of the day. Several months later she was the breeder in the adjoining territory after moving through at least 4 territories. Her former territory contained only the helper male and the breeding male. The breeding male disappeared from the territory during the following winter. The helper attracted a new mate and successfully fledged the first young 3 years after the expulsion. The old female fledged only 1 female on her new territory and then disappeared a year later.

Reproduction

Clutches averaged 2.8 eggs (N=45) with laying dates ranging from April to July. Eggs were usually laid in the male's cavity; however, the female roost tree was used 4 times. Thirty-nine of the 45 group-years in which eggs were observed, nestlings developed. During the 1991 and 1992 nesting season 34 eggs produced 15 fledglings. Eleven of the eggs (32%) failed to hatch, while the other 8 resulted in nestlings that starved.

DeLotelle and Epting (1992) studied group fledging rates over 5 years and found them directly correlated with breeder experience and territory size and inversely correlated with outside intrusion rate. Over the 12 year study the group fledging rate was 0.94 birds/group (N=82). Pairs in which neither bird had breeding experience produced no fledglings (n = 5) in the first year. Experienced nesting pairs without a helper produced significantly ($t = 2.07$, $P=0.05$) more fledglings (1.400 ± 62, n=32) than experienced pairs with a helper (0.940 ± 54, n =18). The fledging rate (0.80, n = 5) for groups with helpers that included one inexperienced breeder was not different from groups with helpers and 2 experienced breeders.

The average population fledging rate ranged from 0.3 to 1.4 fledglings/group per year (**Table 3**). The variability in annual fledging rates was linked to changing breeder experience and fluctuations in outside intrusion rate as discussed above (**Figure 1**).

Mean fledging rate for individual breeders was directly correlated with years of experience on the territory (**Figure 2**). The average rate for males ranged from 0.27 for the first year to 1.33 fledgling/individual for birds residing 5 years on the territory (n = 13 males and n = 60 bird/years). The rate for females increased from 0.57 for the first year to 1.30 fledglings per individual for birds residing 4 years on the territory (n=14 females and 52 bird/years). The rate for females without helpers increased from 0.27 to 1.4 fledglings/individual. The fledging rate in groups with helpers changed little with female nesting experience and averaged about one bird (**Figure 3**).

DISCUSSION

Comparison of Populations

Availability of old trees for cavities and the quality of foraging resources could account for differences in demographics, social dynamics, and reproduction between populations. Territory size in central Florida was nearly twice as large as populations outside Florida because pine stocking levels were about one third that of other areas. The demography of the central Florida population included a high level of breeder experience, high frequency of successful nests, low fledging rate, and high fledgling and adult survivorship as compared to other populations (**Table 3**). The reduced foraging base in the study area may account for the lower fledging rate as compared to northern populations; however, greater social disruption, apparently associated with poor habitat, also had an influence on the lower fledging rate.

Several habitat factors had a positive effect on the frequency of successful nests and adult survival. No observed nest loss at the egg or nestling stage occurred as the result of nest predation or nest competition. Nest predation by rat snakes and flying squirrels, which predate RCW eggs and nestlings (Jackson 1987), occurred infrequently in the central Florida population. Few avian predators are present in the mesic longleaf pine flatwoods, particularly during the spring and summer (Robertson and Kushlan 1974, Emlen 1978). In comparison, predators are more abundant in continental habitats and on the central Florida ridge, which has better pine habitat and more

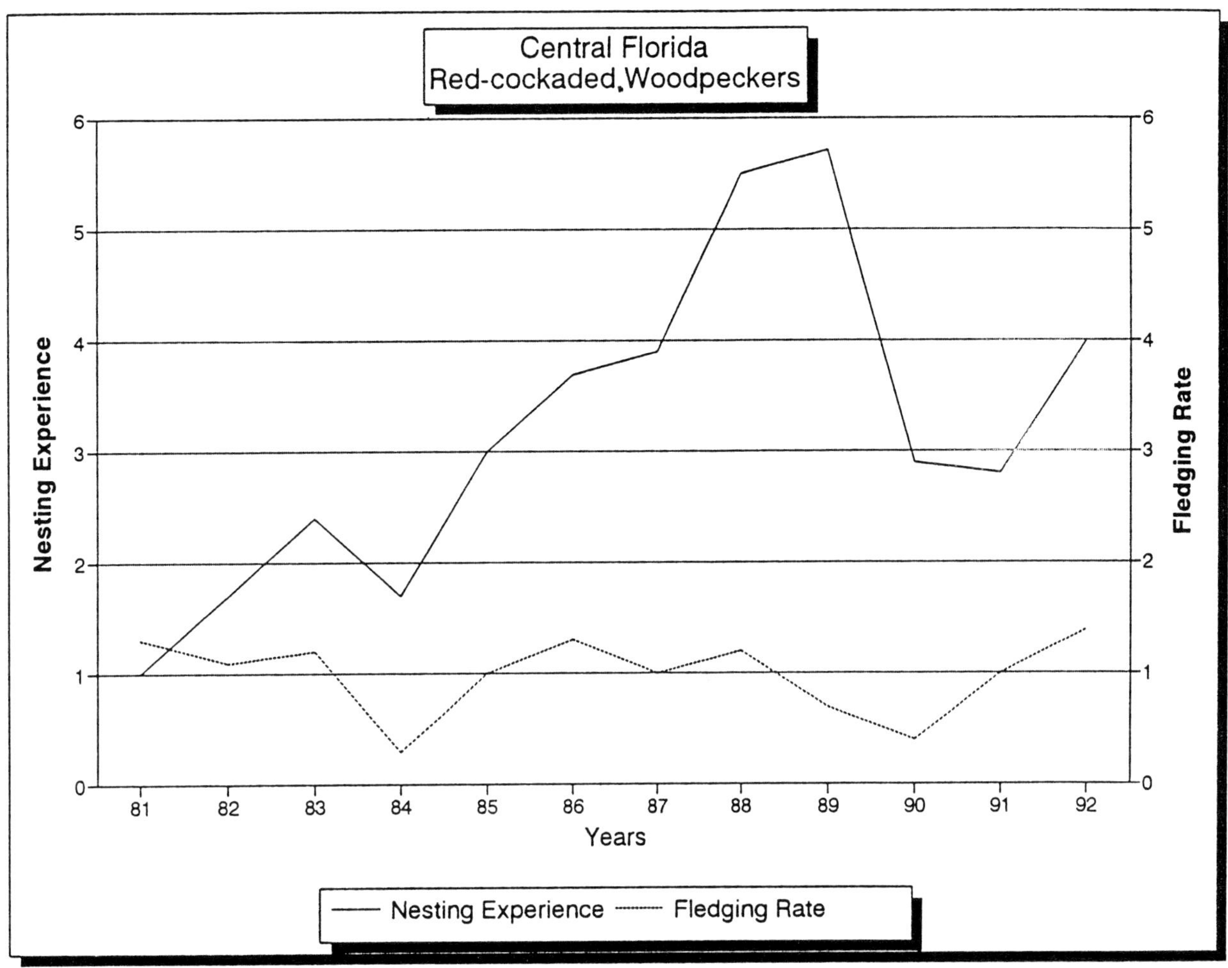

Figure 1. Average population breeding experience for females (solid lines) compared to average fledging rate per group (dashed line for the population). Female experience for this graph was based on a running average beginning from the first year of the study. See methods and Table 2.

hardwoods. This greater predator occurrence could account for the greater nest failures and lower adult survivorship observed by Lennartz and Heckel (1987) and Walters et al. (1988). The density of nest cavity competitors, such as red-headed woodpeckers and flying squirrels, may increase in well-developed hardwood understory (Jackson 1987, Blue 1985). The absence of roost and nest site competitors may lower exposure of RCW to predators due to insufficient roosting cavities. These positive effects probably account, in part, for the high survivorship and higher frequency of successful nests in central Florida.

The differences between fledging rates for central Florida and northern populations was 0.94 and 1.59 fledglings/group, respectively. Many investigators have said that fledgling rate may be influenced by habitat quality (Walsberg 1978, Kodric-Brown and Brown 1978, Repasky 1984). Two aspects of habitat quality that could influence fledgling rate are pine density and pine tree size. Reduced habitat quality may lower fledging rates by creating a greater energy cost for forging and by reducing the food supply for nestlings. Repasky (1984) said that reduçed tree density increases the cost of foraging flight due to greater transient distances. Tree sizes may alter foraging benefits since smaller trees provide less foraging space (Travis 1977, Jackson 1979b, Repasky 1984). DeLotelle and Epting (1992), however, found that habitat quality had only a modest effect on fledging rate within the central Florida population. The mean pine density and dbh for the smaller nesting season

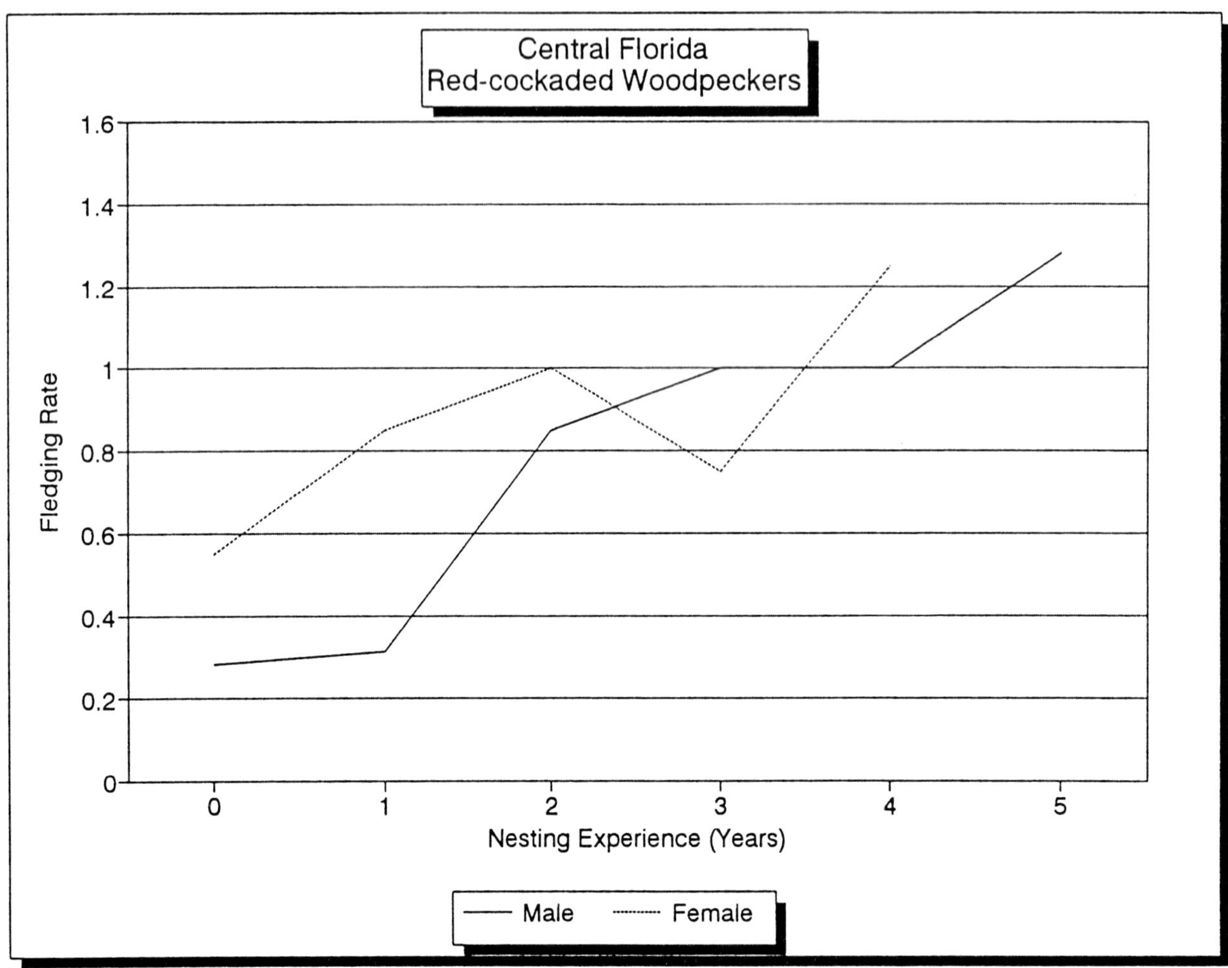

Figure 2. Average fledging rate for male (dashed line) and female (solid line) adult breeders with known years of breeding experience. Fledging rates for birds with more than 5 years for females and 6 years for males were summed.

home ranges in a South Carolina population were 302 stems/ha and 25 cm dbh (Lennartz 1983). Mean tree density and dbh in the larger nesting home range of central Florida habitats were 91 stems/ha and 20 cm dbh.

Despite low fledging rates and the limitations of low pine density and small tree size, the breeding density in the study area (1 group/137 ha) was similar to that throughout the species range as reported in Hovis (1982). This was primarily due to the abundance of old trees. DeLotelle and Epting (1988) reported that old pine stands function as potential nesting sites. These stands of old trees were abundant in most territories (DeLotelle and Newman 1983), while fewer old trees occurred in surrounding habitats that were unoccupied by the woodpecker.

A reduction in extra-territorial habitat is another result of poor foraging resources. Extra-territorial habitat may be much reduced (Zahavi 1974, DeLotelle et al. 1987) if coupled with abundant nesting resources in sparse foraging habitats. Unoccupied habitat between territories in the study area, accounted for only 4% of the pine forest where as in South Carolina extra-territorial habitat accounted for 36% of the area. Compared to other populations, the average territory in central Florida contains about 33% of the foraging resource **(Table 3)** resulting in 70% larger territories and 31% less extra-territorial habitat, (Hooper et al. 1982, Nesbitt et al. 1983, Repasky 1984, Blue 1985, DeLotelle et al. 1987, Epting et al. 1994).

Behavior and Social Structure

Young birds searching or competing for

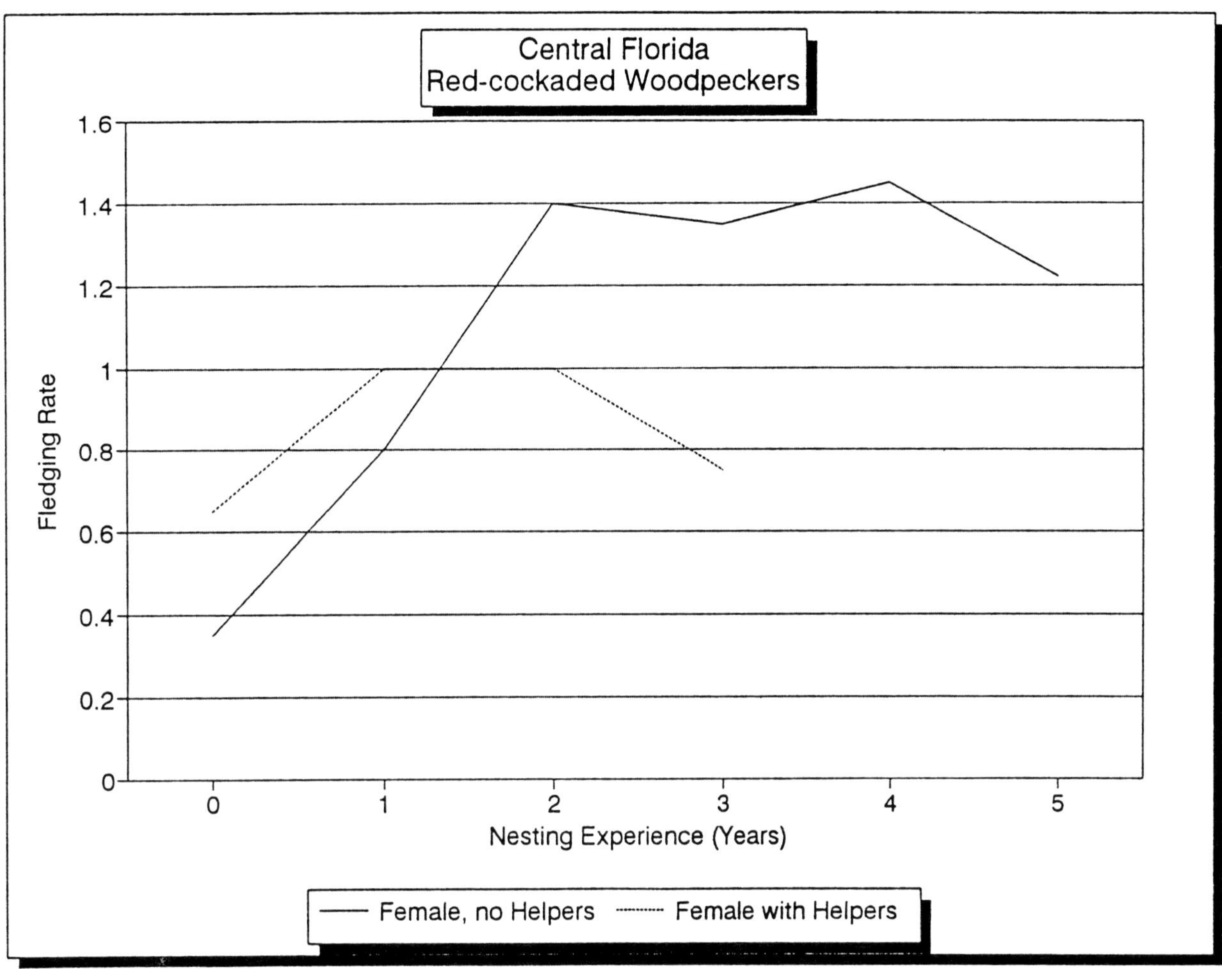

Figure 3. Average fledging rate by years of experience for females in pairs without helpers (solid line, n =37) and in pairs with helpers (dashed line, n = 15).

breeding positions occurred on all territories. Young birds would likely create substantial social interactions during the dispersing process in densely populated areas. Often, the replacement and movement of adult breeding females resulted from competing, young females. These conflicts were caused by female floaters, female helpers, and displaced female breeders (DeLotelle and Epting 1992).

Alternatively, large areas of open habitat between territories may expose dispersing birds to higher predation risk, but may reduce territorial conflicts. Among closely packed territories, young birds can monitor several groups of RCW for the availability of a breeding position. Other social interactions may be generated by the resident birds within the territory.

Aggression by experienced breeding males toward experienced breeding females in Florida (0.52%) was 17 times greater than the South Carolina (0.03%) population (Hooper and Lennartz 1981). This may be a consequence of reduced foraging space on the sparse and smaller pine trees in central Florida. In central Florida, conflicts were even more frequent for first year breeders compared to experienced breeders. Both sources of increased conflict between breeders may be factors contributing to the overall reduced fledging rate of the central Florida population.

When a related breeding female and helper male are older, there is a greater potential for helper-breeding female conflict (Emlen 1982b). In this situation, a helper has a higher probability of achieving breeder status, if he can evict his mother. Also, if the mother or helper leaves, incest can be avoided. In other cases, the male helper may attract, or at least not inhibit, the competitive intrusions of young females (DeLotelle and Epting 1992). These results suggest that the sequence of female breeder movement and level of helper and breeder experience are often linked. This also resulted in a short-term reduced fledging rate within the population. These external and internal behaviors appeared to link demographics and social behavior.

Demographics and Reproduction

The demographics of the central Florida population was influenced by the high survivability of breeders, and the aging and experience of male helpers. The high mean level of survivability, however, declined during years following high turnover of helpers and adult males. The replacement of experienced breeders by experienced helpers did not occur at a constant annual rate over the study years, but often occurred as a simultaneous replacement of breeders by helpers on most of the studied territories. This episodic sequence of male turnover, often occurred before or after several female breeder turnovers. This resulted in reduced level of breeder experience, and reduced fledging rate within the population. This sequence of changes in population demography occurred through 2 complete cycles (4 or more years). Reproduction was at its lowest levels during these turnover periods. Further, long-term study is necessary to confirm the interpretation that a demographic cycle exists in this or other populations.

Interaction of Ecology, Behavior and Demography

The high variability in the pool of potential breeders is apparently the result of an interaction of high breeder and helper survivorship in a constrained habitat. There was only limited opportunity for population increases during years with surplus birds because suitable abandoned cluster sites, other potential cavity tree construction sites, and suitable foraging trees occurred in a sparse or patchy fashion unsuitable for colonization. Surplus birds were probably lost to the population due to the scarcity of these resources. Female helpers were more common in years of high female fledging rates and low breeder female turnover. Alternatively, a few single males occupied territories during periods of female scarcity. A population constrained less by habitat would more likely have a gradual shift in population demographics.

Management Issues

The demographics of a small population varied over time. Surplus birds existed in some years, while in other years there was a reduction in breeders, potential breeders, and fledglings, and the population size declined. These long-term results have implications for enhancing recovery in similar populations of RCW. First, adding young birds to small or moderate sized, habitat-constrained populations may not increase the long-term size of the population, and may contribute to mortality. Second, using small to moderate sized populations to obtain donors, clearly requires at least a recent demographic history of reproduction and breeder turnover to insure that surplus birds are available.

A habitat based approach offers good long-term opportunities for positive recovery results in small to moderate populations. Improvements in foraging habitat such as reducing hardwoods or opening densely stocked pine stands may be required. Reconnecting locally fragmented population centers by providing foraging habitat and newly created cluster sites appears to offer higher probability for both stabilization and long-term survival. If RCW are unable to colonize habitat areas adjoining population centers, population will not expand during years of surplus birds. Cavity and cluster creation can be used to overcome these habitat gaps.

Introducing young birds to newly created cluster sites provides an opportunity to expand and centralize populations. If the problems of habitat limitation are not reduced, transplanting young birds will only provide a short term solution. The population may stabilize or increase for a short time, but will likely decline unless habitat factors are enhanced. Moving young surplus birds from within a population in good reproductive years should provide the best opportunities to increase the number of groups. Following years of high local reproduction, introduction of young females from outside the resident population may increase competition on the better quality cluster sites. This could decrease survivability of resident birds as well as introduced birds. Transplanting young females from outside the local population appears useful where reproduction is low or there are many single males. In years of reduced reproduction, young females or males from outside the population could be added to enhanced or created cluster sites to increase the number of breeding groups.

The current health of a population can best be understood within the perspective of long-term monitoring of demographics. This option is seldom available to most wildlife managers. Management strategies for RCW populations should anticipate year-to-year variability in surplus potential breeders based on limited field survey information. Census of fledging rate is one option, however, mid-winter group size estimates may be a better indicator of surplus birds available for repopulating new or unoccupied habitat.

In other cases, within widely fragmented or unsecured small populations, cavity and cluster creation and enhancement maybe useful in consolidating and protecting some populations. This cluster creation process coupled with the formation of new red-cockaded woodpecker groups by means of introducing young males and females to these created sites may be useful in transferring small populations from unsecured lands to lands with stronger management guidelines.

ACKNOWLEDGMENTS

Financial support for this study was provided by Orlando Utilities Commission, Orlando, Florida. Special thanks are given to Florida Game and Fresh Water Fish Commission and the U.S. Fish and Wildlife Service who assisted in various aspects of the study. Thanks are given to J. Exum and the Florida Department of Correction for providing additional data for the area. Special thanks to G. Lowe and L. Grant for their reviews and comments on earlier drafts of the paper.

Red-cockaded Woodpecker Territory and Habitat Use in Georgia and Florida

Robert J. Epting, Department of Resource Management, St Johns River Water Management District, P.O. Box 1429, Palatka, FL 32178
Roy S. DeLotelle, DeLotelle and Guthrie, Inc., 1220 SW 96th Street, Gainesville, FL 32607
Tim Beaty, Directorate of Engineering and Housing, Fish and Wildlife Section, Fort Stewart, Georgia 31320

ABSTRACT: Red-cockaded woodpecker territory size, habitat types and pine stand selection were studied in Georgia and compared to an earlier study in Florida. Territories in Georgia averaged 67 ha with average stand density of 211 stems/ha > 15 cm dbh and age of 42 years. Territory size and habitat quality for Georgia were more similar to those for northern habitats of the species range than to the Florida site. Variation in Georgia territory size was correlated with habitat quality, including pine density and area of hardwood stands. Foraging use of pine stands was explained by a model of pine age and density, hardwood density, and the distances of the stand to the cluster and average territorial interaction sites. The effect of territoriality on foraging appeared to be less important in the Georgia than the Florida population. Given the entire range of pine stand ages and equal pine densities, red-cockaded woodpeckers prefer older age stands for foraging. Foraging use also increased with pine density and decreased with hardwood density. The correlation of territory size with pine density and hardwood area, and the utility of the foraging use model over a wide range of habitats found in Florida and Georgia suggested that habitat quality was a measurable determinant of territory size and foraging use.
KEYWORDS: red-cockaded woodpecker, territory, foraging use, hardwoods, multi-variate model

The need for suitable foraging habitat for red-cockaded woodpeckers is important to their continued existence (Lennartz and Henry 1986). However, functional models of foraging use by red-cockaded woodpeckers (*Picoides borealis*) have been developed for Florida only (DeLotelle et al. 1987, Porter and Labisky 1988). DeLotelle et al. (1987) reported that given the entire range of pine stand ages and equal pine density, red-cockaded woodpeckers prefer the older age stands. Porter and Labisky (1986) reported a similar positive relationship between average tree size and foraging use. Other studies in South Carolina (Hooper and Harlow 1986) and North Carolina (Repasky, 1984) reported weaker results. A weak selection preference may be an indication that pine stand age and density were poor measures of habitat quality. In addition to differences in pine stocking, hardwoods occur in the understory habitat of Georgia and many populations.

To verify the general utility of the Florida foraging use model, we apply the procedures of DeLotelle et al. (1987) to a Georgia population occupying habitat with higher pine stocking and a hardwood understory. DeLotelle et al. (1987) incorporated the effects of both distance of the stand to the cluster site and to the average territorial interaction site as well as pine age and density in the foraging use model. The analysis procedure can be expanded to include the effects of hardwoods, or other variables of interest, on stand selection. We evaluated factors affecting territory size and foraging use for 18 groups from the south Georgia coastal plain and compared the results to those for Florida.

METHODS

The Georgia study area was the 113,064 ha Ft. Stewart Army Base near Savannah, Georgia, and contained approximately 165 red-cockaded woodpecker groups. Eighteen groups were observed between 1983 and 1984. The Florida study area was the 1328-ha Curtis H. Stanton Energy Center, 21 km southeast of Orlando. Six groups were observed throughout 12 years beginning in the fall 1980. An additional 17 groups occupied surrounding habitat, but were not included in the foraging study.

Forest stand type mapping was conducted for both study areas. The general methods for sampling habitat quality and foraging use, territory size, group size, and population density estimates were described in DeLotelle et al. (1983b, 1987). In both studies, group location, marked with color-coded leg bands, was recorded at 5-minute intervals. Territory was the defended area, and home range the total foraging area (Burt 1943). Territory boundaries were based on the location of group interaction sites and frequency of use. Home ranges were based on modified foraging use polygons.

The habitat in Florida was longleaf pine (*Pinus palustris*) mesic flatwoods with a sparse saw palmetto (*Serenoa repens*) understory interspersed with cypress domes (*Taxodium distichum*), bayheads, and wet prairies. In Georgia, longleaf, slash (*P. elliottii*) and loblolly (*P. taeda*) pine dominated uplands with cypress and mixed hardwood stands in drains and low areas. Stand boundaries were based on differences in tree age and density. Territory shapes were superimposed over the site vegetation maps to obtain the area of each stand in the territory. Forging use was the sum of all observations for a particular pine stand.

The analysis employed the two-stage procedure of DeLotelle et al. (1987) to evaluate the effects of pine stand age and density, hardwood density, and stand distance to both the cluster site and average territorial interaction sites on stand use. In the first step, observed stand use, expressed as a percent of total foraging within the entire territory, was regressed against stand size. The residuals from this regression were percent observed foraging use minus percent foraging use predicted by stand size only. These residuals have several useful statistical properties, including independence from stand size (Afifi and Clark 1984). These residuals were the dependent variable in the second stage of the analysis.

The second analysis stage was multiple regression of stand use (observed minus predicted stand use residuals from the first step) on the first five principal component scores obtained from the transformed variables of pine stand age, pine stand density, hardwood density (Georgia only), and stand distance to both the cluster and average territorial interaction sites. These five principal components represent the uncorrelated information content of the original stand variables. Note that dropping at least one principal component term guarantees statistical independence of those remaining (Afifi and Clark 1984). The location effects were the natural logarithms (Georgia only) of the distances, divided by the diameter of the territory, from the midpoint of each stand to the cluster site and the average territorial interaction site. The pine stand effects were the natural logarithm of the density of hardwoods (stems/ha), pine age, pine stand density (stems/ha), and the interaction (product) term of age and density for each stand. The principal component analysis allows direct interpretation of the principal component loadings (Afifi and Clark 1984). All analyses were preformed using SAS (1985).

Sensitivity analysis is a procedure for investigating the information content of complex regression models (Smith 1970). Two types of sensitivity analysis were used. The numeric analysis solved the regression equation as each variable, in turn, was varied by a one percent increase from mean value with the remaining variables set at mean value. Each change in foraging use, expressed as a percent of the largest change for all variables, is the sensitivity of the model to each variable. The procedure can be extended to examine simultaneously any two of the regression variables over a range of values, and the results presented graphically. The resulting curves are known as lines of indifference, a form of sensitivity analysis well developed in econometrics (Ch. 3, Utility and Preference, Hirshleifer 1976).

RESULTS

The mean territory size for 18 groups in Georgia was 67 (21 - 107) ha and accounted for 84% of the home range (total foraging area). Habitat types within these territories included mixed pine flatwoods (82%), pine/hardwood forest (8%), swamps (2%), and clearcuts (8%). Territories averaged about 12,255 pine stems greater than 15 cm dbh with an additional 15,486 pine stems less than 15 cm dbh. Approximately 19% of the larger pine stems were greater than 25 cm dbh. Mean pine basal area per territory for stems greater than 15 cm dbh was 537±364 m^2, and 771±501 m^2 for all pine stems. Pine stands within the territories consisted mostly of intermediate aged trees, ranging from 30 to 60 years.

Of the 18 territories, seven had been subjected to impacts on foraging resources by shelterwood cutting of pine stands just prior to the study. These impacted territories averaged 84 ha, with a mean home range of 96.7 ha and were significantly ($p < 0.05$) larger than non-impacted territories (57 ha). The average population density (i.e., the number of groups within 2,000 meters) surrounding these groups was 8 and 5, for the impacted and the non-impacted territories, respectively. Population density was greatest around the groups with the largest territories, that is, the groups with the most impact on their territories.

Territory size in Georgia was significantly correlated ($R^2 = 0.80$, $P < 0.0001$, $N = 18$) with the natural logarithms of mean pine basal area per hectare, and the natural logarithms of the total hardwood area (ha) within the territory (territory size = 136.2 + 22.9 x hardwoods - 46.2 x pine density). Other parameters such as group size and population density were not significant.

Table 1. Mean values for habitat and territory variables for south Georgia and central Florida (DeLotelle et al. 1987).

Variable	Population	
	Georgia	Florida
Number of groups	18	6
Territory size (ha)	67	116
No. stands/ territory	5	5
Pine stand age	42	41
Pine stand density[a]	211	121
Total pine basal area (m^2)/territory	771	279
Hardwood Density (stems/ha)	187	0
Mean group size	2.6	3.1

[a] Stems greater than 15 cm dbh for Georgia and 6 cm dbh for Florida.

The quality of pine habitat in Georgia was better than that of Florida. Mean pine stand age was similar for the Florida and Georgia populations (**Table 1**). However, pine stand density in Georgia was nearly twice that for Florida. The difference between Florida and Georgia was even greater when the density of young pines, a size class not in the foraging regression, is considered. The Georgia territories contained variable amounts of hardwood understory while Florida pine stands were virtually hardwood-free.

A subset of 12 of the 18 territories with complete data for the analysis variables was used in the multiple regression. Stand size accounted for 55% of foraging use in Georgia (use = -0.10 + 1.07 x stand size; $P = 0.0001$, $N = 68$) and 63% in Florida (DeLotelle et al. 1987). The multiple regression of stand use on the first five principal components of age, density, age times density, hardwoods, stand distance to the cluster site, stand distance to the average territorial interaction site, and cluster distance times territorial interaction distance terms for the Georgia population (**Table 2**) was significant ($R^2 = 0.52$, $F = 12.8$, $P = 0.0001$, $N = 64$). The regression of stand use on the first five principal components for the Florida population was significant ($R^2 = 80.4$, $F = 21.3$, $P = 0.0001$, $N = 31$). The two-stage model thus accounts for 78% (0.55 + 0.52(1.0 - 0.55)) of stand use variability in Georgia compared to 93% in Florida.

The biological interpretations of the principal component loadings (**Table 2**) for Georgia and Florida were similar (DeLotelle et al. 1987). In assessing stand use, the first principal component was primarily the effect of stand distance to the cluster and to the average interaction site. The second principal component was primarily the habitat quality effect of pine stand density and age. In Georgia, the third principal component was the effect of hardwoods, while in Florida, the third principal component was the effect of age and density and their interaction.

The importance of habitat variables on stand use in Florida and Georgia was compared by the numerical sensitivity analyses (**Table 3**). The importance of stand distance to the cluster site in predicting stand use was nearly identical in Florida and Georgia. The importance of age was similar, with only a 7% difference. Pine stand density was more important in Florida

Table 2. Multiple regression parameters of use on the first five principal components of age (A), density (D), age times density, hardwoods (H), stand distance to the cluster site (C), stand distance to the average territorial interaction site (T), and cluster-site distance times territorial interaction site distance terms for the south Georgia (top) and central Florida (bottom) populations. See text for details of data transformation and standardization.

	Principal Component				
GEORGIA	**1**	**2**	**3**	**4**	**5**
Proportion	0.3430	0.2909	0.1720	0.1254	0.0686
Eigenvectors					
ln D	0.2575	0.5588	0.2308	-0.3322	-0.3016
ln A	0.2120	0.2635	-0.6545	0.4286	0.3920
ln D * ln A	0.3154	0.5999	-0.1286	-0.0791	-0.0691
ln C	0.4747	-0.2349	0.0637	0.4886	-0.5245
ln T	0.4566	-0.2712	-0.0494	-0.5374	0.4408
ln C * ln T	0.5759	-0.3140	0.0064	-0.0528	-0.0305
ln H	0.1542	0.1765	0.7038	0.4116	0.5294
P-value	0.0012	0.0001	0.0061	0.7146	0.6279
Partial r^2	0.1403	0.2741	0.0964	0.0015	0.0027
Slope	-3.6020	5.4681	-4.2184	-0.6391	1.1287
FLORIDA	**1**	**2**	**3**	**4**	**5**
Proportion	0.4740	0.2705	0.1764	0.0726	0.0060
Eigenvectors					
ln D	0.2078	0.7296	-0.0476	-0.1513	-0.0357
ln A	-0.3856	-0.3101	0.6170	0.1984	-0.0433
ln D * ln A	-0.1733	0.5504	0.6300	0.0672	0.0495
C	0.5057	0.0044	0.0560	0.7790	0.3642
T	0.4667	-0.2269	0.3704	-0.5711	0.5162
C * T	0.5519	-0.1307	0.2823	-0.0142	-0.7715
P-value	0.0001	0.0001	0.0571	0.0251	0.0001
Partial r^2	0.2906	0.2856	0.0299	0.0426	0.1548
Slope	-3.6236	4.7554	1.9063	3.5466	-23.4798

Table 3. Relative importance, based on numeric sensitivity analysis, of stand pine age, density, hardwood density, distance to cluster site, and distance to average territorial interaction site, on stand use for the south Georgia and central Florida populations (see text).

	Population	
Stand Effect	**Georgia**	**Florida**
Pine age	100	93
Pine density	30	69
Hardwood density	-9	-
Cluster site distance	-30	-29
Territory site distance	9	-100

than Georgia. The importance of territoriality differs considerably between the two populations. Hardwood density had a modest negative effect on foraging use in Georgia. Comparing habitat variables within each population shows that the effects of territoriality and pine stand age were nearly equally important in Florida, whereas in Georgia, stand age was clearly the single most important foraging parameter.

The simple regression curves of foraging use versus stand distance from the cluster site were linear for Florida and nonlinear Georgia (**Figure 1**). In both populations, stands close to the cluster had disproportionate higher foraging value than stands farther away. The vertical position of the lines at any distance from the cluster site was a direct measure of stand use in the two populations.

The importance of age on foraging use increases with stand distance from the cluster for the Georgia population only (**Figure 1**). That is, the vertical distance between the age curves for Georgia increased with stand distance from the cluster; the rate of increase diminishes at greater distances from the cluster.

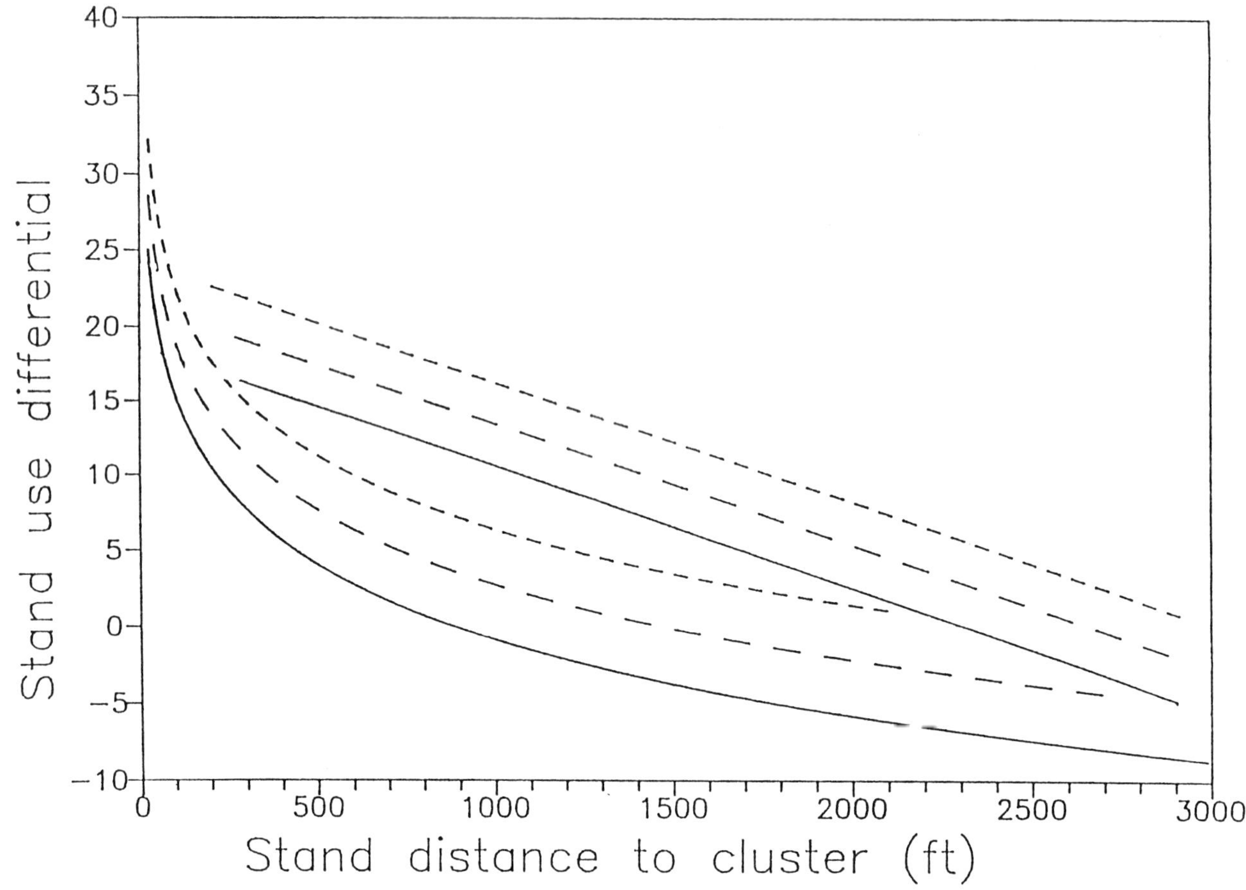

Figure 1. Differential stand use versus stand distance from the cluster site for the central Florida (straight lines) and south Georgia (curved lines) populations for stands of mean age (solid lines), mean age plus one standard deviation (broken lines), and mean age plus two standard deviations (dashed lines). Note that the vertical position of the foraging lines moves upward with each increase in pine age.

Nearer the cluster, the effect of increased stand age was to increase the size of the central area of the territory where stands had positive foraging value. Further from the cluster, the effect of increasing the stand age was to increase the size of the territory over which their was positive foraging value. The lines for Florida and Georgia converge at greater distance from the cluster site.

The negative effect of hardwoods on foraging use increased with distance from the cluster (**Figure 2**). Nearer the cluster, hardwoods had less a pronounced negative effect on forging use. Farther from the cluster, hardwoods had a much greater negative effect on foraging use. The result of increasing hardwoods was a reduction of the size of the territory where pine stands had positive foraging use.

DISCUSSION

Habitat and territories for the Georgia and Florida populations, while generally similar, had several notable differences (**Table 1**). Landscape characteristics, such as wetland drains, affect the shape and spatial distribution of territories. The shape of territories in Georgia was often irregular, some markedly so, corresponding to upland fingers between drains, whereas Florida territories were more circular due to the uniform habitat of the mesic flatwoods community. Cluster site and average territorial interaction site show various patterns of spatial distribution within the territories in Georgia, whereas all Florida study territories generally had a central cluster site and peripheral average territorial interaction site. Territories in Georgia were smaller than territories in Florida (DeLotelle et al. 1987).

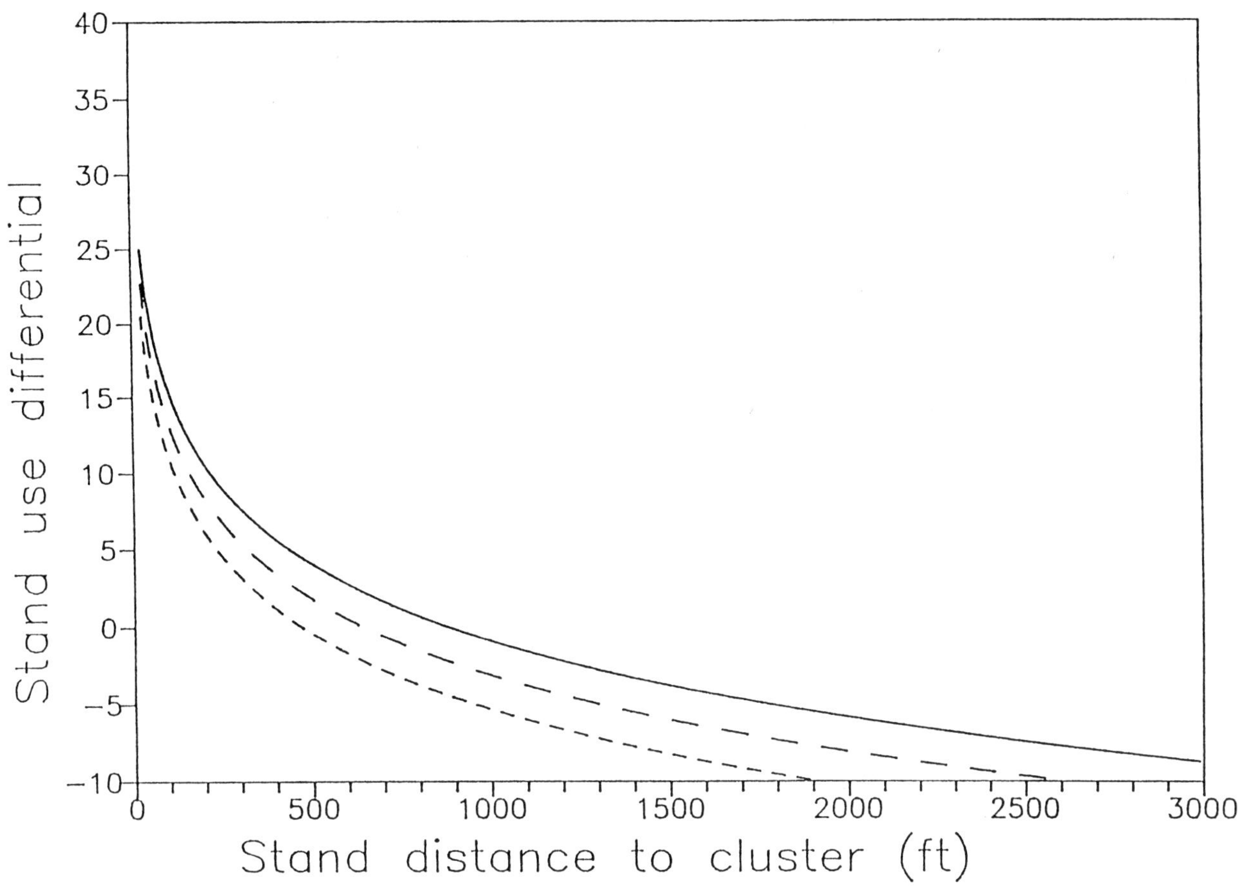

Figure 2. Differential stand use versus stand distance from the cluster site for the south Georgia population for stands with mean hardwood density (solid line), mean hardwood density plus one standard deviation (broken line), and mean hardwood density plus two standard deviations (dashed line). Note that the vertical position of the foraging lines move downward with each increase in hardwood understory.

Previous studies reported a strong relationship between population density and territory size, but only a weak influence of habitat quality and group size on territory size (Hooper et al. 1982, DeLotelle et al. 1987). In contrast to the South Carolina study, pine density was a major factor influencing variation in territory size in the Georgia study. The groups in impacted territories in Georgia were apparently compensating for reduced pine density by incorporating more habitat area during their foraging. In Georgia, population parameters such as group size and population density relative to available habitat had no significant effect on the variation in territory size. The importance of territoriality on foraging use was apparently reduced in the better quality habitat of Georgia (**Table 2**). The higher level of territorial behavior in Florida appears to be an added cost of occupying poor quality habitat (DeLotelle et al. 1987).

In both Georgia and South Carolina, hardwood understory was inversely related to territory size (Hooper et al. 1982). The negative effect of hardwoods on foraging use (**Figure 2, Table 2**) was consistent with the widely held opinion that red-cockaded woodpeckers evolved as a member of a fire sub-climax pine community, where burning maintained hardwoods in restricted areas at low density levels.

The forging preference for older age stands exists in both Florida and Georgia despite the differences in habitat quality. Further, in Georgia, the effect of pine age on foraging use varies with stand distance from the cluster (**Figure 1**). The difference in the lines between populations may be due to landscape differences between the Florida and Georgia sites, since territory size and shape may affect foraging flight times and costs (Repasky 1984) or other causes. Moreover, at a given distance from the cluster, foraging use in both populations was a simultaneous function of pine age and density (**Figure 3**, DeLotelle et al. 1987). In Georgia, foraging use was also a simultaneous function of stand distance from cluster and hardwood density (**Figure 2**).

Foraging use depends on the interaction of several stand variables (**Figures 1, 2**). Older age pine stands in the territory core had greater foraging value than equivalent stands outside this core but within the territory. Conversely, the older the pine stand, the farther it can be from the cluster and still receive a positive foraging use. Hardwoods also affect the size of the central core and its importance to foraging. Therefore, habitat quality is a measurable determinant of territory size and habitat use.

ACKNOWLEDGMENTS

This paper is dedicated to the memory of Don Charbonneau. Special thanks to G. Lowe, L. Grant, and D. Palmer for their reviews and comments on earlier drafts of the paper, and C. Huang for figure preparation. St Johns River Water Management District provided computer support.

The Ecology of Red-cockaded Woodpeckers Associated With Construction and Use of a Multi-Purpose Range Complex at Fort Polk, Louisiana

Jerome A. Jackson, Department of Biological Sciences
Mississippi State University, Mississippi State, Mississippi 39762
Stephen D. Parris, Environmental Division
Directorate of Engineering and Housing, Fort Polk, Louisiana 71459

ABSTRACT: Eight active red-cockaded woodpecker (*Picoides borealis*) cavity tree clusters on and adjacent to the site of a new multi-purpose range complex (MPRC) at Fort Polk, Louisiana, were studied through construction and initial use of the range. Although habitat losses were severe at two sites and some active cavity trees have been lost, only three clusters have been abandoned since construction began. The overall population of red-cockaded woodpeckers has declined, and home ranges of clans within the MPRC have increased in response to habitat losses. Extensive open areas have resulted in increased diurnal raptor presence. Average weight of red-cockaded woodpeckers declined and the decline for females appears greater than for males.
KEYWORDS: red-cockaded woodpecker, Ft. Polk, Louisiana, military lands, home range, reproductive success, foraging ecology, habitat use

The military is always developing new weapons and modes of transportation for troops and those of recent years have vastly increased ranges and mobility over different terrains. To adequately test these and train troops in their use, ever larger, more sophisticated training areas are needed. In 1984 we became involved in the process of selecting the site for a multi-purpose range complex (MPRC) at Ft. Polk, Vernon Parish, Louisiana, to be used by infantry and in training troops in the use of the M-1 tank and other armaments and vehicles. Fort Polk is adjacent to and shares a portion of Kisatchie National Forest in southwest Louisiana. The main part of the Fort is also home to at least 50 active red-cockaded woodpecker (RCW; *Picoides borealis*) cavity tree clusters.

The Army proposed four alternative sites (**Table 1**) which satisfied the stringent safety and operational criteria for a range of this nature. These alternatives included varying numbers of red-cockaded woodpecker clusters and amounts and qualities of associated RCW foraging habitat. Each site included an area of about 1619 ha (4000 acres) that would have to be cleared of trees for range operation. We examined each site with the purpose of recommending the site which would involve the least impact to the red-cockaded woodpecker.

The Army's preferred choice was Alternative 2A, the alternative with the largest number of RCW cavity tree clusters. At this point the Army entered into formal consultation with the U.S. Fish and Wildlife Service. Subsequently, the U.S. Forest Service joined the consultation process because some of the alternatives included cavity tree clusters on shared lands. In November 1984 the Fish and Wildlife Service issued a non-jeopardy opinion for the alternative with the 16-20 RCW cavity tree clusters, granting the Army a take of up to 52 birds.

Although granting permission for the Army to use the preferred alternative, the Fish and Wildlife Service recommended that one of the other alternatives be selected and directed that the impact of the project on the red-cockaded woodpecker be monitored. Ultimately, the Army chose Alternative 2 rather than 2A for several environmental considerations.

Table 1. Alternative sites proposed for the Ft. Polk multi-purpose range complex relative to red-cockaded woodpeckers and their habitat.

	Alternative 1	Alternative 2	Alternative 2A	Alternative 3
RC Clans Present	4-5	2-3	16-20	2-3
Total Cavity Trees	20	26	93	9
Hectares (Acres) to Clear	937 (2315)	1000 (2471)	856 (2115)	484 (1196)
Hectares (Acres) of Pine Sawtimber	761 (1881)	760 (1877)	445 (1100)	0 (0)
Hectares (Acres) of Pine Pole Timber	0 (0)	160 (396)	0 (0)	381 (941)

This paper is a report of our monitoring efforts between 1984 and 1992.

STUDY AREA AND METHODS

The completed multi-purpose range complex involves an area of about 12,141 ha (30,000 acres) including assembly areas, three major lanes of travel for the troops, and an extensive safety fan **(Figure 1)**. Upland forests of the area include longleaf (*Pinus palustris*), loblolly (*P. taeda*), and shortleaf (*P. echinata*) pines. Longleaf was predominant, and all cavity trees were longleaf pines.

There were several scattered long-abandoned cavity trees within the area to be cleared and these were left standing. Only two active RCW cavity tree clusters **(Figure 1, clusters A and B)** were down range and thus had the potential for being hit by fire from the various weapons to be used. The range was laid out specifically to avoid the necessity of removing cavity trees and to allow the active cavity tree clusters and some buffer areas to remain. All active cavity trees within these two clusters were tied in with adjacent pine stands. However, outside of a 61 m (200 ft) buffer and within the range proper, most trees, and all trees bigger than 25.4 cm (10 in) dbh were removed.

Cluster E was located behind firing lines but within the staging area. Broad gravel roads were constructed dissecting this clan's foraging area, but cavity trees, a buffer zone, and an extensive foraging area of mature pines were left.

Cluster F was adjacent to the staging area, but well away from the construction site, although some of the clan's foraging area was cleared. This cluster was abandoned when we first discovered it and as monitoring began, but was reoccupied by a banded bird from Cluster H who ultimately attracted a female.

The remaining clusters are outside of the areas of MPRC construction. Cluster G is located in an area frequently used for bivouacing and in generally poor habitat with much open space, young pine stands, and extensive hardwood encroachment in adjacent stands. Clusters C and D were originally treated as separate, but were ultimately demonstrated to be used alternately by the same clan. Both include older trees surrounded by young foraging habitat. Cluster I is outside of but immediately adjacent to the MPRC area and was monitored as part of this project. It included older cavity trees in a young stand that has a serious understory problem. This cluster and portions of Cluster C-D are on land shared with the Forest Service.

We were able to begin monitoring activities in the fall of 1984. Birds were captured and banded with U.S. Fish and Wildlife Service numbered aluminum bands and color-banded for individual identification. Upon capture and recapture, each bird was weighed to the nearest 0.1 g on a triple beam or an electronic balance. Chicks were removed from the nest for banding between 7 and 10 days of age using the noose technique developed by Jackson (1979a). Adults were captured on the roost using the nets described by Jackson (1977a) or Jackson and Parris (1991). Post-fledging birds were aged using the plumage characteristics described by Jackson (1979a). Nest status was determined by climbing to the cavity and inspecting it with a mirror and light or by monitoring the behavior of adults at the cavity as described by Jackson (1976).

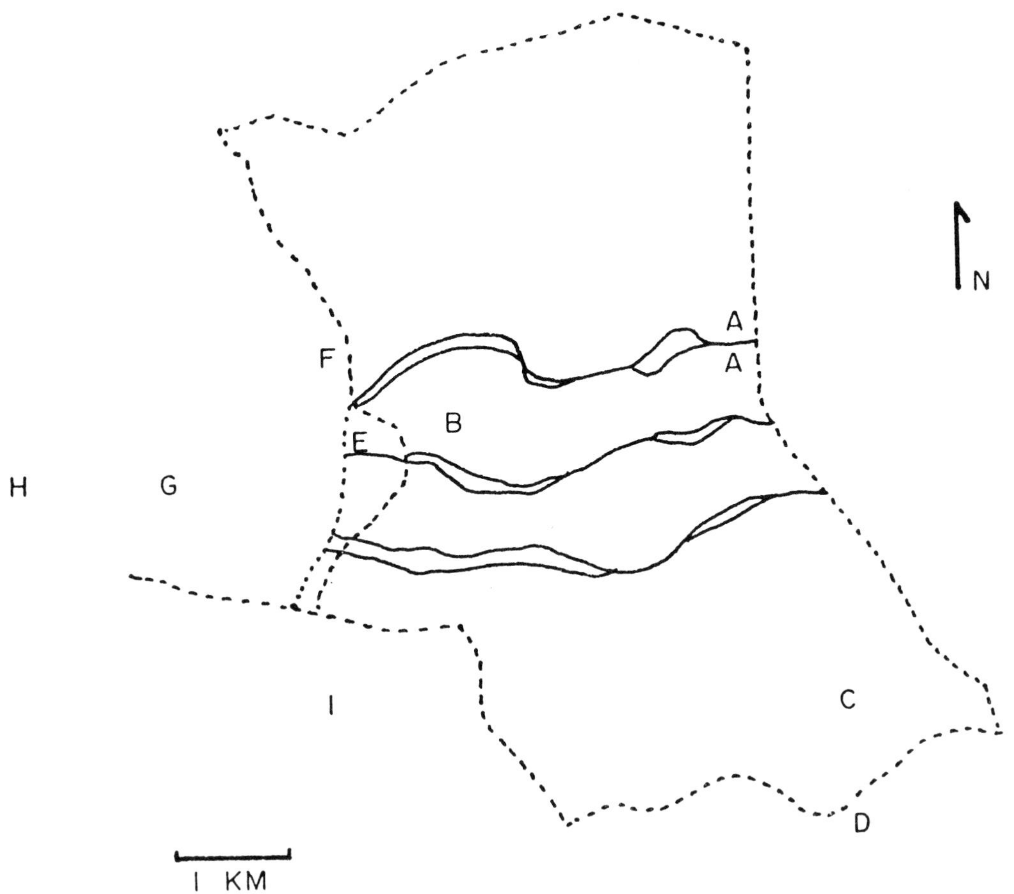

Figure 1. Schematic map of the multi-purpose range complex at Ft. Polk, Louisiana, showing peripheral roads (dotted lines), lanes of travel through the range (solid lines), and locations of red-cockaded woodpeckers cavity tree clusters identified by the letters used in the text. Firing on the range is from west to east.

Minimum home ranges were estimated by following marked birds and plotting locations on maps. As a result of range schedules and other factors, it was not always possible to follow birds from daylight to dark. Furthermore, although we could generally locate foraging birds during each visit, there were dates on which clans could not be located within their previously known range. Thus, we emphasize that ranges determined represent minimum home range. The home range data reported include the areas of the smallest polygons formed by connecting extreme locations of the birds. These include areas of unsuitable foraging habitat (roads, grasslands, and other non-forested areas) habitually crossed by the birds in the course of their daily activities, since crossing these areas involved time and energy expenditures necessary to reach foraging habitat.

Home range data were collected throughout the year and on 22 dates prior to initiation of range clearing, 20 dates during clearing and range construction, and 41 dates after the range became operational.

In the fall of 1986, clearing for the range began. Construction began in 1987, and the range has been in use since 1989. Because of training activities, the range has been closed to our monitoring for as much as 200 days per year since 1989. This made it impossible to check contents of all nests or to band all nestlings, although we are confident of our determinations of successful nesting efforts.

RESULTS AND DISCUSSION

The number of active RCW cavity tree clusters in this area has declined steadily from the original seven to five at the present time

Table 2. Pre-breeding clan sizes for red-cockaded woodpecker cavity tree clusters in and near the multi-purpose range complex, Ft. Polk, Louisiana.

Year	1985	1986	1987	1988	1989	1990	1991	1992
Within the Range								
Cluster A	4	3	2	2	2	2	1	0
Cluster B	3	3	2	2	2	3	2	2
Behind the Range								
Cluster E	3	5	4	4	5	4	4	5
Cluster F	0	1	2	2	2	2	2	3
Range Periphery								
Cluster C-D	2	2	3	3	2	2	3	2
Cluster G	2	2	2	1	0	0	1	0
Cluster H	3	2	2	3	3	2	3	3
Cluster I	2	1	0	0	0	0	0	0
Total Number of Birds	**19**	**19**	**17**	**17**	**16**	**15**	**16**	**15**
Total Number of Active Clusters	**7**	**8**	**7**	**7**	**6**	**6**	**7**	**5**

(**Table 2**). Significantly, only Cluster A within the range has been abandoned, and that abandonment finally came in early 1992. In November 1990, the female's roost tree was toppled by a direct hit; at the same time, the male's roost tree suffered a direct hit and died within a year. The male continued to roost in it but disappeared after the tree snapped off in early 1992. Use of dead cavity trees as roost or nest sites has previously been reported (e.g. Hooper 1982, Patterson and Robertson 1983) and is often symptomatic of inadequate numbers of alternative cavities. In this case no other cavities were available and there were no remaining trees suitable for cavity excavation.

A second cluster that was abandoned was Cluster I. This cluster had a dense hardwood understory reaching cavity height and was abandoned in 1987, prior to range completion and use. Work was being conducted on an existing road passing within about 60 m of the cavity trees at the time the birds disappeared from the area. Abandonment of cavity tree clusters with extensive hardwood encroachment is well known (e.g., Hooper et al. 1980), and we believe this was the major problem rather than activities associated with road work.

The third cluster to be abandoned was Cluster G. Initially active, problems with Cluster G included bivouacing troops at the base of the only active cavity tree, construction of an obstacle course within the 61 m buffer, and generally poor habitat. This cluster was abandoned in 1989, but reactivated in 1991 by a lone male that reactivated a different abandoned cavity tree. That bird disappeared in 1992.

The number of birds per clan just prior to the breeding season ranged from 0 to 5 (**Table 2**). The largest clan size has been in Cluster E which also consistently had the largest number of available suitable cavities (5-8) and the best foraging habitat. Clan size declined in Clusters A and B following range construction. Although young fledged in Cluster A for four years following range clearing, none remained as a helper in any year. Young fledged from Cluster B in six years and only in 1990 did one remain as a helper. None of these young was ever seen after disappearing from its natal cluster although checks of other clusters outside of the Range complex have frequently been made.

Pre-clearing annual home ranges for three clans averaged 135 ha, similar to the 150 ha reported for clans in central Florida pine flatwoods (Delotelle et al. 1987), but much more than the 86.9 ha average home range found for clans in coastal South Carolina (Hooper et al. 1982).

Annual clan ranges increased dramatically where extensive habitat alterations were made, more than doubling for the birds using Cluster A to in excess of 405 ha (1000 acres) (**Table 3**). At this particular cluster, we had predicted that the birds would shift their foraging range to the north, away from the range and into what appears to be good foraging habitat. They did not, and instead extended their range farther into the range, using predominantly 30-40-year-old trees in a very low basal area (<4.6-6.9

Table 3. Minimum annual home ranges (ha (acres)) of selected red-cockaded woodpeckers clans associated with the Multi-Purpose Range Complex, Ft. Polk, Louisiana.

Cluster	Pre-clearing	Post-clearing
A	170 (420)	413 (1020)
B	126 (310)	231 (570)
F	109 (270)	115 (285)
H	91+ (225+)	

Table 4. Numbers of new cavities completed and cavities lost when cavity trees died or were burned out in red-cockaded woodpeckers cavity tree clusters associated with the Multi-Purpose Range Complex, Ft. Polk, Louisiana, 1984-1992.

Cluster	# New Cavities	# Cavities Lost
Within the Range		
A	0	2
B	1	1
Behind the Range		
E	4	0
F	2	1
At Range Periphery		
C-D	2	2
G	1	2
H	2	4
I	0	0
Totals	12	12

m^2/ha (< 20-30 ft^2/acre) environment. We feel that one or both of three factors may have influenced the direction of the birds' home range shift:

(1) an active timber sale to the north might have been a negative factor;

(2) the male's roost tree was the southernmost tree in the cavity tree cluster and was south of a major clearing for target placement; and

(3) temporarily there was an abundance of food in the cleared areas as a result of beetle populations moving in to take advantage of recently downed or damaged trees associated both with range construction and range use.

During the tenure of our monitoring, a number of cavity trees were lost and others were gained (**Table 4**). No new cavities were, however, excavated in Cluster A, and only one new one was excavated in Cluster B.

We believe that the decreased habitat quality associated with clearing for range construction decreased the health of the birds and survivorship of nestlings and fledglings. Indicators of decreased health of individuals included two morphological characteristics of young birds. In 1989, one of the two chicks hatching from a three-egg clutch in Cluster B was half the size of its sibling at about age 10 days. In addition, the only evidence of fault bars on feathers, a known characteristic associated with poor nutrition, was found on fledglings from Clusters A and B (**Figure 2**).

We also believe that the effect of habitat changes, specifically the removal of large trees, had a differentially negative impact on females. In both pre-clearing and post-clearing data we found that females characteristically foraged on the trunks of pines, usually below the lowest branches, while males foraged higher and more often on branches in the crown. Such a sex difference was first reported by Ligon (1968) in Florida and Ramey (1980) and I have also documented it for birds in Mississippi and South Carolina.

Prior to clearing for the range, males and females from Clusters A and B averaged approximately 49 and 48 g at evening roost time, respectively. Following clearing, the average for males dropped to about 48 g and that for females to about 46 g. The lowest adult weights recorded were of females weighing 41.1 and 43.6 g. Our sample sizes are small, and differences within sexes between pre- and post-clearing are not statistically significant. Nonetheless, we feel that there is an obvious problem. Removing big trees differentially removes the foraging substrate most used by females.

As further evidence of a possible sex-specific problem, we note that prior to clearing we captured 16 post-fledging males and 16 post-fledging females in the area. Following clearing of the MPRC, we captured 9 new males, but only 4 new females.

Finally, as if the birds in this area didn't have enough problems, following clearing for the range, we began to notice an increased number of wintering avian predators that might potentially take a red-cockaded woodpecker -- specifically American Kestrels (*Falco sparverius*), Northern Harriers (*Circus cyaneus*), and Loggerhead Shrikes (*Lanius ludovicianus*). Although we did not initially census these species, Jackson kept a daily species list during work at Ft. Polk and only infrequently noted these birds. By the winter of 1989, these species were noted daily and several individuals could easily be found along the range roads. Once, while attempting to mist net an unbanded red-cockaded woodpecker, we were successful in capturing it and an American Kestrel that chased it into the net.

In short, the extensive grassland habitat created by range construction provides optimum habitat for small rodents and wintering sparrows and thus draws in these predators. Red-cockaded woodpeckers forced to fly across the open range to and from foraging areas and cavity trees thus seemed exposed to an increased risk of predation.

Although it has been nearly six years since clearing for the range began and four years since it became operational, we do not feel the total effects of range construction and use on these clusters have been realized. We predict the ultimate loss of Cluster B. We also feel that as a result of the availability of quality habitat and two sets of cavity trees that are now separated by a large road and cleared right-of-way, that Cluster E might ultimately become two clusters.

CONCLUSIONS

1. The impact of habitat modifications on red-cockaded woodpeckers can be a delayed reaction occurring after several years. Impact studies must be long term. We do not feel that we have seen the final chapter in the impact of this range construction on the red-cockaded woodpeckers at Ft. Polk, although the construction was completed four years ago.

2. Large open areas, particularly semi-natural grasslands, with scattered shrubs or small trees provide wonderful habitats for wintering populations of American Kestrels, Loggerhead Shrikes, Northern Harriers, and other avian predators that have the potential of preying on red-cockaded woodpeckers, a bird that flies relatively high and slow. Some of these predators can remain year round.

3. When considering habitat needs of a species, the special needs of all age and sex classes must be considered. E.g., in the case of the RCW, the female's need for large trunks as foraging substrate may not be met by younger trees that might provide for males.

4. When faced with an increasingly adverse habitat, the red-cockaded woodpecker may choose adversity -- to its detriment -- rather than move.

ACKNOWLEDGMENTS

We acknowledge the financial and logistical support of the U.S. Army at Ft. Polk provided through contracts to Eco-Inventory Studies, Inc. Stephanie Young, Jim Grafton, Charles Stagg, and others at Ft. Polk, Emlyn Smith, James Key, Bette Jackson, Peter Jackson, and Jerome A. Jackson, Jr. also assisted with field work.

Short-term Response of a High Density Red-cockaded Woodpecker Population to Loss of Foraging Habitat

Robert G. Hooper, U.S.D.A. Forest Service
Southern Research Station, Charleston, SC 29414
Michael R. Lennartz, U.S.D.A. Forest Service, Washington, DC 20090-6090

ABSTRACT: Removal of 43% of the foraging habitat from a dense aggregation of RCW (*Picoides borealis*) groups had no discernible negative impact on the study groups. The number of groups increased during the period habitat was removed, and group size and reproductive success were similar before and after habitat removal. In conjunction with other studies, the results suggest that RCWs are not sensitive to loss of foraging habitat except at low densities, and that low population density, itself, may be a major factor inhibiting expansion of some small populations.
KEYWORDS: *Picoides borealis,* foraging habitat, Francis Marion National Forest, habitat loss, reproduction, group size, population density

Several studies found that home range size of red-cockaded woodpeckers [RCW] decreased as population density increased (Sherrill and Case 1980, Hooper et al. 1982, DeLotelle et al. 1987). This relationship suggested that some RCW groups may have had more foraging habitat than they apparently needed for satisfactory reproduction and maintenance of group size. Also evident, was that the birds themselves effectively reduced habitat availability as population density increased.

Wood et al. (1985a, 1985b) tested the excess habitat hypothesis in a RCW population of moderate density (mean of 8.2 groups within 1.25 miles of study groups) by removing up to 37% of the foraging habitat in RCW home ranges and observed no adverse effect on the groups. Conner and Rudolph (1991b) found that group size in a sparse population (mean of 2.5 groups within 1.25 miles of study groups) was associated with habitat loss and/or fragmentation, but no effect was observed from even greater habitat loss in a somewhat denser population (4.7 groups within 1.25 miles of study groups on average).

Based on these studies, one might expect groups not to be sensitive to habitat removal, except perhaps at low densities. But if groups were sensitive to habitat loss, perhaps it would be most apparent at high population densities: the denser a population, the less the available habitat to each group, on average, and the more a group would suffer from habitat loss. If groups were not sensitive to habitat loss, then removal of habitat to some threshold would not of course affect them at higher densities.

Here we report the short-term effects of foraging habitat removal on a dense aggregation of RCW groups (mean of 20.6 groups within 1.25 miles of study groups). Unfortunately, the study was prematurely terminated by Hurricane Hugo. Several more years of data would have increased the conclusiveness of the results. However, a similar study is not likely to be repeated in the foreseeable future because of the lack of other high density populations, and because the degree of habitat removal we observed is now prohibited.

METHODS

Study area

The study was conducted from 1979-1989 in the Francis Marion National Forest (FMNF) in South Carolina. The RCW population was large and increasing during the study period: 427 groups were present in 1980-1981 and 477 groups in 1987-1988 (Hooper et al. 1991a). The overall condition and management of the FMNF was described by Hooper et al. (1991a).

We selected a 5000 acre study area of

mostly 90- to 110-year-old longleaf pines (*Pinus palustris*) for the study. The area had a high density of RCW groups (1 group per 100-125 acres of pine stands). In 1980, FMNF staff began plans to regenerate a portion of the extensive area of mature longleaf pine. Because the area had one of the highest densities of RCW groups known, we wanted to determine the effects of habitat removal. Two concentric circles, each encompassing 1500 acres, were located around the areas to be regenerated **(Figures 1, 2)**. Research had been conducted in the area since 1976 (Hooper and Lennartz 1981, Hooper et al. 1982, Lennartz et al. 1987) but 1979 was the first year we had complete data on groups in the core zone. The number of groups and the area of foraging habitat were similar in both the core and peripheral zones before and after removal of the foraging habitat **(Table 1)**. The peripheral zone was intended to keep groups within the core zone from shifting their foraging outside the core zone in response to loss of foraging habitat.

Group performance

We determined reproductive performance of each group in the core zone by climbing nest trees and examining cavity contents with a light and mirror 1-2 times each week from mid-April until nesting ceased in early to mid-June. The number of adults per group was determined by watching banded adults entering cavities to incubate eggs or feed nestlings. If a group did not nest we determined the number of adults by observing their behavior upon leaving roosting cavities in the morning.

All clusters of cavity trees in the peripheral zone were visited 1-5 times during the nesting season each year to determine if they were occupied by groups. A cluster with a nest in a given year obviously represented a group. If at least 2 RCW's roosted in a cluster and did not fly to an adjacent cluster upon leaving their cavity trees in the morning (Hooper and Lennartz 1983), we considered them to represent a group even though we did not find a nest that year.

Removal of foraging habitat

Timber harvests began in 1984 and were completed in 1988. Trees were removed in clearcuts and thinnings. National forest personnel determined the total number of trees removed from the core zone by actual count. Number of pines ≥ 10 inches DBH remaining after cutting was estimated by point sampling with prisms (Dillworth and Bell 1982).

Total trees in the core zone prior to harvesting was determined by adding trees remaining and trees removed. Angular sum, a measure of forest fragmentation devised by Conner and Rudolph (1991b), was determined for each group in both zones before and after loss of foraging habitat. We determined the number of groups within 1.25 miles of each study group in both zones before and after removal of foraging habitat.

RESULTS

In the core zone, 448 acres (32%) of the foraging habitat in pines ≥ 30 years old were clearcut **(Figure 2, Table 1)**. Including the trees removed in thinnings, 43% of the pines ≥ 10 inches DBH were removed from the core zone. Prior to harvesting, 58,000 pines ≥ 10 inches DBH were present in the core zone; after harvesting, 33,000 pines that size remained **(Figure 3)**. Groups averaged 5,800 pines ≥ 10 inches DBH in 1983, and 2,500 pines that size in 1989. Fragmentation, as determined by angular sum (Conner and Rudolph 1991b), increased 10 fold in the core zone as a result of clearcutting **(Table 1)**.

In the peripheral zone, 155 acres of pines ≥ 30 years old were removed after 1983, and the total remaining area of foraging habitat in the 2 zones was similar **(Table 1)**. Considerable habitat removal occurred outside the peripheral zone on national forest and private lands between the beginning and end of the study **(Figures 1, 2)**.

The number of groups in the core zone increased after removal of foraging habitat began in 1984 **(Figure 4)**. One group became defunct in 1983, prior to removal of foraging habitat, and that cluster was without a group until 1987. One colonization (new group and new cluster, see Hooper et al. 1991a) occurred in 1985 and another in 1989.

The number of groups in the peripheral zone also increased after removal of foraging habitat began. The peripheral zone had 13 groups when removal of foraging habitat began in 1984 and 14 in 1989. Two colonizations occurred and 1 group became defunct after habitat removal began.

The mean number of adults in groups within the core zone did not differ significantly before and after removal of foraging habitat

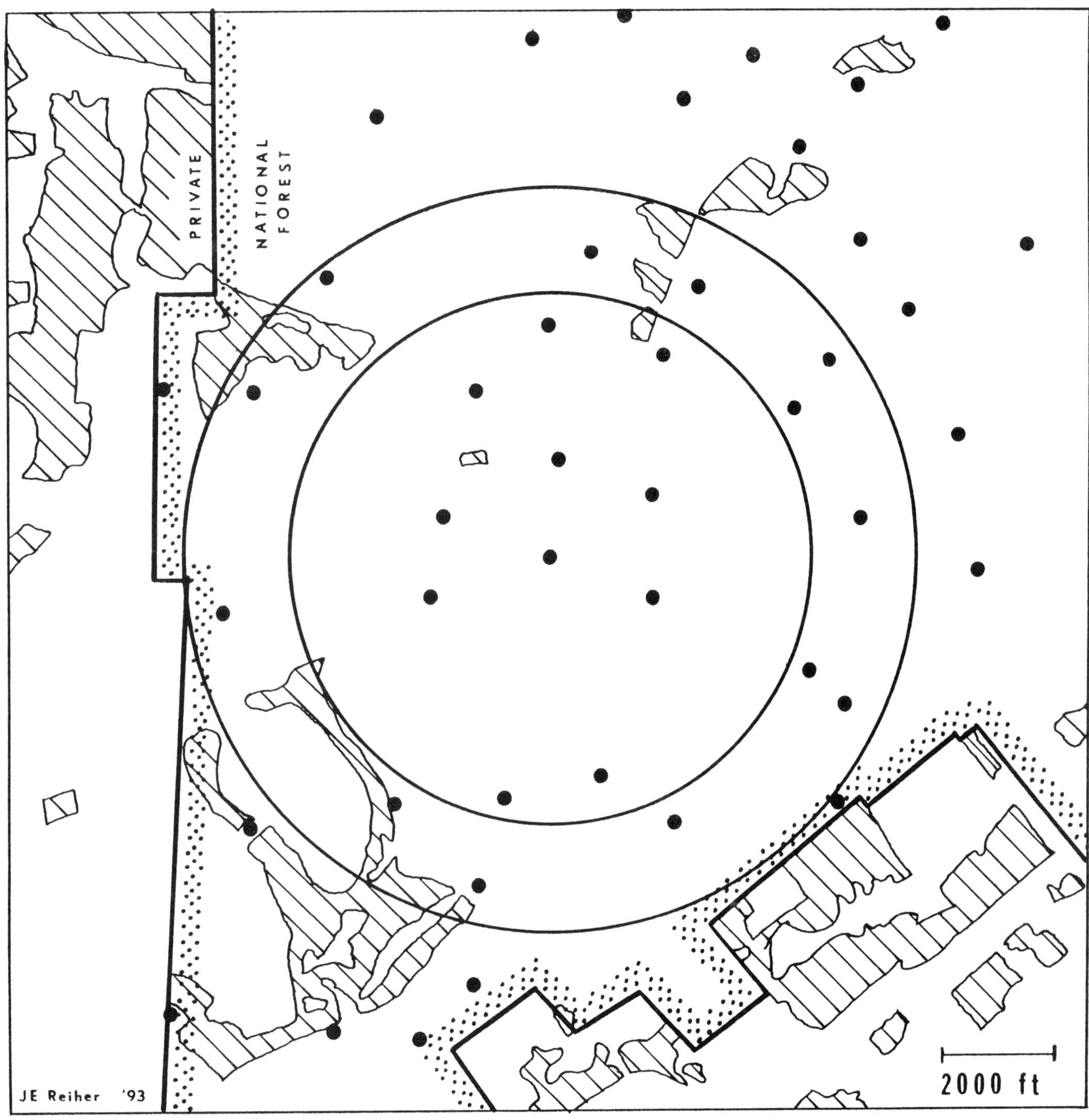

Figure 1. Location of RCW cavity-tree clusters supporting a group of RCWs (solid dots), in and around a 1500 acre core zone (inner circle), and a 1500 acre peripheral zone (outer circle) in 1979, when a study to determine the effects of removal of foraging habitat on the groups in the core zone began. Clearcuts, fields and lakes are indicated by the diagonal lines. The area was part of a high-density subpopulation in Francis Marion National Forest in South Carolina.

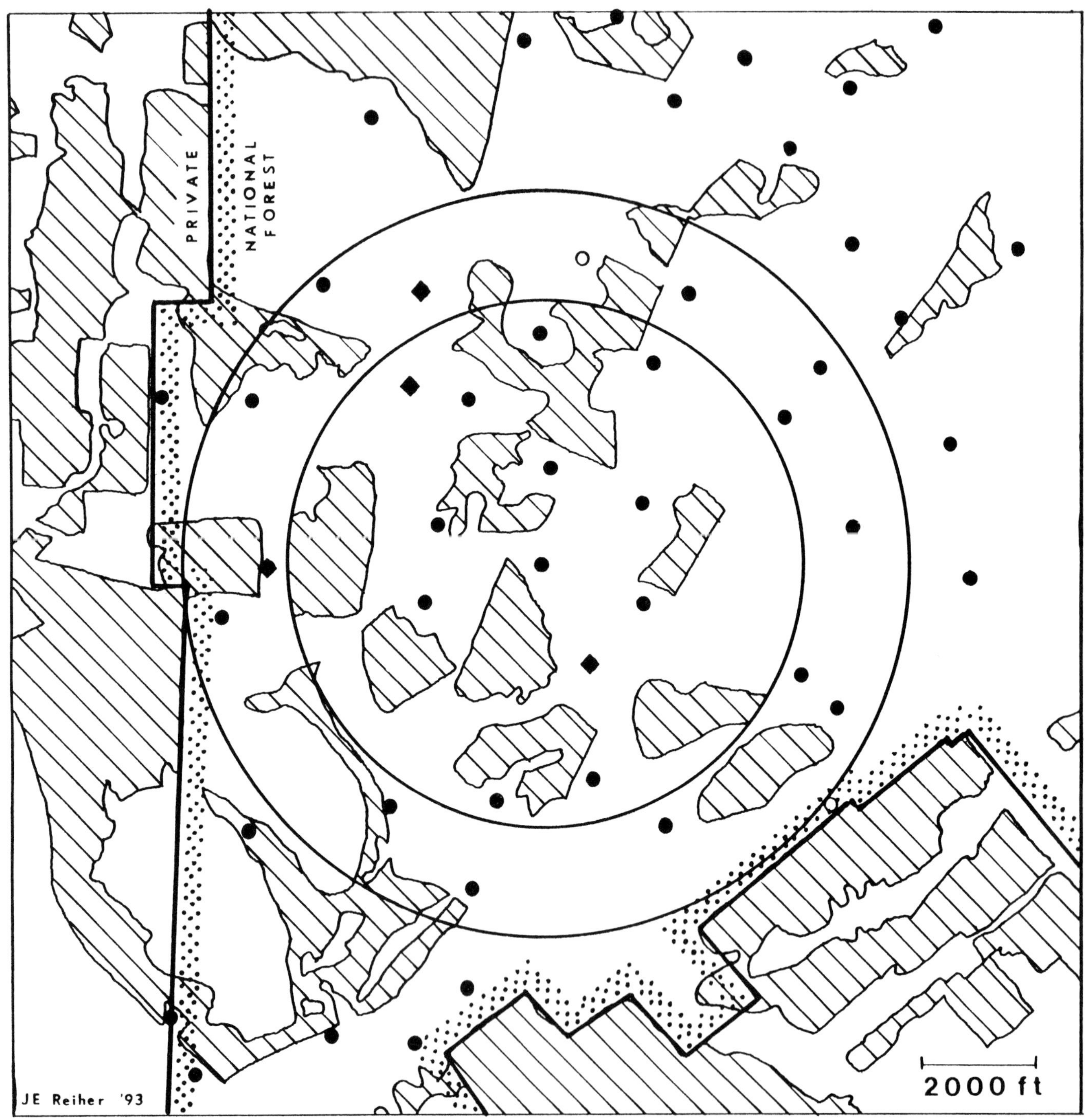

Figure 2. Condition of the area shown in Figure 1, in 1989. Clearcuts, fields and lakes are indicated by diagonal lines. Solid dots are RCW clusters that were present in 1979-1983 and still supported a group of RCWs in 1989. Open dots are clusters that supported groups in 1979-1983, but not in 1989. Solid diamonds are clusters that did not exist in 1979-1983, but supported new groups in 1989.

Table 1. Characteristics of core and peripheral zones of a study area in a high density sub-population of red-cockaded woodpeckers (RCW) in Francis Marion National Forest, South Carolina, before (1979-1983) and during and after (1984-1989)[2] removal of 43% of the pines ≥ 10 inches dbh from the core zone.

Variable	Before Removal Core	Before Removal Peripheral	Removal[2] Core	Removal[2] Peripheral
Total acres	1500	1500	1500	1500
Acres pine ≥30 years old	1404	1337	956	1111
Acres pine < 30 years old	0	71	448	226
Number clearcuts	0	3	8	8
Mean angular sum[1] (range)	22 (0-80)	62 (0-157)	243 (163-317)	182 (21-316)
RCW groups	10-11	13	10-13	13-14
Mean number of	20.6	18.2	23.4	19.7
Mean number of RCW groups in 1.25 miles (range)	20.6 (17-25)	18.2 (12-27)	23.4 (18-27)	19.7 (14-28)

[1]Sum of angles defined by edge of clearcuts, fields and lakes, and the center of each group's cluster of cavity trees. See text.

[2]Removing of foraging habitat was completed in 1988. See figure 2.

Table 2. Mean number of adults in red-cockaded woodpecker groups and reproductive performance of those groups before, and during and after removing 43% of the pines ≥ 10 inches dbh from the 1500 acre core of a 3000 acre study area in Francis Marion National Forest, South Carolina.

Variable	Before Removal[1] mean # (SE)	After Removal[2] mean # (SE)	t	Two-tailed P
Adults	2.4 (0.17)	2.4 (0.08)	0.1	<0.91
Eggs	2.7 (0.23)	2.8 (0.18)	0.2	<0.82
Hatching	1.7 (0.24)	1.5 (0.19)	1.0	<0.33
Fledging	1.2 (0.16)	1.1 (0.12)	0.6	<0.59

[1]1979-1983. Number of groups ranged from 10-11

[2]1984-1989. Number of groups ranged from 10-13. Removal of foraging habitat began in 1984 and was completed in 1988. See figure 2.

began (**Table 2**). The mean number of eggs, hatchlings and fledglings for groups within the core zone, did not differ significantly before and after removal of foraging habitat began (**Table 2**). The mean number of young fledged in 1989, the year after habitat removal was completed, was no lower than in 1980 or 1983, prior to habitat removal (**Figure 5**).

DISCUSSION

The increase in the number of groups in our study area during the period of foraging habitat removal, was associated with an overall increase in the number of groups in FMNF during the same period (Hooper et al. 1991a). Across the FMNF as a whole, foraging habitat was being removed at a rate of 1.5% (areal basis) each year between 1979-1989 and the net increase in number of groups was 1.4% per annum. Clearly the population increases were not prevented by removal of foraging habitat; nor does it seem likely the increases would have been greater had habitat not been removed, given the greater habitat removal (6.4% per year) and the greater increase in groups in our study area (5.0% per year), compared to the FMNF as a whole.

The lack of a negative impact on groups from removal of foraging habitat in our study area was consistent with earlier studies (Wood et al. 1985a, 1985b, Conner and Rudolph 1991b). Except in the case of a low density of groups (Conner and Rudolph 1991b), these

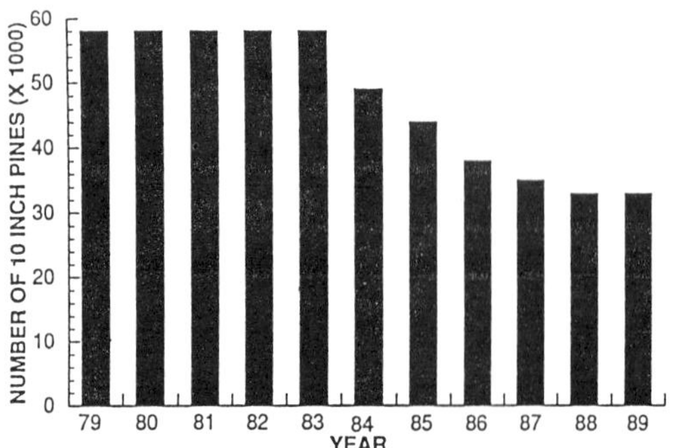

Figure 3. Number of pines ≥ 10 inches DBH in the 1500 acre core zone of a 3000 acre study area within Francis Marion National Forest in South Carolina from 1979-1989. Beginning in 1984 and ending in 1988, pines were removed by clearcutting and thinning.

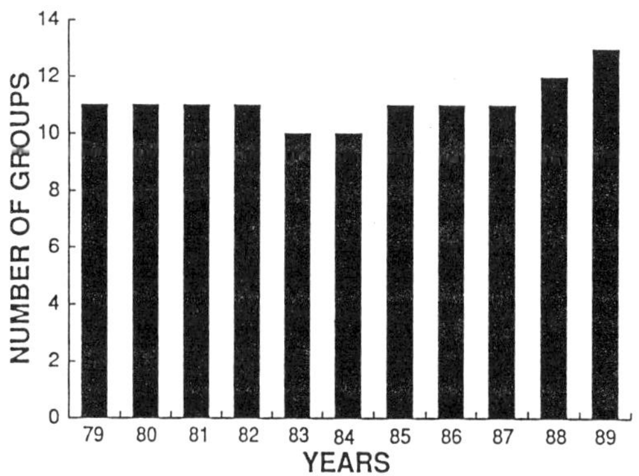

Figure 4. Number of RCW groups in the 1500 acre core zone of a 3000 acre study area within Francis Marion National Forest in South Carolina. Removal of foraging habitat began in 1984 and ended in 1988.

studies suggest that RCW groups are not very sensitive to removal of foraging habitat.

If groups were sensitive to removal of foraging habitat, one might expect their response to be most obvious in populations with a high density of groups. Mean density in our study was 3 times greater than in the Wood et al. (1985b) study, and 5 times greater than in the high density area of the Conner and Rudolph (1991b) study. Yet, the number of groups in our study area increased as total available habitat was being reduced. The increase in the number of groups further decreased the amount of habitat per group, but we detected no change in group performance. A related effect is apparent in the Conner and Rudolph (1991b) study. In their study areas, groups with a sparse density of surrounding groups had twice the available habitat on average as groups with a higher density of surrounding groups. However, it was the groups in the sparse density areas that apparently suffered from a shortage of foraging habitat and from forest fragmentation.

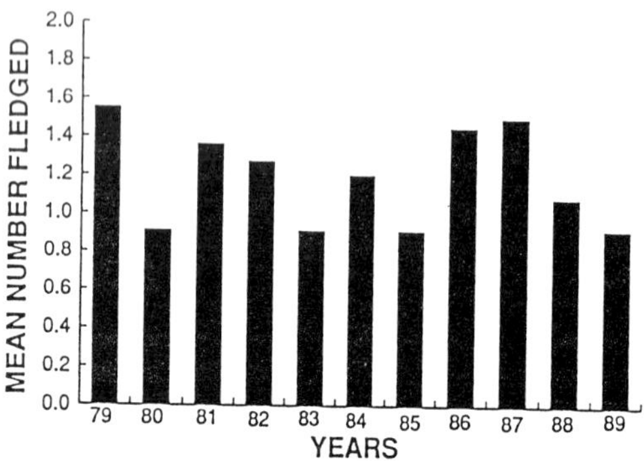

Figure 5. Mean number of young fledged in the 1500 acre core zone of a 3000 acre study area within Francis Marion National Forest in South Carolina. Removal of foraging habitat began in 1984 and ended in 1988.

Groups in dense and moderately dense populations were not shown to be sensitive to loss of up to 43% of their foraging habitat. To the contrary, groups in sparse populations did seem to be sensitive to loss of foraging habitat and forest fragmentation, but yet they had more foraging habitat that groups in the denser populations (Conner and Rudolph 1991b). Thus, one might conclude that it was the density of surrounding groups that was the critical factor in the sparse populations, rather than foraging habitat per se. If so, habitat protection alone may not be effective in increasing small populations. Direct intervention with artificial cavities (Copeyon 1990, Allen 1991, Taylor and Hooper 1991) and translocations of RCW's (Allen et al. 1993; Hess and Costa 1995) may be necessary before low-density populations can increase. These techniques were effective in increasing small populations that had been declining, even when some of them existed in rather poor habitat (Gaines et al. 1995, Reinman 1995, Richardson and Stockie 1995). In contrast, a small population of 11 groups declined to extirpation in seemingly good habitat where no loss of foraging habitat had occurred (Baker 1983b).

What is a critically low population density? A first approximation might lie between the dense and sparse populations studied by Conner and Rudolph (1991b). Their dense and sparse populations averaged 4.7 groups and 2.5 groups within 1.25 miles of study groups, respectively. On a density basis, those numbers of groups are equal to 1 group per 551 and 898 acres, respectively. The 11 groups in Baker's study (1983b) were in a 2,800 acre area equaling a density of 1 group per 254 acres, but this density estimate is not comparable to density estimates based on the mean number of groups within 1.25 miles. If we reduce the 2,800 acres to a circle and then assume the 11 clusters are evenly distributed within that circle, the mean number of groups within 1.25 miles of each of the 11 clusters would be about 4.2, or 1 group per 604 acres.

Current foraging guidelines require 6350 pines $\geq$ 10 inches DBH for each group of RCW (Lennartz and Henry 1985, Henry 1989). Provision of 6350 stems is predicated on reproductive output increasing as number of pine stems increases up to 6350 stems. This guideline was selected to enhance recovery of populations and does not represent the minimum amount of foraging habitat needed for reproduction and long-term site occupancy. If RCW groups are not sensitive to loss of foraging habitat down to the levels discussed here, there may be some circumstances when a RCW population would benefit in the long run by having its foraging habitat reduced below the guideline level. Some examples of such circumstances include:

(1) recovery areas where the risk from hurricanes makes it is especially desirable to have a balance of age classes as soon as practical (Hooper and McAdie 1995);

(2) in the thinning of pine stands to reduce the southern pine beetles hazard (Thatcher et al. 1986);

(3) removal of trees infested with southern pine beetles in order to avoid a major epidemic (Billings and Varner 1986); and

(4) conversion of off-site pine species to longleaf.

Preliminary Fire History of McCurtain County Wilderness Area and Implications for Red-cockaded Woodpecker Management

Ronald E. Masters, Department of Forestry, Oklahoma State University, Stillwater, OK 74078
John E. Skeen and James Whitehead, Oklahoma Department of Wildlife Conservation, 1801 N. Lincoln, Oklahoma City, OK 73105

ABSTRACT: Historical evidence suggests that frequent anthropogenic and lightning fires maintained a mosaic of mixed shortleaf pine-hardwood stands, with an open bluestem or woody understory, among hardwood and hardwood dominated stands on the McCurtain County Wilderness Area (MCWA). Red-cockaded woodpecker populations may be indicative of the extent and quality of the shortleaf pine-bluestem ecosystem in the Ouachita Mountains of Oklahoma and Arkansas. Although the MCWA is forested with virgin old-growth timber, the plant community continues to change. Fire exclusion has been the primary factor advancing hardwood presence into the midstory and lower canopy, and a probable factor in woodpecker population declines. Frequent fires occurred on MCWA until 1956. Presettlement fires occurred with mean fire intervals of 3.5 and 5.6 years respectively, within a 300 and 800 ha area in the wilderness. Fire suppression has extended the mean fire interval from 29.9 to 546.7 years at a dimensionless point. The earliest fire recorded on the MCWA dated to 1710. Fires prior to 1890 alternated from the east to the west side of Mountain Fork River and rarely burned both in a given year, suggesting that Native American fires had habitat management objectives. From 1830-1931, fires exhibited a repeating pattern of short-return intervals followed by 1 or more long-return intervals. From 1939 to 1992, 24% of the fires were from lightning. Lightning-caused fires occur on the MCWA every 3.1 years and occur most frequently from July to October and of lower frequency in March and April. Because large scale, periodic, growing season fires have been important in shaping this ecosystem, prescribed burning should be patterned after the presettlement burning regime.
KEYWORDS: fire history, fire scar, fire suppression, prescribed fire, Oklahoma, red-cockaded woodpecker, wilderness

The McCurtain County Wilderness Area (MCWA) is the largest representative tract of virgin old-growth shortleaf pine (*Pinus echinata* Mill.)-hardwood habitat known to remain in the United States (Stahle et al. 1985). The only known population of red-cockaded woodpeckers (*Picoides borealis*) in Oklahoma exists on the MCWA, with the exception of one cluster site in Pushmataha County. The MCWA population is at the western limit of the species' range. Red-cockaded woodpecker populations have declined precipitously since Wood (1977) first surveyed the woodpeckers status on MCWA (Kelly 1991). From 1974-77, Wood (1977) surveyed 83% of the area and located 29 groups with 86-92 individuals. A more extensive survey in 1989-1990, located 15 active groups with 31 individuals (Kelly 1991). The average number of clusters lost within Wood's (1977) study area was 1.4/year, a 62% decline (Kelly 1991). Since that time, a wind storm destroyed 1 cluster and 3 isolated single bird clusters were abandoned. After banding efforts were initiated, 2 clusters, formerly considered separate, were combined. Currently, only 9 active clusters with 22 individuals are known.

Extensive hardwood intrusion into the midstory and lower canopy layers has been implicated in the decline of red-cockaded woodpeckers on the MCWA (Masters et al. 1989). Elsewhere midstory hardwoods have been related to the abandonment of red-cockaded woodpecker clusters and population declines (Van Balen and Doerr 1978, Conner and Rudolph 1989, Loeb et al. 1992). Fires have been suppressed on the MCWA since 1926 (Masters et al. 1989). Fire suppression allows shifts in community dominance to a greater hardwood component (Bruner 1931, Garren 1943, Komarek 1974, Van Lear 1985, Cain 1987). Where fire is excluded, midstory hardwoods will increase in stem density and percent cover and gradually replace overstory pines (Cain 1987). This happens because shortleaf pines will not reproduce under shade, and as old dominant pines die they will be replaced by midstory hardwoods (Guldin 1986).

The Oklahoma Department of Wildlife Conservation is in early stages of implementing a management plan for the MCWA following formal consultation with the U. S. Fish and Wildlife Service on red-cockaded woodpecker management (Okla. Dep. Wildl. Cons. 1991a,b,c). The MCWA management plan follows a systems approach to ecosystem restoration. Major elements of the plan include reintroduction of fire and limited midstory removal in active, inactive, and replacement clusters. An important part of the plan is reintroduction of fire in a manner that will approximate a presuppression, anthropogenic and lightning fire regimes.

This paper presents data from the first phase of an ongoing study to reconstruct the fire history of MCWA. Fire history was reconstructed from historical accounts, Oklahoma Department of Agriculture, Forestry Services fire records, and from fire scars in tree cross-sections. This information will be used to plan appropriate prescribed burning regimens.

STUDY AREA

The 5,701 ha MCWA is located in the Kiamichi Mountain Range of the Ouachita Highlands in southeast Oklahoma. Broken Bow Reservoir (Mountain Fork River) divides the area into two parts, with most of the area (5,215 ha) on the east side **(Figure 1)**. The area is characterized by steep and narrow, east-west trending ridges, with occasional rock outcrops. Elevation varies from 183 m to 415 m. North and South Linson Creeks are the largest perennial streams that dissect the area (Masters et al. 1989).

The MCWA is covered by virgin old-growth pine-hardwood forest. Shortleaf pine, post oak (*Quercus stellata* Wangenh.), blackjack oak (*Q. marilandica* Muenchh.) and hickory (*Carya* spp.) dominate much of the forest and most of the sites are xeric in nature (Stahle et al. 1985). Mesic sites, such as drainages and lower to mid-north slopes, support white oak (*Q. alba* L.), maple (*Acer* spp.), and hop hornbeam (*Ostrya virginiana* [Mill.] Koch). Mixed pine and hardwood stands are characteristic of this region with a higher frequency of pine on most upper south-facing slopes. This study area has been further described by Carter (1965) and Masters et al. (1989).

The MCWA has been owned by the Oklahoma Department of Wildlife Conservation since 1918. In 1951, the area was designated by the State legislature as a Wilderness Area (Oklahoma Statutes, Title 29, Part 7, p. 56-57) (Okla. Game and Fish Dept 1954:21). Existing statutes require wildfire suppression by the State Division of Forestry (presently Forestry Services) (Oklahoma Statutes, Title 2, Section 1301, Article I, Section 104 and 107). Current management, in compliance with Oklahoma Statutes (Title 29, part 7-701,704, and 7-705) does not allow any vegetation manipulation except by "Department employees pursuant to their duties" (7-705 part 5). Sections in Title 29 indicate that natural processes (e.g., lightning-caused fire) should be allowed unhindered while sections in Title 2 require wildfire suppression. Current state statutes are in conflict.

Until 1991, when the management plan was implemented, no habitat management activities had occurred. Midstory work was completed at 73% of the active and inactive clusters in winter 1991. After surveys indicated no abandonment of treated clusters by late summer 1991, the remaining clusters were treated (Conner and Rudolph 1991a). No woodpecker use was noted at 2 of the untreated clusters during spring 1991.

Since the mid-1920's, wildfires of all origins have been actively suppressed on MCWA, with varying degrees of success (Little and Olmstead 1931). In 1932-1934, construction of some primitive interior roads was begun by the Civilian Conservation Corps for fire control (Okla. State Game and Fish

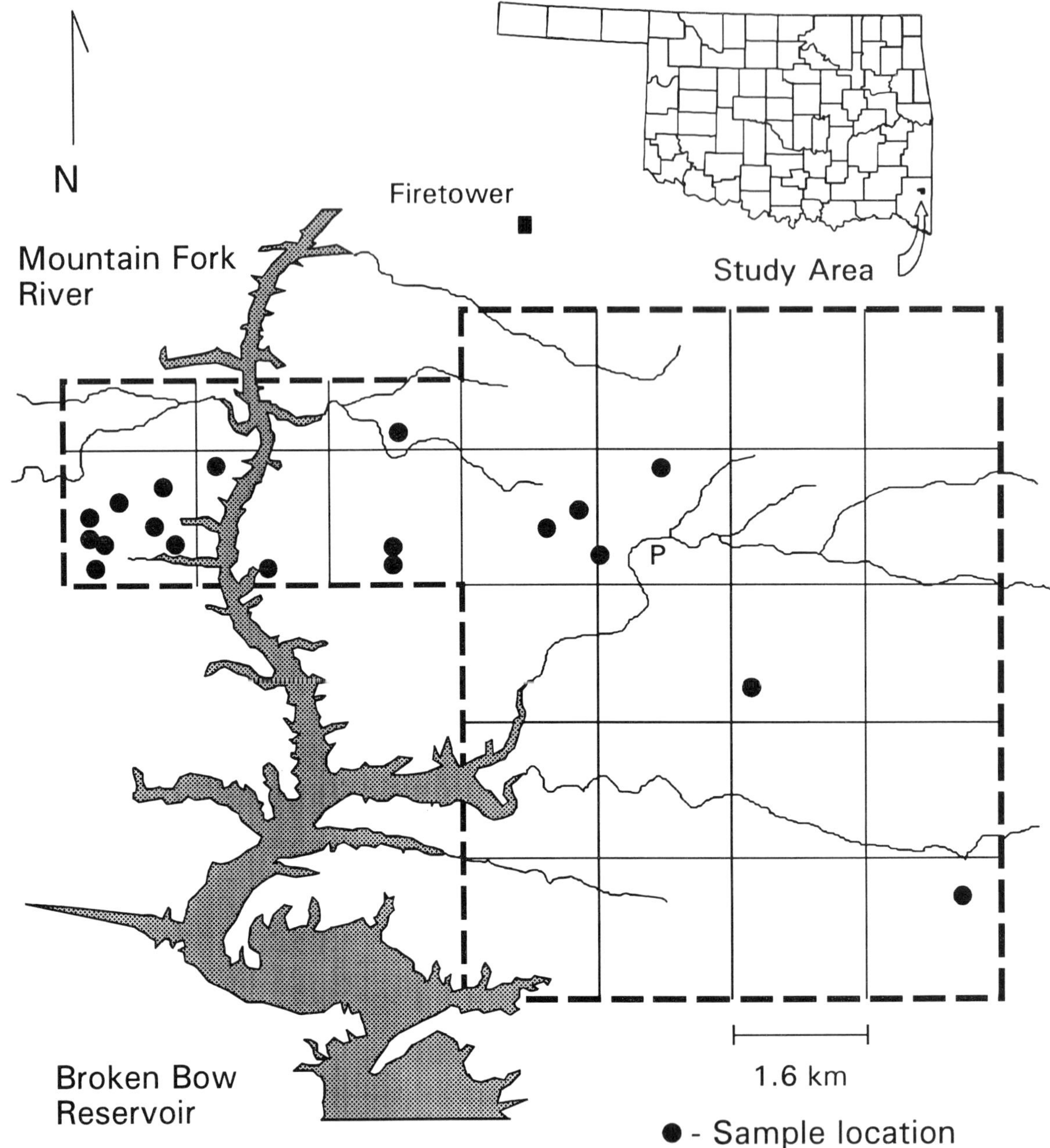

Figure 1. Location of study area and sites that cross-sections were obtained from for fire history reconstruction on McCurtain County Wilderness Area, 1991-1992 and historic photo point (P).

Comm. 1934:27) and completed about 1940. Fire reintroduction began with a prescribed burn of 486 ha on the area in March 1992.

METHODS

Historical Accounts

Historical accounts of travels prior to settlement of the Ouachita Mountain region were reviewed for descriptions of fire occurrence and vegetation cover. Historical literature was searched for earliest photographs depicting forest conditions in McCurtain County, Oklahoma. Settlement of southeast Oklahoma was later than in western Arkansas and eastern Texas, because this area was allotted to the Choctaw Nation in 1820 (Gibson 1965:83). Tribal governments were abolished in 1906 and Oklahoma gained statehood in 1907 (Gibson 1965:324). A time line of major events in human history of southeast Oklahoma and possibly affecting fire return intervals on the Wilderness Area was developed from published accounts in historical and scientific literature.

Forestry Services Data

Fire records for southeast Oklahoma were obtained from Oklahoma Department of Agriculture, Forestry Services files. Fire data from the MCWA and adjacent 1.6 km were tabulated and keypunched. Fire records have been maintained since 1939. These data were supplemented with limited area records.

Tree-ring Data

Site locations for complete stem cross-sections were limited to hardwood stems removed during midstory work and from wind-thrown trees along the path of a June 1991 wind storm. Stems removed were ≤15 m from active, inactive, or replacement cavity trees (Okla. Dep. Wildl. Cons. 1991a,b). This distance was a compromise from early drafts of the plan based upon public input. Replacement cavity trees were ≤ 100 m from active or inactive clusters.

At each of 16 clusters, 4-25 cross-sections were collected. From 1-4 samples were collected at 3 wind-thrown tree sites <400 m from active clusters (**Figure 1**). Samples were collected from February-early March, August-October 1991, and in October 1992. Species sampled included post oak, blackjack oak, white oak, black oak (*Q. velutina* Lam.), southern red oak (*Q. falcata* Michx.), mockernut hickory (*C. tomentosa* Nutt.), black hickory (*C. texana* Buckl.), black gum (*Nyssa sylvatica* Marsh.), black cherry (*Prunus serotina* Ehrh.), eastern red cedar (*Juniperus virginianus* L.), and shortleaf pine.

Sections 2-5 cm in thickness were taken from trees at 5-20 cm above ground line, using a chain saw. Relative crown position occupied by the tree was assigned after Smith (1962). Sections were labeled by location, species, sample number, and diameters were measured and recorded. Site data recorded included location, elevation, slope position, and exposure. Most samples (63.2%) were taken from southwest or south slopes. The remaining samples were taken from northwest, north, and east facing slopes.

Air dried sections were sanded with a belt sander and a orbital finishing sander until a smooth viewing surface was obtained. Cross-sections were then sanded by hand using 220 or 400 grit sandpaper, and polished with steel wool before final viewing. Annual growth rings were counted under magnification to determine tree age. Sections were inspected for presence of charcoal and scar tissue in growth rings to date fire occurrence.

Cross-dating was used to verify ring counts and to determine the presence of false or missing rings (Stokes and Smiley 1968, Baillie 1982). The post oak tree-ring chronology developed from MCWA by Stahle et al. (1985) was used for cross-dating. Phloroglucinol dye (Forestry Suppliers Inc., Jackson, Miss.) was used on samples from diffuse porous species, to assist in counting annual rings. Phloroglucinol was dissolved in 95% ethyl alcohol (1 g/100 cc) and applied to samples for 1 min., then 50% HCL was applied for 1 min. Samples were then washed in water.

Dormant season burns can be problematic to date because of the time sequence in which annual rings are formed. Earlywood, characterized by larger and lighter appearing cells, is formed from March-July. Latewood, smaller and more densely clustered cells, is formed from July until prior to leaf abscission in mid-October (Hawley 1941). Because August-September is drought prone in the Ouachitas, latewood formation may be limited or non-existent in some years. Therefore an apparent dormant season fire scar could have occurred between September of one year and March of the next year. In either instance, scar tissue forms in the later year. Dormant season fires were dated to the year of scar tissue formation unless Forestry Services data indicated that fire events occurred in the previous year (Guyette and Cutter 1991). We define dormant season burns as those occurring from mid-October to March, and growing season burns as those occurring from April to mid-October.

DATA ANALYSIS

Forestry Services and fire scar data were divided into time periods corresponding to settlement and area history (Little and Olmstead 1931, Gibson 1965, Wyckoff and Fisher 1985). Differences in fire frequency and number of sites burned among time segments were examined with analysis of variance. Frequency of trees scarred was not analyzed because frequency may be biased by the size-class of trees sampled (Guyette and Cutter

1991). Protected multiple comparisons were made using LSD ($P < 0.05$). All statistical procedures were conducted with the Statistical Analysis System (SAS Inst., Inc. 1985).

Fire occurrence and distribution were determined by back-dating sections from the time of cutting and through cross-referencing fire scar data from different sites. Mean fire intervals (MFI) were determined by calculating the mean time intervals between fire events (Barrett and Arno 1988). Mean fire intervals were calculated at 3 scales:

1) for the entire MCWA at a dimensionless point;

2) for each side of Broken Bow Reservoir;

3) across all fire scar sample sites (Johnson and Schnell 1985, Guyette and Cutter 1991).

RESULTS

Historical Accounts

Caddo Indians were the earliest inhabitants of the Ouachita Mountains of Oklahoma and Arkansas (Gibson 1965). Earliest published accounts give indication of widespread use of fire by the Caddoans. In 1719, La Harpe traveled through the Ouachita Mountains within 45-55 km of present-day MCWA (Lewis 1924). His account describes forests interspersed with numerous prairies (indicative of frequent fire). La Harpe described crossing a prairie valley, in present Pushmataha County near the Kiamichi River, bounded by a very "thick wood" that extended into the mountains. This wooded area was extremely difficult to travel through on horseback, giving the indication of dense sapling (or larger) stands caused by periodic fire (Lewis 1924:338).

The botanist, Thomas Nuttall traversed this same general vicinity in 1819. South of the area described by La Harpe, Nuttall traveled through extensive "...hilly woods covered with dwarfish post and black oaks, which having been burnt were extremely difficult to penetrate...(Nuttall 1980:170)."

Nuttall (1980) provided considerable detail on forest conditions observed while traveling through the Ouachita Mountains 50-60 km west of MCWA. Sparsely timbered mountains interspersed with a mosaic of numerous prairies and areas of open tall timber were described throughout his account, in addition to references to fire. Pines were apparently relegated to hills and ridges as no mention was made of them in prairie valleys. He described the mountains as "ridges...thinly scattered with pines and oaks" (p. 167), "bare serrated hills...scattered with pine and post-oak" (p. 168), "open woods..hilly prairie land...lofty woods" (p. 169), and "areas overgrown with bushes and trees-half burnt" (p. 180).

Historic accounts of presettlement vegetation and fire evidence in the Arkansas Ouachita Mountains (east of MCWA) was reviewed by Foti and Glenn (1991). They cited early accounts of fire from 1720, 1804, 1819, and 1844. Several of these accounts attribute fires to Indian inhabitants and mention open forest conditions with widely spaced or sparse pine, oak-pine or oak. The open forest conditions were attributed to Indian-set fires (Foti and Glenn 1991). Du Pratz (1774:134, in Foti and Glenn 1991) described September fires that burned for several days, indicating extensive burns. Gerstacker (1854) described frequently burned areas in the Arkansas Ouachitas circa 1840. He also described an area that had not been burned for several years as "thickly overgrown with underwood."

Photographs taken adjacent to or near MCWA, circa 1916-1928, provide evidence that some virgin upland areas were still forested with widely spaced pine trees with little or no hardwood undergrowth or midstory (Honess 1923 [Part I:175,279; Part II:21,71], Gould 1928:5, Smith 1986:7). Bark char and some fire scaring was evident on shortleaf pine in most of these photographs. Photographs taken at points established by Dr. Elbert Little, retired dendrologist U. S. Forest Service, in 1930, 1981, and 1990, on MCWA illustrate dramatic increases in midstory hardwoods **(e.g., Figures 2-4).**

Annual burning was common in southeast Oklahoma, before 1926. From 1/3 to 3/4 of the entire upland pine-oak area in the Oklahoma Ouachita Mountains was burned each year. The Oklahoma Forest Service (presently Forestry Services) established the Southeastern Oklahoma Protective Unit in 1926, for fire prevention and suppression (Little and Olmstead 1931).

Based on historical literature and analysis of recorded fire events, fires were categorized into 6 periods. They were: Caddo (pre 1751), Native American Transition (1751-1833), Choctaw (1834-1889), Settlement (1890-1925), Early Suppression (1926-1956), and Modern Suppression (1957-present) (Little and Olmstead 1931, Gibson 1965, Wyckoff and Fisher 1985).

Figure 2. Photograph taken in summer 1930 near an old cabin. The fire evidence was from a summer 1929 fire. The area at this photo point has not burned from 1939 present. Courtesy of Elbert L. Little, Jr.

The largest single fire event in this period burned approximately 680 ha in March 1942. During March and April of 1942, 5 separate fire events burned approximately 1280 ha in the southern and eastern part of the Wilderness Area. These 5 fire events account for 35% of the total area burned since 1939.

Fire events have occurred during all months since 1939, but occurred more frequently in spring and during the late summer and early fall months (**Figure 6a**). Fires from all causes were more frequent in early spring. However, lightning-caused fires occurred more frequently during late summer than in spring (**Figure 6b**). From 1939-1992, lightning has ignited a detectible fire an average of every 3.1 years (range=2 days to 10.1 years apart) somewhere on the MCWA. Only 3 times during this period has a lightning strike ignited fire in the same section (259 ha).

The MFI at a dimensionless point for the period 1939-1956 was 29.9 years. Through fire suppression this interval was significantly extended (P=0.0001) to 546.7 years at a dimensionless point for the period 1957-1991. The MFI, including both time periods, was 86.0 years at a dimensionless point.

FORESTRY SERVICES DATA

Since 1939, 78 fire events have been recorded on the MCWA, excluding the 1992 prescribed burn (**Table 1**). Frequency (P=0.0001) and areal extent (P=0.0082) of fires have declined significantly since 1956 (**Figure 5**). Incendiary and lightning fires accounted for the most frequent causes and burned the largest area. A total of 41 incendiary fires burned 3,634 ha and 19 lightning fires burned 137 ha.

Tree-ring Data

We examined 184 cross-sections for presence of fire scars. Of these, 94 (51.1%) recorded 224 fire scars (**Figure 7**). Twenty-seven of the fire scars were from 2 fires in 1953, confirmed by Forestry Services fire records. Of the 224 fire scars, 134 were from

Figure 3. Photograph taken in fall 1981 at the same location as Figure 2, illustrating forest midstory change. Courtesy of Elbert L. Little, Jr.

Table 1. Number of fires and area burned by cause on the McCurtain County Wilderness Area, 1939-92.

Cause	Number of Fires	Percent	Total Hectares Burned	Average Hectares Burned/fire
Campfire	2	2.5	3.2	1.6
Debris	2	2.5	2.1	1.0
Incendiary	41	51.9	3,634.2	88.7
Lightning[a]	19	24.1	136.6	7.2
Not found[b]	2	2.5	?	?
Prescribed	1	1.3	485.6	485.6
Smoker	3	3.8	16.5	5.5
Unknown[b]	9	11.4	20.2	10.2
Total	79	100.0	4,298.5	54.4

[a] **One lightning-caused fire was a reignition of a previously suppressed lightning-caused fire.**

[b] **Hectares burned were not reported for some fires.**

samples collected west of Broken Bow Lake and 90 from east of the Lake. One sample recorded 11 fire events and 8 others recorded 5 or more fire events. The earliest recorded fire occurred in 1710.

Fire scars were accurate ±1 year for most samples because of cross-dating and assessment of dormant season burns to the year of callus tissue formation. Some 10 scars were datable to within 4 years because of limited

Table 2. Mean fire intervals (years) between fires on the McCurtain County Wilderness Area, 1834-1992.

Human Influence		Mean Fire Interval (MFI)			
		Fire Scar Data		Forestry Services Data[a]	
Period	Time	East[b]	West[b]	East[c]	West[c]
		MFI (range)	MFI (range)	MFI (range)	MFI (range)
Choctaw	1834-1889	5.6 (2-12)	3.5 (1-7)	--	--
Settlement	1890-1925	3.5 (1-6)	4.4 (2-9)	--	--
Early Supp.[d]	1926-1956	2.1 (1-4)	3.9 (1-8)	0.4 (0.01-2)	2.1 (0.01-5)
Modern Supp.	1957-1992	--	--	2.1 (0.9-8)	11.7 (3-14)

[a]Forestry Services data include all known fires in calculation of mean fire interval.

[b]The minimum area for the east side includes 800 ha and the west side includes 300 ha between sites.

[c]The area for the east side includes 5,215 ha and the west side includes 486 ha.

[d]Forestry Services data for this period include only fires from 1939-1956.

latewood formation for a period of years. These scars were assigned to the nearest fire event (Barrett and Arno 1988, Guyette and Cutter 1991). This gives a more conservative estimate of fire intervals over time (Barrett and Arno 1988).

Fires were frequent on MCWA from the Choctaw Period until Early Suppression (**Table 2**). A fire occurred within the 300 ha area sampled west of the river every 3.5-4.4 years from the Choctaw to the Early Suppression Period (**Table 2**). A fire occurred within the 800 ha area sampled east of the river (excluding the 2 southern sites) every 2.1-5.6 years in the same periods (**Table 2**). The 5.6 MFI may be an overestimate because of limited sample size. Fire intervals prior to 1800 could not be accurately determined because of small sample size of trees in older age-classes.

Generally, a number of sites exhibited fire scars during a given year (**Figure 7**). Given the presence of physical barriers, knowledge of fire behavior in bluestem understories, and historical accounts of aboriginal firing patterns, it is reasonable to conclude that many sites were burned in widespread single fire events. If we assume that all sites that showed fire scars in a given year were burned in a single fire event then fires ranged in size up to a minimum of 400 ha (excluding the 2 southern-most sites on the east side). Inferences cannot be made on fire extent in central and southern parts of the area east of the river, because they were sparsely sampled. At the least in a given year in which fire occurred, multiple sites were burned.

Comparison of fire occurrence between east and west sides of Mountain Fork River (Broken Bow Reservoir) shows an alternating burning pattern (**Figure 7**). Fires alternated from the east to west side of the river and rarely burned both in a given year. Fires occurred in segments of 1-3 year intervals followed by 1-3 longer (4-9 year) periods of fire absence on a given side (**Figure 7**). This periodic pattern was repeated 5X from 1830-1931 on both sides of the river. The periodic alternating pattern persisted after settlement because Choctaws remained in the vicinity and probably influenced early settlers burning patterns.

DISCUSSION

Fire scar chronologies developed from red-cockaded woodpecker clusters, Forestry Services data, and historical fire accounts clearly indicate that periodically, frequent and extensive fires occurred on the Wilderness Area. Frequent fire has occurred on MCWA at least since 1800. Fires continue to occur, but have greatly diminished in extent and in frequency since 1956 (**Figure 5, Table 2**).

Historical literature and data (1719-1859) characterize Ouachita Mountain forest conditions as widely spaced trees with low basal area and stem density, open grassy understories, sparse midstories, and indicate that fire shaped forest conditions (e.g., pine-bluestem community) (Nuttall 1980, Foti and Glenn 1991). Fires were of anthropogenic and lightning origin. Native American-set fires occurred in late summer and fall; were frequent, extensive, and exhibited pattern. This pattern suggests that Native Americans set fires with specific objectives in mind. Similar fire patterns have been observed elsewhere (Jacobs et al. 1985).

Lightning-ignited fire occurrence on the

Figure 4. Photograph taken in July, 1990 at the same location as Figures 2 and 3, illustrating forest midstory change. Courtesy of the senior author.

Wilderness Area corresponds with the greatest monthly frequency of cloud-to-ground strikes for this vicinity (Reap and MacGorman 1989, Orville 1991). The vicinity of MCWA is characterized by maximum frequency of thunderstorm activity during afternoons of summer months (Reap and MacGorman 1989). Our lightning-set fire data closely corresponds with lightning-set fire data for the Ouachita Mountains in Arkansas (Foti and Glenn 1991). Lightning-set fires occur most frequently in the Ouachita Mountains during July-September and less frequently in March and April. Native Americans set fires from September-November, and modified rather than changed the climatic fire regime (Foti and Glenn 1991). The climatic fire regime was responsible for the overall vegetation complex in the Ouachita Mountains (Foti and Glenn 1991). The pattern of burning, rather than specific average burning intervals shaped the mosaic. Regeneration of future stand dominants was probably initiated during longer fire-free periods.

Presettlement aboriginal fires in the Ozarks have averaged 3.2 - 4.3 year intervals (Guyette and McGinnes 1982, Guyette and Cutter 1991). Our mean fire interval data is comparable to presettlement fire intervals in the Ozarks for a similar size area (west side). Fire frequency in the Ozarks decreased with settlement (Arend 1950, Beilmann and Brenner 1951, Guyette and McGinnes 1982, Guyette and Cutter 1991). Johnson and Schnell (1985) speculated that this was true for the eastern Ouachita Highlands.

Inspection of both Forestry Services and fire scar data sets **(Table 2)** indicate that fire frequency increased in the western Ouachita Highlands with settlement, then declined as fire suppression activities became more effective. Others have reported instances of increased fire

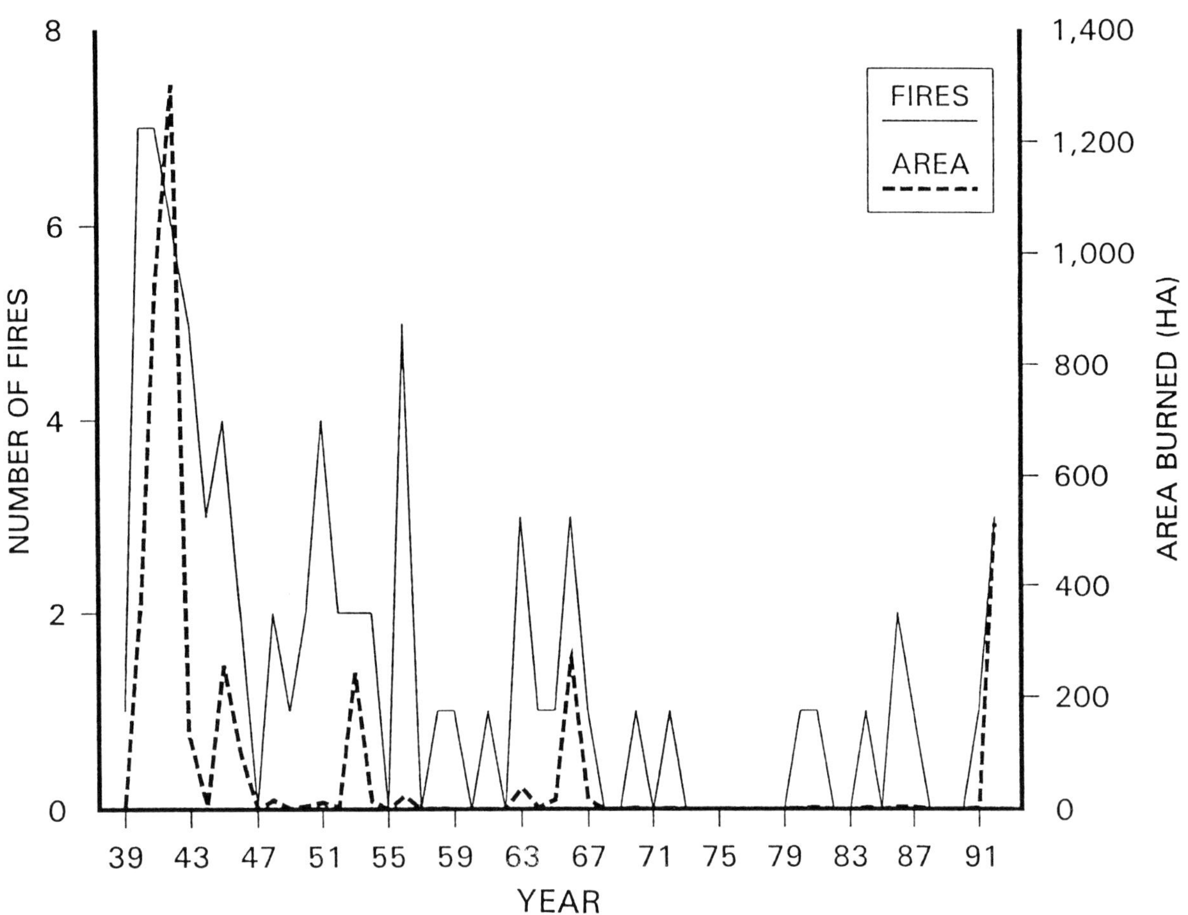

Figure 5. Fire frequency and area burned by year on the McCurtain County Wilderness Area, 1939-1992.

frequency upon settlement (e.g., Heinselman 1973).

Johnson and Schnell (1985) reported fire return intervals to a single point for the eastern Ouachitas as opposed to fire return intervals within a larger area. We calculated comparable intervals at each site and found slightly shorter intervals. We found site specific mean fire intervals on the west side of 24.3 and 16.4 years and on the east side of 18.7 and 26.2 years for the Choctaw and Settlement Periods respectively. Our overall MFI to a single point on MCWA, over the same time period (1780-1938), was 35.5 years compared to their 41.4 years (Johnson and Schnell 1985). However, they used shortleaf pines for fire-dating, and shortleaf pine is less prone to scarring than post oaks (Guyette and Cutter 1991). Johnson and Schnell (1985) did report one shortleaf pine with a MFI of 7.3 years from 1788-1817.

A fire chronology developed from fire scar evidence is conservative because every fire will not cause fire scars in a given stem (Guyette and Cutter 1991). Fire chronologies can be biased by site characteristics. A large-scale fire event will include fire behavior of varying intensity among sites, because of variations in topography (and thus insolation), fuel loads, fuel architecture, and fuel moisture (Andrews 1986) and may not scar trees at a given site. This would lengthen apparent fire intervals considerably, particularly on sheltered sites.

Hot Springs National Park fire suppression records indicate that through fire suppression, fire return intervals had been extended to 1229 years at a dimensionless point for the period 1938-1980, in the eastern Ouachitas (Johnson and Schnell 1985). This is considerably longer than the 546.7 year interval from Forestry Services fire suppression records for the MCWA. However, cultural differences in woods-burning philosophy existed between McCurtain County, Oklahoma and adjacent counties in Arkansas. McCurtain Countians

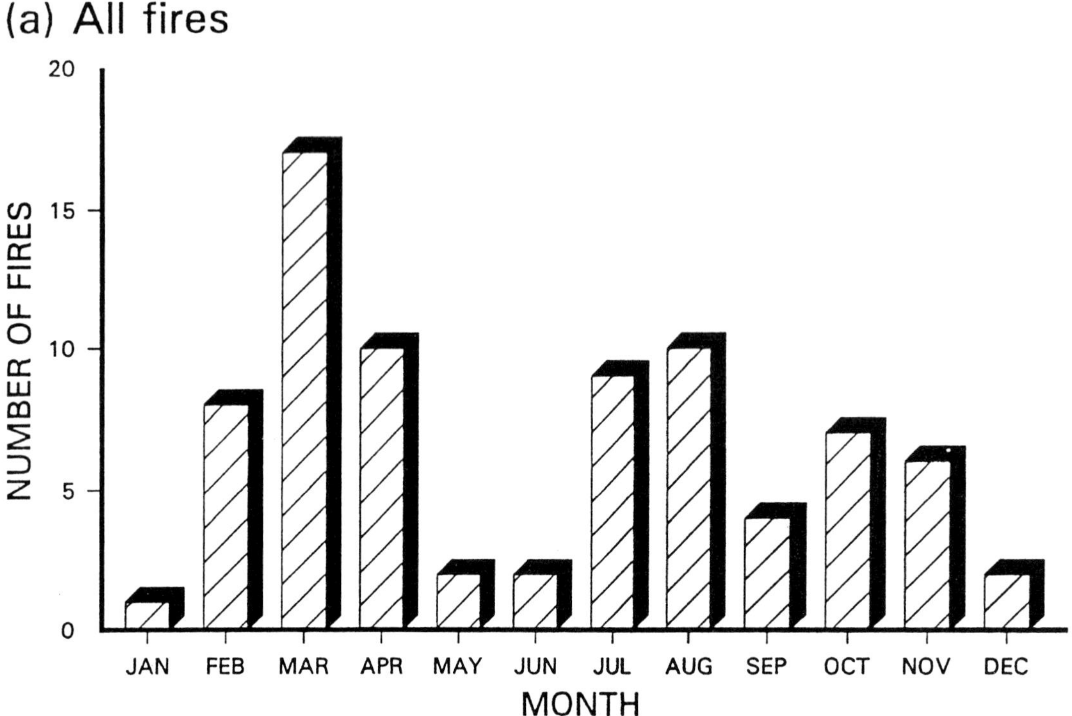

Figure 6 (a). Fire frequency by month from all causes on the McCurtain County Wilderness Area, 1939-1992.

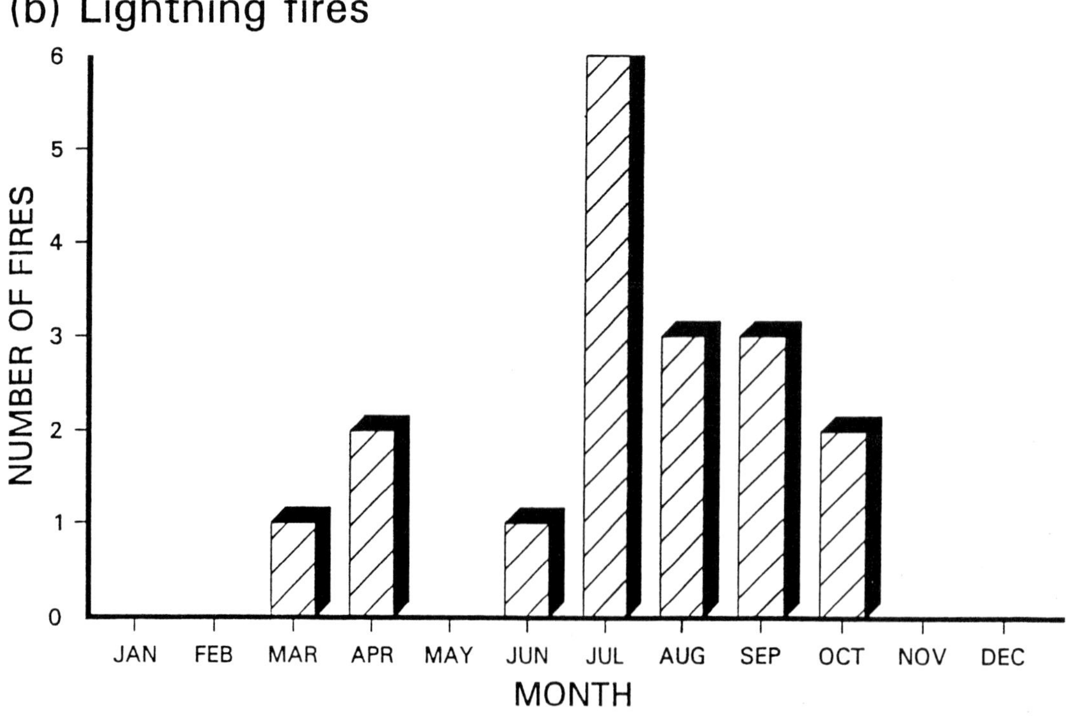

Figure 6 (b). Lightning-caused fire frequency by month from 1939-1992 on the McCurtain County Wilderness Area.

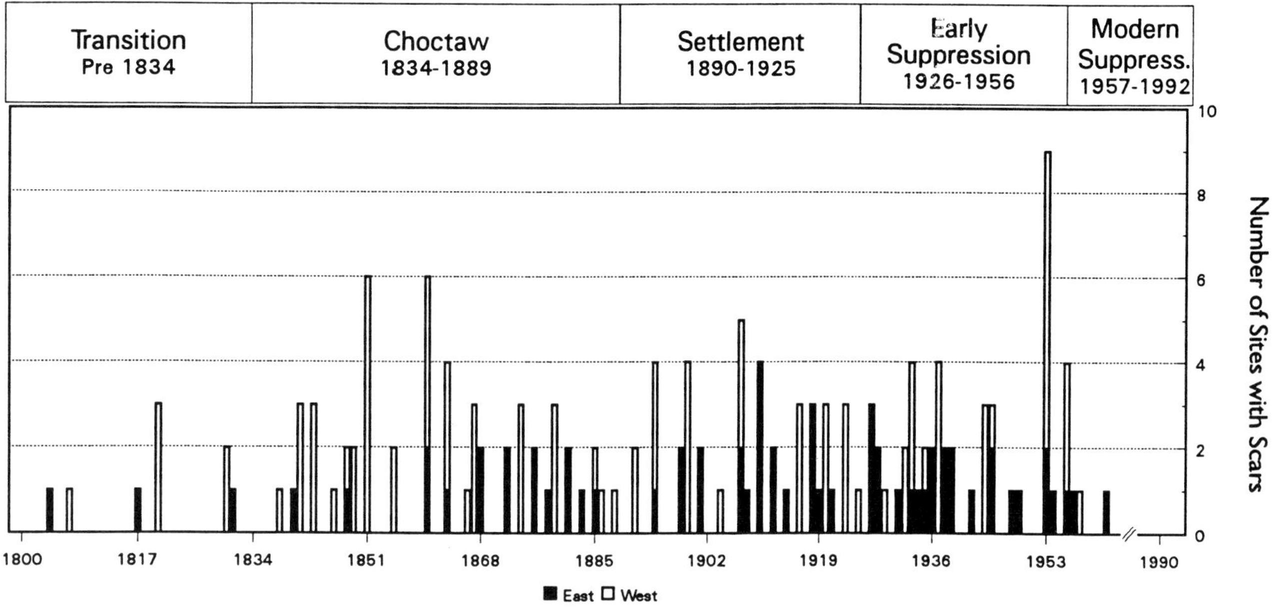

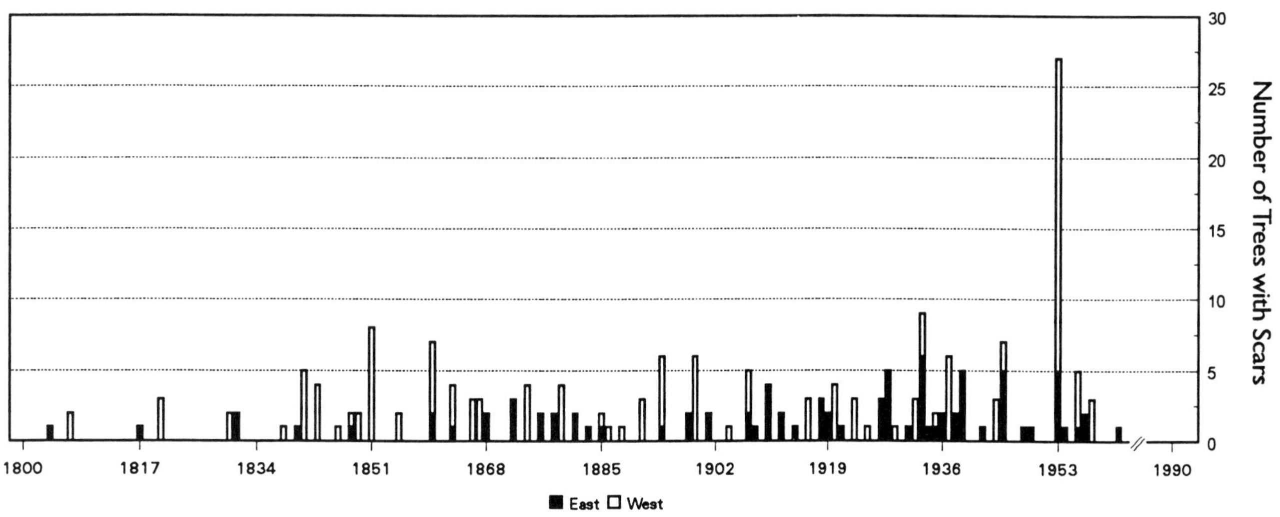

Figure 7. Human influence periods and fire chronology for number of trees and sites exhibiting fire scars on both sides of Broken Bow Lake on the McCurtain County Wilderness Area, 1800-1992.

were more prone to woods burning (circa 1940) than inhabitants of more eastern areas of the Ouachitas (J. Burwell, pers. commun., retired forester, Oklahoma Div. of Forestry). Fire occurrence data presented by Nelson and Zillgitt (1969) support this observation from 1956-1965.

The purpose for reintroduction of fire is to direct vegetation change on the MCWA to conditions prior to fire suppression (Okla. Dep. Wildl. Cons. 1991b). Both anthropogenic and lightning fire combined to shape the unique habitat features required by the red-cockaded woodpecker (Jackson et al. 1986). Red-cockaded woodpecker populations in the Ouachitas may be indicative of the extent and quality of the pine-bluestem community. With the current loss rate of 1.4 clusters/year, the woodpecker may be gone from the area in just over 7 years. Return to a lightning fire regime alone will not return suitable habitat conditions in time to create the red-cockaded woodpecker on MCWA. Lightning fires under current fuel conditions would be less extensive and less likely than in presettlement times.

Under conditions associated with thunderstorms (high relative humidity, high fuel moisture, variable rain) in the Ouachitas, the probability of ignition is less with leaf litter (current conditions) than grass litter (presettlement conditions on some sites). Continuous leaf litter has different fuel architecture, moisture of extinction, and rate-of-spread than grass litter (e.g., fuel model 2 vs. fuel model 8, Andrews 1986). Unlike leaf litter, fuelbeds of bluestem beneath widely spaced trees (basal area of 9 m^2/ha) will carry fires at a moderate rate-of-spread with a light misting rain (Masters 1991).

MANAGEMENT IMPLICATIONS

Red-cockaded woodpecker management efforts have demonstrated a population increase after an aggressive prescribed burning program, and midstory hardwood control where availability of suitable cavity trees was not limiting (Hooper et al. 1991a). Suitable cavity trees will not be limiting on MCWA (Kelly 1991) if limited midstory work is continued around replacement cavity trees.

General Land Office survey data from 1834 in active Arkansas red-cockaded woodpecker cluster locations give presettlement pine basal area of 8.4 m^2/ha (59.1 stems/ha) and hardwood basal area of 6.0 m^2/ha (90.9 stems/ha) (Foti and Glenn 1991). Jackson et al. (1986) recommend a pine basal area of 9.2-13.8 m^2/ha. Similar conditions exist in active and inactive clusters on MCWA, but with a greater number of smaller hardwoods (Kelly 1991). Our goal is to recreate these structural attributes with fire and limited midstory work, not to eliminate hardwoods. Large codominant and dominant hardwoods, particularly oaks, are notably fire resistant to even annual burning (Waldrop et al. 1992). With control of midstory through fire reintroduction, restoration of the pine-bluestem community can be accomplished given enough time (Waldrop et al. 1992).

Compartments with active clusters will be emphasized in prescribed fire planning. Initially, winter burns with high fuel moisture conditions will be used to acclimate old-growth pines to fire (Ferguson et al. 1961). As old-growth pines are acclimated to fire, predominantly growing season burns will be used and frequency increased to less than average intervals to hasten return of an open understory and to control midstory encroachment (Grano 1970, Waldrop et al. 1992). As smaller stems are controlled, a shift should be made to the sequence of short burning intervals followed by 2 longer intervals. Large scale fires will be necessary to mimic the natural fire regime and to hasten return to the "trajectory of presettlement change" (Baker 1992). With small scale burns, even frequent fire has a low probability of landscape level ecosystem restoration (Baker 1992). Site specific information for the remainder of the area is critical to plan a fire management regime for a given site.

ACKNOWLEDGMENTS

This project was supported by the Wildlife Restoration Act Project W-140-M through the Oklahoma Department of Wildlife Conservation, the Cooperative Extension Service, and the Department of Forestry, Oklahoma State University. This is Proceedings Article P-3841 of the Oklahoma Agricultural Experiment Station. We wish to acknowledge the assistance of T. Smith, Forestry Services, for providing fire records of the area. We thank F. James, J. Hemphill, and M. Talley for assistance in collecting samples and C. Wilson for assistance in sample processing and historical reviews.

The Relationship Between Body Size, Trophic Position and Foraging Territory Among Woodpeckers

Robert W. McFarlane
McFarlane & Associates, Houston, Texas 77036

ABSTRACT: I examined the relationship between body weight, trophic position (measured as percent animal matter of stomach contents), and home range or territory size for 15 species of North American woodpeckers. The correlation between body weight and territory was significant but fragile. When the largest, and most poorly categorized species, the ivory-billed woodpecker, was ignored, the correlation lost significance. When the second-largest, but well defined species, the pileated woodpecker, was deleted, the correlation disappeared for the remaining 13 species. The correlation between trophic position and territory was insignificant until I ignored the ivory-billed woodpecker. When body weight and trophic position were both considered, the correlation was significantly improved (r=.821, P=.009). The red-cockaded, black-backed and 3-toed woodpeckers, all highly insectivorous and conifer specialists, defend large territories for their body size. Broad generalizations about woodpecker body size and territories may not be reliable.
KEYWORDS: woodpecker, home range, territory, diet, trophic position

For 30 years ecologists have been enamored with the concept that foraging territory and home range size, and the inverse relationship, population density, are correlated to the body size of birds and mammals (McNab 1963). Factors modifying such relationships include trophic position, latitude, precipitation and habitat quality (Harestad and Bunnell 1979). Mammalogists have concluded that herbivores require separate analysis but carnivores and omnivores do not differ significantly (Peters and Raelson 1984). This allometric approach to ecological problems is considered a good example of predictive ecology (Peters 1991).

Schoener (1968) was the first to broadly apply this concept to birds. **Figure 1** was constructed with Schoener's data to illustrate the general tendency for territory size to increase with body weight. The correlation coefficient was highly significant (r=0.776; P<0.0005) and body weight explained 60% of the variation in territory size. Separate analyses of the 6 diet categories reveal that only 1 category exhibited a statistically significant correlation between body weight and territory size. Small passerines whose diet was more than 90% insects (IP) showed a strong correlation (r=0.715, P<0.0005) and body weight accounted for 51% of the variation in territory size. It appeared that these insectivores were driving the significant correlation for all birds in this sample. Juanes (1986) examined bird density with a larger data set and reached opposite conclusions; body weight accounted for only 9% of the variation in density for 442 insectivores but 58% for 19 raptors.

The red-cockaded woodpecker (*Picoides borealis*) [RCW] appears to occupy a very large foraging territory for its small size. The objective of this study was to examine the relationship between body weight, diet, and foraging territory among North American woodpeckers to determine if the RCW is unique in this respect.

METHODS

This study relied on disparate sources of information. Body weights were based on Dunning (1993). Body weights vary by sex and geographic location, and broad geographic coverage was seldom available. Diet information was based on Beal (1911), an old but useful data set that is unlikely to be replicated. Diet varies by season and habitat, and stomach content analysis fails to account for liquid nourishment, such as tree sap and nectar. The only variable utilized here was the percent of total stomach contents comprised of animal matter. This percentage indicated a relative position on the omnivore-to-insectivore continuum.

Territory size was summarized from numerous sources. Home range (an area habitually traversed by an animal) is usually easier to determine than territory (a defended area used exclusively by an individual, mated pair, or cohesive group of animals). Observation of multiple territorial disputes is required to define a territory. Home range/territory size varies seasonally and geographically in some species. Many home ranges and territories contain patches of unusable habitat. Resolution of the differences between home range and foraging territory, seasonal and geographical variation, and inclusion of low-value habitat appears intractable, and perhaps unnecessary, for the purposes of this study, which must select a single value to represent the species.

RCW (*Picoides borealis*)—Woodpecker territoriality was best known for this non-migratory species. Estimates of home range and territory, both seasonal and annual, were available from Florida (Crosby 1971, Baker 1971, Nesbitt et al. 1978 and 1983, Patterson and Robertson 1981, Porter and Labisky 1986, DeLotelle et al. 1983b, 1987 and 1992, Jerauld et al. 1983), South Carolina (Skorupa and McFarlane 1976, Sherrill and Case 1980, Hooper et al. 1982, Wood et al. 1985a), and Oklahoma (Wood 1983a). The average was 85.0 hectares (SD 48.84; range 14.4 to 213.2 ha). The average weight of 68 birds from Florida, Kentucky, Mississippi, and South Carolina was 46.8 grams (Dunning 1993, Mengel and Jackson 1977, McFarlane data).

Downy woodpecker (*Picoides pubescens*)—Downy woodpeckers appear to occupy home ranges of 10-15 ha but much smaller territories (Schroeder 1983a, Kilham 1983, Bull 1978, Fitch 1958, Lawrence 1967). 7 estimates averaged 3.6 ha.

Hairy woodpecker (*Picoides villosus*)—The average of 11 estimates was 6.0 ha (Sousa 1987, Fitch 1958, Lawrence 1967).

3-toed woodpecker (*Picoides tridactylus*)—3 non-overlapping home ranges of 53, 142, and 304 ha were determined by radio-tagging in Oregon (Goggans et al. 1987).

Black-backed woodpecker (*Picoides arcticus*)—3 non-overlapping home ranges of 72, 123, and 328 ha were determined by radio-tagging in Oregon (Goggans et al. 1987).

White-headed woodpecker (*Picoides albolarvatus*)—A single estimate of 8 ha was located (Bull 1978).

Yellow-bellied sapsucker (*Sphyrapicus varius*)—5 estimates averaged 5.4 ha (Lawrence 1967).

Williamson's sapsucker (*Sphyrapicus thyroides*)—This sapsucker defended a larger territory early in the breeding season but reduced its territory to the vicinity of the nest tree later in the nesting cycle (Sousa 1983b, Bull 1978); 3 estimates averaged 5.7 ha.

Red-bellied woodpecker (*Melanerpes carolinus*)—Fitch (1958) measured 4 home ranges. Reller (1972) studied 3 woodlots and found 5 pairs in a 12.5 ha woods during spring, 2 pairs in a 15.0 ha woods during summer and fall, and 5 individuals overwintering separately on 24.3 ha. Kilham (1983) estimated a 1.7 ha territory. These 8 estimates averaged 4.3 ha.

Red-headed woodpecker (*Melanerpes erythrocephalus*)—Red-headed woodpeckers sometimes migrate to areas of abundant mast during winter. They store numerous acorns and remain nearby to protect them. Reller (1972) reported 8 pairs breeding in a 12.5 ha woods, 13 pairs summering in a 15.0 ha woods. Kilham (1983) found 12 birds wintering in contiguous but separate territories that ranged from 0.03 to 0.28 ha. These 15 areas averaged 0.64 ha.

Lewis' woodpecker (*Melanerpes lewis*)—Some studies indicate that this species defends only a small territory around the nest tree or stored food in winter (Sousa 1983a). A single estimate of 4 ha has been found (Bull 1978).

Acorn woodpecker (*Melanerpes formicivorous)* —In some, but not all, parts of their geographic range these cooperative breeders maintain small territories centered on a common granary where they store acorns gathered from the surrounding area (Koenig and Mumme 1987, Burgess et al. 1982, Kattan 1988). The average size of 31 territories mapped by Koenig and Mumme (1987) was 5.5 ha. Another study indicated that the average distance between territory centers for 112 territories was 127 m (Roberts 1979), which would yield smaller territories. There were differences of opinion concerning aggregated, uniform or random distribution of the territories (Burgess et al., 1982, Mumme et al. 1983, Burgess 1983, Brewer and McCann 1985).

Common flicker (*Colaptes auratus*)—Territories of 7.9 and 11.7 ha were described by Lawrence (1967).

Pileated woodpecker (*Dryocopus pileatus*)—There appeared to be dramatic geographic differences in territory size for this species (Schroeder 1983b). Eleven territories in Missouri averaged 87.5 ha (range 52.9 to 160.1 ha, Renken and Wiggers 1989). Other estimates in the eastern U.S. included 43 ha (Tanner 1942, based on 6 pair/sq mi), 71 ha (Evans and Conner 1979), and the often cited 70 ha (Kilham 1979, 1983) which may be 114 ha, based on the published map. In the coniferous forests of the western U.S., estimates ranged from 130 to 549 ha (Mannan 1984, Bull 1978 and pers. comm., Bull and Meslow 1977). Continent-wide, the average of 23 estimates was 194 ha.

Ivory-billed woodpecker (*Campephilus principalis*)—Tanner (1942) sketched a map of 1 territory which was approximately 375 ha; elsewhere he attributes a territory of 6 sq mi (1554 ha) to this species. Only 2 body weights were found, 454 and 568 g. The contents of 3 stomachs contained 46% animal matter (Tanner 1942).

The data were summarized in **Table 1**. Univariate and multivariate analyses were conducted with the Multivariate General Linear Hypothesis program of Systat.

RESULTS

The correlation between body size and home range/territory size was significant **(Figure 2; r=.613, p=.015)** although only 33% of the variation in territory size was attributed to body size. When the datum for the ivory-billed woodpecker (IBW), based on 1 range estimate and 2 weights, was deleted the significance of the correlation disappeared

Table 1. Body weight, diet and home range/territory of North America woodpeckers.

	Body Wt. (g)			Animal Fraction of Diet		Home Range/Territory (ha)		
Species	N	Ave	Range	N	%	N	Ave	Range
Red-cockaded	63	47	--	76	81.1	68	85	14.4-213.3
Downy	383	27	21-32	723	76.1	7	3.6	0.5-8.9
Hairy	38	68	59-80	382	77.7	11	6.0	0.6-15
3 -toed	15	68	65-74	23	94.1	3	166	53-304
Black-backed	9	70	61-88	28	88.7	3	174	72-328
White-headed	35	61	53-68	44	38.9	1	8	--
Yellow-bellied Sapsucker	52	50	41-62	343	49.3	5	5.4	4.6-6.4
Williamson's Sapsucker	19	48	44-55	47	86.7	3	5.7	4-9
Red-bellied	31	64	--	271	39.9	8	4.3	1.7-8.3
Red-headed	89	72	56-91	443	33.8	15	0.64	0.03-4.9
Lewis'	5	116	--	59	37.5	1	6.1	--
Acorn	86	81	--	84	22.6	31	5.5	1.5-16
Com. Flicker	159	133	106-164	3	60.9	2	9.8	7.9-11.7
Pileated	6	280	250-309	86	72.9	23	194	43-549
Ivory-billed	2	511	454-568	3	46.0	1	375	--

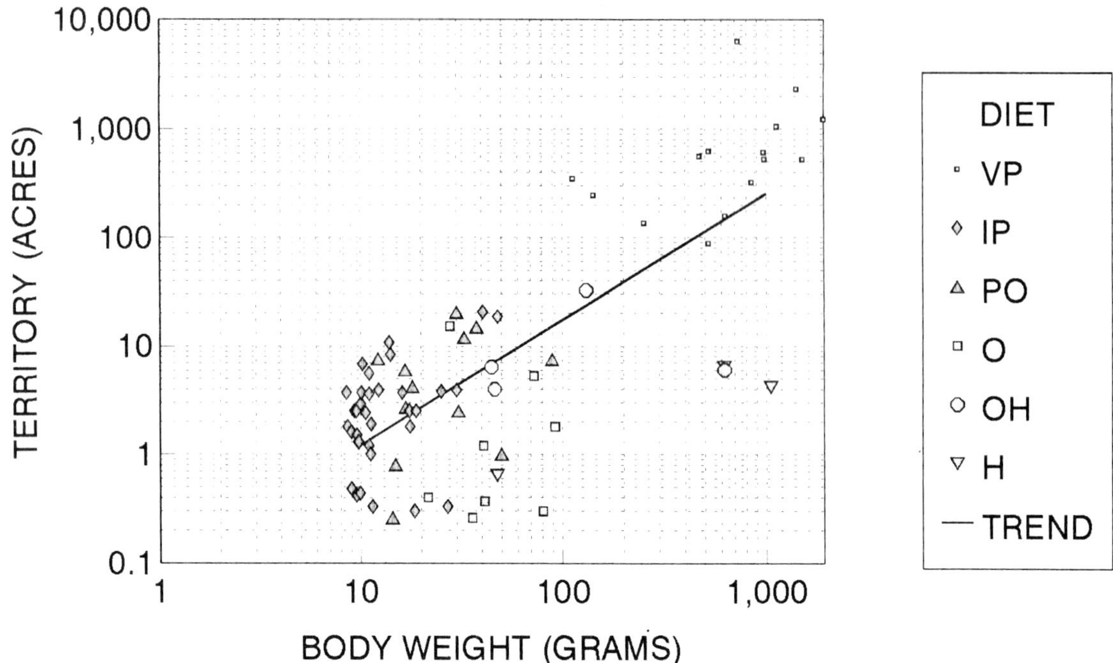

Data source: Schoener 1968
LN(HA) = 1.16 LN(W T) - 2.48 R = 0.776 R - SQ = 0.602 N = 78 spp
Diet: VP>90% vertebrates, IP>90% insects, PO 70-90%, O 30-70%, OH 10-30%, H<10% animal matter

Figure 1. The relationship between body weight and territory size among birds.

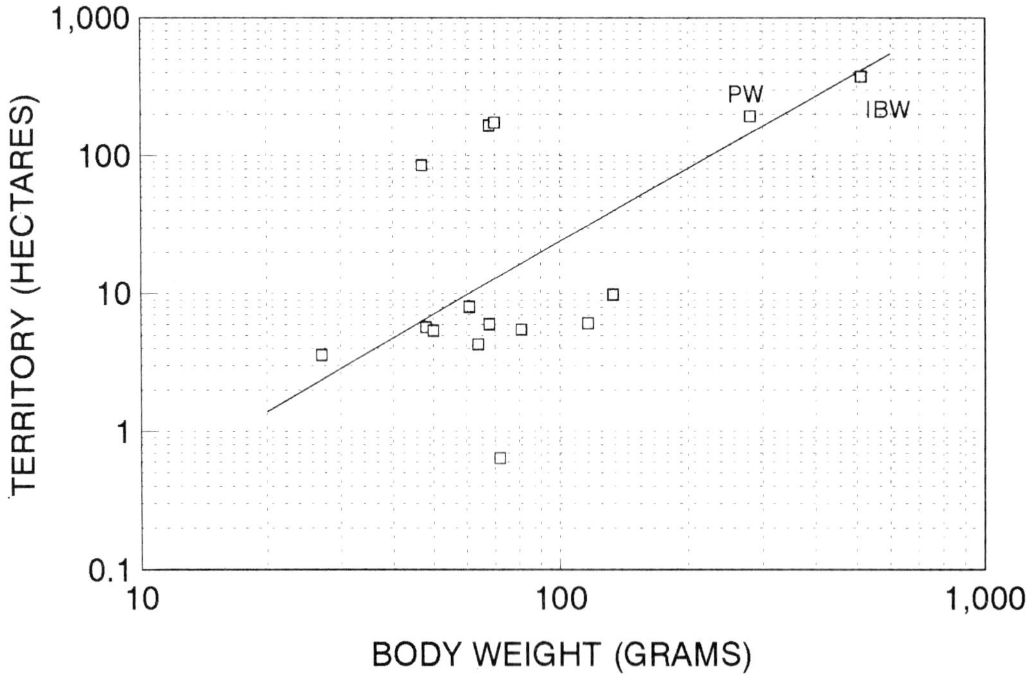

All: LN(HA) = 1.75 LN(WT) - 4.87 R = .613 ADJ R - SQ = .328 P = .015
Minus IBW: = 1.02 LN(WT) - 2.16 R = .346 ADJ R - SQ = .046 P = .226
Minus IBW & PW: = 1.83 + 0.11 LN(WT) R = .028 ADJ R - SQ = .000 P = .927

Figure 2. Body weight versus territory size.

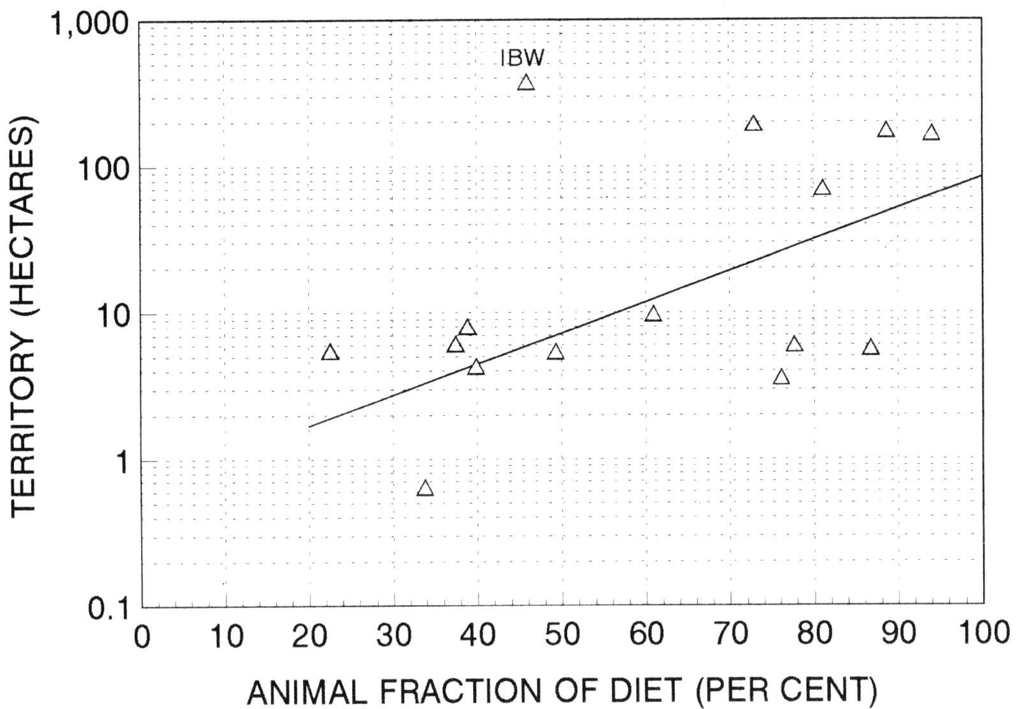

If ivory-billed woodpecker is deleted
LN(HA) = 0.049 (diet) - 0.475 R = .665 ADJ R - SQ = .396 P =.009

Figure 3. Diet versus territory size.

(r=0.346, P=0.226). When the pileated woodpecker (PW) datum based on adequate data was also removed, there was no relationship (r=0.028, P=0.927) for the remaining 13 species. It appeared as if the 2 larger woodpeckers were driving the entire apparent relationship.

The correlation between trophic position, expressed as the percent of animal matter in the diet, and territory size was not significant (**Figure 3; r=0.315, P=0.252**). However, when the IBW datum, based on only 3 stomach samples and unexpectedly low in animal matter (46%), was deleted, a significant correlation (r=0.665, P=0.009) existed for the remaining 14 species and 40% of the variation in territory size was attributable to trophic position.

When body size and trophic position were considered together, the correlation was significantly improved. Interestedly, the relationship was slightly stronger when the IBW was included (n=15 spp, r=0.821, P=0.001) than when it was excluded (n=14 spp, r=0.806, P=0.003).

DISCUSSION

The red-cockaded (47 g, 85 ha), 3-toed (68 g, 166 ha), and black-backed (70 g, 174 ha) woodpeckers have large territories for their body weight. All 3 are highly insectivorous (81, 94 and 89% of diet, respectively) and specialize in conifers (pines, spruce, and pines-firs, respectively)(Bock and Bock 1974). 2 congeners, the downy (27 g, 3.6 ha) and hairy (68 g, 6 ha) woodpeckers, are nearly as insectivorous (76 and 78%), but occupy small territories, forage on both conifers and deciduous trees, and demonstrate substantial foraging niche shifts seasonally (McFarlane 1992). A 6th *Picoides* species, the white-headed woodpecker (61 g, 8 ha), is distinctly omnivorous (39% animal matter in diet) with poorly known but smaller territory requirements; it specializes in conifers but concentrates on their seeds (Short 1982).

At the other end of the spectrum, the red-headed woodpecker (72 g, 0.64 ha) is notable as a generalist (McFarlane 1992) with fly-catching habits; it migrates to areas of mast abundance where it defends very small territories in the winter. The acorn woodpecker (81 g, 5.5 ha) comes closest to being a herbivore (23% animal matter), gathers acorns from a modest territory, and stores them in a cooperatively-defended granary.

CONCLUSIONS

The relationships between body size, trophic position, and home range or territory for North American woodpeckers appeared weak, at best. Statistically significant, but fragile, correlations were found: a single species exerted a substantial effect on the analyses. Better data are needed and broad generalizations based on limited analyses may not be very reliable. While ecologists love a good story, our ecological concepts should be based on a solid foundation, rather than dogma.

ACKNOWLEDGMENTS

I thank Rebecca Goggans for bringing the large territories of 3-toed and black-backed woodpeckers to my attention, and Robert Hooper for sharing his data on RCW territories in the Francis Marion National Forest.

Assessing the Influence of Red-cockaded Woodpecker Colony Site Management on Non-target Forest Vertebrates in Loblolly Pine Forests of Mississippi: Study Design and Preliminary Results

Leonard A. Brennan[1], Jeffery L. Cooper, Kathleen E. Lucas, Bruce D. Leopold, and George A. Hurst, Department of Wildlife and Fisheries, Mississippi State University, MS.

ABSTRACT: We initiated a study to assess the influence of red-cockaded woodpecker (*Picoides borealis*) (RCW) habitat management on non-target forest vertebrates in the loblolly pine forests of east-central Mississippi. Our study design used pair-wise comparisons of presence and relative abundance of vertebrates in **1**) RCW colonies and **2**) similar-aged pine stands that were not managed for nor used by the RCW as a colony site. Study areas were Bienville National Forest (BNF) and Noxubee National Wildlife Refuge (NNWR). Species presence and relative abundance of herpetofauna, small mammals, and resident and breeding birds were sampled. Results from the first year of field sampling indicated that RCW colony site management influenced the distribution and abundance of non-target forest vertebrates. Differences in the way RCW habitat management was conducted on BNF and NNWR also may have influenced the distribution and abundance of forest vertebrates. This type of investigation needs to be replicated for additional years, and with other types of pine-dominated silvicultural systems within targeted recovery populations and habitats of the RCW.

KEYWORDS: amphibians, Bienville NF, birds, loblolly pine, midstory, neotropical migrants, Noxubee NWR, prescribed burning.

[1]Current address: Tall Timbers Research Station, Route 1, Box 678, Tallahassee, FL 32312.

We have, at best, a meager understanding of how red-cockaded woodpecker (RCW) (scientific names of vertebrates are provided in **Appendixes 1-3**) habitat management influences the distribution and abundance of non-target vertebrates. Much of the public land in the South is subject to RCW management. Thus, many of the strongholds of biodiversity in this region are dominated to various degrees by management for a single species. This situation may not necessarily be bad because the RCW is a keystone species in these pine-dominated ecosystems. Nonetheless, informed management demands that we know what effect RCW management is having on other species. In addition, there are competing wildlife issues involved in public land management, such as widespread declines of neotropical migrant birds, that may confound or compliment recovery efforts for the RCW (Block et al., in preparation).

The work described in this paper is an expansion of a study that quantified differences in northern bobwhite abundance between stands that were and were not managed for the RCW (Brennan 1991). Our purpose here is to describe a first approximation of whether RCW habitat management can influence distribution and abundance of terrestrial vertebrates in 2 loblolly pine (*Pinus taeda*) forests. The objectives of this paper are to:

1) summarize the results from the first year of sampling,

2) identify species that are likely to be

impacted positively or negatively by RCW habitat management, and

3) point out the need to replicate this type of study in other pine-dominated silvicultural systems within the geographic range of this endangered woodpecker.

METHODS

Study Areas

We used 2 study areas, Bienville National Forest (BNF) and Noxubee National Wildlife Refuge (NNWR). Both areas are within the mixed loblolly pine-hardwood forest of east-central Mississippi. BNF and NNWR represent different approaches to red-cockaded woodpecker habitat management when considered from the scale of the forest stand. Differences in burning rotation (BNF = 4-5 years, NNWR = 1-3 years), and intensity of physical removal or suppression of hardwood midstory have caused the BNF stands to have a greater basal area, and more biomass of hardwood mid-story than stands at NNWR.

SAMPLING TECHNIQUES

Study Design

We used a pair-wise design with randomized blocks for comparing forest vertebrate community structure in

1) active RCW colony sites; and,

2) similar-aged forest stands that were not managed for, or used by the RCW as a colony site. In this quasi-experimental context, the unmanaged stands can be considered a "control", and the woodpecker stands a "treatment". Forest stands used for sampling were matched according to past burning history, soil type, and average and maximum age of the stand.

Stand size had a great deal of influence on our study design, because forest stands managed as RCW colony sites ranged from 10-100 ha.

Basically, sites managed as RCW colonies were relatively small patches of open, parklike pine stands distributed in a matrix of dense, closed-canopy pine and mixed pine hardwood forest. Thus, the relatively small size of the stands available for sampling limited numbers of replicate survey points that could be located within them. Four replicates per stand type (those managed for red-cockaded woodpeckers, and similar-aged pine stands that were not managed for the woodpecker) were chosen because it represented the maximum amount of sampling that could be accomplished within resources available for this project.

Herpetiles and small mammals

Pitfall trapping was used to sample species presence and relative abundance of reptile, amphibian, and small mammals (Bury and Corn 1987). Four 19-liter (5 gallon) plastic buckets were buried at ground-level. Drift fences (height = 50 cm, length = 5 m) radiated at 3 angles (0, 120 and 240 degrees) from each bucket. After installation, traps were covered for a minimum of 10 days, and were then opened for 10 consecutive days during May and June of 1992.

Birds

Resident and breeding birds were sampled using a variable-radius point count technique (Reynolds et al. 1980). At BNF, 7 count points were located in each stand type (14 count points total). Resident birds were counted 5 times during February 1992 and breeding birds were counted 7 times during May and June 1992. At NNWR, 6 count points were located in each stand type (12 count points total). Resident birds were counted 5 times during February 1992, and breeding birds were counted 8 times during May and June.

Interpretation and Analyses

Data from herpetile, mammal and bird sampling were assessed in terms of species presence, and relative abundance between stand types. Low count frequencies precluded statistical analyses on a species by species basis. Relative abundance of a species between colony and unmanaged stands was considered only for species that were detected >5 times within a particular stand type. Our primary goal is to present a preliminary summary of the first year of data collection for this study. Statistical tests of differential abundance of vertebrate species in particular stand types will be conducted when all data are compiled upon completion of the study. Structural vegetation differences between stands managed and not managed as RCW colony sites were assessed using 25 0.02-ha (15-meter diameter) circular plots from each stand type. Differences in vegetation variables (pine and hardwood basal area, pine and hardwood tree density, and relative amounts of canopy cover) were

Table 1. Number of reptiles and amphibians captured in pitfall traps within managed RCW colonies and unmanaged stands at Noxubee National Wildlife Refuge (NNWR) and Bienville National Forest (BNF), Mississippi, May-June, 1992. Counts are based on 40 trap nights of pitfall trapping in each replicate colony and unmanaged stand (160 trap nights total per stand type). Common names and taxonomic information is provided in Appendix 1.

	Noxubee										Bienville									
	RCW Colony				Unmanaged				Total[a]		RCW Colony				Unmanaged				Total	
Species	1[b]	2	3	4	1	2	3	4	C	U	1	2	3	4	1	2	3	4	C	U
Marbled Salamander			1						1	0		1				1	1		1	2
Mole Salamander															1				0	1
Green Anole	1				1		1		1	1		1	1	2	2				2	4
Fowler's Toad		1	1	2	1	2	4		2	9	19	10	22	7	17	1	12	17	58	47
Midwest Worm Snake													2						2	0
Southern Black Racer				1					1	0										
Southern Ringneck Racer							1		1	0										
Five-lined Skink	4		2						6	0		2			1				2	1
Broadhead Skink	2	1			2				3	2							1	1	0	2
Eastern Narrow-mouth Toad	2	2	1	4	1				9	1	1	5	15	13	6	4	8	10	35	28
Gray Treefrog	1								1	0										
Central Newt													1						0	1
Mississippi Slimy Salamander											3			1		1	4		4	5
Green Frog					8				0	8			1				21		1	21
Southern Leopard Frog			1						1	0										
Ground Skink	3		1	1	2	3	3	2	5	10	8	5	4	1	7	4	1	2	18	14
Fence Lizard	3	1	2	2	3	2		1	8	10		3	1	1		1	5	1	5	1
Midland Brown Snake							1		0	1	1								1	0
Three-toed Box Turtle				1		1			1	1	1		1	1		1	2	2	2	2

[a]Totals represent summation of each species across replicate stands. [b]Numbers represent replicate stands.

examined using two-group discriminant function analysis (Green 1978).

RESULTS

Herpetofauna

Seven species of reptiles and amphibians were captured with sufficient frequency (>5 times in a particular stand type) to make inferences about their relative abundance. At NNWR, species that were most frequently detected in RCW colonies were Fowler's toad, five-lined skink, and eastern narrowmouth toad (**Table 1**). Fowler's toad, and eastern narrowmouth toad, and ground skink were clearly the most abundant herpetiles at BNF, but neither showed a trend to be more abundant in a particular stand type. Two species that were detected most frequently in stands not managed as RCW colonies were the green frog (at both NNWR and BNF), and ground skink (at NNWR only) (**Table 1**).

Mammals

Pitfall trapping indicated that 5 species of mammals were influenced by RCW habitat management. The least shrew and hispid cotton rat were detected most frequently in RCW colonies at NNWR and BNF (**Table 2**). The

Table 2. Number of small mammals captured in pitfall traps within managed RCW colonies and unmanaged stands at Noxubee National Wildlife Refuge (NNWR) and Bienville National Forest (BNF), Mississippi, May-June, 1992. Counts are based on 40 trap nights of pitfall trapping in each replicate colony and unmanaged stand (160 total nights per stand type). Common names and taxonomic information is provided in Appendix 2.

	Noxubee										Bienville									
	RCW Colony				Unmanaged				Total[a]		RCW Colony				Unmanaged				Total	
Species	1[b]	2	3	4	1	2	3	4	C	U	1	2	3	4	1	2	3	4	C	U
Southern Short-tailed shrew	2				1	1			**2**	**2**	3		1	5	1	4	3	1	**9**	**9**
Least Shrew	1	2	5	8					**16**	**0**	2		5	7	1				**14**	**0**
Woodland Vole								1	**0**	**1**					1			1	**0**	**2**
Golden Mouse	1			1	2		1		**2**	**3**				1					**1**	**0**
Marsh Rice Rat				1	1				**1**	**1**				1	1		2		**1**	**3**
White-footed Mouse	4	7	9	3	7	19	9	4	**23**	**29**	3	10	12	28	6	11	12	8	**53**	**37**
Eastern Harvest Mouse	3	5	11				1	3	**19**	**4**	1	4	9		3	2		5	**14**	**10**
Hispid Cotton Rat	1	2	3						**6**	**0**	1	2	3						**6**	**0**
Southeastern Shrew	2	1		3			1	2	**6**	**2**	1		1	2	1	1		2	**4**	**4**

[a]Totals represent summation of each species across replicate stands. [b]Numbers represent replicate stands.

white-footed mouse was detected most frequently in unmanaged stands at NNWR, but in RCW colonies at BNF (**Table 2**). The eastern harvest mouse and southeastern shrew were detected most frequently in RCW colonies at NNWR but not at BNF (**Table 2**).

Resident Birds

At NNWR, resident birds that were detected most frequently in RCW colonies were American goldfinch, Carolina chickadee, Carolina wren, common grackle, northern cardinal, red-winged blackbird, rufous-sided towhee, and white-throated sparrow (**Table 3**). None of the 27 species of resident birds counted at NNWR showed conclusive evidence of being more abundant in the unmanaged stands.

At BNF, the northern cardinal was detected most frequently in the unmanaged stands. All other species of resident birds showed no indication of being more abundant in a particular stand type at BNF (**Table 3**).

Breeding Birds

At NNWR, Bachman's sparrow, blue-gray gnatcatcher, Carolina chickadee, Carolina wren, common yellowthroat, indigo bunting, northern bobwhite, rufous-sided towhee, white-breasted nuthatch, and yellow-breasted chat were significantly more abundant in RCW colonies than they were in unmanaged stands (**Table 3**). At BNF, great-crested flycatcher, and yellow-breasted chat were observed most often in the RCW colonies.

At NNWR, hooded warbler and wood thrush were detected on all counts in unmanaged stands, but never in the RCW colonies (**Table 3**).

Four species of migrants (palm warbler, scarlet tanager, white-throated sparrow, and yellow-rumped warbler) were detected during the spring counting period (**Table 3**). Of these, the white-throated sparrow seemed to have an affinity for unmanaged stands at BNF, but not at NNWR.

Vegetation

Discriminant function analyses indicated that there was more overlap in habitat structure between RCW colonies at BNF than at NNWR (**Figure 1**). At NNWR, 92% of the habitat plots

Table 3. Counts of winter resident birds within red-cockaded woodpecker colony stands at Noxubee National Wildlife Refuge and Bienville National Forest, Mississippi, during February, '1992. Counts based on 5 replicate counts at 6 points at Noxubee, and 5 replicate counts at 7 points at Bienville.

	Noxubee		Bienville	
Species	RCW Colony	Unmanaged Stand	RCW Colony	Unmanaged Stand
American Crow	0'	1	0	2
American Goldfinch	30	6	4	2
American Kestrel	0	0	3	0
American Robin	0	0	1	0
Brown-headed Nuthatch	4	0	2	10
Brown Creeper	1	4	1	0
Carolina Chickadee	12	2	9	17
Carolina Wren	34	12	20	15
Common Flicker	1	0	0	0
Common Grackle	24	0	0	0
Dark-eyed Junco	3	0	0	0
Downy Woodpecker	1	1	2	2
Eastern Bluebird	1	0	0	0
Eastern Phoebe	1	2	1	0
Golden-crowned Kinglet	0	0	4	12
Hermit Thrush	0	2	0	0
Northern Bobwhite	0	0	19	0
Northern Cardinal	13	2	7	17
Pileated Woodpecker	4	1	0	2
Pine Warbler	77	45	78	79
Red-bellied Woodpecker	7	8	1	0
Red-cockaded Woodpecker	19	0	13	1
Red-headed Woodpecker	0	0	1	0
Red-winged Blackbird	18	0	0	0
Ruby-crowned Kinglet	0	0	4	10
Rufous-sided Towhee	15	0	8	6
Song Sparrow	6	0	0	0
Swamp Sparrow	1	0	0	0
Tufted Titmouse	7	6	6	18
White-breasted Nuthatch	7	0	0	0
White-throated Sparrow	35	0	5	0
Winter Wren	5	2	3	5
Wood Duck	7	0	0	0
Yellow-bellied Sapsucker	0	1	0	0
Yellow-rumped Warbler	0	0	5	0

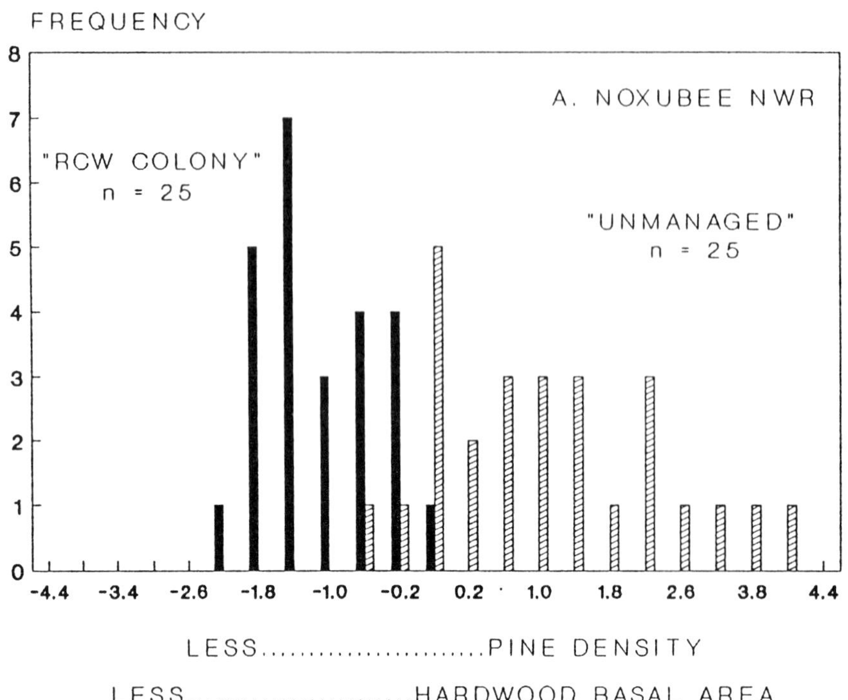

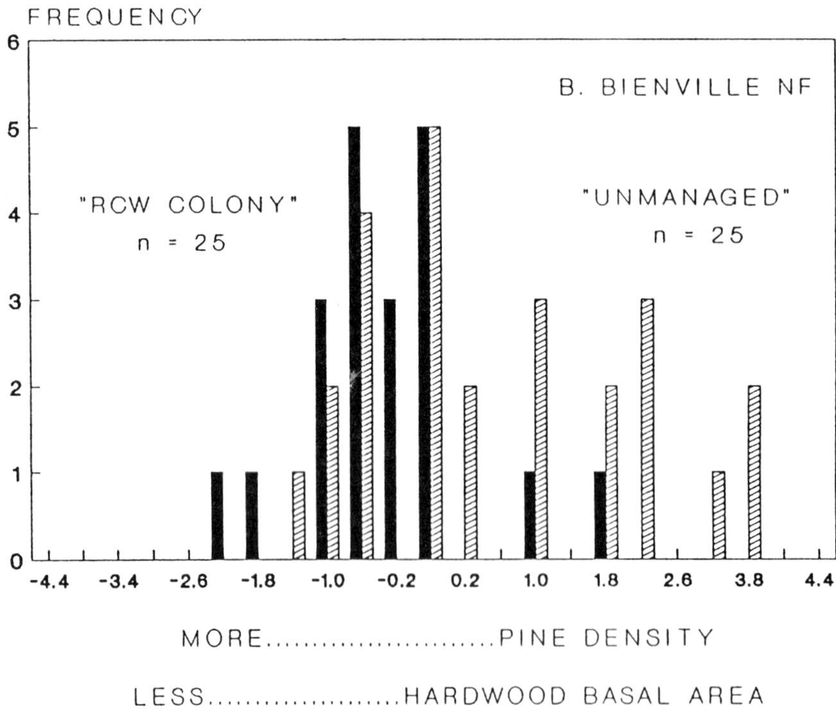

Figure 1. Frequency histograms of discriminant function analysis scores from 0.02 ha habitat plots measured in stands managed for red-cockaded woodpeckers, and unmanaged mature pine stands at Bienville National Forest and Noxubee National Wildlife Refuge in east-central Mississippi. At Bienville, 76% of 50 habitat plots were classified in the correct group, at Noxubee, 92% of 50 habitat plots were classified in the correct group.

were classified in the correct group, whereas at BNF, only 76% were classified in the correct group **(Figure 1).**

DISCUSSION

It is evident that RCW colony site management can influence the distribution and abundance of non-target vertebrates. It is premature to make predictions about how particular species may or may not be impacted by RCW colony site management in loblolly pine silvicultural systems. Our results suggest some relationships that may be of interest to forest managers.

First, it is apparent that the intensity of RCW management influences the vertebrate community. For example, at NNWR, large relative differences in abundance of vertebrates were more frequent than they were at BNF, where vegetation differences between treatments were not as great **(Figure 1)**.

Second, presence of many rarely-detected species makes it difficult to develop clear interpretations. The large number of apparently uncommon species means that it will be difficult to use such information to make predictions about how RCW colony site management will influence a large number of non-target forest vertebrates. Such a situation puts both managers and researchers in a quandary. We pose the following two questions as examples of the trade-offs faced by managers who want to consider issues related to vertebrate biodiversity in the context of RCW colony site management:

1) Would more widespread, and intensive sampling be justified to increase numbers of rare and uncommon animals sampled?; or,

2) Should the manager be concerned with only the vertebrates that have a reliable indication of differential abundance between red-cockaded and unmanaged stands?

Third, a study such as the one described here illustrates the merits and limitations of a survey data set based on a single year. The merits of the study are that it showed there were some obvious differences in the abundance of certain herpetiles, mammals, and birds in relation to red-cockaded woodpecker habitat management. However, the inferential value of such a data set is extremely limited. Although the counts were based on replicate points and replicate stands, and 2 study areas, our effective sample size at present is n=1, because all surveys were conducted during only 1 year (Wiens 1981).

There is clearly a need to expand this type of study to include other pine-dominated ecosystems and geographic regions within the range of the RCW. Unlike the Pacific Northwest where interest in the northern spotted owl (*Strix occidentalis*) inspired federal agencies such as the USDA Forest Service to conduct comprehensive surveys of terrestrial vertebrates in old-growth and other seral stages of Douglas-fir (*Pseudotsuga menziesii*) forests (Ruggiero et al. 1991), wildlife researchers in the South have neglected to make a similar, comprehensive assessment of how RCW habitat management can impact non-target vertebrates. A broad-scale study designed like the one described in this paper that is replicated at a minimum of 5-6 sites across the range of the red-cockaded woodpecker would go a long way towards filling a gaping hole in our knowledge about vertebrate habitat relationships in forest stands managed for this bird.

ACKNOWLEDGMENTS

This study was supported by the USDA Forest Service Southern Forest Experiment Station Ecological Modeling Competition Program. R.N. Conner administered funding. R. Hooper and R. Costa provided editorial advice. At Bienville NF, Dean Elsen, Bruce Davenport and Clint Floyd provided data on stand age and history. Also at Bienville, Edsel Cliburn, Frankie Slade, and Dale Windham of the Mississippi Department of Wildlife, Fisheries and Parks provided housing for field crews and other assistance. At Noxubee, David Richardson, David Ellis, Jim Stockie, Jim Tisdale, and Gwen Cotton provided assistance with virtually all aspects of the project.

Appendix 1. Amphibians and reptiles captured using pitfall traps within RCW colonies and unmanaged stands at Noxubee National Wildlife Refuge (NNWR) and Bienville National Forest (BNF), Mississippi, May-June, 1992.

Scientific Name[a]	Common Name
ORDER CAUDATA	**Salamanders**
FAMILY AMBYSTOMATIDAE	Mole Salamanders
Ambystoma opacum	Marbled Salamander
Ambystoma talpoideum	Mole Salamander
FAMILY SALAMANDRIDAE	Newts
Notophthalmus viridescens	Central Newt
FAMILY PLETHODONTIDAE	Lungless Salamanders
Plethodon mississippi	Mississippi Slimy Salamander
ORDER ANURA	**Toads and Frogs**
FAMILY BUFONIDAE	Toads
Bufo woodhousii	Fowler's Toad
FAMILY HYLIDAE	Treefrogs
Hyla chrysoscelis	Gray Treefrog
FAMILY MICROHYLIDAE	Narrowmouth Toads
Gastrophryne carolinensis	Eastern Narrowmouth Toad
FAMILY RANIDAE	True Frogs
Rana clamitans	Green Frog
Rana utricularia	Southern Leopard Frog
ORDER TESTUDINES	**Turtles**
FAMILY EMYDIDAE	Box Turtles
Terrapene carolina	Three-toed Box Turtle
ORDER SQUAMATA	**Lizards and Snakes**
SUBORDER LACERTILIA	Lizards
FAMILY POLYCHRIDAE	Anoles
Anolis carolinensis	Green Anole
FAMILY PHRYNOSOMATIDAE	Spiny Lizards
Sceloporus undulatus	Fence Lizard
FAMILY SCINCIDAE	Skinks
Eumeces laticeps	Broadhead Skink
Eumices fasciatus	Five-lined Skink
Scincella lateralis	Ground Skink
SUBORDER SERPENTES	Snakes
FAMILY COLUBRIDAE	Colubrids
Storeria dekayi	Midland Brown Snake
Diadophis punctatus	Southern Ringneck Snake
Carphophis amoenus	Midwest Worm Snake
Coluber coluber	Southern Black Racer

[a]Scientific names follow Conant and Collins (1991).

Appendix 2. Mammal species captured and presence recorded within red-cockaded woodpecker colonies and unmanaged stands at Noxubee National Wildlife Refuge (NNWR) and Bienville National Forest (BNF), Mississippi, using pitfall sampling during May-June, 1992.

Scientific Name[a]	Common Name
ORDER INSECTIVORA	**Insectivores**
FAMILY SORICIDAE	Shrews
Sorex longirostris	Southeastern Shrew
Blarina carolinensis	Southern Short-tailed Shrew
Cryptotis parva	Least Shrew
ORDER RODENTIA	**Rodents**
FAMILY CRICETIDAE	Cricetids
Oryzomys palustris	Marsh Rice Rat
Reithrodontomys humulis	Eastern Harvest Mouse
Peromyscus gossipinus	Cotton Mouse
Peromyscus leucopus	White-footed Mouse
Ochrotomys nuttalli	Golden Mouse
Microtus pinetorum	Woodland Vole
Sigmodon hispidus	Hispid Cotton Rat

[a]Scientific names follow Jones et. al. (1986).

Appendix 3. Bird species censused within red-cockaded colony stands and unmanaged stands at Noxubee National Wildlife Refuge and Bienville National Forest from February-June 1992.

Scientific Name[a]	Common Name
ORDER ANSERIFORMES	
FAMILY ANATIDAE	Ducks
Aix sponsa	Wood Duck
ORDER FALCONIFORMES	
FAMILY ACCIPITRIDAE	Hawks
Accipiter cooperii	Cooper's Hawk
Buteo lineatus	Red-shouldered Hawk
Buteo platypterus	Broad-winged Hawk
FAMILY FALCONIDAE	Falcons
Falco sparverius	American Kestrel
ORDER GALLIFORMES	
FAMILY PHASIANIDAE	Quail
Colinus virginianus	Northern Bobwhite
FAMILY MELEAGRIDIDAE	Turkey
Meleagris gallopavo	Wild Turkey
ORDER CUCULIFORMES	
FAMILY CUCULIDAE	Cuckoos
Coccyzus americanus	Yellow-billed Cuckoo
ORDER APODIFORMES	
FAMILY TROCILIDAE	Hummingbirds
Archilochus colubris	Ruby-throated Hummingbird
ORDER PICIFORMES	
FAMILY PICIDAE	Woodpeckers
Colaptes aurates	Common Flicker
Drycopus pileatus	Pileated Woodpecker
Melanerpes carolinus	Red-bellied Woodpecker
Melanerpes erythrocephalus	Red-headed Woodpecker
Sphyrapicus varius	Yellow-bellied Sapsucker
Picoides pubescens	Downy Woodpecker
Picoides borealis	Red-cockaded Woodpecker
ORDER COLUMBIFORMES	
FAMILY COLUMBIDAE	Doves
Zenaida macroura	Mourning Dove
ORDER PASSERIFORMES	
FAMILY TYRANNIDAE	Flycatchers
Myiarchus crinitus	Great-crest Flycatcher
Sayornis phoebe	Eastern Phoebe
Contopus virens	Eastern Pewee
FAMILY CORVIDAE	Jays, Crows
Cyanocitta cristata	Bluejay
Corvus brachyrhyncos	American Crow
FAMILY PARIDAE	Titmice
Parus carolinensis	Carolina Chickadee
Parus bicolor	Tufted Titmouse
FAMILY SITTIDAE	Nuthatches
Sitta carolinensis	White-breasted Nuthatch
Sitta pusilla	Brown-headed Nuthatch

Appendix 3 continued

Scientific Name[a]	Common Name
FAMILY CERTHIIDAE	Creepers
Certhia familiaris	Brown Creeper
FAMILY TROGLODYTIDAE	Wrens
Troglodytes troglodytes	Winter Wren
Thyrothorus ludovicianus	Carolina Wren
FAMILY MIMIDAE	Mimic Thrushes
Dumetella carolinensis	Gray Catbird
Taxostoma rufum	Brown Thrasher
FAMILY TURDIDAE	Thrushes
Turdus migratorius	American Robin
Hylochila mustelina	Wood Thrush
Catharus guttatus	Hermit Thrush
Sialia sialis	Eastern Bluebird
FAMILY SYLVIIDAE	Kinglets
Polioptila caerulea	Blue-gray Gnatcatcher
Regulus satrapa	Golden-crowned Kinglet
Regulus calendula	Ruby-crowned Kinglet
FAMILY VIRONIDAE	Vireos
Vireo griseus	White-eyed Vireo
Vireo flavifrons	Yellow-throated Vireo
Vireo olivaceus	Red-eyed Vireo
FAMILY PARULIDAE	Wood Warblers
Mniotilta varia	Black and White Warbler
Dendroica coronata	Yellow-rumped Warbler
Dendroica pinus	Pine Warbler
Dendroica discolor	Prairie Warbler
Dendroica palmarum	Palm Warbler
Seiurus aurocapillus	Ovenbird
Oporornis formosus	Kentucky Warbler
Geothlypis trichas	Common Yellowthroat
Icteria virens	Yellow-breasted Chat
Wilsonia citrina	Hooded Warbler
FAMILY ICTERIDAE	Blackbirds
Agelaius phoeniceus	Red-winged Blackbird
Quiscalus quiscula	Common Grackle
Molothrus ater	Brown-headed Cowbird
FAMILY THRAUPIDAE	Tanagers
Piranga olivaceus	Scarlet Tanager
Piranga rubra	Summer Tanager
FAMILY FRINGILLIDAE	Finches
Cardinalis cardinalis	Northern Cardinal
Passerina cyanea	Indigo Bunting
Cardeulis tristis	American Goldfinch
Pipilo erythrophthalmus	Rufous-sided Towhee
Aimophila aestivalis	Bachman's Sparrow
Junco hyemalis	Dark-eyed Junco
Zonotrichia albicollis	White-throated Sparrow
Melospiza georgiana	Swamp Sparrow
Melospiza melodia	Song Sparrow

[a]Scientific names follow American Ornithologist's Union (1983).

Red-cockaded Woodpecker Nesting Success in the Sam Houston National Forest in 1989

Brent Ortego, Texas Parks & Wildlife Department, Victoria, Texas 77901
Dawn Carrie, USDA Forest Service, Cleveland, Texas 77327
Kenneth Moore, USDA Soil Conservation Service, Leesville, Louisiana 71446

ABSTRACT: From April through July, 1989, 34 red-cockaded woodpecker (*Picoides borealis*) clans with at least 2 adults were monitored for nesting, and production of eggs, young and fledglings on the Sam Houston National Forest. This population was the westernmost and largest in Texas. Twenty four of 34 clans (71%) successfully fledged at least 46 young (1.35 young/clan). Nesting chronology, production, and causes of nesting failures were discussed in relation to other populations.
KEYWORDS: red-cockaded woodpecker, breeding season, nesting success, southern flying squirrel, predation, Sam Houston National Forest, loblolly pine, longleaf pine, shortleaf pine

As part of a red-cockaded woodpecker (RCW; American Ornithologists' Union 1983) translocation project, the nesting chronology and success of RCW groups in the Sam Houston National Forest (SHNF) were examined. This population was the largest and westernmost in Texas (Ortego et al. 1988). Breeding chronology and nesting success has been studied in other parts of the species' range (Lennartz and Henry 1985), but not on this westernmost population. Most research on this species in Texas focused on the smaller, more eastern populations (Ortego et al. 1988).

Determining breeding chronology and nesting success of this larger population was important in order to schedule and set management practices and priorities. In addition, this information can be compared to populations on private and public lands to the east that are smaller, more fragmented, and declining in Texas (Conner and Rudolph 1989; Ortego and Lay 1988).

From April through July 1989, we studied the reproduction of 34 clans near Lake Conroe in SHNF, about 49 km (30 miles) north of the Houston metropolitan area. This area contains a relatively old loblolly pine (*Pinus taeda*; Hatch et al. 1990) and shortleaf pine (*P. echinata*) forests.

METHODS

All RCW cavity tree clusters listed as active in the SHNF data base in the vicinity of Lake Conroe were visited during the week of 16 April, 1989. We classified all known cavity trees at a cluster as either active or inactive, based on the presence or absence of recently worked resin wells. Cavity tree clusters containing at least 1 cavity tree with active resin wells (Jackson 1977b) were considered active, and based on this technique, each cavity tree cluster was classified as active or abandoned.

From 16 April - 15 May 1989, we used sunrise and sunset roost checks to determine the number of adults in each clan. Several visits to a colony site were some times necessary to determine the number of adults. Weather, juxtaposition of cavity trees, behavior of the birds and the occasional presence of a floater or intruder RCW made determination of clan size difficult with one visit.

Beginning on 24 April (for clans with at least two adults) we checked colonies at approximately 9-day intervals for nesting activity. We visited all active trees and attempted to flush birds from the cavities by tapping and scraping on the trunk and making loud squeaking noises. If we flushed a bird, or observed a bird on or near an active tree, we climbed the tree with Swedish tree climbing ladders and used a mirror and a small halogen light bulb powered by a 6-volt battery to inspect the cavity for the presence of eggs or young. If we did not find eggs in a tree cluster containing two or more adult birds by the time most other clans produced eggs or young, we climbed all active cavity trees in the cluster. If a nest was lost to predation, we continued monitoring for re-nesting. We ceased monitoring for nesting activity in mid-June, after which time we believed it unlikely the nests would be initiated that year.

We assumed a 12-day incubation period beginning with the first egg. Nests found with 1 or 2 eggs were revisited in 17 days, when we estimated the young were 5-9 days old and of the proper size for banding. Nests found with 3 or 4 eggs had an unknown onset of incubation and were visited at 9-day intervals until we found young or the nesting attempt was finished.

We banded young at 5-9 days of age, basing age estimates on Ligon (1970). We used nylon nooses to remove young from cavities (Jackson 1982). In most cases we captured all young from a cavity in a single visit, although 2 visits were needed to capture all young in 2 nests. If all young were not captured after 2 visits, the remaining young were left unbanded.

Each young was banded with a unique combination of 2 colored plastic leg bands, with 1 band being placed on each leg. To facilitate recognition of individuals, no 2 young in a clan had the same color band on the same leg. Ten colors were used.

After banding the young, we did not revisit colonies until 1-9 days after the estimated fledging date, allowing for a 26-day nestling period. We attempted to locate all young from a nest, usually starting at sunrise when adults emerged. Unbanded young were distinguished from adults based on their habit of remaining crouched on tree boles or limbs for up to 65 minutes, their weak flight, and their overall "dingy" coloration. The young were fed frequently by adults for several weeks after fledging, further aiding in their recognition.

When monitoring fledging, we followed a clan until we either located all known young from that nest, or until we were confident that we had located all surviving young. From 2-4 visits were made to each nest site to locate potentially missed young. We sexed all fledglings based on the presence (males) or absence (females) of a red crown patch. We identified color bands and sexed fledglings with binoculars and a spotting scope mounted on a rifle stock.

A Kruskal-Wallis non-parametric ANOVA on ranked data was used to analyze results because the data were not normally distributed (Proc GLM; SAS 1988).

RESULTS

Thirty of 34 RCW clans were observed to nest at least once and four of the clans re-nested after the eggs from the first attempt were believed to be predated. Twenty four of the 34 clans (71%) successfully fledged at least 46 young (1.35 young/clan).

Evidence of predation was observed within 9 days of first discovering eggs in 4 cavities. Evidence consisted of southern flying squirrels (*Glaucomys volans*; Burt and Grossenheider 1976) and egg shell fragments found in 1 cavity, egg shell fragments in another cavity, and 2 empty cavities. In each case, all clan members were observed after the alleged predation, but they no longer roosted in the cavities where nesting had failed.

The 5 4-bird clans were successful in 4 of 6 nesting attempts and they produced 21 eggs (mean = 3.5), at least 10 young (mean = 1.7) and 10 fledglings (mean = 1.7) at a rate of 0.5 young fledged per adult. The 13 3-bird clans were successful in 11 of 14 nesting attempts and produced 41 eggs (mean = 2.9), at least 29 young (mean = 2.1), and at least 19 fledglings (mean = 1.4) at a rate of 0.5 young fledged per adult in clan. The 12 2-bird clans were successful at 9 of 14 nesting attempts and produced 37 eggs (mean = 2.6), at least 26 young (mean = 1.9) and at least 17 fledglings (mean = 1.2) at a rate of 0.7 young fledged per adult in clan.

During the study, we observed fairly synchronized RCW nesting. Clans laid eggs from 23 April through 19 June. The average date of clutch initiation was 3 May, the mode was 28-29 April and the median date was 29 April. RCW fledged young from 31 May through 3 July. The average date of fledging was 9 June, the mode 5 June and the median 6 June.

Clans with 3 or 4 adults averaged 3.1 eggs per nesting attempt compared to 2.6 eggs for 2-bird clans, but the difference was not significant ($P<0.06$). Clans with 3-4 adults averaged 1.5 fledglings and 2-bird clans averaged 1.2 fledglings, and again the difference was not significant (P<0.19).

DISCUSSION

Lay and Swepston (1973) reported nesting occurred from 29 April through 29 July in Angelina and Newton Counties in 1970-1973, in primarily longleaf pine (*Pinus palustris*) forests, about 80 and 110 miles to the northeast of SHNF, respectively. The average date of clutch initiation for 10 nesting attempts in their study was 20 May. Of the 62 nesting attempts observed, 58% were successful. The average clutch initiation date they reported was 17 days later than the 3 May average we observed. However, the reported Angelina and Newton counties nesting dates fall within the range of nesting dates, April through July, reported in other states (Ligon 1970, Lennartz 1983, Lennartz and Henry 1985, Lennartz and Heckel 1988, Walters et al. 1988, DeLotelle and Epting 1992).

The 58% nesting success reported by Lay and Swepston (1973) was lower than the 71% we observed in SHNF, and considerably lower than the 93% in South Carolina, 84% in Florida, 72% in North Carolina, and 69% in Georgia (Lennartz and heckel 1988, Walters et al. 1988, DeLotelle and Epting 1992). It is possible that a combination of smaller population sizes and poorer habitat quality reduced reproduction in Angelina and Newton counties during 1970-1973.

ACKNOWLEDGMENTS

This project was jointly funded with Section 6 monies from the U.S. Fish and Wildlife Service, a Challenge Grant from the National Forests in Texas, and from the Fisheries & Wildlife Division of the Texas Parks & Wildlife Department. Logistical support and cooperation were provided by personnel of the National Forests in Texas. D. Carrie and K. Moore were employed by Texas Parks & Wildlife Department during the project. John Barron with the Texas Parks & Wildlife Department conducted statistical analysis of data.

Evaluation of the Global Positioning System as a Research Tool in the Management of the Red-cockaded Woodpecker

Kathleen E. Franzreb and Haven R. Barnhill[1], Southern Research Station, Department of Forest Resources, Clemson University, Clemson, SC 29634

ABSTRACT: The Global Positioning System (GPS) is a new technology for rapid and accurate determination of positions on the earth that has major applications in resource management. GPS is a satellite-based radio navigation/positioning system developed by the Department of Defense. We compared GPS estimates of the home range sizes for red-cockaded woodpeckers (*Picoides borealis*) at the Savannah River Site in South Carolina with the more traditional, but less precise, method of recording positions on maps developed from aerial photographs that we refer to as the traditional mapping method (TMM). Estimates of 7 home ranges determined from TMM positions varied from 14.5 ha (35.8 ac) to 93.6 ha (231.2 ac) with a mean of 49.7 ha (122.8 ac). Using GPS at these same positions provided estimates ranging from 16.0 ha (39.5 ac) to 94.8 ha (234.2 ac) with a mean of 51.6 ha (127.6 ac). The distance between TMM and equivalent GPS positions varied from 1.3 m (4.3 ft) to 818.6 m (2701.4 ft), and averaged 81.8 m (269.9 ft). There was no significant difference in the mean size of the home range derived by TMM and GPS. As GPS readings were accurate to ± 5 m (16.5 ft), it was estimated that less than 6% of the observations using GPS were incorrectly categorized as to stand type. In contrast, the study suggested that TMM assigned approximately 14 - 20% of the positions to the wrong stand type. The major advantage of GPS was that it determined more accurately positions used to assess stand utilization.

KEYWORDS: GPS, global positioning system, home range, red-cockaded woodpecker, Savannah River Site.

[1]Current address: Georgia Department of Natural Resources, Wildlife Resources Division, 2070 Highway 278 SE, Social Circle, GA 30279.

Determination of home range size of red-cockaded woodpeckers (RCW) has implications for assessing habitat requirements, determining how many groups can be accommodated in suitable habitat, and in developing management guidelines. Previous RCW home range studies were based on following birds and mapping their positions on either aerial photographs or maps (Blue 1985; DeLotelle et al. 1983b, 1987; Hooper et al. 1982; Labisky and Porter 1984; Nesbitt et al. 1978; Patterson and Robertson 1981; Porter and Labisky 1986; Repasky 1984; Skorupa 1979; and Skorupa and McFarlane 1976). Accuracy of this method depended on the photogrammetric and navigational skills of the observer and the limitations of the available maps and photographs. Reliability of these procedures has not been determined.

This paper evaluates the use of the global positioning system (GPS) for determining bird locations. Our objective was to compare the size and configuration of home ranges of 7 RCW groups estimated from group positions determined by GPS and the more traditional mapping method (TMM).

GLOBAL POSITIONING SYSTEM GPS

GPS is an accurate navigational/positional system being developed by the U.S. Department of Defense (DoD). It was developed for the military and only recently became available for civilian use. When complete, GPS will consist of 24 satellites that orbit the earth and transmit unique radio signals, which travel at the speed of light (c). Each radio signal is marked with a code noting the exact time that it was transmitted. GPS determines the position of the receiver by simultaneously measuring the distances between the GPS unit and 3 or more satellites in space, with the satellites acting as precise reference points. The GPS receiver, which is carried in the field by the observer, collects the radio signal and calculates the amount of time (t) required for the signal to travel from the satellite to the receiver. The distance from the satellite to the receiver is calculated by multiplying c by t. To illustrate, if a receiver is "X" km from satellite "A", then the receiver must be somewhere on a sphere with radius "X". If at the exact same time it is determined that the receiver is "Y" km from satellite "B", then the receiver must be somewhere on a sphere with radius "Y" that intersects the sphere with radius "X". Adding input from a third satellite narrows the position even further because there are then only 2 points in space where these 3 distance measurements intersect. Generally only 1 of these positions is logical, as the other position is usually far out in space (or moving at an unrealistic speed). Measuring the distance to a fourth satellite allows for a single position to be determined (however, the fourth satellite is not always necessary as a user can choose to manually input the exact altitude of the receiver antenna to calculate the correct position).

The ability to accurately measure the time required for the signal to travel from the satellite to the receiver is the key to determining the correct position. A 1 millisecond time error may yield a 300 km (187.5 mi) position error (Langley 1991). Numerous potential errors in determining GPS positions exist: the atomic clocks on the satellites are synchronized only to within about 1 millisecond, the GPS receivers are mis-synchronized by an unknown amount of time and environmental conditions such as interference from the ionosphere and troposphere impact the signal. These errors are corrected through various measurements and calculations performed by the GPS receiver software. The DoD controls the accuracy of GPS positions through a process called selective availability (SA). SA is an unknown time error introduced into the satellite radio signals by the DoD to prevent hostile users from instantaneously determining an exact position. SA is the largest error factor in GPS but can be eliminated with differential corrections. To overcome SA effects it is necessary to establish a benchmark GPS receiver (base station) in a known location. The error for each position is determined by continuously collecting data from the base station and comparing each position with the known location. This error factor is used to differentially correct each position calculated by the rover unit. Langley (1991) and Leick (1990) provide a more detailed explanation of the mathematics involved in GPS.

METHODS

The study was conducted at the Savannah River Site (SRS), a nuclear production facility operated by the U.S. Department of Energy (DOE), in Aiken and Barnwell counties in west central South Carolina. Gaines et al. (1994) described the area and the management of its RCW population.

Seven RCW groups were observed for 7 months from May 1992 through November 1992 as part of a 3-year study to determine home range and foraging behavior. In most cases, the group members were followed as they emerged from roost trees in the morning until mid-afternoon. On some days, birds were followed for the entire day or from noon until roosting in the evening. The date, time, observer, leg band colors, and behavioral information were recorded for each observation obtained at 15-minute intervals. A numbered and dated flag was attached to the tree in which the bird was located. The observer continued to follow the birds and take behavioral observations without being delayed by taking a GPS reading. A researcher was then able to return to the flag to coincide with a time period when enough satellites were within range to accurately estimate the positions.

For each group of birds, we determined home range size by the modified minimum area method (Harvey and Barbour 1965), which is based on the minimum convex polygon method (Mohr 1947, Odum and Kuenzler 1955). The modified method differs from the minimum area method in that one-quarter of the maximum range length was used to determine what peripheral positions should delineate the boundary.

Peripheral positions were included if they were less than one-quarter of the maximum length apart. Applying this standard in general would result in smaller home range sizes than using the minimum convex polygon method. A t-test identified significant differences ($P < 0.05$) in home range sizes estimated by GPS and compared to manual mapping procedures. Results herein are preliminary and are mainly presented to illustrate a case history using GPS for wildlife related work.

We used the Trimble Navigation Pathfinder Professional model GPS unit which consisted of a receiver system (6-channel), a lightweight GPS antenna, and a MC-V datalogger. Positions computed by the GPS receiver were recorded on the datalogger, transferred to a personal computer, and manipulated with the PFINDER post-processing software (Trimble Navigation Ltd., 585 North Mary Ave., Sunnyvale, CA. 94086-8000) included with the GPS Pathfinder System. Specifications for the unit are provided in the Appendix.

In general for every day spent following birds in the field, an equivalent amount of time was needed to collect and handle the GPS data. All GPS positions were interfaced with a geographical information system (GIS) (ArcInfo) and then plotted on GIS generated maps.

Logging and data processing procedures followed Trimble Navigation Ltd.'s recommendations, which yielded positions that were accurate to within 2 to 5 m (6.6 to 16.5 ft)(Trimble 1992). Each GPS position was determined by logging a minimum of 180 position fixes (each fix is a single location determination). The position fixes were differentially corrected and averaged to provide a single estimate for each position. The data were formatted to be compatible with the available GIS.

The Savannah River Forest Station produced stand maps at a scale of 1:24,000 from 1:15,840 scale aerial photographs. These maps were incorporated into Environmental Systems Research Institute's ArcInfo (ERSI, 380 New York Street, Redlands, CA. 92376), a GIS. Maps of the area used by each group were generated from this coverage using the GIS and included stand boundaries, roads, and other obvious landmarks such as RCW cavity trees. The maps were plotted on 11" by 17" sheets at scales ranging from 1:2,778 to 1:5,003. The larger maps improved and simplified determining the positions of the birds. After the GIS generated maps were available, we returned to the flagged positions to manually plot them on the maps (TMM). By then some of the flags had been destroyed or lost, so that in some instances the sample size for TMM positions was less than what was available for the GPS positions.

To allow for comparable analysis, the TMM positions were digitized so the GIS and GPS coverages were equivalent. Each TMM position had a matching GPS position. Comparisons were made of the home range sizes estimated by the 2 methods. Also the home range sizes were calculated based on the entire GPS data set for each group. The GIS was used for spatial analysis including calculating the home range areas, determining stand use, and measuring distances from the GPS locations to the equivalent digitized locations.

RESULTS AND DISCUSSION

Home range sizes over a 7 month period for the 7 groups on the SRS estimated by TMM were 14.5 - 93.6 ha (35.8 - 231.2 ac) with a mean of 49.7 ± 27.2 ha (122.8 ± 89.8 ac) (**Table 1**). Based on GPS data of the same positions, home range sizes were 16.0 - 94.8 ha (39.5 - 234.2 ac), and the mean was 51.6 ± 29.6 ha (127.6 ± 97.7 ac). Estimates of the means derived by these 2 methods did not differ significantly ($t = 0.12$, $P > 0.50$). When all GPS positions were included, the mean home range size increased to 54.3 ± 29.3 ha (134.1 ± 96.7 ac).

The number of positions available for estimating the home ranges of the different woodpecker groups ranged from 104 to 254 with the smallest data set yielding the smallest home range estimate. Shapes of individual home ranges were similar for 6 of the 7 groups using TMM positions compared to equivalent GPS data as exemplified by group 16 (**Figure 1**). However, for group 2 (**Figure 1**) the GPS and TMM determined boundaries did not coincide. Some groups used all portions of their home ranges fairly consistently and regularly (i.e., group 2), whereas others seemed to avoid certain portions of the range (group 43) (**Figure 2**).

Results indicated that TMM provided home range estimates similar to those determined by GPS. To further test the accuracy of the traditional mapping method, the distance from each TMM position to the equivalent GPS position was measured. The mean difference in 1,247 equivalent GPS and TMM positions was

81.8 m (269.9 ft), with a minimum difference of 1.3 m (4.3 ft) and a maximum difference of 818.6 m (2701.4 ft) (**Table 2**). Because of the accuracy provided by GPS, these differences may vary by ± 5 m (16.5 ft).

Because the accuracy provided by GPS is within 5 m (Trimble 1992, Jasumback and Luepke 1993), only those GPS locations within 5 m of an edge were considered to be potentially categorized in the wrong stand. The distance from each GPS location to the nearest stand edge was measured to assess the likelihood that the stand information provided by GIS was an accurate reflection of where the bird actually was. Only 80 (5.6%) of 1,411 positions determined by GPS were within 5 m (16.5 ft) of a stand edge. Thus, a maximum of 5.6% of the GPS locations may have been assigned to an incorrect stand.

Although an independent error estimate was not determined for the TMM positions, the mean distance between the TMM and the equivalent GPS positions (81.8 m) was used as the best available estimate of error in the TMM (**Table 2**). A total of 1,082 (79.1%) of 1,368 TMM positions were within 81.8 m of a stand edge, suggesting that a high proportion of TMM positions may have been placed in the wrong stand.

To further identify instances where TMM errors resulted in incorrect stand placement, the number of times both the TMM and equivalent GPS positions were in the same stand was counted. Results indicated that 79.6% (993 of 1,243) of the observations were in the same stand (**Table 3**). Therefore, 20.4% of the observations did not agree as to stand. Assuming that the maximum error attributable to GPS was 5.6%, then the minimum error in stand classification using the TMM data was 14.8% (20.4% total error - 5.6% GPS error). However, it is unlikely that all GPS points falling ≤ 5 m from the stand edge were improperly classified. Therefore, the 14.8% error estimate in the TMM data represented a conservative value. The actual average error rate in determining stand use with the TMM positions probably was between 14.8% and 20.4%.

Estimated error rates in determining stand use were not consistent among groups. Using the most conservative estimates for the TMM data (assuming the maximum GPS error of 5.6%), errors in stand classification ranged from 5.4% in group 3 to 26.6% in group 19. If these estimated misclassification rates were typical of previous habitat use studies based on similar mapping procedures, significant errors may exist in habitat use studies.

The study yielded considerably smaller home ranges than previous studies. For 4 groups in Florida, Porter and Labisky (1986) found mean yearly home ranges to be 129 ha (319 ac) using the harmonic mean method. In an 18-month study of an isolated group in south Florida, Patterson and Robertson (1981) found the home range size to be 159.3 ha (393.5 ac). Other home range estimates include 148.1 ha (365.8 ac) for 4 groups near Orlando, Florida, based on whole or partial days of observation (DeLotelle et al. 1983b). Five groups in south Florida were estimated to have a mean home range of 144.4 ha (356.7 ac) with the largest being 213.2 ha (526.6 ac) (Nesbitt et al. 1978). Two groups studied in the Sandhills in North Carolina had year-long home ranges of 180.3 ha (445.3 ac) and 138.7 ha (342.6 ac) of which 64.0 ha (158.1 ac) and 85.6 ha (34.7 ac), respectively, were maintained as territories (Repasky 1984). Blue (1985) estimated that home range of 3 groups in the Sandhills of North Carolina were 160.8 ha (397.2 ac), 206.2 ha (509.3 ac), and 255.0 ha (629.9 ac) of which 54.3%, 40.7%, and 45.7% of these areas, respectively, were territories as evidenced by intergroup conflicts. Total observed home range of 24 groups in coastal South Carolina ranged from 34 -225 ha (84 ac - 555.8 ac)(average 86.9 ha)(214.6 ac) (Hooper et al. 1982). For 2 groups on the SRS, minimum home range size was estimated at 15.8 ha (39.0 ac) and 16.0 ha (39.5 ac) in the summer, increasing to 33.6 ha (83.0 ac)and 65.8 ha (162.5 ac) in the winter (Skorupal 1979). Previous studies have found that sample size had an effect on home range results. It is likely that final home range estimates, based on a total annual cycle rather than the 7 months described herein, will be somewhat larger than our current estimates and far more accurate as to area needs of the species.

Expected improvements in signal tracking circuitry and signal processing technology will substantially increase the operating efficiency and accuracy in forest habitats (Jasumback and Luepke 1993). GPS was still being developed and the satellite constellation was incomplete. When the study was initiated, there were 16 usable satellites that provided only 13 hours 25 minutes of 3D (positions determined using 4

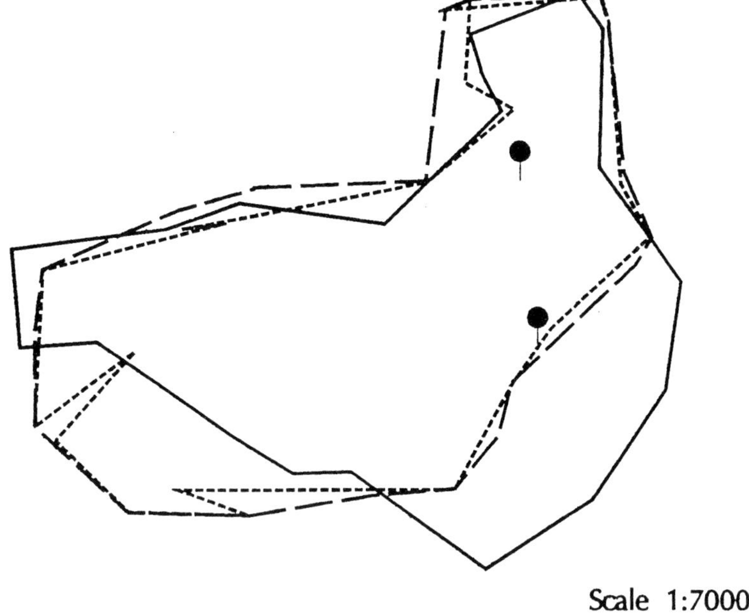

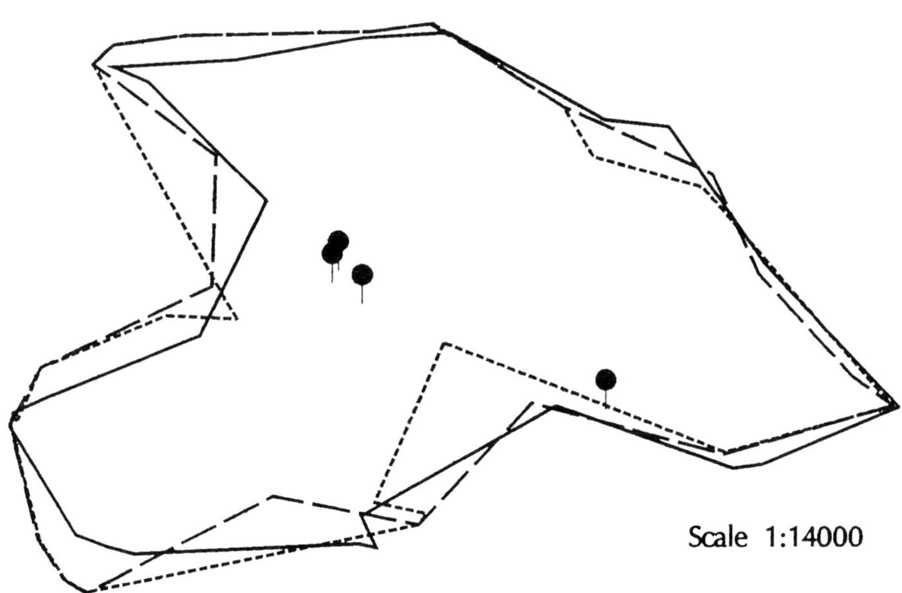

Home Range Determined Using Digitized Observations

Home Range Determined Using Equivalent GPS Observations

Home Range Determined Using All GPS Observations

RCW Cavities

Figure 1. Comparison of home range boundaries derived from traditional mapping method and GPS positions for red-cockaded woodpeckers (group 2 upper, group 16 lower) on the Savannah River Site, South Carolina.

Figure 2. Location of observations obtained from GPS and traditional mapping method positions for red-cockaded woodpeckers (group 2 upper, group 43 lower) on the Savannah River Site, South Carolina.

Table 1. Comparison of home range estimates using GPS and traditional mapping method (TMM) for red-cockaded woodpecker groups at the Savannah River Site.

	Home Range Size (ha) (ac)			
Group #	Traditional Mapping (method)	Equivalent GPS[1]	All GPS[2]	Sample size*
2	14.5 (35.8)	16.0 (39.5)	16.4 (40.5)	104/141
3	28.2 (69.7)	20.2 (49.9)	26.3 (65.0)	154/163
16	93.6 (231.2)	94.8 (234.2)	94.8 (234.2)	210/223
19	44.3 (109.4)	53.9 (133.1)	57.5 (142.0)	230/254
39	76.3 (188.5)	79.3 (195.9)	85.3 (210.7)	193/224
43	56.2 (138.8)	61.8 (152.6)	62.0 (153.1)	170/181
49	33.3 (82.3)	35.5 (87.7)	37.8 (93.4)	182/225
Mean (± SD)**	**49.7±27.2 (122.8±89.8)**	**51.6±29.6 (127.6±97.7)**	**54.3±29.3 (134.1±96.7)**	**1243/1411**

[1]Refers to GPS positions determined at the same location as the traditional mapping method positions.
[2]Using all GPS positions including those under the equivalent GPS column.
*A/B where A is the sample size for traditional mapping method and equivalent GPS data; B total GPS points.
**No significant difference in mean values between TMM vs equivalent GPS positions (t = 0.12 P>0.50).

Table 2. Comparison of distance between each traditional mapping method (TMM) position and its equivalent GPS observation.

		Difference in TMM vs GPS distance in meters (ft)		
Group #	Frequency	Mean Distance	Minimum Distance	Maximum Distance
2	105	94.8 (312.8)	7.2 (23.8)	395.5 (1305.2)
3	156	54.8 (180.8)	7.1 (23.4)	223.6 (737.9)
16	210	92.6 (305.6)	7.0 (23.1)	374.1 (1234.5)
19	230	100.4 (331.3)	1.3 (4.3)	818.6 (2701.4)
39	193	97.3 (321.1)	3.7 (12.2)	367.8 (1213.7)
43	170	79.3 (261.7)	7.6 (25.1)	306.7 (1012.1)
49	183	47.4 (156.4)	4.5 (14.9)	232.3 (766.6)
Weighted mean		**81.8 (269.9)**	**5.5 (18.2)**	**388.4 (1281.7)**

satellites) coverage per day with 11 breaks in usable time. This limited and inconsistent coverage made it difficult to schedule data collection. On December 30, 1992, 20 satellites were providing 19 hours 50 minutes of 3D coverage per day with only 5 breaks in usable time. As the number of satellites increases, GPS data collection should continue to become easier, eventually providing 24 hour 3D coverage.

No special permits were needed to purchase or operate GPS equipment. Various courses were offered in satellite navigation and in the use of GPS equipment. At least 40 hours of intensive hands-on study both in the field and at the computer were needed to become relatively proficient in data collection, storage, retrieval, and manipulation, including interfacing with GIS.

GPS offers many advantages. The system we used allowed us to create a "data dictionary" that logs features and attributes related to the positions. For instance, cavity trees may be identified as a "feature." Various attributes of the cavity tree such as species, dbh, cavity height, active/inactive condition, etc. can be logged and stored with the position information. Attribute values can be logged manually on the keypad or automatically with a user-created barcode chart (each feature, attribute, and attribute value is assigned a unique bar code). Thus, extensive data can be logged while in the field. When accuracy is a substantial consideration, GPS excels. Data can be interfaced with a GIS for extensive spatial analysis and map creation. In this study, the main advantage of GPS was that it provided a more accurate description of stand use.

Table 3. Red-cockaded woodpecker stand use differences between GPS and traditional mapping method (TMM) positions on the Savannah River Site during 1992.

Group	Total # of equivalent obs. in both GPS and TMM coverages	Total # of equivalent obs. in same stand	Percent equivalent obs. in same stand
2	104	81	77.9
3	154	137	89.0
16	210	171	81.4
19	230	156	67.8
39	193	158	81.7
43	170	129	75.9
49	182	157	86.3
Total	**1243**	**989**	--
Mean	**177.6**	**141.3**	**79.6**

The disadvantages of the GPS procedure include the cost. The device we used cost $10,000, although less expensive, less versatile models are available. In addition, extensive training was necessary to learn to operate the device and handle data to become reasonably proficient. Access to a base station was needed to provide the daily correction factors.

GPS technology offers enormous potential for field research on wildlife as well as endangered, threatened, and sensitive species of plants. In RCW research, it can aid in accurately describing locations of active and inactive cavity trees, improving estimates of home range, and accurately determining numerous spatial habitat relationships, including characteristics of preferred foraging habitat, effects of habitat modification, and size of home range in relation to population density and group size. All these can be achieved with TMM, but GPS provides greater accuracy. Although both approaches can be interfaced with GIS technology, GPS can be more time consuming and involves a substantial initial outlay of money. GPS is particularly well suited for relocating points such as previously known cavity trees. Using waypoints, the observer can relocate the location to an accuracy of 2 to 5 m (6.6 to 16.5 ft), an obvious advantage over TMM.

ACKNOWLEDGMENTS

This research was funded by the Department of Energy, Savannah River Site, and its cooperation is gratefully acknowledged. In addition, Patrick Jackson (DOE) and Savannah River Forest Station staff (especially John Irwin, John Blake, Glen Gaines, Rick Chubb, and Harry Park) provided support throughout the study. We are particularly grateful to Susan Steiner for field support. We thank Paul Roberts (Trimble Navigation Ltd.) for providing information on the use of the GPS equipment and Richard Conner for reviewing the manuscript.

APPENDIX

Specifications

The GPS Pathfinder Professional system consisted of a 6-channel receiver, lightweight antenna, and a MC-V datalogger. It operated off a lead acid battery. Measuring 7.0 cm wide x 21.9 cm in diameter x 15.9 cm high and weighing 1.25 kg (2.75 lbs), the receiver was linked to an antenna that was 15.4 cm in diameter x 8.9 cm high and weighed 0.25 kg (0.55 lb.). In this study we used a datalogger (MC-V-T-RD model) that was 10.3 cm wide x 5 cm in diameter x 24 cm high and weighed 0.74 kg (1.63 lbs). The logging memory was 1 M and could store more than 45,000 positions. Internal rechargeable nickel-cadmium batteries permitted the datalogger to operate for 6 to 8 hours per charge. The datalogger allowed entry and storage of complex, user-defined feature and attribute information related to a position. The system, as described, costs $14,950 for a private party or $10,500 under the U.S. Government (GSA) contract with Trimble Navigation Ltd. that was valid until approximately June 1993. Note that prices are subject to change.

Critical Settings

With the Pathfinder Professional various options were available for critical settings, non-critical settings and display options. Critical settings were those that affected the quality of the data collected. An observer could arrange for an audible beep each time a position fix was calculated and recorded to ensure that data were being collected. Using the dynamics code could improve satellite acquisition capabilities depending on whether data were collected from land, sea, or air. The elevation angle mask controlled the minimum elevation above the horizon that satellites could be used. Therefore, the rover elevation mask had to be higher than the base station elevation mask to ensure that the rover did not calculate positions using satellites that were not available to the base station. Logging mode designates whether the datalogger was being used as a rover or base station and whether velocity information was recorded. The dilution of precision (DOP) indicates the quality of the position based on the satellite constellation geometry. The position dilution of precision (PDOP) was generally a good indicator of accuracy, with low PDOP values being more accurate than high values. The PDOP mask controlled the maximum PDOP value that 3D positions (those determined based on 4 satellites) were recorded. The PDOP switch was the maximum PDOP that 3D positions would be determined before switching to the 2D (positions determined using 3 satellites and the altitude of the last 3D position or altitude input by user) mode automatically. Consequently the PDOP switch only applied when the logging mode was auto 2D/3D. The position fix mode controlled whether positions were determined in manual 2D always using 3 satellites, manual 3D always using 4 satellites or auto 2D/3D using 4 satellites when available then automatically switching to 3 satellites when 4 satellites were not available. In general, 2D positions were less accurate than 3D positions unless the exact altitude was manually entered for each 2D position. With the signal strength mask, the observer could set the minimum satellite signal strength used to determine positions. Weaker signals had a smaller signal to noise ratio, yielding less accurate positions. Up to 1 position per second could be obtained using the position logging interval that controlled the rate that position fixes were recorded. The power source allowed the datalogger to draw power from the receiver battery, thus extending the amount of time that data could be collected on a single battery charge. The raw measurement logging interval was only necessary when the datalogger was operating as a base station.

The critical settings for our data collection were beeper on, dynamics code 0 for logging on land, elevation angle mask = 15, logging mode = rover, PDOP mask = 8, PDOP switch = 8, position fix mode = manual 3D, position logging interval = 1, power source = external power, raw measurement logging interval = 10, and signal level mask = 6.

Field Time Required

Once the observer was at the location in the field and ready to obtain a GPS reading, the unit took about 1.5 to 2 minutes to collect sufficient information to determine which satellite constellation would provide the best position. Then an additional 4 to 5 minutes were needed, depending on satellite availability, to obtain the position fix. The unit could collect up to 1 position fix per second under good conditions and a minimum of 180 position fixes were necessary to obtain an accurate average. Once back in the office, it took about 10 minutes to download the approximate 20 positions obtained per day from the datalogger. Several hours were required to retrieve correction data, differentially correct positions, and format data for the GIS.

A Survey for Blood Parasites in Red-cockaded Woodpeckers

M. P. Luttrell, Southeastern Cooperative Wildlife Disease Study, College of Veterinary Medicine, The University of Georgia, Athens, Georgia 30602
P. W. Stangel[1], University of Georgia's Savannah River Ecology Laboratory, Aiken, South Carolina 29802
C. M. Bartlett, University College of Cape Breton, Sydney, Nova Scotia, Canada B1P 6L2.

ABSTRACT: Two types of unidentified microfilariae were found in RCWs (*Picoides borealis*) examined in Texas, Louisiana, Mississippi, and Alabama in 1988. The first type averaged 58.5 µm in length and had a markedly tapered tail that ended bluntly. The second type averaged 78.9 µm and had a lightly tapered tail that ended bluntly. This represents the first report of blood parasites in this endangered species.
KEYWORDS: *Picoides borealis*, red-cockaded woodpecker, survey, microfilariae, hematozoa.
[1]Current address: National Fish and Wildlife Foundation, 1120 Connecticut Avenue, Suite 900, Washington, D.C. 20036.

The RCW is a federally endangered species (U.S. Fish and Wildlife Service, 1970) endemic to the mature pine forests of the southern United States (U.S. Fish and Wildlife Service, 1985). The RCWs distribution has become contracted and fragmented (Jackson 1971; Jackson 1978a; Lennartz et al. 1983a). Many populations are small, isolated, and declining (Costa and Escano 1989).

The U.S. Forest Service has initiated a program for translocating RCWs from larger to smaller populations to help overcome the demographic, genetic, and environmental concerns associated with small population size and insularization (DeFazio et al. 1987, Allen et al. 1993). One potential consequence of moving birds among populations is transmission of disease or parasites. In regard to these concerns, a survey of RCWs was conducted to gather baseline data on blood parasite diversity and abundance.

In 1988, blood from 72 RCWs from 8 populations in Texas, Louisiana, Mississippi, and Alabama (Stangel et al. 1992) were collected for genetic and parasite analyses. Adult birds were captured by placing nets over roost cavity entrances, and 250 µl of blood was taken from the brachial vein (Stangel and Lennartz, 1988). Two blood smears were made from each bird on site, air dried, and shipped to the Southeastern Cooperative Wildlife Disease Study (SCWDS), College of Veterinary Medicine, The University of Georgia, Athens, Georgia 30602, USA. Blood smears were fixed with 100% methanol and stained with Giemsa stain. Initially, smears were scanned at low magnification (100x) of a brightfield microscope over the entire field to search for microfilariae. A second examination was made (1000x) using oil immersion for approximately 10 minutes to look for protozoan parasites.

Unidentified microfilariae were found in 24 of 72 birds (33.3%), representing five of eight populations sampled (**Table 1**). The number of microfilariae detected per bird sampled was few, ranging from a maximum of 2 per positive bird from Louisiana and Alabama to 13 per positive bird from Mississippi. No other blood parasites were observed.

Due to the method of smear preparation,the microfilariae were greatly contracted, and precise identification was not feasible. It was not possible to determine whether sheaths were present. Nevertheless, two microfilarial types

Table 1. Microfilariae in 72 red-cockaded woodpeckers from National Forests in Texas, Louisiana, Mississippi, and Alabama, 1988.

Location	RCWs Sampled	RCWs with Microfilariae	Type of Microfilariae[1]
TEXAS			
Sabine N.F.	6	0	-
Sam Houston N.F.	9	0	-
LOUISIANA			
Kisatchie N.F.			
Vernon District	10	3	B
Kisatchie District	9	3	B
Evangeline District	6	0	-
MISSISSIPPI			
Bienville N.F.	19	11	A,B
DeSoto N.F.	5	4	B
ALABAMA			
Talladega N.F.	8	3	B
Total	72	24	

[1]See text for description of types

were distinguished. Type A averaged 58.5 µm in length with a maximum width of 3 µm (n = 4). Its anterior end was bluntly rounded, and the posterior end tapered markedly to a blunt extremity. Type B averaged 78.9 µm with a maximum width of 3 µm (n = 7). Its anterior end was bluntly rounded, and the posterior end tapered slightly to a blunt extremity.

Greiner et al. (1975) noted that microfilariae had been reported in 6 of 10 species of woodpeckers in North America. These species included the red-bellied woodpecker (*Melanerpes carolinus*), northern flicker (*Colaptes auratus*), downy woodpecker (*P. pubescens*), hairy woodpecker (*P. villosus*), red-headed woodpecker (*M. erythrocephalus*), and yellow-bellied sapsucker (*Sphyrapicus varius*). Of the 42.4% of woodpeckers infected with blood parasites, only 0.3% exhibited microfilariae. Three RCWs examined by Love et al. (1953) were negative for hematozoa as well as microfilariae.

The present paper is the first report of blood parasites in RCWs. Although sample sizes in this study were small, our results suggest there is variation among host populations with respect to the presence and identity of microfilariae.

The significance of filaroid nematode infections in these birds is unknown. Necropsies of infected birds are necessary to identify species of filaroid nematodes present, but the endangered status of the RCW makes surveys for this purpose impractical. However, examination of birds lost to natural mortality could provide additional information on the identification of these parasites and their effect on RCW health. Also, populations designated as donor or recipient sites for translocations could be monitored to further document presence of blood parasites in these birds.

ACKNOWLEDGEMENTS

Out thanks to the many U.S. Forest Service biologists, particularly Emlyn Smith, that assisted with location and capture of RCWs. Research and manuscript preparation were supported by Contract DE-AC09-76SROO-819 between the U.S. Department of Energy and the University of Georgia's Savannah River Ecology Laboratory.

SECTION 5

Cavity Trees as a Resource

Red-cockaded Woodpecker Cavity Trees: An Introduction

Richard N. Conner
Wildlife Habitat and Silviculture Laboratory[1]
Southern Forest Experiment Station, USDA Forest Service
Nacogdoches, Texas 75962

Red-cockaded woodpeckers (*Picoides borealis*) are unique among North American Picidae because they excavate their nest and roost cavities almost exclusively within living pines (*Pinus* spp.)(Steirly 1957a, Hooper 1982). Dead snags, the cavity substrate typically used by most other North American woodpecker species, tend to be rare in the southern pine fire disclimax ecosystem because of frequent fires (Jackson et al. 1986). Red-cockaded woodpecker use of living pines for cavity substrate necessitated adaptations to deal with the pine gum (resin) that flows from wounds made during excavation of living sapwood. Natural selection favored pairs that excavated resin wells, small wounds in the pine's bole, around the cavity entrance. Sticky resin oozed from these wounds, coated the pine's bole creating a barrier that provided protection against climbing rat snakes (*Elaphe* spp.), and enhanced survival of protected nest cavities (Jackson 1974, Rudolph et al. 1990b). Cavities are excavated in the heartwood of pines and have entrances that orient slightly downward **(Figure 1)**.

Red-cockaded woodpeckers prefer old pines for their cavity trees (Jackson and Jackson 1986, Conner and O'Halloran 1987, DeLotelle and Epting 1988, Hooper 1988, Rudolph and Conner 1991). Cavity trees exceeding 360 years-old have been reported (Hedrick 1992). Pines of sufficient diameter are required to assure that enough heartwood is present to contain the cavity chamber (Clark 1993). Additional years beyond the time when sufficient heartwood has formed are necessary for pines to become infected by red heart fungus (*Phellinus pini*). Red heart decays the central heartwood of living pines and it has been frequently assumed that decay reduces the time required for cavity excavation (Jackson 1977c, Conner and Locke 1982, Hooper 1988, Hooper et al. 1991b). The presence and extent of decay are positively correlated to the age of pine trees, an additional reason why old pines are needed for cavity trees. The adaptation of red-cockaded woodpeckers to the southern pine ecosystem has had a significant positive impact on faunal diversity of the forest. Red-cockaded woodpeckers are a keystone species within the ecosystem. They create cavities in a relatively cavity-barren environment, cavities used by numerous other cavity-using vertebrates and invertebrates (Rudolph et al. 1990a). Red-cockaded woodpeckers are the cavity "pathfinders" for species using cavities in living southern pines. Cavities initially made by red-cockaded woodpeckers are often enlarged by pileated woodpeckers (*Dryocopus pileatus*). These enlarged cavities provide nests and roosts for many larger secondary cavity nesters **(Figure 2).** Thus, population declines or the loss of red-cockaded woodpeckers would impact the diversity of the cavity nesting guild within the southern pine ecosystem.

The evolutionary role of cavities as a limited resource for red-cockaded woodpeckers has only recently been understood. As a critical resource that requires extended time periods for construction, cavity availability has likely had a significant role in the evolution of cooperative breeding within red-cockaded woodpeckers (Lennartz et al. 1987, Walters 1990, Walters et al.1988, 1992a). The induction of new woodpecker group formation by providing unoccupied artificial cavities in new sites and abandoned clusters provides strong evidence of the limiting nature of this valuable resource for the woodpecker (Doerr et al. 1989, Copeyon et al. 1991).

Figure 1. Vertical section of a red-cockaded woodpecker cavity showing the cavity chamber in the heartwood and entrance tube through the sapwood.

Figure 2. Enlarged red-cockaded woodpecker cavities are often used by secondary cavity nesters for nesting and roosting sites.

Demand for cavities in southern pine forests is high, as is the potential for competition. Southern flying squirrels (*Glaucomys volans*) are major users of red-cockaded woodpecker cavities and likely compete with them for the best unenlarged cavities (Rudolph et al. 1990a, Loeb 1995). Research on interactions between red-cockaded woodpeckers and southern flying squirrels is essential to improve our management of woodpecker habitat. Anecdotal observation strongly suggests that flying squirrels occasionally prey on young woodpeckers and eggs. Usurpation of cavities by flying squirrels when woodpeckers are absent appears to occur fairly often. Studies investigating flying squirrel excluder devices (SQEDs) (Montegue et al. 1995) are very important to reduce cavity losses to this competitor, particularly in areas where woodpecker populations are small and flying squirrel populations are high. Equally important are studies developing **nonlethal** devices to prevent rat snake predation of eggs and young. Nonlethal snake excluder devices (SNEDs) (Withgott et al. 1995) may be particularly important for cavity trees that have reduced resin producing capacity. Pines selected by biologists for artificial cavity installation and perhaps some senescent pines may not have the resin producing capability of pines selected by woodpeckers.

Losses of red-cockaded woodpecker cavities have a significant negative impact on the woodpecker--both dramatic landscape disturbances (e.g., hurricanes) and smaller scale losses to bark beetles can deplete cavities faster than woodpeckers can excavate new replacements (Hooper et al. 1990, Conner et al. 1991a). Researchers developing new technology to create artificial cavities play a crucial role in attempts to recover the endangered woodpecker. The original idea of inserting a "box" containing an actual woodpecker cavity into a pine tree (Odom et al. 1982) laid the ground work for perfection of a technique using artificial wooden inserts (Allen 1991). Development of drilling techniques (Copeyon 1990) and subsequent modifications (Taylor and Hooper 1991) have perhaps lead to the best ways to artificially create new woodpecker cavities.

The papers presented within this section provide additional exploration and information on the unique characteristics of red-cockaded woodpeckers and some of the necessary management tools for the recovery of the species. The papers expand our knowledge of woodpecker habitat needs and cavity trees, interactions with competitors and predators, and new techniques to aid our management of this endangered woodpecker.

[1]Maintained in cooperation with the College of Forestry, Stephen F. Austin State University, Nacogdoches, Texas 75962.

Red-cockaded Woodpecker Detection of Red Heart Infection

D. Craig Rudolph, Richard N. Conner and Richard R. Schaefer
Wildlife Habitat and Silviculture Laboratory[1]
Southern Research Station, Nacogdoches, Texas 75962

ABSTRACT: Cavity excavation by red-cockaded woodpeckers (*Picoides borealis*) requires substantial time and energy. The presence of decay due to red heart fungus (*Phellinus pini*) allows for easier cavity excavation. Earlier studies support the hypothesis that red-cockaded woodpeckers select trees with a high probability of decay. We analyzed the distribution of woodpecker cavities in relation to the incidence of fungal infection. The results support the above hypothesis, and the additional hypothesis that red-cockaded woodpeckers can preferentially locate their cavities within tree boles to take advantage of more easily excavated decayed heartwood. The cues used by red-cockaded woodpeckers to detect fungal infection are unknown. We tested the hypothesis and rejected that they use the presence of fungal conks as a visual cue.
KEYWORDS: *Picoides borealis,* red-cockaded woodpecker, *Phellinus pini,* cavity excavation, fungal decay

[1]Maintained in cooperation with the College of Forestry, Stephen F. Austin State University, Nacogdoches, Texas, 75962.

Red-cockaded woodpeckers (*Picoides borealis*) (RCW) are unique among woodpeckers in the use of living pines for cavity excavation (Ligon 1970). Cavity excavation requires a large investment in time and energy; mean excavation times in Texas ranged from 1.8-6.3 yr depending on pine species (Conner and Rudolph 1995a). The selection of older trees for cavity excavation sites (DeLotelle and Epting 1988, Hooper 1988, Rudolph and Conner 1991) provides several benefits presumably related to ease of excavation and cavity quality (Jackson and Jackson 1986, Rudolph and Conner 1991).

Red heart fungus decays the heartwood of living pines presumably facilitating the excavation of cavities (Steirly 1957a). Other studies (Jackson 1977c, Conner and Locke 1982, Hooper 1988, Hooper et al. 1991b) have supported the hypothesis that RCWs select for trees with red heart infection. Past studies were not sufficient to determine if RCWs were selecting cavity trees based on direct detection of red heart infection, or other criteria (age, diameter, bark thickness, bole length, stress) that were correlated with red heart infection (Jackson 1977c, Conner and Locke 1982, Hooper 1988, Hooper et al 1991b).

This study examined the distribution of decay in relation to the location of RCW cavities and starts. We tested the hypothesis that RCWs were able to detect decay and thus locate their excavations to take advantage of the more easily excavated decayed tissue. We also tested the hypothesis that RCW used the presence of fungal fruiting structures (conks) to select potential cavity trees.

STUDY AREA AND METHODS

We studied longleaf pine (*Pinus palustris*) cavity trees located on the Angelina National Forest (ANF) in eastern Texas. The 12 RCW groups in longleaf pine habitat excavated cavities almost exclusively in relict pines that were significantly older than the stands they were found in (Rudolph and Conner 1991). For a more complete discussion of the study area see Conner and Rudolph (1989) and Rudolph and Conner (1991).

We selected cavity tree clusters on the basis of ease of access. The majority of cavity trees in each cluster were included; a few trees with difficult access were omitted. We extracted 5 mm increment cores from each tree at breast height (1.5 m), 6 m, 9 m, and 12 m. All cores were visually examined for signs of fungal decay. Fungal infection results in soft, friable and discolored xylem tissue that is clearly visible if decay is advanced (Jackson 1977c, Conner and Locke 1982). Incidence of decay comparisons were based on the G-test (Sokal and Rohlf 1969:591).

Diameter at breast height (DBH) and age were determined for each cavity tree. Age was obtained by counting the number of annual growth increments and adding 5 years to allow for growth to breast height. If the extracted core missed the piths, the core was overlayed on a drawing of concentric circles to estimate the number of missed rings. Decay precluded determining age in several instances. Height to the nearest live branch (bole length) was determined. Cores were also extracted from a point 5 cm above each cavity entrance or cavity start and examined for decay and cavity height was recorded. When cavities or starts were located within the crown of the tree, the number of live branches below the excavation was recorded.

A non-cavity tree was selected from the vicinity of each cavity tree on the basis of similar DBH. Increment cores were extracted from 1.5 m, 6 m, 9 m, and 12 m. Each core was evaluated for decay. Age, DBH and bole length were determined for non-cavity trees.

The experimental evaluation of the use of red heart conks as cues in selection of potential cavity trees was conducted on the Raven District of the Sam Houston National Forest (SHNF). Ten cavity tree clusters, each with a minimum of 6 existing cavity trees, were selected for experimental manipulation. Clusters selected had received midstory treatment within the previous 2 years. Each cluster selected had a minimum of 3 adult RCWs. Existing cavity trees were examined in December 1989.

Eight potential cavity trees lacking existing cavities or cavity starts, were selected in each cluster using the characteristics of cavity trees described by Conner and O'Halloran (1987) as criteria. Selection favored trees that were older, larger, and had larger crown volumes than non-selected trees. Age and DBH were determined for each tree. Four trees in each cluster were randomly designated as experimental and four as control trees. A red heart fungus conk was nailed to each experimental tree. Conks were placed at a height of 10 m on the southwest (225 deg.) side of the bole, similar to typical cavity orientation (Locke and Conner 1983).

Clusters were monitored at 3-6 month intervals between December 1989 and January 1992. During each monitoring experimental and control trees were carefully examined for signs of RCW cavity initiation. All cavity trees in each cluster were examined for signs of cavity initiation in January, 1992.

RESULTS

Cavity trees were significantly different from non-cavity trees in the incidence of decay detected in cores taken at specified heights **(Table 1)**. A total of 32 (60.4%) cavity trees had decay compared to 7 (13.2%) non-cavity trees. Cavity trees were not significantly different from non-cavity trees in DBH, but cavity trees were significantly older than non-cavity trees **(Table 1)**.

Table 1. Comparison of red-cockaded woodpecker cavity trees and non-cavity trees.

	Cavity trees	Non-cavity trees
Age(yrs)[a]	115.1	96.0
DBH(cm)[b]	51.1	48.4
# Decayed (%)[c] @ 4 hts	32 (60.4)	7 (13.2)

[a]Mean ages were significantly different ($t=2.28$, $P<0.05$).
[b]Mean DBH's were not significantly different ($t=0.97$, $P>0.05$).
[c]Incidence of decay was significantly different ($G=24.66$, $P<0.01$).

The frequency of decay detected in cores from specified heights varied with height in cavity trees **(Table 2)**. The lowest incidence of decay was at the 1.5 m height, and the highest incidence was at the 9 m height. The incidence of decay in non-cavity trees was uniformly low.

Decay was detected in cores at 69.8% (37 of 53) of the cavities and 53.5% (46 of 86) of the cavity starts. Decay was detected in 39.0% (62 of 159) of the cores taken at specified heights from 6-12 m in cavity trees. The incidence of decay was significantly greater at the cavities and cavity starts compared to the incidence detected in the cores taken a specified heights within cavity trees ($G = 7.60$, $P < 0.01$).

Table 2. Incidence of decay in red-cockaded woodpecker cavity trees (n=53) detected in cores from 4 heights.

Core ht(m)	1.5	6	9	12
#(%) decayed	14 (26.4)	19 (35.8)	24 (45.3)	19 (35.8)

The mean height of cavities was 9.4 m and the mean height of starts was 9.6 m. The average height of cavities with decay (9.1 m) was lower than the mean height of cavities without decay (10.1 m). The frequency of decay detected in cores from cavities was 69.8% (37 of 53). Cavity starts with decay averaged higher (9.9 m) than starts without decay (9.2 m). The frequency of decay detected in cores from cavity starts was 53.5% (46 of 86). The incidence of decay at cavities above and below mean cavity height (9.4 m) was not significantly different. However, the incidence of decay at the lowest 20% of cavities (< 6.5 m) was significantly greater (G = 6.84, P < 0.01) than at cavities above 6.5 m (**Table 3**). The pattern was reversed for cavity starts (**Table 3**). The incidence of decay was significantly greater at starts above the mean height of 9.6 m than at starts below 9.6 m (G = 6.32, P < 0.05).

Sixteen of the 138 cavities and starts were located above the lowest live branches (**Table 4**). The incidence of decay at cavities and starts located in the crown (i.e. above the lowest live branches) was 62.5% (10 of 16). The incidence of decay at cavities and starts below the crown was 55.7% (68 of 122). The incidence of decay at cavities and starts located above and below the crown was not significantly different (G = 0.08, P > 0.05).

Twenty new excavations were initiated in 10 cavity tree clusters used to evaluate red heart conks as visual cues for RCW selection of potential cavity trees. Nine new excavations were initiated in 8 existing cavity trees. At the time of the final monitoring these excavations consisted of 2 completed cavities and 7 cavity starts. Excavations were also initiated in 11 trees that had no existing excavations at the initiation of the experiment. At the time of the final monitoring these excavations consisted of 6 completed cavities and 5 cavity starts.

None of the initiated excavations were in the 40 experimental trees with artificially attached conks. One excavation was initiated in a control tree and consisted of an active cavity start at the conclusion of the experiment. There was no significant difference between the rate of cavity initiation in the experimental and control trees.

Table 3. Comparison of incidence of decay at red-cockaded woodpecker cavities[a] (n=53) and cavity starts[b] (n=86) above and below specific heights.

	Height			Height	
	<6.5 (m)	>6.5 (m)		<9.6 (m)	>9.6 (m)
Cavities with decay	10	27	Starts with decay	20	26
Cavities without decay	0	16	Starts without decay	29	11

[a]Significant difference with height based on G-test with Yates' correction (G=6.84, P<0.01).
[b]Significant difference with height based on G-test with Yates correction (G=6.32, P<0.05).

Table 4. Comparison of incidence of decay at red-cockaded woodpecker cavities and cavity starts located above and below the base of tree crown.[a]

	Above base of crown	Below base of crown
With decay	10	68
Without decay	6	54

[a]No significant difference based on G-test with Yates correction (G=0.08, P>0.05).

DISCUSSION

Cavity trees and non-cavity trees differed significantly in the incidence of decay. Because the entire heartwood column was not sampled, the actual incidence of decay was presumably somewhat greater in both tree samples. This difference suggests that RCW were able to select trees that have a high probability of containing decay. However, the non-cavity trees were not true control trees precluding a definitive conclusion. The non-cavity trees were significantly younger than the cavity trees, and were from stands in which the woodpeckers had already selected numerous cavity trees.

Older pines on the ANF were relicts that survived the harvesting during the early decades of this century. Due to the rarity of relicts and the relatively high proportion that have been selected as cavity trees by RCWs it would be difficult to select a sample to serve as a true control. This is a general problem that applies to data sets of this type. However, the only other comparable data (Hooper 1988, Hooper et al. 1991b) are less seriously compromised. In these 2 studies, cavity trees and adjacent non-cavity trees were not statistically different for a number of characteristics, including age. This was due to the greater abundance of older trees on those study sites. The prior selection of cavity trees by the woodpeckers had occurred, however. The results reported by Hooper (1988) and Hooper et al. (1991b) indicate that cavity trees had a very significant and much higher incidence of decay than the non-cavity trees (86.4 vs. 9.1%). These results were consistent with the hypothesis that RCWs are able to select trees that have a high probability of containing decay.

The greater incidence of decay at mid-bole were consistent with the biology of red heart fungus. Red heart fungus initiates infection of pines primarily via broken branches of sufficient diameter to expose heartwood (Conner and Locke 1982, 1983). Branches that were once present low on the bole were of insufficient diameter to allow efficient infection. Higher on the bole, branches were of sufficient size but were lost relatively recently, and spread of the decay column had not had sufficient time to become widespread (Conner and Locke 1982). Within the crown, decay was relatively rare since most branches were still intact. Pines typically do not experience significant red heart infection until advanced ages due to the mode of infection, slow growth of the fungus, and presumably the uncertainties of spore dispersal.

The significantly higher incidence of decay associated with cavities at lower heights supports the hypothesis that RCWs can detect the presence of fungal decay. We assume that RCWs would prefer to excavate cavities at greater heights to reduce predation and risk of fire damage. Consequently, the birds might require stronger indications of decay before initiating excavations at low heights. In the absence of available sites with decay the birds would select sites at greater heights. We would predict that the incidence of decay associated with cavities would be greater at lower heights as reported above.

The opposite relationship that we found in the case of cavity starts is also consistent with this scenario. RCWs could be less inclined to complete excavations at lower heights with weaker indications of decay. This assumes, of course, that the ability of the birds to detect decay is not perfect, and that cavities initiated in the absence of decay remain as starts for a significantly longer period of time or are ultimately abandoned due to the difficulty of excavation.

There are obvious limits to the height at which RCWs can excavate cavities. An absolute limit is that a sufficient diameter of heartwood is required to enclose the cavity chamber (Jackson and Jackson 1986), and the diameter of the heartwood decreases with height. The decreasing incidence of decay in the upper portions of the bole, and especially within the crown, also reduces the suitability of these regions.

In current RCW populations a majority of the cavities are excavated in the bole rather than in the trunk within the crown (88% in this study). It has been hypothesized that the birds prefer the more open flight paths present below the crowns (Wood 1983a). The greater incidence of decay and larger heartwood diameters in this region of the bole are also obvious influences. If this were the case, we would predict that cavities and cavity starts located within the crown would have a higher incidence of fungal decay by reasoning analogous to the above consideration of cavities at low heights on the bole. In the present study there was no statistically significant difference between the incidence of decay associated with cavities and starts in the bole compared with those within the crown, although the trend was in the predicted direction. The small sample size of excavations within the crown may have precluded detection of a significant difference.

If, however, we accept the lack of a significant difference between the incidence of decay associated with cavities within the crown and those on the bole, the pattern suggests lack of a strong preference for locating cavities on the bole. The current preponderance of cavities located on boles in most RCW populations may be due to lack of sufficient heartwood and decay within the crown of the relatively young trees available. Observations of cavities excavated by a groups of woodpeckers with access to 250+ year old trees suggests that in these circumstances a higher proportion of the cavities are located within the crown (Conner et al. 1991a, R. T. Engstrom Pers. Comm.).

It is not known what cues RCWs use to locate decay. Two general classes of hypothesis are consistent with current data. They could be selecting trees based on characteristics correlated with fungal infection. Age and correlated characteristics, and high boles indicative of extensive branch prunning are obvious candidates. Available data do not address these questions. Alternatively, they could detect the presence of decay directly. This ability would not only allow them to select trees with a higher incidence of decay, but also to select segments of the tree with a higher probability of decay. External manifestations of fungal infection are one class of potential cues. It is also possible that the birds are able to detect decay by percussing the surface and detecting differences in resonance (Conner et al. 1976).

Pines with advanced red heart infection typically have one or more conks. RCW cavities and cavity starts are frequently located in close proximity to conks. This observation suggested that conks might be used by RCWs to locate decay. Our test rejected this hypothesis.

Data now available (Hooper 1988, Hooper et al. 1991b, this study) support the hypothesis that RCWs are able to select pines that contain fungal decay, and to preferentially locate their cavities within decay columns. This ability allows them to take advantage of the presumably more easily excavated decayed heartwood. The woodpeckers apparently do not use the presence of red heart conks, the most obvious external sign of infection, as a visual cue to locate decay. The possibility remains that they detect hidden decay by percussing the tree and perceiving differences in resonance.

ACKNOWLEDGMENTS

We thank R.G. Hooper for valuable comments on an earlier draft of the manuscript that resulted in substantial improvements.

Excavation Dynamics and Use Patterns of Red-cockaded Woodpecker Cavities: Relationships with Cooperative Breeding

Richard N. Conner and D. Craig Rudolph
Wildlife Habitat and Silviculture Laboratory[1]
Southern Research Station, USDA Forest Service
Nacogdoches, Texas 75962

ABSTRACT: In the fire-climax southern pine ecosystem typical woodpecker cavity sites found in dead trees and snags were likely rare as a result of frequent hot fires. Red-cockaded woodpeckers (*Picoides borealis*) adapted to this ecosystem by using living pines rather than sites in dead wood for cavity excavation. Although old pines ideal for cavities were abundant, cavity excavation time was protracted. Previous studies hypothesized that the extended time required to excavate cavities was a major factor in the evolution of the red-cockaded woodpecker breeding system. We examined the dynamics of cavity excavation and use, and the time required by red-cockaded woodpeckers to excavate cavities in longleaf (*Pinus palustris*), loblolly (*P. taeda*), and shortleaf (*P. echinata*) pines. Cavity excavation time appears to depend on the amount of resin produced by the pine, the presence of red heart fungus (*Phellinus pini*), and the current need for cavities. The average times used to excavate cavities were: 6.3 y in longleaf, 2.4 y in shortleaf, and 1.8 y in loblolly pine. Longleaf pine cavity trees survived longer and were used significantly longer than were loblolly and shortleaf pines. We examine models for the evolution of cooperative breeding relative to cavity excavation dynamics.
KEYWORDS: *Picoides borealis*, cavity excavation, fungi, oleoresins, woodpeckers
[1]In cooperation with the College of Forestry, Stephen F. Austin State Univ., Nacogdoches, Texas 75962.

Red-cockaded woodpeckers (*Picoides borealis*) are unique among woodpeckers in that they nest and roost in cavities they excavate in living pines (Steirly 1957a, Short 1982, Ligon et al. 1986). Their essentially exclusive use of living pines for cavities may be an adaptation to the fire climax pine ecosystem of the South (Ligon 1970, Jackson et al. 1986, Jackson 1989) and possibly avoidance of rat snakes through daily excavation of resin wells (Jackson 1974, 1978d, Rudolph et al. 1990b). Frequent, intense fire, historically characteristic of the southern pine ecosystem, likely eliminated most hardwood trees and snags which are the normal cavity and foraging sites for most other species of southern woodpeckers (Ligon 1970, Conner et al. 1975; Conner 1976, 1980; Kilham 1983). As a result, the main substrate for cavity sites in southern pine ecosystems was living pine trees. Adaptation to excavating cavities in living pines has required the woodpeckers to select old pines (Jackson and Jackson 1986, Conner and O'Halloran 1987, DeLotelle and Epting 1988, Rudolph and Conner 1991), excavate through 10 to 16 cm of live sapwood, cope with pine resin that seeps from the cavity entrance during excavation, and excavate a cavity in sound or decayed heartwood (Jackson 1977c, Hooper et al. 1980, Conner and Locke 1982, Conner and O'Halloran 1987, Hooper 1988, Hooper et al. 1991b). The woodpeckers often excavate cavities in pines in which the heartwood is decayed by red heart fungus (*Phellinus pini*) (Steirly 1957a, Jackson 1974, Conner and Locke 1982). It is well documented that in the absence of infection, red-cockaded woodpeckers can excavate cavities into undecayed heartwood (Beckett 1971, Conner and Locke 1982, Hooper 1988, Hooper et al. 1991b), but it has been inferred that additional time and energy is required to do so.

Red-cockaded woodpeckers are cooperative breeders, where young adults (primarily males) delay breeding, remain with their parents, and aid in territory defense and provisioning of siblings in their parents' nest tree (Ligon 1970, Lennartz and Harlow 1979, Lennartz et al. 1987, Walters et al. 1988, Walters 1990). An ecological constraint model suggests that individuals might delay breeding if the likelihood of breeding success is low because of a limiting habitat resource (Selander 1964, Emlen 1982a). Under such a general model, ecological limitations constrain young individuals from dispersing to form new breeding units. Koenig and Pitelka (1981) proposed a marginal-habitat model where a species cannot survive or reproduce in unsuitable habitat and marginal habitat that is unsuitable for reproduction is minimally available for use between suitable and unsuitable habitat.

Stacey and Ligon (1987, 1991) suggest that cooperative breeding usually evolved based on a benefit from philopatry, often species-specific, rather than solely ecological constraints. They suggest that all species face ecological constraints, and many use habitat that is saturated with breeders and have a nonbreeding subpopulation (floaters) that are free to disperse and breed. Yet, the evolution of cooperative breeding is relatively rare and observed in only 3% of 9,000 bird species (Stacey and Ligon 1987). They suggest that it is the benefits of an already completed cavity that favors natal philopatry and cooperative breeding in red-cockaded woodpeckers.

The evolution of cooperative breeding may be modeled as selection between two life history tactics,

(1) staying on the natal territory to await a breeding vacancy in the vicinity, or

(2) departing after fledging to search for a breeding vacancy. Staying may be selected if territory locations are stable and competition for vacancies is especially intense, due usually to unusual variation in territory quality. Under such conditions birds compete for high quality territories rather than accept low quality ones.

A limited supply of cavities for nesting and roosting as a result of the difficulty (time and effort) of excavating them appears to be associated with the evolution of the cooperative breeding behavior exhibited by red-cockaded woodpeckers (Walters et al. 1988, 1992a, 1992b). Walters (1990) and Walters et al. (1988, 1992a) proposed and explored a critical-resource model for the evolution of cooperative breeding in red-cockaded woodpeckers. They proposed that territories with existing cavities are sufficiently superior to those without cavities that individuals compete only for the former, creating conditions that favor the staying and helping strategy. The value of cavities appears to be closely related to the woodpecker's adaptation to the fire-climax pine ecosystem and use of living pines for cavity trees, resulting from the ecosystem's relative paucity of easily excavated cavity sites in well decayed hardwoods and snags. However, mature pines that were potential cavity trees were likely abundant in the old-growth fire climax pine ecosystem and should not have been a limited resource.

Characteristics of living pines (extensive living sapwood and active transport of pine gum to wound sites) that make them more difficult to excavate than decayed dead snags or hardwoods are a plausible reason for a high value of existing cavities. Lennartz et al. (1987) and Walters et al. (1992b) hypothesized that the extended time required to make cavities may be what makes existing cavities so valuable. Jackson et al. (1979b) suggested that cavity excavation typically takes many years. We examined the dynamics and time requirements of cavity excavation in 3 species of pines, the relative value of each pine species to red-cockaded woodpeckers, and use patterns after cavity completion. We also examined population dynamics of cavity trees to gain insight into cavity excavation rates required to maintain an adequate supply of cavities.

METHODS AND STUDY AREAS

Annual visits were made to red-cockaded woodpecker cavity tree clusters (colonies) on the Angelina National Forest during March and April from 1978 to 1992. Between 1978 and 1982 only selected clusters were visited. Starting in 1983 and continuing through 1992 all cavity trees were visited and checked in detail for status, condition of cavities and existing cavity starts and amount of activity at resin wells, and the cluster area was searched for new cavity starts. A cavity start was considered new if it was < 3 cm deep and showed signs of recent excavation. Cavity trees are primarily longleaf pine (*Pinus palustris*) on the southern portion of the Angelina National Forest, whereas loblolly (*P. taeda*) and shortleaf (*P. echinata*) pines are the only pine species available for use on the portion of the forest north of Lake Sam Rayburn. Cavity tree mortality and cause were determined (see Conner et al. 1991a), cavity enlargement by Pileated woodpeckers (*Dryocopus pileatus*) was recorded, and progress on cavity starts noted. Field records provided by Dan Lay (retired Texas Parks and Wildlife Department Biologist, Nacogdoches, Texas) and published information (Lay and Russell 1970) provided a history of cavity trees in some clusters for as far back as the 1960s. National Forest personnel (Alfredo Sanchez, District Biologist, Angelina National Forest) installed and provided records on cavity inserts and restrictors on the Angelina National Forest for 1991 and 1992.

Between 1987 and 1992 regular visits were made to all active and most inactive cavity tree clusters on the Davy Crockett National Forest where similar data were obtained on the status and condition of shortleaf and loblolly pine cavity trees. Cavity starts that were completed or remained active during the course of our study on both national forests were climbed and cored at cavity height with an increment borer to determine if heartwood decaying fungi were present. Some cavity trees were increment bored up to 5 years after cavity completion, thus in some cases, whether fungi were present or absent prior to cavity excavation was not certain.

Using the null hypothesis that there were no differences in excavation times and woodpecker use among species of pines used for cavity trees, we used analyses of variances (GLM model) and Duncan's Multiple Range Test to evaluate possible differences among tree species (SAS 1988).

STAGES OF RED-COCKADED WOODPECKER CAVITY EXCAVATION

Red-cockaded woodpecker cavity excavation can be divided into 4 separate stages (**Figure 1**). The first stage was the woodpecker's initial wound to the pine which typically resulted in a round 4-6 cm diameter wound that penetrated into xylem tissue 2-6 cm (**Figure 1a**). Excavation of cavity starts during this stage typically produced copious oleoresin (pine gum) flow that resulted in the formation of an icicle-like flow of crystallized resin from the wound to 20-50 cm down the bole. Resin is produced by pines and serves as a first line of defense to "pitch out" infesting bark beetles, or to prevent growth of fungi at wounds. Subjective observation indicated that longleaf pine usually had longer "icicles" of crystallized resin than loblolly or shortleaf pines, but we obtained no direct measure to compare icicles statistically. Concurrent with icicle formation on the bole of the pine was resinosis of the sapwood around the wound caused by excavation (**Figure 1a**). Resinosis occurred when wood tissue became saturated with pine resin. Initial excavation produced copious resin flow in most pines. Often, woodpeckers would abandon excavation at cavity starts while resin was actively being transported to the wound site. When a woodpecker refrained from excavating for 1 or more months, resin would continue to concentrate in the sapwood (xylem) around the wound and eventually harden. Sapwood with resinosis can be excavated by the woodpecker without producing additional resin flow.

The second stage, excavation of the entrance tube through the living xylem tissue (**Figure 1b**), required varying times to complete, and appeared to be related to the ability of pines to produce resin and the possible rate of resinosis around the wound that was penetrating the xylem. As excavation into the sapwood deepened, the extent of sapwood affected by resinosis increased. In pines with minimal resin flow, excavation of the second stage proceeded without pauses for resinosis abandonment, possibly related to more vigorous resin production in this pine species. It is noteworthy that the sapwood thickness of pines (often 10-14 cm) is typically much greater than that encountered by other species of woodpeckers excavating cavities in hardwoods (3-6 cm)(Conner et al. 1976). Thus, the time required for cavity excavation by red-cockaded woodpeckers is increased by the amount of sapwood that must be penetrated.

The time spent in stage 2 may also have related to the "need" for new cavities within the cluster. On some occasions 1 to 3 years of inactivity at a deep start were eventually followed by cavity completion. It could be speculated that such a strategy may be

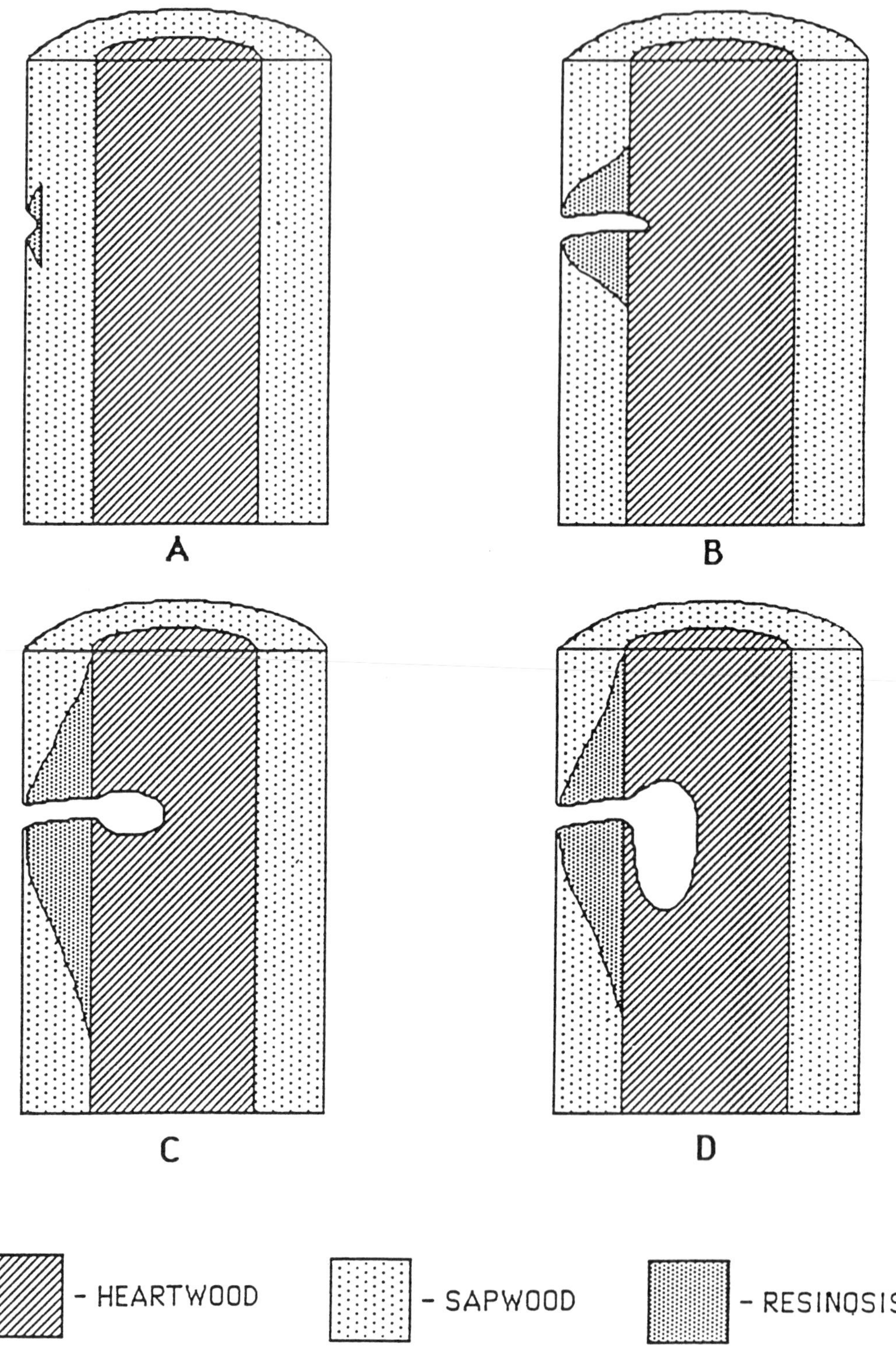

Figure 1. Sequence of red-cockaded woodpecker cavity excavation with gradual saturation of sapwood tissue with pine resin (resinosis), A. Stage 1 occurs when the woodpecker initially wounds the pine, B. Stage 2 involves excavation of the entrance tube to the cavity chamber area which will be located in the heartwood of the pine, C. Stage 3 is the initial cavity chamber excavation that occurs before the woodpecker excavates downward in the heartwood of the pine, and D. Completion of the cavity within the heartwood of the pine.

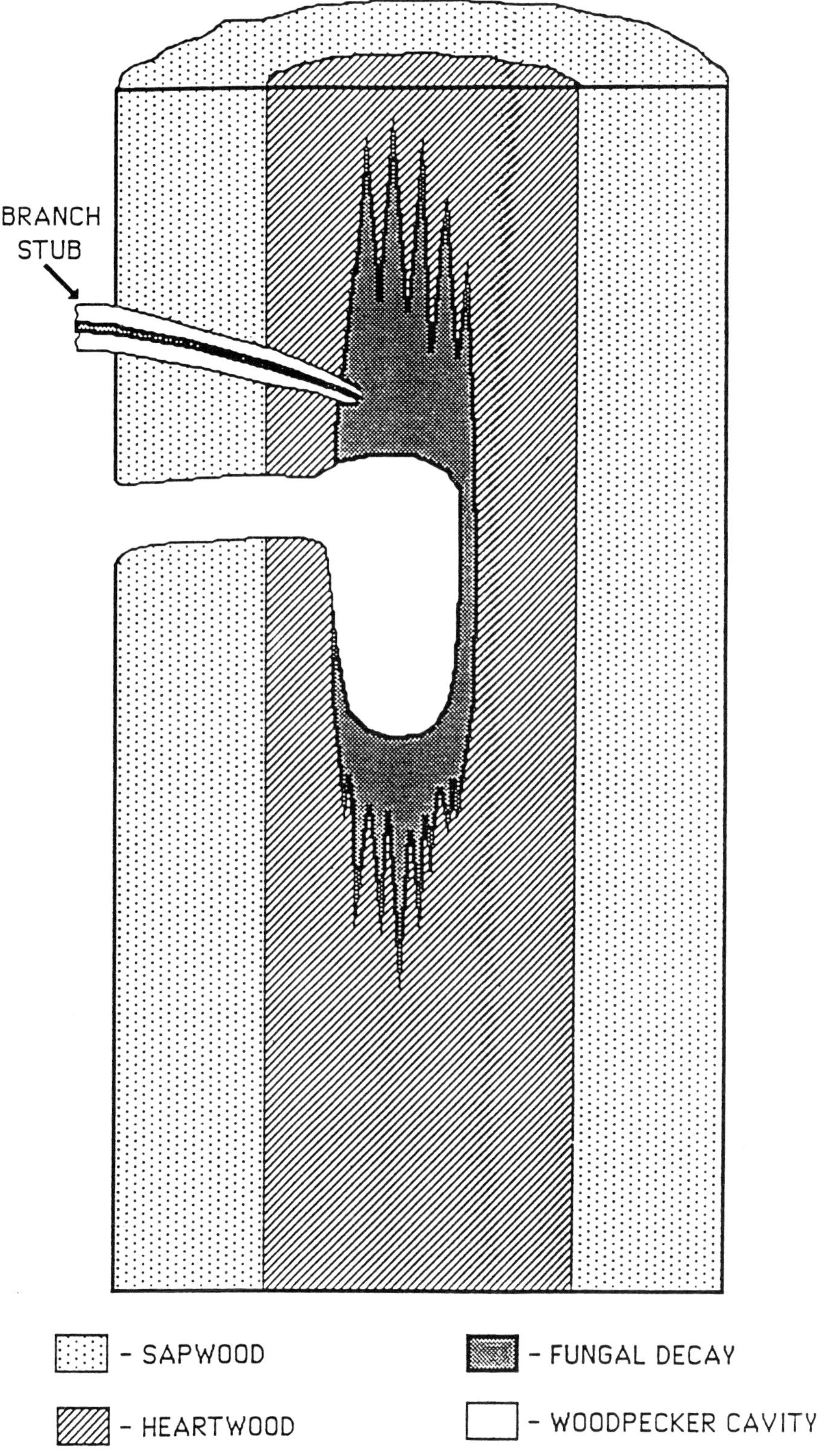

Figure 2. Completed red-cockaded woodpecker cavity showing a broken branch stub heartwood as the path for red heart fungus to infect the heartwood of the pine's bole. A sufficiently decayed heartwood greatly decreases the time and energy requirements of excavating a cavity and cavities are typically excavated in decayed portions of the heartwood.

advantageous if a new cavity for nesting is suddenly needed during the breeding season after competitors have usurped or destroyed the male's roost cavity (Rudolph et al. 1990a). We currently have only indirect observational data to support this hypothesis; alternatively, excavation may have slowed as a result of the woodpeckers reaching sound heartwood in the bole of their intended cavity tree.

The third stage was reached after woodpeckers penetrated the heartwood of the pine (**Figure 1c**). Unlike sapwood that contains living ray and axial parenchyma cells that actively transport resin, the heartwood is dead (Kramer and Kozlowski 1979). Red-cockaded woodpeckers typically must reach the heartwood to excavate their cavity chamber; otherwise the chamber would partially fill with sticky resin and trap the bird. Thus, a sufficient diameter of heartwood must be present to house a cavity chamber (Hooper et al. 1980, Conner and O'Halloran 1987). We examined 26 vertically sectioned red-cockaded woodpecker cavities killed by storms, fires, and bark beetles (see Conner and Locke 1982). We measured the amount of heartwood occupied by cavity chambers by measuring the horizontal distance from 2 cm below the entrance tube to the back of the cavity. Cavity chambers used an average 11.2 cm (STD = 1.9) of the heartwood and measurements ranged from 9.4 to 17.2 cm. Most cavities used between 10 and 12 cm of the heartwood. The larger cavities had been in use for more than 5 years and the cavity chambers had been expanded by red-cockaded woodpeckers during repeated years of use. The expansion of cavity chambers by red-cockaded woodpeckers typically enlarged the back-side of the cavity (horizontally elliptical) and not the lateral portions of the chamber. The chambers had not been enlarged by other species of woodpeckers. Thus, to house the largest cavities, the heartwood must be at least 18 to 20 cm in diameter.

Woodpeckers do not have to wait for resinosis to occur in order to continue excavation of the heartwood (dead xylem tissue). Initially, a small chamber with marginal depth is carved out of the heartwood of the pine. It is during this period that woodpeckers will often begin to roost in the pine and actively peck resin wells around the entrance to their expanding cavity. The pine resin that flows from these wells serves as an effective barrier against climbing rat snakes (*Elaphe obsoleta*) (Jackson 1974, 1978d; Rudolph et al. 1990b). There was a strong association between presence of red heart fungus and red-cockaded woodpecker cavity trees (Steirly 1957a, Jackson 1977c). The rate of chamber enlargement during this stage was affected by the presence and extent of fungal heartwood decay. Heart rots infected the heartwood of pines by growing down the heartwood of large broken branches until the heartwood in the bole of the pine was infected (Conner and Locke 1982) (**Figure 2**). Red-cockaded woodpeckers prefer and appear to be able to actively detect pines with red heart infections of the heartwood for cavity excavation (Conner and Locke 1982, Hooper et al. 1991b, Rudolph et al. 1995).

As with the third stage, the rate of completion of the fourth stage, excavation of the full cavity chamber, was dependent on presence and extent of heartwood decay (**Figure 1d**), and possibly the woodpecker group's need for another cavity. Subjective tests by the authors with a chisel suggest that it is easier to excavate wood while cutting across the grain (as occurs when the woodpecker excavates sapwood and initial parts of the heartwood) than it is to excavate vertically "with" the grain of the wood. Vertical excavation tends to spread wood fibers apart rather than cut them. Thus, presence of well decayed heartwood can greatly facilitate chamber excavation when the woodpecker excavates portions of the cavity chamber that lie below the entrance tube.

DURATION OF CAVITY EXCAVATION AND CAVITY USE PATTERNS

The time used by red-cockaded woodpeckers to excavate cavities was significantly longer in longleaf pines than in either loblolly or shortleaf pines (**Table 1**).

Excavation time averaged 6.3 y in longleaf pines, which was 2 to 3 times greater than the mean times used in loblolly and shortleaf pine. Three cavity starts in longleaf pine were active for more than 9 y; these cavities still had not been completed by the end of the ninth year.

As mentioned above, the presence of red heart fungus can reduce the time required for excavation of the heartwood of pines. Pines with decayed heartwood (n = 15, = 3.7 years) were excavated in less time than pines without decay (n = 10, = 5.0 years)($P = 0.057$).

We compared cavity trees that had been

Table 1. Length of time required for red-cockaded woodpeckers to excavate cavities in longleaf (n=12), loblolly (n=9), and shortleaf (n=12) pines on the Angelina and Davy Crockett National Forests in eastern Texas during a 15-year study.

Tree Species	Mean[1]	S.D.	Min.	Max.
Longleaf pine	6.3 a	2.5	2.0	9.0[2]
Loblolly pine	1.8 b	1.2	1.0	4.0
Shortleaf pine	2.4 b	0.9	1.0	4.0

[1]Common letters indicate non-significant differences, ANOVA with Duncan's Multiple Range Test, P< 0.05.

[2]Three incomplete cavity starts were still being actively excavated in 1992.

used as red-cockaded woodpecker nest trees at least once (**Table 2**). Only cavities that were followed from starts to completed cavities were considered; cavity trees already in use during the early years of the study were not included in the data. Longleaf pines served as red-cockaded woodpecker nest trees for more years than did either loblolly or shortleaf pine. Longleaf pines also were used significantly longer as active roost trees, and total span of utility to the woodpeckers was significantly longer than for either loblolly or shortleaf pines (**Table 2**). The number of years that nest trees were inactive was not significantly different among tree species. Data presented in **Table 2** for loblolly and shortleaf pine accurately represent what occurs over the full span of time that the pines are used by red-cockaded woodpeckers. Such is probably not the case for longleaf pine, and values measured and presented in **Table 2** are minimal figures. Our study was not long enough (even though it spanned 18 years for longleaf pines) to obtain a full view of the entire period longleaf pines were of value to red-cockaded woodpeckers. We do have data obtained during earlier studies of the woodpecker in longleaf pine habitat on the Angelina National Forest (Lay and Russell 1970). Four longleaf pine cavity trees that we began to study in the late 1970s were being used as active cavities for nesting and roosting during the late 1960s. We observed nesting and roosting in all four of these cavity trees into the 1990s. This indicates that longleaf pine cavity trees can be used by red-cockaded woodpeckers as nest and roost sites for at least three decades, more than compensating for the longer time required for initial cavity excavation (**Table 1**). Cavities in longleaf pines would provide a greater benefit of philopatry than those in other pine species.

A general summary of use patterns for cavity trees indicates that longleaf pines were valuable and used over an extended period of time which involved cycles of use and inactivity (**Table 2**). Such was not the case for loblolly and shortleaf pines; both of these species typically were excavated in less time, but were used for fewer years, often dying after being infested by southern pine beetles (*Dendroctonus frontalis*) during the fall (Conner et al. 1991a, Rudolph and Conner 1995).

POPULATION DYNAMICS OF CAVITY TREES OVER A 10-YEAR PERIOD

Red-cockaded woodpeckers were not excavating cavities as fast as they were losing them on the Angelina National Forest between 1983 and 1992 (**Table 3**). The annual mortality of cavity trees over the 10-year period was 3.6 times higher than the excavation of new cavity trees. If losses of red-cockaded woodpecker cavities resulting from Pileated woodpecker enlargement is included the rate of loss exceeded new cavity excavation by 7.0 times. The woodpecker population over this time period ranged from 38 active clusters in 1983 to a low of 19 active in 1988 (Conner and Rudolph 1989).

Annual mortality of loblolly and shortleaf pine cavity trees was more than twice that of longleaf pine cavity trees (t = 2.5, P = 0.03, **Table 3**). However, red-cockaded woodpeckers excavated new cavities in loblolly and shortleaf pines at twice the annual rate of that for longleaf pines (**Table 3**). These values need to be adjusted because of the differences in the number of woodpecker groups in each timber type. An average of about 12 woodpecker groups were present in longleaf pine habitat whereas only about 7 occurred in the loblolly-shortleaf pine habitat. When values are adjusted for number of woodpecker groups, cavity excavation rate per woodpecker group in longleaf pine was about 1/4 the rate of that for those in loblolly-shortleaf pine habitat.

Cavity tree losses of this magnitude are a major concern, and the impact on woodpecker populations was considerable until the

Table 2. Red-cockaded woodpecker use of longleaf, loblolly, and shortleaf pine nest trees over an 18-year span of time on the Angelina National Forest in eastern Texas.

Variable	Mean[1]	S. D.	Min.	Max.
Longleaf pine (n = 27)				
# years used as nest tree	3.1[a]	1.5	1.0	6.0
# years cavity tree was active	8.7[a]	3.3	4.0	17.0
# years cavity tree was inactive	1.6[ns]	2.5	0.0	8.0
Span of years cavity tree was active	10.3[a]	4.3	5.0	18.0
Loblolly pine (n = 17)				
# years used as nest tree	1.8[b]	1.1	1.0	5.0
# years cavity tree was active	4.8[b]	2.3	1.0	10.0
# years cavity tree was inactive	0.4[ns]	0.9	0.0	3.0
Span of years cavity tree was active	5.2[b]	2.7	1.0	10.0
Shortleaf pine (n = 7)				
# years used as nest tree	1.7[b]	1.0	1.0	3.0
# years cavity tree was active	4.6[b]	2.9	2.0	9.0
# years cavity tree was inactive	0.4[ns]	0.8	0.0	2.0
Span of years cavity tree was active	5.0[b]	3.6	2.0	10.0

[1]Means with common letters indicate nonsignificant differences among pine tree species, ANOVA with Duncan's Multiple Range Test, $P<0.05$.

beginning of the 1990s. A cavity drilling technique developed by Copeyon (1990), and later modified by Taylor and Hooper (1991), and cavity inserts (Odom et al. 1982, Allen 1991) revolutionized how cavity losses are approached by forest managers. Use of cavity restrictors (Carter et al. 1989) solved most of the problems with cavity enlargement by other species of woodpeckers. These new techniques, when applied correctly as on the Angelina National Forest **(Table 3)**, can rapidly offset losses due to mortality and repair damage done by Pileated woodpeckers.

DISCUSSION

The time necessary for red-cockaded woodpeckers to excavate cavities in loblolly, shortleaf, and particularly longleaf pines **(Table 1)** appears to have a significant affect on the availability of cavities. Mature pines would have been far more abundant during pre-Columbian times when the woodpecker adapted to the fire climax pine ecosystem than they are during the present time period following the harvest of the old-growth forest. Clearly, cavity excavation for the relatively small red-cockaded woodpecker is a major endeavor, often requiring an entire life span to complete, and in some cases 2 or more generations of woodpeckers. Thus, our data support the benefits-of-philopatry and critical-resources models as proposed by Stacey and Ligon (1987, 1991) and Walters et al. (1992a). Cavity excavation in other species of woodpeckers in the southeastern United States typically requires 2 to 6 weeks from start to finish (Conner et al. 1975, 1976; Kilham 1983; R.N. Conner, Pers. Observation). None of these other species of woodpeckers breeds cooperatively. Cavities are often excavated during the spring of the breeding season and become the male's roost cavity prior to nesting. Although cavities are seldom used again as nest trees, pileated woodpeckers will occasionally use the same cavity during the following year after enlarging the chamber, or excavate a new cavity in the same nest tree (Conner et al. 1975, 1976). Thus, the duration of both cavity excavation and cavity use by red-cockaded woodpeckers are distinctly longer than that for other woodpeckers in the southeastern United States.

In pre-Columbian times the evolution of cooperative breeding in red-cockaded

Table 3. Gains and losses of natural red-cockaded woodpecker cavity/start trees and cavity enlargement by pileated woodpeckers on the Angelina National Forest, Texas over a 10-year period.

	Year										
	1984	1985	1986	1987	1988	1989	1990	1991	1992	Total	Ann. Avg.
					# cavity/start trees[a]						
Loblolly/shortleaf	70	71	72	74	78	79	73	77	72		74
Longleaf	164	169	170	173	175	187	187	187	188		178
Subtotal	234	240	242	247	253	266	260	264	260		252
					# new cavity/start trees[b]						
Loblolly/Shortleaf											
New cavity	3	1	1	2	0	2	1	0	1	11	1.2
New start	4	2	1	4	6	4	0	2	2	25	2.8
Existing cavity/start tree	5	1	0	0	4	2	0	4	2	18	2.0
Longleaf											
New cavity	3	0	0	0	1	1	0	0	0	5	0.6
New start	1	2	0	4	3	3	3	0	0	16	1.8
Existing cavity/start tree	20	3	2	0	1	12	0	1	2	41	4.6
Subtotal	36	9	4	10	14	24	4	7	7	116	12.9
					# cavity trees dying						
Loblolly/shortleaf	1	3	1	4	6	7	7	2	10	41	4.6
Longleaf	2	0	1	1	3	4	3	1	1	16	1.8
Subtotal	3	3	2	5	9	11	10	3	11	57	6.3
					Cavity enlargement[c]						
Loblolly/shortleaf	2	4	1	0	4	2	1	1	0	15	1.7
Longleaf	6	1	8	5	5	6	4	3	2	40	4.4
Subtotal	8	5	9	5	9	8	5	4	2	55	6.1
Net change in useable cavity trees[d]	+3	-2	-1	-3	-8	-8	-9	-3	-10	-41	-4.6
Net change in useable cavity trees all causes[e]	-5	-7	-11	-8	-17	-16	-14	-7	-12	-96	-10.7
# artificial cavities installed[f]								57	50	107	
# cavity restrictors installed[f]							4	31	16	51	

[a]Number reflects newly excavated trees and discovery of existing old cavity trees.
[b]Trees found during spring surveys and preceding 11 months.
[c]All cavities enlarged so extensively that the trees were unuseable for roosting and nesting.
[d]Net change reflects the death of existing cavity trees versus newly excavated trees, but not enlargement.
[e]Net change reflects losses of useable cavities to both mortality and cavity enlargement.
[f]Numbers of installed artificial cavities (primarily inserts) and restrictors are not included as new cavities in the above data.

woodpeckers was likely influenced more by the length of time required to excavate nest cavities than by the limited availability of mature pines in which cavities could be constructed or woodpecker-saturated habitat. Thus, the availability of an extremely valuable resource-an excavated cavity-appears to be the likely basis of natal philopatry, and ultimately, cooperative breeding in red-cockaded woodpeckers. As indicated by Walters (1990) more young red-cockaded woodpeckers disperse than are typically observed in other cooperative breeders. Young woodpeckers are free to disperse to vacant habitat that contains mature pines suitable for cavity excavation, but their inability to rapidly excavate cavities for nesting and roosting appears to decrease the value of vacant habitat lacking existing cavities (Walters 1990). When cavities are provided artificially, red-cockaded woodpeckers readily occupy vacant habitat and form new breeding units (Copeyon et al. 1991, Walters et al. 1992b), suggesting the critical nature of the cavity resource rather than an unsuitable character of the habitat in general.

Red-cockaded woodpecker cavity excavation rates varied significantly among tree species. Excavation times in longleaf pine were much longer than either loblolly or shortleaf pines. Average time to excavate cavities in shortleaf pine was somewhat longer than in loblolly pines, but not significantly longer (**Table 1**). It was perhaps more than coincidental that the pattern of incidence of fungal decay in these 3 species of pines for the eastern Texas area (Conner and Locke 1982) was the direct inverse of the average time periods used by the woodpeckers to excavate cavities. Conner and Locke (1982) detected a 47% infection rate in longleaf pine cavity trees, 87% in shortleaf pines, and 100% in loblolly pines. An average of 6.3 y in longleaf, 2.4 y in shortleaf, and 1.8 y in loblolly pines is in accord with the expectation that longer time periods are required to excavate cavities in an absence of decayed heartwood.

The rate that red-cockaded woodpeckers excavated new cavities was considerably lower than the rate of cavity loss (**Table 3**). Reasons for this disparity may at first not be obvious, but warrant some consideration. The frequency of heartwood decay in pines and the amount of heartwood within the bole is proportional to tree age (Nelson 1931, Kramer and Kozlowski 1979, Clark 1993). Longleaf pine cavity trees are significantly more resistant to infestation by southern pine beetles than are either loblolly or shortleaf pines (Conner et al. 1991a). Older pines have greater diameters of heartwood for cavity excavation in the crowns of pines increasing the heights at which woodpeckers can excavate cavities (Conner et al. 1991a). Such cavities are less likely to be destroyed by the frequent fires that maintained the ecosystem. Historically, the longleaf pine ecosystem that red-cockaded woodpeckers occupy was fifty-fold more common prior to the "Bonanza Era" of timber harvesting that occurred throughout the South between the 1850s and 1930s (Maxwell and Baker 1983). Much of the original longleaf pine forests were replanted with loblolly pine, a species considerably more vulnerable to southern pine beetle infestation. Contemporary forests are also much younger, decreasing the amount of heartwood available at greater heights in the crown, and in many cases failing even to provide sufficient heartwood at heights suitable for cavity excavation. Low cavities in longleaf pine trees are particularly susceptible to ignition by fire, and such cavity trees are regularly lost if not actively protected during prescribed fires (Conner and Locke 1979). All these factors suggest that mortality of cavity trees was considerably lower during pre-Columbian times than it has been since the loss of the South's original pine forest.

Woodpecker populations occupying loblolly and shortleaf pine habitat appear to be more vulnerable to population declines as a result of cavity tree mortality than woodpeckers using longleaf pine habitat. Cavity tree loss rates were much higher in loblolly-shortleaf pine habitat than in longleaf pine habitat. In the past, shortage of cavities in loblolly pine habitat have often been attributed to presence of competitors (Harlow and Lennartz 1983). Particularly vulnerable are areas where cavity tree mortality rates exceed cavity excavation rates and loblolly cavity trees are old, likely having excessive heartwood decay that has destroyed the bottoms of cavities. The unexplained extirpation of the Tall Timbers population where encroachment of hardwood midstory was not a problem (Baker 1983b) might be at least partially explained by an imbalanced excavation/mortality rate ratio and advanced stages of heartwood decay.

ACKNOWLEDGMENTS

We thank R.G. Hooper, J.A. Jackson, J.D. Ligon, and J.R. Walters for comments leading to the improvement of the manuscript. We thank D. Lay for red-cockaded woodpecker cavity tree records from the 1960s through the mid 1970s.

Population Structure and Annual Turnover Rates of Cavities of the Red-cockaded Woodpecker in the Apalachicola National Forest

Frances C. James, Charles A. Hess, Gregory Hagan, and Bowie Kotrla
Department of Biological Science, Florida State University, Tallahassee, Florida 32306

ABSTRACT: The largest remaining population of the red-cockaded woodpecker (*Picoides borealis*) (RCW) is the one in the Apalachicola National Forest (ANF), near Tallahassee, Florida. Current records from the U.S.D.A. Forest Service (FS) list approximately 700 active clusters of cavity trees (sites, colonies) in this area, each of which is assumed to be occupied by a pair of birds, with or without helpers, or a single bird. We report preliminary estimates of the demographic status of the population in 1991 and 1992 for each of the two management districts of the ANF, the Apalachicola Ranger District (ARD) and the Wakulla Ranger District (WRD). The data are based on information from 70 active clusters randomly drawn from FS records of mapped active clusters. Our estimates indicate that the population structure of RCW's in the two districts and the rates of turnover of cavities and cavity trees are different. In 1991 only 22 of the 32 sites in the WRD sample of defended sites were being defended by pairs of birds or pairs with helpers, and ten sites were occupied by single birds. On the other hand, all of the 38 ARD defended sites were being defended by pairs of birds or pairs with helpers. In the ARD, the total numbers of active cavities and active cavity trees were stable between 1991 and 1992, but in the WRD these numbers declined by more than 20%. Comparisons for the WRD with the previous year indicate that 1991 to 1992 was a more stressful period. Habitat for RCW's in the ARD is largely separated from that in the WRD by private land and the Ochlockonee River floodplain. We think that the RCW population of the ANF should be considered two separate populations and managed differently. Our methods of analysis might be useful for making standardized comparisons with RCW populations elsewhere.
KEYWORDS: red-cockaded woodpecker, monitor, Apalachicola National Forest

The red-cockaded woodpecker (RCW) population in the Apalachicola National Forest (ANF) in northern Florida qualifies as officially recovered because it is estimated to have an effective population size of over 250 pairs of birds (USFWS 1985). Since Hurricane Hugo in 1989, the ANF population has been the only population that so qualifies. The actual number of breeding pairs in the ANF is unknown, but the U.S.D.A. Forest Service (FS) estimates that there are close to 700 clusters of active cavity trees (sites, colonies). Of these sites, approximately 500 are west of the Ochlockonee River, in the Apalachicola Ranger District (ARD), and approximately 200 are east of the river, in the Wakulla Ranger District (WRD). James (1991) estimated that more than 25% of clusters considered active by the FS had no birds in 1989 or 1990. The percent of defended sites with single birds was estimated to be 17% in 1989 and 31% in 1990, and the estimated number of pairs for the entire WRD was 90. Here we report the results of further study of the WRD sample of sites for 1991 and 1992, and we compare the results with data for a sample from the ARD.

Our objective is not to make estimates of the sizes of the populations in the two districts or their population trends. Instead we compare the districts using some statistics that reflect the demographic population structure of birds at defended sites and the turnover rates of active cavities and active cavity trees. Because this type of information can be gathered in a shorter period than is necessary to determine overall population trends, we hope that similar information can be used to monitor the health of other RCW populations. In fact, if comparisons were set up carefully, results could be useful for monitoring the effectiveness of management procedures, such as the addition of artificial cavities or the augmentation of sites with birds from source populations. The major source of birds for augmentation thus far has been the ARD.

METHODS

In 1989 and 1990, we randomly drew a sample of 50 sites in the WRD from the 186 mapped sites listed as active in FS files. In 1991, 32 of these sites were being defended by birds and had not received artificial cavities or been augmented by the introduction of birds from elsewhere. Of the other 18, 12 had no birds, three had single birds that were extraterritorial roosters (that is, belonged to groups in adjacent clusters), two had received inserts or been augmented, and one was inaccessible. In 1991 we drew a random sample of 50 sites from the 448 mapped sites in FS files for the ARD. In this sample, 38 sites qualified as above. Of the 12 sites that were excluded, four had no birds, one had an extraterritorial rooster, five were inaccessible in 1991, and two had artificial cavities.

We visited the 70 defended sites on a regular basis, mapped the trees, scored their cavities and start holes for signs of recent activity by birds, and noted where birds were roosting. Our standardized scoring system for cavities and start holes included scoring the number of fresh resin wells near a cavity and the amount of fresh sap on a scale of 1 to 5. We searched the area within one half mile of each site for evidence of new start holes and cavities. In our summary files, if a cavity was judged to be active any time during a year, it was scored as active for that year. If a start hole became an active cavity within a year, it was scored as an active cavity for that year. The objective was to be as inclusive as possible of all cavities and start holes used by the birds in a particular year.

The number of birds in a clan, even excluding the young of the year, varies, because individual birds vary in their site fidelity. The particular cavities being used at any one time for roosting can be even more variable, depending on the season and the circumstances. For the number of adult birds in a cluster, we used the highest estimate of the number of birds that were regularly associated with a site before the breeding season of that year. Young of the year were excluded.

Distributions of birds among sites and of active cavities and active cavity trees among sites were tested with Wilcoxon Mann-Whitney U tests for ordered alternatives among independent observations. We suggest two graphic ways to compare such distributions: as histograms or as quantile plots. The differences between districts in annual rates of the maintenance of active cavities and the rates of transitions from inactive to active cavities or active start holes to active cavities were tested with chi-squared statistics.

Because the WRD sample was drawn more than one year before the ARD sample, and FS records are updated on a regular basis, we were concerned that our WRD sample might not be representative of the sites in the district. Therefore, we drew a second random sample of 50 sites for the WRD in October 1992 and determined how many of these sites were occupied by birds.

RESULTS

Average Differences Between Districts

One way to compare the population structures of two RCW populations, when the ages and sexes of many individuals are unknown, is to compare the numbers of birds that exist in pairs, as helpers, and as single birds at representative samples of occupied sites **(Table 1a)**. When two or more birds were consistently present at a site, we assumed that it supported a mated pair of birds, and the additional birds were scored as helpers. In the ARD, 89 and 94% of such sites had nests in 1991 and 1992. In the WRD, 62 and 89% of sites with two or more birds had nests in 1991 and 1992. Because the monitoring intensity was lower in 1991 than in 1992, we are less confident about its reliability.

The major differences between the samples from the two districts were that the total number of birds in the WRD sample decreased sharply between 1991 and 1992 (67 to 58) and that in both years a substantial fraction (15% or more) of the total number of birds in the WRD sample were single. The average numbers of active cavities and start holes per site (indicators of the level of activity of birds)

Table 1. Summary statistics for birds, cavities, start holes, and trees with signs of activity by birds for samples of defended sites in the Apalachicola Ranger District (ARD, n = 38) and the Wakulla Ranger District (WRD, n = 32) of the Apalachicola National Forest for 1991 and 1992. Table 1a gives totals, percentages, and differences in percentages between districts. Table 1b gives averages by site and differences between averages.

	1991			1992		
	ARD	WRD	%	ARD	WRD	%
Part a.	n (%)	n (%)	Diff.	n (%)	n (%)	Diff.
Birds						
In pairs	76 (78%)	44 (66%)	-12	72 (73%)	40 (69%)	-4
As helpers	22 (22%)	13 (19%)	-3	25 (25%)	8 (14%)	-11
Single	0 (0%)	10 (15%)	15	2 (2%)	10 (17%)	15
	98	67		99	58	

	ARD	WRD	Ave.	ARD	WRD	Ave.
Part b.	n (ave.)	n (ave.)	Diff.	n (ave.)	n (ave.)	Diff.
Birds	98 (2.6)	67 (2.1)	-0.5	99 (2.6)	58 (1.8)	-0.8
Active Cavities	135 (3.6)	94 (2.9)	-0.7	135 (3.6)	71 (2.2)	-1.4
Active Start Holes	74 (1.9)	29 (0.9)	-1.0	54 (1.4)	14 (0.4)	-1.0
Trees with Active Cavities	118 (3.1)	85 (2.7)	-0.4	122 (3.2)	66 (2.1)	-1.1
Trees with Active Start Holes	58 (1.5)	27 (0.8)	-0.7	44 (1.2)	14 (0.4)	-0.6

Table 2. Annual differences in the average number of birds, active cavities, and active cavity trees between 1991 and 1992 in the ARD and from 1990 to 1992 in the WRD.

	ARD	WRD	
	1991-92	**1990-91**	**1991-92**
Birds	0.0	0.3	-0.3
Active Cavities	0.0	0.0	-0.7
Active Cavity Trees	0.1	-0.1	-0.6

were substantially lower in the WRD than in the ARD (**Table 1b**). Differences between districts were generally larger than differences within the WRD between years. The average number of active cavities per site and trees with active cavities per site dropped sharply in the WRD between 1991 and 1992 (94 to 71 and 85 to 66, respectively; **Table 2b**).

Differences in the Distributions of Birds Per Site

Differences between averages in **Tables 1b and 2** do not reveal the differences between districts in the distributions of birds among sites, but these differences can be shown in histograms (**Figures 1a and b**). Because normal curves overlaid on the histograms do not fit the data well, parametric tests are not appropriate, but the differences (e.g. for 1991, see **Table 3**) can be tested reliably by means of the Wilcoxon Mann-Whitney U statistic (**Table 4**). Another way to display such differences in distributions that allows visual comparisons among populations on the same scale is in a quantile plot (Chambers et al. 1983). Here, the fraction of sites that have different numbers of birds is plotted against the number of birds per site (**Figures 2a, b**). One can see in **Figure 2**, for example, that the entire distribution of birds per site is shifted toward the lower end of the scale in the WRD relative to that of the ARD.

Turnover Rates of Active Cavities

Another indicator of the relative health of two RCW populations might be differences in rates of turnover of active cavities. Differences in rates might be correlates of differences in environmental stress experienced by the two populations. Discovery of sources of the differences (**Figure 3**) might suggest where to focus analyses of causes.

The maintenance of a population of cavities

A

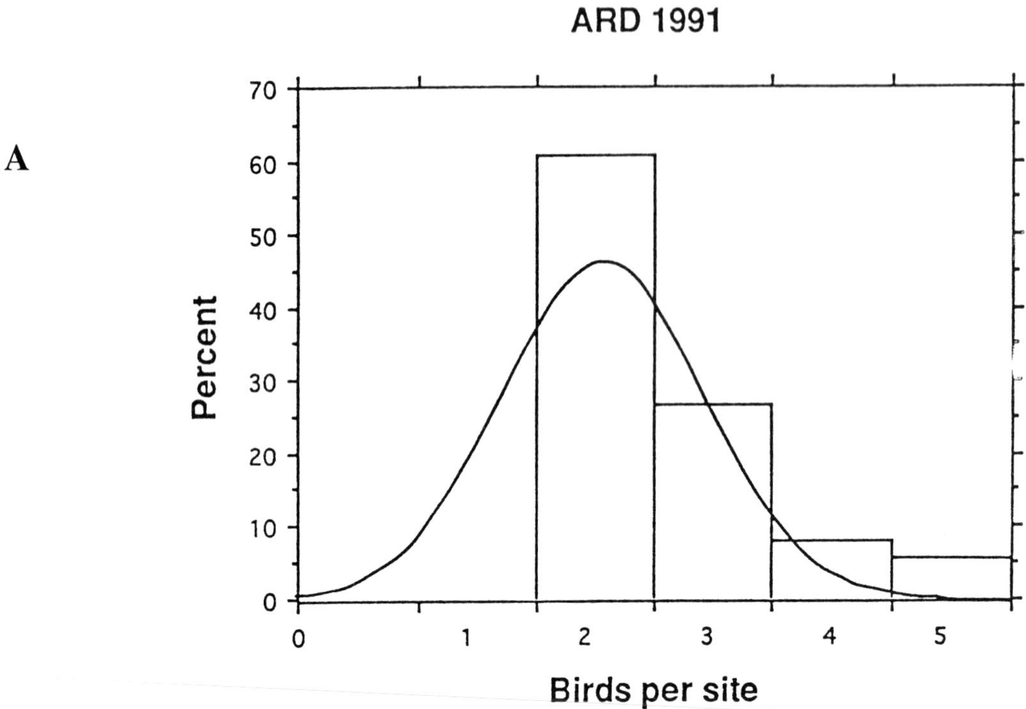

B

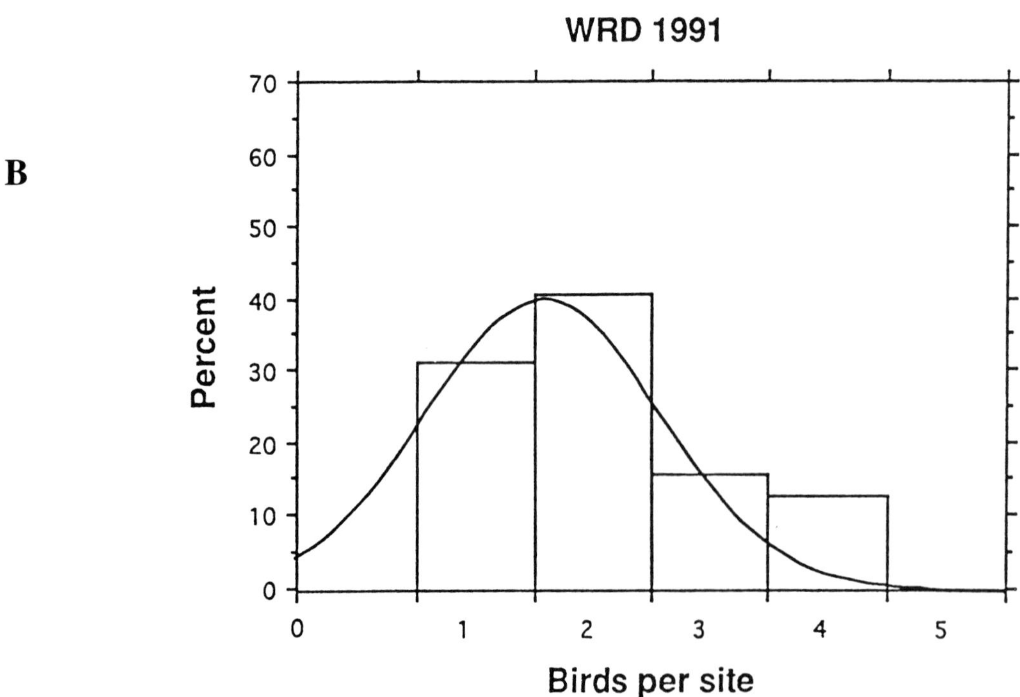

Figure 1a, b. Histogram of the number of sites with different numbers of birds in 1991 for the Apalachicola Ranger District (ARD) and the Wakulla Ranger District (WRD) overlain by ill-fitting normal curves.

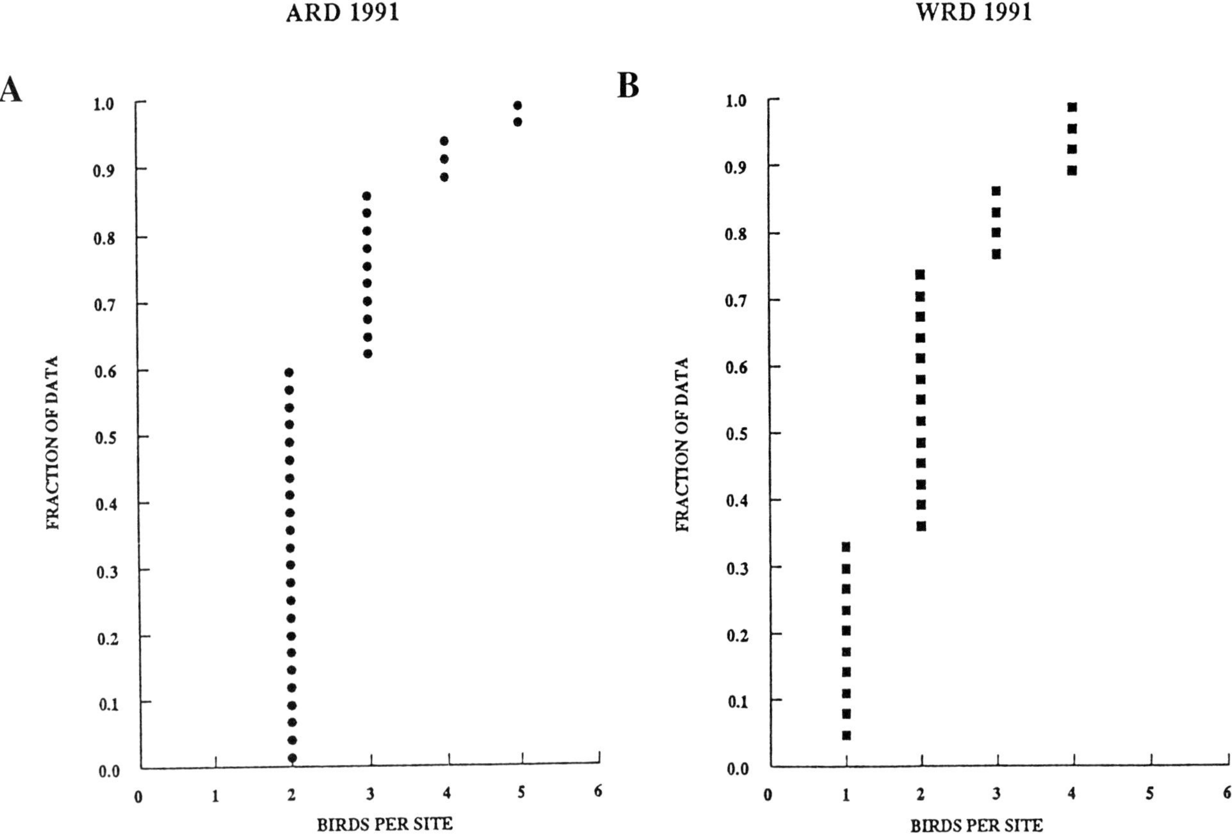

Figure 2a, b. Quantile plots of the cumulative fraction of the sites with different numbers of birds in 1991 (see Chambers et al. 1983).

Table 3. Comparison between districts of the numbers of sites defended by different numbers of birds in 1991.

	District	Number of Birds 1	2	3	4	5	Total
Count	ARD	0	23	10	3	2	38
	WRD	10*	13	5	4	0	32
Percent							
	ARD	0	61	26	8	5	100
	WRD	31	41	15	13	0	100
Difference in Percent		31	-20	-11	5	-5	

***One bird was primarily an extraterritorial rooster in 1991 but defended his site regularly in 1992.**

from one year to the next is a function of four processes: the rate at which active cavities in year 1 remain active, the rate at which inactive cavities are reactivated, the rate at which start holes become active cavities, and the rate at which new cavities are created *de novo* (within the year). An equation for this situation is:

$$a\,(\mathrm{ac}) + b\,(\mathrm{ic}) + c\,(\mathrm{as}) + \mathrm{nc} = \mathrm{x},$$

where ac, ic, and as are the numbers of active cavities, inactive cavities, and active start holes in year 1; nc is the number of *de novo* cavities in year 2; *a*, *b*, and *c* are percentages that were active cavities in year 2; and x is the number of active cavities in year 2. For the ARD between 1991 and 1992,

$$0.78\,(136) + 0.22\,(49) + 0.14\,(75) + 7 = 135,$$

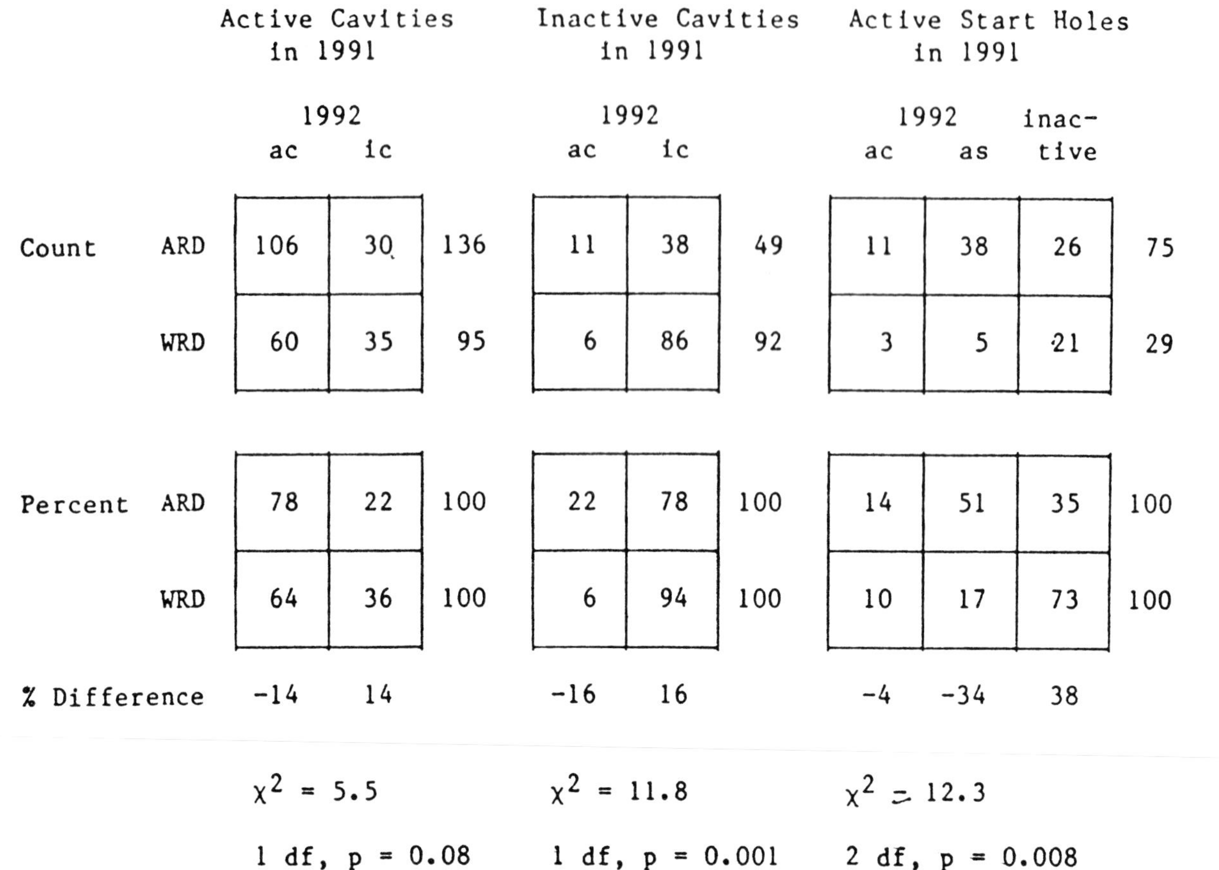

Figure 3. Tests of differences between the ARD and the WRD in changes in the status of active cavities (ac), inactive cavities (ic), and active start holes (as) from 1991 to 1992. For example, of the 136 active cavities in the ARD sample in 1991, 106 (78%) remained active in 1992, and of the 95 active cavities in the WRD sample in 1991, 60 (64%) remained active in 1992. Probability levels have been multiplied by 4 because of 4 simultaneous tests in Figures 3 and 4.

Table 4. Wilcoxon Mann-Whitney U tests of differences between districts in the numbers of birds per site (b), active cavities per site (ac), and active cavity trees per site (act). *Significant at $p < 0.05$, 1 df.

	n	b			ac			act		
		U	z	p	U	z	p	U	z	p
Full sample										
ARD vs. WRD, 1991	38, 32	432	-2.3	*	599	-0.6	ns	498	-1.3	ns
ARD vs. WRD, 1992	38, 32	323	-3.6	*	308	-3.6	*	303	-3.7	*

and for the WRD,

$$0.64\ (95) + 0.06\ (92) + 0.1\ (29) + 2 = 71.$$

Similarly, the number of trees with active cavities in 1992 (y) equals the number in the previous year multiplied by the proportion that remained active (a) + the number of newly activated trees in 1992 (nact):

$$a\ (\text{act}) + \text{nact} = \text{y}.$$

For the ARD it is

$$0.85\ (118) + 22 = 122,$$

and for the WRD it is

$$0.67\ (85) + 9 = 66.$$

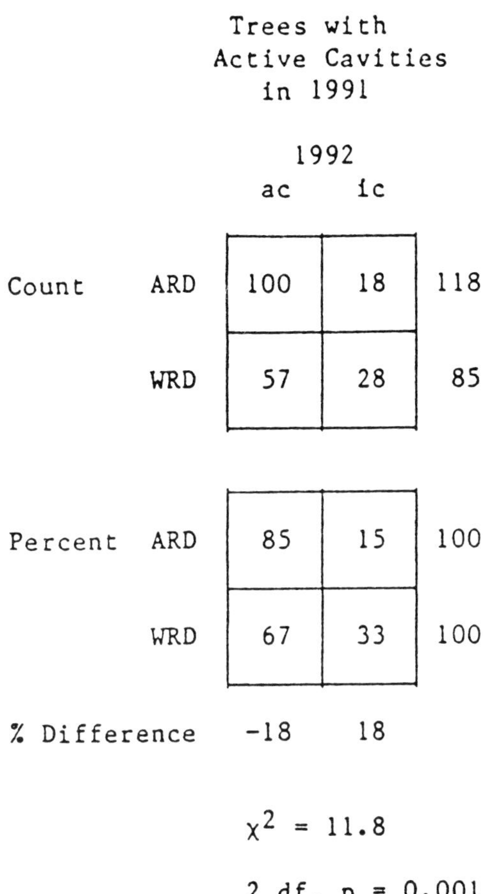
Trees with Active Cavities in 1991

		1992 ac	1992 ic	
Count	ARD	100	18	118
	WRD	57	28	85
Percent	ARD	85	15	100
	WRD	67	33	100
% Difference		-18	18	

$\chi^2 = 11.8$

2 df, $p = 0.001$

Figure 4. Test of the difference between the ARD and the WRD in changes in the status of trees with active cavities from 1991 to 1992. For example, of 118 trees with active cavities in the ARD Sample in 1991, 100 (85%) had active cavities in 1992. The probability level was multiplied by 4 as explained in Figure 3.

Comparisons of the coefficients in the equations for the two districts and chi-squared tests for the data for cavities and trees **(Figures. 3 and 4)** show that all the rates from 1991 to 1992 were lower in the WRD than in the ARD. The largest difference is the 38% higher rate at which start holes became inactive in the WRD. Because the data are for random samples of sites from their respective districts, they indicate major differences in the dynamics of cavity turnover between the populations. The most dramatic difference is the lower rate at which start holes are becoming cavities in the Wakulla Ranger District.

Reliability of Monitoring by Counting Active Cavities

In our samples, rank correlations (r_S) between the numbers of birds (b) regularly present and the number of active cavities (ac) or active cavity trees (act) were not impressively high (b,ac for ARD in 1991, 0.6; in 1992, 0.58; b,ac for WRD in 1991, 0.52; in 1992, 0.69). Also, tests of differences between districts in the distributions of active cavities or active cavity trees among sites were not significant in 1991, although tests of the distribution of birds among sites were significant **(Table 4)**. This result suggests that biologically important differences might be missed if comparisons among populations are based only on information about the numbers of active cavities or active cavity trees per site.

Another Random Sample from the WRD

Because more than 25% of the original 50-site sample in the WRD had no birds, we were worried that our random sample from that district might have been unusual. To allay this concern, in October 1992 we drew a second random sample of 50 sites from updated FS records, which at that time contained records of 170 mapped sites (colonies, clusters). No birds were present at 18 (26%) of these sites. Of these 18 cases, six were due to clerical errors (sites counted twice, etc.), and 12 were truly inactive sites. Our estimate of the percent of inactive sites in the FS list of active sites is thus 12/44 or 27% on the basis of this second sample.

DISCUSSION

Our work indicates that the current monitoring procedure used by the FS to monitor the size of the RCW population in the ANF seriously overestimates the RCW groups (numbers of social units) in the WRD and that it is inadequate for the detection of differences in important factors such as the distribution of birds among sites and the turnover rates of active cavities. See Hooper et al. (1991) for an example of plot-based monitoring applied to the Francis Marion National Forest and Hooper and Muse (1989) for other suggestions of how sampling methods might be improved.

Estimates of population trends based on changes in selected sites are not justified unless they are combined with estimates of the rates at which new sites are being established and inactive sites are being reactivated. A recently completed but unpublished resurvey by the FS of 21 compartments in the WRD and 20 in the ARD (1980-81 and 1989-91) gives numbers that allow these estimates to be made. One new site was established during the 10-year period in the WRD, and four in the ARD (average annual rates per original active site of 0.002 and 0.004 respectively); six and 15 previously inactive sites were reactivated (average annual rates of 0.007 and 0.013). Four cases of possible budding may represent new sites. If previously undetected sites are excluded, this unpublished study indicates a 20% decline in the number of active sites in the WRD sample during the 1980's and at least a stable population in the ARD sample. Because of the findings above, the estimate of James (1991) of 90 pairs for the WRD, and the results reported here, we think that recovered status should be applied only to the ARD population, which is demographically isolated by the Ochlockonee River floodplain from the WRD population. The conservation of the healthy ARD population is especially important because it is serving as a source population for birds that are being translocated to smaller declining populations. The conservation of the WRD population is particularly challenging, because its deterioration is assumed to be recent and potentially reversible. Even so, we do not know the causes of the differences reported here. Contributing factors are most likely to be differences between districts in the foraging habitat, in the burning regimes and their consequences for forest ecology, and/or in the history of the preservation of relict trees.

We hope that the methods of analysis used here (quantile plots, Mann-Whitney U tests of differences in the distribution of birds among sites, and chi-square tests comparing turnover rates at active cavities) will be useful for making comparisons with other populations.

ACKNOWLEDGMENTS

We are very grateful to the following people, who helped with various aspects of the field work, the organization of the files, logistic support, or data analysis: Wilson Baker, Ralph Costa, Todd Engstrom, Michael Evans, Kenny Forhan, Eric Goldman, Michael Kinnison, Dubravka Kufrin, Charles McCulloch, Duane Meeter, Lincoln Moses, Catherine NeSmith, Lora Sylvanima, B.J. Taylor, and Richard West. This research was partially supported by The Nature Conservancy (FLRO Stew.-Gainesvl. 03-26-92) and the National Science Foundation (DEB-9123941).

Turnover of Red-cockaded Woodpecker Nest Cavities on the Piedmont Plateau

Susan C. Loeb and Ernest E. Stevens
Southern Research Station,
USDA Forest Service, Department of Forest Resources,
Clemson University, Clemson, SC 29634-1003

ABSTRACT: Because cavities are a highly valuable resource for the endangered red-cockaded woodpecker (*Picoides borealis*) (RCW), factors leading to cavity loss (e.g., cavity enlargement, interspecific competition, tree mortality) may have a significant impact on populations. We observed fidelity to nest cavities by RCWs in the Piedmont Plateau of Georgia over a 10-year period. We determined the probable causes of nest cavity change and evaluated the effects of changing cavities on the reproductive success. Of the 220 nest attempts that occurred in consecutive years, 50.4% (111) were in a different nest tree than used the previous year. Probable causes of nest cavity change were water in the cavity (7 cases), nest tree mortality (10 cases), cavity enlargement (23 cases), and occupation by other species (33), most commonly southern flying squirrels (*Glaucomys volans*). Nest cavities were used for 1-6 years. Nest cavity change was not related to the status of the breeders or reproductive success in the preceding year, nor did it affect reproductive success in the year that the change occurred. However, continued loss of cavities may induce RCWs to abandon some colony sites.
KEYWORDS: red-cockaded woodpeckers, Piedmont Plateau, cavity use, cavity enlargement, southern flying squirrels.

Red-cockaded woodpeckers (*Picoides borealis*) (RCW) roost in cavities year-round and they use 1 cavity, usually the breeding male's, for nesting (Ligon 1970). Cavities are critical to survival and reproductive success, appropriate cavity trees are often limiting (Lennartz et al. 1983a, Costa and Escano 1989), and excavation of new cavities entails large investments of time and energy (Conner and Rudolph 1995a). Thus, it is not surprising that individual cavities are often used for many years (Ligon 1970, Lay 1973, Jackson 1978b, Harlow 1983). Heavy losses of cavities to competitors, to tree mortality, or to any other causes may have a significant impact on RCW populations (Baker 1983b, DeLotelle and Newman 1983).

In general, populations of RCWs in the Piedmont Plateau are smaller and less numerous than on the Coastal Plain (Lennartz et al. 1983a, 1983b; Costa and Escano 1989). Lowery (1960) and Baker (1982) suggested that longleaf pine (*Pinus palustris*), which is restricted to the Coastal Plain and Sandhills, is preferred over loblolly (*P. taeda*) and shortleaf (*P. echinata*) pines for cavity excavation. After wounding, longleaf pine exudes more resin, allowing the birds to maintain a sticky resin barrier around the cavities for longer periods. Longleaf pine is also less susceptible to southern pine beetle (*Dendroctonus frontalis*) attack and mortality (Coyne and Lott 1976, Hodges et al. 1977). RCW colony sites on the Piedmont Plateau have denser hardwood midstories than those on the Coastal Plain (Lennartz and Heckel 1987). Colony sites with a dense midstory are often abandoned (Conner and Rudolph 1989, Loeb et al. 1992), perhaps because midstories favor competitors and predators. Enlargement of cavities by other woodpeckers, particularly pileated woodpeckers (*Dryocopus pileatus*), also may be greater on the Piedmont Plateau due to moister conditions (Jackson 1978b).

If cavities are lost prior to or at the onset of the breeding season, RCWs may be forced to excavate cavities at a higher rate, utilize suboptimal cavities for nesting, or not nest at all. Thus, information on the duration of nest cavity use and the causes of nest cavity abandonment is important for the protection and management of the species. We examined fidelity to nest cavities by a population of RCWs in the Georgia Piedmont over a 10-year period and determined the probable causes of cavity change. We also evaluated the effects of changing cavities on the reproductive success of the RCWs and compared our results to those of Harlow (1983) from the Coastal Plain to gain more insight into factors impacting RCWs on the Piedmont Plateau.

METHODS

The study was conducted on the Piedmont National Wildlife Refuge (PNWR) and the Hitchiti Experimental Forest (HEF) in central Georgia (Jasper and Jones Counties). These areas are contiguous and have a combined area of approximately 16,000 ha (39,520 ac). Loblolly pine is the dominant species on ridges and upper slopes, but shortleaf pine also is common. On lower slopes and along streams, oaks (*Quercus* spp.), hickories (*Carya* spp.), maples (*Acer* spp.), and blackgum (*Nyssa sylvatica* Marsh.) are dominant. Where fire has been suppressed, these species, dogwood (*Cornus florida* L.), and sweetgum (*Liquidambar styraciflua* L.) are common components of pine stands.

PNWR has been actively managed for RCWs since 1970. Colony sites have been burned every 3-5 years, and, since 1987, hardwood midstories have been removed in active colony sites. Metal restrictor plates (Carter et al. 1989) have been placed on some cavities at PNWR since winter 1991. Until recently, colony sites on the HEF received relatively little active management except for experimental harvesting, thinning, and burning. In summer 1988, Forest Service personnel began installing restrictor plates on selected cavities. Mechanical felling of the midstory within colony sites was begun in 1990.

The study was conducted from 1983 through 1992. In 1983, as many adult RCWs as possible were trapped, banded, and color marked. Since then, trapping of unmarked adults has continued and all nestlings have been banded so that since 1988, essentially all birds in the population are individually identifiable. Throughout the study period the number of active colonies has ranged from 32 to 35 (Lennartz and Heckel 1987, Stevens *unpubl. data*). All RCW cavity trees, including cavity starts, that have been located have been marked with a white band, identified with an aluminum tag indicating colony and tree number, and mapped. Nest cavities were in loblolly pines.

From 1983 through 1987 all active colonies were visited during the nesting season (April-July) and selected trees were climbed and inspected until the nest was found. In addition to inspecting prospective nest cavities, from 1988 through 1992 all cavities in active colony sites were inspected to monitor use of cavities by other species (see Loeb 1993 for details). Contents and status (active or inactive, normal or enlarged) of all cavities that were inspected were recorded. Once a nest was located it was revisited every 2-5 days to determine the numbers of eggs, hatchlings, and nestlings. Nestlings were banded at approximately 5-7 days of age. Fledgling success was determined by observing the clans within 2 weeks of fledging and determining the number of juveniles and their color bands. Clan composition was determined during the nestling period. The nest cavity was observed for at least 2 hours in the early morning or evening with a Questar telescope, and all adults observed feeding young were identified. When possible, the breeders were also identified. Clans were considered to have successfully reproduced if at least 1 young fledged.

Only data for clans that nested for 2 consecutive years were used to determine nest tree turnover. In analyses that considered "no nesting", only clans that had a potential breeding male and breeding female were included. All nest attempts were included in the estimation of the length of time nest trees were used. Probable causes of switching cavities from 1 year to the next were determined by evaluating the contents and status of the previously used tree during the next nesting season. Where several factors could have been responsible for the change, the most conservative cause was assigned. For example, if a cavity was enlarged and also contained a competitor or water, the probable cause for switching was that the cavity was enlarged. If the cavity appeared to be in good condition (not enlarged, still active, no competitor), the cause was "unknown". To further test whether cavity use by other species was related to nest cavity change, data from all cavity tree checks (1988-1992) were used.

RESULTS

Fidelity to a nest cavity tree was not related to the status of the breeders **(Table 1)**. Neither a change in the breeding male ($X^2 = 0.01$, df = 1, $P > 0.05$) nor in the breeding female ($X^2 = 0.93$, df = 1, $P > 0.05$) was more likely to result in a change of the nest cavity tree. Further, there was no relationship between reproductive success during 1 year and the propensity of clans to use the same or different nest cavity the following year ($X^2 = 1.64$, df = 1, $P > 0.05$; **Table 2a**). Clans that changed nest cavities fledged 1.83 ± 0.12 (X ±1 S.E.) young in the year prior to the change. That value was not significantly different ($t = 0.02$, df = 216, $P > 0.05$) from the 1 for clans that used the same nest cavity (1.82 ± 0.11).

Of the 220 nesting attempts that occurred in consecutive years, 50.4% (111) were in a different nest tree than was used the previous year. However, percent change varied from 31.8% in 1988 to 65.5% in 1985 **(Figure 1).** In 73 cases in which birds changed nest cavities, a probable cause could be determined. In 7 cases the nest cavity from the previous year contained water, in 10 cases the nest tree had died, in 23 cases the cavity had been enlarged, and in 33 cases, the cavity contained another species or signs that another species had recently been using the cavity. Southern flying squirrels (*Glaucomys volans*) were in 22 cavities and other species of birds were in 6 cavities. Incidents of birds utilizing nest cavities included 2 red-bellied woodpeckers, *Melanerpes carolinus,* 1 tufted titmouse, *Parus bicolor*, 1 great-crested flycatcher, *Myiarchus crinitus*, and 2 unidentified bird nests with eggs. Nesting material was found in 5 cavities. In addition, there were 12 cases in which a clan contained a principal male and a principal female but no nesting occurred. In 5 of these cases flying squirrels were in the nest cavity of the previous year, in 2 cases other bird species were in the cavity, and in 5 cases the cavities appeared to be in good condition. The frequency of nest cavity change within the population was positively associated with the overall use of cavities by other species **(Figure 2)**. The frequency of annual nest cavity change was significantly correlated with the frequency of cavity use by southern flying squirrels ($r^2 = 0.44$), the frequency of cavity use by all species other than RCWs ($r^2 = 0.65$), and the frequency of cavity use by all species other than RCWs and cavities filled with water($r^2 = 0.83$). When moving to a new nest cavity, RCWs used previously existing cavities in 46 cases and newly excavated cavities (the cavity had not existed or was only a cavity start during the previous nesting season) in 39 cases. Our records did not allow us to determine the previous status of the new nest cavity in the remaining 26 cases.

Of the 171 nest trees that were used during the study, 94 (55.0%) were used for only 1 nesting season, 46 (26.9%) for 2 nesting seasons, 19 (11.1%) for 3 nesting seasons, 7 (4.1%) for 4 nesting seasons, 1 (0.6%) for 5

Table 1. Relationship between nest cavity fidelity and the status of breeding male and breeding female RCWs in the Piedmont Region of Georgia.

	Same Nest Cavity	Different Nest Cavity
Breeding Male		
Same	76	72
Different	9	8
Breeding Female		
Same	55	53
Different	24	16

Table 2. Relationship of reproductive success of RCW clans in the Piedmont Region of Georgia and nest cavity fidelity.

a) Reproductive Success in Previous Year

	Same Nest Cavity	Different Nest Cavity
Reproductive Success		
Successful	91	85
Unsuccessful	18	26

b) Reproductive Success in Current Year

	Same Nest Cavity	Different Nest Cavity
Reproductive Success		
Successful	92	87
Unsuccessful	17	24

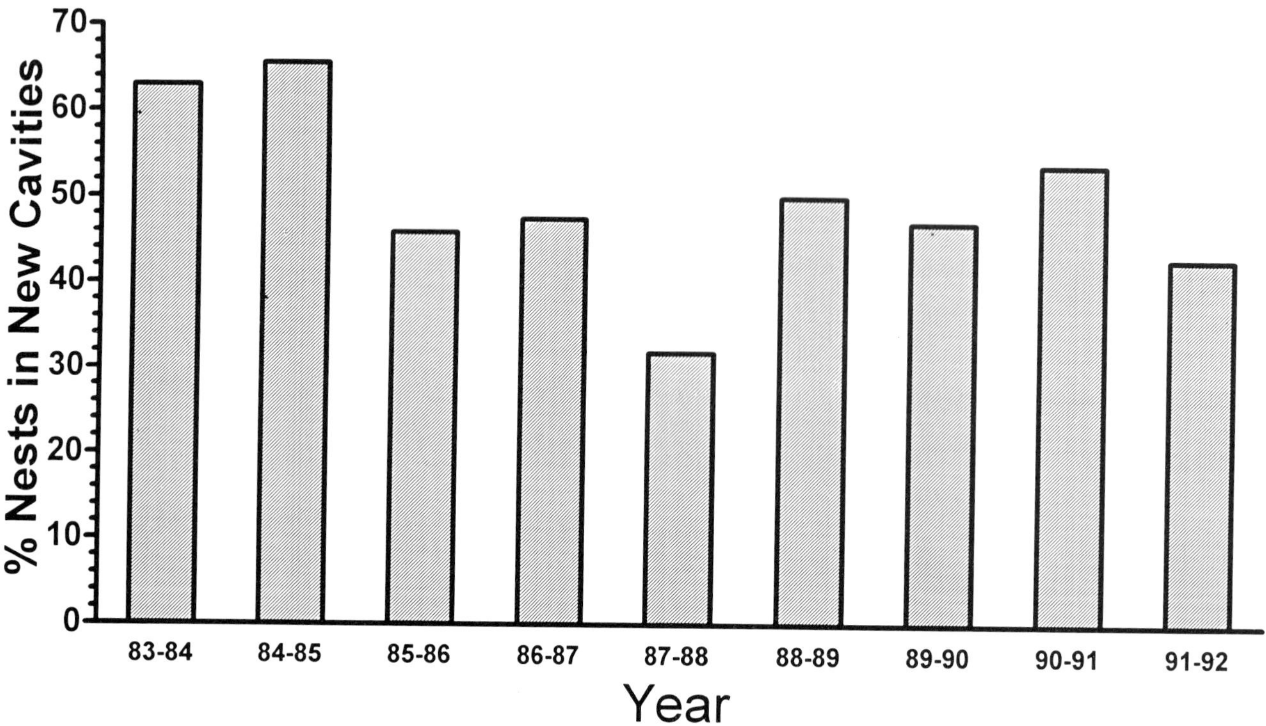

Figure 1. Percent of red-cockaded woodpecker clans that changed nest cavities from the previous years. Data from the Piedmont Region, Georgia.

nesting seasons, and 4 (2.3%) for 6 nesting seasons. Trees that were used during multiple nesting seasons were not always used in consecutive years. For example, of the 4 trees that were used for 6 years, only 2 were used for 6 consecutive years. One of the other trees was used in 1983 and 1984, not used from 1985 through 1988, and used again from 1989 through 1992. The reuse of this tree after 4 nesting seasons was likely due to the addition of restrictors. In 1985 all 3 cavities in this tree were enlarged, probably resulting in the move to a new nest tree. However, in July 1988, restrictors were placed on all 3 cavities and this tree was subsequently used in the following 4 nesting seasons (1989 through 1992). In another colony, Tree #17 was used in 1984, 1985, 1987, 1988, 1990, and 1991. Another tree was used in 1986, even though Tree #17 appeared to be in good condition. No nesting occurred in 1989, perhaps because both the principal male and female were new. Moving to a new nest cavity did not appear to affect the reproductive success of RCWs **(Table 2b)**. Clans that switched nest cavities were just as likely to successfully reproduce as clans that remained in the same nest cavity ($X^2 = 1.22$, df = 1, $P > 0.05$). Furthermore, reproductive success for clans that nested in the same cavity as the previous year (1.95 ± 0.11 fledglings per clan) was not significantly different ($t = 0.91$, df = 218, $P > 0.05$) from that for clans that switched cavities (1.81 ± 0.11 fledglings).

DISCUSSION

The only other study to quantify nest tree fidelity of RCWs was conducted on the Francis Marion National Forest (FMNF) in the Coastal Plain of South Carolina by Harlow (1983). Prior to Hurricane Hugo in 1989, the population of RCWs on the FMNF was one of the largest (Costa and Escano 1989) and had been increasing (Hooper et al. 1991a). The large pine sawtimber volumes, availability of old trees, and an intensive prescribed burning program to control the hardwood midstory were likely responsible for the expansion of this population (Hooper et al. 1991a). In contrast, our study was conducted on a stable but small population in an area which probably has suboptimum conditions.

There were several differences between the Piedmont and FMNF populations in nest tree turnover and the factors associated with it. Fidelity to nest trees was moderately lower in the Piedmont population than in the FMNF population. Of 83 nest attempts in the FMNF population, birds used the same nest tree as in the previous year 59.0% of the time compared to 49.6% of the time in the Piedmont population. Fifty-five percent of the nest trees in the FMNF study were used for more than 1 nesting season whereas only 45.0% of the trees in the Piedmont study were used for more than 1 nesting season. Clans in the FMNF were significantly more likely to change nest trees when reproduction was unsuccessful the previous nesting season whereas past reproductive success was not a significant factor associated with changing nest trees in the Piedmont. Further, when the same male was present for 2 consecutive nesting seasons in the FMNF population, the same cavity was used 79.0% of the time. When the same male was present for 2 consecutive nesting seasons in the Piedmont population, the same cavity was used only 51.4% of the time (**Table 1**). Of the 34 cases in which a different cavity was utilized in the FMNF, there were only 12 cases (35.3%) in which the condition of the cavity or tree (another species occupying the cavity, water in the cavity, or the tree had died) appears to be the cause of the change. In contrast, we found that the condition of the tree or cavity was a probable cause for the switch in 65.8% of the cavity changes. Thus, it appeared that factors internal to the dynamics of the clan (e.g., turnover in breeders and reproductive history) were more important in determining nest cavity change in the FMNF and that factors external to the clan (e.g., condition of the cavity, other species) were more important in determining nest cavity change in the Piedmont population. The 2 most frequent probable causes of nest cavity change were cavity enlargement and occupation of the cavity by other species. Other species commonly use RCW cavities (Baker 1971, Dennis 1971a, Jackson 1978b, Harlow and Lennartz 1983, Rudolph et al. 1990a, Loeb 1993), but their impact on RCW populations is unknown. Because we were not able to follow use of roosting cavities by RCWs throughout the year, we could not determine with certainty whether:

1) the cavities were enlarged or usurped by other species leading to abandonment, or

2) whether the RCWs abandoned the cavities which were later enlarged or occupied by other species.

However, the significant correlations between overall cavity use by species other than RCWs and nest cavity change suggest that there are at least some indirect effects of cavity use and enlargement by other species on RCW nest cavity change. Further, even if cavities were abandoned by the RCWs first and later enlarged or occupied by other species, the destruction or occupation of cavities means that they were no longer suitable or available to the RCWs.

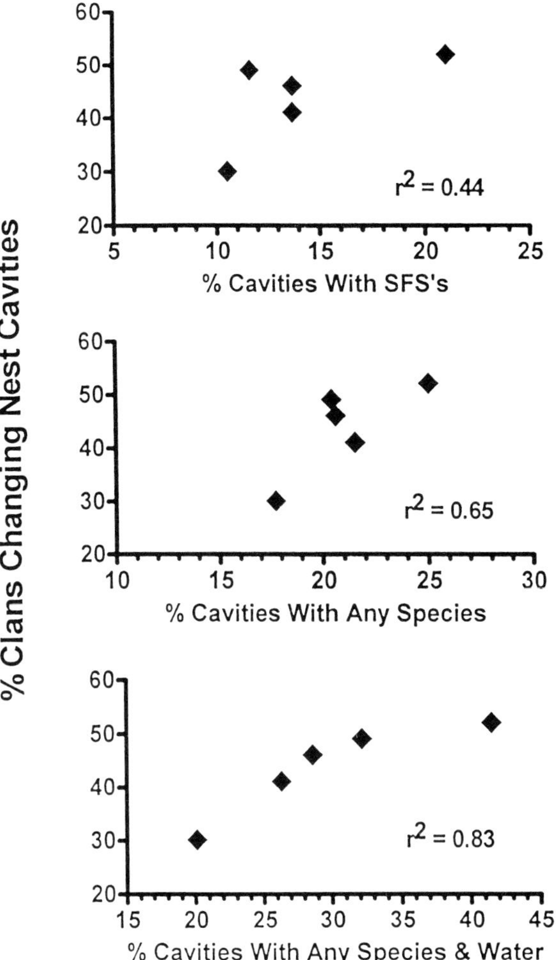

Figure 2. Correlation between the percent of clans that changed nest cavities and the percent of all cavity trees that had: a) southern flying squirrels (SFS's), b) any species of vertebrate, and c) any species of vertebrate or water. Data from the Piedmont Region of Georgia.

Cavity enlargement can be prevented from occurring by placing a restrictor plate to encircle the entire cavity entrance. In some cases, restrictor plates can be used to "rehabilitate" cavities that have been enlarged and extend their use for many years. However, flying squirrels also prefer non-enlarged cavities (Rudolph et al. 1990a, Loeb 1993). Thus, use of full restrictor plates may lead to increased cavity use by southern flying squirrels. On the other hand, the greater availability of suitable cavities for both RCWs and flying squirrels may decrease potential competition.

Because we did not have data on prior use of trees at the beginning of the study and because trees that are presently being used may be used for several more years, length of cavity use is slightly underestimated. Duration of cavity use in our study was similar to that found for RCWs on the FMNF (Harlow 1983). Although the total yield and duration of oleoresin flow from loblolly pine is less than that for longleaf pine (Hodges et al. 1977), loblolly pine appeared to be adequate for nest cavities for RCWs. Several trees were used for 4 or more nesting seasons and in some cases this represented use over 6-8 years. Nest cavity changes due to cavity tree mortality were also similar between the Piedmont and FMNF. 3 of the 34 (8.8%) cavity changes on the FMNF were probably due to death of the previous year's nest cavity tree (Harlow 1983) whereas 10 of the 111 (9.0%) nest cavity changes on the Piedmont probably resulted from death of the cavity tree. In 4 cases, RCWs continued to use the nest cavity from the previous year even though the tree had died in the intervening year. However, 2 of the 4 nests in dead trees were destroyed by black rat snakes (*Elaphe obsoleta*) when the young were approximately 8 and 19 days old. This observation lends support to the hypothesis that RCWs maintain a sticky resin barrier around the cavity to prevent snakes from gaining access (Dennis 1971a, Rudolph et al. 1990b) and suggests that nesting in dead trees can have significant costs for RCWs. Measures to prevent cavity tree mortality from southern pine beetle attack as outlined by Conner et al. (1991a) should be taken.

We detected no significant effects of changing nest cavities on the reproductive success of the RCWs. However, loss of cavities to other species, enlargement, and cavity tree mortality may have effects that we were not able to measure. For example, the increased energy demands of excavating a new cavity and the decreased time available for foraging may affect the RCWs survival. Fewer suitable cavities also may decrease fledgling survival. And, colony sites may be abandoned if alternate cavity trees or suitable potential cavity trees are limited. Several Piedmont colony sites have been abandoned during the 10 years of this study. The population has remained stable because new colonies or clusters have been formed through budding (Hooper 1983). However, had existing colony sites not been abandoned while new colonies were being formed, the population might have increased. Therefore, cavity loss, particularly if it leads to colony abandonment may affect overall population trends.

The Relative Effectiveness of Artificial Cavity Starts and Artificial Cavities in Inducing the Formation of New Groups of Red-cockaded Woodpeckers

Jeffrey R. Walters[1,3], Pamela P. Robinson[1], Warren Starnes[2] and Janice Goodson[1]
[1]Department of Zoology, North Carolina State University,
Box 7617, Raleigh, NC 27695-7617
[2]USDA Forest Service, Croatan Ranger District,
141 E. Fisher Avenue, New Bern, NC 28560

ABSTRACT: Construction of artificial cavities has proved to be an effective way to induce red-cockaded woodpeckers (*Picoides borealis*) to occupy previously unoccupied habitat and thereby form new social groups. Construction techniques are tedious, time-consuming and difficult to learn, however. If simpler techniques could be devised, the labor and fiscal resources necessary to employ such techniques to stimulate population expansion would be greatly reduced. It is much faster and easier to drill cavity starts in live pine trees than to drill complete cavities as has been done in previous studies. To determine whether drilled cavity starts alone would be as effective as drilled completed cavities in attracting red-cockaded woodpeckers to unoccupied habitat, we drilled five cavity starts in each of eight sites that had been abandoned by the birds in Croatan National Forest, North Carolina. We drilled three cavity starts and two complete cavities in each of eight other abandoned control sites. We also constructed five cavity starts in each of nine sites that had never been used by the birds. Only two of eight control sites were occupied two years after drilling, and two of eight experimental sites were. One of nine sites in previously unused habitat was occupied. The low rate of response to drilled cavities may be due to the large number (25) of drilled sites relative to the population size (38 breeding groups in 1990). The experiment indicates that cavity starts alone can induce the formation of new social units, but leaves in question whether they are as effective as complete cavities in this respect.
KEYWORDS: cavities, cavity starts, drilling, North Carolina
[3]Department of Biology, Virginia Polytechnic Institute and State University, Blacksburg, VA 24061-0406

The red-cockaded woodpecker (RCW) is an endangered species endemic to the southeastern United States. It is characterized by several unusual traits, including breeding cooperatively and constructing cavities for roosting and nesting in living pine trees (USFWS 1985; Walters 1990). Its population dynamics also are unusual, due largely to behavior linked to the cooperative breeding system (Walters 1991; Walters et al. 1992a). One example important to conservation efforts is that birds rarely occupy habitat that does not already contain usable existing cavities. The birds prefer competing for existing clusters of cavity trees to constructing new clusters in previously unoccupied areas. Thus new groups form at very low rates in most populations, primarily through the process of budding, in which one group splits into two. Typically the rate of budding is too low to counter rates of loss of existing groups, let alone allow for population expansion (Walters et al. 1988). There is circumstantial evidence that populations occasionally may expand at higher rates (Hooper et al. 1991a), but the conditions under which this may occur are unclear. Therefore efforts to expand populations currently depend heavily on artificial cavity construction.

In 1989 construction of artificial cavities using a drilling technique (Copeyon 1990) was employed successfully in the North Carolina Sandhills to induce red-cockaded woodpeckers to occupy previously unused habitat and thereby form new social groups (Copeyon et al. 1991; Walters et al. 1992a). This technique and others developed subsequently (Taylor and Hooper 1991; Allen 1991) have since been used to induce birds to remain in territories in which previously existing woodpecker-excavated cavities had been destroyed (Hooper et al. 1990). Additionally the widespread construction of artificial cavities in previously unoccupied habitat during the past few years across the Southeast attests to the promise these techniques hold for stimulating population expansion and inhibiting population decline (Gaines et al. 1995; Reinman 1995; Richardson and Stockie 1995; Watson et al. 1995).

These techniques are not without their problems, however. If in drilling the vertical chamber of the cavity in the heartwood of the pine the surrounding sapwood is breached, birds can be killed by becoming trapped in the sap that leaks into the cavity (Copeyon 1990). To learn to drill cavities properly so that such problems can be avoided requires extensive training. Existing techniques are tedious and time-consuming, which often limits the number of cavities a manager has the resources to construct. If simpler techniques could be devised, artificial cavity construction could be used more widely and more rapidly in meeting management objectives. Currently there are no known alternatives to cavity construction that have proven successful in stimulating population expansion through formation of new RCW groups (but see Hooper et al. 1991a).

It is much faster and easier to drill cavity starts than to drill complete cavities that include the vertical cavity chamber. The purpose of this study was to determine if drilled cavity starts alone could induce occupation of previously unoccupied habitat, and if so, if they were as effective as drilled completed cavities in this respect. To do this we constructed cavity starts in a set of experimental sites in the Croatan National Forest in North Carolina, and compared the response of red-cockaded woodpeckers to these sites to their response to a set of control sites in which we drilled complete cavities. We determined that the birds did indeed occupy sites in which only cavity starts were constructed, but a low overall rate of response precluded us from determining if sites with only cavity starts were as effective as sites with complete cavities.

METHODS

The Croatan National Forest is located in the Coastal Plain of North Carolina, and includes 157,851 acres. These acres are managed for a variety of natural resources. The plant communities of the Croatan National Forest range from xeric pine / scrub oak to pocosin wetlands (Wells and Shunk 1931; Snyder 1978; Christensen 1988). The pocosin wetlands, which cover more acres than any other habitat type, are characterized by a diverse, dense shrub understory and a sparse overstory of pond pine (*Pinus serotina*) (Christensen 1981b). The pocosin wetlands grade into mesic and wet pine flatwoods characterized by an overstory of longleaf pine (*P. palustris*) and/or loblolly pine (*P. taeda*), a diverse understory of small shrubs and a diverse ground cover dominated by wire grass (*Aristida stricta*). Pine / scrub oak communities occur on sandy soils in the southern portion of the Forest. These communities have an open canopy of longleaf pine, a sparse understory dominated by several oak species (*Quercus* sp.) and a ground cover dominated by wire grass. In both flatwoods and pine / scrub oak communities, the understory layer varies from being high and dense in areas from which fire has been excluded to being low and sparse in other areas, most of which have been burned in recent years.

RCWs on the Croatan are concentrated in the flatwoods and pine/scrub oak communities, although they also occur in pocosins. About 10,000 acres were being managed for the woodpecker, reflecting the amount of habitat the woodpeckers currently used. Much of the habitat that was not used was pocosin located in the center of the Forest. The population consisted of four solitary males and 38 breeding groups for a total of 110 adults in the 1990 breeding season (**Fig. 1**).

Cavities were constructed in the fall and winter of 1990-1991 by Forest Service personnel. A few cavities were constructed using the technique described by Copeyon (1990), and the remainder by a modified version of this technique described by Taylor and Hooper (1991). Briefly, the horizontal entrance tunnel was drilled first, and then another tunnel was drilled vertically beginning

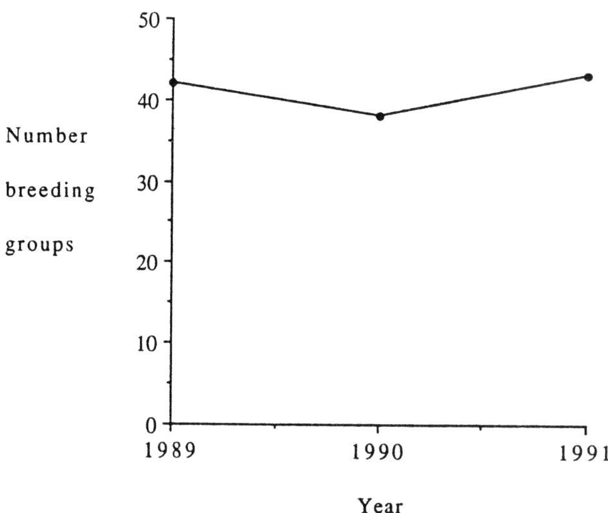

Figure 1. The number of breeding groups (groups containing an adult female) of red-cockaded woodpeckers on the Croatan National Forest, 1989-1991.

above the entrance tunnel, intercepting the entrance tunnel in the heartwood. The accesstunnel was then extended and reamed out to form the cavity chamber. The portion of the access tunnel above the entrance tunnel was then sealed off. The technique we used for drilling cavity starts was described in Copeyon (1990). This consisted simply of drilling a horizontal tunnel that angles slightly upward through the sapwood and into the heartwood, and reaming out a wider area at the end of the tunnel to represent the beginning of the cavity chamber. Thus cavity starts essentially corresponded to completed entrance tunnels.

We selected 16 existing inactive cavity clusters in which to drill. These clusters contained old woodpecker cavities that were no longer used by the birds. No birds were roosting in these sites prior to drilling. We paired the clusters according to characteristics such as habitat type and location within the Forest, and determined which member of each pair was to be an experimental site and which a control site by flipping a coin. We drilled five cavity starts in each of the experimental sites, and two complete cavities and three cavity starts in each of the control sites. We also selected nine areas of suitable habitat that had never been used by RCWs, that is, contained neither birds nor old cavities. These sites were located 1-4 km from the nearest existing cavity cluster. In each of these vacant sites we also drilled five cavity starts. In all sites, whether experimental, control or vacant, we cleared any existing understory vegetation within 10 m of the trees in which cavities were drilled.

Trees in which cavities and cavity starts were constructed were selected based on the criteria outlined by Copeyon (1990). All but a few of the trees that contained sufficient heartwood to meet the selection criteria were over 100 years of age. Nearly all cavities were constructed in longleaf pine, a few in loblolly pine. Only one cavity start in one site was in a loblolly pine among the vacant sites (of 45 total cavities), and only one complete cavity and one cavity start (in different sites) among control sites (of 40 total cavities and cavity starts). All of the cavities and cavity starts in each of two experimental sites were in loblolly pine, and four of five in a third experimental site (of 40 total cavities). The cavities were screened initially until we determined that any sap flow within them had ceased. All were made available to the birds simultaneously by removing the screens just prior to the 1991 breeding season. All sites were then checked regularly for active resin wells, an accurate indicator of use of cavities by RCWs (Jackson 1977b). Sites were checked during both the breeding and non-breeding seasons through the fall of 1992. Active sites were monitored during the breeding season using the protocol described in Walters et al. (1988). Using this protocol we monitored reproductive activity and censused the birds using the site. Nearly all birds in this population were individually identifiable from color bands at the time the experiment began.

RESULTS

In the breeding season of 1992, the second breeding season since the cavities and cavity starts were constructed, two of the eight control sites were active, and two of eight experimental sites were active **(Table 1)**. Two additional control sites and one additional experimental site were used in the 1991 breeding season, but were abandoned by the 1992 breeding season. Thus a total of four control sites and three experimental sites were used over the two years of the experiment **(Table 1)**. The low level of overall response to drilled cavities precludes detecting any difference in effectiveness of cavity starts compared to complete cavities.

Only one vacant site was used, and it was occupied in the 1992 breeding season **(Table**

1). Overall, combining vacant and experimental sites, four of 17 sites containing only cavity starts were used by the birds in the two years of the experiment.

Most of the sites were used by unpaired males. One site, an experimental site, was occupied by a male-female pair during the 1992 breeding season, but they did not attempt nesting. Two of the control sites, one each in 1991 and 1992, were occupied by floater females (Walters et al. 1988), that is, by females that had not yet settled in a territory with a male and were using the site temporarily for roosting. Birds that occupied sites with only cavity starts continued excavation, in three cases completing construction of the cavity chamber. All of the cavities used were in longleaf pine.

Response to constructed cavities and cavity starts was by no means complete by the 1992 breeding season. Two additional experimental sites were first used in the fall of 1992. Whether this use proves temporary or persists into the 1993 breeding season remains to be seen. In all other cases, use first detected in the non-breeding season persisted into the following breeding season.

Table 1. Response of red-cockaded woodpeckers to drilled completed cavities and drilled cavity starts that were installed at various sites in the Croatan National Forest, North Carolina.

	1992 Breeding Season Only		Calendar Years 1991 and 1992	
Treatments	Used	Not Used	Used	Not Used
Experimental[1]	2	6	3	5
Controls[2]	2	6	4	4
Vacants[3]	1	8	1	8

[1]Initially inactive cavity tree clusters to which five drilled cavity starts were added.

[2]Initially inactive cavity tree clusters to which two drilled completed cavities and three drilled cavity starts were added.

[3]Sites without cavity trees to which five drilled cavity starts were added.

DISCUSSION

The low rate of use of drilled cavities contrasts with previous results from the North Carolina Sandhills. In the Sandhills, 90% of sites in which cavities were drilled were occupied in the first breeding season following drilling (Walters et al. 1992a), whereas in the Croatan National Forest only 50% of the clusters to which complete cavities were added were used within two years. One possible explanation is that the drilling techniques used were slightly different. The cavities drilled in the Croatan National Forest were among the first constructed by the Forest Service using their technique (Taylor and Hooper 1991) and we experienced more problems with sap leakage than occurred in the Sandhills study in which Copeyon's (1990) technique was used. However, since cavities remained screened until the leakage stopped, this seems unlikely to account for the difference in cavity use.

The difference in techniques seems an unlikely explanation given that the rate of response to cavity starts was also low and that the differences between the techniques are small. A more likely explanation lies in the difference between the Croatan National Forest and the Sandhills in the ratio between previously existing groups of birds and drilled sites. In the Sandhills 20 sites were drilled in an area inhabited by 217 groups of birds (Copeyon et al. 1991; Walters et al. 1992a). In the Croatan National Forest the ratio was 25 drilled sites to 38 groups of birds. Perhaps in the Forest there were too few non-breeding individuals (helpers, floaters, dispersing fledglings) seeking breeding positions to ensure that drilled sites would be located by appropriate individuals. In the Sandhills new groups were formed by birds from these non-breeding classes that moved to drilled sites (Walters et al. 1992a).

If occupancy depended primarily on the chances that drilled sites were encountered by non-breeding birds, one might expect drilled sites in portions of the Forest containing few previously existing groups to be less likely to be occupied than those in parts of the Forest with many existing groups. It was not clear that this was the case. Three drilled sites were located in the southeastern corner of the Forest where only one solitary male lived at the start of the experiment. The closest other birds were more than 10 km away. Yet one of these three sites was occupied. Eight drilled sites were located within the area of highest woodpecker density, containing 19 groups and one solitary male. Only three of these eight sites were occupied. Overall, if one sums all portions of the Forest in which the ratio of previously existing groups to drilled sites is less than one, two of ten drilled sites were occupied. In areas where the ratio was greater than one, five of 15 sites were occupied. It seems that degree of isolation of a drilled site is less important in determining probability of occupancy than the size of the nonbreeding classes relative to the number of drilled sites.

Another possible explanation of the difference in rate of response to constructed cavities between Croatan National Forest and the Sandhills is that the sites in which cavities were placed may have been superior in some respect in the Sandhills. For example, the quality of foraging habitat might differ. This, too, seems an unlikely explanation of our results. All of the sites in the Forest met USFWS foraging habitat guidelines, both in terms of quantity and quality.

In the Sandhills several drilled sites were occupied by previously existing groups who shifted from their previous cluster to a new drilled cluster (Walters et al. 1992a). This did not occur at all in this study. Generally drilled sites were farther removed from existing occupied clusters on Croatan Forest than in the Sandhills, but not invariably. Perhaps on the Forest existing groups had more suitable cavities than in the Sandhills, and thus were less attracted to new cavities removed from their existing ones.

That drilled cavity clusters were initially occupied by unpaired males rather than breeding pairs is not surprising. This was also observed in the Sandhills (Walters et al. 1992a). In the Sandhills unpaired males eventually obtained mates, and clusters initially occupied by unpaired males eventually became productive breeding sites (Walters et al. 1995a). Presumably this same process will be repeated in the Croatan National Forest.

It is difficult to evaluate the effectiveness of cavity starts in inducing birds to occupy new or previously abandoned areas due to the low level of response to both control and experimental sites. However, it is at least clear that cavity starts alone can induce the formation of new groups since some sites containing only starts were occupied. The trends in the data suggested that cavity starts may not be as attractive to RCWs as complete cavities. However, given that there was some response, coupled with the many advantages of drilling starts rather than complete cavities described earlier, further use of this technique is warranted. Some loss in efficiency is acceptable due to gains in the number of sites that can be drilled and reduction of risk to the birds. Further studies in larger populations will be necessary to determine if a difference in efficiency exists, and if so, if it is sufficiently small that drilling of cavity starts instead of complete cavities is justified.

CONCLUSIONS

Although the overall rate of response to drilled cavities and drilled cavity starts was low, we were able to demonstrate that cavity starts alone can induce the formation of new groups of red-cockaded woodpeckers in previously unoccupied areas. Thus even this minimum level of effort can produce some success. Clusters with only cavity starts may prove less attractive than clusters with complete cavities, although we did not have sufficient data to demonstrate this. If the difference in effectiveness is sufficiently small, given that more drilling can be accomplished using cavity starts with a given amount of labor and money than can be accomplished using complete cavities, using cavity starts may be a more efficient use of resources. Further research in areas of higher population density is needed to explore this possibility.

ACKNOWLEDGMENTS

Funding was provided by the USDA Forest Service and the North Carolina Agricultural Research Service. W. E. Taylor and C. S. Bachler constructed the cavities, and T. Bohannon and K. Brust assisted with data collection. We thank the editors for providing comments on an earlier version of the paper.

Use of Artificial Cavities for Red-cockaded Woodpecker Mitigation: Two Studies

J.H. Carter III[1], R. Todd Engstrom[2], and Patricia M. Purcell[1],
[1]Dr. J.H. Carter III & Associates, Southern Pines, NC 28388,
[2]Tall Timbers Research Station, Tallahassee, FL 32312

ABSTRACT: Provisioning inactive and new red-cockaded woodpecker (*Picoides borealis*) (RCW) colony sites with artificial starts and cavities was used as mitigation for highway construction impacts to existing RCW colony sites in Georgia and North Carolina. Three sites were provisioned in Georgia and 4 were provisioned in North Carolina. After 2 breeding seasons, 2 new breeding groups and a solitary territorial male were established in Georgia. Two solitary territorial males had occupied provisioned sites in North Carolina after 1 year. Careful implementation of the mitigation techniques described herein show promise for offsetting unavoidable adverse impacts within RCW populations.
KEYWORDS: red-cockaded woodpecker, mitigation, artificial cavity, provisioning

In December 1990, the Georgia Department of Transportation initiated a mitigation project for impacts caused by highway construction to a red-cockaded woodpecker (*Picoides borealis*) (hereafter, RCW) colony site in southwestern Georgia. The impacts resulted from the widening of U.S. Highway 319 between Thomasville and the Florida state line in Grady and Thomas Counties. The mitigation project was in response to a Jeopardy Opinion issued by the U.S. Fish and Wildlife Service (USFWS) for adverse impacts to the foraging habitat of 1 active RCW colony site. Mitigation involved the construction of 3 artificial RCW colony sites, with the goal of establishing 1 new territorial male RCW.

The North Carolina Department of Transportation (NCDOT) began a mitigation project in August 1991 to compensate for adverse impacts to 2 RCW groups caused by the widening of U.S. Highway 15-501 in Moore County, N.C. The 15-501 project had resulted in a Jeopardy Opinion which was issued by the USFWS on 15 February 1991. As a Reasonable and Prudent Alternative, NCDOT agreed to create or rehabilitate 4 RCW colony sites on State-owned land at McCain, Hoke County, N.C. This paper summarizes the results of these efforts.

PROJECT AREAS

The Georgia project area is within the "Plantation Country" or "Red Hills" of southwest Georgia. The landscape is characterized by very large quail plantations, some of which are thousands of acres in size. These plantations are managed for hunting, forest products, and agriculture. There are large expanses of mature pine forest, which has a long history of prescribed burning. This habitat is excellent for RCWs, and though this species is known to occur throughout the area, the exact distribution and status of colony sites have been poorly documented.

The McCain property is located in western Hoke County in the southern North Carolina Sandhills. It consists of approximately 1700 acres owned by the State of North Carolina (Department of Agriculture and Division of Prisons). McCain is bordered by the Fort Bragg (FB) Military Reservation on the east, and private lands on the other sides. Surrounding land uses include military training, agriculture, forestry, and rural residential housing.

MITIGATION SITE DESCRIPTIONS

Georgia

The 3 areas chosen for artificial colonies in Georgia were Pebble Hill Plantation,

Springwood Plantation, and Elsoma Plantation (**Figures 1 and 2**). The Pebble Hill site was located southwest of Thomasville about 1 mile west of U.S. 319. It was forested with mature longleaf (*Pinus palustris*), shortleaf (*P. echinata*), and loblolly pines (*P. taeda*), and there were some midstory and overstory hardwoods. One active RCW colony (PH1) was located approximately 4660 feet southeast of the artificial colony (PH2).

The Springwood site was located southwest of Thomasville, approximately 1 mile east of U.S. 319. It was about 2 miles from the Pebble Hill site. The area was forested with mature longleaf pine, with some slash pine (*P. elliottii*) in flatwoods and around sinks. Some stands of mixed pines also occurred onsite. Three active RCW colonies were found in the immediate area. One (S3) was located 2900 feet north of the artificial colony (S4), another (S2) was 4365 feet to the south-southwest, and the third (S1) was about 6000 feet to the southwest (**Figure 1**). The latter 2 colonies were separated by a large swamp from the habitat around the artificial colony.

Elsoma Plantation was located south of Thomasville and was several miles east of the other 2 mitigation sites. It was forested with mature stands of mixed pines, with considerable hardwood midstory in some areas. Elsoma was reported to have only a single abandoned RCW colony (E1), but surveys located 4 additional colonies. Three colonies were active (1 had been recently reactivated-E2), and 1 was recently abandoned (E4). The recently abandoned colony was chosen as a mitigation site (see below). The 2 nearest colonies, both active, were located 4290 feet to the southeast (E5) and 1650 feet to the northeast (E3), respectively (**Figure 2**). Several other active colonies occurred on nearby Arcadia Plantation, but were more than 1 mile away (**Figure 2**). The use of Elsoma was critical to the fulfillment of the mitigation project, primarily because insufficient time was available to seek another location by the time it was determined that there was not room to add a completely new colony. The USFWS agreed that the recently abandoned colony (E4) could be rehabilitated with new cavities and starts and substituted for 1 of the new colonies.

North Carolina

McCain was a heavily forested tract, and had been managed in recent years primarily for the production of pine straw. There were 2 extensive stand types on the tract: longleaf pine-scrub oak (*Quercus spp.*) and loblolly pine plantation. Longleaf pine was found on undisturbed sites. Stands averaged approximately 68 years old and varied in density from quite sparse to very dense. Typical longleaf community types were Xeric Sandhill Scrub, Pine-Scrub Oak Sandhill (most common), and Longleaf Pine Woodland. The latter type supported the densest stands of timber. Loblolly pine plantations occurred on old field sites, averaged 52 years old, and were heavily stocked. Other community types present at McCain included Xeric Upland Hardwood, Streamhead Pocosin, beaver pond, and a pond cypress (*Taxodium ascendens*)-shrub Carolina Bay.

There were 5 active RCW colonies on McCain in 1981 (3 breeding groups and 2 solitary males) (**Figure 3**). In addition there were 2 breeding groups (FB 37 and 53) and 2 abandoned colonies (FB 46 and 52) on FB immediately adjacent to the McCain boundary, and other FB colonies were located slightly further east (**Figure 3**). In 1989, an experimental colony site with artificial cavities (MCCA E01) was constructed by personnel from N.C. State University (NCSU). This colony was occupied by a breeding group in 1990, a solitary male in 1991, and a breeding pair in 1992. The 5 original McCain colonies have not fared well. There was considerable abandonment in the early to mid 1980's, and by 1990, only a single breeding group remained (MCCA 04). The male from this group moved to MCCA 03 in 1991, where he was a solitary nonbreeder. All of the other McCain colonies were inactive. There have been 2 status changes in the adjacent FB colonies. FB 53 has become inactive, and FB 46 was briefly reactivated before becoming inactive again.

METHODS

Georgia

Several potential artificial colony locations were located in the Georgia project area by W. Wilson Baker of the Red Hills Conservation Association during the winter of 1990-91. Three sites were chosen for more intense ground surveys in late February 1991. These sites were surveyed on foot to document the presence or absence of RCW cavity trees, and to locate adjacent RCW colonies. Information from these surveys was pooled with existing data supplied by local foresters and biologists to produce a map of all RCW colonies in the immediate mitigation area **(Figures 1 and 2).**

Foraging habitat analyses were conducted at all 3 mitigation sites. Procedures contained in "*Guidelines for the preparation of biological assessments and evaluations for the Red-cockaded Woodpecker*" (Henry 1989) were followed. One-tenth acre plots were set up at 10 chain intervals on linear transects 10 chains apart. Data collected included the age of a representative pine tree, pine basal area (BA), and the number of pine stems >10" diameter at breast height (dbh). Since nearly all sampled pine stems were over 40 years old, and most pine stands contained mixed age classes, pine stands were stratified by density or not at all (all stand data pooled). Data from like stand classes were averaged and the average pine BA and stems >10" dbh were then multiplied times the acreage for that stand class, which was determined with a dot grid. The pine BA and stems >10" dbh were then added across classes to get the total pine BA and total stems >10" dbh for each site.

Although one-half mile radius foraging circles were utilized to some extent, the sampled habitats were primarily determined by the location of adjacent active colonies and non-contiguous habitat boundaries (mostly hardwood swamps and lakes). Thus some areas slightly more than one-half mile from artificial colonies were counted as foraging habitat, while some areas within one-half mile were not. In areas where there was considerable circle overlap, all available habitat in the general area was pooled, then divided by the number of colonies present. The goal was to document adequate habitat for an artificial colony, not to fully characterize the amount of foraging substrate within one-half mile of such a site.

Color-banding of adult RCWs was initiated in colonies adjacent to the proposed mitigation sites in late February 1991. Adults were captured at roost time using a hoop net attached to a telescopic pole. An attempt was made to band all RCWs in colonies within 1 mile of the proposed mitigation sites. This effort involved 1 colony at Pebble Hill (PH1), 2 at Springwood (S2 and S3), and 7 colonies in and around Elsoma (3 on Elsoma- E2, E3, and E5, and 4 on nearby Arcadia Plantation- A5, A6, A10, and A13) **(Figures 1 and 2)**. An attempt was also made to band all nestling RCWs in these colonies. Nestlings were removed from a cavity with a flexible noose, banded and returned to the cavity.

Between 11 and 14 March 1991, artificial starts and cavities were placed in the 3 chosen colonies (PH2, S4, and E4) using techniques described in Copeyon (1990). Three starts and 2 completed cavities were placed in each colony, with the exception of S4, where 4 starts and 2 cavities were constructed. Each start/cavity was temporarily screened to prevent RCWs from entering the sappy entrance tunnels. Most screens were removed by 21 April 1991, and all were removed by 18 May 1991.

North Carolina

During the late summer and fall of 1991, McCain and a one-half mile radius around it were surveyed for RCW cavity trees and colonies. Though a few previously unknown cavity trees were detected, no new colonies were found. Four colonies were chosen for the placement of artificial cavities. These were the existing inactive colonies MCCA 01, 02, and 04, and a new site (MCCA E02) located between MCCA 03 and 05 **(Figure 3)**. These sites were chosen because of the presence of suitable numbers of old-growth pines that were needed for artificial cavity construction. MCCA 05 was not used because it lacked old-growth pines.

One-half mile radius foraging circles were set-up around the center of each mitigation colony and around all adjacent active colonies **(Figure 3)**. Plots were set-up and data collected as were described previously. The methodology was sufficient to obtain at least several sample points in each of the major stand types. For analysis purposes natural longleaf stands were separated into 3 density classes (dense, moderate, and sparse). Other stand/habitat types were mixed longleaf-loblolly pine, longleaf plantation, loblolly plantation, and pocosin. Nonforested areas, hardwood and hardwood-pine stands, and pine plantations less than 30 years old were considered nonforaging habitat. Data from plots in like stands were pooled, and averages for BA and stems >10" dbh were obtained for each type. Separate averages were obtained for stands in the MCCA 02-03-04-E02 circles and the MCCA E01-01/FB 37-38-39-47 circles.

The total pine BA and total stems >10" dbh for a type were calculated as described above.

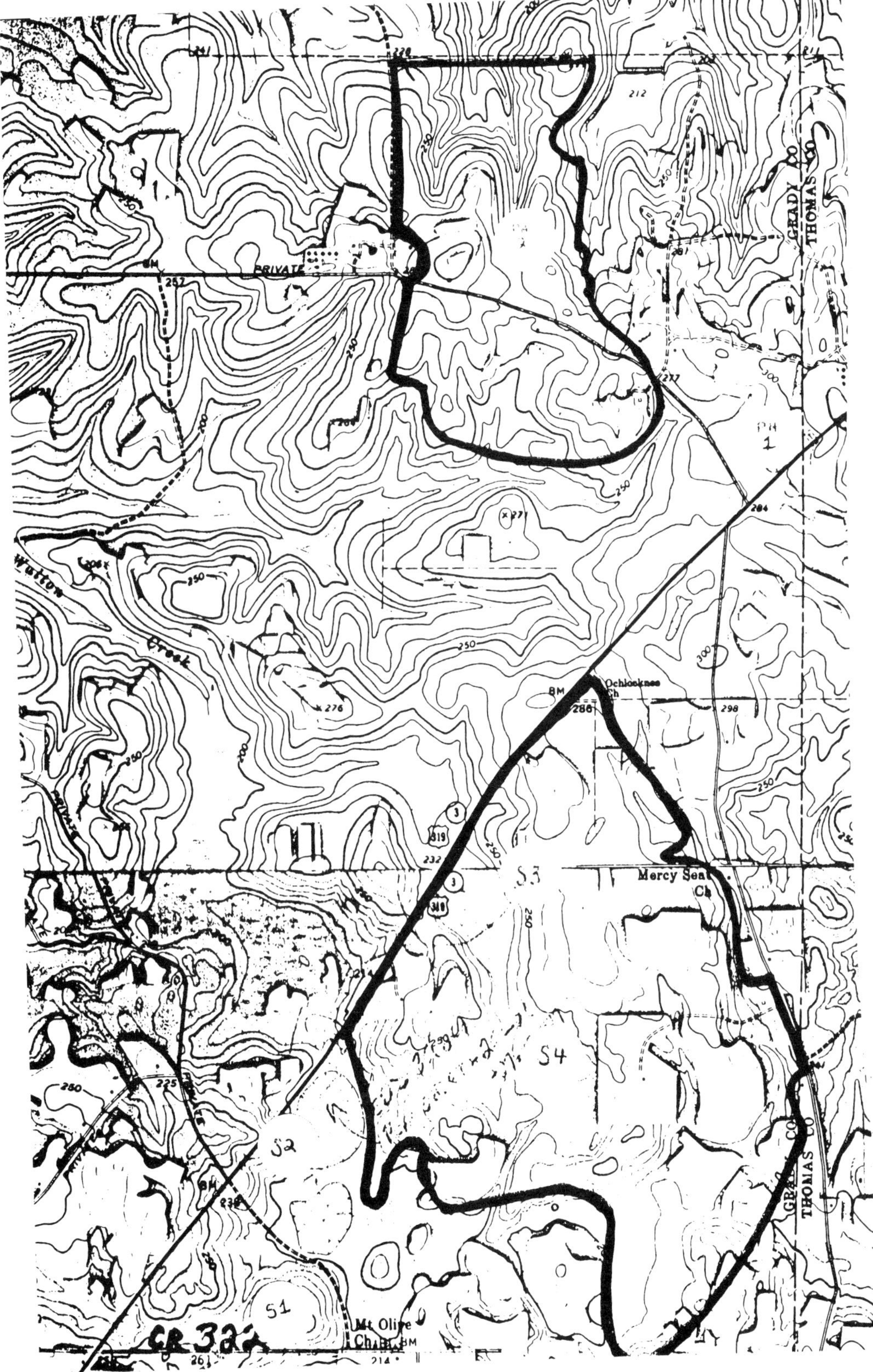

Figure 1. Colony site locations and foraging habitats evaluated at Pebble Hill and Springwood Plantations. Heavy lines outline areas included in foraging habitat analyses. PH=Pebble Hill; S=Springwood; colonies with artificial cavities are PH2 and S4.

Figure 2. Colony site locations and foraging habitat evaluated at Elsoma Plantation. Heavy line outlines area included in foraging habitat analysis. E=Elsoma; A=Arcadia; colony with artificial cavities is E4.

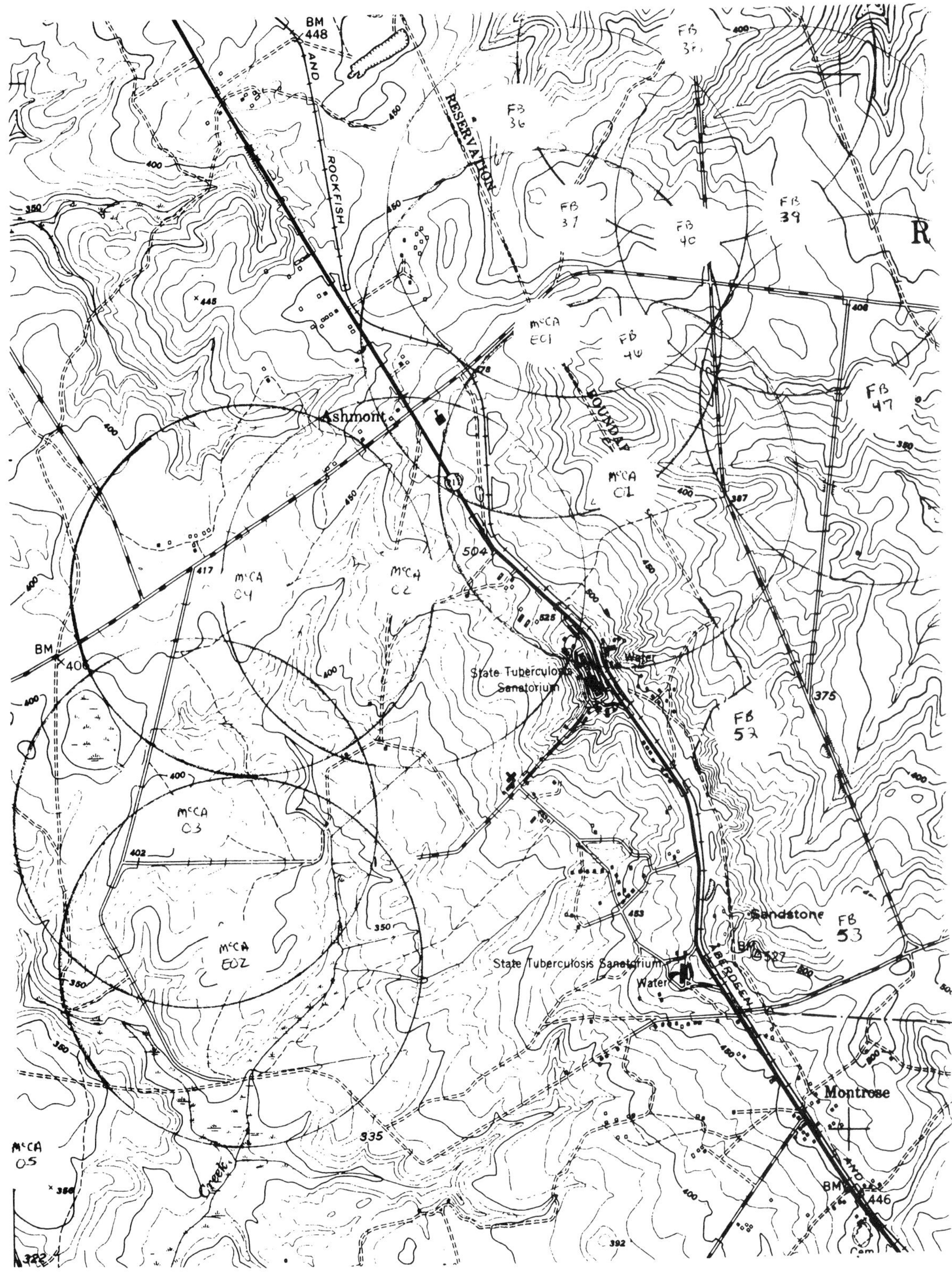

Figure 3. Locations of red-cockaded woodpecker foraging circles and abandoned colony sites at and adjacent to McCain, Hoke County, N.C.

These stand totals were then added for each circle to get the circle totals. Foraging habitat in areas of circle overlap was divided by the number of overlapping circles. A foraging circle must have at least 8490 sq. ft. of pine BA and 6350 pine stems >10" dbh in stands contiguous to the colony site in order to have sufficient foraging habitat (Henry 1989).

RCWs in the McCain area were already color-banded since McCain was within the area of the long-term NCSU RCW demographic study. Unbanded adults and nestlings were captured and color-banded during this study as the need arose.

In October and November 1991, MCCA 01, 02, 04 , and E02 were provisioned with 3 starts and 2 cavities each. One constructed cavity and 2 starts were placed in MCCA 03 in order to stabilize that active site. Cavities were temporarily screened to prevent access, but most screens were removed in January 1992, and all screens were removed by March 1992.

MITIGATION RESULTS AND DISCUSSION

Georgia

Adequate foraging habitat was documented at the Springwood (S4) and Elsoma (E-4) colonies, with borderline sufficiency at the Pebble Hill Colony (PH2). The decision was made to proceed with PH2, because only 1 foraging circle overlapped the PH2 circle, and it had over 475 acres of habitat not counted in the analysis.

Most adult RCWs in colonies adjoining the Georgia mitigation sites were color-banded by early April 1991. A few birds could not be caught because their cavities were too high or roost locations were unknown. RCW nestlings were banded in 5 of the adjacent colonies. No nests were found at the existing Pebble Hill colony (PH1) and at 1 of the 3 active Elsoma colonies (E2) during the 1991 breeding season.

RCW activity was first seen at an artificial cavity on 12 April 1991 at the Elsoma colony (E4), and a single RCW was subsequently seen on 2 occasions. Both artificial cavities were active on 1 June 1991, and 2 RCWs were seen on 4 June 1991. One of these birds had been banded along with another male on 26 February 1991 in the colony across the pond to the northeast (E3). The other bird, a probable female, was unbanded. The banded male was not part of the E3 group during the 1991 breeding season. It is possible that he was a territorial bird in the E4 site all along, and was cross-roosting in the adjacent colony (E3). A single RCW had been heard in E4 during the construction of cavities in mid-March. At that time, 2 existing RCW cavities had very small amounts of fresh RCW chipping, but there was no direct evidence of a RCW roosting in the site based upon morning and evening checks. However, even if this RCW does not represent a new territorial male, the artificial cavities have restored breeding potential to this colony by providing quality nest sites. Unfortunately, 1 of the artificial cavity trees died during the summer of 1991, and the unbanded bird disappeared. However, 1 of the artificial starts was completed by the RCWs. On 14 April 1992, an unmarked female RCW was captured and banded in this colony, and on 1 May, a nest was found in tree #5E. The nest failed, but the attempt proved that a breeding pair had become established.

RCW activity was first seen in the Springwood colony (S4) on 18 May 1991, and an adult male was captured and banded on 28 May 1991. Both artificial cavities were active on 1 June 1991. Observations of this bird indicated that he was a territorial male. He did not join any other RCW group, and had an aggressive encounter with the RCW group (S3) to the north. An unbanded RCW, probably a female, was seen in this colony during the fall. Four eggs were found in tree #15E on 28 April 1992, and a single young was fledged. The adult female was banded in July 1992, but by that time the breeding male had been replaced by a year old male from colony S2. The fledgling was not seen after July.

RCW activity was detected in the Pebble Hill colony (PH2) on 22 May 1991, and an adult male was captured and banded on 12 June 1991. Two attempts to follow this bird failed when it disappeared to the south and southeast. No other colonies are known from the immediate area in which this bird headed. He was not part of the other RCW group at Pebble Hill. All available evidence indicates that this bird is also a new territorial male. This bird continued to be a solitary male during the 1992 breeding season.

North Carolina

There was twice the minimal foraging habitat, allowing for circle overlaps, in the MCCA 02, 03, 04 and E02 circles. Foraging habitat was also sufficient, though less so, in the MCCA E01 circle. The MCCA 01 circle had a deficiency of 863.5 square ft. BA, but an excess of 1217.1 stems >10" dbh. However, there was ample pine BA in the overlap area between MCCA E01 and 01 to allow a shift of the needed BA from E01 to 01. With this "paper transfer" accomplished, all the mitigation circles had adequate foraging habitat.

During the late winter of 1992, NCDOT personnel and contractors cleared at least 10 acres in and around each of the provisioned colonies of nearly all understory hardwoods. Most of the understory was removed by mechanical means. During May and June 1992, the N.C. Forest Service burned most of these cleared areas, plus much of the surrounding foraging habitat.

A single provisioned cavity had become active in MCCA 01 by April 1992, though the RCW responsible for this activity was never identified. Single active cavities were found in MCCA 04 and E02 in July, and there was an active start in MCCA 04. Activity continued in these colonies through December 1992, when 2 cavities and 1 start were active in MCCA 04 and 2 cavities were active in MCCA E02. One provisioned start tree was killed in a prescribed burn in MCCA 04 in June. MCCA 01 had become inactive by December, and MCCA 02 has not been active yet.

Two new RCW territories with solitary males have been established at McCain to date. A second-year male from a Camp Mackall site (MACK 17) was seen in MCCA 04 on 24 June 1992, and a hatching year male from MACK 11 was seen in the same colony on 25 August 1992. Those colonies are 6.4 and 9.3 miles southwest of McCain, respectively. The latter bird was still in MCCA 04 in early January 1993, but it was not clear if the former individual was still present (its cavity tree was active). An unbanded male of unknown age was banded in MCCA E02 on 17 November 1992, and he was still there in early January 1993.

The single male RCW in MCCA 03 attracted a mate and they fledged 2 young in 1992. It is not known if the provisioned cavities and understory suppression helped facilitate this success.

CONCLUSIONS

The mitigation goal in Georgia was to establish a single new territorial male in 1 of 3 provisioned colonies. In the 2 breeding seasons since provisioning, 2 new breeding groups and a single territorial male have been established. No negative impacts on surrounding groups have been detected. The goal in North Carolina was to create 2 new breeding groups. In the year since provisioning, 2 new territorial males have been established in the provisioned colonies, with no impacts on adjacent groups. It is expected that 1 or more of these colonies will achieve breeding status in the 1993 breeding season.

The mitigation process described herein has proved highly successful in achieving short-term, local population growth within existing RCW populations. However, it should be noted that in each case, the provisioned colony was placed next to existing occupied territories within major RCW populations. This process is not likely to be as successful in areas with low RCW densities, a highly fragmented population, or no population at all.

Following is a brief step-down list of procedures followed.

1. Locate provisioning sites with suitable potential cavity trees.
2. Survey surrounding habitat for a 1 mile radius in order to document other RCW colonies and foraging circle overlaps.
3. Conduct foraging habitat analyses to document adequacy of foraging habitat.
4. Evaluate mid story conditions and treat if too tall/dense.
5. Construct artificial cavities and starts; monitor for leakage.
6. Color band adult and nestling RCWs to document that new birds in provisioned colonies did not abandon adjacent colonies.
7. Monitor provisioned colonies for RCW activity.

Response to Drilled Artificial Cavities by Red-cockaded Woodpeckers in the North Carolina Sandhills: 4-Year Assessment

Jeffrey R. Walters[1], J. H. Carter, III, Phillip D. Doerr and Carole K. Copeyon[2]
Department of Zoology, North Carolina State University, Box 7617, Raleigh, NC 27695

ABSTRACT: Drilled cavities were first used to induce formation of new groups of red-cockaded woodpeckers in the North Carolina Sandhills in the 1989 breeding season. We initially reported that 18 of 20 sites with drilled cavities were used, resulting in the formation of 12 new social units consisting of 7 pairs and 5 unpaired males. Data from 3 subsequent breeding seasons indicate that the new groups were able to coexist with previously existing groups. One site has never been occupied, another was first occupied after 1989 and a third that was initially occupied in 1989 was later abandoned. By 1992, the number of groups added to the population by cavity construction had increased to 13. There was no obvious effect of the new groups on previously existing groups other than neighboring ones. The number of sites producing young increased from 4 in 1989 to 14 in 1992. The number of sites occupied by solitary males decreased from 5 to 0, and the number occupied by groups that included helpers increased from 0 to 7.

KEYWORDS: artificial nest cavities, North Carolina, population dynamics, productivity

[1]Department of Biology, Virginia Polytechnic Institute and State University, Blacksburg, VA 24061-0406

[2]U.S. Fish and Wildlife Service, Pennsylvania Field Office, 315 South Allen Street, State College, PA.

The red-cockaded woodpecker (*Picoides borealis*) is an endangered species endemic to the southeastern United States, characterized by a highly unusual biology. It is unique in excavating cavities for roosting and nesting in living pine trees. Due to difficulties inherent in excavating through living sapwood, cavity excavation takes months to years to complete (Jackson et al. 1979b; Conner and Rudolph 1995a). The red-cockaded woodpecker also is unusual among North American birds in being a cooperative breeder (USFWS 1985; Walters 1990). Many young males remain in their natal groups as helpers, sometimes for several years (Ligon 1970; Lennartz et al. 1987; Walters et al. 1988). Walters (1990; 1991; Walters et al. 1992a) has linked the cooperative breeding system to the highly unusual population dynamics characterizing this species. He has proposed that because of the high cost of constructing cavities, individuals compete for breeding vacancies on territories with existing cavities rather than move into unused habitat and construct a new set of cavities. This could explain the very low rates at which new groups form in most populations of this species (Walters 1991; but see Hooper et al. 1991a).

In 1989, artificial cavities were constructed using a drilling technique (Copeyon 1990) in order to test this hypothesis in the North Carolina Sandhills. Eighteen of 20 sites in which drilled cavities were placed were occupied in the first breeding season following completion of drilling (Copeyon et al. 1991; Walters et al. 1992a). Some sites were used by previously existing groups that shifted to the drilled cavities from their previous cavity cluster. Twelve new groups formed in response to drilling, including seven pairs and five unpaired males. Another dramatically successful employment of artificial cavities followed in the Francis Marion National Forest. After Hurricane Hugo destroyed 87% of the cavity trees in the Forest, cavities were constructed to prevent the remaining birds from abandoning territories that no longer contained enough suitable cavities (Hooper et al. 1990).

Construction of artificial cavities is now

widely employed as a management tool to stimulate population expansion and halt population decline by reducing abandonment of territories. Although the basis for predicting that the technique will be successful is well grounded in theory (Walters 1991), as well as initial successes, it has not yet been shown that population responses to cavity construction are stable. That is, it has not been demonstrated that new groups that form in drilled sites can coexist with previously existing groups over the long term. Possibly foraging habitat or some other resource is insufficient for long-term coexistence where population density is increased by response to drilled cavities. In this paper we report on the dynamics of the Sandhills population in the 3 breeding seasons subsequent to 1989. We show that the initial population increase stimulated by cavity construction was stable over this period, and that group size and productivity in drilled sites increased considerably. Our data suggest that cavity construction can result in productive new groups that permanently increase population size.

METHODS

The research was conducted within a study area of over 110,000 ha within the Sandhills of North Carolina. The area is forested predominantly with second-growth longleaf pine (*Pinus palustris*), but is diverse in terms of land use and associated vegetation. Habitat is described in detail in Walters et al. (1988) and Carter et al. (1995b). The woodpecker population inhabiting this area has been completely marked since the early 1980s (Carter et al. 1983; 1995b). Each year the population is censused based on individual identifications of color-marked birds, and the reproductive activity of each woodpecker group is monitored. Methods of population monitoring are described in detail in Walters et al. (1988). The final product of population monitoring each breeding season are determinations of the:

(1) activity status of all cavity tree clusters;

(2) identity and status of all adults in all active clusters; and,

(3) number and identity of all young fledged in each group.

The process by which sites were chosen for drilling is described in detail in Copeyon et al. (1991). Briefly we selected 20 vacant sites, each consisting of unoccupied forested areas that appeared suitable for red-cockaded woodpeckers. Each of the vacant sites was located at least 0.5 km from existing cavity tree clusters. We paired the sites according to habitat features and selected one member of each pair for cavity drilling by flipping a coin. The second member of each pair served as a control site. We also selected 20 inactive cavity tree clusters, paired them, and selected an experimental and control site from each pair in the same way. To each experimental site, whether an inactive or vacant one, we added two complete drilled cavities and three cavity starts.

Cavities and cavity starts were drilled in old, live pines using the technique described in Copeyon (1990). Cavity starts consist of an entrance tunnel through the sapwood into the heartwood, with a widened area at the end of the tunnel in which a bird can roost. Complete cavities consist of the entrance tunnel plus a vertical cavity chamber in the heartwood.

We constructed cavities during February-March 1988 and November 1988-February 1989, and reported on occupancy of cavities in the 1989 breeding season (Copeyon et al. 1991; Walters et al. 1992a). Since that time we have included both experimental and control sites in our regular population monitoring procedures (see above). These procedures generated data on activity status of all sites, identity and status of birds using the sites, and reproductive success within the sites. In this paper we report on use of experimental and control sites during the 1990, 1991 and 1992 breeding seasons.

RESULTS

No cavities were lost to breakage of the tree at or below cavity height during the 4 years since cavity construction. One cavity tree was killed by fire, but there was no tree mortality that can be linked to cavity construction. As of the 1992 breeding season, no control sites, either vacant or inactive, were occupied by red-cockaded woodpeckers. Thus the dramatic difference in response to control and experimental sites reported in 1989 (Copeyon et al. 1991; Walters et al. 1992a) persisted in 1992.

We previously reported that 18 experimental sites, including 9 vacant sites and 9 inactive sites, were occupied in 1989. During the subsequent 3 years, 1 additional inactive site was occupied, so that only 1 vacant site was not used. Only 1 site, a vacant one, that was used in 1989 was subsequently abandoned.

Some sites originally were used by previously existing groups and others by new groups, so that the total number of new social units added to the population in response to cavity construction was 12 in 1989 (**Table 1**). In some cases, groups neighboring the newly formed ones were subsequently lost, so that by 1992 there was no net increase in the number of social units in these areas. Thus, where new social units were reported in 1989, there was a net loss of 3 social units in subsequent years (**Table 1**). However, this loss was balanced by formation of additional new groups in areas where in 1989 previously existing groups had captured experimental sites (**Table 1**). None of the previously existing groups that had captured experimental sites in 1989 (that is, were using both their original cluster and an experimental cluster), were able to hold both clusters over the long term. In one case a new group seized the experimental cluster, and in two cases the previously existing group moved to the experimental cluster, but their original cluster was then seized by a new group. Thus, in 1992 the number of new social units added to the population in response to cavity construction was 13. The number of experimental clusters in use in 1992 was still 18, as the additional cluster occupied was balanced by the abandonment of another that had been used in 1989.

Generally there were changes in status of both experimental cavity tree clusters and previously existing clusters neighboring experimental clusters. Clusters shifted between being inactive, captured and used by a woodpecker group, as is characteristic of cavity tree clusters generally in this population (Doerr et al. 1989). Thus the net number of new groups (compared to 1988) in these areas fluctuated around the initial number of 12, varying from 10-13, with the locations of the added groups shifting from year to year.

We reported that in 1989 many of the new social units consisted of males that had not yet obtained a mate, and we suggested that such units would eventually develop into breeding pairs (Copeyon et al. 1991; Walters et al. 1992a). Indeed this proved to be the case. As shown in **Figure 1**, the proportion of new social units consisting of solitary males declined to 0 by 1992, whereas the proportion consisting of groups with helpers has increased from 0 in 1989 to 39% in 1992. The typical developmental sequence characterizing these groups was that first an experimental site would be occupied by a male, who subsequently would attract a mate. The pairs then began to produce young, and some male offspring remained as helpers.

The development of new social units was also evident in reproductive data. In 1989 only a few of the experimental sites produced young, primarily those used by previously existing groups (**Fig. 2**). However, the number of successful nests, and fledglings produced from those nests, has steadily increased (**Fig. 2**), so that by 1992 the experimental sites were highly productive. The number of fledglings produced per breeding pair in our population has ranged from 1.19 to 1.74 during the years 1980-1992, whereas it was 1.78 among the 18 pairs using experimental sites in 1992.

Table 1. Responses of red-cockaded woodpeckers to experimental cavity tree clusters, 1989 and 1992.

1989 Occupant (#)	# New Social Units 1989	# New Social Units 1992
New group (6)	6	5
New unpaired male (4)	4	2
Existing pair		
Previous cluster used (2)	2	1
Previous cluster abandoned (1)	0	1
Existing unpaired male		
Previous cluster abandoned (1)	0	0
Captured existing pair (4)	0	3
None (2)	0	1

One concern about the experimental clusters was that the number of cavities contained within them might be insufficient to support a woodpecker group over the long term. Many of the groups not only have used the complete cavities we constructed, but also have converted drilled cavity starts into complete cavities by excavating a cavity chamber. Interestingly, many have also begun constructing new cavities in additional trees. To

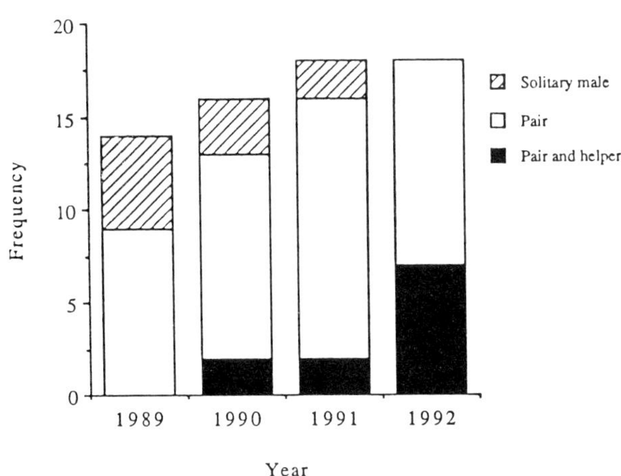

Figure 1. Composition of social units in experimental cavity tree clusters. The number of solitary males, pairs and pairs with one or more helpers during the breeding seasons of 1989-1992.

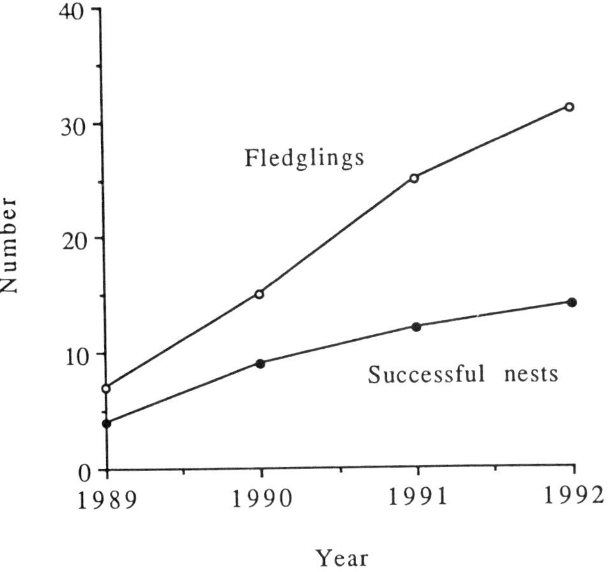

Figure 2. The number of successful nests and the number of fledglings produced in experimental cavity tree clusters, during the breeding seasons of 1989-1992.

date no bird-initiated cavities have been completed, which is not surprising considering the length of time required for cavity construction (Conner and Rudolph 1995a). It appears that the birds will be able to add to their set of cavity trees without additional drilling being required. This suggests that trees suitable for cavity excavation existed in these sites, but the birds chose not to move into these areas and initiate excavation until drilled cavities were added.

DISCUSSION

Our results indicate that new groups may be added to populations by cavity construction without causing losses of neighboring existing groups, at least over a period of four years. It will be necessary to monitor population dynamics in the vicinity of experimental cavity tree clusters for additional years, as responses to perturbations are often delayed in this species, due primarily to the resiliency of older breeding males. A key observation will be whether stability persists when breeding males die and are replaced by new, younger males. To this point, there are no hints of widespread instability triggered by addition of drilled cavity tree clusters.

Our estimate of 13 new breeding groups added to the population in response to cavity construction is a conservative one, since if any groups neighboring the experimental clusters were lost we concluded there was no net addition. That is, we considered every loss adjacent to the experimental clusters to be due to the addition of the new social unit, when in fact some or all losses could have been due to other factors. There were in fact losses of groups during the study period in parts of the population removed from drilled clusters.

Overall the population declined slightly during the study period despite the addition of new groups in response to cavity construction. Furthermore, the experimental clusters experiencing neighboring losses shifted over time. In some cases a neighboring group was lost immediately, but reformed later. In other cases there was no immediate loss, but in subsequent years a loss occurred. This suggests that some of the losses observed may not have been caused by the addition of new groups in experimental clusters.

We attempted to evaluate the extent to which losses of neighboring groups may have been caused by addition of new groups in experimental sites rather than general population trends by comparing changes among neighboring groups to changes among more distant groups in the general vicinity of experimental sites. The average number of clusters considered to be neighbors of an experimental cluster was 3, whereas the average number of additional clusters considered to be in the general vicinity was 8. The latter were those non-neighboring clusters within 3-4 km (depending on density) of the experimental cluster.

Among groups neighboring the experimental clusters there was a loss in 5 cases and a gain in only 1, whereas among non-neighboring groups in the greater vicinities of these sites there were 6 cases of loss and 3 cases of gain. These results are inconclusive, but suggestive of a higher rate of loss among neighbors. We conclude that most additions of new groups did not cause a loss of a previously existing group, but a few (best estimate of 2-3) did. More research will be necessary to determine what distinguishes cases where there is a compensating loss from ones in which the new group is able to coexist with all previously existing groups in the vicinity. There was no obvious relationship between distance between neighboring groups and drilled clusters and whether neighboring groups were lost in this study, using a minimum distance between experimental clusters and previously existing clusters of 0.5 km.

One encouraging result is that experimental sites did not remain captured for any length of time. Previously existing groups that initially captured experimental clusters 0.5 km or more from their own cluster were not able to hold both sites over the study period. At least in our study population, clusters this distant from one another are independent in terms of use by the birds except for short periods. The fear that constructed cavity tree clusters would simply be incorporated into existing territories has not materialized.

Our results certainly offer encouragement that new social units may be permanently added to populations by cavity construction. Our data also indicate that new units can be highly productive, and come to function in every way like groups in woodpecker-excavated clusters. Artificial cavity tree clusters, if sufficiently distant from existing clusters, appear to function like woodpecker-excavated clusters as well.

Results through four years continue to justify the conclusions drawn from initial results in 1989 about the role of cavity trees in population dynamics of red-cockaded woodpeckers (Copeyon et al. 1991; Walters 1991; Walters et al. 1992a). The birds appear to compete for territories with existing cavity trees rather than occupy unused space and construct new cavity tree clusters. It is particularly telling that birds in experimental sites are now excavating additional cavities on their own, as this indicates the habitat was suitable for occupancy and construction of new cavity tree clusters before cavities were drilled. That is, the birds could have occupied these vacant areas and inactive clusters, but they chose not to. This further reinforces the notion that cavity construction will be a much more effective means of increasing population size than will be providing high quality vacant habitat that lacks cavities (Walters 1991). It seems appropriate based on our results to attempt to expand populations using cavity construction to meet recovery objectives for the species.

ACKNOWLEDGMENTS

This research was supported by NSF grants BSR-8307090 and BSR- 8717683, the Department of the Army, Fort Bragg, and the North Carolina Agricultural Research Service. We thank Fort Bragg, North Carolina State Parks, the North Carolina State Museum of Natural Science and the North Carolina Wildlife Resources Commission for allowing us to construct cavities on lands they manage. We also thank J. Hammond for assistance in cavity construction, and the many graduate students, technicians and summer interns who assisted in collecting population data.

Augmentation from the Apalachicola National Forest: The Development of a New Management Technique

Charles A. Hess, USDA Forest Service, P.O. Box 579, Bristol, Florida, 32321
Ralph Costa, U.S. Fish and Wildlife Service, College of Forestry and Recreation Resources, Clemson, South Carolina 29634

ABSTRACT: Active measures are now needed to recover many small populations of the red-cockaded woodpecker (*Picoides borealis*) left on public lands. One of these measures used recently by the US Forest Service is called augmentation, the process of moving juvenile females from large healthy populations to single males in isolated declining populations. Because the population on the Apalachicola National Forest has the largest and healthiest remaining population, eighteen birds were moved from this forest in the first three years of the program, with 11 showing some measure of success in their new location. In the course of this program, information has been gathered that can be used to increase the probability of a positive result. First, the availability of a suitable cavity in the receiving cluster is essential to the retention of the new bird in that cluster. Second, transport of birds that have already dispersed from their natal clusters is unlikely to be successful.
KEYWORDS: augmentation, transport, *Picoides borealis*

The development of single male clusters is part of the natural evolution of a new red-cockaded woodpecker (RCW) group, because males disperse to new areas and then recruit females to join them (Walters 1989; Hooper et al. 1991a). In small isolated populations, however, well established clusters may be reduced to single males and dispersing females have a low probability of finding them (Conner and Rudolph 1989). Thus, one of the signs of a declining population is a high percentage of single male clusters (Walters et al. 1988). One way to alleviate this problem is to move young females to isolated males (Defazio et al. 1987). This process, "augmentation", can help sustain small populations in isolated areas.

The population of RCW's in the Apalachicola National Forest (ANF) is the largest remaining population (Costa and Escano 1989). For the past several years it has been used as a source of young females for the recovery of small populations elsewhere. This paper describes the first eighteen cases of transporting young female RCWs from the ANF to other populations. We also offer some recommendations for the best procedures to follow when conducting augmentation.

METHODS

In 1989 a banding program was initiated on the Wakulla Ranger District of the ANF. Starting in the last week of April, each of about 100 clusters was visited at least once every ten days until nesting was detected. Once nesting was detected, the tree was climbed and a count of eggs was made. If eggs were detected, the cluster was visited eight days later to obtain an estimate of the age of the chicks and to schedule the final nest visit to band 6-7 day old chicks. If chicks were four or more days old, either on the second trip or when the nest was initially detected, they were banded.

Chicks were removed from the nest with a monofilament noose attached to a plastic tube (Jackson 1982) and banded with an aluminum U.S. Fish and Wildlife Service band and three color bands, each leg receiving two bands. The three color system allowed each bird to receive unique markings. The information was recorded and birds were returned to their cavities.

Three days following the expected fledging date, cluster sites were visited early in the morning. Birds from each cluster were followed for 30 minutes or until a total count of the group had been made and all birds in the group had been identified. The sexes of juveniles were determined by the presence or absence of the red crown patch characteristic of the top of the head of juvenile males.

If a fledgling female RCW was detected in the cluster, the group received regular visits to monitor her presence and determine where she roosted. In the autumn, when a request came from another national forest, where a single male had been identified and its cluster prepared for receipt of a young female bird, transport was arranged. Transport was conducted in two different fashions. The first was at night. The young female was captured after she had gone to roost and she was placed in a transport box. The box was a wooden box 25 cm x 25 cm x 37 cm with the inside screened on one side, carpeted on the other sides, and topped with cloth. In this box the bird was transported by truck to the receiving population. Upon arrival, the female was placed in a cavity near the recipient male's cavity. A screen, attached to a fifteen meter cord, was placed over the entrance to the female's cavity to hold her there until morning. In the morning, when the male began calling, the screen was pulled off and she was released. Daytime transport followed the same procedure, except that the bird was caught at dawn and transported during the day to the receiving cluster. In these cases the bird was fed one or two crickets every 45 minutes. Upon arrival at the new cluster site, the female was held until just before the male came in to roost, then placed in her cavity. When travel time did not allow arrival before the male went to roost, the female was put in her cavity after dark. Morning release procedures were identical to those following night transport.

This program began in 1989 on the Wakulla Ranger District of the ANF. After the first year, augmentation activities were moved to the Apalachicola Ranger District. This decision was made because data being collected on the Wakulla District indicated that the sub-population there was probably declining; there was such a high proportion of single male birds there (James 1990) that even a few young females could not be spared.

RESULTS

In he first three years of the program, between November of 1989 and February of 1992, 18 young females were transported to five national forests (Cherokee NF in Tennessee; Desoto and Homochitto NFs in Mississippi; Ocala NF in Florida; Sabine NF in Texas) and to the Department of Energy Savannah River Site in South Carolina (**Table 1**). The measure of success was whether the new female remained in the receiving cluster through the subsequent breeding season. In that case she was attempting to fill a breeding position in the population and could be considered to be part of the host population. In 11 of the 18 cases (61%), the transported birds remained with their new mates through the following breeding season. The number of successful cases increased from 1 of 4 (25%) in the first year to 3 of 6 (50%) in the second, and 7 of 8 (87%) in the third year.

DISCUSSION

The augmentation program is a management technique that is still in the development phase. Each year the cases should be evaluated so that the process can be improved in subsequent years. The four birds moved in the first year had a poor success rate, with only one of them remaining in the new cluster. In retrospect, it was apparent that these failures were mainly due to the poor condition of the receiving cavities. The bird moved to the Ocala National Forest was placed in an abandoned cavity with a restrictor plate. Two birds were moved to the DeSoto National Forest in December of 1989. One bird was placed in an artificial cavity made of a section of plastic PVC pipe attached to the side of a tree. The second bird had to be returned to the ANF because the selected recipient cavity had a wasp nest in it and there were no other suitable unoccupied cavities in the cluster area. Because these Ocala and DeSoto cases failed, we decided that in the 1990-91 season, we would be especially careful to place females in clusters with cavities considered to be in good condition. Because new technology for artificial cavity construction was being developed at this time (Copeyon 1990; Allen 1991), this criterion could be easily met.

In 1990-91, six young female RCWs were moved from the ANF, with a 50% success rate. In these cases all cavities were considered to be in good condition as they all had artificial cavities available in the cluster. In reviewing the cases, two of the three failures were attributable to problems either during transport or at the time of release. The first bird, which

Table 1. List of female red-cockaded woodpeckers transported from the Apalachicola National Forest (ANF) in the first three years of the program.

Date	Age*	From Cluster	Receiving Population	Condition of Receiving cluster	Outcome**
Year 1					
11-89	HY	232-4	Ocala Nat. Forest.	Fair	Failed
12-89	HY	232-11	DeSoto Nat. Forest	Poor	Failed
12-89	HY	232-1	DeSoto Nat. Forest	Poor	Returned
1-90	SY	209-1	DeSoto Nat. Forest	Good	Success
1-90	AHY	232-1	Homochitto Nat. Forest	Good	Failed
Year 2					
11-90	HY	11-5	Ocala Nat. Forest	Good	Success
12-90	HY	22-7	Cherokee Nat. Forest	Good	Failed
12-90	HY	69-4	Homochitto Nat. Forest	Good	Failed
12-90	AHY	73-5	Homochitto Nat. Forest	Good	Failed
1-91	SY	23-1	DeSoto Nat. Forest	Good	Success
1-91	SY	67-7	DeSoto Nat. Forest	Good	Success
Year 3					
10-91	HY	17-1	Savannah River Site	Good	Success
10-91	HY	23-8	Savannah River Site	Good	Success
11-91	HY	29-1	DeSoto Nat. Forest	Good	Success
12-91	HY	11-11	Homochitto Nat. Forest	Good	Success
1-92	SY	44-11	DeSoto Nat. Forest	Good	Success
1-92	SY	20-13	Wakulla District ANF	Good	Success
2-92	SY	22-6	Sabine Nat. Forest	Good	Success
2-92	SY	72-4	Sabine Nat. Forest	Good	Failed

*** HY, hatching year hatched that calendar year; SY, second year, hatched in the spring of the previous calendar year; AHY, after hatching year, hatched some previous calendar year but actual year unknown.**
**** Success, transported bird remained in new cluster through at least one breeding season; Failed, transported bird left the area within a few days of transport and was not seen again; Returned, transported bird was returned to the ANF and released at it's capture site.**

was moved to the Cherokee National Forest, was transported by airplane and the temperature at the time of release was much lower than the bird had experienced in Florida. Either the air travel or the temperature, or both, may have had a negative effect on the bird. After emerging from her cavity, she did not call, as expected, but climbed the tree to a sunny spot, fluffed her feathers, and stayed there for more than an hour. She did not respond to the calls of the resident male, though she must have heard them. She disappeared later that day and was not seen again.

The two failed cases of moving birds to the Homochitto NF occurred on the same day. In one case the male at the receiving cluster flew out of his cavity and directly out of the cluster area. The female, upon release, came out and called, but, because the male was not in the area, she never made contact with him. Having no male to entice her to stay, she left the cluster and was not seen again in the area. The second bird that was moved to the Homochitto was a dispersing female, captured in a cluster where the breeding female was not her mother. It was assumed that this young bird was an "extra" female and would be an good candidate for the program. This same assumption had been made with two birds moved in 1989, from clusters 232-4 and 232-1. The juvenile female hatched in 232-4 was captured after she had dispersed to another cluster; the bird from 232-1 was an extra female where the breeding female was still present. The assumption that they were extra females may have been valid but, after three failed attempts to move dispersing females, it became clear that such birds should not be used in augmentation.

With this experience, our plans for transport of birds in the 1991-92 season had two restrictions. First, the receiving cluster must have cavities in good condition for receipt of the bird, and second, the bird must come from her natal cluster prior to her dispersal. Eight birds were moved in the 1991-92 season, of which seven were successful. The single failure was caused by mistakenly releasing a juvenile female into a cluster that already had a mated pair of birds.

CONCLUSION

The results of the 1991-92 program show that this program can be successful in contributing to the maintenance of small populations of RCWs. Although the augmentation program is very labor intensive, its development and refinement have been progressing rapidly and important elements have been identified for making the program successful. Managers at receiving populations must make sure that their cluster sites are in the best possible condition (for example, midstory free) before birds are transported. Habitat conditions at the receiving site must be improved to acceptable standards prior to the transport of the female bird. The cluster should have at least three suitable cavities, before managers request that a female be brought to a single male. Typically, cavities that have not been used and maintained by RCW's for several years are no longer suitable as roost cavities (Walters et al. 1992b). Therefore, artificial cavities, either inserts or drilled cavities, will normally be required in recipient clusters. Qualified personnel must assure that the cluster site is indeed occupied only by a single male bird. Disturbance to the site after delivery of the new bird should be minimized. This work requires personnel that are experienced in working with RCWs and who are familiar with the cluster under consideration. These criteria are important parts of a successful augmentation program.

ACKNOWLEDGMENTS

The work could not have been completed without the help and support of Ron Smith, District Ranger on the Wakulla R.D.; Andy Colaninno, District Ranger, on the Apalachicola R.D.; and the following field personnel: S. Fitzgerald, F. Sanders, G. Titus, J. Kappes, O. Rivera, D. Quinn and M. Donaldson. The project was financially supported by the National Forests of Florida of the U.S. D.A. Forest Service and the National Fish and Wildlife Foundation. We thank F.C. James and R.G. Hooper for comments on the manuscript.

Interspecific Competition for Red-cockaded Woodpecker Cavities in the Apalachicola National Forest

John J. Kappes, Jr., University Florida, Gainesville, FL.
Kisatchie National Forest, Pineville, LA.
Larry D. Harris, University Florida, Gainesville, FL.

ABSTRACT: Use of red-cockaded woodpecker (*Picoides borealis*) cavities by various species was investigated in longleaf pine habitat in Apalachicola National Forest. The hypothesis that availability of snags within 250m of red-cockaded woodpecker (RCW) clusters influences the rate of RCW cavity occupation by other species was tested. Results suggest that snag availability may reduce competitive pressure from all potential competitors combined, but more work with larger sample sizes is needed to fully examine this management issue. Red-bellied woodpeckers (*Melanerpes carolinus*) occupied an average of 27% (range 7-62%) of the RCW-suitable cavities per cluster. Geographic variation in RCW cavity occupancy rates by red-bellied woodpeckers may be explained in part by variations in site productivity. In areas where the soils are better for longleaf growth, snags tend to be larger and this greater availability of the larger snags that are preferred by red-bellied woodpeckers may reduce red-bellied woodpecker competitive pressure for RCW cavities.
KEYWORDS: red-cockaded woodpecker, red-bellied woodpecker, red-cockaded woodpecker cavities, snags, interspecific competition

Interspecific competition is recognized as a potential threat to red-cockaded woodpecker (*Picoides borealis*) colony stability and reproductive success (Dennis 1971a,b; Jackson 1978b, Lennartz and Harlow 1979, Baker 1983b, Harlow and Lennartz 1983, Hovis 1984, Walters 1990, Rudolph et al. 1990a). Effects of interspecific competition was among the Recommended Research Initiatives of the recent "RCW Summit" in 1990 and the RCW Recovery Plan (USFWS 1985) indicate the same need for research. At issue is " whether retention of snags within colonies increases or decreases competitive pressure".

Due to the present scarcity of stands of trees suitable for cavity, competition for existing cavities is intense (Jackson 1974, 1978b). At least 24 species of vertebrates are documented to use RCW cavities (Dennis 1971a; Baker 1971; Beckett 1971; Jackson 1978b; Harlow and Lennartz 1983; Rudolph et al. 1990a; Kappes, pers. obs.). Although most species tend to use abandoned RCW cavities, others usurp active ones. Pileated woodpeckers (*Dryocopus pileatus*) are known to enlarge RCW cavities making them available to larger cavity-nesting species such as the wood duck (*Aix sponsa*), kestrel (*Falco sparverius*) and screech owl (*Otus asio*) (Dennis 1971a,b; Jackson 1978b, Rudolph et al. 1990a).

Pileated woodpeckers seem to enlarge red-cockaded woodpecker cavities primarily while foraging (Harlow and Lennartz 1983) inasmuch as they rarely roost or nest in the cavities they enlarge (Dennis 1971a,b; Jackson 1978b). Indeed, Dennis (1971a) noted that the entrance diameter of many cavities enlarged by pileateds actually exceeds the diameter of pileated nest cavities. Cavities that have been enlarged by pileateds are rarely used by RCW, (Dennis 1971a, Jackson 1978b), and even abandon their cluster sites when all or most suitable cavities are impacted (Walters 1991).

Different researchers have drawn differing

conclusions on the use of red-cockaded woodpecker cavities by red-bellied (*Melanerpes carolinus*) and red-headed woodpeckers (*M. erythrocephalus*). Beckett (1971) and Jackson (1978b) erroneously concluded that neither red-bellied nor red-headed woodpeckers are capable of entering a RCW cavity without first enlarging it. But Ligon (1970), Baker (1971), and we have observed that red-headed and red-bellied woodpeckers may use at least the larger diameter red-cockaded woodpecker cavities without further enlargement.

Nonetheless, Jackson (1978b) noted that on Noxubee National Wildlife Refuge in Mississippi, red-bellied woodpeckers are so common that few red-cockaded woodpecker cavities escape at least some enlargement by that species. Red-cockaded woodpecker cavities that have been used by red-bellies or red-headeds for nesting may subsequently be reactivated by RCWs (Baker 1971; Jackson 1978b). Whether or not a red-belly or red-headed woodpecker enlarges a RCW cavity it usurps may depend on the cavity's history (usually unknown to researchers), or perhaps the original diameter of the cavity upon its completion by a red-cockaded woodpecker.

Most researchers conclude that flying squirrels and red-bellied woodpeckers are the principal competitors for RCW cavities (Dennis 1971a, Harlow and Lennartz 1983, Rudolph et al. 1990a, Ligon 1970, Jackson 1978b). In addition, red-bellied woodpeckers and flying squirrels are known to destroy RCW eggs and young (Ligon 1970, Jackson 1978b, Walters 1991). In some areas, defending cavity sites from red-bellied woodpeckers is an important daily activity of RCWs (Ligon 1970, J.K., pers. obs.). Both Ligon (1970) and Jackson (1978b) found the red-bellied woodpecker to be a principal competitor for RCW cavities and Jackson (1978b) noted that of two cluster sites he studied on Noxubee Refuge, red-bellied woodpeckers usurped the only active cavities and may have caused the abandonment of both clusters.

Most interactions between flying squirrels and red-cockaded woodpeckers undoubtedly occur at night, and are thus more difficult to study. Rudolph et al. (1990a) found that flying squirrels were the most common potential competitor using "optimal" RCW cavities, but concluded that they were not actually usurping cavities; they hypothesized that neither a RCW nor a flying squirrel is capable of evicting the other species if it already occupies the cavity.

Underlying factors regulating the dynamics of interspecific competition for RCW cavities are complex with most hypotheses implicating some aspect of forest structure. Harris and Scheck (1991) suggested that as RCW habitat has become increasingly fragmented, red-bellied woodpeckers have increased in abundance that results in increased competition between the two species. In areas where cluster sites are not burned regularly, hardwood midstory development may lead to increased cavity competition with red-bellied woodpeckers and flying squirrels, and increase the rate of cavity enlargement by pileated woodpeckers (Ligon et al. 1986, Costa and Escano 1989, Conner and Rudolph 1989).

The abundance of snags in the vicinity surely influences the intensity of interspecific competition for RCW cavities (Wood 1983b, USFWS 1985, Harlow and Lennartz 1983, Rudolph et al. 1990a). For example, Wood (1983b) and Rudolph et al. (1990a) attributed the low intensity of interspecific competition on their study sites to an abundance of alternative cavity sites in the form of snags.

A major objective of this study was to test the hypothesis that red-bellied woodpecker competitive pressure for red-cockaded woodpecker cavities, is directly related to the number of snags suitable for cavity excavation by red-bellied woodpeckers.

STUDY AREA AND METHODS

The study was conducted in Leon County, Florida on the Wakulla Ranger District of the Apalachicola National Forest. The red-cockaded woodpecker population on the Wakulla district appears to be in decline (James 1991; U.S. Forest Service, unpublished data). Several factors may be contributing to this decline. The uplands of this area were historically dominated by longleaf pine but much of this (approximately 40%) has been converted to slash pine plantation (R. Costa, pers. comm.). In many parts of the district, encroachment by titi (*Cyrilla* and *Cliftonia* spp.) has inundated formerly suitable RCW habitat. The current fire regime has apparently failed to curb this encroachment. As recently as 1990, unmerchantable relict pines were regularly removed as cull trees (Kappes, pers. obs.). These factors have undoubtedly reduced the acreage and quality of RCW foraging habitat, and diminished the availability of potential cavity trees.

Covering approximately 2800 ha, the study area includes a group of ten currently active clusters in four compartments centered around U.S. Forest Service compartment 232. Compartment 232 contained six of the ten study clusters. Contiguous clusters were chosen to maximize the likelihood of documenting the effect of intercluster movements, dispersal, or extraterritorial roosting, on cavity usage.

The ten study colonies are all located in mature, open longleaf pine forest. The study site had been burned approximately every five years during the winter. This regime has encouraged the development and maintenance of an understory dominated by saw palmetto (*Serenoa repens*), wiregrass (*Aristida stricta*), and runner oak (*Quercus pumilla*). The regular fire and poor site conditions have prevented development of a significant hardwood midstory. Approximately half of the sites have a few scattered turkey oaks (*Quercus laevis*), but removal for firewood had drastically reduced this species prior to the initiation of this study.

In each study cluster all cavity trees were marked and numbered. The number of cavity trees per cluster ranged from three to eight. All RCW group members, and nearly all red-bellied woodpeckers occupying cavities within the ten study clusters were captured and banded with distinctive colored plastic leg bands and aluminum U.S. Fish and Wildlife Service leg bands.

Data on cavity usage were collected from April to August, 1991. The small sample size (n=10 clusters) was necessary to gather regular, detailed documentation of cavity usage. The use of RCW cavities by various species was documented by conducting roost checks and by visual examination of the cavity chambers. During roost checks, 1 to 3 observers were strategically positioned at locations where birds could be seen entering the cavities to roost. Any observed interspecific or intraspecific interactions associated with cavities were recorded. Roost checks were conducted in each cluster 2-3 times per month.

Because roost checks may fail to detect the presence of non-avian occupants of RCW cavities (e.g. flying squirrels, rat snakes, etc.) cavities were also accessed using Swedish sectional climbing ladders and examined using a droplight and dentist's mirror. Cavity contents (e.g. nest material, eggs, flying squirrels , rat snakes, water, etc.) were recorded. The cavity chambers were examined weekly during the breeding season (April to August) and monthly from September to March. Trees were also climbed when roost checks were inconclusive.

Cavity diameters were measured using a set of calipers. The horizontal diameter of the entrance tunnel was measured, following Jackson's (1978) suggestion that this was the best indicator of the body size limit of animals capable of entering a cavity. The measurement at the narrowest point of the entrance tunnel was recorded.

DEVELOPING A SNAG INDEX

In order to test the hypothesis that red-bellied woodpecker competitive pressure, for red-cockaded woodpecker cavities, is directly related to the number of snags suitable for cavity excavation by red-bellies, it was necessary to:

(1) develop a site-specific index of the suitability of snags for red-belly cavity excavation in the mature longleaf pine habitat of the study area, and

(2) develop indices of red-belly competition for RCW cavities.

At this study site the principal competitors for RCW cavities are red-bellied woodpeckers and flying squirrels (pers. obs.). Red-bellies are primary cavity nesters capable of excavating their own cavities in snags, whereas flying squirrels require cavities previously excavated by woodpeckers or those resulting from fungal rot. Therefore red-bellies are more likely to respond to the availability of snags, with or without cavities, than flying squirrels. The red-bellied woodpecker was the most abundant medium-sized primary cavity-nester in the study area (Kappes, pers. obs) and probably the most important cxcavator of snag cavities used by medium-sized cavity-dependent species (e. g. flying squirrels, Eastern bluebirds, and great-crested flycatchers). We therefore concluded that in order to study the influence of snag availability on interspecific competition for red-cockaded woodpecker cavities, it was most important to estimate the availability of snags suitable for cavity excavation by red-bellied woodpeckers.

Based on a review of the literature, McComb et al. (1986, citing Hardin and Evans 1977, and Conner et al. 1983) concluded that a snag had to be at least 25 cm in DBH and at least 9 m in height to be suitable for cavity excavation by red-bellied woodpeckers. Using this as a reference point, and allowing for deviation, we concluded that recording the top condition, % bark cover, number of limbs > 1m, DBH, and total height of each snag > 21 cm DBH and > 5 m height, and within 250 m of each of the 10 study clusters, was sufficient to include all potential red-belly nest snags. An inventory of snags within 250 m of each cluster was conducted from July to September, 1991. Snags in advanced stages of decay (e.g. large splits in sapwood, large portions of sapwood gone) and those that had decayed little (e.g. twigs or needles still present, most branches still intact) were deleted from the sample since in our study area such snags were not observed to be used by red-bellies for nest cavity excavation (Kappes, pers. obs.).

In order to develop an index of the suitability of potential red-bellied woodpecker nest snags, searches for red-bellied nests were conducted in the ten study clusters and in similar areas of open, mature, longleaf pine habitat within 5 km of the study area. These searches were conducted during the month of July in 1991 and 1992. A total sample of 23 pine snags used for red-bellied nests was collected. For each of these nest snags, the top condition (presence or absence of top), percent bark cover, number of limbs > 1 m in length, DBH, and total height were recorded. These data on snags used by red-bellieds were used to characterize, quantitatively, snags that are suitable for nest cavity excavation by this species. Wilcoxon's rank sum tests were used to compare the percent bark cover, number of limbs > 1 m in length, DBH, and total height of the sample of red-bellied nest snags (n=23) to the snags in the cluster sites that were not known to be used by red-bellies (n=138). A Chi-squared test was used to test for differences between the two samples for top condition.

INDICES OF "COMPETITION"

Three indices of interspecific competition for RCW cavities were used:

(1) the number of RCW cavities usurped by red-bellieds (#RBUSRPS),

(2) the average proportion of RCW cavities occupied by each species of potential competitor, and

(3) the average proportion of RCW woodpecker cavities occupied by all potential competitors combined.

The movement of a red-belly to a red-cockaded cavity was categorized as an usurpation if the newly occupied cavity had been:

(1) active,

(2) occupied by a RCW up to the time of red-belly occupation and

(3) the displaced RCW subsequently roosted outside. In some instances, observed aggressive interactions provided additional evidence. I was able to accurately identify cavity usurpations by red-bellies using the above criteria.

Red-cockaded woodpeckers prefer cavities with entrance tunnel diameters of less than 60 mm (Rudolph et al. 1990a, Kappes and Harris, in prep.). All such cavities were considered "suitable" cavities. Cavities exceeding 60 mm in diameter were excluded from this analysis.

Variables representing the estimated occupancy rates of "suitable" RCW cavities are defined in **Table 1**.

The estimated rate of occupation of suitable cavities was based on the average proportion of the "suitable" RCW cavities that were occupied by each species at the time that roost checks were conducted. Between nine and eleven roost checks were conducted, at each study cluster, during the 1991 breeding season (April to August).

RESULTS AND DISCUSSION

Cavity occupation by red-bellied woodpeckers

During the 1991 breeding season, red-bellied woodpeckers usurped red-cockaded woodpecker cavities a total of seven times in the 10 study clusters. The number of these usurpations ranged from zero to two per cluster. The average proportion of "suitable" RCW cavities per cluster occupied by red-bellied woodpeckers was 0.27 with a range of 0.07 to 0.62 (Kappes and Harris, in prep.). The average proportion of the cavities per cluster occupied by flying squirrels was 0.09 with a range of zero to 0.42 (Kappes and Harris, in prep.).

Snag Index

Red-bellied nest snags were of significantly greater DBH and total height than snags not used for nests (Kappes and Harris, in prep). Red-bellied nest snags did not differ from other snags in the percent bark cover, number of limbs, or top condition (Kappes and Harris, in

Table 1. Spearman's rank correlation coeficients (r_S) between variables* representing indices of interspecific competition, and the number of snags considered suitable for cavity excavation by red-bellied woodpeckers (N=10). Kappes and Harris, unpublished.

	#RBUSRPS	RBOCC	FSOCC	RBFSOCC	CMPOCC	UNOCC
#SNAGS	-0.02	-0.31	-0.46	-0.49	-0.52	+0.43

*** #RBUSRPS = the number of times red-bellied woodpeckers usurped active RCW cavities, RBOCC = the average proportion of "suitable" red-cockaded woodpecker cavities (suitable being defined as having an entrance tunnel diameter < 60 mm) occupied by red-bellied woodpeckers, FSOCC = the average proportion of suitable RCW cavities occupied by flying squirrels, RBFSOCC = the average proportion of suitable RCW cavities occupied by either red-bellies or flying squirrels, CMPOCC = the average proportion of suitable red-cockaded woodpecker cavities occupied by potential interspecific competitors (any species, including wasps, snakes, birds, etc. that exclusively occupied cavities), UNOCC = the average proportion of suitable RCW cavities that were unoccupied. A Spearman's rank correlation coefficient of 0.56 is significant at the 10% level of significance.**

prep). Thus, only snag DBH and height were used to evaluate the suitability of snags. A snag was considered "suitable" if it's measurements fell within the range of DBH and height measured for the 23 red-belly nest snags (22.2-40.8 cm and 7-22 m, respectively).

Snags and competition

The cavity occupancy rates of the two most frequent interspecific occupants of red-cockaded woodpecker cavities, red-bellied woodpeckers and flying squirrels, were negatively related to the number of suitable snags in the vicinity of red-cockaded woodpecker clusters, but these relations were not statistically significant **(Table 1).** The correlation between the number of snags and the occupancy rate of all potential competitors combined was stronger and approached significance at the 10% level **(Table 1)**. The consistently negative relation between the occupancy rates of potential competitors and the number of snags suggests a relation, albeit not statistically significant. Consistent with this pattern is the positive correlation between the number of snags and the average proportion of of red-cockaded woodpecker cavities that were unoccupied. The latter relation suggests that the availability of alternative cavity sites for potential competitors may lead to a greater availability of cavities for red-cockaded woodpeckers. In total, these results indicate that increased availability of snags may alleviate interspecific competition for red-cockaded woodpecker cavities by providing alternative cavity sites for potential competitors. However, more work, with larger sample sizes, is needed to fully examine these relations.

Rudolph et al. (1990a), working in longleaf pine habitat in eastern Texas, did not observe red-bellied woodpeckers roosting or nesting in RCW cavities. Rudolph et al. (1990a) concluded that while red-bellies were "moderately common" in the study area, ample lightning and beetle-killed pines appeared to provide for suitable alternative nest and roost sites. At least two factors may explain the discrepancies between our study and that of Rudolph et al. (1990a). The first relates to possible differences in red-bellied abundance between the two areas. Red-bellied woodpeckers were "abundant" in our northwestern Florida site. More quantitative estimates of red-belly abundance in longleaf habitat in both areas would be useful.

A second factor relates to the differences in site productivity between the two sites. The heavier soils at the western end of the longleaf belt are superior to the eastern areas for longleaf growth (Wahlenberg 1946). Wahlenberg (1946) added that in the deep, coarse sands of northwestern Florida, the growth of pine timber is too slow to be commercially feasible (Wahlenberg 1946:12). The longleaf pines growing in the eastern Texas study area are noticeably larger in DBH than those of similar age at the Florida site (Kappes pers. obs.); this results in a greater availability of the larger snags red-bellied woodpeckers appear to prefer.

ACKNOWLEDGMENTS

Thanks to F. Sanders, K. Bettinger, and S. Carr for assistance with field work; and to R. Costa for providing essential field equipment.

A Technique to Deter Rat Snakes from Climbing Red-cockaded Woodpecker Cavity Trees

James H. Withgott, Department of Biological Sciences, University of Arkansas, Fayetteville, AR 72701
Joseph C. Neal, Arkansas Cooperative Fish and Wildlife Research Unit University of Arkansas, Fayetteville, AR 72701
Warren G. Montague, Poteau Ranger District, USDA Forest Service, Waldron, AR 72058

ABSTRACT: Throughout its range, the red-cockaded woodpecker (*Picoides borealis*) faces the threat of nest and roost predation by rat snakes (*Elaphe obsoleta)*, which are skilled tree-climbers. We experimentally tested a technique to prevent snakes from reaching woodpecker cavities. The snake excluder device (SNED) we tested consisted of a 60-cm-wide band of light-weight aluminum flashing wrapped around and stapled to the trunk of a cavity tree at breast height. The smooth surface of the flashing deprives snakes of features to grip with their ventral scales. We tested 17 wild-caught black rat snakes (*E. o. obsoleta*) for ability to climb past SNEDs. Thirteen snakes attempted 41 climbs on 5 trees fitted with SNEDs; none were successful in crossing SNEDs. The largest snakes could not be induced to climb, and so could not be tested. However, large snakes may represent less of a threat to red-cockaded woodpeckers than smaller snakes, which are superior climbers. We suggest several measures that should make the SNED fully effective in preventing snake predation.
KEYWORDS: Arkansas, black rat snake, endangered species, Ouachita National Forest, predation, rat snake, red-cockaded woodpecker, snake excluder device

Rat snakes are known to attempt to access nest and roost cavities of red-cockaded woodpeckers (RCWs) (Dennis 1971a, Jackson 1978d, Neal 1992, Richardson and Stockie 1995). The woodpecker excavates resin wells and maintains constant flows of resin on active cavity trees. This behavior may represent an adaptation which protects against predatory attacks by climbing snakes (Jackson 1974). While the resin barrier may deter most snake climbs (Jackson 1974, Rudolph et al. 1990b), some snakes are successful in reaching woodpecker cavities (Jackson 1978d; Neal 1992). In the Ouachita National Forest of Arkansas, snakes attempted to climb 50% (n=12) of woodpecker nest trees in 1991 and 18% (n=38) of nest and roost trees in 1992 (Neal 1992). At Noxubee National Wildlife Refuge in Mississippi, gray rat snakes (*E. o. spiloides*) attempted to climb 64% (n=22) of nest and roost trees of red-cockaded woodpeckers in 1992 (Richardson and Stockie 1995). Even if only a small percentage of rat snakes are able to cross resin barriers, the high frequency of climbing attempts makes them a threat to conservation of the RCW. woodpeckers may be especially vulnerable to snake predation in recently installed artificial cavity inserts (Allen 1991) or drilled cavities (Copeyon 1990) if these cavities lack substantial accumulations of resin.

Nylon monofilament garden mesh (3/4" and 1/2") wrapped around trunks of cavity trees has been used to capture climbing snakes, thus documenting climbs while preventing snakes from reaching cavities (Eichholz and Koenig 1992, Neal 1992, Richardson and Stockie 1995). Climbing snakes pass their heads and necks through the mesh, but their thicker midbodies become stuck and their overlapping scales prevent them from backing out. However, use of mesh snake traps as a management technique is inappropriate, since even frequent checking of traps may not be sufficient to prevent high rates of snake mortality during hot weather.

In this paper, we introduce a nonlethal technique to inhibit predatory snake climbs. We present results of experimental tests to assess the effectiveness of our method using captive black rat snakes.

STUDY AREA

Experiments were conducted in the Ouachita National Forest in Scott County of western Arkansas. The area included mixed stands of mature shortleaf pine (*Pinus echinata*) and oak (*Quercus* spp.) managed under U.S. Forest Service guidelines including prescribed fire and midstory reduction. Climbing trials were conducted on 5 active RCW cavity trees in 3 cavity tree clusters, as well as on 5 control trees near 1 of the cavity tree clusters.

MATERIALS AND METHODS

The Device

Bands of thin-gauge aluminum flashing were attached to cavity trees in the Ouachita N.F. in 1991 and 1992 to deter climbing by black rat snakes on RCW nest and roost trees. The bands of flashing, which we named snake excluder devices (SNEDs), were 60 cm wide and were secured tightly around the trunk at breast height with a staple gun (**Figure 1**). Prior to attachment of each SNED, the surface of the trunk was scraped with a limb saw to allow for a tighter fit. Bark was also scraped approximately 0.3 m above and below each SNED. We predicted that the smooth surface of the flashing would deny snakes features with which to brace themselves as they climbed. Predator guards operating on the same general principle have been used to reduce nest predation on eastern bluebirds (*Sialia sialis*) (Kingston 1991, Krueger 1991), seaside sparrows (*Ammodramus maritimus*) (Post and Greenlaw 1989), and wood ducks (*Aix sponsa*) (Bellrose 1955, Hester and Dermid 1973).

To assure that these highly reflective aluminum strips did not deter woodpeckers from occupying roost cavities, each SNED was monitored on the evening after its installation. In no case did a woodpecker refuse to enter its cavity. In only one case did a woodpecker hesitate briefly before entering, and this occurred at an exceptionally low cavity at which the SNED was attached only 1.1 m below the cavity entrance.

During 1992, 38 RCW nest and roost trees received a SNED. However, because mesh traps (see above) were attached to these trees beneath the SNEDs in order to document snake climbs, no field testing of the effectiveness of the SNEDs was possible.

The Experiments

Trials were run with 17 black rat snakes on 5 consecutive days between 14 and 18 July 1992 to test the effectiveness of SNEDs in deterring climbs by rat snakes. Eleven of the snakes were captured in the wild during the spring and summer of 1992, and 6 snakes were caught during the summer of 1991. All snakes were captured in western Arkansas and eastern Oklahoma. Snakes were fed and cared for in captivity and appeared healthy at the time of the climbing trials. Eight snakes were judged to be male and 4 to be female, while the sex of 5 was undetermined. Snakes ranged in snout-vent length from 64 to 183 cm and in total length from 76 to 201 cm. No snakes in the process of shedding their skins were used in climbing trials.

We tested snakes individually for climbing ability on 5 woodpecker cavity trees with SNEDs (SNED trees) and on 5 control trees lacking SNEDs (control trees). Selection of control trees was based on similarity to the 5 SNED trees in DBH, bark roughness, and degree of lean (see below). These parameters were characterized for both sets of trees (**Table 1**). Temperature and relative humidity were measured, and control tree trials were conducted under similar weather conditions as SNED tree trials. Loose bark was scraped from control trees to simulate the limb saw treatment of SNED trees. All trees on which snakes were tested were shortleaf pines. Bark roughness was measured by wrapping a string around the trunk at three levels - 5, 15, and 30 cm below the lower edge of the SNED.

Two sets of measurements were taken on each tree, one before and one after the climbing trials. For each crevice extending $\geq$ 5 mm in depth inward from the string, area of a plane slicing through the crevice inward from the string was estimated using measurements of width and depth. Mean planar area of crevices $\geq$ 5 mm in depth was determined. The number of crevices $\geq$ 5 mm deep per cm of trunk diameter at each level of each tree was recorded as "frequency" of crevices. Lean of

Figure 1. Black rat snake attempting to climb past a snake excluder device (SNED) attached to the trunk of a nest tree of red-cockaded woodpeckers in the Ouachita National Forest, Arkansas. Photo: J.C. Neal. Public domain, USFS.

Table 1. Characteristics of SNED trees and control trees used in climbing trials of Black Rat Snakes, Ouachita National Forest, Arkansas. *t*-tests revealed no significant difference between tree types.

Tree char.	SNED	Cont.	*t*	*P*
DBH[1]	43.2	42.7	0.31	NS
BR (f)[2]	0.33	0.32	0.96	NS
BR(m)[3]	32.8	27.3	0.38	NS
Lean(°)	5.4	3.7	1.28	NS

n of SNED trees = 5
n of control trees = 5
df for all t-tests = 8
NS = nonsignificant; $P>.10$
[1]Diameter at breast height
[2]Frequency = number of crevices ≥ 5 mm deep per cm of trunk circumference beneath string (see text for full explanation).
[3]Mean area = mean planar area of crevices ≥ 5 mm deep beneath string (see text for full explanation).

each tree was measured as degrees of departure from the vertical between ground level and the base of the SNED.

Tests were conducted 1 tree at a time, with all snakes being given the opportunity to climb a particular tree before tests on another tree were run. Snakes were presented 1 at a time, in random sequence, to each tree. Each snake was removed from its cage and placed gently on the bark near the tree's base, with head oriented upward 60-100 cm above ground level (*sensu* Rudolph et al. 1990b). If a snake refused to climb voluntarily, or after repeated gentle prodding, it was put back into its cage and the next snake in the sequence was tested. Snakes that refused to climb a tree were given a second opportunity after all snakes had been tested once.

Snakes that began climbing after placement on a tree were observed from several meters away with no further interference. During the climbs, we recorded:

1) amount of time snakes spent stretching their bodies upward over the surface of the SNED in an initial search for a way across;

2) amount of time snakes spent exploring the base of the SNED after a failed initial attempt to cross it.

We also recorded, by visual estimation, the maximum distance across the SNED that each snake reached during each trial. These three measurements quantified the effort snakes expended in climbing.

For control trees, small pieces of tape were placed on the bark at heights equivalent to the average heights of the top and bottom of the 5 SNEDs tested. Bark roughness was measured in reference to the lower piece of tape. We recorded **1)** time spent climbing while heads of snakes were between the pieces of tape, and **2)** time spent climbing while snakes moved from snout to vent past the top piece of tape (equivalent to successful crossing of the SNED area).

The percentage of trunk circumference used by snakes during each attempt was visually estimated for climbs on both types of trees. This served as a fourth measure of climbing effort.

Success rates for climbing snakes were compared for SNED trees and control trees using Chi-square analysis (with "success" defined as a snake's vent reaching the top of a SNED or the top piece of tape). SNED and control trees were compared for DBH, bark roughness, and degree of lean using t-tests of means.

RESULTS

Thirteen of our 17 snakes voluntarily climbed trees. On SNED trees, snakes climbed on 38 of 74 first opportunities and on 3 of 33 retrials. On control trees, snakes climbed on 37 of 65 first opportunities and on 6 of 28 retrials.

No snakes were able to climb past SNEDs. Of 41 climbs on SNED trees, none were successful. By contrast, on control trees, snakes climbed above the tape simulating the top of a SNED in 40 of 43 climbing trials. The difference in success rates between SNED and control trees was highly significant (Chi-square, $X^2 = 69.13$, df = 1, $P < 0.0001$).

Only one snake was able to stretch its head past the top of a SNED. This snake (a female, 109 cm in snout-vent length) spent 250 seconds attempting to cross the SNED, before descending. In only 3 other trials did snakes manage to extend their heads three-quarters of the way up a SNED. Mean maximum distance reached by snakes in the 41 SNED trials was 24 cm beyond the lower edge of the 60-cm-wide SNED.

Snakes on SNED trees covered an average of 55 percent of each tree's circumference in attempting to cross the SNEDs, while snakes on control trees covered only 17 percent of the circumference. This difference was statistically significant (Kolmogorov-Smirnov one-tailed two-sample test, $X^2 = 19.48$, $P < .0001$).

On SNED trees, snakes spent an average of 72 seconds (range = 0-364) in initial attempts to cross SNEDs. This includes only time during which heads and forebodies of snakes were stretched over the metal flashing. Snakes spent an average of 77 seconds (range = 0-791) exploring the lower edge of the flashing and making subsequent climbing attempts from this position. During these episodes, snakes moved laterally around the base of the SNED, with head and forebody extended alternately above and below its lower edge.

On control trees, snakes spent an average of 61 seconds (range = 23-188) moving their snouts from the level of the bottom piece of tape to the level of the top piece of tape. They then spent an average of 80 seconds (range = 28-182) moving their bodies from snout to vent past the top piece of tape.

Snakes showed no apparent difficulty climbing control trees. They ascended by gripping bark crevices with their ventral scales, moving upward at a constant pace. On SNED trees, however, snakes were unable to grip the smooth surface of the metal flashing. Typically, snakes climbed upward, flicking their tongues, until roughly 15 cm of their bodies were extended over the metal. They then slowed their upward progress and began waving their forebodies back and forth, searching for a grip. Many continued swinging back and forth as they explored, inching upward with the middle and posterior portions of their bodies. Unable to find purchase points, some snakes descended, but in 71 percent of the trials they began climbing laterally around the tree at the lower edge of the SNED, flicking their tongues and trying repeatedly to move upward over the flashing **(Figure 1).** Many snakes made multiple climbing attempts from successive positions along the lower edge of the SNED.

Snakes sometimes used small objects on the surface of SNEDs to their advantage in climbing. Several snakes used horizontally-oriented staples to support themselves while climbing, and 1 snake used a loose and projecting vertical staple. Along the bole of a SNED tree, a thin nylon cord (used for pulling a bird trap frame up to the cavity) extended from the ground up to the cavity. Two small snakes used this cord to successfully cross the SNED. These 2 climbs were not included in the data analysis.

The 4 snakes that refused to climb in all trials (SNED and control) ranked as the first, second, third, and ninth longest individuals among our sample of 17 snakes. These non-climbing snakes were significantly longer than snakes which climbed (Wilcoxon test, $z = 2.32$, $P = 0.01$). Snakes that did climb successfully ranged in size from 64 to 117 cm in snout-vent length and from 76 to 142 cm in total length.

In most of their trials, the 3 largest snakes that did not climb repeatedly oriented downward after being placed on the trunk. On the few occasions when they attempted to climb upward they struggled laboriously, making slow progress, knocking off flakes of loose bark, and slipping frequently before descending. The smaller non-climbing snake was able to grip the bark well, but never attempted to proceed upward. (This individual climbed upward readily when placed on a nearby black oak (*Quercus velutina*) trunk with deeply fissured bark and a post oak (*Q. stellata*) with vines running up the trunk).

Snakes held in captivity since 1991 did not differ significantly in climbing performance from snakes captured in 1992. Snakes caught in 1991 climbed control trees slightly faster than 1992 snakes, and showed greater effort in climbing SNED trees, according to 3 of 4 measures of effort, but these differences were not significant. (All t-values were less than 1.08, with 37-39 degrees of freedom, giving P-values above 0.10).

SNED trees did not differ significantly from control trees in DBH, bark roughness, or lean **(Table 1)**. Weather conditions were also similar during SNED and control trials. Temperatures ranged from 21.4 to 31.1 degrees C during SNED trials and from 22.9 to 27.3 degrees C during control trials. Relative humidity ranged from 56 to 90 percent for SNED trials and 54 to 96 percent for control trials. Skies were cloudy the majority of the time. A few SNED trials and several control trials were run in very light rain, which did not seem to affect the climbing surface. During sunny periods we placed snakes on shaded sides of trees.

DISCUSSION

SNEDs successfully deterred climbs of black rat snakes of various sizes despite apparent determination of the snakes to climb upward past the flashing. Employed under certain guidelines (see below), we predict the

SNED will be an effective management tool for eliminating snake predation at RCW nest and roost cavities. Use of SNEDs should prove especially beneficial on RCW trees with recently installed artificial cavities, where sparse resin accumulations may be inadequate for deterring snakes.

Snakes spent an average of more than 1 minute in initial attempts to cross SNEDs. They spent well over 1 additional minute exploring the lower edges of SNEDs and making subsequent attempts to cross them from alternate positions. On average, snakes extended their forebodies over 40 percent of the height of SNEDs. They also covered over half the circumference of each SNED tree during their climbs, with several snakes crawling completely around the trunk before descending. These measures quantify and confirm our observations during the trials that snakes were actively attempting to traverse the SNEDs.

Climbing rates on control trees were distinctly faster than on SNED trees, and snakes climbed more or less straight upward, covering much less of the trunk circumference. This supports our conclusion that SNEDs were responsible for inhibiting upward progress of snakes.

While most snakes climbed trees voluntarily, the motivation behind their behavior was unclear to us. Both escape behavior and exploratory behavior may have played major roles. Whatever factors caused them to climb, snakes in this experiment likely differed in behavior from snakes in the wild climbing toward active woodpecker nests, motivated by pursuit of prey. Snakes motivated by hunger may expend greater effort climbing than did snakes in our trials. As such, our experiment does not completely rule out the possibility that snakes in the wild may occasionally be able to cross SNEDs. Field tests of the effectiveness of SNEDs against snakes in the wild will be run in the future.

A second limitation of our experiment was that only snakes of small and moderate size attempted to climb SNED trees. The largest snakes generally refused to climb and appeared unable to grip bark effectively. This tendency follows that noted by Rudolph et al. (1990b), who reported that small rat snakes showed greater climbing ability than large ones. Thus, large rat snakes may represent less of a threat to woodpecker nests and roosts than small snakes. However, it is possible that large snakes motivated by hunger might attempt to climb cavity trees. A large snake might be able to extend its forebody past the top of a SNED while still gripping the bark below the SNED, and successfully cross it. Ample evidence indicates that large snakes do climb trees. Black rat snakes caught in mesh traps on the trunks of RCW trees in the Ouachita N.F. measured up to 163 cm in total length (Neal 1992), while those caught on nest trees of other bird species measured up to 170 cm (J.H. Withgott, unpubl. data). Two radio-tagged black rat snakes each measuring 171 cm in total length were found in cavities high in oak trees in the Ouachita N.F. (J.H. Withgott and C.J. Amlaner, unpubl. data).

We expected that snakes used in our experiments might climb less effectively than wild snakes if living in captivity weakened their muscles through lack of exercise. Our results ruled out this potential problem. The snakes kept in captivity for 12 months climbed as effectively as those captured within 3 months of the trials. The results strongly suggested that captivity for prolonged periods did not diminish the health and strength of our snakes. From this observation we predict that snakes held in captivity for up to a year probably are not substantially weaker than snakes in the wild.

Wild snakes may exhibit greater climbing ability than our experimental snakes due to the motivation of prey pursuit. Therefore, managers may want to consider using flashing of widths greater than 60 cm. We feel 60 cm should be adequate to deter climbing snakes, but have no data on the relative effectiveness of various widths of flashing. While rat snakes normally do not exceed 183 cm in snout-vent length, records of 250 cm do exist (Conant and Collins 1991).

Regardless of the width of flashing used, successful employment of the SNED technique necessitates that:

1) Flashing should be stapled onto trees with staples oriented vertically and flush against the surface of the flashing. Staples oriented horizontally or loose vertical staples may provide small but significant purchase points for climbing snakes;

2) SNEDs should be kept free of a buildup of resin droplets, which may dry and provide a rough surface to aid climbing snakes;

3) Bark should be kept as smooth as possible above and below SNEDs, to deprive snakes of deep crevices with which to brace themselves.

4) For RCW cavity trees whose branches contact or substantially overlap those of adjacent trees, SNEDs should be attached to the adjacent trees as well as the cavity tree. In 1991, Neal (1992) captured a rat snake climbing a tree whose limbs overlapped those of an RCW nest tree.

Installation of SNEDs represents the best available strategy for countering the widespread threat of rat snake predation on RCW nests and roosts. Mesh snake traps can be effective in deterring snake climbs, but may result in high rates of snake mortality unless checked on a very frequent basis. The objective should be to protect the endangered RCW while minimizing harm to non-target animals. We urge the use of SNEDs rather than mesh traps not only for ethical reasons, but for an ecologically practical one. Rat snakes prey on competitors for RCW cavities such as southern flying squirrels (*Glaucomys volans*) and other woodpecker species (Hoyt 1957, Nolan 1959, Fitch 1963, Jackson 1970, Dennis 1971a, Stickel et al. 1980, Ernst and Barbour 1989). Reduction of rat snake densities near RCW cavity tree clusters could lead to higher densities of cavity competitors.

The SNEDs we recommend:

1) do not harm snakes

2) are constructed of inexpensive materials,

3) require only a few minutes to install, and,

4) need virtually no maintenance.

This simple technique provides wildlife managers with an effective tool for reducing snake predation on RCW nests and roosts with a minimum of effort.

ACKNOWLEDGEMENTS

The Ouachita National Forest has provided generous support for graduate research projects involving two of us (Withgott and Neal), for which we are grateful. We also thank the Southern Forest Experiment Station, Nacogdoches, Texas, and the Arkansas Audubon Society Trust for additional support.

We thank those who captured snakes for use in this study and related studies: G. Bukenhofer, T. Dunn, L. Garner, S. Garner, H. Hicks, P. Li, P. Moore, K. Piles, J. Ray, M. Revels, L. Shores, E. Stewart, J. Walters, C. Warriner, E. Wilson, K. Zyskowski. S.B. Isenberg, D.A. James, and D.C. Rudolph made helpful comments on a draft of the manuscript.

Techniques for Excluding Southern Flying Squirrels from Cavities of Red-cockaded Woodpeckers

Warren G. Montague, Poteau Ranger District, USDA Forest Service, P.O. Box 2255, Waldron, AR 72958
Joseph C. Neal, Arkansas Cooperative Fish and Wildlife Research Unit, University of Arkansas, Fayetteville, AR 72701[1]
James E. Johnson, U.S. Fish and Wildlife Service, Arkansas Cooperative Fish and Wildlife Research Unit, University of Arkansas, Fayetteville, AR 72701
Douglas A. James, Department of Biological Sciences, University of Arkansas, Fayetteville, AR 72701

ABSTRACT: Between 1990 and 1993, a population of red-cockaded woodpeckers (*Picoides borealis*) (RCWs) in 13 to 15 clusters of cavity trees was studied in the Ouachita National Forest of west-central Arkansas. Part of this study involved potential cavity conflicts between RCWs and southern flying squirrels (*Glaucomys volans*) (SFS). These squirrels occurred in all clusters. Inexpensive squirrel excluder devices (SQEDs) were installed on 11 inactive cavities of 10 trees. SQED treatments included **1)** installation of twin bands of aluminum flashing above and below cavity entrances to impede squirrel access to cavities, and **2)** thorough cleaning of cavities to make them habitable by RCWs after verified use of cavities by squirrels. In some cases, steps 1 and 2 were applied, and squirrels occupying cavities were captured and translocated from cavity tree clusters.

Prior to SQED treatments, 10 of these cavities were occupied by squirrels; the eleventh cavity had been abandoned by RCWs, but was not occupied by squirrels. Squirrels abandoned use of six cavities after SQED treatments. With the addition of squirrel translocations, 10 of 11 SQED-treated cavities were eventually reoccupied by RCWs.

RCWs were not deterred from using cleaned cavities with SQEDs and reinitiated use of such cavities which were formerly occupied by squirrels. Translocation of squirrels occupying a given cavity at the time of SQED treatment appeared to reduce the chances that other squirrels would reoccupy cavities. SQEDs appeared to work best in clusters of cavity trees where RCWs had fledged young. SQED treatments resulted in making more natural cavities available for use by RCWs.

KEYWORDS: southern flying squirrel, red-cockaded woodpecker, cavity use, Ouachita National Forest, endangered species management

[1]Present address: Poteau Ranger District, U.S.D.A. Forest Service, P.O. Box 2255, Waldron, AR 72958.

A key element in group formation in cooperatively breeding red-cockaded woodpeckers (*Picoides borealis*) (RCWs) is availability of suitable cavities (Walters et al. 1992a). However, numerous other species of vertebrates use these cavities (Dennis 1971a, Jackson 1978b, Rudolph et al. 1990a). Competition for cavities is sometimes intense, especially during nesting seasons (Harlow and Lennartz 1983, Neal et al. 1992). In North Carolina, southern flying squirrels (*Glaucomys volans*) were more frequent occupants of RCW cavities than any avian species encountered (Everhart 1986). Flying squirrels are more likely than other cavity-users to use the cavities preferred by RCWs (Loeb 1993). In Georgia, 7 of 51 RCW nest failures were attributed to cavity usurpation by squirrels (Lennartz and Heckel 1987), but in Texas there was little evidence that squirrels forced RCWs to roost in the open or to use suboptimal cavities (Rudolph et al. 1990a).

Flying squirrels have been mentioned as incidental predators of bird eggs and nestlings (Muul 1968, Harlow and Doyle 1990) and have been known to sometimes crush or possibly eat eggs of RCWs (M. LaBranche, pers. comm.) and destroy eggs of other cavity nesting birds (Stabb et al. 1989). Stomachs from flying squirrels collected in clusters of cavity trees of RCWs in South Carolina mostly contained acorn starch grains and seed coat parts (Harlow and Doyle 1990). These results suggest that habitat degradation resulting from hardwood encroachment and consequent production of hardwood mast has favored competitors such as squirrels for RCW cavities.

Techniques described in this paper reduced squirrel use of RCW cavities and promoted reoccupation of cavities by the woodpecker. Wildlife management techniques that favor this endangered species in interspecific conflicts for cavities have potential to speed recovery of this woodpecker.

STUDY AREA

Studies were conducted in the Ouachita National Forest (Ouachita NF) in Scott and Polk counties of west-central Arkansas. Cavity trees of RCWs in the Ouachita NF áre shortleaf pines (*Pinus echinata*) that occurred in maturing stands of second-growth pine or mixed pine-hardwood (Neal and Montague 1991). Flying squirrels are common in the Ouachita NF (Heidt 1977, pers. obs.).

METHODS

During three breeding seasons, 1990 through 1992, 13 to 15 RCW groups were monitored. Cavity trees were checked for signs of RCW use (Jackson 1977b, 1978c). During the nesting season, mid-April to early July, ladders were used to climb nest and roost cavity trees. Cavities were inspected with a light and mirror. Nests were checked every 7-10 days from 29 May to 27 June in 1990 and 1 to 2 times weekly from 23 April to 24 July in 1991 and 12 April to 24 July in 1992. Additional observations were made throughout the year including the period through April 1993.

In 1990, all active cavity trees (n=26) were inspected for the presence of squirrels. Squirrels were removed with a flexible retrieving tool 60 cm long (e.g., NAPA service tool). In 1991, active (n=31) and recently active (n=8) cavity trees (e.g., categories 1 and 2 in Rudolph et al. 1990a) were checked for presence of squirrels. When squirrels were found, the cavity entrance was blocked. A small piece of sponge attached to the end of the retrieving tool was soaked with a purified grade of ethyl ether (e.g., Fisher Chemical E 134-1) and inserted into the sealed cavity for 1-2 minutes to sedate the animals. Anesthetic treatment facilitated extraction and reduced potential injuries to the squirrels from struggling. Use of ether to simplify handling of flying squirrels had been employed successfully elsewhere on the Ouachita NF (G. Heidt, pers. comm.). Squirrels sometimes suffered minor injuries as a result of removal due to design of the retrieving tool.

Beginning on 15 May 1991, squirrels were extracted from active or recently active RCW cavities (**Table 1, Numbers 1-11**). The retrieving tool and a battery-powered portable vacuum (e.g., Black & Decker Power Pro, DB6000 with the AK10 accessory kit) equipped with a 75-cm-long flexible hose (e.g., Hoover Elite upright vacuum hose) were used to clean the cavity of shredded bark, mosses, and other materials carried there by squirrels. A heavy-duty staple gun was used to attach squirrel excluder devices (SQEDs) to the cavity tree. SQEDs were not installed on active cavity trees.

SQEDs (**Figure 1**) consisted of paired strips of 20.3-cm-wide lightweight aluminum flashing stapled tightly to the bark above (1 strip) and below (1 strip) the cavity entrance. Points of attachment of SQED strips above and below the cavity entrance varied due to **1)** irregular tree trunk surfaces and **2)** varied patterns of resin well excavations. The objective was to minimize interference with the function of resin wells, while keeping the strips as close as possible to the cavity entrance. SQEDs were not painted or otherwise disguised.

After initial SQED installations, sap flows from resin wells often formed a "bridge" of resin across some SQED strips. This resulted in a textured surface that sometimes allowed squirrels to traverse the previously smooth surface of the SQED. For this reason, the original flat SQED design was modified into a sap-deflecting SQED (**Figure 2**). This was done by cutting and bending down the top two cm of each SQED strip at an angle away from

Table 1. Field tests of Squirrel Excluder Devices (SQEDs) in the Ouachita National Forest, Arkansas, 1991 to 1993.

Number and location[a]	Results and narrative[b]
Number 1 **c 1252** **s 26** **tree 6**	Result: SFS abandoned cavity, RCW reoccupied cavity. Two SFS present when SQEDs installed 15 May 1991. RCW in cavity by 21 May. SFS crossed sap "bridge" on upper SQED strip on 28 Aug. Sap-deflector SQED installed and RCW in cavity again by Nov. 1991. Tree killed by southern pine beetles in early 1992; RCW abandoned cavity. (RCW nesting in this cluster successful in 1991, but not 1992.)
Number 2 **c 1257** **s 20** **tree 13**	Result: SFS removed, RCW reoccupied cavity. SQEDs installed on recently active RCW cavity on 22 May 1991, at which time 1-3 SFSs occupied cavity. SQED did not exclude SFS, so SFS removed from cluster 31 May. Cavity empty until 28 August 1991 when fledgling RCW roosted until autumn 1992. When juvenile RCW moved to new cluster, breeding male RCW used cavity in Oct. 1992. Small amount of SFS material removed spring 1993; flat SQED heavily covered by sap bridge. (RCW nest in this cluster successful in 1991 and 1992.)
Number 3 **c 1253** **s 5** **tree 4**	Result: SFS did not abandon cavity. Two SFS in recently active RCW cavity on 22 May 1991 when flat SQED installed. Thereafter, cavity repeatedly filled with SFS material. One of three SFS removed from cavity 23 March 1993. (RCW nesting in this cluster unsuccessful in 1991 and 1992.)
Number 4 **c 1274** **s 9** **tree 7**	Result: SFS abandoned cavity; RCW did not reoccupy cavity until early 1993. SQED installed 23 May 1991 with one SFS present. No SFS on 28 May and thereafter. Some fresh RCW work on resin wells in 1992, but no RCW roosting until between mid-October 1992 and mid-February 1993. (RCW nesting in this cluster unsuccessful in 1991, but successful in 1992.)
Number 5 **c 1252** **s 26** **tree 4**	Result: SFS abandoned cavity, RCW reoccupied cavity. RCW nest cavity in 1990 season and still appeared active when SFS observed in cavity on 28 May 1991. SQED installed June 1991, with one SFS present. Renewed RCW activity by 8 July and still active in early 1992. Sap "bridge" had formed on lower flat SQED strip and SFS in cavity on 21 April 1992 when sap-deflecting SQED installed. One SFS present on 4 May 1992 escaped. Evidence of RCW activity 10 June 1992 and RCW roosting observed 8 August 1992. (RCW nesting in this cluster successful in 1991, but not 1992.)
Number 6 **c 1257** **s 28** **tree 6**	Result: SFS abandoned cavity, RCW reoccupied cavity. No evidence of RCW use in spring 1991. Four SFS in cavity on 1 May 1991. SQED installed 29 May. Tree reactivated by RCW in Nov. and Dec. 1991 and cavity still active 23 Jan. 1992. This RCW disappeared, but cavity subsequently used by another RCW[c]. (No RCW nesting in this cluster in 1991 or 1992.)
Number 7 **c 1257** **s 28** **tree 1** **(upper cavity)**	Result: SFS abandoned upper cavity, RCW reoccupied cavity. RCW nest tree in 1990. Upper and lower cavities (see Number 8) used alternately by RBW, SFS and RCW during 1991. Sap-deflecting SQEDs installed on both cavities 23 Oct. 1991 and RBW evicted by a readjusted restrictor. Thereafter, different RCW roosted in cavity. (RCW nesting in this cluster unsuccessful in 1991 and cluster inactive in 1992.)

Table 1 continued

Number and location[a]	Results and narrative[b]
Number 8 **c 1257** **s 28** **tree 1 (lower cavity)**	Result: SFS removed from cavity, RCW reoccupied cavity. SFS in RCW cavity during RCW nest season, at least from 25 April to 3 May 1991. On 4 Aug. 1992, four male SFSs were removed from this cavity. No RCW activity noted until 14 March 1993, when lower cavity was occupied by an RCW. This RCW had roosted previously in another tree (see Number 6, above).
Number 9 **c 326** **s 14** **tree 1**	Result: SFS abandoned cavity, RCW reoccupied cavity. RBW occupied RCW cavity in 1990. RBW excluded by restrictor. No RCW use until late summer 1991. SFS present 25 April 1991. Sap-deflector SQEDs installed 25 June 1991. Juvenile female RCW roosting by 23 July 1991, and cavity still in use in autumn 1992. Cavity clean and empty in March 1993, but not occupied by RCW. (RCW nesting in this cluster successful in 1991 and 1992.)
Number 10 **c 326** **s 14** **tree 2**	Result: No SFS present, but RCW reoccupied cavity. No history of SFS use of cavity. RCW roost cavity in 1991, but inactive during June 1991. Pileated woodpecker hole in the back of the cavity chamber patched and a sap-deflecting SQED installed 25 June 1991. RCW roosting confirmed 7 Oct. and still active 25 March 1993. (RCW nesting in this cluster successful in 1991 and 1992.)
Number 11 **c 326** **s 14** **tree 8**	Result: SFS removed, RCW reoccupied cavity. At least one SFS in cavity on 22 April 1992 when SQEDs with sap-deflectors installed. SFS present on 23 April 1992, at which time it was removed. Cavity clean and in use by an RCW by 16 July 1992; still used by RCW on 25 March 1993.

[a]Includes example number compartment (c) and stand (s) locations of cavity tree clusters in the Ouachita NF, and tree number within the cluster.

[b]Abbreviations include RCW (red-cockaded woodpecker), RBW (red-bellied woodpecker) and SFS (southern flying squirrel).

[c]The cavity was recleaned of SFS materials on 18 Nov. 1992 and remained clean thereafter. On 19 Dec. 1992 an RCW from the Kisatchie NF in Louisiana was moved to the Ouachita NF and released near this tree; it chose the SQED-treated tree as a roost and remained until 14 March 1993, when it moved to another SQED-treated cavity tree (Table 1, Number 8).

the tree. Sap flowing from above the SQED would then be deflected away from the surface of the SQED by these sap-deflecting flaps. The bottom edge of each SQED strip was still fitted tightly against the bark, but a space was left along the top edge of each SQED strip so that sap flow would continue along the tree bole behind the SQED, rather than over it. In a further modification, small flaps of flashing were stapled behind the SQED strips between each sap deflector flap, forming a continuous surface or rim around the entire tree bole **(Figure 3).** These two modifications prevented sap from forming a textured surface across the SQED.

Initially, squirrels were returned to cavities after attachment of SQEDs. We predicted that squirrels would abandon treated cavities because SQEDs would make it more difficult to access cavities by climbing up or down the tree trunk. In those cases where squirrels were not excluded by the SQEDs **(e.g. Table 1, Number 2)**, squirrels were translocated (or removed) from the cavity tree cluster. We predicted that with SQEDs in place, squirrels previously unfamiliar with the cavity would find it difficult to either locate or access treated cavities.

In 1991 and 1992, 11 cavities in 10 trees were treated with SQEDs, including one inactive cavity without a history of squirrel use. Cavities with SQEDs were inspected to verify

Figure 1. Squirrel excluder device (SQED) with sap-deflecting modification installed on RCW cavity tree. Ouachita NF, 1991.

whether;

1) squirrels had been excluded from the cavity, and,

2) RCWs had reoccupied the SQED-treated cavity.

RESULTS

In 1990, 35 squirrels were found in seven RCW clusters; 13 were in a single cluster of inactive cavity trees. Almost all squirrels were in cavities that had not been active recently (e.g., categories 3-5 in Rudolph et al. 1990a).

In 1990, no squirrels were observed occupying cavities having evidence of recent RCW activity. In 1991, however, a squirrel occupied a woodpecker cavity that had recently been the roost of a one-year-old male RCW **(Table 1, Number 8)**. This squirrel occupation

Figure 2. Close view of sap-deflecting modification of squirrel excluder device (SQED) installed on RCW cavity tree. Ouachita NF, 1991.

lasted at least from 25 April to 3 May 1991, a period when RCWs were laying eggs in the Ouachita N.F.

In 1992, there were four instances in which squirrels reoccupied SQED-treated cavities (**Table 1, Numbers 1, 3, 5, 8**). In one such reoccupation (**Table 1, Number 5**) squirrels had apparently crossed a sap "bridge" that had formed on the SQED. Installation of a sap-deflector SQED and recleaning the cavity was followed by squirrel abandonment.

From 1991 to 1993, squirrels abandoned use of nine cavities after SQED treatment; all of these cavities were eventually occupied by RCWs (**Table 1**). The first SQED treatment of a cavity (**Table 1, Number 1**) resulted in rapid abandonment of the cavity by squirrels and an almost immediate RCW reoccupation. Before

Figure 3. Sap-deflecting squirrel excluder device (SQED) modified by addition of gap flaps which formed a continuous surface or ring around the tree. Ouachita NF, 1993.

the end of 1991, five more cavities previously occupied by squirrels were reoccupied by woodpeckers (**Numbers 2, 5, 6, 7, 8**). Squirrels did not abandon one cavity that received SQED treatments (**Number 3**). During 1992 and early 1993, most cavities with SQEDs were still occupied by the woodpeckers (**Numbers 4, 5, 7, 8, 10, 11**). Four cavities were unoccupied by either squirrels or RCWs and one tree died.

One SQED tree (**Table 1, Number 2**) with a recently active RCW cavity was occupied by three squirrels. This cavity tree was adjacent to a stream-course with other trees in such close proximity that they could easily be used by squirrels to glide directly to the cavity between the SQED strips. When squirrels did not abandon this cavity after SQED treatment, they were captured and translocated. The cavity was later occupied by a juvenile male RCW that fledged from a nest tree approximately 400 m away. The cavity remained in use until late 1992. In September 1992, this juvenile male briefly joined another woodpecker group, then moved to a cluster of cavity trees unoccupied by RCWs. The cavity with SQEDs was then occupied by the breeding male of this juvenile male's original group.

In 1991, reoccupation of SQED-treated cavities occurred in or adjacent to home ranges of groups of RCWs that had successfully fledged young (**Table 1, Numbers 1, 2, 5, 6, 7, 8, 9**). Cavities that received SQED treatments but were not reoccupied by woodpeckers during 1991 (**Numbers 3, 4**) were in territories of woodpecker groups that suffered nest failures or did not nest in 1991.

Examples Numbers 7 and 8 (**Table 1**) illustrate the complex pattern of interspecific use of cavities originally excavated by RCWs. In late 1991, these two cavities in the same tree received SQEDs (**Figure 4**). RCWs had nested in this tree in 1990, but after the 1990 nesting season, the cavities had been alternately occupied by squirrels, red-bellied woodpeckers

Figure 4. Two squirrel excluder devices (SQED) installed on one RCW cavity tree. Metal flashing near the base of the tree prevents climbing by snakes. Ouachita NF, 1992.

(*Melanerpes carolinus*), and two different RCWs. No nesting by RCWs occurred in the cluster in 1991 and the cluster was inactive during the nesting season in 1992. However, when SQEDs were installed on 23 October 1991, the upper cavity became a roost for an RCW and the lower cavity was occupied by a squirrel. On 19 November 1991, a red-bellied woodpecker was roosting in the lower cavity; no RCW roosted in the upper cavity. Through January 1992, a red-bellied woodpecker occupied the upper cavity while the lower cavity remained vacant. Installation of restrictors (sensu Carter et al. 1989) excluded the red-bellied woodpecker. In summer of 1992, following a period in which the cluster was inactive, the upper cavity was reoccupied by an RCW; this new occupant was not the same bird that had formerly roosted in the tree. It has since paired with a female that was translocated from the Kisatchie National Forest **(Table 1, c)**.

Some occupations of high-quality RCW cavities by squirrels may be a natural consequence of RCW reproductive failure. At the end of the 1992 nesting season, the former nest cavity of a woodpecker group that suffered nest failure during 1992 was occupied by a squirrel. The woodpecker nest had failed by 2 June, when remains of dead young were found in the cavity. The cavity was empty on 11 June, but on 22 June the entrance tunnel was obstructed by mud nests of sphecid wasps, which were removed from the cavity. On 6 July, a squirrel was in the cavity. During the same period, RCWs had activated a cavity in another area of the cluster.

As of April 1993, after more than two years of monitoring and SQED treatments, **1)** 6 of 11 cavities were roosts for RCWs, **2)** 4 of 11 cavities were clean and habitable but unoccupied by either RCWs or squirrels, and **3)** 1 of 11 cavities was in a dead tree and therefore unlikely to be used by RCWs.

DISCUSSION

Monitoring efforts confirmed high rates of cavity use by flying squirrels in the Ouachita NF. It is probable that none of the 11 cavities that received SQED treatments would have been reoccupied by RCWs without some combination of treatments described in this paper. The importance of maintaining an abundant supply of serviceable cavities to promote woodpecker recruitment has been demonstrated (Walters et al. 1992a). Developing an effective technique to discourage or prevent such occupancy could preserve the integrity of scarce natural cavities, which seem to be preferred by the woodpeckers, and could defer the need to supplement existing clusters with expensive, artificial drilled or insert cavities. In the Ouachita NF, the number of habitable natural cavities increased with translocation of squirrels and cleaning of cavities.

SQED installations including thorough cleaning of cavities and or translocation of squirrels usually resulted in reoccupation of cavities by RCWs. RCWs were not deterred from using cleaned cavities with SQEDs and reinitiated use of such cavities formerly occupied by flying squirrels. Translocation of squirrels in conjunction with the SQED treatment appeared to further reduce the chances that squirrels would reoccupy cavities suitable for use by RCWs.

The effectiveness of SQED treatments also appeared related to tree spacing. Thus, cavity tree clusters with well-spaced trees could enhance the effectiveness of SQED treatments. We recommend that paired SQED strips be installed as close to the cavity entrance tunnels as possible (while minimizing interference with resin wells) to decrease chances that squirrels will be able to glide directly to the cavity entrance from adjacent trees.

SQED strips, cavity cleaning, and translocation of squirrels provided nonlethal methods of excluding squirrels from specific cavities. This resulted in increased occupancy rates of these cavities by RCWs for indeterminant periods of time.

Caution is urged in interpreting these results. First, cavity trees with SQEDs were not paired with control trees which might have allowed a more precise understanding of natural mechanisms leading to abandonment of cavities by flying squirrels. Secondly, ecological relationships that exist between various users of cavities originally excavated by RCWs are unclear; therefore, subtle effects of removing squirrels on this community of cavity users are also unclear. Similarly, the potential effects that routine removals of squirrels from cavities in 1990 had on treatments during 1991 and 1992 are unclear.

ACKNOWLEDGMENTS

Support was provided by the U.S.FS, with additional assistance from the Arkansas Cooperative Fish and Wildlife Research Unit and the University of Arkansas. The management style of G. Landrum of the Poteau Ranger District, which encourages innovation and initative, has been a special inspiration to the senior author. This project and others involving RCWs on the Ouachita N.F. have benefited from the able field assistance of K. Piles and others on the Poteau Ranger District. G. Heidt and his students from the University of Arkansas-Little Rock shared with us their techniques for handling flying squirrels. Leadership by C. Annett and D. James during a seminar on community ecology at the University of Arkansas strengthened Neal's appreciation for the complex interactions involving the community of cavity users.

Vulnerability and Resistance of Red-cockaded Woodpecker Cavity Trees to Southern Pine Beetles in Texas

William G. Ross and David L. Kulhavy, College of Forestry
Stephen F. Austin State University, Nacogdoches, Texas 75962
Richard N. Conner, Southern Research Station, Nacogdoches, Texas 75962

ABSTRACT: Southern pine beetles (*Dendroctonus frontalis* Zimmermann) are responsible for over half of the mortality of loblolly (*Pinus taeda* L.) and shortleaf (*Pinus echinata* Mill.) pine trees used by red-cockaded woodpeckers, *Picoides borealis* (Vieillot) for nesting and roosting cavities over much of eastern Texas. Resin flow and xylem moisture potential, often used as indicators of pine susceptibility to bark beetle mortality, were measured in fourteen red-cockaded woodpecker cavity tree clusters in the Angelina and Davy Crockett National Forests. No differences in xylem moisture potential were found, although resin flow varied by site, tree species, and cavity tree type. These cavity trees and the stands in which they occur have characteristics associated with enhanced risk to southern pine beetle. Traditional southern pine beetle control tactics, along with newer strategies involving pheromones and artificial cavities, are useful and necessary in reducing the loss of cavity trees, but forest management that produces healthier habitat is the best solution.
KEYWORDS: red-cockaded woodpecker, southern pine beetle, resin flow

Southern pine beetles (*Dendroctonus frontalis* Zimmermann) (SPB) are the most important forest insects in the southern states from the standpoint of economic impact and potential for rapid destruction of large areas of mature pine forest. They are also a major cause of death to red-cockaded woodpecker *Picoides borealis* (Vieillot) (RCW) cavity trees (Conner et al. 1991a, Mitchell et al. 1991). Loss of red-cockaded woodpecker cavity trees occurs for a variety of reasons. In Texas loblolly and shortleaf stands in the Angelina, Davy Crockett, and Sam Houston National Forests, Conner et al. (1991a) found that fifty-three percent of cavity tree mortality was due to attack by the SPB. This was followed by windsnap (29.6%), fire (7.1%), windthrow/root rot (4%), old age/suppression (1.6%), and herbicide (accidental) (0.8%). They believed that most of the cavity trees that died of unknown reasons were killed by SPB. Black turpentine beetles *Dendroctonus terebrans* (Olivier) and the engraver beetles, *Ips calligraphus* (Germar), *Ips grandicollis* (Eichhoff) and *Ips avulsus* (Eichhoff) are ubiquitous in southern pine forests, but do not approach the SPB in importance with respect to impact (Wahlenberg 1946).

Adult SPBs are small insects, about the size of a grain of rice (2 to 4 mm), cylindrical in shape, and light brown to black (usually) in color. Their life cycle begins when mature adults emerge from infested trees. After emergence, the beetles search for a new host tree. Initial attack begins when female SPBs land on a suitable host, chew into the bark, defecate, and begin to release the pheromone frontalin. This pheromone, mixed with host odors, particularly alpha-pinene, attracts males seeking to mate. Led by the females, the beetles bore into the phloem where they build winding S-shaped galleries for oviposition. After hatching, the larvae go through 4 stages of development (instars), pupate, then emerge as adults. Depending on the season, the life cycle takes from 3 to 8 weeks to complete (Payne 1980). SPB are capable of several annual generations under ideal circumstances. Pines colonized by SPB are almost invariably killed by girdling by egg gallery construction and effects of a blue stain fungus (*Ophiostoma minus*) carried by the insects which physically plugs the xylem.

A potentially important area of interaction between RCWs, SPBs and southern pines is the resin system of the pine trees (Ross et al. 1991). Both RCWs and SPBs require these trees as hosts. RCWs peck small holes, called resin wells, around cavity entrances that cause a cascade of pine resin on the bole around and beneath the cavity entrances. This resin barrier serves as a shield against rat snakes, *Elaphe obsoleta* (Say), a major RCW predator (Jackson 1974, Rudolph et al. 1990a). It seems to have little effect, however, on cavity competitors (Rudolph et al. 1990b). Whether pine resin alone is an attractant or arrestant for SPB is still debated by forest entomologists (Heikkenen 1977, Payne 1980). Newly excavated RCW cavity trees do seem to be unusually susceptible to SPB attack (R. N. Conner pers. obs.) as do freshly wounded mature pines in general. A primary defense against bark beetle attack is preformed resin flow (Hodges et al. 1979, Payne 1980, Paine et al. 1985, Nebeker et al. 1988). Preformed resin is resin in the tree's resin ducts at the time of wounding or beetle attack rather than resin produced as a response to these stimuli. Bark beetles, particularly during endemic periods, are often unable to colonize and kill pines with high resin flow (Lorio et al. 1990).

Objectives of this study were to: **(1)** measure effects of red-cockaded woodpecker cavity excavation and resin well pecking of preformed resin flow in shortleaf, and loblolly pine trees in Texas, **(2)** measure effects of RCW activity on tree moisture stress, and **(3)** evaluate relative susceptibility of red-cockaded cavity trees to southern pine beetles.

METHODS

Study Areas

Resin and xylem moisture potential data were collected periodically during the growing seasons of 1986 through 1989 in RCW cavity tree clusters in the Bannister Wildlife Management Area (BWMA) of the Angelina National Forest (ANF) in eastern Texas, and 1989 through 1990 in the Neches District of the Davy Crockett National Forest (DCNF) near Crockett, Texas.

RCW trees in the BWMA and DCNF are loblolly and shortleaf. Trees in RCW stands used for sampling were divided into the following categories, with 50 to 100 total trees sampled in each RCW study area:

1. Trees currently used for RCW nesting and roosting (Active);

2. Trees previously used for nesting and roosting, but currently not used by RCW (Inactive);

3. Trees having external characteristics associated with RCW trees, such as age, crown condition, and evidence of heart rot. (Potential);

4. Beginning in 1988 in the BWMA newly excavated trees were added as a separate category.

Resin Flow

Resin flow was measured by driving a 2.54 cm diameter circular arch punch (after Lorio et al. 1990) to the interface of xylem and phloem at approximately 1.4 m (DBH) on the bole. Holes were punched between 0700 and 1000h to minimize effects of diurnal variation in resin flow (Nebeker et al. 1988). Triangular metal funnels were placed under the wounds to divert exuded resin into clear plastic graduated tubes. Resin flow was recorded 8 and 24 hours after wounding. After 24 hour readings were taken, funnels and tubes were removed and the bark plug replaced. On woodpecker trees in both the BWMA and DCNF, only one sample was taken during any one sampling period to avoid placing undue stress on the trees.

Xylem Moisture Potential

Xylem moisture potential was evaluated in RCW cavity clusters using the pressure chamber technique (Scholander et al. 1965). Twig samples were taken from the upper crowns of cavity and non-cavity trees selected from among the trees sampled for resin flow. Sampling was performed between 1300 and 1500 h at the same time resin sampling was being conducted. Twig samples were collected using a 12 gauge shotgun, and moisture status evaluated within 60 sec of collection.

ANALYSIS

Data were analyzed using the SPSS^X statistical software package (Norusis 1985). Resin flow at 8 and 24 hours was analyzed separately for each species and by each forest. Resin flow by species was analyzed using the Mann-Whitney U-Wilcoxon Rank Sum test (Norusis 1985). Kruskal-Wallis non-parametric rank analysis was used to evaluate resin flow by cavity tree type. When differences were significant at $P \leq 0.05$, the non-

Table 1. Overall resin flow for all trees (both cavity and non-cavity) in milliliters at 8 and 24 hours by species and forest.[1] N refers to total number of samples.

	Angelina National Forest			Davy Crockett National Forest		
Species	N	8 hr ml (SD)	24 hr ml (SD)	N	8 hr ml (SD)	24 hr ml (SD)
Loblolly	598	3.75 a (4.05)	5.37 a (5.77)	126	5.89 b (5.21)	9.21b (8.66)
Shortleaf	391	4.36 b (4.44)	6.57 b (6.65)	331	3.53 a (3.89)	5.93 a (6.31)

[1]Resin flow differs significantly between species for 8 and 24 hour measurements (a = 0.05, Mann-Whitney U-Wilcoxon Rank Sum Test [Norusis 1985]).

Table 2. Eight and twenty-four hour resin flow in milliliters by cavity tree type, Angelina National Forest, 1986 and 1987.[1] N refers to total number of samples.

	Loblolly Pine			Shortleaf Pine		
Cavity Tree Type	N	8 hr ml (SD)	24 hr ml (SD)	N	8 hr ml (SD)	24 hr ml (SD)
Active	86	3.08 a (3.10)	4.13 a (4.28)	91	2.35 a (2.07)	3.55 a (2.70)
Inactive	112	4.08 a (5.65)	5.74 a (7.49)	52	4.87 b (5.31)	6.92 b (7.02)
Potential	128	2.87 a (2.72)	4.39 a (3.99)	59	3.31 a (4.06)	6.15 ab (7.32)

[1]Within columns, means followed by the same letter are not significantly different (a = 0.05; Kruskal-Wallis non-parametric rank analysis [Norusis 1985]; non-parametric multiple comparison procedure [Daniel1990])

parametric multiple comparison procedure described by Daniel (1990) using an experimentwise error rate of 0.20 was used. The same procedures were used to analyze xylem moisture potential.

RESULTS

Overall 8 and 24 hour resin flow, combining all cavity tree types, showed significant differences in resin flow by species (**Table 1**), but with the species exhibiting highest resin flow differing by forest. In the Angelina National Forest, shortleaf pine had higher resin flow, whereas loblolly pine had the highest resin flow in the DCNF.

Resin flow in the Bannister Wildlife Management Area prior to 1988 was analyzed separately from that afterwards because of the addition of newly excavated cavity trees as a fourth category. In the loblolly pines, no differences in resin flow were discovered, either at eight or twenty-four hours (**Table 2**). More variation was observed between shortleaf cavity trees. At eight hours, inactive cavity trees had higher resin flow than active or potential trees. At 24 hours, inactive trees still had higher resin flow than active trees.

Analysis of sample trees by cavity tree type in 1988 and 1989 showed similar difference by species. In the BWMA trees, no significant differences in resin flow by cavity tree type in loblolly pine (**Table 3**) were found. In shortleaf pine however, newly activated cavity trees had much higher resin flow than old active, inactive or potential. Inactive trees still showed a trend toward higher resin flow compared with active and inactive trees, but differences were no longer statistically significant.

Results from the Davy Crockett National Forest were different from the Angelina National Forest (**Table 4**). Active loblolly pine cavity-trees had significantly higher resin flow than the potential trees. For shortleaf cavity trees, resin flow was highest in potential trees and lowest in inactive trees.

No significant differences were found in xylem moisture potential between cavity tree types in either forest (**Tables 5 and 6**). It should be emphasized, however, that these are results taken only during hours of peak stress and do not include newly excavated cavity trees.

Table 3. Eight and twenty-four hour resin flow in milliliters by cavity tree type, Angelina National Forest, 1988 and 1989.[1] N refers to total number of samples.

		Loblolly Pine			Shortleaf Pine	
Cavity Tree Type	N	8 hr ml (SD)	24 hr ml (SD)	N	8 hr ml (SD)	24 hr ml (SD)
Active	81	4.23 a (4.15)	6.12 a (6.58)	68	4.27 a (3.44)	6.04 a (5.20)
Inactive	65	4.78 a (4.08)	6.20 a (5.60)	23	5.64 a (5.88)	8.03 a (8.48)
Potential	95	3.75 a (3.81)	5.59 a (5.57)	40	4.59 a (4.06)	7.09 a (6.77)
New Active	29	4.57 a (3.84)	7.64 a (6.26)	29	10.07 b (4.50)	14.70 b (6.68)

[1]Within columns, means followed by the same letter are not significantly different (a = 0.05; Kruskal-Wallis non-parametric rank analysis [Norusis 1985]); non-parametric multiple comparison procedure [Daniel 1990]).

Table 4. Eight and twenty-four hour resin flow in milliliters by cavity tree type, Davy Crockett National Forest, 1989 and 1990.[1] N refers to total number of samples.

		Loblolly Pine			Shortleaf Pine	
Cavity Tree Type	N	8 hr ml (SD)	24 hr ml (SD)	N	8 hr ml (SD)	24 hr ml (SD)
Active	22	9.27 b (5.38)	13.82 b (9.21)	82	3.70 ab (4.50)	5.90 ab (7.28)
Inactive	48	5.67 ab (4.89)	9.21 ab (8.27)	111	2.59 a (3.11)	4.27 a (4.78)
Potential	56	4.75 a (4.97)	7.39 a (8.24)	136	4.25 b (3.85)	7.39 b (6.49)

[1]Within columns, means followed by the same letter are not significantly different (a = 0.05; Kruskal-Wallis non-parametric rank analysis [Norusis 1985]); non-parametric multiple comparison procedure [Daniel 1990]).

Table 5. Xylem moisture potential in megapascals (MPa) by species and cavity tree type, Angelina National Forest, 1986 through 1989.

		Loblolly Pine		Shortleaf Pine
Cavity Tree Type	No. Samples	(MPa) (SD)	No. Samples	(MPa) (SD)
Active	58	-1.64 (0.30)	60	-1.65 (0.31)
Inactive	62	-1.66 (0.30)	34	-1.64 (0.34)
Potential	80	-1.66 (0.30)	43	-1.60 (0.32)

Table 6. Xylem moisture potential in megapascals (MPa) by species and cavity tree type, Davy Crockett National Forest, 1989 and 1990.

		Loblolly Pine		Shortleaf Pine
Cavity Tree Type	No. Samples	(MPa) (SD)	No. Samples	(MPa) (SD)
Active	10	-1.77 (0.15)	16	-1.82 (0.15)
Inactive	8	-1.69 (0.30)	11	-1.78 (0.10)
Potential	18	-1.73 (0.22)	10	-1.74 (0.17)

DISCUSSION

Cavity excavation and continual resin-well pecking by red-cockaded woodpeckers on loblolly and shortleaf pine trees appears to affect preformed resin flow. Direction and magnitude of these effects vary considerably with tree species and site conditions, however. Also, variance in resin flow between cavity tree types is significant, hence the use of non-parametric statistics in the analysis. Because of this variation it is difficult to generalize about effects of RCW activity on cavity tree vulnerability to bark beetles in terms of preformed resin flow. Given their general senescence, low vigor, and infection by pathogens, RCW cavity trees, especially shortleaf and loblolly, are naturally at a stage of life where vulnerability to mortality from all sorts of factors, including bark beetles, may be high.

Traditional approaches to suppression of bark beetle infestations include salvage cutting, cutting infested trees and applying insecticide, cutting infested trees plus a buffer zone and leaving these trees in the woods, and piling infested material and burning (Swain and Remion, 1981). All of these strategies except piling and burning may be appropriate under suitable conditions to prevent SPB infestations from reaching RCW cavity trees. Behavioral chemicals, such as the SPB anti-aggregation pheromone Verbenone, are currently being evaluated experimentally as a method of diverting attack away from high-value areas such as RCW colonies (Billings and Upton 1993). Artificial nest cavity construction, most commonly involving insertion of a prefabricated nest box into a suitable living pine, is showing great promise in mitigating the impact of natural cavity tree loss. As important as these tactics are in the effort to prevent further decline in RCW populations and habitat, they are not substitutes for long-range management that provides a forest environment less favorable to the SPB and more favorable to the RCW.

Long-term pro-active management strategies to favor the woodpeckers in loblolly/shortleaf stands must include reducing risk of bark beetle attack by increasing overall forest health. Longleaf pine is more resistant to SPB attack (Wahlenberg 1946), and should be restored on suitable sites. Timely thinnings to maintain low to moderate pine basal area will provide space for trees to maintain radial growth and vigor resulting in increased resistance to SPB. Prescribed fire at natural frequencies and seasons will help maintain an open, SPB resistant stand (Conner and Rudolph 1989, Conner et al. 1991a). When beetle populations are epidemic in an area, activities that might wound or otherwise stress trees, such as logging or even prescribed fire under some conditions, should be avoided in or near RCW colonies.

Forest fragmentation, in addition to creating dispersal and habitat problems for the RCW (Conner and Rudolph 1991b), can also lead to increased bark beetle activity. Poorly designed clearcuts may result in suddenly and substantially increased wind velocity in adjacent stands (Spurr and Barnes 1980, Chrismer et al. 1995, Conner and Rudolph 1995b) resulting in wind-thrown and wind-snapped trees. These damaged trees are utilized by *Ips* engraver beetles and black turpentine beetles which may spread to adjacent healthy trees and either kill them outright or weaken them to the point that vulnerability to SPB is greatly increased. Swaying caused by sudden changes in wind speed also results in damaged root systems, making trees more susceptible to insects and diseases (Spurr and Barnes 1980). Seed-tree and shelterwood regeneration, especially the irregular variations (Smith 1986) reduce fragmentation, retaining RCW habitat and reducing wind-related trouble (Conner et al. 1991b).

Prompt response to SPB outbreaks is critical to reducing impact, but is inadequate unless accompanied by pro-active management. Devising innovative approaches to managing forest ecosystems for optimum health of their native constituents in conjunction with providing society with the forest products and values it demands is the continuing challenge facing foresters and other resource management professionals.

ACKNOWLEDGEMENTS

This work was supported through a cooperative research grant with the Southern Forest Experiment Station, USDA Forest Service and Stephen F. Austin State University (agreement no. 19-86-068) and McIntire-Stennis funds, College of Forestry.

Tree Characteristics, Resin Flow, and Heartwood Rot in Pines (*Pinus palustris, Pinus elliottii*), with Respect to Red-cockaded Woodpecker Cavity Excavation, in Two Hydrologically-Distinct Florida Flatwood Communities

Reed Bowman[1] and Christopher Huh
Archbold Biological Station, Lake Placid, FL 33852

ABSTRACT: In south-central Florida, where longleaf pines (*Pinus palustris*) and slash pines (*Pinus elliottii*) co-occur, red-cockaded woodpeckers (RCWs) excavate cavities only in longleaf pines. However, south of the longleaf pine range, in southwest Florida, RCWs excavate cavities in slash pines. These slash pines grow in hydric flatwoods. In south-central Florida, slash and longleaf pines tend to occur in mesic flatwoods. We compared heartwood fungal infection, resin flow rates, and other tree characteristics between hydric slash pines in southwest Florida and mesic slash and longleaf pines in south-central Florida. Hydric slash pines greater than 60 years old were significantly smaller in dbh and tree height and tended to have lower crown-bole ratios than either mesic slash or longleaf pines of similar age. Hydric slash pines had more heartwood rot. Longleaf pines had the greatest resin flow. Resin flow was influenced by crown-bole ratio in both longleaf and slash pines, but the relationship differed by habitat. In hydric habitats peak resin flow was associated with lower crown-bole ratios than in mesic habitats. In both habitats, RCWs excavated cavities in trees with the crown-bole ratios associated with maximum resin flow.
KEYWORDS: resin flow, flatwoods, slash pine, cavity excavation, longleaf pine, red heart.
[1]Author to whom correspondence should be addressed.

The red-cockaded woodpecker (*Picoides borealis*), a federally endangered species, is endemic to old-growth pine forests of the southeastern United States. Its historic range occurred from southern Virginia, south through Florida and west to Texas (Hooper et al. 1980). Red-cockaded woodpeckers are unique among North American piciformes in that they excavate their cavities in living pine trees (Ligon 1970). Throughout their range, red-cockaded woodpeckers excavate cavities in a variety of pine species: longleaf pine (*Pinus palustris*), slash pine (*P. elliottii*), shortleaf pine (*P. echinata*), pond pine (*P. serotina*), loblolly pine (*P. taeda*), pitch pine (*P. rigida*), and Virginia pine (*P. virginiana*) (Hooper et al. 1980). Ecological and demographic patterns likely vary among red-cockaded woodpecker populations in different regions and using different pine species (Jackson 1995).

In the southern half of peninsular Florida, red-cockaded woodpeckers excavate cavities in both longleaf pine and slash pine (Shapiro 1983). The southern edge of the range of longleaf pine, however, occurs in south-central Florida, thus red-cockaded woodpecker populations in south Florida do not have longleaf pine available for cavity excavation. Longleaf pine and slash pine occur sympatrically in south-central Florida. At the Avon Park Air Force Range (APAFR), active red-cockaded woodpecker clusters occur in many mixed longleaf/slash pine stands, but cavities are excavated only in longleaf pine (Bowman and Fitzpatrick 1993). At three large red-cockaded woodpecker populations south of the southern range-limit of longleaf pine, all cavities occur in slash pines (Patterson and

Robertson 1981, Shapiro 1983, Beever and Dryden 1992). Slash pine occurs in a variety of hydrologically-different habitats (Abrahamson and Hartnett 1990), yet in southwest Florida, 89 of 112 known red-cockaded woodpecker clusters and 91 of the 92 active groups occur in hydric slash pine flatwoods (Beever and Dryden 1992).

Territories of red-cockaded woodpeckers in hydric slash pine flatwoods appear to be larger than birds nesting in upland mesic and xeric longleaf forests (Patterson and Robertson 1981, Nesbitt et al. 1983). Typically, stocking rates are lower in hydric versus mesic slash pine flatwoods and under existing federal guidelines, much of the preferred red-cockaded woodpecker habitat in southwest Florida would be considered poor or even unsuitable (Beever and Dryden 1992).

Red-cockaded woodpecker populations at APAFR appear to be declining (Bowman and Fitzpatrick 1993). Intensive management, including artificial cavity provisioning and red-cockaded woodpecker translocations, may be necessary to ensure the long-term persistence of this population. Red-cockaded woodpecker groups inhabit most of the large, healthy longleaf pine stands at APAFR, but many mesic slash pine stands with old, large trees that appear morphologically similar to longleaf cavity trees, are unused. To effectively manage this population it is important to understand why woodpeckers do not use mesic slash pine in south-central Florida, but prefer hydric slash pines in southwest Florida. Hydric slash pine flatwoods in southwest Florida are under considerable development pressure, threatening regional woodpecker populations. Understanding how ecological differences in nesting habitat influences demographic patterns may play a critical role in the conservation of red-cockaded woodpeckers in Florida. This led us to pose the question -what differences exist between mesic longleaf pine, mesic slash pine and hydric slash pine in characteristics that may influence red-cockaded woodpecker cavity excavation.

We compared tree characteristics (diameter-at-breast-height (dbh), tree height, and crown-bole ratio), heartwood fungal infection, and resin flow rates among three groups; hydric slash pine in southwest Florida and mesic slash pine and longleaf pine in south-central Florida. We chose these parameters because each has a potential influence on red-cockaded woodpecker demographic patterns (Steirly 1957, Conner et al. 1976, Hooper et al. 1980).

METHODS

We measured mesic longleaf pine and slash pine at two sites in south-central Florida: APAFR in Polk and Highlands counties and Archbold Biological Station (ABS) in Highlands county. Over 20 active red-cockaded woodpecker clusters occur at APAFR. Although red-cockaded woodpeckers no longer occur at Archbold Biological Station, several large, old, and relatively undisturbed stands of mesic slash pine and longleaf pine occur on or near the station. We measured hydric slash pine at the Cecil M. Webb Wildlife Management Area (CMW) in Charlotte county. Over 20 active red-cockaded woodpecker clusters occur at CMW and most cavities are excavated in hydric slash pine.

Tree characteristics

It was not possible to compare cavity trees among all three groups since no red-cockaded woodpecker cavities occur in mesic slash pine. Thus we sampled trees that appeared morphologically similar to cavity trees at each site. We sampled 9 longleaf pine and 28 slash pine at ABS, 38 longleaf pine and 39 slash pine at APAFR, and 40 slash pine at CMW. We estimated tree age by coring at breast height with an increment borer, staining the core with phloroglucinol, and counting the annuli. This method of age determination often underestimates tree age, thus we added five years to each estimate (K. Olsen, APAFR, pers. comm.). The age distribution of sampled trees, however, differed among the three sites, thus we based our analyses only on those trees 60 years or older. The age of cavity trees at APAFR ranged from 35 to 128 years old (avg. = 78) (Bowman and Fitzpatrick 1993), thus our sample reflects an age distribution similar to that of cavity trees in the region. This reduced our samples to 26 mesic longleaf pines, 33 mesic slash pines, and 23 hydric slash pines.

We measured dbh with a standard diameter tape. We determined tree height with a clinometer. Crown-bole ratio was calculated by determining the height of the lowest living branch then dividing the distance between the lowest living branch and the top of the tree by total tree height.

We also measured tree characteristics of 13 cavity trees in hydric slash pine at CMW. Tree characteristics of longleaf cavity trees at APAFR were collected as part of a larger study (Bowman and Fitzpatrick 1993).

Heartwood fungal infection

To determine the presence or absence of heartwood fungal infection we cored all trees at dbh and at 6 m (mean cavity height at APAFR). We examined cores from all trees for visible evidence of heartwood decay. Decay resulted in visible darkening of the heartwood and a marked softening of the wood. Cores from trees with advanced decay often were impossible to age accurately since the heartwood usually disintegrated. Presence or absence of visible decay was noted.

We found slightly higher rates of fungus infection from cores taken at 6 m than those sampled at dbh for all groups. In no case did a tree testing negative at 6 m, either visibly or by chemical detection, test positive at dbh. Thus for all comparisons between groups we used results from cores taken at 6 m.

We removed heartwood chips from each sample and cultured them following procedures described by Nobles (1965) and Gilbertson and Ryvarden (1987). Heartwood chips were removed aseptically, placed on malt agar, and cultured in darkness at 25° C. After one week, inocula from each culture were removed and placed on two new plates. These plates were incubated for five to eight weeks. During this period, macroscopic characteristics (mycelial color and texture) and micromorphology (hyphal septation) of the cultures were noted. *Phellinus pini* is the only heartwood rot known to occur in red-cockaded cavity trees (Jackson 1977c), thus cultures from cavity tree cores were grown and used for comparison during fungal identification.

We used two chemical tests to determine the presence of heartwood fungus. The presence of extracellular oxidase in culture is indicative of *Phellinus* (Gilbertson and Ryvarden 1987). We added a solution of gum guaiac and ethanol to the mycelial mat. The resulting color change indicated the presence or absence of extracellular oxidase. We also transferred inocula from each plate to gallic acid agar and cultured the inocula for approximately two weeks. The appearance of a dark reaction zone indicated presence of extracellular oxidase while no reaction indicated absence of the enzyme.

Resin flow

We used the procedure described by Ross et al. (1991) to measure resin flow rates. We drove a 2.5 cm diameter arch punch (Osborne Co.) into the interface of phloem and xylem on the southwest side of the tree bole. Holes were cut at eye level (~ 1.7 m). A 10 cm vertical groove was excavated in the bark below the hole and a 25 ml clear plastic graduated cylinder was placed in the groove below the hole. We recorded the volume of resin at 12 and 24 h post-tapping. After 24 h, we replaced the bark plug formed by the arch punch. To minimize the diurnal effect on resin flow rate described by Nebeker et al. (1988), all holes at ABS and APAFR were punched during the hours of 0700 - 0900 and 1900 - 2100. Sampling dates were July 30, August 5, 11, 18, and 21 for ABS trees and August 6, September 5, 14, and October 2 for APAFR trees. Resin flow data were collected for all study trees at these locations. Approximately equal numbers of longleaf and slash were sampled on each sample date. At CMW, resin flow was determined for 23 slash trees on December 10. Weather data (high and low temperatures) for these dates were obtained from ABS records.

Statistical analyses

Analyses were performed using the SYSTAT statistical package (Wilkinson 1990). We used univariate analysis of variance (ANOVA) to test for differences in tree characteristics among the three groups. Where differences occurred we used the Tukey test for pair-wise comparisons. We used multivariate analysis of variance (MANOVA) to test for the effects of time of tree punching and sampling date on resin flow. After pooling data, ANOVAs were run to examine variation in resin flow among the three groups. Categorical variables (visible core decay, presence/absence of *P. pini* in culture) were compared among groups using Chi-square tests. We used Spearman rank correlation to test for relationships between resin flow and continuous variables such as tree height, dbh, age, and crown-bole ratio. We used a quadratic regression model (NONLIN, Wilkinson 1990) to further test the relationship between crown-bole ratio and resin flow.

RESULTS

We found significant differences among the three groups in tree height, dbh, and crown-bole ratio (ANOVA, P<0.05 for all three comparisons) **(Table 1)**. Height of hydric slash pines was significantly lower than either mesic slash pine or longleaf pine. Hydric slash pines also had the smallest dbh and lowest crown-bole ratio of the three groups. Mesic slash pines had significantly larger dbh than did similarly-aged hydric slash pines. Hydric and mesic slash pines did not differ statistically in crown-bole ratio, but longleaf pine had a larger ratio than did hydric slash pine.

The proportion of trees in each group that had heartwood fungus varied significantly for both visible detection (X^2 = 14.70, df = 2, P = 0.006) and chemical detection (X^2 = 8.72, df = 2, P = 0.013) **(Figure 1, A and B)**. We found significantly higher amounts of visible decay in hydric slash pine than in mesic slash pine (2x2 contingency table, $P < 0.05$). Over 40% of all hydric slash pines had visible heartwood decay. Longleaf pine did not differ statistically from either hydric or mesic slash pine in the percent of trees with visible decay **(Figure 1A)**. All three groups had higher fungal infection rates based on chemical detection. Both hydric and mesic slash pines, however, had high rates (> 50%) of chemically-detected fungal infection and both were significantly higher (2x2 contingency table, $P < 0.05$) than longleaf pine **(Figure 1B)**.

Both time and date of sampling had a significant effect on resin flow rates, but the effects differed by group. Resin flow in longleaf pine after 24 h was significantly greater in trees tapped between 1900 - 2100 than those tapped between 0700 - 0900. We found no diurnal variation in resin flow rates in mesic slash pines. Resin flow also varied significantly by sampling date for both longleaf and mesic slash pines. Because we sampled approximately equal numbers of longleaf and slash pines during each sampling date and tapped an equal number during both a.m. and p.m. sampling times, we pooled dates and times for inter-group comparisons. Hydric slash pines at CMW were tapped during the late afternoon.

Resin flow varied significantly between groups at both 12 and 24 h **(Figure 2)**. During both sampling intervals, longleaf pines produced significantly more resin than either slash pine groups. Resin flow did not differ between hydric and mesic slash pine. For all three species, most resin flow occurred within 12 h of tapping. In longleaf pine, 87% of the total resin flow occurred during the first 12 h. In mesic and hydric slash pine, this percentage was 71% and 73%, respectively.

Resin flow varied greatly for all three groups. In longleaf pine, 24 h resin flow ranged from 0 ml to 26 ml. Similar amounts of variation, though the range was not quite as high, occurred in both mesic and hydric slash pine. Neither tree age, height, nor dbh was

Table 1. Characteristics of non-cavity trees > 60 years old among longleaf pine and slash pine in mesic flatwoods and slash pine in hydric flatwoods.

Species	N	Tree dbh (cm)	Tree height (m)	Crown-bole ratio
Longleaf pine	26	35.5 ± 3.4 ab[1]	17.4 ± 3.3 a	.50 ± 0.07 a
Mesic slash pine	33	36.3 ± 2.3 a	20.4 ± 3.1 b	.47 ± 0.08 ab
Hydric slash pine	23	31.4 ± 2.2 b	14.5 ± 2.9 c	.42 ± 0.07 b

[1] **Within columns, means assigned the same letter are not significantly different (P<.05; Tukey test).**

Table 2. Characteristics of red-cockaded woodpecker cavity trees in mesic longleaf pine and hydric slash pine.

	Mesic longleaf pine (N = 294)	Hydric slash pine (N = 13)	Significance[1]
Tree age[2]	78.4 ± 17.2	103.6 ± 22.6	$P < 0.05$
Tree dbh (cm)	30.7 ± 5.4	28.9 ± 4.1	n.s.
Tree height (m)	14.6 ± 2.7	14.9 ± 2.6	n.s.
Crown-bole ratio	0.44 ± 0.1	0.33 ± 0.1	$P < 0.05$
Cavity height (m)	5.7 ± 1.7	8.1 ± 1.9	$P < 0.05$

[1] **Independent T-test**

[2] **Sample size for tree ages was 120 trees**

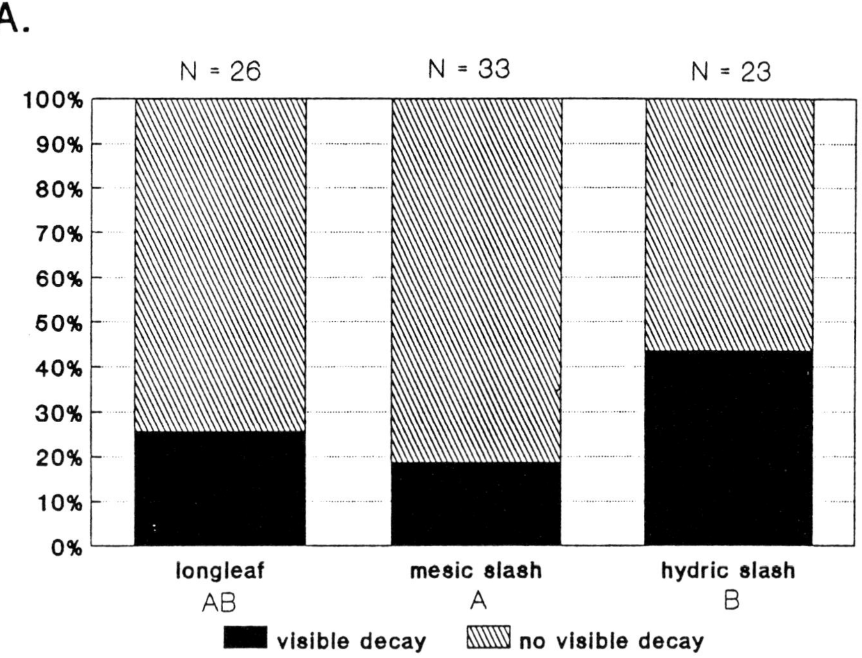

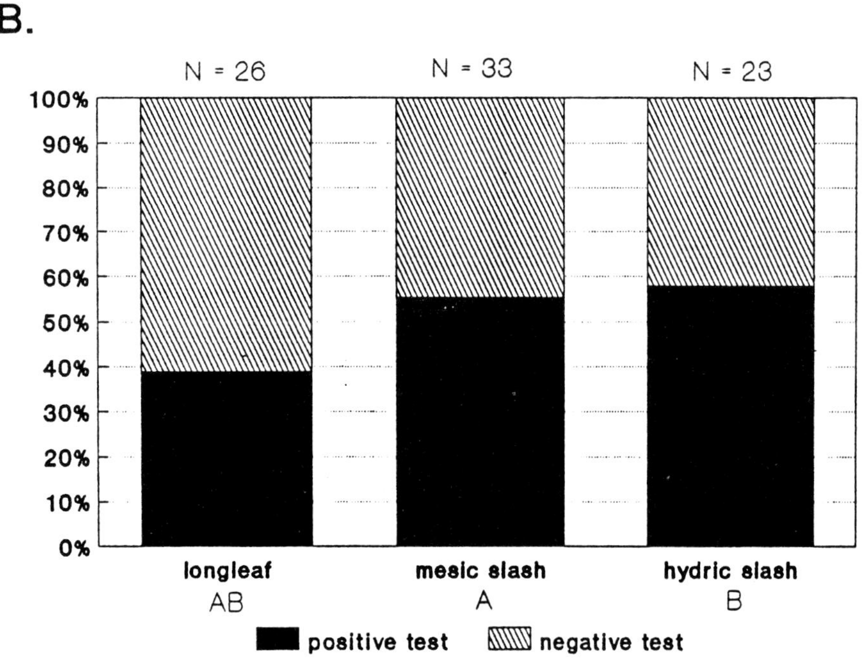

Figure 1. Proportion of randomly-selected non-cavity trees with visibly (A) or chemically (B) detectable heartwood fungus infection. Columns assigned the same letter are not significantly different.

related to 24 h resin flow. We found no linear correlation between crown-bole ratio and resin flow; however, the data scatter suggested that maximum resin flow rates were associated with median crown-bole ratios. A quadratic regression model with crown-bole ratio as the independent variable explained 47% of the variation in resin flow rates in longleaf pine **(Figure 3A)**. A similar pattern existed in mesic slash pine, the quadratic regression model explaining 41% of the variation. Resin flow rates in hydric slash pine were less variable, but when plotted against crown-bole ratio showed a similar tendency for resin flow to peak with a certain crown-bole ratio. This ratio was somewhat smaller, however, than the ratio associated with peak resin flow in longleaf and mesic slash pine (0.35 vs. 0.50). A quadratic regression model with crown-bole ratio as the independent variable explained only 32% of the variation in resin flow in hydric slash pines **(Figure 3B)**.

We transformed crown-bole ratio into a categorical variable and examined the relationship with resin flow using ANOVA. Resin flow varied significantly with crown-bole ratio for both longleaf pine **(Figure 4A)** and hydric slash pine **(Figure 4B)**. Maximum resin flow was associated with crown-bole ratios of 0.4-0.6 and 0.2-0.4 for each group, respectively.

Crown-bole ratios of hydric slash pine cavity trees were significantly lower ($T = 5.71$, $P < 0.001$) than longleaf pine cavity trees. In longleaf pine, 78% of all red-cockaded woodpecker cavity trees had a crown-bole ratio between 0.4 and 0.6, the range associated with maximum resin flow. Virtually all of the cavity trees in hydric slash pine (12 of 13, 92%) had crown-bole ratios between 0.2 and 0.4, the range associated with maximum resin flow in hydric slash pine **(Figure 5)**.

To test whether these distributions differed from random, we compared crown-bole ratios of cavity trees with unused trees greater than 60 years old. Cavity trees in both longleaf pine and hydric slash pine had a significantly lower crown-bole ratio than the population of similar-aged, but unused trees in that habitat **(Figure 6)** ($T = 3.79$, $P < 0.001$ and $T = 2.7$, $P = 0.013$, respectively, Independent T-test).

We found several differences between cavity trees in hydric slash pine and mesic longleaf pine **(Table 2)**. Although hydric slash pine red-cockaded woodpecker cavity trees were significantly older than longleaf pine cavity trees, they were similar in height and dbh, though having a smaller crown-bole ratio. Cavities were significantly higher ($P<0.05$) in hydric slash pines than in longleaf pines **(Table 2)**.

DISCUSSION

Slash pines in hydric flatwoods are shorter, smaller in dbh, and tend to have smaller crown-bole ratios than similarly-aged slash pines or longleaf pines in mesic flatwoods. In both mesic and hydric flatwoods, south Florida slash pine (var. *densa*) is the dominant canopy tree. South Florida slash pine may have evolved a large degree of phenotypic plasticity, enabling it to grow in both mesic and hydric conditions, as a response to annual and supra-annual variation in the drought-flood cycles typical to south Florida (Squillace 1966). Langdon (1963) reported that mature south Florida slash pine attained a height of 33.5 m and a dbh of 41 cm, however, in undisturbed southwest Florida hydric pine flatwoods mature slash pines typically are shorter (18.3 - 22.9 m) and have a smaller dbh (25.6 - 30.8 cm) (Beever and Dryden 1992), suggesting that hydric conditions impose growth limitations on pines.

Shapiro (1983) reported that red-cockaded woodpecker cavity trees in south Florida tend to be older than elsewhere in the southeastern United States (but see Bowman and Fitzpatrick 1993); however, she made this comparison based on a limited number of other studies. We found that cavity trees in hydric and mesic habitats were similar in size, but hydric slash pine cavity trees were significantly older than longleaf pine cavity trees. The mean ages of longleaf cavity trees in our study were younger than those reported by Shapiro (1983) for south-central Florida, but similar to the mean for other studies. Our slash pine age data did not differ from Shapiro's data. It may simply take trees, either slash pine or longleaf pine, in poor-quality sites longer to attain a diameter large enough to permit cavity excavation (Thompson and Baker 1971). However, in both mesic and hydric sites, cavity tree size was, in general, smaller than that reported for cavity trees elsewhere over the range of red-cockaded woodpeckers (Conner and O'Halloran 1987). At several sites in south Florida, the average height and dbh of the overstory was lower than that of cavity trees, as found in hydric habitats during this study, suggesting that woodpeckers

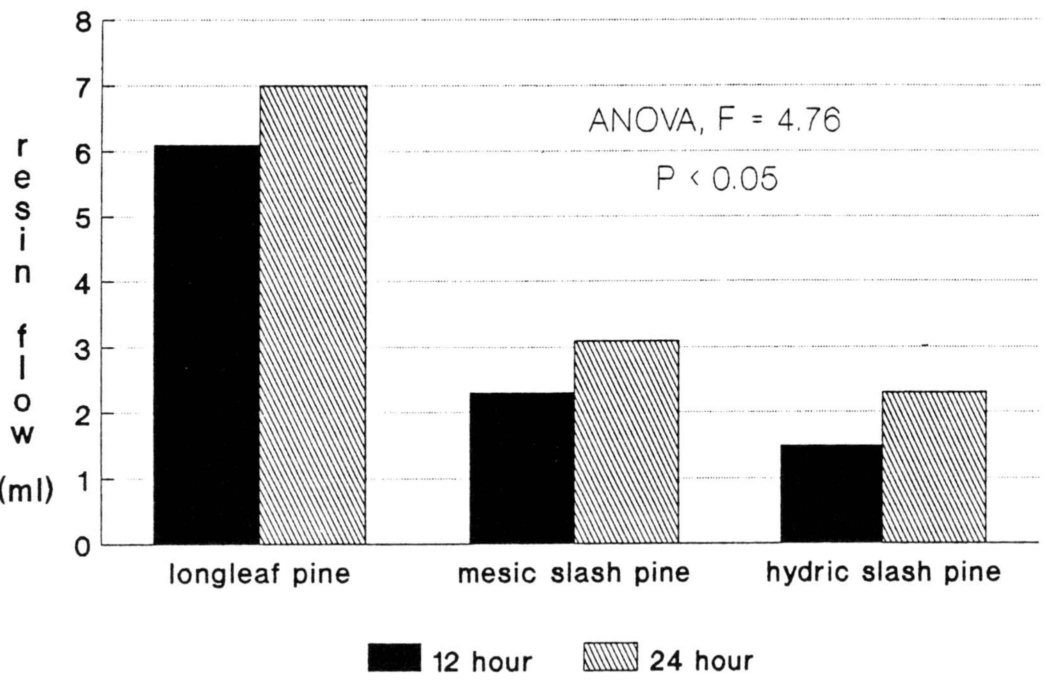

Figure 2. Total resin flow over a 24 h interval following wounding for three groups of randomly - selected non-cavity trees.

choose dominant trees for cavity excavation (Shapiro 1983) and that these trees may be relatively uncommon.

Hydric pine flatwoods display other characteristics consistent with poor growing conditions. Hydric slash pine stands typically have lower basal areas and stem densities than mesic slash pine stands farther north (Shapiro 1983, Beever and Dryden 1992). Within red-cockaded woodpecker clusters, distances between cavity trees are greater and territories are larger (Patterson and Robertson 1981, Nesbitt et al. 1983, Shapiro 1983, Beever and Dryden 1992), probably because of the lower tree densities. Red-cockaded woodpeckers may prefer hydric flatwoods to mesic flatwoods for a variety of reasons. Mesic slash pine flatwoods in southwest Florida typically have a dense midstory vegetation. Stress induced by poor growing conditions, fire, and periodic flooding may make trees in hydric flatwoods more susceptible to insect attack, providing an abundant food source for woodpeckers (Beever and Dryden 1992). Beever and Dryden (1992) also suggested that mature trees stressed by conditions of hydric pine flatwoods may prove more suitable for cavity excavation.

Conner and O'Halloran (1987) reported that red-cockaded woodpecker cavity trees in east Texas exhibited a phenomena of growth suppression and then release, more frequently than unused pines. They suggested that growth suppression may lead to characteristics conducive to cavity excavation that are independent of tree age. Suppression may cause lower limbs to be dropped, increase the heartwood-sapwood ratio, or make the tree more susceptible to fungal heartwood rot. Our results are consistent with growth suppression in hydric slash pines. Hydric slash pines had smaller crown-bole ratios than trees in mesic flatwoods, suggesting that lower limbs were more likely to be absent. Hydric slash pines had a much higher incidence of visible heartwood rot than mesic slash pines or longleaf pines. Fungal infections may occur through wounds left from fallen branches (Conner et al. 1976, Conner and Locke 1982). We found that the incidence of cultures with *Phellinus pini* diagnostic characters was higher than the incidence of visible decay in mesic slash pines. *Phellinus pini* infections may be ubiquitous in slash pine, but the higher incidence of visible decay in hydric slash pines may result from weakened resistance to fungal attack caused by the stressed conditions of hydric flatwoods.

A.

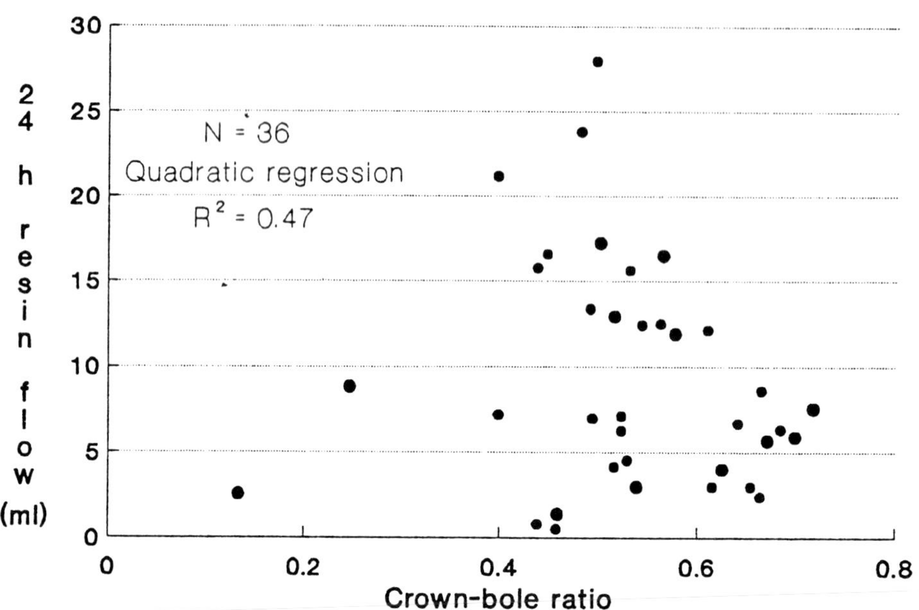

B.

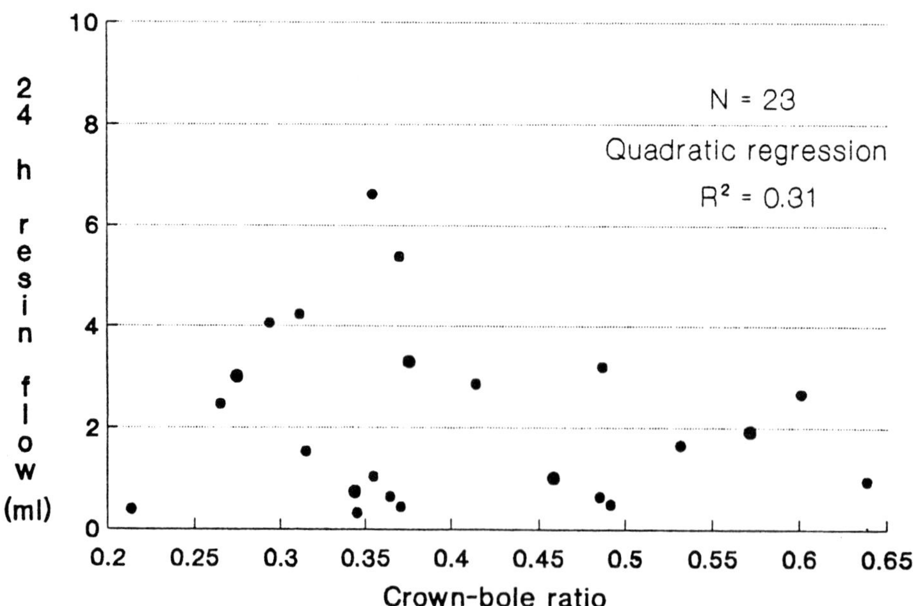

Figure 3. Relationship between crown-bole ratio and total resin flow 24 h after wounding for randomly-selected non-cavity longleaf pines in mesic flatwoods (A) and slash pines in hydric flatwoods (B). Quadratic regression based on 30 iterations with resin flow as the dependent variable.

A.

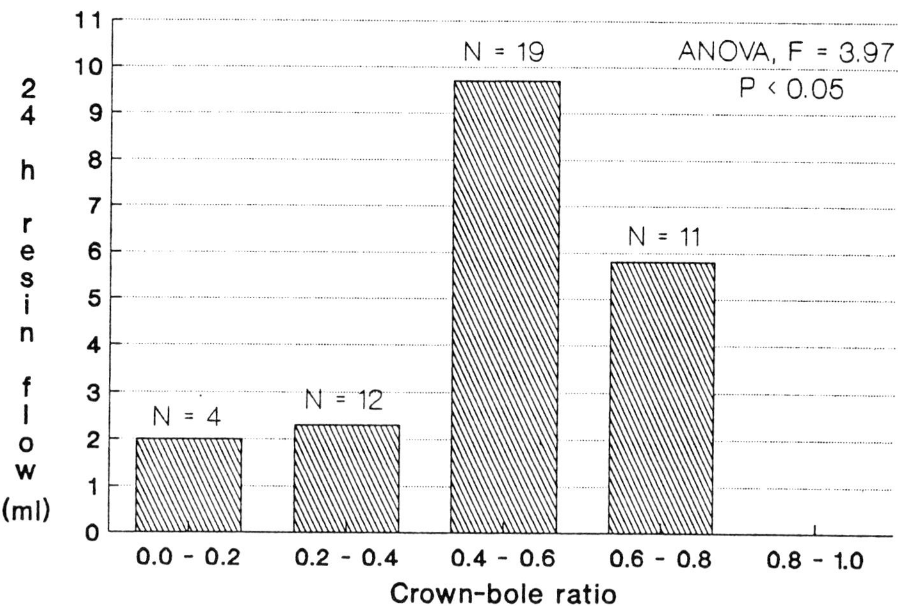

B.

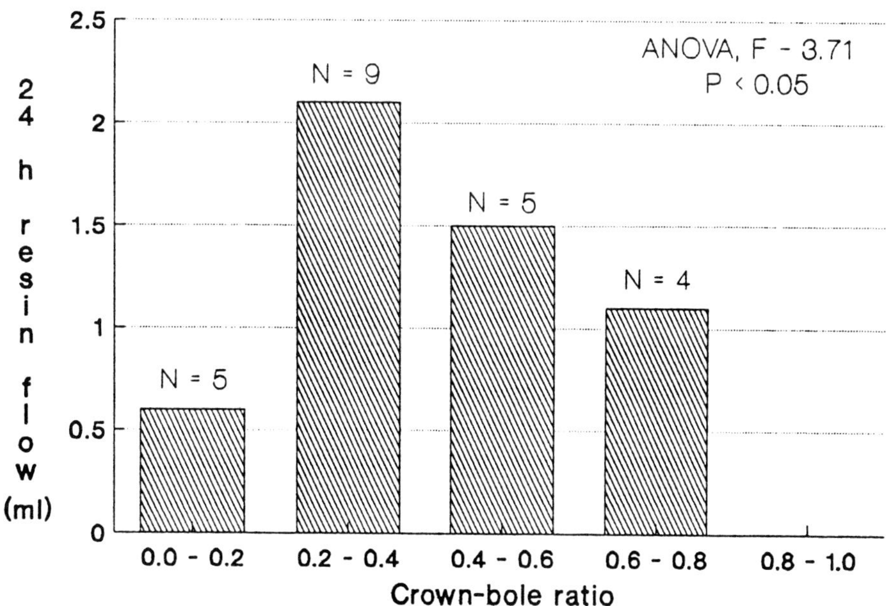

Figure 4. Resin flow associated with different crown-bole ratios for randomly-selected non-cavity longleaf in mesic flatwoods (A) and slash pines in hydric flatwoods (B).

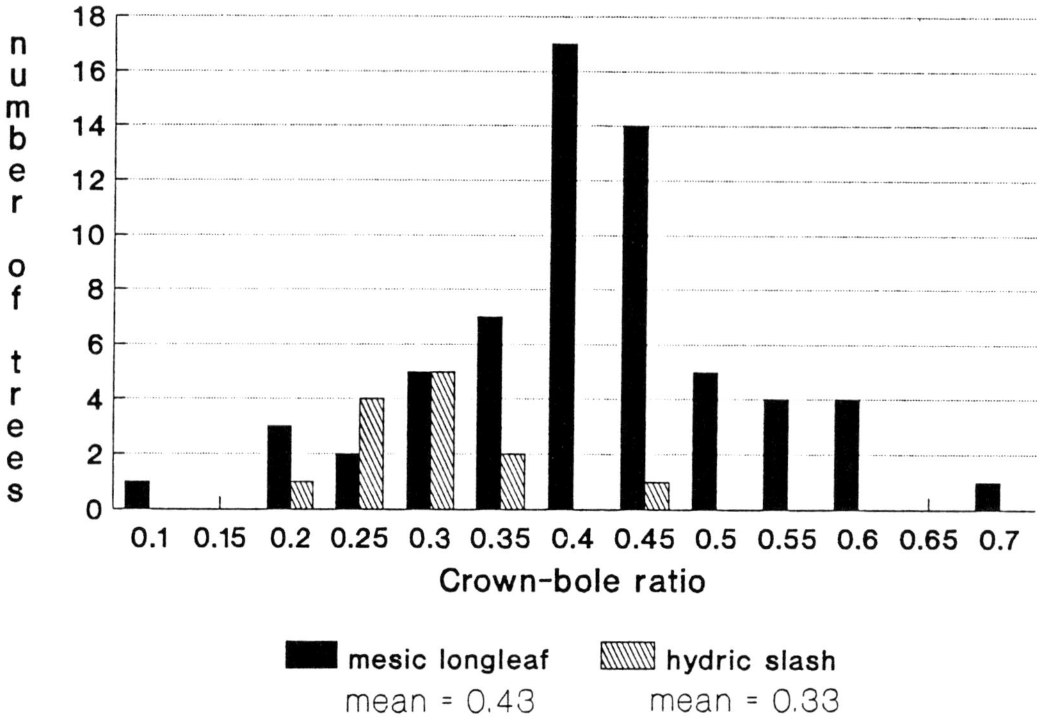

Figure 5. Frequency distribution of crown-bole ratios of red-cockaded woodpecker cavity trees in mesic longleaf and hydric slash pine.

Slow growth increases the susceptibility to heartwood rot in loblolly pines (Gruschow and Trousdell 1958). We did not examine cores for evidence of growth suppression, however, hydric slash pines had significantly smaller dbh than similarly-aged mesic slash pine, suggesting, at least, slower growth rates. We also did not examine heartwood-sapwood ratios, however, cavities excavated in hydric slash pines were higher than those in mesic longleaf pines. The parsimonious explanation is that cavity height is negatively proportional to crown depth. Hydric slash pines have small crowns thus red-cockaded woodpeckers can excavate cavities higher in the tree. Heartwood diameter, however, decreases with height but increases with age. It is possible that high cavity heights reflect large heartwood-sapwood ratios higher in the bole, also consistent with growth suppression in hydric slash pines.

Longleaf pines yielded more resin than either mesic or hydric slash pines, as has been reported by others (Schopmeyer and Larson 1955, Hodges et al. 1977). Lower total flow in slash pine results from increased viscosity and slower flow rates (Hodges et al. 1977). We observed no differences in resin flow between hydric and mesic slash pines, however, we measured mesic slash pine resin flow rates during the late-summer and early-fall and hydric slash pine rates in early December. Average temperatures in December were 10° C cooler than during the earlier sampling period and these differences may have influenced our results.

Resin production occurs in pine needles, thus the size of a pine's crown influences resin flow rates (Wahlenberg 1946, Schopmeyer and Larson 1955); however, Hodges et al. (1979) reported that physical properties of resin (viscosity, crystallization rate, flow rate, and total yield) were not strongly related to morphological characteristics (dbh, growth rate, crown-bole ratio, or age). Red-cockaded woodpecker cavity trees in eastern Texas had significantly higher crown-bole ratios than unused trees and Conner and O'Halloran (1987) suggested this apparent selection may relate to resin production. However, Locke et al. (1983) reported that mean ratio of bole length to height of pines was greater in red-cockaded woodpecker colony sites than at randomly-selected control sites, suggesting woodpeckers selected trees with relatively small crown-bole ratios. They suggested that forest stands occupied by red-cockaded woodpeckers had

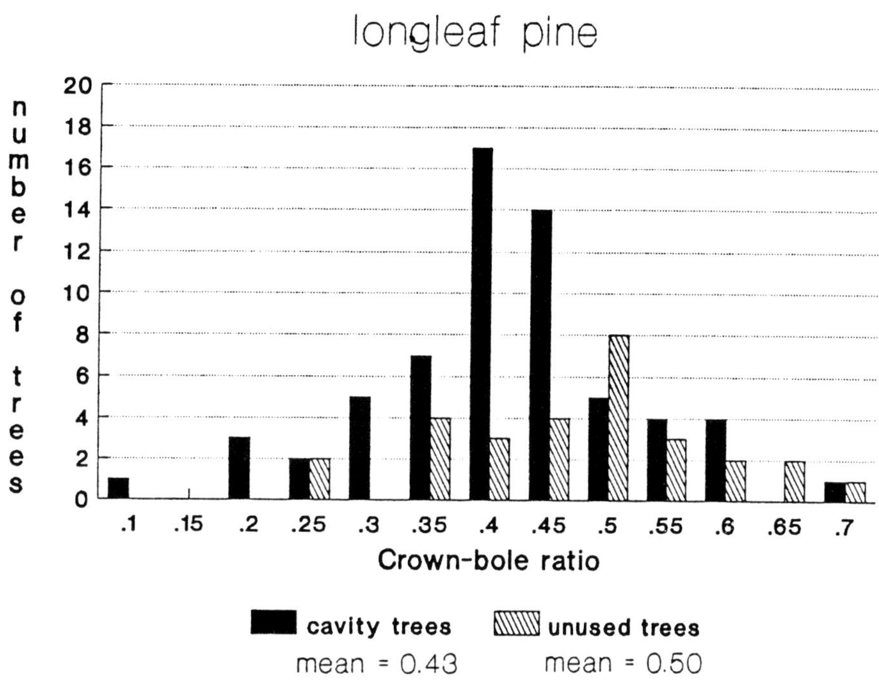

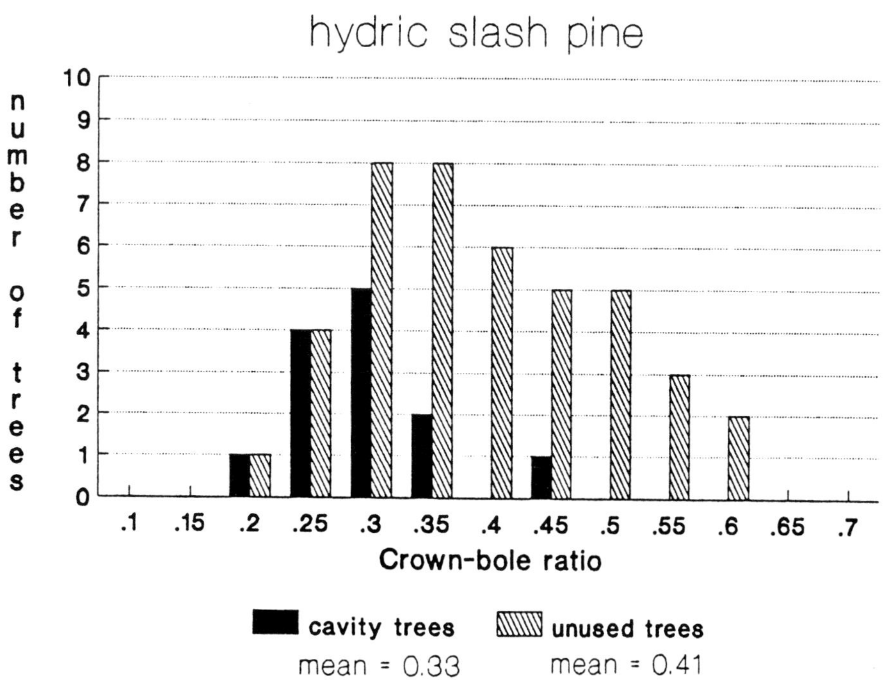

Figure 6. Frequency distribution of the crown-bole ratios of red-cockaded woodpecker cavity trees vs. randomly-selected non-cavity trees > 60 years old in mesic longleaf pine flatwoods (A) and hydric slash pine flatwoods (B).

once been stocked at greater pine densities and experienced greater crown competition than unused stands. These conditions would also have resulted in growth suppression and release. Our findings are different than those of Conner and O'Halloran (1987) but may not contradict their conclusions. Crown-bole ratio was the only tree morphology characteristic that related to resin production, however, the relationship was not linear. Peak resin flow was associated with median crown-bole ratios in both longleaf and slash pine. For both tree species, red-cockaded woodpeckers appeared to utilize trees for cavity excavation with a crown-bole ratio different than randomly selected unused trees of similar age and utilize trees with a crown-bole ratio associated with maximum resin flow. Resin production generally increases in response to cavity excavation, but our results and those of Conner and O'Halloran (1987) suggest that red-cockaded woodpeckers may select trees for maximum resin potential.

Beever and Dryden (1992) noted that good quality, mature hydric slash pine flatwoods have approximately 54 trees per acre (133 trees/ha), 5 to 8 pine stems of 10 inches (25.8 cm) or larger in dbh (12 to 20 trees/ha) and a basal area of approximately 20 square feet per acre (4.6 sq. m/ha). Existing federal guidelines for red-cockaded woodpecker habitat would consider these stocking rates marginal or inadequate as foraging habitat. However, more than 80 active red-cockaded woodpecker clusters occur in hydric slash pine flatwoods in southwest Florida (James 1993). Hooper (1988) proposed a model of relative preference by red-cockaded woodpeckers for trees to excavate cavities among longleaf pines of different ages, growth rates, and heartwood conditions. Red-cockaded woodpeckers should prefer decayed trees regardless of growth rates. Clearly many hydric slash pines, though slow growing, have decayed heartwood, providing preferred sites for cavity excavation. Individual tree characteristics may be of greater importance in red-cockaded woodpecker habitat selection than general forest stand characteristics (Conner and O'Halloran 1987).

Our data suggest that where longleaf pine and mesic slash pine co-occur, red-cockaded woodpeckers prefer longleaf pine. This may be because slash pines in mesic flatwoods have a relatively low incidence of heartwood rot and often occur in areas of dense midstory. Where longleaf pine has been extensively logged, midstory control and implementation of an artificial cavity program in suitable mesic slash pine stands might increase the amount of potential nesting habitat available for red-cockaded woodpeckers.

On peninsular Florida, a large proportion of the woodpecker population occurs in southwest Florida and inhabits hydric slash pine flatwoods. Our results and those of others (Jansen and Patterson 1983, Shapiro 1983, Beever and Dryden 1992) indicate this community may be ecologically different from what is generally recognized as suitable red-cockaded woodpecker habitat. These differences pose a strong argument for the development of different management guidelines for red-cockaded woodpeckers in south Florida (Beever and Dryden 1992). These differences also have potential impacts on demographic patterns and are worth studying further.

ACKNOWLEDGMENTS

We thank Paul Ebersbach, Robert Progulske, Kurt Olsen, and Sam Van Hook at the Avon Park Air Force Range Natural Resources Office for their assistance. Larry Campbell, of the Florida Game and Fresh Water Fish Commission, allowed us access to Cecil M. Webb Wildlife Management Area. Discussions with K. Dryden, M. Poole, and D. Jansen improved this manuscript.

Hardwood Midstory Control in Red-cockaded Woodpecker Colonies: Analysis of Change in Stand Structure and Species Composition

R. R. Cahal, III, D. L. Kulhavy, W. G. Ross and J. G. Gage
College of Forestry, Stephen F. Austin State University, Nacogdoches, TX, 75962

ABSTRACT: Four red-cockaded woodpecker (*Picoides borealis*) colonies totaling 75.3 acres in the Bannister Wildlife Management Area, Angelina National Forest, were cruised following midstory reduction operations. All trees, both standing and felled, were either measured for diameter at 4.5 ft. (DBH) for trees ≥ 5 inches or tallied into 2-inch diameter classes for smaller trees. Total basal area (BA) was reduced by 24%, with essentially all of the hardwoods removed. Stand structure and species composition were changed from a mid-successional loblolly and shortleaf pine stand with abundant hardwood and pine understory and midstory to an even-aged stand consisting almost entirely of large (13.8 inches average DBH) pines.
KEYWORDS: midstory, hardwoods, management

The red-cockaded woodpecker (RCW) is a non-migratory picid endemic to the southeastern United States. RCWs are unique among the North American woodpeckers because of their need to excavate nests in living pines and their decided preference for pine habitat with sparse hardwood midstory (Ligon 1971, USDA 1979, Hooper et al. 1980).

Hardwood midstory within RCW colony sites has been associated with cavity and colony abandonment (Hopkins and Lynn 1971, Van Balen and Doerr 1978, USDA 1979, Hooper et al. 1980, Locke et al. 1983, Hovis and Labisky 1985, Conner and Rudolph 1989). RCW response to the presence of hardwoods is not fully understood, but indications of increased interspecific competition and nest predation accelerates as hardwood stand components increase (Lennartz 1988). Hardwood midstory has been controlled in East Texas via prescribed fire, cutting, mechanical tree chipping, pushing over with skidders, and herbicide applications (Conner and Rudolph 1989, Conner and Rudolph 1991a). Conner and Rudolph (1991a), investigating the effects of midstory removal on colony status, determined that sudden hardwood reduction caused no short-term detrimental impact to the RCW when conducted during the non-breeding season.

The purpose of this research was to quantify the effects of hardwood midstory reduction on stand structure and species composition in loblolly (*Pinus taeda*) and shortleaf (*Pinus echinata*) pine stands in eastern Texas.

METHODS

Stand structure and species composition following control of midstory pines and hardwoods, accomplished between 1990 and 1991, were examined in four RCW colonies, totaling 75.3 acres, on the Bannister Wildlife Management Area, Angelina National Forest. Pine and hardwood saplings (< 5 inches DBH) were counted and recorded in two-inch diameter classes. Larger (≥ 5 inches) hardwoods and pines were measured for diameter at 4.5 ft. Frequency tables were constructed using the SPSS statistical software package (Norusis 1985).

RESULTS

In the four red-cockaded woodpecker colonies 52.6% (5,071 of 9,644) of the pines (mostly understory and midstory) and 99.5%

Table 1. Trees cut and trees remaining in four red-cockaded woodpecker colonies on the Bannister Wildlife Management Area, Angelina National Forest.

Species	Total Trees	Number Cut	Number Remaining
All Pines	9644	5071	4573
All Hardwoods	4224	4205	19
Red Oaks ≥ 5 in.	143	140	3
White Oaks ≥ 5 in.	229	222	7
Oaks and Other Hardwoods < 5 in.	3852	3843	9
Total	13,868	9,276	4,592

Table 2. Mean diameters (inches) and standard deviations (SD) of trees cut and trees remaining in four red-cockaded woodpecker colonies on the Bannister Wildlife Management Area, Angelina National Forest.

Species	Pre-treatment Mean Diameter (SD)	Mean Diameter Cut (SD)	Mean Diameter Remaining (SD)
All Pines	9.0 (6.1)	4.7 (3.1)	13.8 (4.9)
All Hardwoods	4.8 (3.2)	4.8 (3.2)	7.8 (4.0)
Red Oaks ≥ 5 in.	9.7 (4.7)	9.7 (4.7)	11.2 (4.4)
White Oaks ≥ 5 in.	10.6 (5.4)	10.6 (5.5)	8.2 (4.6)
Oaks and Other Hardwoods < 5 in.	4.3 (2.4)	4.2 (2.4)	6.3 (2.8)
Total Mean Diameter	7.7 (5.7)	4.7 (3.2)	13.7 (4.9)

(4,205 of 4,226) of the hardwoods were cut **(Table 1)**. Stand structure and species composition were changed from a mid-successional loblolly and shortleaf pine stand with abundant understory and midstory hardwoods and pines to an even-aged stand consisting almost entirely of large pines **(Figures 1 and 2)**. In addition to almost completely removing the hardwoods, the midstory reduction treatment increased the mean pine diameter from 9.0 inches to 13.8 inches **(Table 2)**. Before treatment, 11.0% (10.2 of 92.6) of the total basal area (BA) was composed of hardwood species. Less than 1% of the post treatment BA is hardwood **(Table 3)**. Of the total pre-treatment basal area, 23.5% (21.8 of 92.6) was cut in the four colonies.

DISCUSSION

Midstory control significantly altered stand structure and species composition in the four red-cockaded woodpecker colonies examined. Before midstory control, the stand overstory was dominated by pines with relatively few large hardwoods except along intermittent streams. Pine and hardwood midstory occurred throughout the stands. Although hardwood basal area was only eleven percent of the total before treatment **(Table 3)**, hardwoods comprised 30.4% (4,224 of 13,868) of the total stems **(Table 1)**. This stand structure is characteristic of southeastern "old-field" secondary succession at about a mid-successional stage. Without disturbance, the process of succession would have culminated in predominantly oak-hickory stands (Spurr and Barnes 1980).

Following midstory removal, the stands were composed primarily of large loblolly and shortleaf pines with an even-aged diameter distribution (Smith 1986) **(Figures. 1 and 2)**. Midstory control created stands much more even-aged than before by almost completely eliminating the smaller (less than 10 inch) diameter classes. At the residual basal area continuous intensive management in the form of prescribed fire, herbicides, brush-hogging or

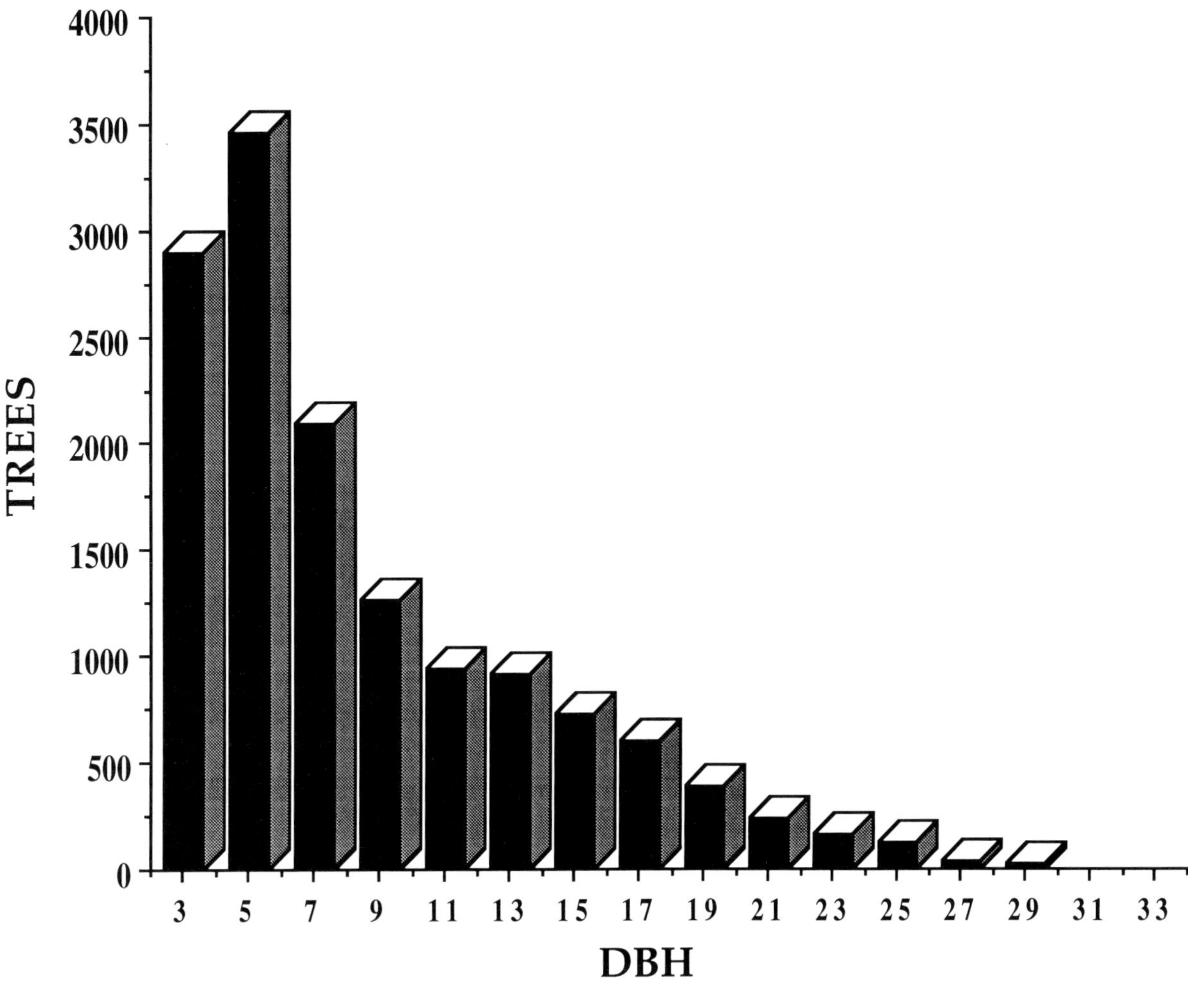

Figure 1. Diameter distribution (inches) for all trees before midstory removal in four red-cockaded woodpecker colonies, Bannister Wildlife Management Area, Angelina National Forest.

Table 3. Basal area per acre (BA/A) of trees cut and trees remaining in four colonies on the Banister Wildlife Management Area, Angelina National Forest.

Species	BA/A Remaining	BA/A cut	Pre-treatment BA/A
All Pines	70.7	11.7	82.4
All Hardwoods	0.1	10.1	10.2
Red Oaks ≥ 5 in.	< 0.1	1.2	1.2
White Oaks ≥ 5 in.	< 0.1	2.3	2.3
Oaks and Other Hardwoods (< 5 in.)	< 0.1	6.7	6.7

other woody vegetation control will be imperative or hardwood and pine understory and midstory will inevitably return (Spurr and Barnes 1980, Conner and Rudolph 1989). Future management practices must balance the RCW's need for sparce midstory with the need for pine regeneration to maintain RCW habitat. Complete elimination of midstory pines from a

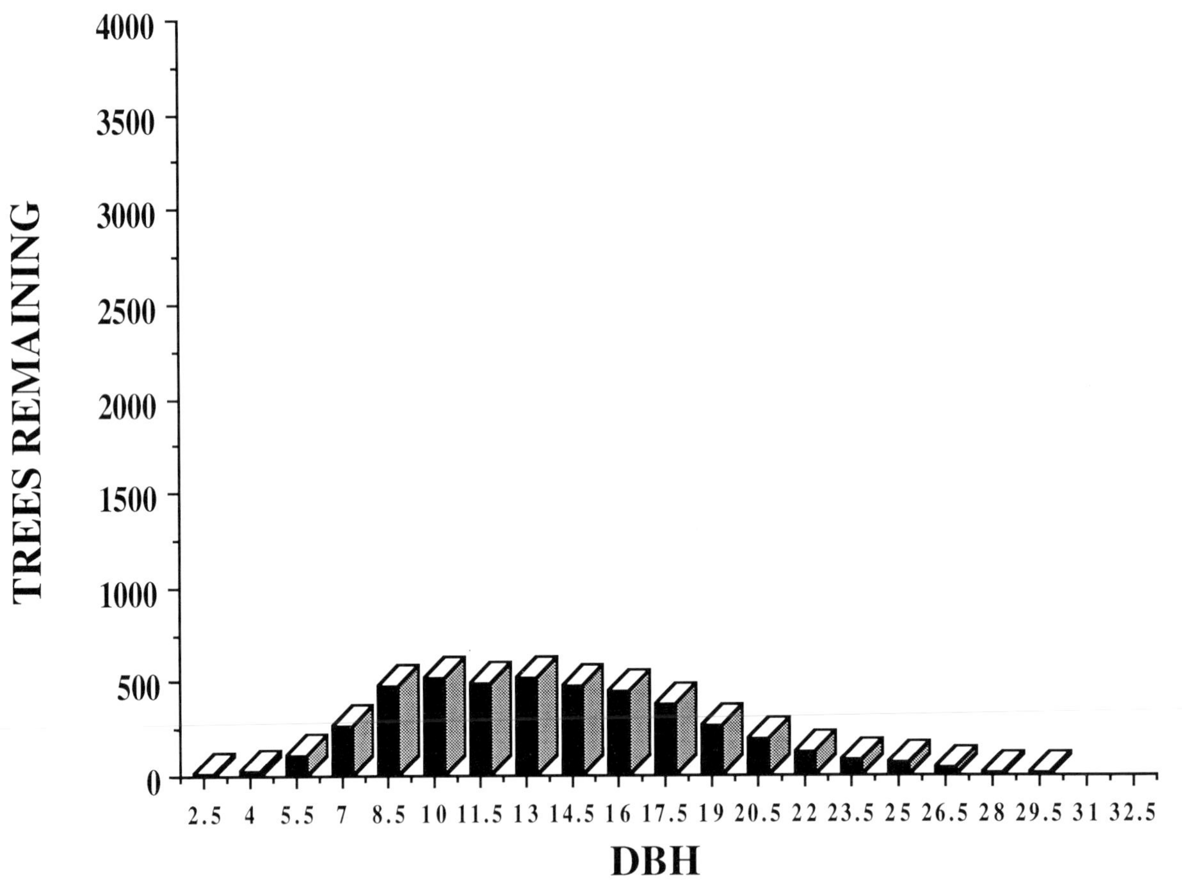

Figure 2. Diameter distribution (inches) for all trees after midstory removal in four red-cockaded woodpecker colonies, Bannister Wildlife Management Area, Angelina National Forest.

stand may ultimately lead to a deficit of suitable cavity trees because of the large discontinuity between diameter classes within the stand.

Although red-cockaded woodpeckers are favored by judicious midstory control in cluster areas, large scale hardwood removal may have a profound effect on other wildlife species requiring hardwoods for roosting, nesting, and denning. Squirrels, turkeys, and white-tail deer would be influenced by loss of acorns and hickory nuts and soft mast. The extent of replacement of these wildlife foods by the early-successional grasses, forbes, and browse favored by fire and open forest canopies will vary by species. Long-term monitoring is needed to determine the effects of midstory removal on red-cockaded woodpecker cavity tree cluster areas in loblolly and shortleaf pine stands.

Site/Stand Factors Associated with Red-cockaded Woodpecker Colonies on the Noxubee National Wildlife Refuge, Mississippi

T. Evan Nebeker[1], Matthew Pelligrine[2], Robert A. Tisdale[1] and John D. Hodges[2]
[1]Department of Entomology, [2]Department of Forestry
Mississippi State University, Mississippi State, MS 39762

ABSTRACT: Site, stand and tree characteristics and the incidence of red heart disease in 16 active red-cockaded woodpecker (*Picoides borealis*) colonies at the Noxubee National Wildlife Refuge, Mississippi were determined. Colonies were found mostly on flat upland sites with a mean site index (base age=50) of 92. Pine basal area averaged 16 m^2/ha with loblolly pine (*Pinus taeda*) comprising 92% and shortleaf pine (*P. echinata*) comprising 8% of pine species present. Various hardwoods were also present and had an average basal area of 4 m^2/ha. Average density of pines was found to be 171 trees/ha and that of hardwoods 44 trees/ha. Dominant and codominant pines averaged 31 m in height, with a mean DBH of 41 cm, and mean age of 83 years. Fifty-one cavity (completed) trees were sampled for red heart disease and it was found that only 30% were infected. This is in contrast to an earlier study that reported 90% of trees with completed cavities had obvious signs of red heart disease, while 67% of the trees without red heart contained only cavity starts.
KEYWORDS: red-cockaded woodpecker, *Picoides borealis*, southern pine beetle, red heart disease, Noxubee NWR.

Red-cockaded woodpeckers (RCW) are generally associated with mature pine stands. Jackson and Jackson (1986) have suggested several reasons why the RCW require mature timber:

(1) older trees are prone to red heart disease, which weakens the heartwood facilitating cavity excavation;

(2) older trees are taller and have few lower branches, allowing the woodpeckers to build their cavities out of reach of tree climbing predators;

(3) older trees have a higher percent heartwood; cavities built in sapwood would fill with resin;

(4) older trees have the proper diameter for the construction of the cavity; and

(5) older trees have a lower resin flow than younger trees. Resin, flowing out of the resin wells, is just enough to discourage climbing predators, particularly snakes, but not enough to block cavity entrances.

Researchers in Georgia and Texas (Belanger et al. 1988; Mitchell et al. 1991) have examined potential interactions of woodpeckers and beetles in their respective states, but comparative results have not been published for Mississippi. The purpose of this study was to help alleviate this gap. Specifically this study was designed to determine site, stand and tree characteristics and the incidence of red heart disease in active RCW colonies at the Noxubee NWR.

METHODS AND MATERIALS

For comparative purposes, the procedures outlined by Belanger, Hedden and Lennartz (1988) for the Georgia Piedmont were followed. Cavity trees were mapped for each of the 16 active colonies. The center of each colony was determined as follows. A line was first drawn from the northern-most tree to the southern-most tree, and then a second line from the eastern-most tree to the western-most. The center was designated as the intersection of the two lines. Since the colonies were of irregular shape, in some cases the center point did not reflect the actual colony center and was adjusted visually.

The colony sites and surrounding forest were inventoried using point sampling (Avery and Burkhart 1983). Five sample points were chosen in each colony: 1 at the center and 1 at 46 m from the center in each of the 4 cardinal directions. Each point was the center of a variable radius prism plot. Point samples did not overlap. Sample trees were designated with a 5-factor prism. Trees with a diameter at breast height (dbh) of 10 cm or greater were tallied. Field data of interest for hazard and risk rating models included species, diameter, and crown class of each tree. These measurements were used to determine stand density and basal area of pines and hardwoods.

Four dominant or codominant trees were selected at the central sample point of each colony. Measurements on these trees included total height and height to the live crown. Increment cores were taken on these trees at dbh to determine age and radial growth for the last 10-year period (1981-1991). These measurements were used to determine site index, as calculated, from age-height relationships (Schumacher and Coile 1960).

Site variables examined included landform, percent slope and depth of the A horizon. Samples of surface soil (0-15 cm) were collected and taken to the laboratory to determine their sand, silt and clay content.

We determined the occurrence of red heart by visually examining increment cores (5mm diameter) taken at 3, 6, 9 and 12 m above the ground in 301 trees on 9 different sites. We were unable to culture the suspected red heart fungus in the laboratory for positive identification.

RESULTS AND DISCUSSION

Noxubee NWR Mississippi

Tree, site and stand parameters were collected to characterize active RCW colonies on the Noxubee NWR and to provide a basis of comparison with colonies studied in the Piedmont (Belanger et al. 1988) and the Texas Lower Coastal Plain (Mitchell et al. 1991) **(Table 1)**. Stand characteristics in all locations were the result of silvicultural treatments such as thinning and prescribed burning.

The colony sites at Noxubee were found mostly on flat upland sites, typical of the interior flatwoods of Mississippi. Colony 12 was the only colony located on a major slope (over 15%). Colonies, 9 and 11, although mostly on flat uplands, included a gentle slope which ended in a narrow bottomland bordering Bluff Lake. Colonies 13 and 14 were on terraces. Upland soils were various loams of the Stough-Prentiss-Myatt associations. Soils of the Mathiston-Urbo association were present in the bottomlands. These loams included silt loams, silt clay loams, sandy loams and clay loams. Due to logging disturbance and frequent prescribed fires, A horizons were rarely more than 2 centimeters deep and were often non-existent. Site indices (base age 50) for pine were fairly high at 80-100 except for colony 1 which had a lower site index of 70.

Loblolly and shortleaf pines were found in the colony sites. The pine composition of the colonies was 39 to 100% loblolly with an average of 92%. Only colony 5 had a greater percentage of shortleaf than loblolly. Predominant hardwoods were hickories (*Carya sp.*), southern red oak (*Quercus falcata* Michx.) and sweetgum (*Liquidambar styraciflua L.*). The understory in the colonies consisted of shrubs, such as sumac (*Rhus copallina L.*), ferns, various hardwood saplings, and pine regeneration. The height of most of the understory was between 0.5 - 2.0 m.

The colony sites were fairly open with characteristically low basal areas averaging 16 m^2/ha^{-1} for pine and 4 m^2/ha^{-1} for hardwoods. Total trees per hectare averaged 171 for pine and 44 for hardwoods.

Pines in the colony sites tended to be tall and averaged 41 cm dbh. The average height of dominant and codominant colony trees was 31 m with a height to live crown of 15.9 m. Thus, dominants and codominants had mean percent live crowns of 51.3%. The average age of dominants and codominants was 83 years with a range of 60-106 years. Dominants and codominants put on an average of 10.5 mm of radial growth over the past five year period (1985-90) and 14 mm over the preceding five year period (1980-85), indicating that the growth of the trees had slowed in recent years.

Piedmont NWR and Hitchiti Experimental Forest, Georgia

Trees in piedmont colonies at the Piedmont NWR and the Hitchiti Experimental Forest

Table 1. Comparison of mean site, stand and tree characteristics of red-cockaded woodpecker clusters at the Noxubee National Wildlife Refuge, Piedmont and Texas study sites.

Variable	Noxubee	Piedmont	Texas
% Slope	3.0	9.0	NA*
Depth A Horizon (cm)	0.9	7.0	NA
% Surface Clay	18.0	18.0	NA
Site Index Base Age = 50	92.0	81.0	NA
Age	83.0	87.0	87.0
Pine Basal Area (m^2/ha)	16.0	15.0	15.0
Hardwood Basal Area (m^2/ha)	4.0	5.0	1.0
Pines/ha	171.0	205.0	NA
Hardwoods/ha	44.0	94.0	NA
% Loblolly (Total Pine)	92.0	94.0	NA
% Shortleaf (Total Pine)	8.0	6.0	NA
Pine DBH (cm)	41.0	40.0	53.0
Pine Height (m)	31.0	23.0	29.0
Pine % Live Crown	49.0	45.0	NA
Radial Growth (mm Last 5 Yrs)	10.5	5.6	NA
Radial Growth (mm Previous 5 Yrs)	14.0	8.1	NA

* NA: data not available.

were growing on good pine sites. Soils were various loams, especially sandy loams, while silt loams were more common on the Noxubee NWR. Mean site indices were lower for Piedmont sites (81) than the Noxubee NWR sites (92). The average age of dominant and codominant trees was 87 years, while trees on the Noxubee averaged 83 years. Basal areas of pines and hardwoods were practically identical at both sites, however density of trees was higher on Piedmont than Noxubee sites. Pine density on piedmont sites (205 trees/ha) was higher than Noxubee sites (171 trees/ha) while hardwood density was considerably higher (94 vs 44). Pine composition on the sites was similar.

Mean diameter of pines on Piedmont and Noxubee sites was similar, however, Noxubee pines were taller (31 m vs 23 m) and had greater percent live crown (49% vs 45%). Noxubee pines also put on better radial growth over the past 10 years (24.5 mm vs 13.7 mm).

Angelina National Forest, Texas

Mitchell et al. (1991) did not cruise the Texas colonies, but obtained their information using aerial photos of the Angelina National Forest. However, Conner and O'Halloran (1987) had collected some tree and stand characteristics. Pine basal area at the Texas sites was similar to the other two locations, but hardwood basal area was very low (1 m^2/ha^{-1}). Mean cavity tree age was 87 years. If this figure also represents mean age of pines in the stands, then the Texas pines are the same age as those in the Piedmont and slightly older than those on the Noxubee. Texas cavity trees were also of larger dbh (53 cm) than pines on the other two locations, and were taller (29 m) than Piedmont pines, but of similar height to Noxubee pines.

Incidence of Red Heart

Red heart disease has not been widely studied in the South, perhaps because most southern pine is harvested before red heart has a chance to develop. The incidence of red heart disease increases with age. Hepting (1971) described one study of a representative pine stand in the Atlantic Coastal Plain. Incidence was less than 10% in stands younger than 140 years but greater than 50% in stands older than 140 years.

Several authors maintain that the presence of red heart disease is essential for the construction of RCW cavities (Steirly 1957a, Ligon 1970). Others believe that the relationship is coincidental, a function of the tree's age (Beckett 1971). Jackson (1977c) located 265 cavity trees on the Noxubee NWR and the adjacent John W. Starr Memorial Forest (Mississippi State University) in Mississippi and found that 90% of the trees with completed cavities had obvious signs of red heart disease. Sixty-seven percent of the trees containing unfinished cavities showed no sign of the disease. Jackson found it difficult to age the trees in his study since red heart had destroyed the inner annual rings. He estimated the age of the trees to be between 40 and 116 years.

Hooper (1988), working on colony sites in the Francis Marion National Forest in South Carolina, found that young (≤ 80 yr) longleaf pine with cavities had a significantly higher frequency of red heart disease than adjacent trees of similar size and age (86% vs 9%). Later, working in the same area, Hooper et al. (1991b) compared the frequency of red heart in cavity start and control trees in loblolly and longleaf pine and found that 57% of start trees had red heart as opposed to 7% of controls. The authors concluded that RCW's on the Francis Marion selected trees infected with red heart disease.

Conner and Locke (1982) looked at the association of RCW and red heart disease on the Angelina and Davy Crockett National Forests and the I.D. Fairchild State Forest in east Texas. Their methods were unique in that they examined cross sections of recently killed cavity trees and cultured suspected fungi, as opposed to visually examining samples removed with increment bores. The authors found heart rot in 63% of cavity trees and 44% of cavity-starts. Although most heart rot was red heart disease, 6 other species of fungi were identified. Longleaf had a lower incidence of decay than loblolly and shortleaf pine, possibly because longleaf produces more resin than other species (Hodges et al. 1979), and thus may slow or halt infection. This is interesting, since longleaf is generally considered the preferred species for cavity construction (Hooper et al. 1980). Conner and Locke speculated that the birds may prefer infected trees, but could not make this conclusion since they lacked estimates of heartwood decay in the stands.

If RCWs prefer trees infected with red heart, this would seem to explain why they select old trees. Heart rot does not become prevalent in stands of southern pine until the trees reach a relatively old age, for example, 75 years for loblolly pine and 100 years for longleaf (Wahlenberg 1946). Attempts have been made to artificially inoculate younger trees, age 26 to 28, with red heart, to create potential habitat for the birds (Conner and Locke 1983).

The difficulty of excavating a cavity in living pine seems to support a preference for trees infected with red heart disease. According to Hooper et al. (1980), most woodpecker species take as little as two weeks to hollow out a cavity, while the red-cockaded may take several months or even years to finish. Rudolph et al (1995) suggest that the birds are quite capable of making cavities in trees without red heart disease, however infected trees are definitely preferred and are much easier to excavate.

Another hypothesis would be that red heart colonizes the tree after the woodpeckers excavate their cavities. This hypothesis is supported by wind dissemination of fungal spores, and by the tendency of the spores to colonize branch stubs in the upper trunk. Conner and Locke (1982) examined the possibilities of red heart spores infecting the tree once excavation had begun. They found that once RCW's began excavation, resin. saturated the sapwood preventing infection. The heartwood, however, did not saturate and was susceptible to infection, although this was not apparent in visually examined cross sections of 32 recently killed cavity trees. Based on these visual examinations, the authors concluded that red heart was present throughout the entire cavity site prior to excavation.

We examined over 400 trees on the Noxubee NWR colonies for red heart disease, but only 20 of the 301 that were sampled at different heights were found to be infected **(Table 2)**. There is of course the possibility that red heart occurred in small, scattered patches in the trunks of the trees, patches that the sampling technique would have likely missed. Of the 51 cavity trees sampled, only 15 (30%) were found to have red heart **(Table 2)**. These results do not necessarily indicate that the birds do not need red heart. Woodpecker cavities are not very large and would only require a small strip of red heart for construction. Although the trunk around inactive cavities was intensively sampled, it was not possible to sample directly above or below the hole for fear of damaging

Table 2. Incidence of red heart trees in red-cockaded woodpecker colonies on the Noxubee National Wildlife Refuge.

	NON CAVITY TREES			CAVITY TREES		
COLONY	SAMPLED	YES	NO	SAMPLED	YES	NO
3	30	0	30	6	2	4
4	30	0	30	5	2	3
5	30	0	30	8	2	6
6	30	1	29	3	1	2
7	30	1	29	4	1	3
8	30	0	30	11	2	9
10	30	0	30	6	3	3
12	30	3	27	6	2	4
15	10	0	10	2	0	2
Total	**250**	**5**	**245**	**51**	**15**	**36**

the cavity and possibly making it less attractive to the birds for future use. In fact, examination of a few cross sections of cavity trees revealed red heart directly above and below the cavity. A reasonable hypothesis may be that RCWs prefer red heart trees for cavity construction, and may even seek them out, but do not actually require red heart. Rudolph et al. (1995) have expressed similar sentiments.

Our results are contrary to what Jackson (1977c) found on the Noxubee NWR and the adjacent John W. Starr Memorial Forest. Jackson found that 90% of the cavity trees he examined had "obvious signs" of red heart disease. This dilemma would be best solved by additional attempts of culturing samples from RCW cavities (Conner et al. 1976) and examining cross sections of recently killed cavity trees as in Conner and Locke (1982). It may also be explained by the changes in stand structure through time.

ACKNOWLEDGMENT

We would like to thank the personnel of the Noxubee National Wildlife Refuge for their assistance in completing this study. We also acknowledge the assistance of numerous undergraduates. This study was supported by the Mississippi Agriculture and Forestry Experiment Station, Mississippi State, MS.

SECTION 6

Status

Status of the Red-cockaded Woodpecker and Its Habitat

Frances C. James

Biologists and managers agree that the overall long-term decline in the number of Red-cockaded Woodpeckers (RCW's) continued through the 1980's. Although no coordinated range-wide sampling program has ever assessed the status of the species, Jackson (1971), Lennartz et al. (1983), and Costa and Escano (1989) have reported the results of responses to questionnaires. Summaries are also available for individual populations and for statewide surveys. Six of the papers in this section document the results of the most recent such efforts by states (Smith and Martin for Louisiana; Cox and Baker for Florida; Baker for Georgia; Cely and Ferral for South Carolina; Ortego, Krueger, and Barron for Texas; Bradshaw for Virginia). Because the authors of all six papers estimate seriously declining numbers during a period when sampling effort has been increasing, they probably underestimate true rates of decline. The conclusion of all these papers is that management at least until 1990 was insufficient to allow even stable populations to persist.

The other four papers in this section are a summary of a recent comprehensive survey of RCW's in the Red Hills Hunting Plantations of northern Florida and southern Georgia (Engstrom and Baker); a report of the results of the first three years of study of the RCW population at Eglin Air Force Base in northwestern Florida (Hardesty and coworkers); a report of the results of a pilot project to study the habitats of RCW's on private industry-owned land in Arkansas, Georgia, Louisiana, and Texas (Wigley and Sweeney); and the results of a range-wide summary that compares information about the status of the RCW in the early 1980's and with that about its status in 1990 (James).

In the range-wide summary, the data are for the numbers of known active sites (clusters of cavity trees, each representing the center of a territory defended by a pair of birds with or without helpers or by a single bird). In 1990 approximately 4000 such sites were occupied by an estimated 9300 individual RCW's, a decline of 23% from the early 1980's. An estimated 3400 breeding pairs occupied known sites in 1990, a decrease of about 1000 pairs since the early 1980's. Between Virginia and Texas, 34 populations had at least 10 active sites, and 50% of all known sites were in six populations. Even so, a few unsurveyed areas are likely to harbor additional birds. Examples are plantations in the outer coastal plain of South Carolina (Cely) and private lands in central and southern Florida (Cox and Baker) and northern Louisiana (Smith), where some land owners are not inclined to cooperate with either surveys or the recovery process.

The news of continuing declines in a species that has been on the endangered species list for more than 20 years is discouraging. What is the problem? What is the real prospect for recovery, as defined in the Recovery Plan (500 active sites in each of 15 widely distributed populations)? Is any real progress being made at resolving the apparent conflict between timber harvest and conservation? These questions must be addressed, especially now, as the U.S.D.A. Forest Service, the manager of 12 of the 15 recovery populations **(see James, Table 4)**, is finalizing its long-awaited new RCW management guidelines. Even larger questions involve how the new directive of the administration to use ecosystem management should be implemented and whether endangered-species programs should be subsumed under ecosystem approaches to land management.

Clearly two of the specific problems that contribute to the decline of the RCW are the limited availability of relict trees for the construction of cavities and the increasing isolation of ever-smaller populations. According to Costa and Escano (1989), emergency measures are needed to sustain RCW populations through what they view as a demographic bottleneck in the age structure of potential cavity trees. In that case the question becomes, "How can managers keep populations from failing altogether while presently degraded forests are being restored to conditions that will support healthy populations of birds?" Recent successes in moving birds between populations (Hess and

Costa; Rudolph et al. 1992) and in enticing birds to form new clusters by providing them with artificial cavities (see papers in this volume) have been so successful that the draft guidelines prepared by the Forest Service (USDA 1993) advocated widespread use of these intensive management techniques.

Probably more controversial, because it impinges on the management of saw timber, is the issue of how to restore the longleaf pine (*Pinus palustris*) ecosystem so that it can support its full diversity of animals and plants. In this system, if even remnants of the original grassy ground cover remain, restoration involves primarily a vigorous program of prescribed burning that emphasizes summer burns and time. A reasonable harvest of saw timber during this period can be compatible with the restoration program. Restoration of longleaf pine in areas that were site prepared and planted with loblolly and slash pine is more difficult and takes longer, and methods for ecosystem management in areas that were originally loblolly or shortleaf pine must be developed.

We are unlikely to see again the ancient longleaf pine forests of the southeastern coastal plain that Wells and Shunk (1931) described as unquestionably "one of the most wonderful forests in the world . . . with millions of trees measuring a yard or more in diameter." Even after its original harvest, management has involved unsustainable exploitation (Myers 1990). Many longleaf pine forests were replaced with plantations of loblolly (*Pinus taeda*) (Smith), slash (*P. elliottii*), and sand (*P. clausa*) pine. Past policies of fire suppression (Cely) have led to massive hardwood encroachment (Conner and Rudolph 1989). It took legal pressure to get these policies stopped.

Even now, however, one district on one national forest supports a population of RCW's that qualifies as recovered. That district is the Apalachicola Ranger District of the Apalachicola National Forest in northern Florida. The best sections of this district, those with a self-sustaining, all-age forest and the full biodiversity of the longleaf pine ecosystem, are pine savannas that are frequently burned in summer (**Figure 1**). Managers elsewhere who want to implement ecosystem management should set into motion long-term plans to provide resources at least equal in quality to those available to this population. What else needs to be done? Private landowners should be eligible for financial incentives, cooperative management agreements, or conservation easements to encourage them to manage their land in naturally regenerating forests from which saw timber can be harvested in a sustainable manner (Ortego and Lay 1988; Engstrom and Baker).

Figure 1. Longleaf pine savanna in the Post Office Bay area of the Apalachicola National Forest, the area with the highest density of red-cockaded woodpecker clusters in the forest (photograph by C. A. Hess).

The fact that RCW's persist on some private-industry lands that do not meet the requirements of the Recovery Plan (Wigley and Sweeney) shows that those recommendations (USDI 1985) are not the lower limit of what the birds can tolerate. Sometimes we learn in indirect ways. Consider the irony of the situation on Eglin Air Force Base, where the RCW survived several decades of unsustainable management of longleaf pine saw timber partly because for 40 years accidental fires in buffer zones around test ranges promoted a healthy longleaf pine forest there.

Clearly, we need a new long-range plan for ecosystem restoration in the southeastern pine forests, one that will eventually provide long-term self-sustaining forests in which it will not be necessary to provide boxes for the RCW's or to move the birds around. All new guidelines, including the Forest Service guidelines (USDA 1993), should be judged in light of this objective. Ecosystem management for the longleaf pine system will require far more prescribed burning than in the past, better protection for relict trees, and the establishment of naturally regenerating forests.

The Status of the Red-cockaded Woodpecker in 1990 and the Prospect for Recovery

Frances C. James, Department of Biological Science, Florida State University, Tallahassee, Florida 32306

ABSTRACT: Primarily on the basis of responses to a questionnaire and information in the literature, red-cockaded woodpeckers were known to occur at approximately 4000 sites (clusters of cavity trees with signs of activity, formerly called "active colonies") in 1990, a decline of 23% from the early 1980's. Because the number of birds at a site varies from one to about six, even excluding birds of the year, it is difficult to estimate the number of birds that occupied these sites. If the average number of sites per site were 2.3, the number of birds in 1990 would have been about 9300. If 85% of the sites supported a mated pair of birds, there would have been 3400 pairs, a decrease of about 1000 pairs at known sites in 10 years. The monitoring intensity was much greater and the area monitored was larger in 1990 than in the early 1980's, so the actual decline in number of sites was undoubtedly higher than 23%. In 1990 over 50% of all known sites were in six populations: Apalachicola National Forest, North Carolina Sandhills, Francis Marion National Forest, Vernon and Kisatchie Districts of the Kisatchie National Forest, Eglin Air Force Base and adjacent land, and Red Hills Hunting Plantations.

With a few exceptions, the management of the southern pine ecosystem on both public and private lands between the early 1980's and 1990 was not conducted so that red-cockaded woodpeckers could maintain their numbers. Even in populations designated as recovery populations by the U.S. Fish and Wildlife Service (U.S.D.I. 1985), there were more than 300 fewer active sites in 1990 than in the early 1980's. In terms of mimicking the natural processes of the longleaf pine ecosystem, the population in the Red Hills of southwestern Georgia, which is on private land and is managed by long rotation and selective cutting of trees, had the most ecologically sound management regime (Engstrom and Baker 1995).

Since 1990 the response of the birds to several intensive management techniques specifically designed to alleviate the isolation of single males (transport of juvenile females to them) or to alleviate cavity limitation (placement of cavity restrictors to keep out other species, the addition of drilled artificial cavities or cavity inserts) has been encouraging. However, for the establishment of self-sustaining populations, these practices will have to be accompanied by a greater commitment to ecosystem-level management than has been seen in the past, including less disturbance to habitats from the harvest of trees, more protection for old trees, better control of hardwood encroachment, and burning regimes that are more frequent than every five years and that emphasize burning in the summer. In spite of some recent gains (Escano 1995, Richardson and Stockie 1995), long-term recovery of the southern pine ecosystem, and the red-cockaded woodpecker in particular, will require a far larger commitment to ecosystem-level management than has been the case until now.
KEYWORDS: rangewide survey, inventory, decline, prognosis, *Picoides borealis*

This paper gives the results of a project conducted by the American Ornithologists' Union Conservation Committee's Subcommittee for the Conservation of the Red-cockaded Woodpecker. The project was designed to assess the rangewide status of the red-cockaded woodpecker in 1990 and, to the extent possible, to make comparisons with its status in the early 1980's. The information is based on responses to a questionnaire sent to more than 65 biologists and administrators, subsequent correspondence, published literature, and personal contacts.

The questionnaire asked for information

about the locations of populations and distances to the nearest other populations. Sites with birds (active colonies, clusters of cavity trees at least one of which had freshly worked resin wells) within 32 kilometers (20 miles) of one another were considered to belong to the same population of birds, regardless of land ownership category (federal, state, private). Each site was presumed to have been occupied by a mated pair of birds, a pair with one or more helpers, a single unmated bird, or rarely two unmated birds. Although sites were assumed to have been within defended territories, a small percentage of sites were probably occupied only by single birds who were roosting in a cavity tree there but were actually members of adjacent clans.

The objective of the present project was not to make estimates of red-cockaded woodpecker populations or their trends based on a sampling regime, but rather to respond to the stated need (Ligon et al. 1986, Anonymous 1990) for a rangewide summary of what was known about the distribution and abundance of the species. The organization of the information about sites into presumably functional natural populations, regardless of land ownership, was intended to help the reader see where cooperative agreements among land owners might best promote conservation.

Inaccuracies are surely present in the information in the tables, and some places have still not been surveyed, especially on private lands. In a few cases, when data from direct counts of sites with birds were not available for the early 1980's, numbers were taken from estimates based on the sample field surveys reported by Lennartz et al. (1983a). The numbers of abandoned sites (clusters of cavity trees without signs of recent work by birds, "inactive colonies") are not reported.

Two sources of bias in the type of data reported here make them unsuitable for making reliable estimates of actual population trends. First, if broad areas are not checked and the observer visits only previously known sites, new sites will not be found, and the results will be biased toward indicating a decline. Second, if the same area is checked but monitoring intensity is higher in the later period and more sites are found, bias will be toward indicating an increase. Participants in this project reported that much larger areas were searched and that monitoring intensity was either the same or higher in 1990 than in the early 1980's. Therefore, if the numbers indicate declines, the actual declines are likely to have been larger than the numbers suggest.

RESULTS

For 1990 the project identified approximately 4000 sites in 13 states **(Tables 1-3)**. Compared with the early 1980's, this was an overall decline in the number of known sites of 23% **(Table 2)**. Even with the two sources of bias toward finding more active sites in the latter period, the figures show that states with fewer than 100 sites in the early 1980's (Oklahoma, Virginia, Kentucky, and Tennessee) had declines in the numbers of known sites of 50% or more between the early 1980's and 1990 **(Table 1, Figure 1)**. The numbers of known sites were fewer in 1990 in all states except Georgia, where the increases were clearly due to more extensive surveys in two populations **(see footnotes to Table 3)**.

Nearly 90% of the sites in 1990 were in Florida, Georgia, North and South Carolina, Louisiana, and Texas. More than 50% were in six populations: Apalachicola National Forest in Florida, with 15% of all known sites; the North Carolina Sandhills and the Francis Marion National Forest in South Carolina, with 9% each; the Vernon and Kisatchie Ranger Districts of the Kisatchie National Forest and adjacent land in Louisiana, with 8%; Eglin Air Force Base and adjacent land in Florida, with 6%; and the Red Hills Hunting Plantations, primarily in Georgia, with 5% **(Tables 3, 4)**.

It is difficult to estimate how many birds there were from the numbers of sites because the number of birds at a site can vary from one to about six. Using the value 2.3 (close to the average for the two districts of the Apalachicola National Forest in the early 1990's, James et al. 1995), the estimate of the total number of birds for the early 1980's is approximately 12,000, and for 1990 it is 9300 **(Table 2)**. It is also difficult to estimate the number of breeding pairs of birds. An estimate of 85% as the number of active sites having a pair or a pair with one or more helpers leads to an estimate of approximately 3400 pairs in 1990, a decrease of about 1000 pairs since the early 1980's **(Table 2)**.

As defined here, populations were separated by at least 32 kilometers (20 miles) from other known sites where RCW's were present. Excluding places with only one active

Table 1. The total numbers of known clusters of active cavity trees (active colonies) in the early 1980's and in 1990. States are listed in decreasing order of the numbers for the early 1980's. Relevant papers are listed in footnotes.

State	Early 1980's	1990	Change	% Change	% of all sites in 1990
Florida[a]	1136	1116	-20	-2%	28
Louisiana	866	464	-402	-46%	11
South Carolina	861	615	-246	-29%	15
North Carolina[b]	701	465	-236	-34%	11
Georgia[c]	581	636	+55	+9%	16
Arkansas	289	132	-157	-54%	3
Texas[d]	288	265	-23	-8%	7
Mississippi	230	154	-76	-33%	4
Alabama	201	157	-44	-22%	4
Oklahoma[e]	31	15	-16	-52%	+
Virginia	12	5	-7	-58%	+
Kentucky[f]	8	4	-4	-50%	+
Tennessee[g]	6	1	-5	-83%	+
Total	5210	4029	-1181		

[a]Baker et al. (1980), Cox et al. (1995), Wood and Wenner (1983).
[b]Carter et al. (1983).
[c]Baker (1982), Engstrom and Baker (1995).
[d]Ortego and Lay (1988), Ortego at al. (1988), Conner and Rudolph (1989).
[e]Wood and Lewis (1977), Masters et al. (1989).
[f]Jackson et al. (1976), Kalisz and Boettcher (1991).
[g]Nicholson (1977).

Table 2. Number of known sites with red-cockaded woodpeckers, estimated total number of birds, and estimated number of pairs of birds in the early 1980's and in 1990. The third number in each category is the difference between the first two. Numbers are rounded to the nearest 10.

Number of sites			Estimated number of birds (sites x 2.3)			Estimated number of pairs of birds (sites x 0.85)		
Early 1980's	1990	Difference	Early 1980's	1990	Difference	Early 1980's	1990	Difference
5210	4030	-1180	11,980	9270	-2710	4430	3420	-1010

site and scattered records not assignable to any population, the project identified 59 such populations in 1990 (**Table 3**). Of these populations, only 34 had at least 10 active sites (**Figure 2**). Most populations were isolated from one another by far more than 32 km, judged to be the maximum dispersal distance of individual birds (Walters 1989).

The primary responsibility for the recovery of the red-cockaded woodpecker lies with agencies that manage federal land. Table 3 includes data both for the early 1980's and for 1990 for 28 such properties (15 national forests, eight military bases, and five national wildlife refuges). The only major increases in the 1980's on these lands were due to more complete surveys (Apalachicola National Forest, Ft. Benning). Excluding these two cases, the change in number of sites on these 28 properties was an overall loss of 464 sites (345 in national forests, 97 on military bases, and 22 in national wildlife refuges). These numbers exclude the populations in the Carolina Sandhills and the separate districts of the Kisatchie National Forest, where numbers were not available for the early 1980's.

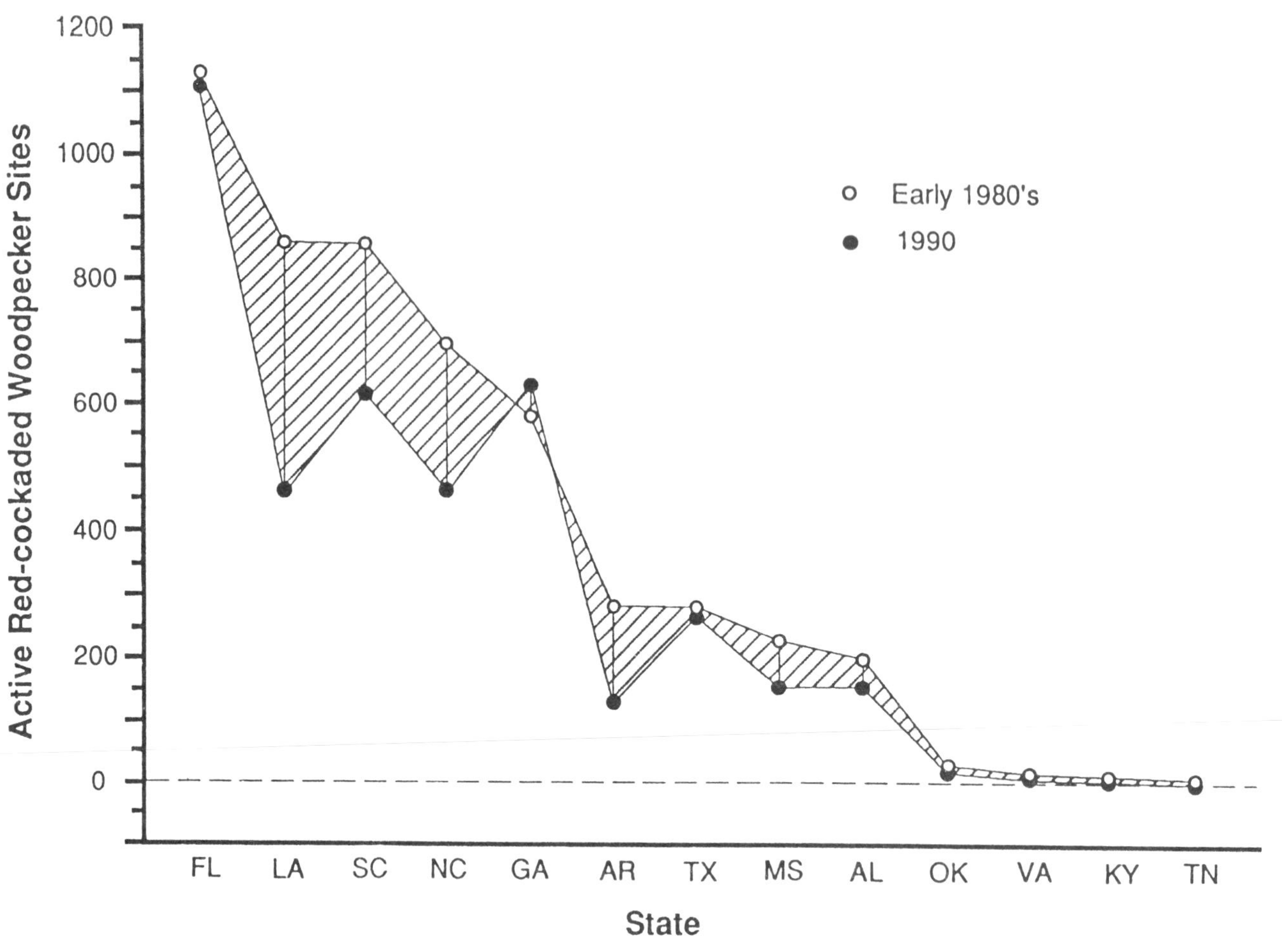

Figure 1. The number of active red-cockaded woodpecker sites ("colonies," clusters of cavity trees) in the early 1980's and in 1990 by state. States are in descending order of their numbers of sites in the early 1980's.

DISCUSSION

Although information about the status of the red-cockaded woodpecker is more complete now than ever before, it has never been fully adequate (Hooper and Muse 1989, U.S.D.I. 1985, Ligon et al. 1986). Jackson's (1971) initial estimate of the number of birds (2939) at the time of the first symposium (Thompson 1971) was based on 23 localities. Later Jackson (1978a) estimated that there was a maximum of 3473 sites (clusters, "active colonies") rangewide. In addition, several summaries of records of sites in individual states have been published (Jackson et al. 1976, Nicholson 1977, Wood and Lewis 1977, Baker et al. 1980, Baker 1982, Carter et al. 1983b, Wood and Winner 1983, Ortego and Lay 1988, Masters et al. 1989), and information from Florida and Louisiana is updated in this volume (Cox et al. 1995, Smith and Martin 1995). At the time of the second symposium (Wood 1983c), Lennartz et al. (1983a) reported the results of a comprehensive systematic field survey of red-cockaded woodpeckers on federal lands. Using the 460-meter circular scale method of Harlow et al. (1983) to estimate the number of sites ("colonies") from the distribution of cavity trees in 17 national forests, nine military bases, and 11 national wildlife refuges, Lennartz et al. (1983a) estimated that there were 2121 sites on these properties and more than 3000 active sites on federal land in the early 1980's. Although this number was higher than they had expected, they cautioned that "very few local populations could be considered reasonably secure." Escano (1995) gives the results of the most recent surveys by the U.S.D.A. Forest Service.

In 1989, the year that Hurricane Hugo

Figure 2. The location of the 34 populations of red-cockaded woodpeckers with more than 10 active sites ("colonies," clusters of cavity trees) in 1990. Each population is separated by at least 32 km (20 miles) from the nearest adjacent population. Populations are listed below alphabetically by states and from west to east within state, and the number of sites is given in parentheses. Abbreviations are as in Table 4. Alabama: Talladega NF (127); Arkansas: Ouachita NF (16), southeastern Arkansas, commercial timberland and Felsenthal NWR (105); Florida: Eglin Air Force Base, Blackwater State Forest, Conecuh NF in Alabama (237), Apalachicola National Forest (590), Osceola National Forest (44), Withlacoochee State Forest (39), Camp Blanding (15), Orange County, private land (24), Big Cypress National Preserve and adjacent land (51), Three Lakes Wildlife Management Area (15); Georgia: Ft. Benning (176), Red Hills Hunting Plantations (180), Piedmont NWR, Oconee NF and adjacent land (35), Okefenokee NWR (45), Ft. Stewart (128); Louisiana: Vernon and Kisatchie RD's of Kisatchie NF and adjacent land (316), Winn RD of Kisatchie NF and adjacent land (29), Catahoula RD of Kisatchie NF (31), Evangeline RD of Kisatchie NF and adjacent land (56); Mississippi: Homochitto NF and adjacent land (29), Bienville NF and adjacent land (91), DeSoto NF (16), Noxubee NWR and adjacent land (18); North Carolina: North Carolina Sandhills (371), Croatan NF and Camp LeJeune (81); Oklahoma: McCurtain County Wilderness Area and adjacent land (15); South Carolina: Ft. Jackson (15), Francis Marion National Forest and adjacent land (370), Carolina Sandhills NWR and Sandhills State Forest (122); Texas: Sam Houston NF and adjacent land (166), I.D. Fairchild State Forest (12), Davy Crockett NF and adjacent land (30), Angelina NF, Sabine NF, and adjacent land (39).

Table 3. Records of the number of sites (clusters of cavity trees)[a] being used by the red-cockaded woodpecker in the early 1980's and in 1990, arranged in descending order within states. Subtotals are in parentheses below totals for populations separated by at least 35 km (20 miles). Each site included one or more pine trees with signs of recent work by birds and was the center of activity for a social unit consisting of a single bird, a pair, or a pair with one or more helpers. Abbreviations: NF, National Forest; NWR, National Wildlife Refuge; DOD, Department of Defense; WMA, Wildlife Management Area (private owner, state manager); RD, Ranger District. Managers: F, federal government; S, state government; C, city or county government; P, private (timber company, plantation, etc.).

Locality	Management	County (or Parish)	Early 1980's	1990
ALABAMA				
1. Talladega NF	F		175	127
a. Oakmulgee RD	F	Bibb, Chilton, Dallas, Hale, Perry, Tuscaloosa	(175)	(120)
b. Shoal Creek RD	F	Calhoun, Clay, Cleburne, Talladega	(?)	(7)
2. Conecuh NF	F	Covington, Escambia	15	13
3. Tuskegee NF	F	Macon	?	1
4. Bankhead NF	F	Winston	8	0
5. Geneva SF	S	Geneva	?	5
6. WMA's	S	Coosa, Covington	1	2
7. Gulf State Park	S	Baldwin	1	?
8. Interstate Rest Area	S	Conecuh	?	1
9. Lake Purdy	C	Shelby	?	1
10. Scattered additional records	P	Butler, Escambia, Lowndes, Washington	1	7
ARKANSAS				
1. Ashley, Bradley, Calhoun, Union Counties			248	105
a. Georgia-Pacific Corp.	P	Ashley	(92)	(27)
b. Potlatch Lumber Co.	P	Bradley, Calhoun	(131)	(50)
c. Felsenthal NWR	F	Ashley, Bradley, Union	(25)	(28)
2. Ouachita NF, Poteau and Cold Springs RD's	F	Scott	25	16
3. Ross Foundation	P	Clark	2	6
4. Pine City Natural Area and adjacent forest	S, P	Monroe	3	3
5. International Paper Co.	P	Lafayette	1	2
6. Scattered additional records	P	Dallas, Drew, Grant, Perry, Pulaski, Saline	10	?
FLORIDA				
1. Apalachicola NF	F	Franklin, Leon, Liberty, Wakulla	510	590[b]
2. Eglin Air Force Base/ Blackwater State Forest			293	237
a. Eglin Air Force Base (DOD)	F	Okaloosa, Santa Rosa, Walton	(243)	(208)[c]
b. Blackwater State Forest	S	Santa Rosa	(50)	(29)
c. add Conecuh NF in Alabama (see above)				
3. Osceola NF	F	Baker, Columbia	44	44
4. Withlacoochee State Forest	S	Citrus, Hernando	21	39

Table 3 continued

Locality	Management	County (or Parish)	Early 1980's	1990
5. Big Cypress National Preserve and adjacent land	F, P		18	51[d]
a. Big Cypress Nat. Preserve	F	Collier	(18)	(36)
b. Adjacent land	P	Collier	(?)	(15)
6. Orange County, including C.H. Stanton Energy Center, Orlando Utilities Commission	P	Orange	50	24
7. Camp Blanding (DOD)	F	Clay	23	15
8. Three Lakes WMA	S	Osceola	15	15
9. Ocala NF	F	Marion	41	5
10. Avon Park Bombing Range (DOD)	F	Polk	25	8
11. Scattered Wildlife Management Areas (J.M. Corbett, J.W. Webb, Pt. Washington, Ed Ball, Bull Creek, Croom, Kicco, Fisheating Creek, Three Lakes) and Cary State Forest	S	Bay, Charlotte, Gulf, Nassau, Osceola, Palm Beach	50	44
12. St. Marks NWR	F	Wakulla	4	4
13. Old Venus	P	Highlands	1	4
14. Ochlockonee River State Park	S	Wakulla	1	1
15. Additional records	S, P	Alachua, Brevard, Calhoun, Duval, Glades, Lee, Leon, Levy, Putnam, Volusia	40	35
GEORGIA				
1. Fort Benning (DOD)[e]	F	Chattahoochee, Muscogee	111	176
2. Fort Stewart (DOD)	F	Bryan, Liberty, Long, Tattnal	138	128
3. Red Hills Hunting Plantations	P	Grady, Thomas	144	180[f]
4. Okefenokee NWR	F	Charlton, Clinch, Ware	48	45
5. Piedmont NWR/Hitchiti Exper. Forest/B.F. Grant Memorial Forest, Oconee NF	F		46	35
a. Piedmont NWR	F	Jones	(24)	(23)
b. Hitchiti Exper. Forest	F	Jones	(10)	(11)
c. Oconee NF	F	Jasper, Putnam	(12)[g]	(1)
6. Ft. Gordon (DOD)	F	Columbia, Jefferson, McDuffie, Richmond	35	3
7. Laura S. Walker State Park	S	Ware	1	1
8. Scattered additional records[h]	P	Seminole, Decatur, Baker, Montgomery, and others	30	28
KENTUCKY				
1. Daniel Boone NF, London Ranger District	F	Laurel, Whitley	8	4

Table 3 continued

Locality	Management	County (or Parish)	Early 1980's	1990
LOUISIANA[i]				
1. **Vernon and Kisatchie RDs of Kisatchie NF and adjacent land**				316
a. **Vernon RD**	F	Vernon		(186)
b. **Fort Polk and Peason Ridge**	F	Vernon, Natchitoches		(63)
c. **Kisatchie RD**	F	Natchitoches		(59)
d. **Adjacent land**	P	Beauregard, Vernon, Natchitoches		(8)
2. **Evangeline RD of Kisatchie NF, Alexander NF, and adjacent land**				56
a. **Evangline RD**	F	Rapides		(46)
b. **Alexander State Forest**	S	Rapides		(5)
c. **Adjacent land**	P	Rapides		(5)
3. **Catahoula RD of Kisatchie NF**	F	Grant		31
4. **Winn RD of Kisatchie NF and adjacent land**				29
a. **Winn RD**	F	Winn		(18)
b. **Adjacent land**	P	Winn		(11)
5. **Morehouse, Union, and Ouachita Parishes**	P	Morehouse, Ouachita, Union	78	9
6. **Fontainbleau State Park and adjacent land**	S, P	St. Tammany	7	6
7. **River Pines Plantation**	P	Livingston		5
8. **D'Arbonne NWR**	F	Ouachita, Union	5	5
9. **Minden Munitions Plant (DOD) and Caney RD of Kisatchie NF**	F	Claiborne, Webster	6	0
10. **Scattered additional records**	P	Catahoula, Tangipahoa, Scott, Bossier		7
MISSISSIPPI				
1. **Bienville NF and adjacent land**		Scott, Smith	120	91
a. **Bienville NF**	F		(120)	(86)
b. **Adjacent land**	P		(?)	(5)
2. **Homochitto NF and adjacent land**		Amite, Franklin, Wilkinson	50	29
a. **Homochitto NF**	F		(50)	(27)
b. **Adjacent land**	P		(?)	(2)
3. **Noxubee NWR and adjacent land**			30	18
a. **Noxubee NWR**	F	Noxubee, Oktibbeha, Winston	(23)	(17)
b. **Georgia-Pacific Corp.**	P	Noxubee, Winston	(7)	(1)
4. **Desoto NF**	F		30	16
a. **Biloxi RD**	F	Harrison, Stone	(9)	(7)
b. **Black Creek RD**	F	Forrest, Green, Perry	(10)	(1)
c. **Chickasawhay RD**	F	Green, Jones, Wayne	(11)	(8)

Table 3 continued

Locality	Management	County (or Parish)	Early 1980's	1990
NORTH CAROLINA				
1. North Carolina Sandhills	F, S, P	Cumberland, Harnett, Hoke, Lee, Moore, Richmond, Scotland	593	371
a. Ft. Bragg Military Reservation (DOD) and Sandhills Game Land (S) (both N.C. State Study Areas)			(259)	(219)
b. Outside N.C. State Study Areas			(est. 250)	(152)
c. Private (P)			(84)	(?)
2. Croatan NF/Camp LeJeune Marine Base (DOD)			69	81
a. Croatan NF	F	Carteret, Craven, Jones	(43)	(47)
b. Camp LeJeune (DOD)	F	Onslow	(26)	(34)
3. Sunny Point Military Ocean Terminal	F	Brunswick	7	5
4. Pee Dee NWR	F	Anson, Richmond	3	1
5. Alligator River NWR	F	Dare, Hyde, Tyrell	4	1
6. Scattered additional records	S, P	12 counties	25	6
SOUTH CAROLINA				
1. Francis Marion NF and adjacent land			514	370
a. Francis Marion NF[j]	F	Berkeley, Charleston	(434)	(315)
b. adjacent land	F, S, P		(80)	(55)
2. Carolina Sandhills NWR/Sandhills State Forest	F, S	Chesterfield	127	122
3. Ft. Jackson Military Reservation (DOD)	F	Richland	18	15
4. Santee NWR	F	Clarendon	5	1
5. Savannah River Plant (DOE)/USFS	F	Aiken, Barnwell	17	7
6. Scattered additional records	P		180	100
OKLAHOMA				
1. McCurtain County Wilderness Area/ Pushmataha County	S, P	McCurtain, Pushmataha	31	15
TENNESSEE				
1. Cherokee NF	F	Polk	?	1
2. Great Smoky Mountains NP	F	Blount	1	0
3. Catoosa WMA	S	Cumberland, Morgan	4	0
4. Koopers WMA	S	Campbell	1	0
TEXAS[k]				
1. Sam Houston NF vicinity	F		172	166
a. Sam Houston NF	F	Montgomery, San Jacinto, Walker	(172)	(133)

Table 3 continued

Locality	Management	County (or Parish)	Early 1980's	1990
b. Jones State Forest	S	Montgomery	(?)	(15)
c. Private land	P	Grimes, Montgomery, Walker	(?)	(18)
2. Angelina and Sabine NF and vicinity	F, P		66	39
a. Angelina NF (south)	F	Angelina, Jasper	(26)	(13)
b. Sabine NF	F	Newton, Sabine, Shelby	(28)	(8)
c. Angelina NF (north)	F	San Augustine	(12)	(7)
d. Rayburn area	P	Jasper	(?)	(6)
e. Scrappin Valley	P	Newton	(?)	(5)
3. Davy Crockett NF and vicinity	F		46	30
a. Davy Crockett NF	F	Houston, Trinity	(46)	(26)
b. Private land	P		(?)	(4)
4. I.D. Fairchild State Forest	S	Cherokee	(?)	12
5. Buna	P	Newton	(?)	3
6. Brushy Creek	P	Trinity	(?)	3
7. Longleaf Trail	P	Polk	4	2
8. Indian Reservation	F	Polk	?	5
9. Scattered additional records	P	Hardin, Jasper, Newton, Sabine, Tyler	?	5
VIRGINIA				
1. Gray Lumber Company	P	Sussex	9	5
2. Suffolk	P		2	0
3. Grafton	P	York	1	0

[a]Often recorded as "active colonies" or "active colony sites."

[b]Reduced from 684 reported by USFS to 590 because of colonies counted twice, colonies not on maps, and percentage of abandoned colonies found by James (1991) in the Wakulla Ranger District. The increase for the forest as a whole is due to more extensive survey and a reorganization of the records.

[c]Number is for April 1993 of an ongoing survey; estimated total is 235 (J. Hardesty pers. comm.).

[d]See Patterson and Robertson (1981). Increase is due mainly to better survey (D. Jansen pers. comm.).

[e]The figures reported are from Baker (1989). The later increase is due to a more extensive survey.

[f]A recent and ongoing intensive survey gives 180 active sites for March 1993 (Engstrom and Baker pers. comm.). See also Engstrom and Baker (1995).

[g]Figure for 1986 (Costa and Escano 1989).

[h]In each of 21 counties in Georgia, there were one or more active colonies in the late 1970's and none in 1989 (Baker 1989).

[i]Data for the early 1980's are incomplete, but R. Martin estimates 770 active sites for the entire Kisatchie NF, Ft. Polk, Peason Ridge, and adjacent lands in Vernon and nine other parishes. Of these, 603 were in the national forest, 123 at Ft. Polk and Peason Ridge, and 44 on adjacent private land. Lennartz et al. (1983a) estimated 430 active sites in the entire Kisatchie National Forest for 1980-2. By 1992, there were no active clusters in Union, Claiborne, or Webster Parishes. As separated here into populations separated by at least 20 miles, the Kisatchie National Forest had 245 active clusters in the Vernon and Kisatchie RD's, 56 in the Evangeline RD, 31 in the Catahoula RD, and 18 in the Winn RD for a total of 350 in 1992 (E. Smith pers. comm.).

[j]The Francis Marion NF supported 427 clans with at least two adults in the breeding season (groups) in 1980-81 and 477 in 1987-88 (Hooper et al. 1991a). Hurricane Hugo struck in September 1989, destroying 87% of the cavity trees and killing 63% of the birds. Following installation of 537 artificial cavities, there were 238 groups in the 1990 nesting season. With the installation of an additional 291 cavities, the number of groups increased to 320 by the 1992 nesting season (Watson et al. 1993). These numbers do not include sites defended by single birds, but the numbers in the table do include them.

[k]For locations in Texas where status was known and coverage equivalent, there was a 34% loss of active colonies in the 1980's (B. Ortego pers. comm.).

Table 4. Number of active sites (clusters, active colonies) in 1990 in the 20 largest populations. Numbers are rounded to the nearest five. The numbers of sites are for the entire natural population, not just the federal land.[a]

		Number of sites	% of all sites
1.	Apalachicola National Forest, FL	590	15
2.	North Carolina Sandhills, NC	370	9
3.	Francis Marion NF, SC	370	9
4.	Vernon and Kisatchie Ranger Districts of Kisatchie National Forest and adjacent land, LA	315	8
5.	Eglin Air Force Base, Conecuh NF and adjacent land, FL and AL	235	6
6.	Red Hills Hunting Plantations, GA	180	5
7.	Ft. Benning, GA	175	4
8.	Sam Houston National Forest and adjacent land, TX	165	4
9.	Ft. Stewart, GA	130	3
10.	Talladega National Forest, AL	125	3
11.	Carolina Sandhills National Wildlife Refuge adjacent land, SC	120	3
12.	Bienville National Forest, MS	90	2
13.	Croatan National Forest and Camp Le Jeune, NC	80	2
14.	Evangeline Ranger District of Kisatchie National Forest and adjacent land, LA	55	1
15.	Big Cypress National Preserve and adjacent land	50	1
16.	Okefenokee NWR, GA	45	1
17.	Osceola National Forest, FL	45	1
18.	Withlacoochee State Forest, FL	40	
19.	Angelina National Forest, Sabine National Forest, and adjacent land, GA	40	1
20	Piedmont NWR, Oconee NF, and adjacent land, GA	35	1

[a]According to specialists (Anonymous 1990) and the federal recovery plan (USFWS 1985), the species could be declared to have recovered if it had 500 active sites in each of 15 populations distributed widely through its original geographic range. The twelve recovery populations in national forests, as listed in the Draft Environmental Impact Statement (USDA 1993), include the Conecuh NF; separate recovery populations in the Oakmulgee and Shoal Creek RD's of the Talladega NF; Osceola NF; only the Apalachicola RD of the Apalachicola NF, the combined Oconee NF, Hitchiti Experimental Forest, and Piedmont NWR; only the Vernon RD of the Kisatchie NF; Bienville NF; Chickasawhay RD of the DeSoto NF, Croatan NF, Francis Marion NF and Sam Houston NF. The other three recovery populations are the North Carolina Sandhills; Ft. Benning, Georgia; and Ft. Stewart, Georgia.

decimated the population in the Francis Marion National Forest, Conner and Rudolph (1989) documented severe declines in three national forests in Texas, and Costa and Escano (1989) reported recent declines in all but four national forests in the South. Then a special meeting of experts, called the "Scientific Summit" (Anonymous 1990), called for a plan to stabilize declining populations. The participants agreed with Hooper and Muse (1989) that better data about the status of the species were urgently needed, with Costa and Escano (1989) that in several populations the present age structure of trees in red-cockaded woodpecker habitat constitutes at least a 20-year bottleneck in the availability of potential cavity trees for the birds, and with Conner and Rudolph (1991b) that the isolation of small populations is causing a demographic crisis. They declared the situation an emergency for managers.

In the last few years, conservation efforts have increased dramatically. Walters and coworkers (Copeyon 1990, Walters et al. 1995) at North Carolina State University showed that birds could be induced to establish new sites by the provision of artificial cavities in living trees. Refinements of this management technique (Allen 1991, Taylor and Hooper 1991), often in combination with the use of cavity restrictors to exclude competing species, transport of birds between sites (Hess and Costa 1995), and predator control, have been applied with dramatic success in four national forests (Escano 1995), at the Savannah River Site (Gaines et al. 1995), and at the Noxubee National Wildlife Refuge (Richardson and Stockie 1995). The most impressive example has been the ongoing recovery of the population in the Francis Marion National Forest since Hurricane Hugo (Hooper et al. 1990, Hooper et al. 1991a, Watson et al. 1995). The long-term consequences of these intensive practices will not be known for several more years, and some caution seems warranted (Rudolph and Conner 1995, Walters et al. 1995b).

Implications of This Survey and Prospects for Recovery

The participants in the 1990 meeting (Anonymous 1990) agreed that fulfillment of the intent of the criteria for recovery in the federal recovery plan (U.S.D.I. 1985) would require the existence of at least 15 healthy populations each with 400 breeding pairs. Because some sites support only single birds and some pairs do not breed every year, the target number of active sites per population was judged to be 500 (Anonymous 1990). Only the population in the Apalachicola National Forest so qualifies, and even there the population in the eastern Wakulla District is, not stable (James 1991, James et al. 1995). Although the population in the Francis Marion National Forest was increasing before Hugo (Hooper et al. 1991a), the only overall increases reported in recovery populations **(Table 4, footnote)** of the red-cockaded woodpecker for the 1980's (Apalachicola National Forest, Ft. Benning, and the Croatan National Forest plus Camp LeJeune, **Table 3**) are due to better monitoring, not to real population increases. In spite of greatly increased monitoring effort, the number of known sites in recovery populations was more than 300 less in 1990 than in the early 1980's **(Tables 3 and 4)**. Two of the recovery populations listed by Costa and Escano (1989) had 16 and five active sites in 1990 and showed declines of 47% and 88% since the early 1980's respectively (DeSoto and Ocala National Forests). The cautious optimism suggested by increases in the numbers of active sites since 1990 in the six populations mentioned above must be tempered with the recognition of the simultaneous losses in maintenance of midstory and adequate burning regimes in many areas (Escano 1995).

Because private owners of small properties and large timber companies control 90% of the pine timberland in the South (Lennartz et al. 1983a), they should be encouraged and even given incentives to manage that land in a more ecologically sound manner. The fact that the Red Hills Hunting Plantations in Georgia support the sixth largest population of birds, while also having aesthetically pleasing all-age forests, protection of the native ground cover, and a program of selective cutting that has produced millions of board feet of timber over the past several decades (Neel 1971, Engstrom and Baker 1995) is strong support for this type of unevenage management. Even when timber need not be harvested, populations of red-cockaded woodpeckers can be supported (see, for example, Carter et al. 1983b, on suburban habitats).

If the red-cockaded woodpecker is to recover, its management will have to combine the application of the short-term intensive measures mentioned above and a much stronger commitment to ecosystem-level management (Krusac 1995, Dabney 1994, for U.S.D.A. Forest Service approaches). Such management must include less disturbance from the harvest of trees, more protection of old trees, better control of hardwood encroachment, and more frequent burning regimes, preferably in summer. For the pinelands in the South, regardless of ownership of the land, such management is most feasible if it is coordinated with a program of sustainable harvest of pine timber. Red-cockaded woodpecker conservation could become a premier example of simultaneous management for the protection of a major southern ecosystem and for economic gain. Until now, the bird has been a victim of an unbalanced compromise between provision for its needs and forestry practices locked into the pine plantation resource (Knight 1978). The future of the species will depend on how serious managers and foresters are in implementing their stated commitment to the preservation of biological diversity on both public and private land. It will also depend on how priorities are set at all levels of government for the management of endangered species. In Florida, for example, where 28% of all known active sites occurred in 1990, the red-cockaded woodpecker is ranked very low on the priority list of endangered species (Millsap et al. 1990). Because ground was lost in the 1980's, the challenge for the 1990's is even greater.

ACKNOWLEDGMENTS

Sincere thanks are due to all contributors to the survey. Ronald Escano and Ralph Costa were particularly helpful. Other important contributors were: Alabama: Mark Bailey; Arkansas: Tom Foti, Larry Hedrick, Douglas James, John McLemore, Warren Montague, William Shepherd; Florida: Wilson Baker, James Cox, Roy Delotelle, Todd Engstrom, John Fitzpatrick, Jeff Hardesty, Bruce Hagedorn, Deborah Jansen, Nancy Joiner, Wayne Marion, Catherine NeSmith, Donald Wood; Georgia: Wilson Baker, Todd Engstrom, Dwight Harley; Kentucky: Floyd Gibbs, John MacGregor; Louisiana: Joe Hogan, Richard Martin, Emlyn Smith; Mississippi: Chris Frye, Mike Hurst, Jerome Jackson, Timothy Mersman, David Smith; North Carolina: Jay Carter, Jeffrey Walters; Oklahoma: John McLemore; South Carolina: John Cely, Glen Gaines, Robert Hooper; Tennessee: Paul Hamel; Texas: Richard Conner, Brent Ortego, Craig Rudolph; Virginia: Ruth Beck.

Red-cockaded Woodpecker Distribution and Status in Louisiana

Emlyn B. Smith, U.S. Forest Service, Catahoula Ranger District, 5325 Highway 8, Bently, Louisiana 71407-9725
Richard Martin, The Nature Conservancy of Louisiana, P.O. Box 4125, Baton Rouge, Louisiana 70821

ABSTRACT: Red-cockaded woodpeckers (*Picoides borealis*) are scattered from the northern parishes of Louisiana to the southern third of the state. The greatest concentration of active colonies (97%) is located in the central portion of the state in 5-6 separate populations. Public lands are the stronghold for the red-cockaded woodpecker in Louisiana and support over 90% of the known active colonies in the state. Kisatchie National Forest is the largest segment of public land in the state, and supports 340 active colonies. On private lands, red-cockaded woodpeckers are found in isolated colonies or small clusters of five or fewer colonies. The current distribution of red-cockaded woodpeckers in Louisiana is related to land use patterns, fragmented historic ownership and habitat alterations. Recovery of the red-cockaded woodpecker in Louisiana depends on cooperation and participation between public and private landowners. The greatest challenge lies with the U.S. Forest Service, which manages over 471,000 acres of suitable red-cockaded habitat with a hypothetical potential to support 1884 colonies. Currently, 92% of the remaining longleaf is on public land, 76% of which is on Kisatchie National Forest. Restoration of natural forest communities on public land offers the best hope for red-cockaded woodpecker recovery in Louisiana.
KEYWORDS: Louisiana, Kisatchie National Forest, private land, public land

When most people think of Louisiana, they conjure up images of wetlands, dark water bayous and coastal habitats, but half of the state is forested in upland pine (Calhoun 1992). There are two distinct upland pine regions in Louisiana, separated by the Mississippi River and its floodplain. The eastern area is referred to as the Florida Parishes and we will refer to the area west of the Mississippi river as Central Louisiana. There is strong evidence to suggest that longleaf pine (*P. palustris*), shortleaf pine (*P. echinata*), Oak (*Quercus* spp) and Hickory (*Carya* spp) were the forest types that dominated the pine forests in Louisiana prior to European settlement. Longleaf was predominant in the Florida Parishes, and the shortleaf-oak-hickory complex was equally abundant as longleaf in central Louisiana.

The Louisiana Natural Heritage Program (unpublished data) estimated pre-European forest cover after examining geology and soil data and consulting historic accounts of vegetative cover such as Lockett (1969). The Florida parishes were predominantly longleaf (1,512,105 acres) with a mix of shortleaf-oak-hickory (400,930 acres), slash (*P. elliottii*) (36,090 acres) and mixed hardwood - loblolly (*P. taeda*) (44,800 acres) yielding a total of close to two million acres of suitable red-cockaded woodpecker habitat. The central portion of the state comprised a much larger land base, the majority of upland pine sites were forested in shortleaf pine-oak-hickory (4,415,662 acres) and the rest in longleaf (3,785,010 acres).

Predictions concerning the red-cockaded population prior to European settlement can be derived from recovery plan density figures: one clan per 200 acres in longleaf and one clan per 300 acres for shortleaf pine oak hickory (Lennartz and Henry 1985). These density factors and the estimated composition of pre-

European forest can be used to estimate a theoretical historical red-cockaded population in Louisiana. In the Florida Parishes, a red-cockaded population of approximately 9,165 clans was possible. In central Louisiana, there may have been 33,643 red-cockaded clans prior to European settlement. Thus, these figures hypothesize a possible historical population of 42,808 clans statewide.

Historical records acknowledge the occurrence of red-cockaded woodpeckers (*Picoides borealis*) in Louisiana. Vernon Bailey reported in 1899 that the red-cockaded was the most common species of Picidae in Louisiana pine forests (Oberholser 1938). Beyer (1908) and Bent (1939) found red-cockaded woodpeckers extremely common in open longleaf pine forests of Louisiana during the first few decades of the 20th century. However, by 1938, Oberholser already considered the red-cockaded woodpecker to be a rare permanent resident. Information concerning the status of the species in Louisiana since the initial historical records is extremely limited. Prior to the late 1980s, data collected on the red-cockaded in Louisiana were restricted to incomplete information on the location and status of colonies on public land and even sketchier reports from private land.

CURRENT SITUATION

Of the 5.25 million acres of longleaf estimated to exist prior to European settlement, only 3 percent (171,325 acres) is still forested in longleaf (Vissage et al. 1992). Ninety-two percent of the remaining longleaf (156,790 acres), occurs on public land, seventy-six percent (118,443 acres) of which is on U.S. Forest Service land (unpublished U.S. Forest Service data). Because of difficulties in interpreting forest cover data, the current extent of shortleaf-oak-hickory is unclear; however shortleaf pine cover is now apparently less than 10% of pre-European conditions (Vissage et al. 1992)

Although suitable red-cockaded woodpecker habitat has declined markedly in the past 100 years, the pine forests of Louisiana support the fifth largest number of red-cockaded in the country (James 1995). All data of red-cockaded groups were either supplied by landowners, public land biologists' records or the Louisiana Natural Heritage Program. There are gaps in the data because some landowners are unwilling to release information. Most of the known gaps are in the northern portion of the state, which if disclosed, could have significant management implications for existing colonies in Arkansas.

PUBLIC LAND

At one time, suitable red-cockaded woodpecker habitat was distributed across the upland environments of the state. Today, suitable habitat in Louisiana is limited to isolated refugia, with the larger blocks owned and managed as public holdings. The largest of these are managed by the U.S. Forest Service, which administers a total of 601,650 acres in 6 disjunct blocks. The 6 units range from the 165,334-acre Winn Ranger District to the 32,000-acre Caney Ranger District, but only the 5 largest districts still support red-cockaded woodpeckers. Based upon a total of 471,000 acres of potential habitat, the 6 Kisatchie National Forest districts could support approximately 1,884 red-cockaded clans (**Table 1**). However, because of fragmented ownership patterns, current forest composition and age distribution, and the mandate for multiple-use management, the number of active colonies is far below the theoretical maximum. In 1992, the 5 districts of Kisatchie National Forest supported 340 active colonies.

The Vernon Ranger District is the smallest district with red-cockaded woodpeckers; however 63,820 of it's 85,113 acres are potentially suitable red-cockaded habitat. Fire has been an important feature of the historical and contemporary landscape for this parish, creating and sustaining near ideal red-cockaded habitat. The Vernon District currently supports 186 active and 56 inactive colonies, which compares favorably to our theoretical pre-European population of 255 active colonies (**Table 1**). The number of active colonies on the remaining districts that still support red-cockaded woodpeckers is clearly far below our estimates of pre-European population size. The Caney district contains only 3 inactive colonies and is not managed for red-cockaded.

Forestwide records indicate that the number of active red-cockaded colonies had increased on all districts between 1988 and 1992 except the Kisatchie which decreased twenty-one percent since 1988, and now supports only 59 active colonies. While data regarding the decline are incomplete, the 87 inactive colonies on the district suggest a downward trend in

Table 1. Estimated number of active red-cockaded colonies in Louisiana.

	PRE-EUROPEAN SETTLEMENT	1992	% OF PRE-EUROPEAN
FLORIDA PARISHES	9,165	13	<1%
CENTRAL LOUISIANA	33,643	451	1%
	FEDERAL LAND		
USDA FS			
Vernon R.D.	255	186	73%
Kisatchie R.D.	325	59	18%
Evangeline R.D	281	46	16%
Catahoula R.D.	387	31	8%
Winn R.D.	464	18	4%
DOD			
Ft. Polk	330	40	12%
Peason Ridge	165	23	14%
USFWS			
D'Arbonne	10	5	50%
	OTHER OWNERSHIPS		
Private and State	40,591	43	<1%

habitat suitability from earlier conditions.

Of the Kisatchie National Forests' 471,200 acres typed as pine and pine-hardwood, 25 percent is longleaf. The suitable habitat for the five districts supporting red-cockaded woodpeckers is (all except Caney) 436,200 acres, which could theoretically support 1744 clans. If the Forest Service trend towards "ecosystem management" continues, we could look forward to a significant percentage of the 277,144 acres of offsite loblolly being restored to longleaf. The representation of shortleaf-oak-hickory on the forest is less than 4 percent of the total forest area and should also increase if restoration forestry is emphasized. The 2 largest districts have the potential to support many more red-cockaded colonies than are currently present, but fire suppression and past timber management practices have altered the landscape and decreased habitat suitability. Intensive forest management will be required on all districts before red-cockaded numbers will increase significantly, and even approach our estimates of pre-European densities.

The manner in which colony data had been collected and tracked prior to 1988 does not lend itself to accurate interpretation of population trends. Now that inventories have been completed for each district, and monitoring programs established it will soon be possible to determine current trends in red-cockaded abundance on Forest Service lands in Louisiana. Although the number of active colonies known from most districts increased between 1988 and 1992, the increase is likely due to locating previously unknown active colonies.

The only other public lands in Louisiana with known active red-cockaded colonies are Fort Polk Military Reservation (two disjunct tracts), Fontainbleau State Park, Alexander State Forest and D'Arbonne National Wildlife Refuge. The Fort Polk reservation contains almost 66,000 acres of suitable red-cockaded habitat immediately adjacent to the Vernon Ranger District. As a block, these two federal holdings support 226 active colonies, which constitutes the largest red-cockaded population in the state. The military mission for the Fort causes only infrequent conflict with management responsibilities for red-cockaded recovery. The second area managed by Fort Polk is Peason Ridge Bombing Range, which consists of 33,000 acres adjoining the Kisatchie Ranger District. The 23 active colonies on Peason Ridge combined with those on the Kisatchie District increases the total number of active colonies in that population to 82.

The D'Arbonne National Wildlife Refuge in northeast Louisiana, Fontainbleau State Park in southeast Louisiana and Alexander State Forest in central Louisiana support about five active colonies each. The genetic exchange between these two sites and clans on adjacent private land is presently unknown since information about private land colonies is incomplete.

PRIVATE LAND

All recent data indicate that the trend in red-cockaded woodpecker abundance on private lands in Louisiana is downward and that local extirpations continue (unpublished Louisiana Natural Heritage data). In 1992, it was estimated that there were about 40 active colonies on private lands, or less than 10 percent of the known Louisiana population. About 4.4 million acres of Louisiana's 5 million acres of pine-dominated timberland were owned by private individuals or timber companies (Vissage 1992). Thus, 90 percent of the known active RCW colonies are located on 10 percent of the potentially suitable habitat in Louisiana.

There are only 3 tracts of private land in Louisiana that support significant numbers of active colonies. Those "populations" consist of 5-30 active colonies and are more-or-less managed for red-cockaded woodpeckers; one landowner has recently submitted a species protection plan to the U.S. Fish and Wildlife Service. Most of the remaining Louisiana colonies on private land are isolated from other active groups and are surrounded by fragmented forest lands that are, at best, only minimally managed for red-cockaded. However, several colonies exist near Kisatchie National Forest boundaries and should receive immediate attention.

Although red-cockaded woodpeckers are not faring well on Louisiana private lands, management of private lands is instrumental for achieving the 1985 recovery plan goals of establishing a viable population in the state. To date, there has been little concerted effort toward management of private timberlands for red-cockaded woodpeckers, nor have many private landowners participated in attempts to assess the true status of the species in this state. For numerous reasons, landowners are reluctant to release information pertaining to the existence of active colonies on their land.

Recent discussions about using isolated clans to provide birds for augmentation on public lands have increased dialogue between agencies responsible for red-cockaded woodpecker management and private landowners. However, this dialogue is only likely to continue if we assure landowners that releasing information will help them. It is imperative that we encourage private landowners to participate in management and recovery efforts because the true status and distribution of red-cockaded woodpeckers in Louisiana will never be known without their assistance. Further, the observed decline on private lands will likely continue unless landowners are granted financial incentive for colony maintenance. If such incentives are not forthcoming, augmentation and relocation programs using birds produced on private lands should be pursued.

FUTURE OUTLOOK AND RECOMMENDATIONS

Red-cockaded woodpeckers have been eliminated from over 90 percent of their historic range in Louisiana. Statewide distribution continues to shrink and populations continue to decrease, thus extirpation on most, if not all private land in the state is probable. We still have time to incorporate colonies on private land into restoration programs; however, quick action is required if we hope to do more than document the continued decline and ultimate extirpation of red-cockaded outside of federal lands in Louisiana.

The pattern of red-cockaded woodpecker distribution in Louisiana is similar to that across the entire region. The majority of suitable habitat is on public land, therefore the future of the red-cockaded in Louisiana as well as across the range relies on public land management. In Louisiana the greatest burden and challenge lies with the largest landowner - the U.S. Forest Service. Although it is possible to encourage growth of our current population to a level much more suited to recovery of the species, it is unrealistic to think the state could ever support the same number of red-cockaded woodpeckers it did prior to European settlement and development. Habitat restoration, including intensive prescribed burning programs, is the primary means to insure the survival and recovery of the red-cockaded woodpecker on public land in Louisiana.

Protection and management of corridors between red-cockaded colonies on Forest Service lands and colonies on military land should be a priority. All available tools should be used to bring these critical areas under long term protection, including acquisition, stewardship, management agreements and land exchanges. Secondly, we must complete comprehensive red-cockaded woodpecker surveys of private lands adjacent to federal tracts and all large blocks of suitable habitat on private land; active colonies adjacent to federal land should be incorporated into the public land management plans. Third, methods of linking the various Kisatchie National Forest districts should be explored. Although Stangel et al (1992) found "normal" heterozygosity levels for 3 Kisatchie National Forest districts, encouraging gene flow among districts should be a priority because the districts will only become more isolated over time. Fourth, management of the 5 Kisatchie districts that support red-cockaded should continue toward landscape management for multiple resources with emphasis on recovery of the listed species. Finally, we must work to establish incentives for private landowners that encourage compliance with red-cockaded woodpecker management guidelines.

ACKNOWLEDGEMENTS

We are grateful to many individuals with the Kisatchie National Forest, including those that assisted with data collection, and provided access to data. Appreciation is extended to managers at Ft. Polk and D'arbonne National Wildlife Refuge for their disclosure of information. Very special thanks to all private landowners who contributed to our knowledge and the Louisiana Natural Heritage Program for collecting and compiling data on private land. Special thanks to Latimore Smith for his help estimating pre-European settlement forest conditions in Louisiana and to Alan Dorian for his thoughtful review of earlier draft manuscripts.

Status, Distribution, and Conservation of the Red-cockaded Woodpecker in Florida: a 1992 Update.

James Cox, Florida Game and Fresh Water Fish Commission
Tallahassee, Florida 32399-1600
W. Wilson Baker, The Nature Conservancy, Tallahassee, Florida 32301
Don Wood, Florida Game and Fresh Water Fish Commission
Tallahassee, Florida 32399-1600

ABSTRACT: Data gathered from a variety of sources were used to estimate the status and distribution of red-cockaded woodpeckers (*Picoides borealis*) in Florida. Locations for 1150 active sites and 690 inactive sites were obtained and processed using geographic information systems technology. Other data sets were also referenced to identify additional areas with potentially active sites. Most (80%) active sites occurred on federal lands, but several significant populations also occurred on state-owned lands. Although private lands support fewer than 7% of the known active sites, additional active sites likely occur on private land holdings in south and central Florida. Conservation efforts need to focus on habitat management on existing public lands and acquisition of private lands that support larger populations.
KEYWORDS: Apalachicola National Forest, Avon Park Air Force Range, Big Cypress National Preserve, Blackwater State Forest, Corbett Wildlife Management Area, corridor, dispersal, distribution, Eglin Air Force Base, Fakahatchee Strand State Preserve, Florida, Florida Natural Areas Inventory, geographic information system, hurricane, Ocala National Forest, Osceola National Forest, private land, rotation, Three Lakes Wildlife Management Area, Webb Wildlife Management Area, Withlacoochee State Forest.

Surveys for red-cockaded woodpeckers (*Picoides borealis*) (RCW) have been initiated on several public lands in Florida since the statewide status reports of Baker et al. (1980) and Wood and Wenner (1983). Results from Florida's breeding bird atlas project (Kale et al. 1992) and surveys of proposed development projects have also produced records of RCW from new areas. We consolidated information collected by these various sources to provide an update on the recent status of the species in Florida. Data were compiled using geographic information systems technology (GIS), which facilitated additional analyses useful in developing conservation strategies for this species. The computerized database will also enable future status reports to be prepared with greater ease and accuracy.

METHODS

Requests for **(a)** maps of sites where RCW had been recorded recently, or **(b)** geographically referenced coordinates for sites where RCW had been recorded, were distributed to land managers, research scientists, and environmental consultants in Florida. Sites were defined as clusters of cavity trees or individual trees separated from other clusters. An "active" site was broadly defined using the criteria of Jackson (1977b) and Hooper et al. (1980). We did not ask respondents to determine whether active sites supported breeding pairs or successfully produced young. The status of the species as reported here thus should be viewed primarily as a detailed description of the distribution of recently active sites from which some coarse estimate of population size might be made. Jackson (1977b) and James (1991) discussed additional caveats regarding the use of "active" sites in estimating population status.

Florida Department of Transportation county road maps (1:126,720) and 7.5-minute topographic maps (1:24,000) were used by most respondents who plotted active sites by hand. We digitized active sites from these maps, and although there is variation in the scale of the maps, we feel the accuracy is sufficiently precise for the statewide analyses performed. A listing of data sources is provided in **Appendix 1**.

Two additional sources of information were referenced that differed substantially from the data described above. The first source was Florida's breeding bird atlas project (Kale et al. 1992). From 1984-1990, volunteer bird watchers collected information on breeding birds in well defined areas of Florida. The areas surveyed corresponded to one-sixth of a 7.5-minute topographic quadrangle map, an area of approximately 3000 ha. Within these blocks, breeding status was categorized as "possible," "probable," or "confirmed" based on behavioral observations.

The second source consisted of data processed by the Florida Natural Areas Inventory (FNAI). FNAI maintains records of RCW collected as part of numerous surveys. We accessed the FNAI records and determined activity through site visits, comparison with breeding bird atlas records, and consultation with local biologists.

We also reviewed breeding bird survey data (see Robbins et al. 1986) for additional RCW records. However, we found no information that was not represented in the previously described data sets.

RESULTS AND DISCUSSION

The distribution of active sites **(Figure 1)** is primarily restricted to public lands in Florida. Fewer than 7% of the known active sites occurred on private lands, and most populations on private lands were relatively small. **Table 1** presents our estimated number of active and inactive sites within 4 regions of Florida. The 1150 active sites represent a slight increase over the 1139 active sites reported by Wood and Wenner (1983) and 943 active sites reported by Baker et al. (1980). This increase in numbers undoubtedly reflects the discovery of previously unknown sites rather than an increase in population size. For example, approximately 40 sites have been located in southwestern Florida since the publication of previous status reports. These sites were most likely active during the period covered by earlier reports.

RCW records from the breeding bird atlas **(Figure 2)** closely mirror sites shown in **Figure 1**. Even though breeding bird atlas records were less precise than the maps of active sites we received, atlas records included some potentially active sites not identified in other data sets. Atlas records denoted by the letters A-F **(Figure 2)** represent unique locations not included in the FNAI database nor in our survey of biologists. Additional work is needed to pinpoint the locations of these potentially active sites. In addition, several FNAI records that had not been updated by site visits within the last 20 years occurred in atlas blocks where RCW were recorded. This result indicates that these FNAI records were likely active within the last 5-7 years.

Not surprisingly, Florida's largest 2 populations occurred on the Apalachicola National Forest and Eglin Air Force Base. Although these populations were quite large and ostensibly secure, recent studies suggested both populations were declining (James 1991, Hardesty 1995). Given that 70% of the active sites reported for Florida occur on these 2 areas, these apparent declines are alarming. Two other federally managed areas, Big Cypress National Preserve and Osceola National Forest, had sizeable populations with > 30 active sites. In addition, Avon Park Air Force Range had approximately 20 active sites with at least 8 active sites known to occur on private lands to the north (see below).

Although most (80%) active sites occurred on federally managed lands, state-owned lands in Florida also supported some significant populations. The 3 largest such populations were on Corbett Wildlife Management Area, Webb Wildlife Management Area, and Blackwater State Forest. These and other smaller populations help to maintain a broad distribution of protected populations, which is essential to safeguarding against catastrophic events such as hurricanes (Cely 1995). However, no state-owned area contains more than 25 active sites, and these small populations may face relatively high chances of extinction over the next several decades without proper habitat management (see Baker 1983b, Cox et al. 1995).

Many public areas have the capacity to support much larger populations if better habitat conditions were maintained. Engstrom and Baker (1995) described a population of RCW consisting of approximately 170 active sites and occupying about 12,350 ha. Timber

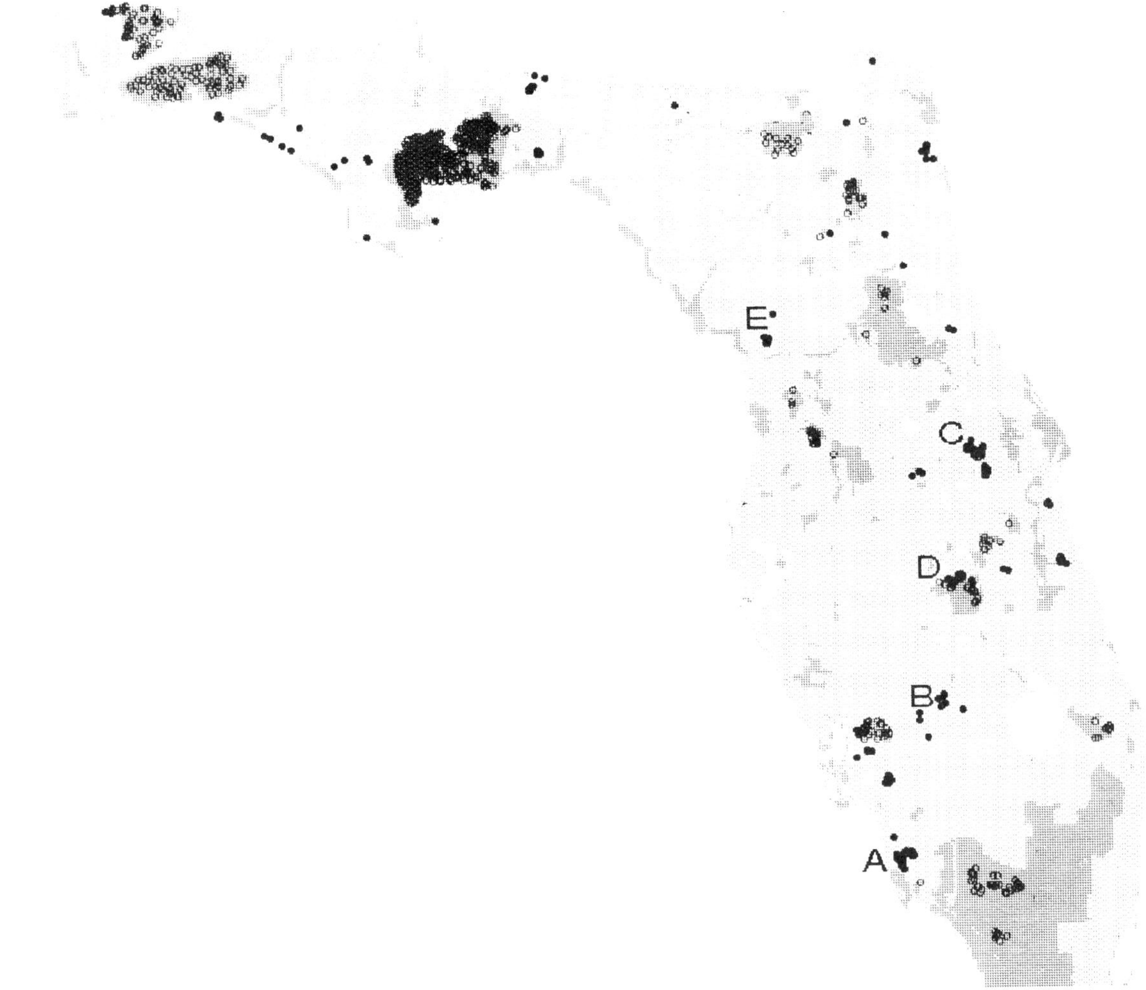

Figure 1. Distribution of reported active sites of red-cockaded woodpeckers in Florida. Open circles represent active sites on public lands; closed circles are active sites on private lands. Dark gray areas represent existing conservation areas. Lettered areas are referenced in the text.

production and recreational hunting were carried out over most of that area. Public lands twice the size of the area described by Engstrom and Baker (1994), or larger, should be capable of supporting large RCW populations and still allow timber production and other multiple-use requirements. Blackwater State Forest, Osceola National Forest, Camp Blanding Military Reserve, Corbett and Webb wildlife management areas, and Citrus and Croom tracts of the Withlacoochee State Forest represent large public landholdings that support relatively small woodpecker populations. The number of inactive sites reported for those areas indicates room for growth, and the security of RCW in Florida would be enhanced significantly by establishing more appropriate habitat conditions on those areas.

Private lands supported at least 106 active sites, which approximated the estimate of Baker et al. (1980). The large number of active sites recently found on private lands in southwestern and central Florida **(Table 1)** suggests that more sites might be located if greater access to private lands were allowed. Most of the recent records from these areas resulted from surveys required for proposed development projects. More complete RCW surveys on private lands are essential to developing conservation strategies for this species and assessing the effects of proposed land-use changes and development projects in this region.

One of the largest known clusters of active sites on private lands in Florida was west of the

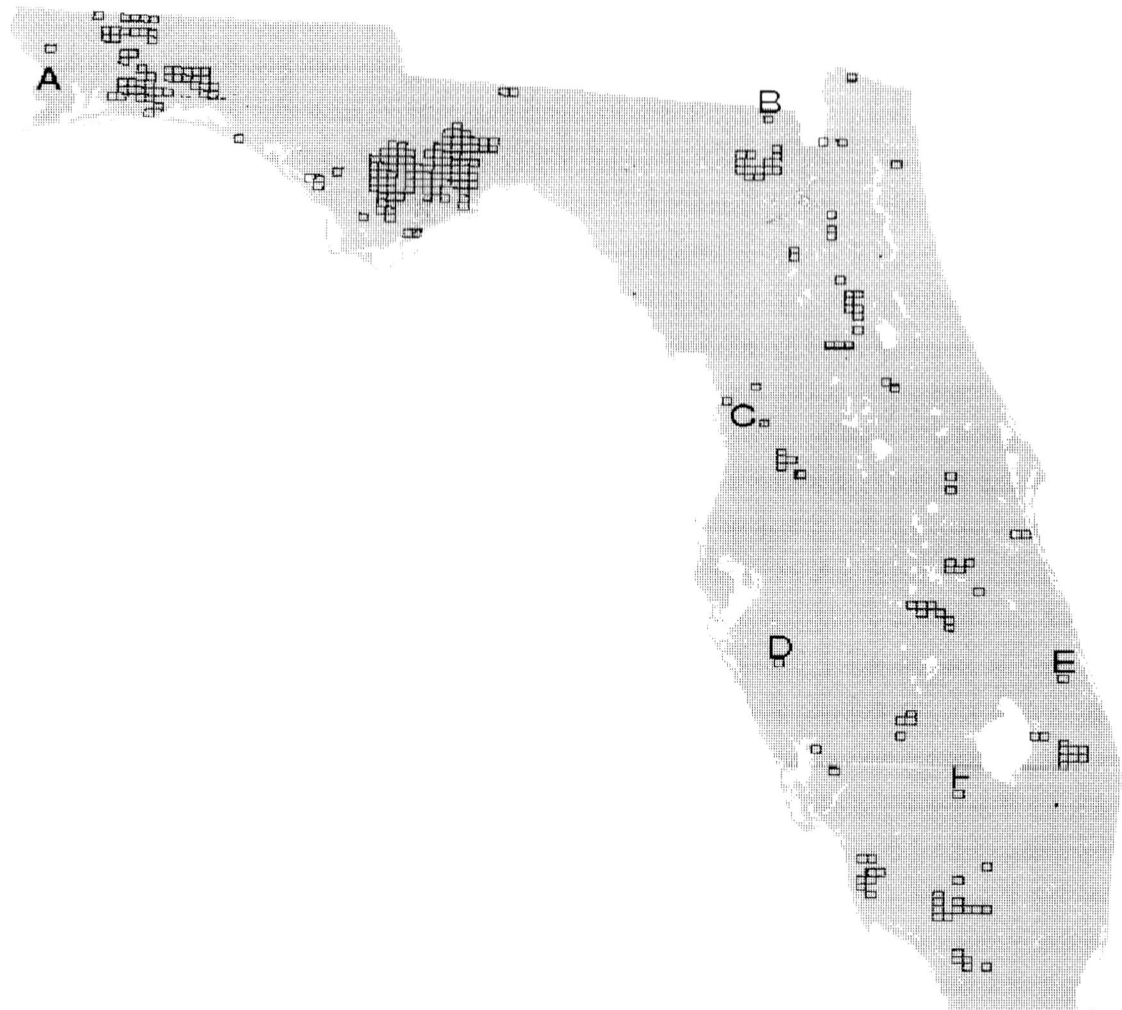

Figure 2. Distribution (dark rectangles) of breeding bird atlas records (Kale et al. 1992) for red-cockaded woodpeckers. Lettered areas indicate unique records of red-cockaded woodpeckers that warrant more thorough investigation.

Big Cypress National Preserve and Fakahatchee State Preserve in Collier County (**Figure 1, area A**). The 14 active sites located here are > 50 km from the population on BCNP, which means that frequent interchange is unlikely. However, this isolated group is sufficiently large to sustain itself for many decades with appropriate habitat management (Cox et al. 1994). Furthermore, occasional translocations from nearby populations could ameliorate other threats potentially facing this small population (Chesser 1981, Simberloff et al. 1992). An effort to acquire a portion of this area was undertaken by the Florida Conservation and Recreational Lands Committee, but local opposition surfaced because state acquisition would reduce ad valorem taxes in the county (D. Bailey, pers. comm.).

Another region with several unprotected active sites occurred in Lee, Glades, and Highlands counties in southwestern Florida (**Figure 1, area B**). Protection of areas south and southwest of Webb Wildlife Management Area (the large public area shown near B) could increase the size of the protected population in this region to approximately 40 active sites. The distance between the Webb Wildlife Management Area and the cluster of active sites on private lands is 10 km, and aerial photography indicates that much of the intervening area consists of apparently suitable habitat. The protection of additional habitat in this area could create a fifth protected population in the state consisting of > 30 active sites. Active sites on private lands in Orange

Table 1. Estimated number of active and inactive red-cockaded woodpecker sites in Florida. Atlas records are reports in Kale et al. (1992) that have not been confirmed. Acronyms are NWR National wildlife refuge; SP State park; and WMA Wildlife management area.

Region	Reported Active	Reported Inactive
FLORIDA PANHANDLE		
Public Lands		
Eglin Air Force Base	208	180
Blackwater State Forest	29	25
Apalachicola National Forest	590	212
St. Marks NWR/Ochlockonee SP	7	1
Subtotal	834	418
Private Lands		
Bay County	3	?
Escambia County	Atlas Record	
Franklin County	1	?
	Atlas Record	
Gulf County	3	?
Leon County	3	18
Wakulla County	6	12
Walton County	3	13
Subtotal	19	43
NORTHERN PENINSULA		
Public Lands		
Osceola National Forest	43	89
Ocala National Forest	8	54
Camp Blanding Military Reserve	15	8
Austin Cary Memorial Forest	1	1
Goethe State Forest	6	3
Subtotal	73	155
Private Lands		
Alachua County	2	?
Duval County	5	2
Hamilton County	1	?
Putnam County	Atlas record	
Volusia County	2	?
Nassau County	Atlas record	
Subtotal	10	2
CENTRAL PENINSULA		
Public Lands		
Avon Park Air Force Range	21	13
Bull Creek WMA	1	?
Chinsegut Nature Center	2	?
Kicco WMA	1?	?
Three Lakes WMA	15	?
Withlacoochee State Forest		
Citrus Tract	4	?
Croom Tract	15?	15?
Richloam Tract	Atlas record	
Subtotal	59	28
Private Lands		
Brevard County	7	2
Highlands County	5-7	?
Orange County	24	17
Polk County	8	10?
St. Lucie County	Atlas record	
Subtotal	44-46	19

Table 1 continued

Region	Reported Active	Reported Inactive
SOUTHERN PENINSULA		
Public Lands		
Webb WMA	17	?
Corbett WMA	23	5
Big Cypress National Preserve*	38	?
Subtotal	8	5
Private Lands		
Charlotte County	7	?
Collier County	14	10
Glades County	6	?
Hendry County	Atlas record	
Lee County	6	?
Subtotal	33	10
Total	1142-50	690

***prior to the passage of Hurricane Andrew**

County (**Figure 1, area C**) also represent a large population that has good chances of long-term persistence. Conservation of habitat in this region could establish a sixth protected population > 30 active sites and further enhance the geographic distribution of large, protected populations.

A third important cluster of active sites exists on private lands north of Avon Park Air Force Range (**Figure 1, area D**). At least 8 active sites have been located here (B. Progulske, pers. comm.) and more likely occur in the area. This area is extremely important for at least two reasons in addition to the presence of RCW. First, conservation of habitat on private lands would significantly enlarge the size of the protected population in the area and create a seventh protected population consisting of > 30 active sites. Second, conservation of habitat north of Avon Park may also provide greater continuity with the RCW population located on Webb Wildlife Management Area just northeast of Avon Park Air Force Range. The boundaries of Avon Park Air Force Range and Webb Wildlife Management Area are separated by approximately 14 km of private lands. Conservation of habitat north of Avon Park could bring the two management areas to within 5-8 km of each other, which could enable more frequent exchanges to occur between populations (see below).

Some private lands that formerly supported RCW apparently no longer do. We found no evidence of activity for formerly active sites in Union, Jefferson, and Hardee counties, and the records for private lands in Bay and Volusia counties in Wood and Wenner (1983) apparently have declined to only a few pairs or lone individuals. Newly discovered active sites in Levy County (**Figure 1, area E**) were recently acquired through Florida's Conservation and Recreational Land program, but this area is now threatened by a proposed high-speed road project (T. Gilbert, pers. comm.). The prospects for active sites on other private lands are not good. During the preparation of this report we received reports of timber operations on at least 3 private areas containing active sites and not discussed above.

The compilation of site records using GIS allows for some additional exploratory analyses. For example, the establishment of corridors and "landscape linkages" has become a strategy in some conservation efforts (Simberloff et al. 1992). However, the effectiveness of such linkages is affected by several chance events that become increasingly improbable over larger distances. For corridors to be effective for RCW, a dispersing bird must find and use a corridor, survive while traversing the corridor, find and establish a territory in the new conservation area, acquire a mate, and then successfully reproduce young. The progeny must then also survive and reproduce. In addition, such an event may need to occur fairly frequently to influence some problems facing small populations (Lande and Barrowclough 1987). The benefits of interchange also critically depend on the size of the populations in question. For example, no amount of interchange can totally alleviate inbreeding if the populations exchanging individuals are themselves small (Chesser 1981).

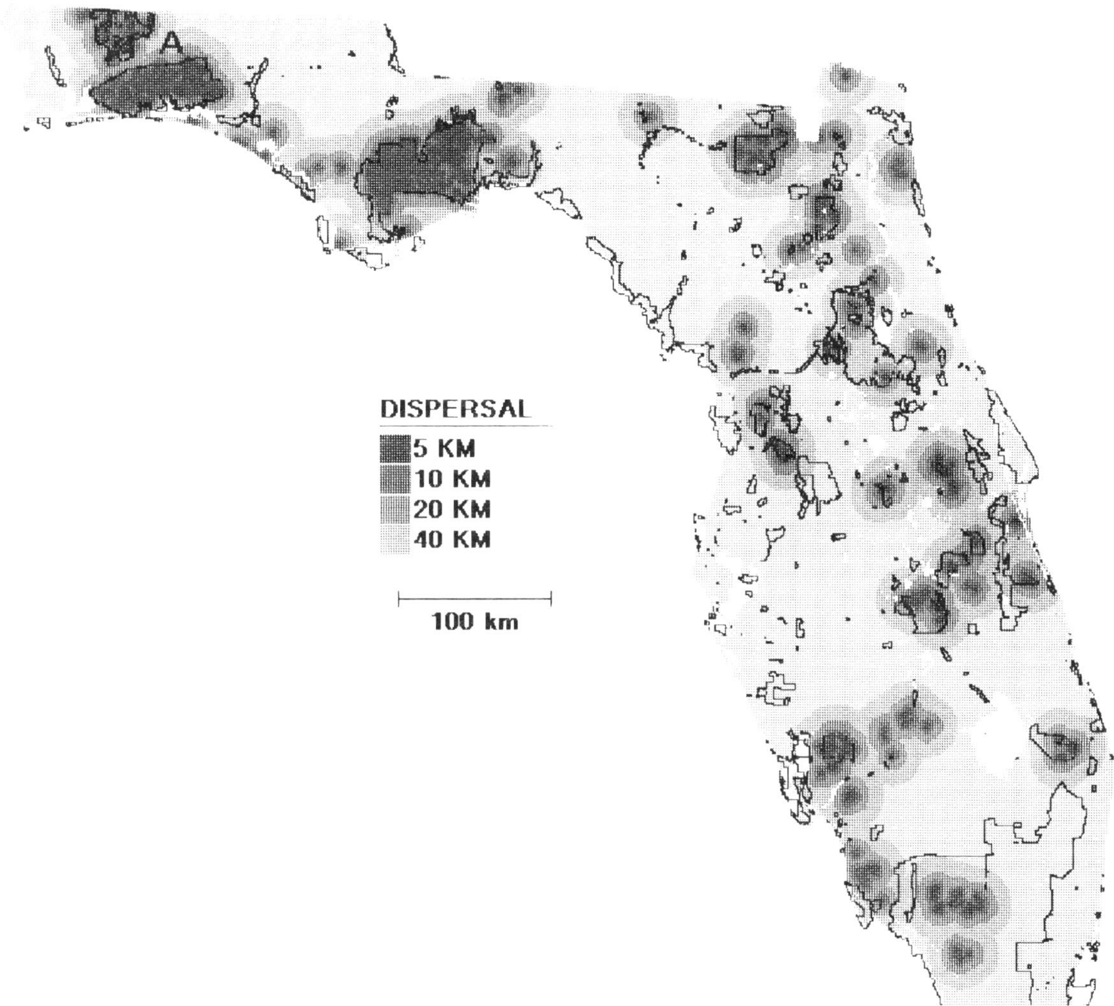

Figure 3. Potential dispersal distances around known active sites. Existing conservation areas are shown as dark lines. The lettered area indicates two large populations on existing conservation areas that appear to be sufficiently close to allow frequent exchanges to occur.

Based on Walter's (1989) estimate of RCW dispersal, we created buffers of 2.5, 5, 10, and 20 km around each active site **(Figure 3)**. We consider these intervals representative of **(1)** frequent, **(2)** infrequent, **(3)** very infrequent, and **(4)** accidental dispersal distances. Note that 2 sites that occurred within 40 km of each other would appear to be linked using the maximum distance interval, while 2 sites that occurred within 5 km would appear to be linked by the shortest dispersal interval used.

The Eglin Air Force Base and Blackwater State Forest **(Figure 3, area A)** populations are the only ones on existing conservation areas that lie close enough to each other for frequent exchanges to occur. A portion of the area between these public lands was proposed to the Florida Conservation and Recreation Lands Committee for acquisition (Anon. 1991). Appropriate habitat conditions need to be restored in the intervening area since it consists largely of short-rotation, commercial-timber land uses (Simberloff et al. 1992). In addition, better management of habitat within existing conservation areas is needed since the current populations are either small or declining (Hardesty 1994). Neither population may currently produce a sufficient number of dispersing birds for frequent interchanges. When contrasted with the area described by Engstrom and Baker (1994), Eglin and Blackwater are certainly large enough (ca. 140,000 ha and 70,000 ha, respectively) to sustain much larger RCW populations. The number of inactive sites reported for both Eglin and Blackwater also suggests much larger populations in the recent past.

Linking other populations on existing public lands shown in **Figure 3** will require infrequent or accidental dispersal events, unless the corridors provide nesting and foraging habitat. Unfortunately, most of the habitat in the intervening areas shown in **Figure 3** is not suitable. Exceptions to this may be some of the private lands in southwest and east central Florida.

Although corridors may have some value in RCW conservation efforts, the security of RCW populations in Florida would probably be most enhanced by other activities. First, greater attention must be given to habitat management on public lands. The Osceola and Ocala national forests, Withlacoochee and Blackwater state forests, and Three Lakes, Webb, and Corbett wildlife management areas appear to have the capacity to support much larger RCW populations, while declines reported for populations on Eglin Air Force Base and Apalachicola National Forest need to be addressed immediately. A second priority is conservation of habitat supporting populations on private lands in central and southwest Florida. Maintenance of these distinctive populations will increase the size of some of the smaller of Florida's manageable populations. Habitat conservation in these areas will also broaden the geographic distribution of manageable populations and lessen the chances of RCW populations becoming extinct as a result of catastrophic events such as hurricanes and disease epidemics.

APPENDIX 1.

Locations of active and inactive sites in the Apalachicola National Forest were digitized from a 1:126,720 forest service map.

Geographically referenced locations of active and inactive sites in Blackwater State Forest were obtained from D. Hardin.

Locations of active and inactive colonies in Camp Blanding Military Reserve were digitized from 1:24,000 maps prepared by J. Garrison.

Geographically referenced locations of active sites in Avon Park Air Force Range were obtained from B. Progulske and R. Bowman.

Geographically referenced locations of active sites in Big Cypress National Preserves were obtained from D. Jansen.

Active and inactive sites on private lands in northeast Florida were digitized from 1:126,720 maps prepared by M. Allen.

Active and inactive sites on private lands in central Florida were digitized from 1:126,720 maps prepared by R. DeLotelle.

Geographically referenced active and inactive sites on the Citrus Tract, Withlacoochee State Forest, were obtained from C. Smith.

Active and inactive sites on private lands in southwest Florida were obtained from K. Dryden and J. Beever.

Active and inactive sites in Walton County were derived from field surveys conducted by the Florida Game and Fresh Water Fish Commission.

The Distribution and Status of the Red-cockaded Woodpecker (*Picoides borealis*) in Georgia, 1992

W. Wilson Baker, The Nature Conservancy
625 N. Adams St., Tallahassee, FL 32301

ABSTRACT: In 1992, 891 red-cockaded woodpecker (*Picoides borealis*) clusters (649 considered active) were listed in 35 counties of Georgia. There are three population centers in Georgia which comprise 80 percent of the state-wide number of active clusters. Georgia differs from many states in that the major population centers are on Department of Defense lands, and one on private lands. Thirty percent of the state's clusters are on private lands. Some of the private tracts of land are valuable high quality longleaf pine forest ecosystems, a very diverse and endangered biological system.
KEYWORDS: red-cockaded woodpecker, *Picoides borealis*, Georgia, distribution, status, longleaf pine

Georgia has such a large forested land base in the coastal plain that it should be an important center for red-cockaded woodpeckers (*Picoides borealis* Vieillot). Georgia is also centrally located in the range of this southern pine forest specialist.

In order to ascertain the distribution and status of this bird in Georgia, I compiled a state-wide survey on data collected from 1966-1980. In 1979-1980 I resurveyed some of the clusters known previously (Baker 1982). Approximately a decade later Baker and Thompson (1990) conducted an updated survey. This paper is a revised and updated version of the 1990 unpublished report.

METHODS

An attempt was made to revisit all red-cockaded woodpecker clusters on private lands that were known to exist at the end of the 1980 study. All but three clusters were revisited and a more complete inventory of the Red Hills Plantation area was initiated. All red-cockaded woodpecker clusters reported on private land since the 1980 study were visited. On public lands we relied on the natural resource personnel with those agencies to supply us with updated information. The determination of clusters as active versus inactive was determined by using the now standard methods (Jackson 1977b, Hooper et al. 1980).

On private land I excluded clusters in the inactive status if the cavity trees stayed in that category for ten or more years. I didn't follow this policy on public lands because some still lack a complete inventory and individual cluster status is often unknown or based on old data.

RESULTS AND DISCUSSION

Comparison with 1980 Survey

The apparent increase in total number of clusters (active and inactive) from 1980 to 1992 (674 to 891 clusters) is mainly a reflection of updated and improved inventories of the three red-cockaded woodpecker population centers: Ft. Benning, Ft. Stewart and Red Hills Plantations. In fact, declines on public land have occurred at Ft. Gordon and Piedmont areas (U.S.F.S. Memo and S. Willard, T. Johnson, and R. Brooks, Pers. comm.).

Twenty-one of 54 counties that had active or inactive clusters on private land in 1980 were not included in this summary because the clusters were inactive for ten or more years. Clusters in 2 of the 54 counties listed (Dodge and Hancock) were actually destroyed during the original study period. Since 1980, clusters have been found in two new counties: Baker and Ben Hill. **(Figure 1)**.

There has been a steady decrease in active clusters on private lands other than a few areas such as the Red Hills Plantations. Seventy-two percent of the private clusters, excluding the Red Hills, went from active to inactive in the last decade. The decline is the result of a combination of: an elimination of present and future cavity trees, inadequate foraging habitat, inadequate burning, and population fragmentation.

1992 Survey

A total of 891 clusters were located in 35 counties, all in the coastal plain or lower Piedmont regions. (**Figure 1 and Table 1**). There are three larger population centers: Ft. Benning, Ft. Stewart and Red Hills Plantations. Two smaller population aggregations occur; the Piedmont and Okefenokee National Wildlife Refuge. (**Figure 1 and Table 1**). Isolated clusters on private lands are scattered throughout the state, mostly in the coastal plain. Sixty-six percent of all active clusters known in Georgia are on public land; 53% being on two military bases: Ft. Benning and Ft. Stewart. As mentioned previously, (Baker 1982) Department of Defense lands provide a very important red-cockaded natural resource base in Georgia (80% of public lands active clusters).

A significant difference between Georgia and most other states is that one of the red-cockaded woodpecker population centers is on private lands. The Red Hills Plantation area where red-cockaded woodpecker clusters occur are mostly tracts under approximately twenty ownerships. This population is discussed in more detail by Engstrom and Baker (1995). During the 1980 survey, I lacked access to a number of these properties and underestimated the number of clans.

No red-cockaded woodpeckers are known to occur near the Georgia coast, or on the coastal islands. In 1981, in a relocation experiment, twelve birds were moved from Ft. Stewart to a barrier island approximately 35 km to the southeast (Odom et al. 1982). By 1983, only two birds were accounted for: one still at the St. Catherines Island release site and one approximately 16 km south of the release site on Sapelo Island (Odom 1983). Based on annually conducted Christmas Bird Counts, the last sighting of the red-cockaded woodpecker on St. Catherines Island was 15 Dec 1990 and the last sighting on Sapelo Island was 31 Dec 1988 (Urban 1991, Dopson 1989). This means that the woodpeckers were at least 10 and 8 years old respectively.

As I mentioned in the 1982 report, the longleaf pine ecosystem is very important to red-cockaded woodpeckers. The loss of quality longleaf forest sites noticed during the 1992 resurvey was very disturbing. During this survey work, 35 tracts in 15 counties and under 32 ownerships contained acreage of what I considered quality longleaf pine forest community. Some of these parcels are extensive enough to be very important and the best of what's left in Georgia and the southeast.

RECOMMENDATIONS

1) Thorough inventories should be completed on public lands and followed by periodic systematic monitoring. We cannot adequately assess the importance of these properties for red-cockaded woodpecker recovery without complete and accurate inventories.

2) Land managers of public lands, especially Ft. Benning and Ft. Stewart, should initiate an aggressive prescribed burning program. The inadequate intensity and frequency of fires on public lands is steadily degrading the pine forest community. In fact, the natural longleaf pine ecosystem is endangered partly because of the lack of frequent fires.

3) Immediate attention needs to be given red-cockaded woodpecker groups adjacent to public land populations. This is especially true for land adjacent to Okefenokee NWR. There could be management plans and cooperative agreements developed between Osceola National Forest in Florida and the Okefenokee National Wildlife Refuge, Dixon Memorial State Forest, Laura Walker State Park in Georgia and selected privately owned tracts similar to the one developed between Piedmont NWR, Hitchiti Experimental Forest and Oconee National Forest. There are still red-cockaded woodpecker clusters and adequate habitat surrounding these public lands. Cooperative agreements should include incentives for private landowners to manage for the benefit of the woodpeckers.

4) Develop a statewide plan for the management of isolated red-cockaded woodpecker clusters on private lands.

5) Recognize and help in the conservation of the natural resources of the unique Red Hills Plantation region. This is one of the largest red-cockaded woodpecker population centers in

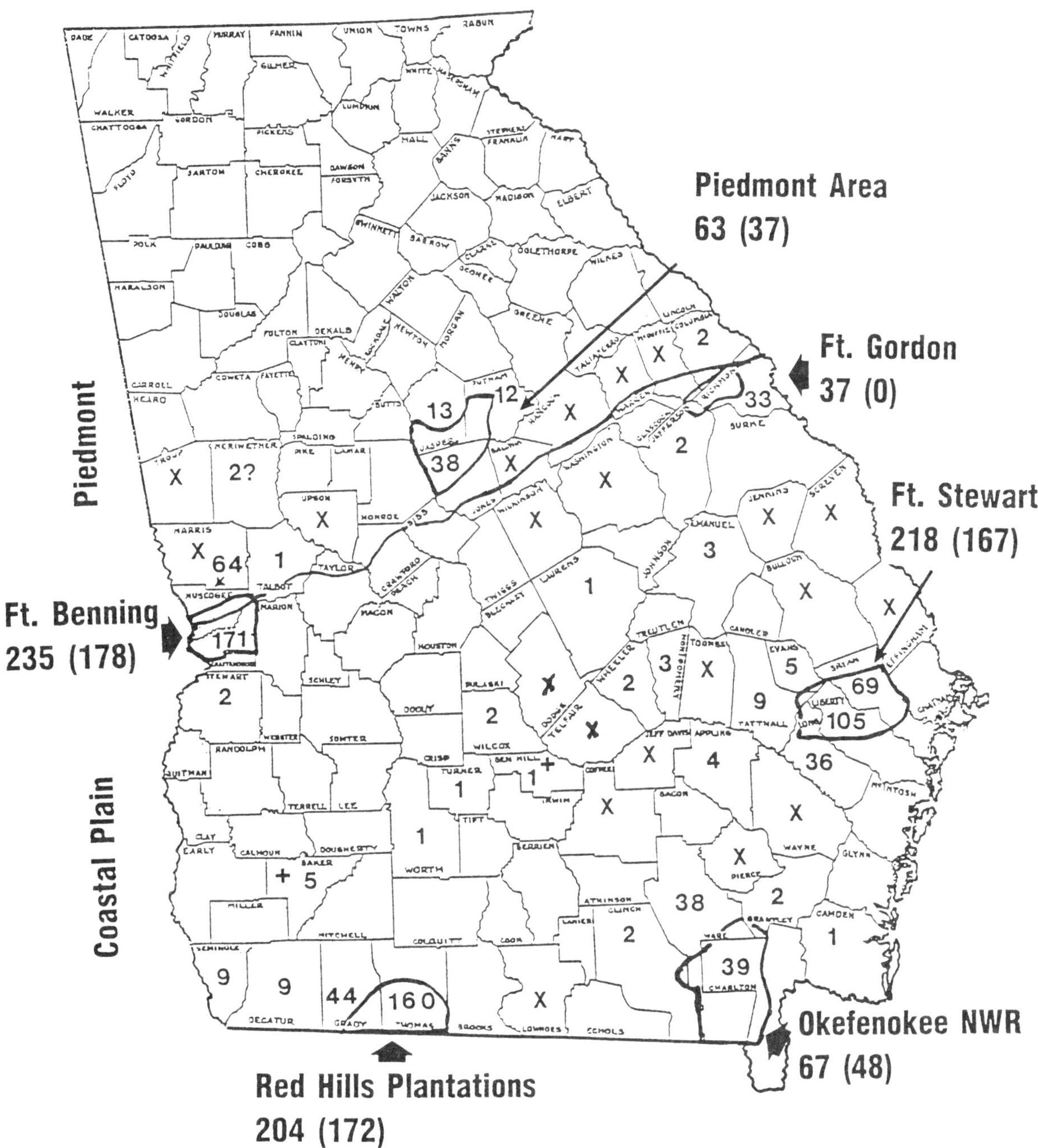

Figure 1. Red-cockaded woodpecker clusters in Georgia: 1992. Numbers represent total clusters with active clusters in parentheses. Population centers are indicated. Counties that have clusters that have been inactive for more than 10 years are indicated by an 'X'. Counties where clusters have been discovered since 1980 are indicated by a '+'.

Table 1. Status of red-cockaded woodpecker clusters in Georgia: 1992

County	No. of clusters*	Status of clusters	
		Active	Inactive
Appling	4	2	2
Baker	5	2	3
Ben Hill	1	1	0
Brantley	2	2	0
Bryan	69	51	18
Camden	1	1	0
Charlton	39	27	12
Chattahoochee	171	124	47
Clinch	2	0	2
Columbia	2	0	2
Decatur	9	6	3
Emanuel	3	2	1
Evans	5	3	2
Grady	44	38	6
Jasper	13	9	4
Jefferson	2	0	2
Jones	38	27	11
Laurens	1	1	0
Liberty	105	75	30
Long	36	32	4
Meriwether	2 ?	2 ?	0
Montgomery	3	3	0
Muscogee	64	54	10
Putnam	12	1	11
Richmond	33	0	33
Seminole	9	6	3
Stewart	2	0	2
Talbot	1	1	0
Tattnall	9	9	0
Thomas	160	134	26
Turner	1	1	0
Ware	38	31	7
Wheeler	2	1	1
Wilcox	2	2	0
Worth	1	1	0
TOTAL Counties 35	**891**	**649**	**242**

*** Clusters considered abandoned or destroyed were not counted**

Georgia and the largest private land population in the Southeast. Under its present management it is also probably the most viable population in the state. The Red Hills Conservation Association has been formed to help protect all the special qualities of the Red Hills region. Red-cockaded woodpecker clusters under the same management patterns continue south into Leon and Jefferson counties, Florida.

6) Many privately owned pinelands with Red-cockaded woodpeckers are also tracts of high importance for regional biological diversity. The red-cockaded woodpecker is just one inhabitant of this diverse system. Longleaf pine forests with high quality natural understory vegetation are rare, especially in large tracts. Georgia is fortunate in having some very high quality sites, some even with old growth longleaf pine. Every effort should be made to encourage landowners to protect and manage such areas for their total biotic diversity. Indeed some of these tracts are of national significance, the best that we have left.

ACKNOWLEDGEMENTS

Most of the survey work on private lands was done by the author, the late John W.

Table 2. Land ownership of red-cockaded woodpecker clusters in Georgia: 1992

	Number of clusters				
Land Ownership	**Active**	**% of State Total**	**Inactive**	**Total**	**% of State Total**
Publicly Owned					
Dept. of Defense					
Ft. Benning	178	27.4	57	235	26.4
Ft. Gordon	0	0.0	37	37	4.1
Ft. Stewart	167	25.7	51	218	24.5
Subtotal	**345**	**53.1**	**145**	**490**	**55.0**
Natl. Wildl. Refuges					
Okefenokee	48	7.4	19	67	7.5
Piedmont	22	3.4	8	30	3.4
Subtotal	**70**	**10.8**	**27**	**97**	**10.9**
Natl. Forests					
Oconee	1	0.1	11	12	1.3
Hitchiti Exp.For.	13	2.0	2	15	1.7
Subtotal	**14**	**2.1**	**13**	**27**	**3.0**
State Park (L. Walker)	1	0.1	0	1	0.1
Other (B.F.Grant Mem.For.)	1	0.2	5	6	0.7
Total	**431**	**66.4**	**190**	**621**	**69.7**
Privately Owned					
Industrial Forests	7	1.1	3	10	1.1
Red Hills Plantations	172	26.5	32	204	22.9
Other	39	6.0	17	56	6.3
Total	**218**	**33.6**	**52**	**270**	**30.3**

Thompson, C. William Dopson and Donald G. Lynch. On public lands we relied on agency biologists and resource people with those agencies for updated information. On the Red Hills plantation population, I was assisted by R. T. Engstrom, J. McKinley, P. Massey, J. H. Moore, L. Neel, owners and managers. Partial funding for the 1989-1990 survey work was covered by a grant from Georgia DNR (Project 619-990284). I am grateful to R. Costa, R.T. Engstrom, R.G. Hooper and D.L. Kulhavy for comments on this manuscript.

Status and Distribution of the Red-cockaded Woodpecker in South Carolina

J. E. Cely and D. P. Ferral, Nongame-Heritage Trust Section
South Carolina Wildlife and Marine Resources Department
Columbia, South Carolina 29202

ABSTRACT: Over a 12-year period from 1977-1989, 95 of 168 red-cockaded woodpecker (RCW) colonies (56.5%) were lost in South Carolina. Hardwood encroachment was the leading cause of loss (32.6%), followed by Hurricane Hugo (27.4%), timbering (21.0%), unknown (15.8%), and other (5.0%). Total loss without Hurricane Hugo was 41%. Woodpecker losses were greater on private than public lands (63.1% versus 36.8%). Hurricane Hugo was the leading cause of loss on private land (30.0%), followed by timbering (28.3%), hardwood encroachment (25.0%), unknown (11.6%), and other (5.0%). For public lands, hardwood encroachment caused the most woodpecker losses (45.7%), followed by Hurricane Hugo and unknown (both 22.8%), and timbering (8.5%). About 60% of the estimated 1000 RCW colonies in South Carolina occurred on public land while the other 40% were on private land. Most RCW colonies were found in the outer coastal plain; a smaller concentration was found in the sandhills region and few were found in the inner coastal plain. It is expected that RCWs will continue to decline in the state, especially those from small colony tracts isolated from other woodpecker populations. Private quail plantations may harbor 25% of all RCW colonies in South Carolina; because of their large size and management practices that benefit RCWs, special consideration should be given to the role these plantations can have in RCW conservation.
KEYWORDS: South Carolina, distribution, abundance, Hurricane Hugo, quail plantations, burning, timbering, RCW status

In 1977 K.B. Stansell of the South Carolina Wildlife and Marine Resources Department and R.M. Hendrick of Clemson University initiated a state-wide survey of RCW *(Picoides borealis)* (RCW) colonies on public and private lands in South Carolina. This survey was done to collect status and distribution information and to provide management assistance and technical guidance to land owners and managers. By 1988 more than 150 RCW colony locations had been surveyed, mapped, and compiled in the Nongame-Heritage Trust data base. Since some colony sites had not been visited in > 10 years, it was felt that a resurvey of all known RCW colonies was needed. This paper gives a status update of these colonies, categorizes colony losses, shows the current distribution of RCWs in South Carolina, and provides an estimate of RCW abundance in the state.

METHODS

Woodpecker colonies were initially located via questionnaires mailed to public land managers, industrial forest owners, and managers of large private land holdings. Properties where the presence of RCWs was indicated were visited to determine the number of active woodpecker colonies; these locations were mapped on USGS topographic maps and filed with the Nongame-Heritage Trust Section. The majority of colonies were located during 1977-78 but new colonies continued to be discovered through 1983. A reassessment of all RCW colonies began in 1988 and continued through 1989; new colonies were also found during the reassessment period but were not used for the status assessment.

Revisited colonies were recorded as active, inactive, or missing; if inactive or missing, colony loss was attributed to: hardwood understory/midstory encroachment; timbering; unknown; or other. A fifth category of colony loss was necessary after the passage of Hurricane Hugo through the state in September, 1989. An effort was also made to survey adjoining lands for any missing colonies that had perhaps been displaced by the above activities.

Abundance estimates were based on our surveys as well as reports from managers and biologists of public lands, the distribution and amount of pine forest within the state, and the land use of these forests were also used for abundance estimation.

RESULTS

Between 1977-1989, 95 of 168 RCW colonies (56.5%) were lost in South Carolina. Hardwood encroachment was considered the leading cause of loss(32.6%), followed by Hurricane Hugo (27.4%), timbering (21.0%), unknown (15.8%), and other (3.1%)(included wildfire and residential development) **(Table 1)**. Colony losses were greater on private than public lands (63.1% versus 36.8%) and varied by category between private and public land: Hurricane Hugo was the leading cause of loss on private land (30.0%), followed by timbering (28.3%), hardwood encroachment (25.0%), unknown (11.6%), and other (5.0%). For public lands, hardwood encroachment caused the most losses (45.7%), followed by Hurricane Hugo and unknown (both 22.8%), and timbering (8.5%) **(Table 1)**.

Colony loss also varied geographically between the 18-county inner coastal plain which had 98 colonies (60.1% of the total), but lost a greater percentage (71.4%) than the 10-county outer coastal plain which had 65 colonies (39.8%) with a 28.5% loss. Marginal habitat for 5 RCW colonies in 4 counties occurred on federal lands in the lower piedmont of South Carolina but only 1 is currently active.

In addition to the remaining 23 active RCW colonies on private land found during the survey, another 188 active colonies have been found or reported from private lands (some reports are estimates), and 590 colonies are known from public lands within the state **(Table 2)**, making a total of about 800 active RCW colonies known to occur in South Carolina. We feel that public lands in the state have been well inventoried for RCWs, and based on the distribution and extent of pine forests in the coastal plain, and management practices on certain private lands beneficial to RCWs, another 200 colonies may exist on private lands within the state, resulting in a total estimate of 1000 RCW colonies in South Carolina. This figure is split about 60% - 40% between public and private occurrence, respectively.

Table 1. Red-cockaded woodpecker colony loss, South Carolina, 1977-1989

Reason for Loss	Private (%)	Public (%)	Total (%)
Hardwood Encroachment	15 (25.0)	16 (45.7)	32 (32.6)
Hurricane Hugo	18 (30.0)	8 (22.8)	26 (27.4)
Timbering	17 (28.3)	3 (8.5)	20 (21.0)
Unknown	7 (11.6)	8 (22.8)	15 (15.8)
Development and Wildfire	3 (5.0)	0	3 (3.1)
Total Loss	60 (63.1)	35 (36.8)	95

The majority of public RCW colonies occurred in the outer coastal plain (68%, n = 401), primarily on the Francis Marion National Forest (376 colonies). The fall line sandhills region supported 29.7% (n = 175) of RCW colonies at 6 different public locations, notably the Sandhills National Wildlife Refuge (about 120 colonies) while only 13 colonies (2.2%) were from 5 public lands in the inner coastal plain. The distribution of RCWs on private lands was similar to those on public lands.

The distribution of known RCW colonies by tract ownership was skewed toward most properties having few colonies. Of 52 private ownerships, 46 (88.3%) tracts had 5 or fewer colonies and the majority (57.6%) had only 1 active RCW colony; no private tract had more than 50 colonies. For the 20 public land areas, 12 had 5 or fewer woodpecker colonies and 7 had only 1. Only 2 public lands, the Francis Marion National Forest and Sandhills National Wildlife Refuge, supported more than 50 colonies each **(Figure 1)**.

The current distribution of active RCW colonies in South Carolina is centered in the northern coastal plain **(Figure 2)** roughly bordered by the Great Pee Dee River on the north and the Cooper Rivers to the south. This region shares characteristics favorable for

RCWs such as large public land holdings, extensive pine forests, and large numbers of private quail plantations where management practices benefit RCWs. The southern coastal plain has little land in public ownership, but has an extensive network of quail plantations, especially in Allendale, Hampton, Jasper, and portions of Beaufort and Colleton Counties. Private lands were often difficult to survey because of lack of access and more RCWs undoubtedly occurred in the southern coastal plain than indicated in **Figure 2.**

Currently 9 of 28 coastal counties have no known active RCW colonies: Bamberg, Beaufort, Calhoun, Colleton, Dillon, Lee, Lexington, Marion, and Marlboro. Only Bamberg, Beaufort, and Calhoun had no previous records of RCWs within the last 15 years.

Table 2. RCWs Colonies on Public Lands in South Carolina, 1992.

Federal	
U. S. Forest Service	
Francis Marion	376
U. S. Fish and Wildlife Service	
Sandhills National Wildlife Refuge	120
Santee National Wildlife Refuge	1
Department of Defense	
Ft. Jackson	15
Charleston Naval Weapons Station	5
Hawe Creek Corps of Engineers	1
National Park Service	
Congaree Swamp National Monument	1
Department of Energy	
Savannah River Site	6
Total Federal	525
State	
Forestry Commission	
Sandhills State Forest	35
Manchester St. Forest	2
Wildlife and Marine Res. Dept.	
Webb Wildlife Center	8
Yawkey Wildlife Cent.	7
Santee Coastal Reserve	3
Cheraw Fish Hatchery	1
Parks, Recreation and Tourism	
Cheraw State Park	3
Santee State Park	1
Hampton Plantation Park	1
Public Service Authority	
Persanti Island	3
University of South Carolina	
Wedge Plantation	1
Total State	65
Total Public	590

DISCUSSION

Despite being a listed endangered species for > 20 years, our findings showed that the RCW has suffered rather widespread declines on both public and private lands in South Carolina. This decline, based on the woodpecker's habitat needs and trends in pine forest age, land-use, and forest fire suppression efforts in the state, appears to be a long-term process that has been underway for some time, and will probably continue into the immediate future. It was noteworthy that at no location did we find an increase in RCWs; the only location in South Carolina in recent years to report an expanding population of woodpeckers was the Francis Marion National Forest prior to Hurricane Hugo (Hooper et al. 1991a). The SRS RCW population has increased as a result of very intensive management using "artificial" means (cavity inserts and translocation) and should not be considered a representative sample.

Declines have also been reported in other parts of the woodpecker's range in the southeast. James and Neal (1989) found a loss of 65% for 92 RCW colonies in Arkansas over approximately 10 years; Ortego and Lay (1988) reported a 71% loss of 31 colonies on private lands in east Texas over a 16-year period. In Georgia, Baker (1981) found a 35% colony loss over a 15-year period from 1966-1980, and Thompson (1976) reported a 23% colony loss in 4 years for 10 southeastern states.

Most of the RCW declines in South Carolina occurred on private lands (63.1%), especially those tracts with single-colony birds. Typical characteristics of these small, isolated colony sites were fragmented, multi-owner woodlands interspersed with agricultural lands, highways, commercial and residential development, wetlands, and other habitats unsuitable for RCWs. Even under ideal management, these isolated colonies could have eventually disappeared because of low woodpecker numbers, inbreeding and lack of recruitment.

Although the status of RCWs in South Carolina is not encouraging, colony declines may level off somewhat in the future, since

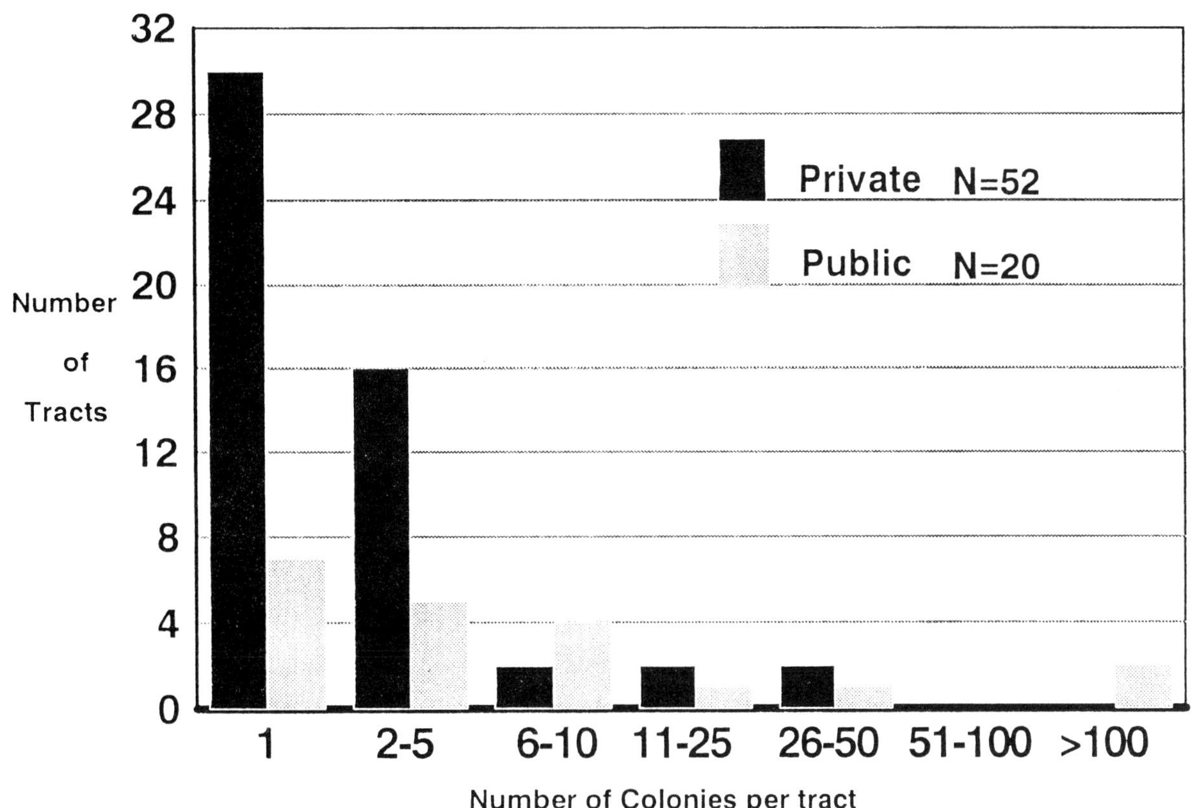

Figure 1. Distribution of red-cockaded woodpecker colonies in South Carolina by private and public ownership tracts, 1992.

those colonies at the most risk have already, or will probably be, lost in the near future, and multiple-colonies on the larger tracts have a greater chance for long-term survival. Trends in RCW habitat, however, is not promising. Lennartz et al. (1983b) found a 13% loss of old-growth pine forest in the South over a 25-year period, and only 2.5% of the commercial pine acreage in this region is considered suitable nesting habitat for RCWs. In South Carolina, about 5.4 million acres, or 45% of the state's timberland, is in softwood, mostly yellow pine, but only about 40% is >30 years old and less than 100,000 acres or 1.8% percent, is >70 years old (Tansey and Hutchins 1988).

Hurricane Hugo's passage through South Carolina in September, 1989, struck the center of RCW abundance in the state and destroyed about 63% of the woodpeckers on the Francis Marion National Forest which had 1 of the world's largest estimated populations at 477 colonies (Hooper et al. 1990). Many large private plantations with favorable habitat for RCWs near the Francis Marion National Forest in Charleston, Berkeley, Williamsburg, and Clarendon County suffered very high timber losses with canopy damage often > 70%. The exact woodpecker loss on private lands in this area will never be known, but it could have well exceeded the 30% loss found in our survey.

It could be argued that a storm of Hurricane Hugo's magnitude, considered to be a 100-year event (Hooper et al. 1990), was a natural and relatively uncommon occurrence within a short-term cycle of RCW fluctuation. When looking at primarily anthropogenic reasons for RCW declines, and disregarding hurricane losses (n = 26), RCWs declined 41% over the 12-year period. Hardwood encroachment accounted for almost half the RCW loss (44.9%), followed by timbering (28.9%), unknown (21.7%) and other (4.3%). Hardwood encroachment was even more noticeable on public lands, when not counting Hugo losses (n = 8), with a 59.2% loss, followed by unknown (29.6%), and timbering (11.1%). When disregarding hurricane loss (n = 18) for private lands, timbering caused the biggest declines (40.4%), closely followed by hardwood encroachment (35.7%), unknown (16.6%), and

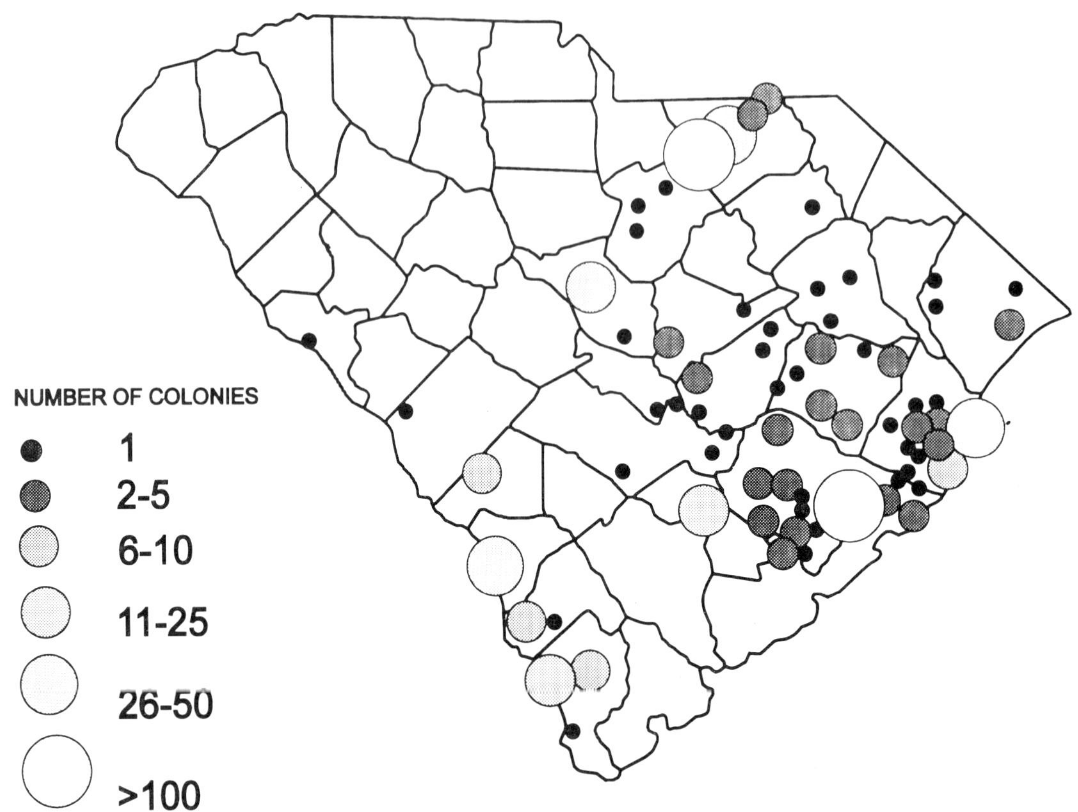

Figure 2. Distribution of active red-cockaded woodpecker colonies in South Carolina, 1992.

other (7.1%).

Because of infrequent visits to most woodpecker sites, it was not possible in 15.8% of the cases (unknown) to identify a reason for loss of a woodpecker colony. We tried to be conservative and list a reason only if we felt reasonably certain of the cause. In some cases there also appeared to be multiple reasons for a loss, e.g.colonies with hardwood encroachment and timbered foraging habitat; we tried to select the most obvious cause in these instances but had to list some as unknown. It was also possible that, due to infreqent visits, the wrong reason for a loss was given, as in cases of hardwood encroachment causing colony lost after which timber was cut; timber harvesting was the apparent reason when hardwood encroachment was the actual reason.

Hardwood encroachment was the leading cause of RCW colony loss in this survey and has been identified as a serious threat to the woodpecker in other parts of its range (Conner and Rudolph 1991a). The long growing season in South Carolina promotes rapid growth in hardwood understories; without a fire frequency of 2-4 years, these understories can quickly take over a stand and make it unsuitable for woodpeckers. The state park system, which supported approximately 16 RCW colonies at 6 parks in the mid to late 1970's, has currently only 5 active colonies, primarily due to an administrative reluctance to burn state park woodlands. Even on properties with active burning programs, such as the state forests, manpower shortages and increased burning regulations have resulted in excess hardwood growth at some colony sites. Burning on private lands, especially for the smaller tracts, poses even greater challenges: small parcels of multiple ownerships, hostility to burning, and liability concerns are just some of the obstacles to overcome for effective private-land burning. The large private plantations however, have a long-standing tradition of fire use which should continue to benefit wildlife, including RCWs.

Past efforts to document the range-wide numbers and distribution of RCWs have

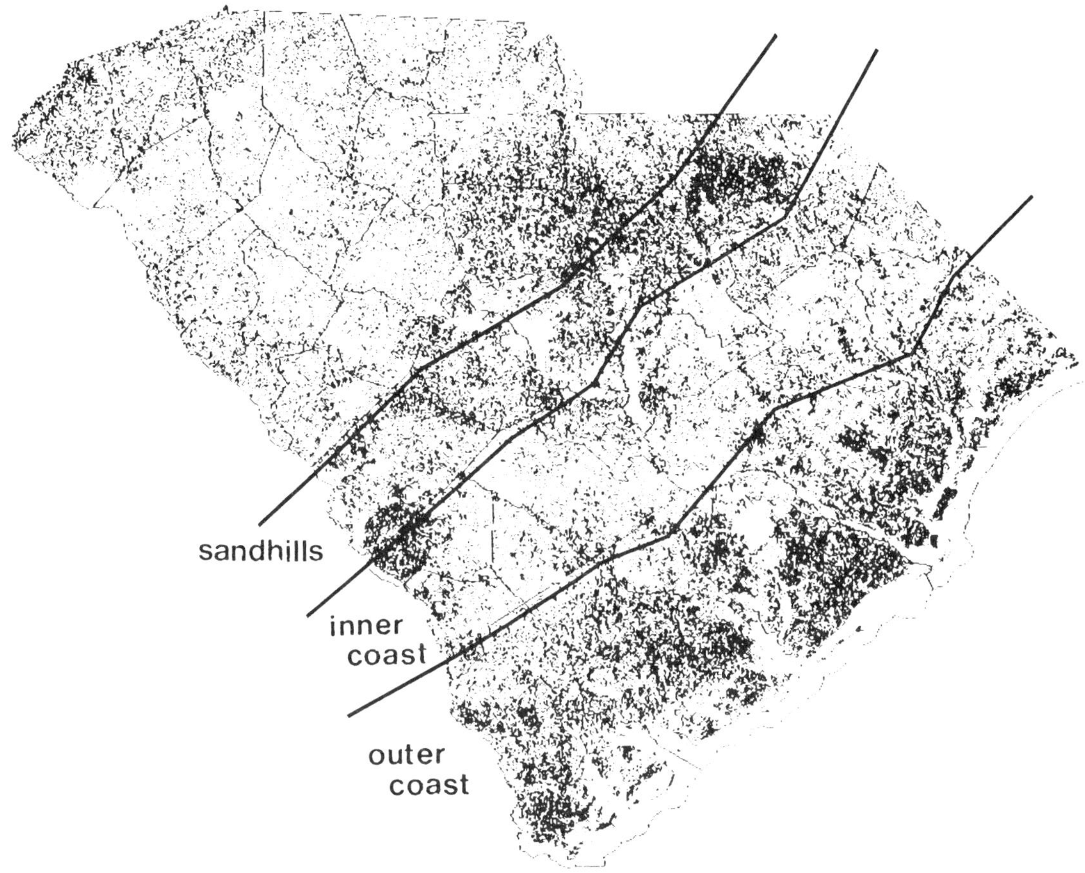

Figure 3. Distribution of evergreen forest in South Carolina; except for live oak forest along the coast line, the remainder is pine cover. Based on SPOT images, 1989-90. Courtesy S. C. Land Resources Conserv. Comm. and Univ. South Carolina.

resulted in estimates of >3000 colonies on federal lands only (Lennartz et al. 1983a) while an earlier estimate by Jackson (1978a) for total numbers of colonies, including those on private land, was 1500-3500 colonies. Using our estimate for South Carolina of 1000 RCW colonies, plus those of Baker (1981) for Georgia (674 colonies), Carter et al. (1983a) for North Carolina (909 colonies), and Wood and Wenner (1983) for Florida (1,139 colonies), nearly 4000 RCW colonies alone are estimated to occur in this 4-state region (data from other states, however, is 10 years old; more recent information is available in this symposium).

The distributional pattern of RCW colonies in South Carolina is influenced by at least several factors: the presence of large public land holdings; extent of pine forest cover; and large plantations that manage for bobwhite quail. The concentration of RCW's in the outer coastal plain reflects all 3 factors, while the concentration of woodpeckers in the sandhills physiographic province, which has few quail plantations, reflects the first 2. The inner coastal plain of South Carolina is the state's farm belt and features counties with some of the smallest percentages of forest cover - typically in the range of 45-55% - and has 11.9% of its total land area in natural pine forest; counties in the sandhills have forest cover in the range of 65-75%, with 17.6% of its land area in pine forests (excluding plantations), while the outer coastal plain has forest cover and natural pine acreage similar to the sandhills region (Tansey and Hutchins 1988) **(Figure 3)**.

Although the southern coastal plain has few public lands compared to the northern coast, more RCW's probably occur here than indicated **(Figure 2)**. This region and much of the South Carolina outer coastal plain supports numerous quail plantations totaling > 400,000 acres. The use of prescribed fire and longer timber rotations on quail plantations benefit plantation is estimated to have 50 RCW colonies, and as many as 250 colonies, or 25% of all RCW colonies in South Carolina, may occur on quail plantations. The association of RCWs with private quail plantations has been noted before (Baker 1981).

Conventional wisdom suggests that the future of RCW conservation is on public lands, but we feel more emphasis should be placed on the role that large, private, non-industrial woodlands, such as quail plantations (some of which exceed 20,000 acres in South Carolina), can have in woodpecker recovery. Jackson (1978a) and Stangel et al. (1992) have noted the importance that small woodpecker populations and their habitats can have in contributing to genetic variability in the species and to serve as corridors for dispersal and gene flow. The current range-wide RCW distribution has resulted in a few more or less isolated large populations on public lands; Hurricane Hugo has shown us that even these heretofore "secure" populations may be at risk from a natural catastrophe. As Stangel et al. (1992) recently noted, the RCW recovery goal of 15 large populations (250 breeding pairs) is far from being met; a prudent RCW conservation strategy may be one that considers the management of buffer zones and corridors of private land near these large public populations as part of a total recovery effort. Such an effort will require economic incentives in order to work.

CONCLUSIONS

RCW declines in South Carolina, and the likelihood of continued declines for the immediate future, are discouraging. Although the rate of colony loss is large, this rate hopefully will begin to decrease since highest colony losses were also those at greatest risk. The management and recovery of this species within the state and throughout its range will be a daunting task due to its specific habitat requirements.

However, recent advances in RCW management techniques, including artificial cavity construction and translocation of woodpeckers to new locations, has proven effective in restoring declining or depleted populations. New management technology has proven successful at several localities in South Carolina including Savannah River Site, Fort Jackson Military Reservation, Charleston Naval Weapons Station, and the Francis Marion National Forest; other sites are also under consideration.

Although timber harvesting is often the most dramatic and obvious cause of RCW decline, the growth of hardwood understory in the colony site is the leading cause of loss in South Carolina. Because of biases against prescribed burning, and more stringent regulations for its use, control of the hardwood understory will be a significant management challenge on private as well as some public lands.

The distribution of RCWs in South Carolina features concentrations of woodpeckers in the northern coastal plain, specifically the outer coastal plain and sandhills region, with a gap of few birds from the inner coast. The southern coast also has fewer RCWs and little public land but has much potential woodpecker habitat in private ownership. The large quail plantations, many of which are contiguous with one another and represent holdings > 20,000 acres in some cases, should have a greater role in RCW recovery. These plantations and other select private lands could be useful as buffer zones around woodpecker populations on public lands, could also serve as connective corridors between populations, and reduce the fragmentation of woodpecker populations. Plantation ownership, along with management goals and objectives, is undergoing change; economics is playing an increasing role in the future of these large tracts and some incentives should be found to encourage management for RCWs on these properties.

ACKNOWLEDGEMENTS

K.B. Stansell and R.M. Hendrick initiated the woodpecker survey; persons who contributed to the questionnaire and allowed access to their property are too numerous to mention but we thank them all. We also acknowledge the assistance of the managers, biologists, and foresters on state and federal lands for making this survey possible. R. Lacy of the S.C. Land Resources Conservation Commission provided the state map of evergreen forest cover and S. Bennett made the other figures. Funding for this survey was made possible through Section 6 aid from the U.S. Fish and Wildlife Service and the Check for Wildlife Program, S.C. Wildlife and Marine Resources Dept. This paper benefited from the reviews of R. Costa and R.G. Hooper.

Status and Management Needs of Red-cockaded Woodpeckers on State and Private Land in Texas

Brent Ortego, Texas Parks & Wildlife Department, 2601 N. Azalea, Victoria, TX
Mike Krueger, Texas Parks & Wildlife Department, P.O. Box 207, Lampassas, TX
Ed Barron, Texas Forest Service, College Station, TX

ABSTRACT: From 1988 to 1992, surveys were conducted for RCWs (*Picoides borealis*; RCW) on all Texas Parks and Wildlife Department and Texas Forest Service lands that had potential to support this species. All known active RCW sites on private lands in 1987 were re-visited at least once to determine status, and all newly reported sites after 1987 were inspected when permission was obtained from the landowner. A total of 26 clans were confirmed on state and 50 on private lands in Texas. Needs for public education, clarification of regulations, financial assistance to private landowners, RCW surveys, monitoring of clan status, habitat management, cooperative agreements between landowners, corridor development, augmentation and cavity construction, and priorities are discussed.
KEYWORDS: Texas, status, private land, state land, management, national forest

Pine (*Pinus* spp.) forests have existed in East Texas for about 4,000 years (Webb 1987). Almost all of the 6,400,000-ha of Pineywoods (Hatch et al. 1990) was heavily cut-over or cleared of forests by 1924 (McWilliams and Lord 1988).

Because the red-cockaded woodpecker (*Picoides borealis*;) (RCW) is adapted to old-growth, open pine woodlands (Ligon et al. 1986), it was believed to have been a common species across the Southeast; including East Texas (Hooper et al. 1980). Little data is available on the extent of RCW populations before the major logging era, but there is little doubt that current populations of this species are significantly lower.

The Texas Parks and Wildlife Department (TPWD) began searching for RCW populations in Texas in 1968. By the end of the year, the species was found in only 9 of the original 40 counties of its range (Lay 1969). Periodic surveys continued on public and private lands in older stands of pine, and observations of this species were solicited from biologists and the general public to locate additional populations. By 1977, 31 colonies were known on private lands. Intermittent surveys continued until 1985. From 1985-1987 intensive re-surveys on private lands were conducted at all known RCW colony sites and adjacent lands. Analysis of the results indicated that colonies on private lands had declined at a rate of 4.4% per year during the 16 prior years (Ortego and Lay 1988). Although alarming, the rate of decline was lower than on 3 of the 4 Texas national forests during roughly the same period (Conner and Rudolph 1989).

During the 1985-1987 re-surveys, reports of newly discovered RCW colonies were also verified and surveys of potential habitat on private and public land continued where possible. At the end of 1987, data from all prior TPWD RCW investigations were compiled and combined with available data from the National Forests in Texas. Ortego et al. (1988) reported a total of 212 active clans in Texas; 25 on private, 18 on state and 169 on federal lands.

METHODS

From 1988-1992, all TPWD and Texas Forest Service (TFS) property with a potential for supporting RCW were systematically surveyed by staff of both agencies. In addition, all known RCW colonies in 1987 on state and private lands were reinspected at least once during this study, with known sites on state land being monitored annually. Newly reported RCW colonies were inspected and verified as in the prior studies (Lay 1969, Ortego and Lay 1988).

Data was compiled by population. A populations was considered any RCW clan or cluster of clans that was farther than 8-km (5 miles) from another clan. Populations were tallied by 8-km spacing increments.

RESULTS AND DISCUSSION

The 1988-1992 surveys confirmed the presence of a total of 15 populations and 50 clans on private lands, and 4 populations and 26 clans on state lands (**Table 1**). We believe TPWD and TFS lands have been accurately surveyed, but private lands have only been thoroughly surveyed in a relatively few places. Industrial timber companies were the predominant private landowner supporting RCW clans. Their properties supported 22.5 clans. Non-industrial private landowners supported 19.5 clans and 2 residential subdivisions supported 8 clans.

In 1968, Lay (1969) spent considerable amount of time searching and only located clans in 9 counties (at a time when the species was believed to be more abundant than it is today). Twenty-three years of additional surveys by TPWD, TFS and the U.S. Forest Service (USFS) confirmed the presence of clans in 16 counties on federal, state and private lands in Texas; a gain of 7 counties. Intensive survey efforts since 1987 have discovered at least 30 additional clans on private and 8 on state property. These "new" clans are not thought to be the result of population expansion, but new discoveries of long established populations. The only population expansion observed during this study was when a clan on a state tract increased to 2 clans.

Even including extensive searches for RCW on private and state lands by TPWD and TFS, and the 267,000-ha National Forests in Texas by the USFS, only about 5% of the East Texas Pineywoods region has been surveyed. National forest and state lands have been thoroughly surveyed, but private lands have been thoroughly surveyed in only a relatively few places. However, only a small percentage of the unsurveyed habitat on private land is currently suitable for the RCW because of short

Table 1. Distribution of RCW Populations on Private and State Lands by distance classes between populations and by distance classes from populations on Federal lands in Texas

	km Distance Apart				
km	**<8**	**8-16**	**16-24**	**>24**	**Total**
	< 24 km of Federal Lands				
	Private Lands				
Pops.	6	1	1	0	8
No. Clans	1-4	6	1	0	0
Total	15	6	1	0	22
	State Lands				
Pops.	0	2	0	0	2
No. Clans	0	1-2	0	0	
Total	0	3	0	0	3
	> 24 km of Federal Lands				
	Private Lands				
Pops.	0	0	3	4	7
No. Clans	0	0	1-16	2-3	
Total	0	0	18	10	28
	State Lands				
Pops.	0	0	1	1	2
No. Clans	0	0	15	8	
Total	0	0	15	8	23

rotation pine management and conversion of pines to non-pine (McMahan et al. 1984). The presence of additional RCW clans on private lands needs to be examined further. We believe there may be several small populations, or portions of known populations, still undetected by agency surveys.

MANAGEMENT NEEDS

RCW clans on private and state lands represent 17.7 and 9.2 %, respectively, of the known RCW clans in Texas. They collectively can play an important role in the recovery of this species.

Regulatory agencies need to: make regulations as clear as possible and broadly available; educate the public on the population status and guidelines needed for recovery; provide technical assistance from their staff or through contracts with appropriate organizations in a prompt manner; make allowances for incidental take if landowners are following an approved plan which provides for a net benefit to the species; as much as possible, allow the landowners to meet their personal goals and budgets for their land; develop cooperative agreements between landowners in order to manage populations, rather than just single clans; and provide financial assistance to cover much of the cost of RCW management.

EDUCATION

Education needs for the red-cockaded woodpecker include information and technical assistance on:

(1) recognition of RCW cavities and habitat and the bird itself;

(2) knowledge of limiting factors and management needs for species recovery, ecology and management;

(3) the function and importance of the cavity tree, stand structure, midstory, and how to maintain forest and site characteristics conducive to red-cockaded woodpecker. The recently released Final Environmental Impact Statement for National Forests in the Southern Region pertains to these lands only but contains information of the RCW. This document will be useful as an information source.

(4) the potential impact of southern pine beetles (*Dendroctonus frontalis*;) on RCW cavity trees and habitat, especially in loblolly and shortleaf pine stands. This issue is addressed in the natural disturbances section of this volume including Coulson et al., Nebeker et al., Rudolph and Conner and forest canopy gaps in RCW cavity tree clusters (Chrismer et al.) ; and

(5) where to seek assistance when questions on the RCW are raised. Assistance includes USDI Fish and Wildlife Service, Red-cockaded Woodpecker Recovery Coordinator and field personnel with RCW assignments; publications available in a soon-to-be-released bibliography of the red-cockaded woodpecker; and other qualified personnel.

Education of the public needs to be addressed in a general manner on a broad scale to get the support needed for funding of RCW recovery; in a more detailed manner at a regional scale to address the status and needs of the RCW versus local land use practices; and in a very specific manner at local levels to deal with management needs in the immediate vicinity of RCW clans and populations.

Clarification of Regulations

Species specific endangered species regulations are complex. Landowners and regulatory agencies have had difficulty in interpreting them. Many private landowners do not understand the regulations, do not want regulatory agency personnel on their property, nor do they want agencies to know about the presence of endangered resources. Regulations need to be made clearer, and should be broadly distributed to the public and to all managers of RCW.

Financial Assistance

Some landowners that have managed their property to created habitat conditions preferred by RCWs, still harbor small RCW populations. Economic incentives and other incentives that support RCW management should be encouraged. Currently, most landowners harboring RCWs are bearing the financial burden of having endangered species on their property. Some landowners harboring RCWs on their property would benefit from various forms of financial assistance to maintain populations. Retaining and/or managing habitat for RCWs can be expensive, and few landowners can afford to so without some assistance or compensation.

Survey Needs

Thorough surveys within 8 km of known RCW sites should be conducted before management plans are developed. There are a number of 1-3 clan populations in Texas that are >16 km from other RCW populations that have been surviving for many years and in all likelihood there may be others. These cavity tree clusters may be important in long-term genetics and population viability (Stevens, this volume, Costa, pp. 67-69, this volume). Surveys should determine population size and demographics.

Current habitat conditions should be used to help prioritize areas to be searched, but young pine habitats should not be disregarded. Our surveys have discovered active cavities in older-aged property boundary trees and in relic pine trees located within younger forests.

Known populations need to be monitored at least annually population size, structure and reproductive success. Plans to augment single male clans should be part of the annual monitoring and management program. Hess and Costa (this volume) discuss augmentation techniques and results in Florida.

Previously unsurveyed RCW habitat should

continue to be searched. During the last 5 years, we have discovered 30 additional clans, suggesting that there are probably still a number of undiscovered populations.

Habitat Management

Short-term and long-range habitat management plans need to be developed to include both immediate actions needed to maintain the survival of a population and long-term solutions to ensure population survival while complying with state and federal guidelines. They should also be designed to fit the goals of the landowner and their budget.

Short-term management plans developed by us have been of the highest priority because they were needed for immediate survival of clans. The plans we have developed with private and state landowners focused primarily on mid-story control by chemical, mechanical and fire treatments, and adjusting pine basal to reduce southern pine beetle hazard. Other actions implemented were installation of metal cavity restrictors (Carter 1989) designed to minimize or eliminate cavity competition.

Long-term habitat management and planning is more complicated and should include planned forest regeneration and development to replace foraging and nesting habitat, maintenance and development of corridors or chains of islands of high quality habitat between clans, regular burning and herbicide treatments to control hardwood mid-stories in colonies. Contingency plans to address short-term cavity shortages with artificial cavities (Allen 1991, Copeyon 1990), cooperative agreements with landowners to manage habitats in a complementary manner, and exchange of RCWs for augmentation if the situation dictates, may be warranted. Costa (this volume) addresses these issues. Periodic habitat monitoring, annual population monitoring and planned revisions and/or amendments should be part of any long-term RCW management plan.

Cooperative Agreements

Few private landowners own enough land to contain an entire clan, much less an entire population. Each landowner's situation is unique and innovative agreements are needed to meet the basic requirements for RCW. Many times clans have been observed nesting on 1 ownership, but depending on another ownership for at least part of their foraging needs. Many landowners recognize the need to protect the nest site, but fewer are able to accept the concept of providing foraging habitat.

Costa (this volume) develops the concept of the Memorandum of Agreement and Habitat Conservation plans.

Corridor Development

Maintenance and creation of continuous habitat corridors or chains of high quality habitat islands between and within populations is desirable. These habitat linkages would facilitate bird movement which is desirable to enhance opportunities for reproduction, colonization and exchange of genetic material.

Cost of corridor/island acquisition and management between populations at distances > 8 km across private land could be very expensive. We suggest that development and management of linkages between distant populations be a secondary priority to maintaining the habitat within a colony and between colonies at distances <8 km.

If longer linkages are to be attempted in Texas, we suggest looking at ways to link existing relatively large populations on the Jones State Forests and on private lands in Montgomery County with larger populations on the Sam Houston National Forests, and populations on the Fairchild State Forest with those on the Davy Crockett National Forest. Also, populations on private land in northern Jasper and Newton counties may be benefited by shorter corridors to populations on the Angelina and Sabine National Forests.

Augmentation

During the early 1980's, data indicated steady declines on private land (Ortego & Lay 1988), and on 3 of 4 national forests in Texas (Conner & Rudolph 1989). No solutions were available to address serious short-term problems of population isolation and fragmentation, and cavity loss. The outlook for recovery was very poor.

Since then, Ortego has been involved with augmentation of bachelor colonies (a technique developed by DeFazio et al. 1987), and installation of metal cavity restrictors (Carter 1989) on enlarged cavities. Augmentation has shown great promise in Texas and cavity restrictors are a useful technique. Two techniques (artificial drilled cavities, Copeyon 1990; cavity inserts, Allen 1991) were recently developed for the creation of man-made cavities where suitable natural cavities are a potential limiting factor. Both of these techniques, along with augmentation, show great promise for improvement of demographics problems associated with isolated populations. The cavity insert has been successfully used to provide cavities in trees that are much younger than the typical 70-120 year old cavity tree in Texas. Augmentation and cavity inserts have been used on the National Forests in Texas and state forests, and have been instrumental in recent population increases on federal land (Larry Bonner, personal comm.). The possibility for recovery is greatly improved.

Priorities

There are currently 76 known clans on private and state lands (this study), 205 on the National Forests in Texas (Larry Bonner, personal communications), and 1 on the Big Thicket National Preserve. Prioritization of management efforts is needed because current federal and state funding is not adequate for the management of all RCW. Several major choices are available to integrate private/state populations with federal holdings, but the 2 most obvious extremes are:

(1) placing most efforts into managing the larger populations because they have a higher probability of being saved; or

(2) placing most efforts into managing small isolated populations because they will soon be lost without any management.

Prior to the development of augmentation and cavity development techniques, option 2 was not viable because management capability for most small, isolated populations on private lands were limited. Application of these techniques, along with the several management needs discussed above, could perpetuate populations on state and private lands where they represent 27% of the RCW population in Texas.

ACKNOWLEDGMENTS

Funding for this project was provided by Section 6 of the U.S. Endangered Species Act, Federal Aid in Wildlife Restoration Act, TPWD Fisheries & Wildlife Division, and TFS.

Habitat Use by a Relict Population of Red-cockaded Woodpeckers in Southeastern Virginia

Dana Bradshaw, Center for Conservation Biology,
The College of William and Mary, Williamsburg, Virginia 23185

ABSTRACT: Red-cockaded woodpecker (*Picoides borealis*)(RCW) habitat in Virginia is relegated to isolated patches of poor quality stands of mature loblolly pine (*Pinus taeda*). Six clans of RCWs were studied in southeastern Virginia from 1980 to 1983. Foraging observations were separated into breeding season and nonbreeding season observations. Quantity and quality of all foraging substrate types were characterized within identified foraging stands.

Breeding season foraging ranges were 46 to 75 ha. Nonbreeding season foraging ranges were 84 to 167 ha. RCWs selected pines > 30 cm dbh as a foraging substrate disproportionate to their availability. The continued decline in older pine stands and the trend away from prescribed burning suggests a foreseeable end to this species in Virginia.

KEYWORDS: red-cockaded woodpecker, *Picoides borealis*, foraging ecology, population status, loblolly pine, Pinus taeda, Coastal Plain, Virginia, private land, foraging habitat, home range

With the apparent extirpation of red-cockaded woodpeckers (*Picoides borealis*) (RCW) from Maryland in the 1970's, Virginia shares the northern terminus of the species' range with Kentucky (Kalisz and Boettcher 1991). Unfortunately, all of Virginia's RCWs are located on private lands, placing them in a precarious position relative to their potential longevity. Just 16 years ago, there were at least 23 active colony sites in the state (Miller 1978) with evidence of at least 24 inactive sites. Presently there are only 5 active sites **(Table 1)**. Losses were due to habitat degradation from lack of management or habitat destruction from the cutting down of colony sites and foraging areas. In addition, the RCW population in Virginia appears to be biologically isolated. The future of this species in Virginia has already been determined by past landuse history and current landowner decisions. This paper extracts one of the few contributions that the Virginia population has to offer: knowledge of peripheral population behavior.

STUDY AREA AND METHODS

In the early 1980's an effort was undertaken to study the year-round foraging ecology of the red-cockaded woodpecker in Virginia (Bradshaw 1990). At least 12 colonies existed at that time. Six colonies were selected for study due to their accessibility and the presence of a breeding pair. All study sites were located in the southeastern corner of the state in Sussex County. Each site and its associated foraging areas comprised a habitat "island",

Table 1. Red-cockaded woodpecker distribution and population trends in Virginia.

Year	Occupied Counties	Active Sites	Number Adults
1977[1]	8	23	47
1980	4	12	34
1983	3	10	22
1986	2	7	14
1989	1	6	14
1992	1	5	12

[1] **Data derived from Miller (1978); and from minimum estimates backdated from post-1977 colony discoveries.**

representing remnant tracts of mature loblolly pine (*Pinus taeda*) in a region of the state otherwise dominated by pine plantations, bottomland hardwoods, or agriculture. All 6 sites occurred on land owned by lumber companies for timber production.

Data were collected year round on foraging birds from December 1980 through December 1983 totalling over 1100 hours of observations. Foraging observations were pooled into breeding season (April 1 - Sept 30) or nonbreeding season (October 1 - March 31). Tree species, diameter, and time spent on each tree were recorded for each bird observed. Diameter classes used were: small (5-15 cm), medium (15.1-30 cm), and large (>30 cm).

Observations began at sunrise. Each study clan had 2-5 birds. None of the birds at these sites were marked, therefore sexual differences among foraging birds could not be determined. Thus, foraging position was not recorded because resource partitioning between sexes is well documented (Ligon 1968, Ramey 1980, Hooper and Lennartz 1981).

Foraging observations were mapped to document foraging habitats, and birds were followed where possible until at least early afternoon to record maximum distance from colony site (Nesbitt et al. 1978). Daily movement boundaries were consolidated to provide measures of range usage year round. Contiguous foraging stands were mapped collectively to assess total foraging range. Peripheral foraging stands not contiguous with the colony site or other foraging stands were mapped separately with their acreage added to the collective range total. None of the colonies were close enough to induce territorial interactions.

Forest inventory data were collected on timber stands known to provide foraging substrate. Basal area values of pine and hardwood, stand age, stand size, and distance from colony were recorded (**Table 2**).

RESULTS

Within each season, G-tests comparing expected with observed foraging time per pine size class for all 6 clans combined showed significant differences across size classes (breeding $p<0.05$, nonbreeding $p<0.001$) (**Figures 1. & 2.**). During the breeding season RCW's foraged less on pines <30 cm dbh and more on pines >30 cm dbh than expected based on size class availability. Nonbreeding season data for all clans combined produced the same results, but selection for pines >30 cm dbh was more pronounced.

Across seasons, nonbreeding season foraging time per pine size class differed significantly from breeding season foraging time per size class ($p<0.001$). Birds spent more time foraging on pines >30 cm dbh in the nonbreeding season than in the breeding season.

A multiple regression equation initially used the following independent variables: distance from colony site to the foraging stand, stand age, stem density of each of the pine and hardwood classes, and basal areas of pine and hardwoods. The dependent variable, foraging time, was expressed as the percent of the clan's foraging time spent in the stand. Regression results ($r^2 = 0.71$, $p<0.001$) suggested that the percent of a clan's foraging time allocated to any particular stand during the breeding season was a negative function of a stand's distance from the colony site, stem density of large hardwoods, and pine basal area. Breeding season foraging time was a positive function of both medium and large pine stem densities.

During the nonbreeding season, similar relationships existed. Regression results ($r^2 = 0.70$, $p<0.001$) suggested that the percent of a clan's foraging time allocated to any particular stand was a negative function of the stand's distance from the colony site and the stand's small pine stem density. Foraging time was a positive function of the density of pines >30 cm dbh.

Foraging range was defined as the total area utilized by a clan year-round. Breeding season foraging ranges for the 6 sites averaged 62 ha (range 47-75 ha). Nonbreeding season foraging ranges averaged 120 ha, almost twice that of breeding season ranges (**Figure 3**). However, there was greater variability in nonbreeding seasons ranges, 84-167 ha. Breeding season observations were always bound by the nonbreeding season foraging range boundaries.

DISCUSSION

In this study, similar to most other studies on the RCW, birds foraged almost exclusively on live pines (Lennartz and Henry 1985). Non-pine foraging observations comprised less than 1% of the total. Unlike Skorupa and McFarlane (1976), who recorded 10% use of hardwoods in

Table 2. General forest stand data for nesting and foraging stands.

Clan	Stand[1]	Size (ha)	Age (yrs)	Distance from Colony Site (meters)	Basal Area (m^2/ha)	
					Pine	Hardwood
One	A	40	80	150	6.7	12.2
	B	43	40	610	9.0	3.7
	C	25	50	690	4.8	3.9
	X	13	60	1140	12.9	3.4
Two	A	26	80	80	15.4	6.4
	P	5	20	80	24.3	1.1
	B	16	60	300	15.4	6.4
	C	17	50	490	14.0	3.2
	D	19	60	990	15.4	4.8
Three	A	5	50	150	8.7	2.5
	B	15	40	230	9.6	5.0
	P	16	20	300	22.7	0
	D	21	60	300	12.6	3.7
	E	21	50	380	14.5	9.4
	F	16	50	610	4.4	5.5
	G	3	60	910	7.1	3.9
Four	A	33	75	80	13.3	0.7
	B	34	70	460	10.3	0.5
	C	53	70	760	5.7	17.7
	D	12	70	760	12.4	5.0
Five	A	18	130	80	12.9	2.1
	B	19	130	150	15.6	6.7
	C	17	130	300	14.0	6.3
	D	21	130	450	13.3	4.1
	E	8	130	450	11.7	4.8
	P	22	20	680	34.7	3.0
Six	A	35	60	80	13.8	0
	P	11	20	230	25.9	0.7
	C	57	60	300	15.1	6.9
	D	23	40	380	14.5	8.3
	E	24	40	530	13.3	6.0

[1]A=colony stand; P=plantation pine; X=noncontiguous stand; others according to distance.

winter as opposed to none in summer, no substrate shift was recognized in the present study. It should be noted however that of those hardwood foraging observations recorded, almost all occurred in the first few weeks after fledging of young. Where it could be discerned, most of these observations involved young of that year, suggesting possible inexperience in selecting foraging substrata.

Within all foraging stands and in both seasons, pines >30 cm dbh were selected as a foraging substrate disproportionate to their availability. Although observed use of such pines was greater than expected for breeding and nonbreeding seasons, the difference was much more pronounced for the nonbreeding season. Selection for larger or taller pines as preferred foraging substrate has also been demonstrated by others (Skorupa 1979, Hooper and Lennartz 1981, Porter and Labisky 1986).

Several factors have been suggested as to why larger or taller pines might be selected for foraging. Jackson and Jackson (1986) suggested that the greater height and larger lateral branches of large pines offer greater structural diversity to bark foragers. This diversity could facilitate spatial separation of the sexes. An advantage of older pines is that the thicker more fissured bark of this size class may provide more food, or larger food items, per unit area (Hooper and Lennartz 1981, Jackson 1986). Jackson thought that the bark

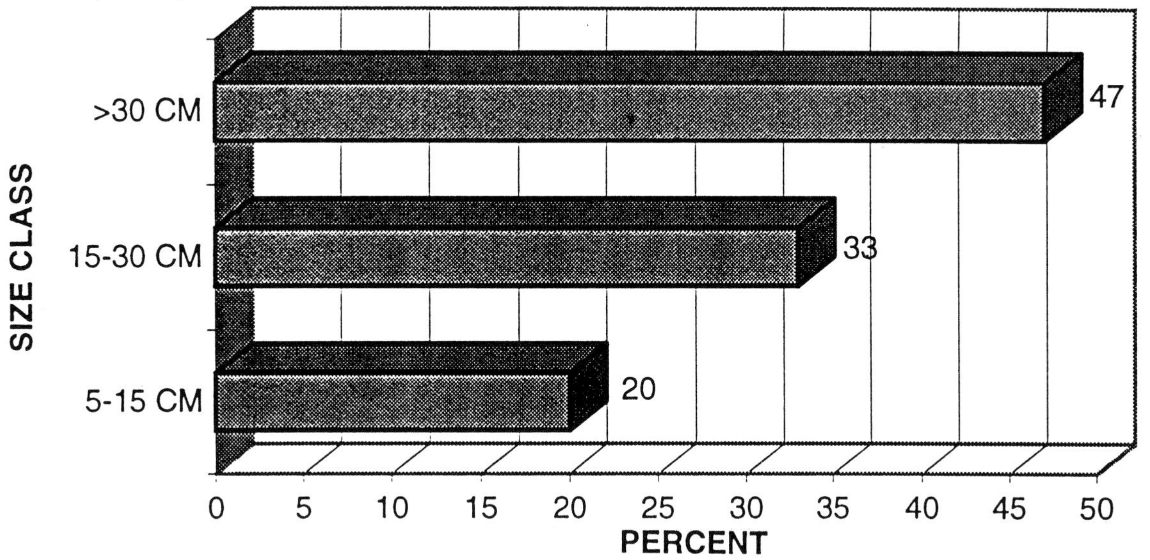

Figure 1. Proportion of total pine foraging habitat by size class.

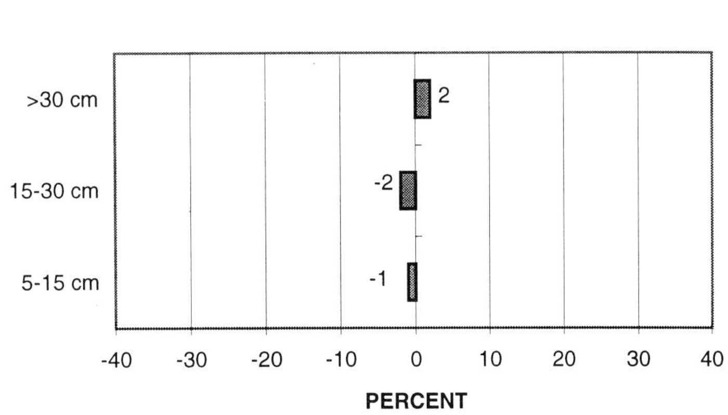

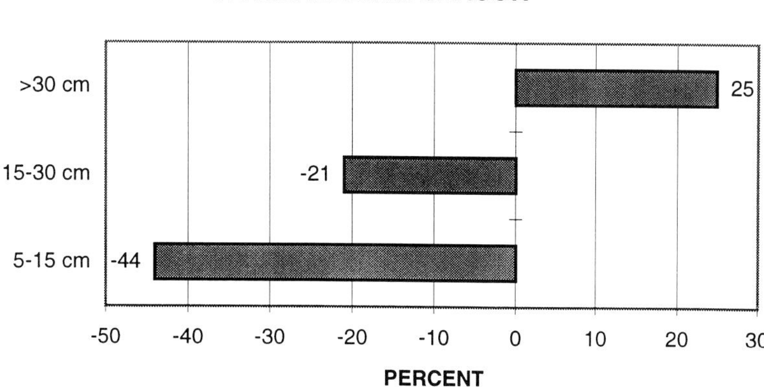

Figure 2. Observed pine size class use as percent deviation from expected based on size class availability.

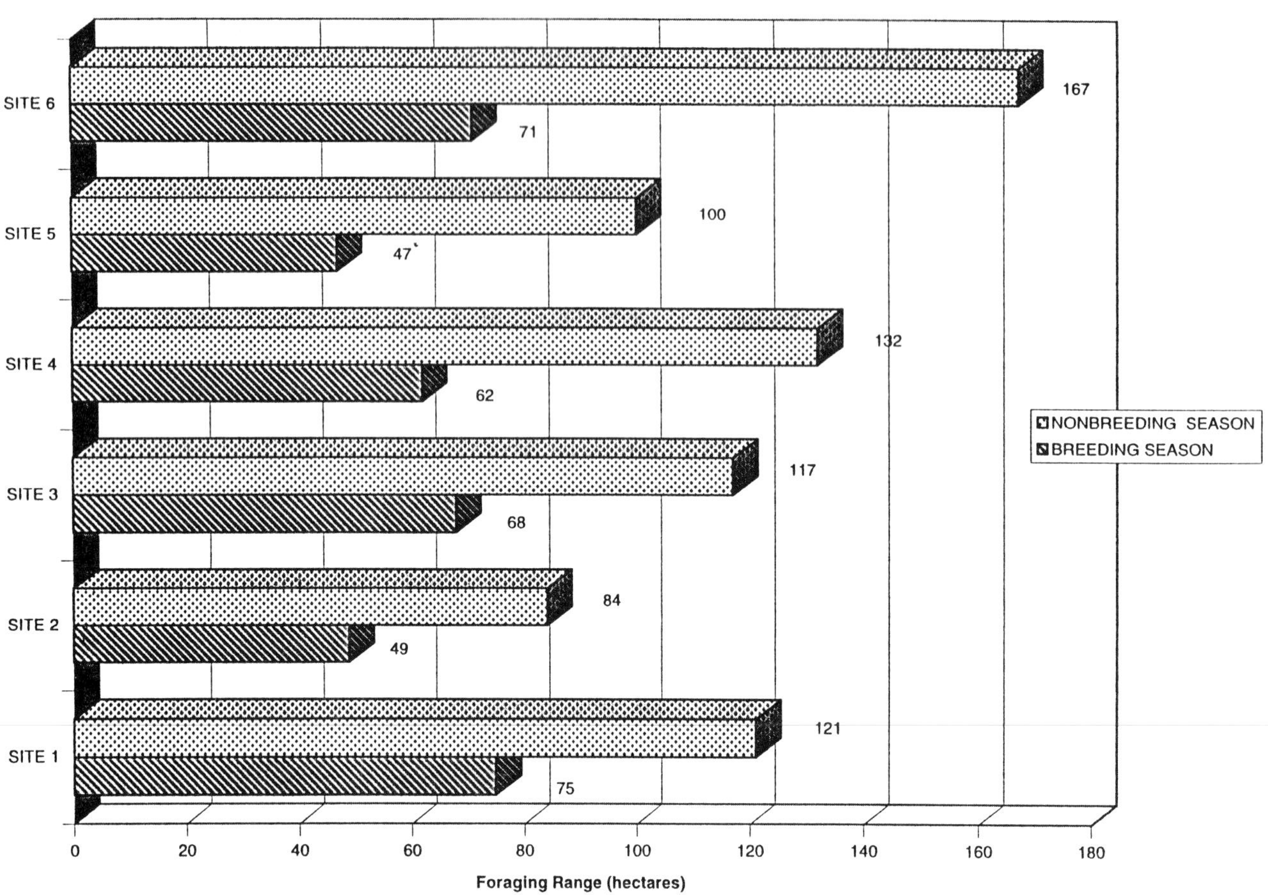

Figure 3. Foraging ranges for breeding and nonbreeding seasons for each clan.

plate size and ease of removal increased with tree age to at least 60 years. Hooper (pers. comm.) found that prey biomass on individual trees increased with tree age up to 85 years and then decreased as bark plates increasingly slough off with advancing age. With a diet almost exclusively composed of arthropods, it could be hypothesized that older or larger trees with thicker bark may in fact harbor larger numbers of insects in more northerly latitudes like Virginia. More severe winters may impose greater demands on arthropods to locate more protected overwintering quarters. It also follows from Grubb's work (1978) that woodpeckers may select larger surface areas as protection from windchill during thermally stressful conditions, suggesting a foraging site selection independent of prey densities (Skorupa, 1978).

Although small pines were used less than expected based on availability, they were selected more often as a foraging substrate in the breeding season than the nonbreeding season. In this study, young plantation pine stands were located directly adjacent to 3 of the colony sites. Each of these sites produced breeding season foraging times on small pines of at least 5 times that amount spent on small pines during the nonbreeding season for the same stand. By virtue of the birds' close association with the colony site during the breeding season, it is reasonable to expect them to utilize resources as available, relative to their distance from the colony.

During the nonbreeding season, less than 5% of the total foraging time was spent in pine stands less than 15 cm in diameter. Hooper and Lennartz (1981) found that red-cockaded woodpeckers tended to avoid stands of trees with diameters less than 11 cm year round. Ligon's (1971) work in Florida suggested that food resources were scarcer in young forest habitat than in older pinelands. An extension of this finding is an interpretation by DeLotelle et al (1983) that foraging on stems as small as 5 cm dbh is a compensatory effect for poor quality habitat. These findings notwithstanding, it did appear that small pines contributed

Table 3. Proportion of foraging time spent in each stand by season.

Clan Site	Stand[1]	Foraging Time: Breeding (%)	Foraging Time: Nonbreeding (%)
One	A	45	51
	B	43	21
	C	11	16
	X	0	12
Two	A	55	53
	P	10	1
	B	23	22
	C	12	12
	D	0	11
Three	A	35	21
	B	15	6
	P	19	4
	D	13	28
	E	10	16
	F	8	11
	G	0	14
Four	A	50	46
	B	29	31
	C	9	6
	D	12	17
Five	A	16	32
	B	32	21
	C	18	25
	D	16	15
	E	14	7
	P	5	1
Six	A	54	25
	P	9	2
	C	26	38
	D	12	20
	E	0	16

[1]A=colony stand; P=plantation pine; X=noncontiguous stand; others according to distance.

Table 4. Ownership of pine timberland by selected age classes: Coastal Plain of Virginia, 1991[1]

Age class	Public	Forest Industry	Private	All Ownerships
		(hectares)		
All Classes	26,552	182,124	330,495	539,171
61-70	---	926	20,795	21,722
71-80	---	2,610	4,256	6,866
80+	---	3,633	6,689	10,322

[1]Derived from Thompson, M.T. 1991. Includes both plantation and natural pine stands comprised of species as follows: 94% loblolly, 5% Virginia, and 1% shortleaf/pond.

significantly to the summer food resource base, although they did not ever appear to be a "preferred" foraging substrate. However, in the winter small pines were clearly neglected in favor of larger and older pine stands even at the cost of having to travel greater distances (**Tables 2 and 3**). DeLotelle et al. (1987) also documented woodpeckers foraging selectively on trees of significantly greater height and greater diameters than were randomly available within stands.

Distance emerged as one of the best predictors of foraging time allocation during the breeding season. This was not unexpected given the constraints placed on a clan in summer to maintain fidelity to the colony site for much of the time. Plotting proportion of foraging time per stand over distance of stand from colony shows an inverse relationship. Foraging time falls off rapidly over distance, terminating altogether beyond 760 m for all clans in this study. For the nonbreeding season, proportion of time spent foraging at stands between 460 and 1220 m did not change significantly. Preference for larger pines in the winter seemed to negate the significance of distance if larger or older pine stands were available. In this study, 1220 m was the greatest distance away from a colony site that birds were observed foraging.

Foraging ranges for the RCW in Virginia were similar to those further south (Hooper et al. 1982, DeLotelle et al. 1987, Porter and Labisky 1986). Nesbitt et al. (1983) suggested that foraging range is inversely proportional to the relative quality and amount of available foraging habitat. They recorded an average summer foraging range of 145 ha for 5 clans in southwest Florida, all occurring in less than optimal habitats. Other studies found seasonal ranges varied from 12 to 405 ha (Nesbitt et al. 1978, Sherrill and Case 1980, Hooper et al. 1982, Jackson and Jackson 1986). Thus Virginia's RCWs did not exhibit range requirements unique to the species. The 2 smallest total foraging ranges, 84 and 100 ha respectively, were documented in habitats with oldest stand ages and highest ratios of pine basal area to hardwood basal area. The largest total foraging range, 167 ha was characterized by the youngest overall foraging stands.

The red-cockaded woodpecker benefits from pine sawtimber management. However, in Virginia these habitats are in short supply. The late 1970's marked the beginning of an apparent crash in the already small RCW population in Virginia. This is best viewed in relation to the 46 year trend in forest resources in Virginia which culminated in a 72 percent decrease in the acreage of softwood sawtimber from 1940-1986 (Cushwa 1988). During the 10 year period from 1977 to 1986, there was a 72 percent increase in area in pine plantations in the state, exceeding a third of the total forest cover classified as softwood (Bechtold et al. 1987). In the Coastal Plain, between 1985 and 1991 there was an 11 percent increase in pine plantations which now account for over half of all pine stands in that physiographic area (Thompson 1991). None of these plantations exceed 40 years in age.

Of the 540,000 ha of pine timberland in the Virginia Coastal Plain, less than 8 percent exceeds 60 years of age **(Table 4)**. Furthermore, none of this older pine sawtimber is in public ownership. Unfortunately, there has been no incentive to manage these properties in a manner consistent with RCW conservation. Prescribed fire was formerly used throughout the Coastal Plain in pine forest management, but there has been little use of fire there since the early 1980's. Therefore, hardwoods are allowed to grow unchecked to the point that most of these areas no longer constitute suitable habitat. Bringing these stands into a useful condition would require intensive management and cost. There is no incentive for landowners to do this at this time.

In conclusion, the red-cockaded woodpecker in Virginia appears hardy, having endured decades of isolation as a population, and as individual clans in some cases. However, the continued decline in suitable habitat may have sealed the fate of the species in the Commonwealth. Given the size of the population, its remoteness from other populations, and the condition of the remaining habitat, it will be increasingly difficult to maintain these birds on private lands in Virginia.

ACKNOWLEDGEMENTS

I am grateful to Dr. Bryan Watts for his review and comments, to Julie Bradshaw for editorial assistance and to Toni Harrison for assistance in producing the final manuscript. I would also like to thank Gray Lumber Company and Union Camp Corporation for permission to access their lands, without which none of this data could have been collected.

Red-cockaded Woodpeckers on Red Hills Hunting Plantations: Inventory, Management and Conservation

R. Todd Engstrom, W. Wilson Baker, Tall Timbers Research, Inc., Tallahassee, Florida 32312

ABSTRACT: Hunting plantations in the Red Hills physiographic region of North Florida and South Georgia contain the largest population of red-cockaded woodpeckers on private lands. With more than 179 active clusters, the Red Hills population is the sixth largest anywhere. As an index of the population density and health, the median number of active clusters within 5 km of each active cluster is 34 (range 0-62). Forest management practices developed by Herbert L. Stoddard, Sr. and W. Leon Neel provide excellent red-cockaded woodpecker habitat. Some of the important practices are protection of native groundcover, frequent prescribed fires, maintenance of a standing forest with single-tree silviculture, and retention of old trees infected with red heart disease. A strategy for long-term management for this woodpecker population is an important part of the Red Hills Conservation Initiative developed at Tall Timbers Research Station.
KEYWORDS: red-cockaded woodpecker, *Picoides borealis*, hunting plantations, Red Hills, distribution, Florida, Georgia, status, conservation, management

The largest existing population of red-cockaded woodpeckers (*Picoides borealis*) [RCW] on private lands is found on hunting plantations in the Red Hills physiographic region of Georgia and Florida. This region contains some of the best remaining examples of native plant communities in the southeastern United States, including excellent old-growth longleaf pine (*Pinus palustris*) forests. Longleaf pine forest types, the preferred habitat of the RCW covered 28 million ha of the Atlantic and Gulf Coastal Plains (Wahlenberg 1946) before the original forests were destroyed in the early part of this century (Williams 1989).

The purposes of this paper are:

(1) to report on a preliminary survey of the RCW population in the Red Hills region;

(2) describe the land uses and management practices that have influenced woodpecker distribution and abundance in the region; and

(3) discuss the conservation strategy that will insure the long-term persistence of the RCW in the region.

The Red Hills

The region surveyed for RCWs is bounded by the Ochlockonee River to the northwest, the Aucilla River to the east and the Cody Scarp to the south (**Figure 1**). Soils are characterized by a surface layer of loamy sand over red clay. The drainage pattern include several large lakes that resulted from collapse of karst sinkholes in ancient streambeds. The rich soils and prominent lakes first attracted Indians and later cotton growers to the region (Brueckheimer 1988). After the Civil War, wealthy Northerners were drawn to the Red Hills because it had good railroad connections and a highly acclaimed health retreat for treatment of tuberculosis. The balsamic vapors emanating from the virgin pine forests were thought to be beneficial to patients with the disease.

Although originally attracted to the Red Hills for health reasons, people quickly discovered the excellent sport hunting in the area. Former cotton plantations with a patchwork of share-cropper fields supported large populations of quail, dove and turkey (Brueckheimer 1988). Large acreages bought by northern industrialists, including the Whitney, Vanderbilt, Mellon and Hanna families, were devoted to sport hunting.

In 1992, approximately 121,500 ha were maintained by 50 families as hunting plantations in the Red Hills (Brueckheimer

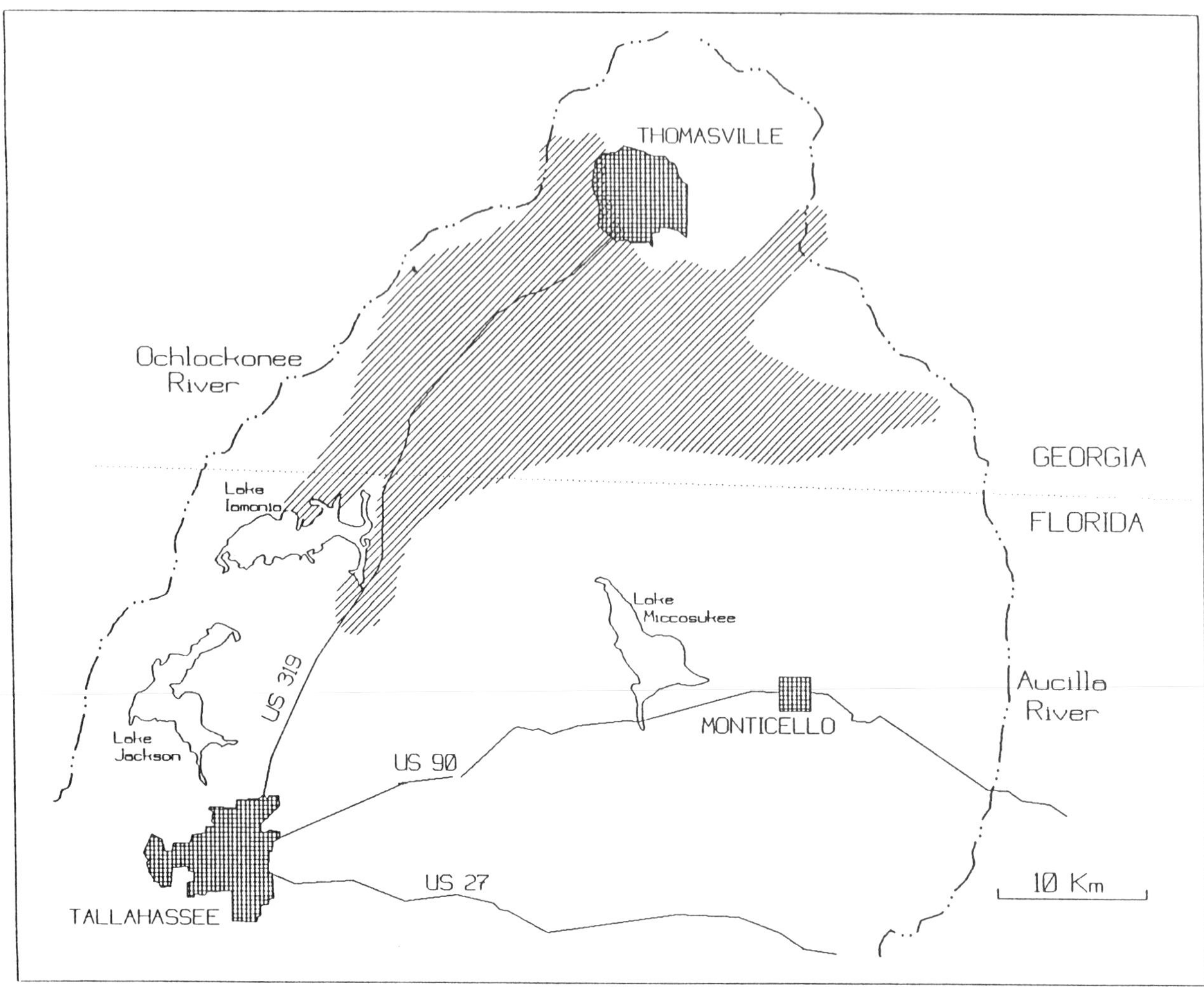

Figure 1. General distribution of RCW cavity tree clusters within the Red Hills region.

1988). The forested areas are a mixture of loblolly (*P. taeda*), shortleaf (*P. echinata*) and longleaf pines in oldfield stands and pure longleaf pine in virgin stands. The mature longleaf stands tend to support a high density of RCW clusters with more woodpeckers per group than the second growth stands. Comparing cavity tree characteristics and woodpecker group size on undisturbed longleaf pine to oldfield sites has been one of our long-term projects.

METHODS

Our inventory of RCWs in the Red Hills was conducted in 1992. Mr. Leon Neel's knowledge of RCW cavity tree cluster locations, accumulated from 40 years as a consulting forester in the region, served as the foundation for the inventory. After examining clusters known to Mr. Neel, we searched potential habitat from woods roads for RCW cavity trees. Cluster locations were plotted on 7.5 minute USGS topographic maps. Clusters were determined to be active if 1 or more cavity trees had signs of woodpecker activity (Jackson 1977b, Hooper et al. 1980). All clusters were visited in 1992, except for clusters known to be long abandoned. We did not conduct a systematic survey of all suitable habitat in all of the plantations, but our survey was 90-95% complete. As such, this paper must be viewed as a preliminary survey and an underestimate of the number of woodpecker clusters in the region.

We determined if neighboring clusters were separate by observing groups of woodpeckers within clusters and by using a rule-of-thumb that clusters be separated by at least 0.3 km. We did not use the circular scale technique (Harlow et al. 1983) because we did

Table 1. Classification of red-cockaded woodpecker clusters by county.

	Active	Inactive	Long Inactive*	Total
Thomas Co., GA	134	27	7	168
Grady Co., GA	38	6	4	48
Leon Co., FL	7	5	16	28
Total	179	38	27	244

*These clusters have been inactive for at least 10 years.

not plot all cavity trees in each cluster. The circular scale technique probably would have underestimated the number of clusters, because the clusters were very close together in the Red Hills (Reed et al. 1988a).

After all clusters were located on topographic maps, latitude-longitude coordinates were determined by digitizing cluster locations. We used these coordinates in ARC/INFO (Anonymous 1991a) to determine the number of active clusters within circles having radii of 3.2 and 5 km around each active cluster. These numbers represent respectively female and male (excluding males that stay in their natal territories) median dispersal distances in a North Carolina population (Walters et al. 1992b). We calculated the number of clusters within these circles because the long-term use of a cluster may depend in part on the density of woodpeckers in the neighborhood (Conner and Rudolph 1991b).

RESULTS AND DISCUSSION

RCW populations in the Red Hills

We located 244 cavity tree clusters in the Red Hills region **(Table 1; Figure 1)**. Some 179 of the clusters (73.3%) were active. Twenty-seven of the 65 inactive clusters (42%) had been inactive for at least 10 years. We believed our inventory was approximately 90 to 95% complete.

Determining trends in this population was not possible, because access to some of these private lands had been limited previously. The apparent increase from 147 active and inactive clusters in 1980 (Baker 1982) to 216 in the Georgia portion of the Red Hills in 1992 was the result of improved access to the hunting plantations during the past year through the efforts of the Red Hills Conservation Association. Recently, owners permitted ecological surveys of their hunting plantations partly out of concern over highway expansion, proposed gas and power lines and increasing urbanization.

The median number of active clusters within 5 km of each active cluster in the Red Hills was 34 (range 0 to 62). The median number of active clusters within 3.2 km of each active cluster in the Red Hills was 17 (range 0 to 38). These numbers provide an index of density and reflected the probability of immigration from other clusters of males (5 km) and females (3.2 km) respectively.

The high density probably indicate a healthy population especially on plantations with large acreages of old timber. However, significant subpopulations of woodpeckers in the Red Hills have been extirpated (Neel 1971), including the subpopulation at Tall Timbers Research Station (Baker 1983b).

With 179 active clusters, the Red Hills is the sixth largest extant RCW population (Costa and Escano 1989, Hardesty 1992, Baker pers. comm., F.C. James pers. comm.) and certainly the largest RCW population on private lands. The next largest population on private land is 110 clusters in the towns of Southern Pines and Pinehurst, North Carolina (Carter et al. 1983a).

Red Hills Land Management Practices

The Red Hills support a large RCW population primarily because land management in the Red Hills is based on an ecosystem management philosophy. In the best examples of this type of management, a single-tree selection silvicultural system produces upland pine forests that retain some of the structure of undisturbed longleaf pine forests. Methods developed by Mr. Herbert L. Stoddard, Sr. and refined by Mr. W. Leon Neel (Neel 1971) have been used on some of these plantations for 60 years. Increased production of northern bobwhite (*Colinus virginianus*) and other game species, timber production, an aesthetically pleasing appearance, and protection of rare and endangered biota, including the RCW, are results of this type of management.

Important elements of the management practices from an ecological point of view are:

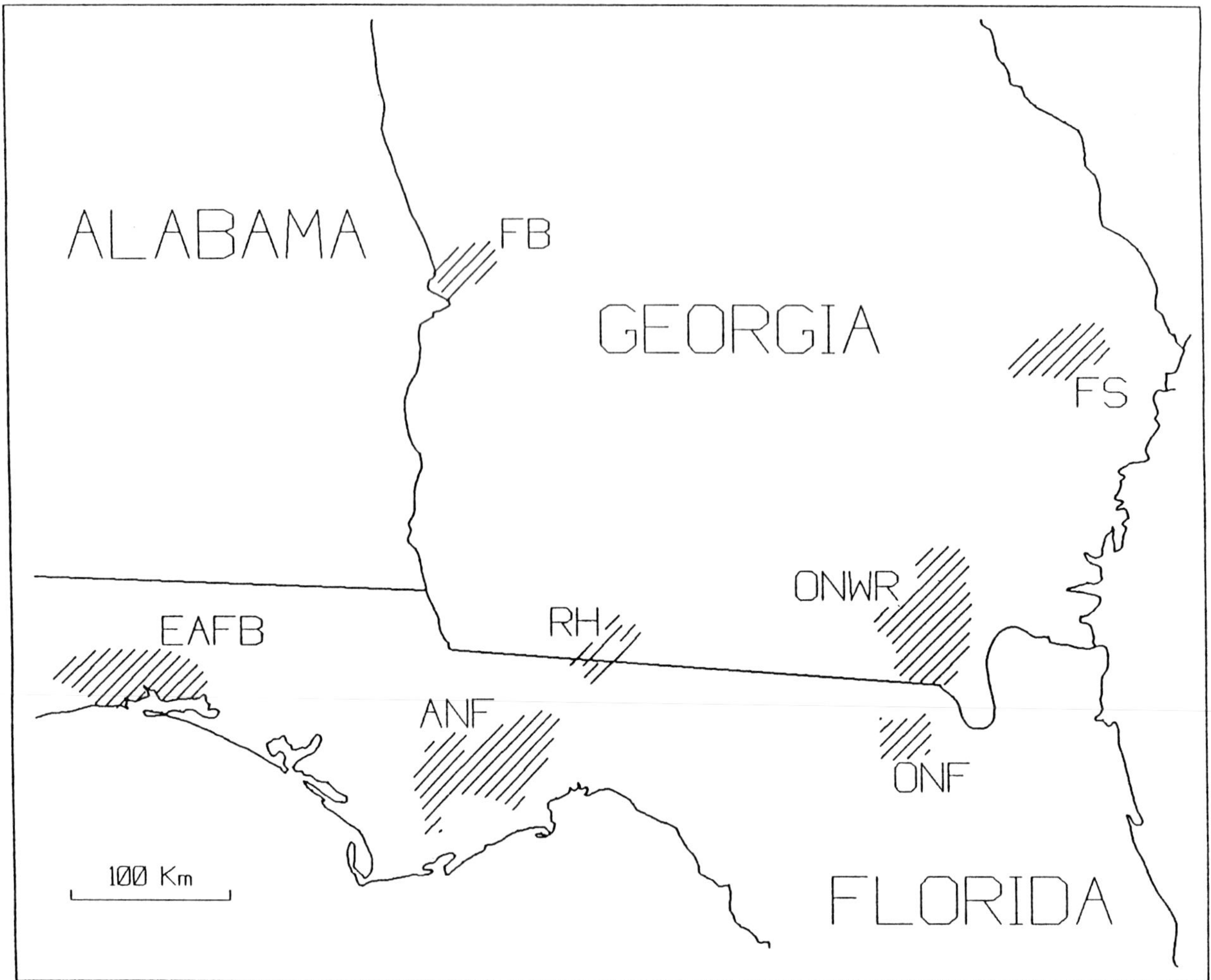

Figure 2. The Red Hills RCW population in relation to other regional population centers. EAFB=Eglin Air Force Base; ANF=Apalachicola National Forest; RH=Red Hills; FB=Fort Benning; FS=Fort Stuart; ONWR=Okefenokee National Wildlife Refuge; ONF=Osceola National Forest.

(**1**) minimized mechanical disturbance or removal of native ground cover,

(**2**) maintenance of a standing multi-aged forest with natural regeneration, and

(**3**) maintenance of patches of old trees within a matrix of younger forest.

Composition and quality of the ground cover strongly influences fire behavior and frequent fire is essential for the long-term perpetuation of ecosystem function. A standing forest that is never clearcut provides constant foraging territory for woodpeckers. The component of old-growth trees retained within the forest are likely to have red-heart and thus provide a supply of potential cavity trees.

From a landscape perspective, it is important to minimize loss of large tracts of forest. Good management of an isolated plantation will be insufficient to sustain the RCW population if surrounding plantations are converted to housing developments or the habitat is degraded. Urbanization not only fragments or eliminates high quality woodpecker habitat, but makes prescribed burning more difficult because of complaints about smoke and the liability of escaped fires.

Future Conservation Efforts

A unique opportunity exists to maintain a significant RCW population in the relatively wild lands of the Red Hills hunting plantations. RCW cavity tree clusters are currently spread over 37 different tracts that are owned by 35 different families. Many of these hunting plantations are contiguous and form a large region of high quality habitat. Coordinated conservation action will be necessary to maintain the current population. It may even be possible to recover the Red Hills population by increasing the size to 400 breeding pairs (Anonymous 1990).

The Red Hills population is 39 km from the large population in Apalachicola National Forest (**Figure 2**) and unsuitable habitat (Tallahassee) separates the two. Thus, the Red Hills population is isolated from the Apalachicola population (Anonymous 1990).

The Red Hills Conservation Initiative, sponsored by Tall Timbers Research, Inc. and the Red Hills Conservation Association, is a multifaceted effort to preserve historical, cultural and ecological resources within the region. The RCW is just one species that will be well-served by protection and wise management of mature upland pine forests.

The strategy for conservation and management of the RCW in the Red Hills requires the cooperation and involvement of landowners and managers. It includes the following steps. The first step, which is largely complete, is to conduct an inventory of clusters. This will be followed by an assessment of critical areas within the region and development of conservation priorities, possibly using gap analysis (Scott et al. 1993). The next step will be to produce educational materials and hold gatherings to inform landowners, foresters and land managers about the life history and legal aspects of the woodpecker. Landowner concerns regarding consequences of the Endangered Species Act need to be addressed. Next, best management practices for the woodpecker need to be identified specifically for the Red Hills. The goal is to promote existing land uses and management practices that allow the landowner to maintain the woodpecker population at little or no cost. We believe that management practices over the past 60 years as developed by Mr. Herbert L. Stoddard, Sr. and Mr. W. Leon Neel have been beneficial for the woodpecker. We hope to document and refine the Stoddard-Neel technique as a model for other areas.

We hope to insure RCW conservation by forming cooperative management agreements and conservation easements with individual landowners. Federal habitat management incentives to landowners would be an invaluable investment. A private cooperative conservation plan in the Red Hills can be a model for integration of ecological considerations with current land uses.

ACKNOWLEDGMENTS

This paper is dedicated to Mr. Herbert L. Stoddard, Sr. and Mr. W. Leon Neel. Their careful management of forests on Red Hills hunting plantations has maintained much of the native biological diversity of the region. The Red Hills stands today as an important population center for red-cockaded woodpeckers largely because of their wisdom and attention.

We thank Larry Carlile, Robert Hooper, Paul Massey, Jim McKinley, and Felicia Sanders for assisting with the field survey. Jim Cox of the Florida Game and Fresh Water Fish Commission and Dean Jue of Florida State University provided invaluable assistance with the digitizing process and ARC/INFO analysis. Jim McKinley, Janisse Ray, Larry Carlile and Felicia Sanders provided useful comments on the manuscript.

Status and Distribution of the Fourth Largest Population of Red-cockaded Woodpeckers: Preliminary Results From Eglin AFB, Florida

Jeffrey L. Hardesty[1,2] and Ruthe J. Smith[1] [1]Florida Cooperative Fish and Wildlife Research Unit, Dept. Wildlife, Ecology and Conservation Univ. of FL, Gainesville, FL 32611
[2]The Nature Conservancy, Gainesville, FL 32611
Carl J. Petrick and Bruce W. Hagedorn, Natural Resources Division, Eglin Air Force Base, FL. 32578
H. Franklin Percival, Florida Cooperative Fish and Wildlife Research Unit, Dept. of Wildlife, Ecology and Conservation, Univ. of FL, Gainesville, FL 32611

ABSTRACT: Despite extensive habitat disturbance and loss of longleaf pine (*Pinus palustris*), observations suggested that Eglin Air Force Base, Florida, contained a significant population of red-cockaded woodpeckers (*Picoides borealis*). Surveys were begun in 1990 and population monitoring started in 1992. Preliminary results indicated that 235 (210-262) active clusters and 232 (205-254) inactive clusters were present in 1992. The population was apparently stable during 1990-1992, based on changes in activity of a random sample of clusters. Further evidence for stability was offered by the relatively low percentage (9.8%) of active clusters occupied by single males, similar to that of a stable population in North Carolina, and secondly, by the stability in the number of normal cavities found among active territories. However, rapid loss of cavities (est. >20% annually) among inactive clusters due to enlargement by other species suggested that options for natural population increase through reoccupation are becoming fewer. An ecosystem-based approach to management that focuses on landscape-scale restoration is expected to halt further declines in the short-term and lead to recovery in the long-term.
KEYWORDS: red-cockaded woodpecker, longleaf pine, sandhills, line transect survey, population monitoring, population dynamics, Department of Defense, cavity dynamics, population stability, prescribed fire, ecosystem management, Eglin Air Force Base

That Eglin Air Force Base in Florida's panhandle once contained a large population of red-cockaded.woodpeckers (*Picoides borealis*) (RCW) had been assumed for some time (Baker et al. 1980); however, estimates of its status were based on anecdotal information and partial surveys. This paper examines data from an extensive recent survey of the distribution and status of the RCW on Eglin, and discusses first-year findings from longer-term studies of population dynamics and habitat relationships. Also discussed are Eglin managers' efforts to integrate RCW recovery efforts into a broader context of ecosystem management.

STUDY AREA

Eglin Air Force Base is located in the Florida panhandle on the Gulf of Mexico approximately 72 km east of Pensacola, Florida (**Figure 1 inset**). It is the largest U.S. Air Force installation and encompasses approximately 188,185 ha (464,000 ac). The primary mission of Eglin Air Force Base is the development and testing of conventional munitions and sensor tracking systems (DoD 1993).

The Eglin landscape ranges from barrier beaches, coastal lagoons, and estuarine wetlands on the Gulf Coast, to the rolling, highly xeric dunal ridge uplands comprised of well-drained deep sands. Elevations range from sea level on the coast to nearly 90 m in the uplands. Principal soil types in the uplands are generally classified as belonging to the Lakeland Association (SCS 1969). Annual rainfall averages 160-168 cm (DoD 1993). The sandhills ecological association (FNAI & FLDNR 1990) is the predominant ecotype, comprising 78% of the landbase (DoD 1993). This association is characterized by an overstory of scattered to dense longleaf pine (*Pinus palustris*), a midstory of mostly turkey oak (*Quercus laevis*), as well as other hardwoods, and a rich ground cover of various herbaceous species. Dense midstories of oak in some areas, a co-dominant overstory of sand pine (*Pinus clausa* var. *immuginata*) in others, and a limited distribution of wiregrass (*Aristida sp.*) throughout are variously related to edaphic conditions and past management practices, most notably fire suppression and overharvest of longleaf pine. The distribution of old-growth longleaf (>100 years of age) reflects past management practices. Other land use areas and ecological associations included administrative areas (12%), sand pine (3%), and pine/mixed hardwood associations (2%), wetlands, riparian areas, flatwoods, and barrier islands (5%) (DoD 1993).

HISTORY

Eglin Air Force Base is variously comprised of former public and private lands. The Choctawhatchee National Forest was created in 1908 and consisted of 53,825 ha interspersed with private landholdings in the western two-thirds of what is now Eglin Air Force Base (DoD 1993). The National Forest had acquired an additional 84,133 ha by the time the area was converted to military purposes, becoming Eglin Air Force Base in 1940. In 1943, an additional 49,778 ha of private lands were acquired.

Eglin's longleaf pine systems have been exploited for more than a century, as were similar areas throughout Florida and the Southeast (e.g., Myers 1990). Cattle grazing occurred from the 1800s-1930s. During the time that Eglin was owned by the U.S. Forest Service, the land was initially managed for naval stores, and later, following the industry's collapse, for timber production. The Forest's 1939 management plan (USDA 1939) classified 133,551 ha as "longleaf pine ridge type," 61% of which had been "cut over." Eglin managers continued the logging of longleaf pine through the 1980s. On former Forest Service lands, minimum leave standards dictated that between 6 and 15 trees ≥25.4 cm dbh were left per ha, with not more than 50 trees of lesser diameters (DoD 1993). Timber harvests on formerly private lands varied from minor in select areas (e.g., Patterson Natural Area) to complete denudation on the majority (DoD 1993). The removal of longleaf stumps to accommodate the distillate wood industry resulted in the removal of 400,000 tons across 74,000 ha from 1961 to 1990. In addition, areas cleared for the operation of the military's mission more than doubled, from 12,950 ha in 1949 to 26,612 ha in 1978 (DoD 1993). Plantations of off-site slash (*Pinus elliottii*) and sand pine occur on 29,353 ha.

Extensive fire suppression began in 1927 in Choctawhatchee N.F., and early records indicate almost total elimination of natural fire (USDA 1931). The later military fire control program was equally successful, and prior to 1989, growing season fires primarily occurred adjacent to test ranges. As weapons testing began, accidental fires increased and, as a result, some areas adjacent to bombing ranges have been continuously exposed to fire since the mid-1940s. Eglin managers began an extensive controlled burning program in 1976, including the introduction of growing season fire in 1989 (DoD 1993).

METHODS

Survey

Beginning in June 1990, the study area was systematically surveyed for RCW cavity trees using line transect techniques described in Wood and Lewis (1977). Some areas had been surveyed prior to 1990. Criteria developed by Jackson (1977b, 1982) and Hooper et al. (1980) were used to determine cavity status and activity. An approximation of the number and distribution of active clusters (i.e., the collective set of cavity trees defended by a group) was determined through use of the circular scale technique (CST) (Harlow et al. 1983, Lennartz and Metteauer 1986), whereby each delineated cluster estimated the presence of one group. A group could include a single male or a breeding pair with or without helpers. We tested the accuracy of CST by comparing the mapped number of clusters (average; 46) determined by applying the CST to a known number of active clusters (42) derived from behavioral observations. Our accuracy (91.3%) was similar to that observed (92%) by Lennartz and Metteauer (1986). In addition, the mean (268.5±19 m SE) and 90th percentile (444.1 m) of furthest distances between active cavity trees within clusters were similar to those obtained by Harlow et al. (1983) (282 m, and 460 m, respectively). On the basis of a comparison of means (one-sample *t*-tests; $P > 0.49$) and the relative accuracy of estimation, we concluded that there was insufficient evidence to believe that the distributions of distances were sufficiently different to warrant rejecting the use of the CST, or of using a different diameter circle.

Abandoned cavity trees and clusters also may be important resources (Doerr et al. 1989, Walters et al. 1988, Walters 1990). Groups or individuals will roost in, reoccupy, capture, or form new groups in abandoned clusters (e.g., Doerr et al. 1989). As noted by Walters et al. (1988:278), however, if the CST is used, which requires the presence of one or more active trees, "abandoned or unoccupied clusters do not exist." Thus, inactive tree groups were delimited by the CST using inactive trees, and using criteria from Walters et al. (1988), whereby both behavioral and spatial characteristics were considered when available. Approximately 15% of inactive trees were assigned to inactive clusters through the use of behavioral data (e.g., cavity trees adjacent to a closely monitored sample of clusters) and another 15% using spatial criteria (groups of ≥2 cavity trees >0.5 km from other groups of ≥2 cavity trees). The remaining 70% were assigned to non-overlapping groups that encompassed the largest number of inactive trees in groups ≥ 2 as delineated by a circle of 460 m.

Resurvey and Population Monitoring

Sets of active and inactive clusters were chosen as part of a long-term monitoring program to document population trends, group size, productivity, breeder turnover rates, foraging habitat use, home range, and response to management. Using a random numbers table (Zar 1984), 53 active clusters were chosen from a total of 160 known (as of March 1992), and 50 inactive clusters were chosen from a total of 201 known (as of November 1992).

Active clusters were monitored extensively at the beginning of the breeding season (March and April) and adults were banded to determine group size and affiliation using methods similar to Walters et al. (1988). Afterwards, clusters were visited approximately every 7-9 days to record the number of eggs, nestlings, and incidentally, adults (Walters et al. 1988). Nestlings, approximately 5-7 days old, were captured and uniquely color banded using techniques described in Nesbitt et al. (1978, 1982) and Jackson (1982). Clusters were visited 20-25 days after nestlings were banded to determine the number and sex of fledglings (Walters et al. 1988). Productivity was estimated using the mean ± SE for the number of fledglings produced among all groups, all breeding groups, and successful groups.

Because of a late start, inactive clusters were first visited after the breeding season. To determine activity, all previously known cavity trees in the delineated cluster were checked once, as were selected nearby trees adjacent to the cluster, and a cursory search was made for unrecorded cavity trees within the 460 m perimeter of the cluster.

A preliminary estimate of a short-term population trend was determined using methods described in James (1991) that combined and compared total percent changes among samples of active and inactive clusters between initial survey and resurvey. While this method does not yield a direct estimate of longer-term population stability, increase, or decrease, it does allow an approximate estimate of a trend during the period of the survey (1990-1992). We used a hypergeometric distribution to estimate variance, appropriate when sample n is a substantial portion of the finite population of size N (Cochran 1977, Zar 1984). Confidence limits (95% C.L.) for finite populations were calculated following Burstein (1975). Population stability was estimated using the following formulae:

1. Estimated no. of active clusters = (% previously active clusters active x total est. no. active) + (% previously inactive clusters active x total est. no. inactive clusters);

2. Estimated no. inactive clusters = (% previously active clusters inactive x estimated no. active clusters) + (% previously inactive clusters inactive x estimated no. of inactive clusters);

3. Estimated no. of active clusters with pairs = (% of active clusters occupied by pairs x estimated no. of active clusters) + [(% of previously inactive clusters active x est. no. of inactive clusters] x % of active clusters occupied by pairs.

Ninety-five percent C.L. for trend estimates were calculated by substituting the extremes of the confidence limits, as calculated for each parameter estimate (e.g., percent of previously active clusters active), in the above formulae.

Cavity loss and recruitment

Cavity trees in all sample clusters were reexamined to determine if their status had changed (active or inactive, start or complete, normal or enlarged, or tree mortality) since the initial survey. Each cluster received a cursory resurvey to determine the presence of previously unidentified cavity trees; however, newly found cavity trees were not included in analyses. The number of exposure days (the number of days elapsed from the date of the original survey to first resurvey) were calculated for each cluster. Per cluster calculations were as follows:

1. Percent cavity loss due to tree mortality = [(no. of complete normal cavities lost due to tree mortality/no. of previously normal complete cavities)/no. of exposure days] x 365 days x 100.

2. Percent cavity loss due to enlargement = [(no. of previously normal complete cavities enlarged upon resurvey/total no. of previously normal complete cavities)/no. of exposure days] x 365 days x 100.

3. Percent new cavity excavation = [(no. of new complete normal cavities upon resurvey/no. of previously normal complete cavities)/no. of exposure days] x 365 days x 100.

4. Percent net change in complete normal cavities = percent new excavation – percent loss due to tree mortality + percent loss due to enlargement.

Population-wide estimates for the above parameters were the means and standard errors of per cluster means.

RESULTS

Survey

From 1990 through 1992, 65,157 ha were surveyed during 3,570 person-days, and as of December 1992, the survey was estimated to be 85% complete. A total of 3,532 cavity trees had been identified, including 814 judged to be active. Active and inactive trees were grouped by spatial arrangement and behavioral observation into 208 active and 205 inactive clusters (1.02 active:1.00 inactive) **(Table 1)**. The total number of clusters was estimated to be approximately 467, including an estimated 235 active and 232 inactive clusters; estimates were based on researcher knowledge of the unsurveyed remaining suitable habitat and the observed ratio of active to inactive clusters. The number of active clusters and the proportion of clusters active were higher in the western portion of the base **(Figure 1)**. Survey error (rate of missed cavity trees or misjudged tree activity) was not estimated.

Resurvey and Population Monitoring

Resurveys of randomly chosen, previously active clusters indicated that 51 of 53 (96.2%) were still active, and 2 of 50 (4%) previously inactive clusters had been reoccupied between 1990-1992 **(Table 1)**. Time intervals between initial and resurveys of active and inactive clusters (422 ± 19 days and 475 ± 36 SE days, respectively) were judged to be not different (Mann-Whitney *U*-test, $Z = 1.01$, $P = 0.31$). Of the 51 active clusters, 46 (90.2%) were occupied by pairs (with or without helpers) and 5 (9.8%) by solitary males; pre-breeding season average adult group size was 2.55 ± 0.13 SE **(Table 2)**. Eighty-eight percent of groups attempted to breed, and 89% were successful **(Figure 2)**. On average, RCWs on Eglin produced 1.57 ± 0.13 SE fledglings per group in 1992 (excluding 5 single male groups) **(Table 3)**.

Analysis suggested that the number of active clusters had remained approximately stable throughout the survey (1990-1992); the initial survey estimated 235 clusters *vs.* 235.4 clusters from the extrapolated results of the resurvey **(Table 1)**.

Cavity Loss and Recruitment Rates

Preliminary annualized rates of cavity loss and recruitment suggested that among active clusters, cavity loss was compensated by new excavation **(Table 4)**. Inactive clusters, however, lost an average of >20% of complete normal cavities annually, primarily due to enlargement by other species. In neither case was tree mortality a significant factor, though similar in each.

DISCUSSION

Red-cockaded woodpeckers underwent apparently large declines after 1900 on lands presently incorporated within Eglin AFB. Moderate to severe habitat disturbance

Table 1. Summary results from a survey (85% complete) of RCW habitat, a resurvey of a random sample of active and inactive clusters, and an estimate of the total number of clusters by activity on Eglin Air Force Base, Florida, 1990-1992.

	# of clusters	Percentage	*95% confidence limits
A. Total estimated clusters on Eglin AFB based on surveys and available habitat			
Active clusters (observed 208)	est. 235		
Inactive clusters (observed 205)	est. 232		
B. Resurvey of active clusters (*n* = 53)			
With birds	51/53	96.2%	88.7 - 99.3%
Without birds	2/53	3.8%	0.7 - 11.3%
With pairs (with and without helpers)	46/51	90.2%	80.0 - 96.1%
With single males	5/51	9.8%	3.9 - 20.0%
C. Resurvey of inactive clusters (*n* = 50)			
With birds	2/50	4.0%	0.8 - 12.4%
Without birds	48/50	96.0%	87.6 - 98.2%
D. Estimates for Eglin AFB in 1992			
Active clusters (0.962 x 235) + (0.04 x 232) =	235.4		210.3 - 262.1
Inactive clusters (0.038 x 235) + (0.96 x 232) =	231.6		204.9 - 254.4
Active clusters w/ pairs (0.902 x 235.4) + [(0.04 x 231.6) x 0.902)] =	220.6		189.8 - 253.8
Active clusters w/ single males (0.098 x 235.4) + [(0.04 x 231.6) x 0.098)] =	24.0		9.3 - 52.8

***Hypergeometric distribution.**

Table 2. Adult RCW group sizes associated with a random sample of active groups (*n* = 51) in 1992; mean = 2.55 ± 0.13 SE adults. Data are from Eglin Air Force Base, Florida.

	Group Size			
	Solitary males	Pair	Pair + 1 helper	Pair + 2 helpers
%	9.8%	43.1%	29.4%	17.6%
(no.)	(5)	(22)	(15)	(9)

occurred during this period, both before and after Eglin came into public ownership, including grazing, turpentining, widespread high-grade logging, the near deforestation of once privately held lands, and over-harvest of longleaf pine on the Choctawhatchee National Forest. Under Air Force management additional impacts have included fire suppression, limited intensive site-prep, road building, development of off-site pine plantations, and conversion of lands to military and civilian facilities (DoD 1993).

During 1990-1992, RCW concentrations occurred primarily on lands acquired in 1908 (53,825 ha) by the federal government as part of the original National Forest. Lands acquired from private owners (133,960 ha) in subsequent decades were largely deforested, with some exceptions, prior to federal ownership, as evidenced by only widely scattered old-growth longleaf. Generally, National Forest minimum leave standards insured that some old-growth trees were retained as seed stock on federal lands, although this resulted in leaving only 6-15 trees ≥25 cm dbh per ha. The Eglin forestry program continued active logging and plantation development, especially in the 1970s. The Air Force created 2 large environmental zones around major test ranges in 1973, indirectly protecting large numbers of old-growth trees. A number of exceptional stands of old-growth are present, including at least one relatively natural stand of >1000 ha (Patterson Natural Area) with numerous

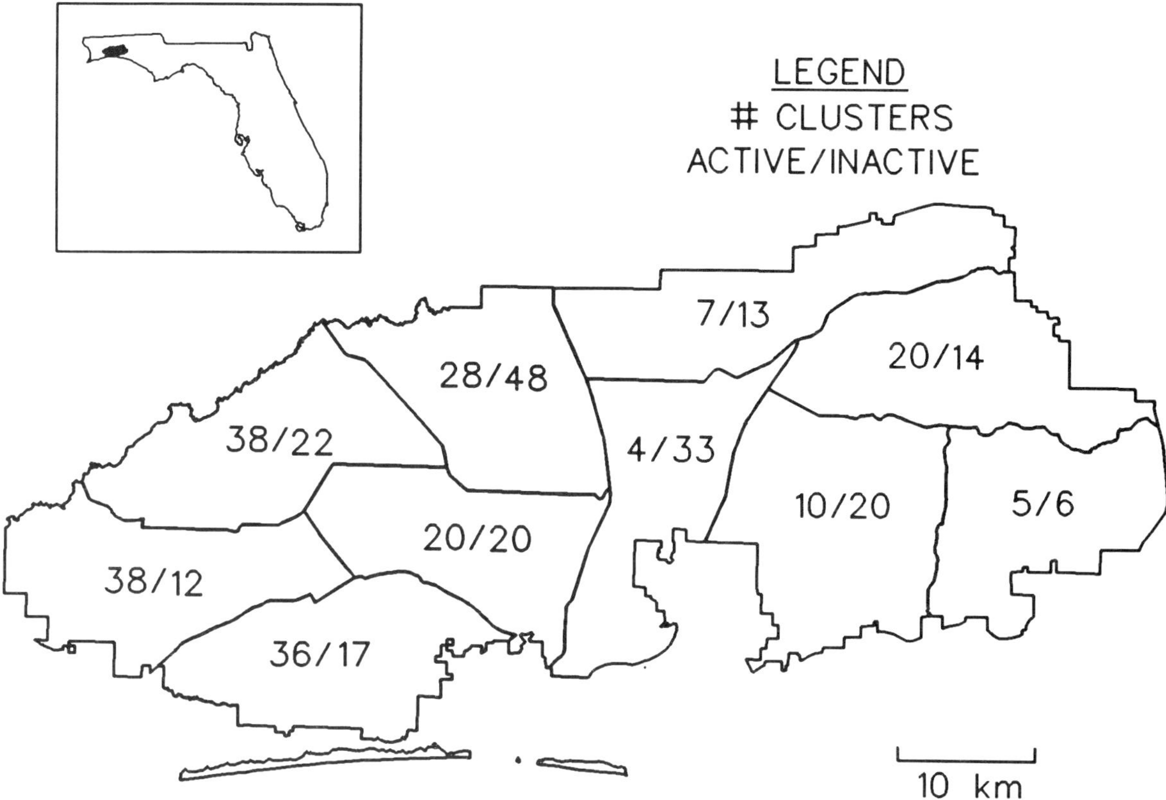

Figure 1. Distribution of RCW clusters (active/inactive) on Eglin Air Force Base, FL.

Table 3. Summary of 1992 reproduction for 46* of 51 randomly chosen red-cockaded woodpecker groups on Eglin AFB, Florida.

Group Status	Mean ± SE (fledglings/grp)
All groups	1.61±0.14 (74/46)
Pairs	1.55±0.19 (34/22)
Pairs with ≥1 helper	1.67±0.20 (40/24)
All breeding groups	1.64±0.14 (74/45)
Breeding pairs	1.62±0.19 (34/21)
Breeding pairs with ≥1 helper	1.67±0.20 (40/24)
All successful breeding groups	1.85±0.12 (74/40)
Successful pairs	1.89±0.14 (34/18)
Successful pairs ≥1 helper	1.82±0.18 (40/22)

***Excluding 5 groups occupied by solitary males.**

individual trees >200 years old.

After 1927, all fires were rigorously suppressed, especially on federal lands, as fire was thought to be detrimental to longleaf pine (USDA 1931). Only in recent years has prescribed fire been reintroduced on any appreciable scale, including growing season fire in 1989. Wildfires and fires associated with test ranges are normally still suppressed. Fire suppression has had severe consequences, perhaps second only to removal of the original forest. Suppression has destabilized largely non-successional sandhill plant communities dominated by an overstory of longleaf pine and an understory of perennial grasses and herbs. In the absence of growing season fire (or in many cases, of any fire), hardwoods and non-serotinous sand pine have moved from protected bayheads, steepheads, creeks, and wetlands onto sandhill ridges, and one or the other are either dominant or co-dominant with longleaf pine in many former longleaf sites. In particular, closed stands of sand pine dominate approximately 25,000 ha of former longleaf/scrub habitat in the southeastern portion of the reservation. Notable exceptions to this trend occur adjacent to some test ranges, where weapons testing and training have led to numerous accidental fires over the past 40 years.

The present-day RCW distribution appears to mirror past disturbance on at least two spatial scales, although data from quantitative

Table 4. Mean percent cavity loss and recruitment rates based on a resurvey of cavity trees and cavities in a random sample of previously active (n=53) and inactive clusters (n=53) on Eglin Air Force Base, Florida, 1990-1992. Analysis included only previously normal complete cavities or new normal complete cavities.

	Cavity loss		Cavity recruitment	Net change
	Tree mortality %±SE	Enlargement %±SE	New excavation %±SE	% ±SE
Active clust*	1.5%+0.7	6.1%+1.8	15.5%+4.8	+7.9%+4.4
Inact.clust*	0.7%±0.5	32.9%±7.0	13.1%±6.3	-20.5%±7.1

***Average time intervals, initial survey to resurvey: Active:421.5 days±18.6 SE; Inactive:474.7 days±36.5 SE.**

studies are not yet available. At the landscape scale encompassing the entire Eglin metapopulation, active clusters tend to occur in clumps, with the majority of active clusters apparently associated with lands obtained in the original Choctawhatchee National Forest, some former railroad lands, and lands lying adjacent to test ranges with high fire frequencies. Inactive clusters occur most regularly in the highly fragmented and cutover formerly private lands in the northwest and central portions of the reservation **(Figure 1)**.

At the scale of individual trees, cavity trees tend to be widely, but thinly dispersed; many hectares have few to no old-growth stems. Many inactive clusters consist of 2-4 old-growth trees (unpubl. data) associated with young longleaf stands adjacent to one or more sand pine or slash pine plantations, whereas most active clusters tend to occur in less fragmented areas of more continuous longleaf pine forest.

Early information on the distribution of RCWs in this landscape is lacking. Review of historical Eglin records (DoD 1949) indicates that 160,000 ha of suitable RCW habitat probably existed prior to 1900, including approximately 155,000 ha of sandhill. The present day density is approximately 1.0 cluster/157 ha (413 active and inactive clusters observed/65,160 ha surveyed). Given that Eglin's current population density has been influenced by past disturbance, historical densities were no doubt higher. Densities of approximately 2 to 4 times those measured at Eglin have been recorded in extant populations (cited in USFWS). The ratio of active to inactive clusters presently is 1.01/1.0 (208/205); however, ratios elsewhere were also higher, e.g., the nearby Apalachicola Ranger District (U.S. Forest Service) exhibits a ratio of 3.18/1.0 (R. Costa, pers. comm.). All else being equal, these ratios and densities suggest that prior to 1900, a minimum of 513 active clusters [(235 active/467 total clusters) x (160,000 ha/157 ha cluster^{-1})], and possibly >1,000, were present. Current numbers (est. 235 active clusters) probably represent one-half to one-fourth the original population size.

At present, the number of RCW clusters appears to be stable; however, this assessment is based on only 1 year of a 5-year monitoring program, and hence should be viewed as preliminary. Evidence supports, but does not prove, this contention. A decrease of 2 among a sample of 53 previously active clusters was balanced by an increase of 2 observed among a sample of 50 previously inactive clusters **(Table 1)**. In order to be 95% confident that a decline had occurred, >6 (>11.7% of 53) previously active clusters would have been found vacant, or no (<1% of 50) previously inactive clusters would have become reoccupied.

However, the rate of cluster occupation by single males (9.8%) **(Table 2)**, while lower than in other populations, is suggestive of further declines. Single male occupation is thought to be a prelude to abandonment following desertion by the breeding female, a shortage of females, or inability of the breeding male to attract a female, usually because of a decline in habitat quality or quantity (Walters 1991). Ninety-five percent confidence limits of single male occupation estimates (3.9 - 20.0%) include or approach estimates from other populations; 11% of clusters were occupied by single males in a stable North Carolina Sandhills population (Walters et al. 1988), 23% in a declining Texas population (Conner and Rudolph 1989), and 37.8% in one district of the Apalachicola NF where evidence suggests a decline is underway (James 1991).

Competition for a limited number of

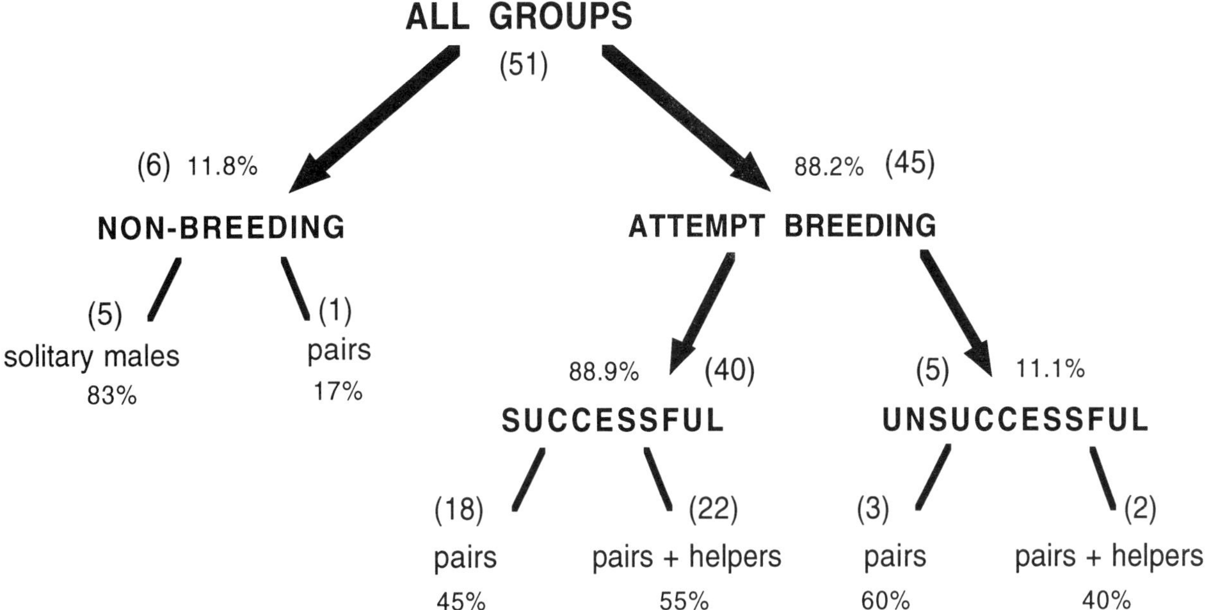

Figure 2. Summary of 1992 reproductive outcomes among a random sample (n = 51) of RCW groups on Eglin AFB, FL.

suitable cavities appears to be the critical factor leading to evolution of cooperative breeding in RCWs (Walters et al. 1988, Walters 1990, 1991). Population growth is limited by the preferred strategy of male RCWs to inherit their natal territory or to replace a breeder within an existing nearby territory. Colonization of unoccupied habitat is rare (Walters et al. 1988), but can occur (Hooper et al. 1991a).

Loss of individual cavities can result from enlargement by other species (e.g., Kalisz and Boettcher 1991) or loss of cavity trees. One might expect that in a declining population, cavity loss would occur at a greater rate than cavity gain, even among active clusters. The preliminary results of an examination of cavity loss/recruitment among active clusters at Eglin suggested that within-cluster cavity loss was compensated by recruitment (**Table 4**). However, an annual net loss of >20% of complete normal cavities among abandoned clusters appears to have occurred during the same period. Taken separately, these data suggest stability among active clusters. However, if the proportion of single male clusters is predictive of imminent cluster abandonment, then net cavity loss among inactive clusters suggests that the present population may be poised for further declines unless immediate remedial steps are taken.

The decline in RCWs at Eglin appears to be due to systemic landscape-wide problems principally related to past management. Eglin's population of 210-262 active clusters is the fourth largest, and has been designated as a Recovery Population by USFWS (R. Costa, pers. comm.). As such, Eglin will play a critical role in eventual recovery and delisting. Further, it appears that weapons testing and RCW recovery can occur in the same landscape. Eglin managers recognize, however, that a narrow focus on RCW recovery may detrimentally impact other systems or species, some of which are of national significance, and may limit, rather than increase, future management options. For example, aggressive mechanical removal of hardwoods in the poor soils of the sandhills appears to negatively impact understory vegetation, the most important functional (fire-carrying) and species-rich component (e.g., Noss 1989) of the sandhills ecosystem.

CONCLUSIONS

In the opinion of consulting scientists, the future of RCW recovery on Eglin is inextricably linked to the restoration of a functioning sandhills ecosystem as defined by process, structure, and composition (Hardesty 1992). To accomplish this, Eglin managers developed an ecologically-based Natural

Resource Management Plan (NRMP) (DoD 1993), with the principal goals being long-term restoration of the sandhills and associated biodiversity, continued military use, and significant public use. Managers invited wide participation in the planning process, and included outside cooperators from agencies, universities, user groups, and non-governmental organizations. The NRMP calls for extensive and intensive reintroduction of fire, especially growing season fire; inventories of biological diversity; research and development of ecological and economical habitat restoration and silvicultural practices; preservation and restoration of fire-dependent and fire-carrying understory plant species; reclamation of slash and sand pine plantations and areas where sand pine has encroached into uplands; creation of a sustainable forestry program based on uneven-aged management; development of an integrated public use program; and development of a landscape-level and adaptive management decision making process. If managers are successful, and if DoD continues to support enlightened ecosystem management, then Eglin may well serve as a model for how the Defense Department can co-exist with nature — including RCWs — in a working military landscape.

ACKNOWLEDGEMENTS

The study was funded by the Dept. of Defense and the U.S. Fish and Wildlife Service (RWO-80). The authors would like to thank the present Command of Eglin AFB, the staff and leadership of the Natural Resources Branch, especially R. McWhite, S. Seiber, D. Green, and D. Teague, and members of the Fish and Wildlife Cooperative Research Unit, Dept. of Wildlife and Range Sciences, University of Florida. Special thanks are due the many members of the survey team for their persistence and untiring efforts. We thank C. Hess, W. Baker, W. Platt, S. Hermann, B. Bartush, J. Walters, T. Engstrom, F. James, J. Dabney, R. Costa, D. Wood, J. Gore, J. Troxel, K. Meyer, M. Huffman, J. Young, and J. Smallwood for their invaluable advice. We also thank R. Hooper and R. Costa for their review of our paper.

Habitat Characteristics Associated with Persistent Red-cockaded Woodpecker Colonies on Industry Lands

T. Bently Wigley and Sheron W. Sweeney
National Council of the Paper Industry for Air and Stream Improvement, Inc.
Department of Aquaculture, Fisheries and Wildlife
Clemson University, Clemson, SC. 29634-0362

ABSTRACT: We reported the initial results of a pilot project on red-cockaded woodpecker (*Picoides borealis*) habitat relationships on lands managed by the forest products industry. Data were summarized for 22 colonies known to have existed at least 5 years; 21 were known to be at least 9 years old. Lands around the colonies usually were in multiple ownerships with 87% owned by industry. All 22 colonies had >60 acres of suitable foraging habitat within 0.5 mile of the colony center and 21 had >125 acres. On average, 32% of foraging habitat within the 0.5-mile circle was within 0.25 mile of colony center. Other habitats within 0.5 mile were: forested but not suitable for foraging (103 acres), young forest regeneration (69 acres), and non-forested (10 acres). Most (19) colonies had >8,490 ft^2/acre of pine basal area and all colonies had >3,000 ft^2/acre in pine trees >10 inches DBH. Fragmentation indices suggested that landscapes generally contained multiple habitats clumped into relatively large, contiguous patches with simple shapes. Red-cockaded woodpeckers were found persisting in a broad range of conditions for most habitat variables, some below recommendations presented in the recovery plan and draft guidelines for private lands.
KEYWORDS: forest products industry, foraging habitat, fragmentation indices, private lands, red-cockaded woodpecker

Although private landowners are not charged with recovery of red-cockaded woodpeckers (RCW), the birds persist and, in places, thrive on privately owned lands. There may be large amounts of potential RCW habitat on private lands. Lennartz et al. (1983b) has suggested that as much as 75% of suitable nesting habitat was located on private lands. To date, however, no comprehensive survey has been conducted (Ortego and Lay 1988), and little is known about RCW numbers on private lands.

Private landowners are interested in the status of RCWs on their lands for several reasons, including ethical and fiduciary considerations. For example, member companies of the American Paper Institute have adopted a comprehensive set of environmental and forestry principles that require a commitment to integrating the growing, nurturing, and harvesting of trees with conservation of habitat for wildlife (McMahon 1992), including endangered species. Thus, for many private corporate and noncorporate landowners, protecting occupied habitats for endangered and threatened species is the "right thing to do."

Private landowners also should be interested in providing for RCWs in order to comply with the Endangered Species Act (ESA). Despite the popular misconception that the ESA does not apply to private lands (e.g., Ortego and Lay 1988), private landowners are required by the ESA to avoid "taking" endangered species. The definition of "take" encompasses a number of activities, including those that might "harm" or "harass" endangered species. In its regulations (50 C.F.R. § 17.3), the U.S. Fish and Wildlife Service (USFWS) has stated that "harm" includes "significant habitat modification or degradation where it actually kills or injures wildlife by significantly impairing essential behavioral patterns, including breeding, feeding, or sheltering."

Thus, it is important that private landowners understand how their management will affect the immediate and long-term viability of the RCW.

Despite the importance of RCW management to private landowners, few studies have been conducted on privately owned lands. Data from private lands are important because management objectives on private lands usually are different from those on public lands. Forest management activities on private lands and ownership patterns often result in stands and landscapes structured very differently from those on public lands. And, private landowners are held under the ESA and USFWS regulations to different standards than federal agencies. Thus, research across a broad spectrum of habitat conditions such as those often found on private lands is important to identify the full range of variation in habitat and landscape configurations that can successfully sustain RCWs while also meeting landowner objectives. Identifying the full range of variation should promote operational flexibility for accommodating RCWs and expand opportunities for private landowners to provide suitable habitat.

The purpose of this study was to describe current habitat conditions surrounding persistent RCW colonies (5 or more years since discovery) found on privately owned lands. This research was the pilot study in a long-term project to monitor RCW performance on privately owned lands. We thank the private landowners who donated data and other information for this paper. We also extend our thanks to A.A. Lucier, L.L. Irwin, and J.M. Sweeney for reviewing this manuscript. This research was funded by the National Council of the Paper Industry for Air and Stream Improvement, Inc.

METHODS

We identified RCW colonies on lands owned by private, forest products companies in Arkansas, Louisiana, Georgia, and Texas. A map showing all forested stands, other habitats, and geographic features within 0.5 mile of each colony's geometric center (geometric center of all active cavity trees) was digitized, and area and perimeter length of each stand or landscape feature calculated. Landowners provided the most recent forest inventory data available. These data were collected as a part of each company's routine forest inventory program and included, for pines and hardwoods, the number of trees by 2-inch diameter classes (DBH) and basal area (BA). For pines and separately for hardwoods within 0.5 mile of the colony center, we summed the estimated number of stems $\leq$10 inches DBH, number of stems >10 inches DBH, basal area in trees $\leq$10 inches DBH, and basal area in trees >10 inches DBH.

Forested stands, other habitats, and geographic features within 0.5 mile of each colony center were classified into 1 of 4 categories:

(1) suitable foraging,
(2) forested but not suitable for foraging,
(3) young forest regeneration, and
(4) non-forested.

Stands were termed suitable for foraging if >50% of BA was pine, 10 to 80 ft^2 BA was present in pine trees >10 inches DBH, and the stand was not separated by >300 feet from other suitable foraging habitat (Costa 1992). Costa (1992) also suggested that stands have "open" characteristics. We were unable to examine every stand to evaluate this recommended characteristic. Therefore, these estimates of suitable foraging habitat may be liberal.

Forested habitat not suitable for foraging included stands that did not meet the definition of suitable foraging habitat. Pine plantations were placed in the "forested habitat not suitable for foraging" category once they reached the 4-inch DBH class. Newly regenerated stands that had not reached the 4-inch DBH class were categorized as "young forest regeneration." Non-forested habitats included landscape features such as lakes, pastures, oil-well sites, and residential sites.

Using maps (1:1,320 scale) of habitats around the colonies, we calculated the angular sum of non-forested and young forest regeneration (or open) habitats as an index of fragmentation (Conner and Rudolph 1991b). To calculate angular sum, we identified the lateral edges of non-forested and young regeneration habitats as viewed from the colony center. The sum of these angles was then measured. Angular sum can range from 0 to 360; low values indicate small amounts of open habitat and high values indicate high amounts. We calculated separate angular sums for habitats within 0.5 and 0.25 mile of the colonies.

We also placed a 40 X 40 matrix over the

habitat maps; each grid cell in the matrix represented a square of 132 X 132 feet (0.4 acre). In each cell within the matrix we entered 1, 2, 3, or 4, to represent the 4 habitat categories previously described. Cells containing multiple habitats were classified according to the dominant habitat. Zeros were entered in cells outside the 0.5 mile circle. A spatial analysis program, modified from the program SPAN described by Turner (1990), was used with these matrices to calculate other indices of habitat distribution and fragmentation. We calculated 4 indices: dominance, contagion, fractal dimension, and the Taylor index to patchiness (O'Neill et al. 1988, Taylor 1977).

The dominance index (D) is based on the Shannon-Weaver diversity index (Shannon and Weaver 1962) and measures the extent to which specific land uses, i.e., habitat categories previously described, dominate the landscape. Because dominance is sensitive to variation in the number of habitat types within landscapes, we used a D value scaled for number of habitats (Turner 1990). Scaled D will vary from 0 to 1. At high values, 1 or 2 land uses dominate the landscape; at low values many uses are present in approximately equal proportions.

The contagion index (**C**) measures the amount of fragmentation or the degree to which land uses are fragmented or clumped. It is the probability of a grid cell dominated by land use i being found adjacent to a grid cell dominated by land use j (O'Neill et al. 1988). We also used a scaled C index, which may range from 0 to 1. Thus, landscapes with large, contiguous patches will have large values of C and landscapes dissected into many small patches will have lower values of C.

The Taylor index of patchiness (**T**) is another index to fragmentation. It is the probability of encountering a different habitat type while traveling from one grid point to the next along the central vertical and horizontal axes of the grid matrix. Again, a scaled T was used which ranges from 0 to 1; 0 represents homogenous habitats and 1 represents very heterogeneous habitats.

Fractal dimension (**F**) is a measure of the fractal geometry of a landscape (O'Neill et al. 1988, Mandelbrot 1983) and is estimated by regressing polygon area against perimeter length for each landscape patch. Fractal dimension ranges from 1 to 2. Landscapes dominated by simple patterns, e.g., squares, will have low values of F. Landscapes dominated by complex or convoluted patterns will have high values of F; a value of 2 would correspond to shapes drawn by Brownian movement (O'Neill et al. 1988).

RESULTS AND DISCUSSION

Data were collected for 34 colonies on lands owned by 5 forest products companies. Data for 12 colonies, however, were unacceptable because the colonies had not been discovered until after 1987 or forest inventory data were unavailable for some stands. Thus, 22 colonies were used in analyses. Because our sample was limited to 5 landowners, our results probably are not representative of all colonies on industry lands. However, these data are the first characterization of RCW habitats on industry lands, and serve to identify a range of habitat conditions apparently suitable for the RCW.

Colony age

The average year of discovery for the colonies was 1979 (SD = 6 years). One colony was discovered in 1987, and the remainder were discovered prior to 1983. Six colonies were known to have existed before 1973. Therefore, most (21) of these colonies had persisted >9 years under management regimes used by some forest products companies.

Active cavities and trees

Although no data are yet available for number of woodpeckers in each clan, the number of active cavities and starts suggest that most of the clans consisted of multiple birds. Most colonies (20) had >1 active cavity and 10 had >3 active cavities (**Table 1**). Most (16) also had ≥1 start. The average colony had about 5.5 active cavities and starts (**Table 1**).

Ownership pattern

Land around the colonies usually was in multiple ownerships; 16 of the 22 colonies had ≥2 landowners within 0.5 mile. Within 0.5 mile, an average of 438 acres (SD = 62) was in industry ownership, 51.8 acres (SD = 63) was in private nonindustrial ownership, 9 acres (SD = 34) was in federal ownership, and 5 acres (SD = 17) was in state ownership. At the 0.5-mile scale, therefore, most of the colonies probably were persisting in landscapes more highly fragmented by ownership than colonies found in recovery populations. A higher proportion of land within 0.25 mile of the average colony was in industry ownership. Within a 0.25 mile radius, an average of 122 acres (SD = 7) was in industry ownership, 3.5 acres (SD = 7) was in private nonindustrial ownership, 0.5 acre (SD = 1.2) was in federal ownership, and no land was in state ownership.

Table 1. Number of active cavities, starts, and trees with cavities or starts for 22 red-cockaded woodpecker colonies on forest products industry lands in Arkansas, Georgia, Louisiana, and Texas, 1992.

Habitat Variable	Mean	S.D.	Min.	Max.
# active cavities	3.7	2.4	1	10
# starts	1.8	1.6	0	5
# trees with active cavities or starts	3.7	2.2	1	9

Habitat types

Within 0.5 mile, 21 colonies had >125 acres of suitable foraging habitat, exceeding recommendations by the recovery plan (U.S. Fish and Wildlife Service 1985). All 22 colonies had >60 acres of foraging habitat and surpassed the minimum proposed by Costa (1992). One colony had 60-100 acres, 4 had 151-200 acres, 3 had 201-300 acres, and 14 had >300 acres. About 63% of the area within 0.5 mile of the colonies was suitable foraging habitat **(Table 2)**.

About 75% of the area within 0.25 mile of the colonies was suitable foraging habitat **(Table 2)**. On average, 32% of foraging habitat within the full 0.5 mile circle was within 0.25 miles of the colony center. Colonies with >200 acres of total foraging habitat (n = 17) had 29% of it within 0.25 miles of the colony center. The 4 colonies with 151-200 acres foraging habitat had 37% of it within 0.25 mile. The colony with 60-100 acres had 64% of its foraging habitat within 0.25 mile.

Most (20) colonies had forested habitats not suitable for foraging within 0.5 mile of their center. For all 22 colonies, forested habitats not suitable for foraging constituted about 22% of area within 0.5 mile of the colonies **(Table 2)**. The 20 colonies with this habitat type had an average of 113 acres (SD = 104); 5 colonies had <25 acres, 1 had 26-50 acres, 2 had 51-100 acres, 10 had >100 acres. Within 0.25 mile, 11 colonies had forested habitats not suitable for foraging. Those 11 colonies had an average of 19 acres (SD = 25 acres) in this type; 9 had 1-25 acres, 1 had 26-50 acres, and 1 had 51-100 acres.

Table 2. Acres of habitats within 0.5 and 0.25 mile of the geometric center of 22 red-cockaded woodpecker colonies on forest products industry lands in Arkansas, Georgia, Louisiana, and Texas, 1992.

Habitat Variable	Mean	S.D.	Min.	Max.
0.5-mile radius				
Suitable foraging habitat	319.1	117.8	61.7	489.0
Forested but not suitable foraging	102.6	104.3	0	403.1
Young forest regeneration	68.6	84.6	0	300.6
Non-forested	10.3	23.5	0	100.0
0.25-mile radius				
Suitable foraging habitat	93.6	26.6	37.5	125.6
Forested but not suitable foraging	9.4	19.5	0	83.1
Young forest regeneration	22.0	24.8	0	90.0
Non-forested	0.9	2.3	0	9.4

For all 22 colonies, about 22% of forested habitats not suitable for foraging were pine plantations 4- to 10-inches DBH. Studies such as Hooper and Lennartz (1981) have shown that red-cockaded woodpeckers will forage (in some cases proportional to availability) in stands with merchantable pine stems ≤10 inches DBH. The majority of the colonies (14), however, had no plantations with 4- to 10-inch-DBH trees within 0.5 mile. The 8 colonies with such plantations had an average of 63 acres (SD=106) or about 12% of the area within 0.5 mile in this type. Only 2 colonies had 4-10-inch plantations within 0.25 mile; 1 was <1 acre and the other was 64 ac.

Young regeneration (<4 inches DBH) made up about 14% of the area within 0.5 mile of all colonies **(Table 2)**. For the 15 colonies with young plantations, this type encompassed an average of 101 acres (SD = 85) or 20% of the area within 0.5 mile. Two had 1-25 acres of young regeneration, 4 had 26-50 acres, 2 had 51-100 acres, and 7 had >100 acres. One colony had about 60% of the area within 0.5 mile in young regeneration.

Within 0.25 mile, an average of 18% of area was in young forest regeneration (**Table 2**). The majority of colonies (13) had young regeneration within 0.25 mile; 3 had 1-25 acres, 8 had 26-50 acres, and 2 had 51-100 acres. The 13 colonies with young regeneration within 0.25 mile, averaged 37 acres (SD = 21) or about 30% in this habitat type.

An average of 2% of the area within 0.5 mile of the colonies was in non-forested habitats (**Table 2**). Eleven colonies had no non-forested habitat, 9 had 1-25 acres, and 2 had 51-100 acres. The eleven colonies with non-forested habitat had an average of 21 acres (SD = 30) or about 4% in this type. Within 0.25 mile, only 4 colonies had non-forested habitat; all 4 had 1-25 acres. Those 4 colonies had an average of 5 acres (SD = 3) or about 4% in non-forested habitat.

Foraging substrate

Within 0.5 mile, the colonies averaged 55 ft^2 of pine BA per acre of suitable foraging habitat (**Table 3**). Most (19) colonies had >8,490 ft^2 pine basal area as recommended in the recovery plan (U.S. Fish and Wildlife Service 1985); the remaining 3, however, had <6,600 ft^2 BA. About 30% of total pine basal area in the 0.5-mile circle was located within 0.25 mile of the colony centers.

All of the colonies had >3000 ft^2 BA of pine trees >10 inches DBH within 0.5 mile, as proposed for private lands by Costa (1992). The majority of the colonies (18) exceeded the recommendation by at least 2,000 ft^2 BA. The average colony had 32 ft^2 BA in pine trees >10 inches DBH per acre of suitable foraging habitat (**Table 3**). About 29% of basal area in >10-inch-pine trees was located within 0.25 mile of the colonies.

Within 0.5 mile, colonies averaged 31 pine stems >10 inches DBH per acre of suitable foraging habitat (**Table 3**). Although many colonies (15) met or exceeded recovery recommendations (U.S. Fish and Wildlife Service 1985) with ≥6,350 pine stems >10 inches DBH, 7 did not. Most of the colonies (17) had ≥5,000 stems >10 inches DBH, as proposed by Costa (1992). Five colonies, however, did not meet this draft recommendation for private lands. About 28% of pine stems >10 inches DBH were located within 0.25 mile of the colony centers.

Within 0.25 mile, the colonies averaged 133 stems and 24 ft^2 BA in pine trees ≤10 inches DBH per acre of suitable foraging habitat (**Table 3**). About 31% of basal area and stems of pine trees ≤10 inches in the full 0.5 mile circle was within 0.25 mile of the colony centers.

Landscape patterns

Most (19) colonies had 3 or 4 habitat types present within 0.5 mile. Three colonies, however, had only 2 habitat types. Habitats within 0.5 mile were dissected into an average of 6.9 patches (SD = 4.1). Suitable foraging habitat was divided into an average of 2.2 patches (SD = 2.4, range = 1-11). Eighteen colonies had foraging habitat apportioned among only 1 or 2 patches.

Within 0.25 mile, an average of 2.2 habitats (SD = 0.7, range = 1-3) were apportioned among 3.7 patches (SD = 4.0, range = 1-19). Suitable foraging habitat within 0.25 mile was dissected into 1.4 patches (SD = 1.7, range = 1-9). Foraging habitat for 21 colonies was in only 1 or 2 patches.

Values for both C and T also suggested that habitats surrounding the colonies were clumped into relatively large, contiguous patches (**Table 4**). Scaled contagion values for habitats within the 0.5- and 0.25-mile circles, tended to be high; C for both circles averaged about 75% of the maximum value. Variability in C within the 0.5-mile circles was low, ranging only about 0.30. Within 0.25 mile of the colonies, however, C had a range of about 0.80. Values of T at both the 0.5-mile and 0.25-mile scales were low, averaging <0.1, and did not exceed 0.20.

Because the number of patches within 0.5 mile of the colonies was low (<10 for 15 colonies), fractal dimension was not calculated for each colony. Rather, fractal dimension was calculated for all 22 colonies combined. For all habitats within 0.5 mile of the colonies, fractal dimensions were 1.111 for suitable foraging habitat, 1.146 for forested habitats not suitable for foraging, 1.175 for young forest regeneration, 1.204 for non-forested habitats, and 1.134 for all habitats. Thus, patch shapes

Table 3. Structure of suitable foraging habitat within 0.5 and 0.25 mile of the geometric center of 22 red-cockaded woodpecker colonies on forest products industry lands in Arkansas, Georgia, Louisiana, and Texas, 1992.

Habitat Variable	Mean	S. D.	Minimum	Maximum
	0.5-mile radius			
BA[1] of pine trees >10in dbh	10,254	6,162	3,098	26,367
BA of pine trees ≤10 in dbh	7,193	4,600	968	15,893
Total pine BA	17,446	7,909	4,820	35,502
No. pine stems >10 in dbh	9,920	5,870	3,183	25,964
No. pine stems ≤10 in dbh	39,049	30,518	6,263	114,145
	0.25-mile radius			
BA of pine trees >10 in dbh	2,944	1,569	1,003	7,369
BA of pine trees ≤10 in dbh	2,247	1,490	286	5,837
Total pine BA	5,192	2,136	1,653	10,547
No. pine stems >10 in dbh	2,779	1,528	813	7,463
No. pine stems ≤10 in dbh	12,488	10,139	1,573	42,294

[1]Basal area is square feet.

Table 4. Indices to landscape pattern within 0.5 and 0.25 mile of the geometric center of 22 red-cockaded woodpecker colonies on forest products industry lands in Arkansas, Georgia, Louisiana, and Texas, 1992.

Habitat Variable	Mean	S.D.	Min.	Max.
	0.5-mile radius			
Scaled dominance	0.371	0.234	0.000	0.825
Scaled contagion	0.758	0.082	0.619	0.930
Scaled Taylor patchiness index	0.059	0.042	0.013	0.179
Angular sum	134	112	0	360
	0.25-mile radius			
Scaled dominance	0.434	0.339	0.013	1.000
Scaled contagion	0.759	0.120	0.197	1.000
Scaled Taylor patchiness index	0.027	0.026	0.000	0.115
Angular sum	102	99	0	360

for all habitat types within 0.5 mile, including regeneration, tended to be simple.

Within 0.25 mile, F values were very similar to those for the full 0.5 mile circle. F values were 1.072 for suitable foraging habitat, 1.196 for forested habitats not suitable for foraging, 1.159 for young forest regeneration, and 1.107 for all habitats. There were not enough non-forested habitats within 0.25 mile of the colony to allow computation of F for that habitat type.

Within 0.5 mile, scaled dominance averaged <40% of its possible maximum value (**Table 4**). Seventeen colonies had D values <0.5, 3 had D values of 0.5-0.75, and 2 had D values >0.75. Thus, the 0.5-mile landscapes tended not to be dominated by only 1 habitat type. Within 0.25 mile, foraging habitat was more dominant but D still averaged about 40% of its possible maximum value. Twelve colonies, however, had D values >0.5.

Within 0.5 mile, the angular sum of non-forested and young regeneration patches averaged 37% of its maximum possible value (**Table 4**). Seven colonies had A values <25, 1 had an A value of 51-100, 9 had A values of 101-200, and 5 had A values >200. Within 0.25 mile, A averaged 28% of the maximum possible value. Seven colonies had A values <25 within 0.25 mile, 2 had A values of 51-100, 7 had A values of 101-200, and 4 had A values >200.

The angular sum of a habitat patch may be affected by its size, distance from the center of the colony, shape, and orientation. Unfortunately, it is difficult to identify how much each of these factors makes to the overall value of A. Indices such as D, C, F, and T provide alternative measures of landscape pattern, and also may be useful predictors of RCW habitat quality, particularly when combined with other stand-level habitat descriptors.

CONCLUSIONS

The RCW colonies we examined were persisting in a broad range of conditions for most habitat variables. Although conditions for many of the colonies met or exceeded recommendations found in the recovery plan and proposed guidelines for private lands, some did not. These results suggest that RCW can persist on private lands in forests managed for timber and other products under even-aged and uneven-aged management systems. It is reasonable to hypothesize, on the basis of these data, that habitat adequacy is a complex multivariate issue and that some habitat variables may function in a compensatory manner. Future research on industry lands by NCASI will attempt to associate reproductive performance and long-term persistence of these and other RCW colonies with habitat and landscape characteristics.

Literature Cited

Abrahamson, W.G. and D.C. Hartnett. 1990. Pine flatwoods and dry prairies. pp. 103-149 *In* R.L. Myers and J.J. Ewel (eds.). Ecosystems of Florida, Univ. Central Florida Press, Orlando, FL.

Adams, L.W. and L.F. Dove. 1989. Wildlife reserves and corridors in the urban environment. Natl. Inst. Urban Wildl., Columbia, MD.

Affeltranger, C. 1971. The red heart disease of southern pines. pp. 96-99 *In* R.L. Thompson (ed.). The Ecology and Management of the Red-cockaded Woodpecker. Bureau of Sport Fisheries and Wildlife, U. S. Dept. Int., and Tall Timbers Res. Sta., Tallahassee, FL.

Afifi, A.A. and V. Clark. 1984. Computer-aided multivariate analysis. Lifetime Learning Pubs., Belmont, CA.

Alexander, R.R. 1964. Minimizing windfall around clear cuttings in spruce-fir forests. Forest Sci. 10:130-142.

Allen, D.H. 1991. An insert technique for constructing artificial red-cockaded woodpecker cavities. U.S. Dept. Agr. For. Serv. Gen Tech Rep. SE-73, Southeastern For. Exp. Sta., Asheville, NC.

Allen, D.H., K.E. Franzreb and R.E.F. Escano. 1993. Efficacy of translocation strategies for red-cockaded woodpeckers. Wildl. Soc. Bull. 21:155-159.

Allendorf, F.W. and R.F. Leary. 1986. Heterozygosity and fitness in natural populations of animals. pp. 57-76 *In* M.E. Soulé (ed.). Conservation biology: the science of scarcity and diversity. Sinauer Associates, Sunderland, MA.

American Ornithologists' Union. 1983. Check-list of North American Birds. American Ornithologists' Union, Washington, DC. 6th ed. Allen Press, Inc., Lawrence, KS..

Andrews, P.L. 1986. BEHAVE: Fire behavior prediction and fuel modeling system—burn subsystem, Part 1. U.S. Dept. Agr. For. Serv. Gen. Tech. Rep. INT-194, Intermtn. Res. Sta., Ogden, UT.

Anonymous. 1928. Third storm of the season sweeps over the turpentine country. Savannah Weekly Naval Stores Review and J. Trade. 38(25):3, 30.

Anonymous. 1935. Severe losses in Perry, Florida, section result of storm. Savannah Weekly Naval Stores Review and J. Trade. 45(26):15.

Anonymous. 1952. Mean number of thunderstorm days in the United States. U.S. Dept. Commerce, Weather Bureau Tech. Pap. No. 19.

Anonymous. 1965. Coordinated campaign to salvage timber downed by Betsy. pp. 32 In Oct. 15 issue, Southern Lumberman.

Anonymous. 1990. Scientific summit on the red-cockaded woodpecker: summary report. Southeast Negotiation Network, Georgia Institute of Technology, Atlanta, GA.

Anonymous. 1991a. ARC/INFO Version 6.0. Environmental Systems Research Institute, Inc., CA.

Anonymous. 1991b. Blackwater-Eglin Connector. File No. 880131-57-1, CARL Program. Florida Dept. Natural Resources. Tallahassee, FL.

Anthes, R.A. 1982. Tropical cyclones: their evolution, structure and effects. Meteorological Monogr. 19.

Arend, J.L. 1950. Influence of fire and soil on the distribution of eastern red cedar in the Ozarks. J. For. 48:129-130.

Assmann, E. 1970. The principles of forest yield study. Pergamon Press, Oxford.

Audubon, J.J. 1849. Ornithological biography. Vol. 5. Adam and Charles Black, Edinburgh, Scotland.

Avery, T.E. and H.E. Burkhart 1983. Forest measurements, 3rd ed. McGraw-Hill Book Co., NY.

Baillie, M. G. L. 1982. Tree-ring dating and archaeology. The Univ. of Chicago Press, Chicago, IL.

Baker, C.K., A.E. Fullwood and J.J. Collins. 1990. Lodging of winter barley (*Hordeum vulgare* L.) in relation to its degree of exposure to sulphur dioxide. New Phytologist 114:191-197.

Baker, D. 1989. Industry sues Forest Service over endangered bird policy. Region 8 For. Serv. Timber Purchasers Council, Atlanta, GA.

Baker, J.B. 1987. Silvicultural systems and natural regeneration methods for southern pines in the United States. pp. 175-191 *In* Proc. of the seminar on forest productivity and site evaluation. (ROC). Taipei, Taiwan. Taipei: Council of Agr.

Baker, J.B. and W.E. Balmer 1983. Loblolly pine. pp. 148-152 *In* Burns, R.M. comp. Silvicultural systems for the major forest types of the U.S. Dept. Agr, For. Serv. Handbook 445. Washington, D.C.

Baker, J.B. and O.G. Langdon. 1990. Loblolly pine. pp. 497-512 In R.M. Burns and B.H. Honkala (eds.). Silvics of North America. Vol. 1, U.S. Dept. Agr., For. Serv., Conifers. Agr. Handb. 654. Washington, D.C.

Baker, W.L. 1992. Effects of settlement and fire suppression on landscape structure. Ecology 73:1879-1887.

Baker, W.W. 1971. Progress report on life history studies of the red-cockaded woodpecker at Tall Timbers Research Station. pp. 44-59 *In* R.L. Thompson (ed.). The Ecology and Management of the Red-cockaded Woodpecker. Bureau of Sport Fisheries and Wildlife, U.S. Dept. Int., and Tall Timbers Res. Stat., Tallahassee FL.

Baker, W.W. 1982. The distribution, status and future of the red-cockaded woodpecker in Georgia. pp. 82-87 *In* R.R. Odom and J.W. Guthrie (eds.). Proceedings Nongame and Endangered Wildlife Symposium, Georgia Dept. Natural Resources, Game and Fish Division, Tech. Bull. WL 5.

Baker, W.W. 1983a. A non-clamp patagial tag for use on

red-cockaded woodpeckers. pp. 110-111. *In* D.A. Wood, (ed.). Red-cockaded Woodpecker Symposium II Proc., Florida Game and Fresh Water Fish Comm., Tallahassee FL.

Baker, W.W. 1983b. Decline and extirpation of a population of red-cockaded woodpeckers in northwest Florida. pp. 44-45 *In* D.A. Wood (ed.). Red-cockaded Woodpecker Symposium II Proceedings. Florida Game and Fresh Water Fish Commission, Tallahassee, FL.

Baker, W.W. and J.W. Thompson 1990. The distribution and status of the red-cockaded woodpecker (*Picoides borealis*) in Georgia, 1989. Unpub. Rept. to GA DNR on project 619-990284.

Baker, W.W., R.L. Thompson and R.T. Engstrom. 1980. The distribution and status of red-cockaded woodpecker colonies in Florida: 1969-1978. Florida Field Naturalist. 8(2):41-45.

Balboni, M.L. 1985. Storm damaged red-cockaded woodpecker colonies. Unpublished letter to District Ranger, Apalachicola R.D., Bristol, FL. Dated 11 Dec 1985.

Barrett, S.W. and S.F. Arno. 1988. Increment-borer methods for determining fire history in coniferous forests. U.S. Dept. Agr. For. Serv. Gen. Tech. Rep. INT-244, Intermtn. Res. Station, Ogden, UT.

Beal, F.E.L. 1911. Food of the woodpeckers of the United States. U.S. Dept. Agr. Biol. Survey, Bull. No. 37.

Bechtold, W. A., M. J. Brown and John B. Tansey. 1987. Virginia's forests. U.S. Dept. Agr. For. Serv. Res. Bull. SE-95, Southeastern For. Exp. Sta., Asheville, NC.

Beckett, T. 1971. A summary of red-cockaded woodpecker observations in South Carolina. pp. 87-95 *In* R.L. Thompson (ed.). Proc. Symp. The Ecology and Management of the Red-cockaded Woodpecker. Bureau of Sport Fisheries and Wildlife, U.S. Dept. Int., and Tall Timbers Res. Sta., Tallahassee FL.

Beever, J.W. III and K.A. Dryden. 1992. Red-cockaded woodpeckers and hydric slash pine flatwoods. Trans. 57th No. Amer. Wildl. and Nat. Res. Conf. 57:693-700.

Beier, P. and S. Loe. 1992. A checklist for evaluating impacts to wildlife movement corridors. Wildl. Soc. Bull. 20:434-440.

Beilmann, A.P. and L.G. Brenner. 1951. The recent intrusion of forests in the Ozarks. Ann. Missouri Bot. Garden 38:261-282.

Belanger, R.P. 1980. Silvicultural guidelines for reducing losses to the southern pine beetle. pp. 165-177 *In* R.C. Thatcher, J.L. Searcy, J.E. Coster and G.D. Hertel (eds.). The Southern Pine Beetle. U.S. Dept. Agr., For. Serv. Tech. Bull. 1631.

Belanger, R.P. and E.V. Brender. 1968. Influence of site index and thinning on the growth of planted loblolly pine. Georgia For. Res. Pap. 57, Georgia For. Res. Coun., Macon, GA.

Belanger, R.P. and B.F. Malac. 1980. Silviculture can reduce losses from the southern pine beetle. U.S. Dept. Agr. Comb. For. Pest Res. and Dev. Prog., Agr. Handbook No.576.

Belanger, R.P., G.E. Hatchell and G.E. Moore. 1977. Soil and stand characteristics related to southern pine beetle infestations: a progress report for Georgia and North Carolina. pp. 99-107 *In* Proc. Sixth South. For. Soils Workshop. U.S. Dept. Agr. For. Serv. Southeastern. Area State and Private Forestry, Atlanta, GA.

Belanger, R.P., R.L. Hedden and M. Lennartz. 1988. Potential impact of the southern pine beetle on red-cockaded woodpecker colonies in the Georgia Piedmont. So. J. Appl. For. 12:194-199.

Belanger, R.P., R.L. Hedden and F.H. Tainter 1986. Managing Piedmont forests to reduce losses from the littleleaf disease-southern pine beetle colmplex. U.S. Dept. Agr. For. Serv., Agr. Handbk. 649, Washington, D.C.

Belanger, R.P., R.L. Porterfield and C.E. Rowell. 1981. Development and validation of systems for rating susceptibility of natural stands in the Piedmont of Georgia to attack by the southern pine beetle. pp. 79-86 *In* R.L. Hedden, S.J. Barras and J.E. Coster (eds.). Proc., Symp. on Hazard Rating Syst. For. Insect Pest Manage. U.S. Dept. Agr. For. Serv. Gen. Tech. Rep. WO-27, Washington, D.C.

Bellingham, P.J. 1991. Landforms influence patterns of hurricane damage: evidence from Jamaican montane forests. Biotropica 23:427-433.

Bellrose, F.C. 1955. Housing for wood ducks. Illinois Nat. Hist. Surv. Circ. 45. Second printing, rev.

Bennett, W.H. 1968. Timber management and southern pine beetle research. For. Farmer 27:12-13.

Bennett, W.H. 1971. Silvicultural techniques will help control bark beetles. pp. 289-295. *In* Proc. 1971 Southern Regional Tech. Conf. Soc. Amer. Foresters.

Bent, A.C. 1939. Life histories of North American woodpeckers. 1964 Dover publications reprint of U.S. National Museums Bull., No. 174, 334 pp.

Bergstrom, J.C. and J.R. Stoll. 1987. A test of contingent market bid elicitation procedures for piecewise valuation. Western J. Agr. Economics 12:104-108.

Beyer, G.E., A. Allison and H.H. Kapman. 1908. List of birds of Louisiana. Part V. Auk 15:439-448.

Billings, R.F. 1980. Direct Control. pp. 179-192 *In* R.C. Thatcher, J.L. Searcy, J.E. Coster and G.D. Hertel (eds.). The southern pine beetle. U.S. Dept. Agr. For. Serv. Tech. Bull 1631.

Billings, R.F. and B.G. Hynum. 1980. Southern pine beetle: guide for predicting timber losses from expanding spots in east Texas. Circ. 249, Texas For. Serv.

Billings, R.F. and W.W. Upton. 1993. Effectiveness of synthetic behavioral chemicals for manipulation and control of southern pine beetle infestations in East Texas. pp. 555-563 *In* J.C. Brissette (ed.). Proceedings Seventh Bienniel Southern Silvicultural Res. Conference. U.S. Dept. Agr. For. Serv. Gen. Tech. Rep. SO-93, New Orleans, LA.

Billings, R.F. and F.E. Varner. 1986. Why control

southern pine beetle infestations in wilderness areas? The Four Notch and Huntsville State Park experiences. pp. 129-134 *In* Kulhavy, D.L. and R.N. Conner (eds.). Wilderness and natural areas in the Eastern United States: A management challenge. Center for Applied Studies, School of Forestry, Stephen F. Austin State Univ., Nacogdoches, TX.

Billings, R.F., C.M. Bryant and K.H. Wilson. 1985. Development, implementation and validation of a large area hazard- and risk-rating system for southern pine beetle. pp. 226-232 *In* S.J. Branham and R.C. Thatcher (eds.). Integrated Pest Management Research Symposium: The Proceedings. U.S. Dept. Agr. For. Serv. Gen. Tech. Rep. SO-56, New Orleans, LA.

Blanche, C.A., J.D. Hodges, T.E. Nebeker and D.M. Moehring. 1983. Southern pine beetle: the host dimension. Bull. 917, Miss. Agr. and For. Exp. Sta., Mississippi State, MS.

Blanche, C.A., T.E. Nebeker, J.D. Hodges, B.L. Karr and J.J. Schmitt. 1985. Effect of thinning damage on bark beetle susceptibility indicators in loblolly pine. Proc. Third Biennial Southern Silvicultural Conference. U.S. Dept. Agr. For. Serv. Gen. Tech. Rep. SO-54, New Orleans, LA.

Block, W.M., D.M. Finch and L.A. Brennan. in preparation. Single-species versus multiple species approaches for the management of neotropical migrant birds. *In* Management of Neotropical Migrant Birds, Oxford Univ. Press.

Blue, R.J. 1985. Home range and territory of red-cockaded woodpeckers utilizing residential habitat in North Carolina. M.S. Thesis, North Carolina State Univ., Raleigh, NC.

Bock, C.E. and J.H. Bock. 1974. On the geographical ecology and evolution of the 3-toed woodpeckers, *Picoides tridactylus* and *P. arcticus*. Amer. Midl. Nat. 92:397-405.

Bodie, W.C. 1985. Hurricane Kate damage to red-cockaded woodpecker habitat. Unpublished letter to Regional Forester, U.S. Dept. Agr. For. Serv., Atlanta, GA. Dated 16 Dec 1985.

Bowman, R. and J.W. Fitzpatrick. 1993. Florida scrub jay and red-cockaded woodpecker populations at the Avon Park Air Force Range. Final Rep., Department of Defense, Avon Park, FL.

Boyce, J.S. 1961. Forest pathology. McGraw-Hill Book Company, Inc. New York.

Boyer, W.D. 1963. Development of longleaf pine seedlings under parent trees. U.S. Dept. Agr. For. Serv., South. For. Exp. Sta. Res. Pap. S0-4, New Orleans, LA.

Boyer, W.D. 1990. Longleaf pine. pp. 405-412 *In* R.M. Burns and B.H. Honkala (eds.). Silvics of North America. Vol. 1, Conifers. U.S. Dept. Agr., For. Serv., Agr. Handbook 654, Washington, D.C.

Boyer, W.D. 1991. Manuscript review, restoration of longleaf pine. Unpublished letter to G. Bengston, Center for Forested Wetlands Research, Charleston, SC. Dated 7 May 1991.

Boyer, W.D. 1993. Long-term development of regeneration under longleaf pine seedtree and shelterwood stands. So. J. Appl. For. 17: 10-15.

Boyer, W.D. and D.W. Peterson. 1983. Longleaf pine. pp. 153-156 *In* R.M. Burns (ed.). Silvicultural systems for the major forest types of the United States. U.S. Dept. Agr. For. Serv. Agr. Handb. 445, Washington, D.C.

Boyer, W.D. and J.B. White, 1990. Natural regeneration in longleaf pine. pp. 94-113 *In* R.M. Farrar (ed.) U.S. Dept. Agr., For. Serv., Gen. Tech. Rep. S0-75, Southern For. Exp. Sta., New Orleans, LA.

Bradshaw, D.S. 1990. Habitat quality and seasonal foraging patterns of the red-cockaded woodpecker (*Picoides borealis*) in southeastern Virginia. Unpublished M.A. thesis, College of William and Mary, Williamsburg, VA.

Bradwell, J. 1947. Here and there in the gum belt. Savannah Weekly Naval Stores Review and J. Trade. 57(29):22.

Bramlett, D.L. 1990. Pond pine. pp. 470-475 *In* R.M. Burns and B.H. Honkala (eds.). Silvics of North America. Vol. 1, Conifers. U.S. Dept. Agr., For. Serv., Agr. Handbook 654. Washington, D.C.

Bramlett, D.L. and R.N. Kitchens 1983. Virginia pine. pp. 167-169 *In*: R. M. Burns, (comp.). Silvicultural systems for the major forest types of the U.S. Agr. For. Serv. Handb 445. Washington, D.C.

Branham, S.J. and R.C. Thatcher (eds.). 1985. Proc. Integrated Pest Management Symposium. U.S. Dept. Agr. For. Serv. Gen. Tech. Rep. SO-56, New Orleans, LA.

Bray, W.L. 1901. Destruction of timber by the Galveston storm. The Forester. 7:52-56.

Brennan, L.A. 1991. How can we reverse the northern bobwhite population decline? Wildlife Soc. Bull. 19:544-555.

Brewer, R. and M.T. McCann. 1985. Spacing in acorn woodpeckers. Ecology 66:307-308.

Bridges, E.L. and S.L. Orzell. 1989. Longleaf pine communities of the West Gulf Coastal Plain. Natur. Areas J. 9:246-263.

Brown, C.A. 1945. Louisiana trees and shrubs. Louisiana Forestry Commission, Bull. No. 1.

Brown, J.L. 1974. Alternate routes to sociality in jays—with a theory for the evolution of altruism and communal breeding. Amer. Zool. 14:63-80.

Brown, J.L. 1978. Avian communal breeding systems. Ann. Review Ecol. Syst. 9:123-155.

Brown, L.R. and J.J. Jacobson. 1987. The future of urbanization: facing the ecological and economic constraints. Worldwatch Paper 77, Worldwatch Institute, Washington, D.C.

Brueckheimer, W.R. 1988. Leon County hunting plantations an historical and architectural survey. Volume I. Final Report, Historic Tallahassee Preservation Board, Tallahassee, FL.

Bruner, W.E. 1931. The vegetation of Oklahoma. Ecol. Mono. 1:99-188.

Bull, E.L. 1978. Specialized habitat requirements of

birds: snag management, old growth, and riparian habitat. pp.74-82 *In* R.M. DeGraaf (ed.). Proc. workshop on nongame bird habitat management in the coniferous forests of the western United States. U.S. Dept. Agr. For. Serv. Gen. Tech. Rep. PNW-64, Portland, OR.

Bull, E.L. and E.C. Meslow. 1977. Habitat requirements of the pileated woodpecker in northeastern Oregon. J. For. 75:335-337.

Burgess, J.W. 1983. Reply to a comment by R.L. Mumme et al. Ecology 64:1307-1308.

Burgess, J.W., D. Roulston and E. Shaw. 1982. Territorial aggregation: an ecological spacing strategy in acorn woodpeckers. Ecology 63:575-578.

Burns, R.M. 1983. Introduction. pp. 1-2 *In* R.M. Burns (comp.). Silvicultural systems for the major forest types of the United States. U.S. Dept. Agr. For. Serv. Agr. Handbook 445. Washington:

Burstein, H. 1975. Finite population correction for binomial confidence limits. J. Amer. Statist. Assoc. 70:67-69.

Burt, W.H. 1943. Territoriality and home range concepts as applied to mammals. J. Mammal. 24:346-352.

Burt, W.H. and R.P. Grossenheider. 1976. A field guide to the mammals. Houghton Mifflin Co., Boston, MA.

Burton, J.D. and D.M. Smith. 1972. Guying to prevent wind sway influences loblolly pine growth and wood properties. U.S. Dept. Agr. For. Serv., Res. Paper SO-80,. Southern Forest Exp. Sta., New Orleans, LA.

Bury, R.B. and P.S. Corn. 1987. Evaluation of pitfall trapping in northwestern forests: trap arrays with drift fences. J. Wildlife Manage. 51:112-119.

Cahal, R.R., D.L. Kulhavy, W.G. Ross and J.G. Gage. 1995. Hardwood midstory control in red-cockaded woodpecker colonies: analysis of change in stand structure and species composition. *In* D.L. Kulhavy, R.G. Hooper and R. Costa (eds.). Red-cockaded Woodpecker: Recovery, Ecology and Management, Center for Applied Studies, College of Forestry, Stephen F. Austin State Univ., Nacogdoches, TX.

Cain, M.D. 1987. Survival patterns of understory woody species in a pine-hardwood forest during 28 years without timber management. pp 141-147 *In* R. L. Hay, F. W. Woods and H. Deselm (eds.). Proc. Central Hardwood Conf. VI, Knoxville, TN.

Calhoun, M. (ed) 1992. Louisiana Almanac, Pelican Pub. Co. Gretna, LA.

Carney, C.B. and A.V. Hardy. 1962. North Carolina hurricanes. U.S. Weather Bureau, Raleigh, NC.

Carter, J.H., III and R.T. Engstrom. 1995. Use of artificial cavities for red-cockaded woodpecker mitigation: two studies. *In* D.L. Kulhavy, R.G. Hooper and R. Costa (eds.). Red-cockaded Woodpecker: Recovery, Ecology and Management, Center for Applied Studies, College of Forestry, Stephen F. Austin State Univ., Nacogdoches, TX.

Carter, J.H.,III, R.T. Stamps and P.D. Doerr. 1983a. Red-cockaded woodpecker distribution in North Carolina. pp. 20-23 *In* D.A. Wood (ed.). Proc. Red-cockaded Woodpecker Symp. II, Florida Game and Fresh Water Fish Comm., Tallahassee, FL.

Carter, J.H. III, R.T. Stamps and P.D. Doerr. 1983b. Status of the red-cockaded woodpecker in the North Carolina Sandhills. pp. 24-29 *In* D.A. Wood (ed.).. Proc. Red-cockaded Woodpecker symp. II FL Game and Fresh Water Fish Comm., Tallahassee, FL.

Carter, J.H., III, J.R. Walters and P.D. Doerr. 1995. Red-cockaded woodpeckers in the North Carolina Sandhills: A 12-year population study. *In* D.L. Kulhavy, R.G. Hooper and R. Costa (eds.). Red-cockaded Woodpecker: Recovery, Ecology and Management, Center for Applied Studies, College of Forestry, Stephen F. Austin State Univ., Nacogdoches, TX.

Carter, J.H., III, J.R. Walters, S.H. Everhart and P.D. Doerr. 1989. Restrictors for red-cockaded woodpecker cavities. Wildl. Soc. Bull. 17: 68-72.

Carter, K.K. and A.G. Snow Jr. 1990. Virginia pine. pp. 513-519. *In*: R.M. Burns and B.H. Honkala (eds.). Silvics of North America, Vol. 1, conifers. U.S. Dept. Agr.: For. Serv., Agr. Handb. 654, Washington, DC.

Carter, P.R. and K.D. Hudelson. 1988. Influence of simulated wind lodging on corn growth and grain yield. J. Prod. Agr. 1:295-299.

Carter, W.A. 1965. Ecology of the summer nesting birds of the McCurtain Game Preserve. Ph.D. Thesis, Oklahoma State Univ., Stillwater.

Caswell, H. 1989. Matrix Population Models. Sinauer Associates, Sunderland, MA.

Cely, J.E. 1991. Wildlife effects of Hurricane Hugo. J. Coastal Res. SI No. 8, 319-326.

Cely, J.E. and D.P. Ferral. 1995. Status and distribution of the red-cockaded woodpecker in South Carolina. *In* D.L. Kulhavy, R.G. Hooper and R. Costa (eds.). Red-cockaded Woodpecker: Recovery, Ecology and Management, Center for Applied Studies, College of Forestry, Stephen F. Austin State Univ., Nacogdoches, TX.

Chambers, J.M., W.S. Cleveland, B. Kleiner and P.A. Tukey. 1983. Graphical methods for data analysis. Duxbury Press, Boston, MA.

Chapman, C.S. 1905. A working plan for forest lands in Berkeley County, South Carolina. U.S. Dept. Agr., Bureau For., Bull. 56, Washington, DC.

Chapman, H.H. 1923. The causes and rate of decadence in stands of virgin longleaf pine. The Lumber Trade J. 84:11,16-17.

Chesser, R. 1981. Chapter 4. Isolation by distance: relationship to the management of genetic resources. pp. 66-77 *In* Schoenwald-Cox, C., S. Chambers, B. MacBryde and L. Thomas (eds.). Genetics and Conservation. A reference for managing wild animals and plant populations. Benjamin/Cummings Pub. Co. Menlo Park, CA.

Chrismer, G.M., W.G Ross and D.L. Kulhavy. 1995. Disturbance frequency and impact in red-cockaded woodpecker habitat in East Texas. *In* D.L. Kulhavy, R.G. Hooper and R. Costa (eds.). Red-cockaded Woodpecker: Recovery, Ecology and Management, Center for Applied Studies, College of Forestry,

Stephen F. Austin State Univ., Nacogdoches, TX.

Christensen, N.L. 1981a. Fire regimes in southeastern ecosystems. pp. 112-136 *In* H.A. Mooney, T.M. Bonnickson, N.L. Christensen, J.E. Lotan and W.A. Reiners (eds.). Fire regimes and ecosystem properties. U.S. Dept. Agr. For. Serv. Gen. Tech. Rep. WO-26, Washington, D.C.

Christensen, N.L. 1981b. The structure and development of pocosin vegetation. pp. 43-61 *In* Richardson, C.J. (ed.). Pocosin Wetlands. Hutchinson Ross Publishing Co.

Christensen, N.L. 1988. Vegetation of the Southeastern Coastal Plain. pp. 317-363 *In* Barbour, M.G. and W.D. Billings (eds.). North American Terrestrial Vegetation. Cambridge: Cambridge Univ. Press.

Clark, A. 1992. Influence of tree factors and site on formation of heartwood in loblolly and longleaf pine for red-cockaded woodpecker colonization in the southeast. Proc. Ann. Conf. Southeastern. Assoc. of Fish and Wildl. Agencies 46:79-87.

Clark, A. 1993. Characteristics of timber stands containing sufficient heartwood for cavity excavation by red-cockaded woodpecker clans. pp. 621-626 *In* J.C. Brissette (ed.). Proc. 7th Biennial Southern Silv. Res. Conf., Southern For. Exp. Sta., New Orleans, LA.

Clark, T.D. 1986. Ships to nowhere. The southern yellow pine fleet of World War I. J. For. Hist. 29:4-16.

Clutton-Brock, T.H. (ed.) 1988. Reproductive Success. Univ. Chicago Press, Chicago.

Cochran, W.G. 1977. Sampling Techniques. 3rd. ed. John Wiley, New York.

Conant, R. and J.T. Collins. 1991. A field guide to reptiles and amphibians of eastern/central North America. 3rd ed. Houghton Mifflin Co. Boston.

Conner, R.N. 1976. Nesting habitat for red-headed woodpeckers in southwestern Virginia. Bird-Banding 47:40-43.

Conner, R.N. 1980. Foraging habitats of woodpeckers in southwestern Virginia. J. Field Ornithol. 51:119-127.

Conner, R.N. and B.A. Locke. 1982. Fungi and red-cockaded woodpecker cavity trees. Wilson Bull. 94:64-70.

Conner, R.N. and B.A. Locke. 1983. Artificial Inoculation of red heart fungus into loblolly pines pp. 81-82 *In* D.A. Wood, (ed.). Red-cockaded Woodpecker Symp. II, Florida Game and Fresh Water Comm. and U.S. Dept. Int. Fish and Wildl. Serv. Tallahassee, FL.

Conner, R.N. and K.A. O'Halloran. 1987. Cavity tree selection by red-cockaded woodpeckers as related to growth dynamics of southern pines. Wilson Bull. 99:398-412.

Conner, R.N. and D.C. Rudolph. 1989. Red-cockaded woodpecker colony status and trends on the Angelina, Davy Crockett and Sabine National Forests. U.S. Dept. Agr., For. Serv., Res. Pap SO-250, New Orleans, LA.

Conner, R.N. and D.C. Rudolph. 1991a. Effects of midstory reduction and thinning in red-cockaded woodpecker cavity tree clusters. Wildl. Soc. Bull. 19:63-66.

Conner, R.N. and D.C. Rudolph. 1991b. Forest habitat loss, fragmentation and red-cockaded woodpecker populations. Wilson Bull. 103: 446-457.

Conner, R.N. and D.C. Rudolph 1995a. Excavation dynamics and use patterns of red-cockaded woodpecker cavities: relationships with cooperative breeding. *In* D.L. Kulhavy, R.G. Hooper and R. Costa (eds.). Red-cockaded Woodpecker: Recovery, Ecology and Management, Center for Applied Studies, College of Forestry, Stephen F. Austin State Univ., Nacogdoches, TX.

Conner, R.N. and D.C. Rudolph. 1995b. Wind damage to red-cockaded woodpecker cavity trees on eastern Texas National Forests. *In* D.L. Kulhavy, R.G. Hooper and R. Costa (eds.). Red-cockaded Woodpecker: Recovery, Ecology and Management, Center for Applied Studies, College of Forestry, Stephen F. Austin State Univ., Nacogdoches, TX.

Conner, R.N., R.G. Hooper, H.S. Crawford and H.S. Mosby. 1975. Woodpecker nesting habitat in cut and uncut woodlands in Virginia. J. Wildl. Manage. 39:144-150.

Conner, R.N., J.C. Kroll and D.L. Kulhavy. 1983. The potential of girdled and 2,4-D-injected southern red oaks as woodpecker nesting and foraging sites. South. J. Appl. For. 7:125-128.

Conner, R.N., O.K. Miller, Jr. and C.S. Adkisson. 1976. Woodpecker dependance on trees infected by fungal heart rots. Wilson Bull. 88:575-581.

Conner, R.N., D.C. Rudolph, D.L. Kulhavy and A.E. Snow. 1991a. Causes of mortality of red-cockaded Woodpecker cavity trees. J. Wildl. Manage. 55:531-537.

Conner, R.N., A.E. Snow and K.A. O'Halloran. 1991b. Red-cockaded woodpecker use of seed-tree/shelterwood cuts in eastern Texas. Wild. Soc. Bull. 19: 67-73.

Connor, M.D. and H.N. Wallace. 1982. Biological evaluation of southern pine beetle on Noxubee National Wildlife in Mississippi *In*: A.C. Mangini, M.C. Hicks and R.C. Kertz (eds). 1988. Biological evaluation of southern pine beetle on the Noxubee National Wildlife, Mississippi. U. S. Dept. Agr. For. Serv. Alexandria Field Office Rep. No. 89-2-3.

Copeyon, C.K. 1990. A technique for constructing cavities for the red-cockaded woodpecker. Wildlife Soc. Bull. 18:303-311.

Copeyon, C.K., J.R. Walters and J.H. Carter. 1991. Induction of red-cockaded woodpecker group formation by artificial cavity construction. J. Wildl. Manage. 55:549-556.

Costa, R. 1992. Draft red-cockaded woodpecker procedures manual for private lands. U.S. Fish and Wildlife Serv.. Atlanta, GA.

Costa, R. 1995. Red-cockaded woodpecker species recovery: challenges, responsibilities, and opportunities for the agencies and individuals involved. *In* D.L. Kulhavy, R.G. Hooper and R. Costa (eds.). Red-cockaded Woodpecker: Recovery, Ecology and Management, Center for Applied Studies, College of Forestry, Stephen F. Austin State Univ., Nacogdoches, TX.

Costa, R. and R.E.F. Escano. 1989. Red-cockaded woodpecker: status and management in the southern region in 1986. U.S. Dept. Agr., For. Serv., Tech. Pub. R8-TP 12, Southern Region, Atlanta, GA.

Coster, J.E., T.L. Payne, E.R. Hart and L.J. Edson. 1977. Aggregation of the southern pine beetle in response to attractive host trees. Environ. Entomol. 6:725-731.

Coulson, R.N. 1980. Population dynamics. pp. 71-105 *In* R.C. Thatcher, J.L. Searcy, J.E. Coster and G.D. Hertel (eds.). The southern pine beetle. U. S. Dept. Agr. For. Serv. Sci. and Educ. Admin. Tech. Bull. 1631.

Coulson, R.N., P.B. Hennier, R.O. Flamm, E.J. Rykiel, L.C. Hu and T.L. Payne. 1983. The role of lightning in the epidemiology of the southern pine beetle. Zeit. ang. Entomol. 96:182-193.

Coulson, R. N., E. J. Rykiel, M. C. Saunders, T. L. Payne, R. O. Flamm and P. B. Hennier. 1985. A conceptual model of the role of lightning in the epidemiology of the southern pine beetle. *In* L. Safranyik (ed.).The role of the host in population dynamics of forest insects. Proc. IUFRO Conf. Banff, Alberta, Canada.

Coulson, R.N., R.O. Flamm, P.E. Pulley, T.L. Payne, E.J. Rykiel and T.L. Wagner. 1986. Response of the southern pine bark beetle guild to host disturbance. Environ. Entomol. 15: 859-68.

Coutts, M.P. 1983. Root architecture and tree stability. Plant and Soil 71:171-188.

Cox, J., R. Kautz, T. Gilbert and M. MacLaughlin. 1994. Closing the gaps in Florida's wildlife habitat conservation system. Nongame Wildlife Program Tech. Rep., Florida Game and Fresh Water Fish Commission. Tallahassee, FL.

Coyne, J.F. and L.H. Lott. 1976. Toxicity of substances in pine oleoresin to southern pine beetles. J. Georgia Entomol. Soc. 11:301-305.

Croker, T.C. 1958. Soil depth affects windfirmness of longleaf pine. J. For. 56:432.

Crosby, G.T. 1971. Home range characteristics of the red-cockaded woodpecker in Florida. pp.60-70 *In* R.L. Thompson (ed.). The ecology and management of the Red-cockaded Woodpecker. Bur. Sport Fish. Wildl. and Tall Timbers Res. Sta.

Curtis, J.D. 1943. Some observations on wind damage. J. For. 41:877-882.

Cushwa, Charles T. 1988. A preliminary assessment of forest wildlife habitat in Virginia. Final report, Multi-State Fish and Wildlife Information Systems Project, Blacksburg, VA.

Cutler, D.F., P.E. Gasson and M.C. Farmer. 1990. The wind blown tree survey: analysis of results. Arborcult. J. 14:265-286.

Daniel, W.W. 1990. Procedures that utilize data from three or more independent samples, Ch. 6 *In* Applied Nonparametric Statistics. 2nd ed. PWS-KENT Pub. Co., Boston, MA.

Davis, K.P. 1966. Forest management: regulation and valuation. McGraw-Hill Book Co.. New York. 519 p.

Dawson, W.R., J.D. Ligon, J.R. Murphy, J.P. Myers, D. Simberloff and J. Verner. 1986. Report of the advisory panel on the Spotted Owl. Audubon Conservation Rep. No. 7.

Day, W.R. 1950. The soil conditions which determine wind-throw in forests. Forestry 23:90-95.

DeFazio, J.T. and M.R. Lennartz. 1987. Establishment of viable red-cockaded woodpeckers at the Savannah River Plant—Progress Report (Unpublished). U. S. Dept. Agr. For. Serv., Southeastern For. Exp. Sta.

DeFazio, J. T., Jr., M. A. Hunnicutt, M. R. Lennartz, G. L. Chapman and J. A. Jackson. 1987. Red-cockaded woodpecker translocation experiments in South Carolina. Proc. Ann. Conf. Southeastern. Assoc. Fish and Wildl. Agencies 41: 311-317.

DeLotelle, R.S. and R.J. Epting. 1988. Selection of old trees for cavity excavation by red-cockaded woodpeckers. Wildl. Soc. Bull. 16(1): 48-52.

DeLotelle, R.S. and R.J. Epting. 1992. Reproduction of the red-cockaded woodpecker in central Florida. Wilson Bull. 104 (2):285-294.

DeLotelle, R.S. and J.R. Newman. 1983. Possible factors influencing red-cockaded woodpecker colony abandonment: a case study. pp. 104-106 *In* D.A. Wood (ed.). Red-cockaded Woodpecker Symposium II Proc. Florida Game Fresh Water Fish Comm., Tallahassee.

DeLotelle, R.S., R.J. Epting and J.R. Newman. 1983a. Preconstruction monitoring of red-cockaded woodpeckers at the Orlando Utilities Commission Curtis H. Stanton Energy Center. Environ. Sci. and Eng. Rpt. 80-309-701.

DeLotelle, R. S., J. R. Newman and A. E. Jerauld. 1983b. Habitat use by red-cockaded woodpeckers in central Florida. Pages 59-67 *In*, D.A. Wood (ed.). Red-cockaded Woodpecker Symp. II Proc. Florida Game and Freshwater Fish Commission, Tallahassee, FL.

DeLotelle, R.S., R.J. Epting and J.R. Newman. 1987. Habitat use and territory characteristics of red-cockaded woodpeckers in central Florida. Wilson Bull. 99:202-217.

Dennis, J.V. 1971b. Species using red-cockaded woodpecker holes in northeastern South Carolina. Bird-Banding 42: 79-87.

Dennis, J.V. 1971c. Utilization of pine resin by the red-cockaded woodpecker and its effectiveness in protecting roosting and nest sites. pp. 78-86 *In* R.L. Thompson (ed.). The Ecology and Management of the Red-cockaded Woodpecker. Bureau of Sport Fisheries and Wildlife, U. S. Dept. Int. and Tall Timbers Res. Stat., Tallahassee FL.

Dennis, J.V. 1971a, Map on file at The Belle W. Baruch Forest Science Institute. Georgetown, S.C.

Department of Defense. 1949. Eglin AFB Management Plan. Eglin AFB, FL.

Department of Defense. 1993. Eglin AFB natural resource management plan: Final draft. Eglin AFB,

FL.

Derr, H.J. and H. Enghardt. 1957. Some forestry lessons from Hurricane Audrey. Southern Lumberman 195:142-144.

DeWalle, D.R. 1983. Wind damage around clearcuts in the Ridge and Valley Province of Pennsylvania. J. For. 81:158-159, 172.

Dillman, D.A. 1978. Mail and telephone surveys: the total design method. New York: John Wiley & Sons, Inc.

Dillworth, J.R. and J.F. Bell. 1982. Variable probability sampling. Oregon State Univ. Book Stores. Corvallis.

Doerr, P.D., J.R. Walters and J.H. Carter, III. 1989. Reoccupation of abandoned clusters of cavity trees (colonies) by red-cockaded woodpeckers. Proc. Ann. Conf. Southeastern. Assoc. Fish Wildl. Agencies 43:326-336.

Doggett, C.A. 1989. North Carolina forest damage appraisal—Hurricane Hugo. Unpublished Report, North Carolina For. Serv., Raleigh.

Dopson, C.W. 1989. 89th Christmas Bird Count, Sapelo Island, Georgia Amer. Birds 43(4) 826.

Du Pratz, L. P. 1774. The history of Louisiana. English ed. T. Becket, London. Facsimile reproduction 1975, Louisiana State Univ. Press, Baton Rouge.

Dunning, J. B., Jr. (ed.). 1993. CRC Handbook of Avian Body Masses. CRC Press, Boca Raton, FL.

Dunston, C.E. 1910. Preliminary examination of the forest conditions of Mississippi. Bull. No. 7. Mississippi State Geological Survey.

Eichholz, M.W. and W.D. Koenig. 1992. Gopher snake attraction to birds' nests. Southwest. Nat. 37: 293-298.

Emigh, T.H. and E. Pollak. 1979. Fixation probabilities and effective population numbers in diploid populations with overlapping generations. Theor. Pop. Biol. 15:86-107.

Emlen, J.T. 1978. Density anomalies and regulatory mechanisms in land bird populations on the Florida peninsula. Amer. Nat. 112:265-286.

Emlen, S.T. 1982a. The evolution of helpers I. An ecological constraints model. Amer. Nat. 119:29-39.

Emlen, S.T. 1982b. The evolution of helping. II. The role of behavioral conflict. Amer. Nat. 119:40-53.

Emlen S.T. 1991. The evolution of cooperative breeding in birds and mammals. *In* J.R. Krebbs and N.B. Davies (eds.). Behavioral Ecology: an Evolutionary Approach. 3rd ed. Blackwell, Oxford.

Engstrom, R.T. and W.W. Baker. 1995. Red-cockaded woodpecker and Red Hills hunting plantations: inventory, management, and conservation. *In* D.L. Kulhavy, R.G. Hooper and R. Costa (eds.). Red-cockaded Woodpecker: Recovery, Ecology and Management, Center for Applied Studies, College of Forestry, Stephen F. Austin State Univ., Nacogdoches, TX.

Engstrom, R.T. and G.W. Evans. 1990. Hurricane damage to red-cockaded woodpecker (*Picoides borealis*) cavity trees. Auk 107:608-610.

Epting, R.J., R.S. DeLotelle and T. Beaty. 1995. Red-cockaded woodpecker territory and habitat use in Georgia and Florida. *In* D.L. Kulhavy, R.G. Hooper and R. Costa (eds.). Red-cockaded Woodpecker: Recovery, Ecology and Management, Center for Applied Studies, College of Forestry, Stephen F. Austin State Univ., Nacogdoches, TX.

Ernst, C.H. and R.W Barbour. 1989. Snakes of eastern North America. Geo. Mason Univ. Press. Fairfax, VA

Escano, R.E.F. 1995. Red-cockaded woodpecker, extinction or recovery: summary of status and management on our national forest. *In* D.L. Kulhavy, R.G. Hooper and R. Costa (eds.). Red-cockaded Woodpecker: Recovery, Ecology and Management, Center for Applied Studies, College of Forestry, Stephen F. Austin State Univ., Nacogdoches, TX.

Evans, K.E. and R.N. Conner. 1979. Snag management. pp.215-225 *In* R.M. DeGraaf, (tech. coord.). Management of north central and northeastern forests for nongame birds. U.S. For. Serv. Gen. Tech. Rep. NC-51, St. Paul, MN.

Everhart, S.H. 1986. Interspecific utilization of red-cockaded woodpecker cavities. Ph.D. thesis, N.C. State Univ., Raleigh.

Fares, Y., P.J.H. Sharpe and C.E. Magnuson. 1980. Pheromone dispersion in forests. J. Theor. Biol. 84: 335-59.

Fargo, W.S., R.N. Coulson, J.A. Gagne and J.L. Foltz. 1979. Correlation of southern pine beetle attack density, oviposition, and generation survival with host tree characteristics and preceeding beetle life stage within the host. Environ. Entomol. 8:624-628.

Farrar, R.M. Jr. 1984. Density control— natural stands. pp. 129-154 *In* B.L. Karr, J.B. Baker and T. Monaghan (eds.). Proc. symp. on the loblolly pine ecosystem (west region). Ext. Serv. Mississippi State Univ. . Pub. 1454. Mississippi State, MS.

Farrar, R.M. and W.D. Boyer. 1991. Managing longleaf pine under the selection system—promises and problems. pp. 357-368 *In* S.S. Coleman and D.G. Neary (eds.). Proceedings of the 6th biennial southern silvicultural research conference, U.S. Dept. Agr. For. Serv., Vol. 1. Gen. Tech. Rep. SE-70, Asheville, NC.

Farrar, R.M. and P.A. Murphy. 1987. Taper functions for predicting product volumes in natural shortleaf pines. U. S. Dept. Agr., For. Serv., Res. Pap. SO-234. Southern For. Exp. Sta., New Orleans, LA.

Farrar, R.M. Jr. and P.A. Murphy 1989. Objective regulation of selection-managed stands of southern pine—a progress report. pp 231-241. *In* J.H. Miller (eds.). Proc. of the fifth biennial southern silvicultural research conference. U.S. Dept. Agr., For. Serv., Gen. Tech. Rep. S0-74, New Orleans, LA.

Ferguson, E. R., C. B. Gibbs and R. C. Thatcher. 1961. "Cool" burns and pine mortality. Fire Contr. Notes 21(1):27-29.

Fickle, J.E. 1978. Defense mobilization in the southern pine industry: the experience of World War I. J. For. Hist. 22:206-223.

Fitch, H.S. 1958. Home ranges, territories, and seasonal movements of vertebrates of the Natural History Reservation. Univ. Kansas Publ., Mus. Nat. Hist.

11:63-326.

Fitch, H.S. 1963. Natural history of the black rat snake (*Elaphe o. obsoleta*) in Kansas. Copeia 1963: 649-658.

Flader, S.L. 1974. Thinking like a mountain. Univ. Nebraska Press, Lincoln, NE.

Flamm, R.O., R.N. Coulson and T.L. Payne. 1988. The southern pine beetle. pp. 531-553 *In* A.A. Berryman (ed.). Dynamics of Forest Insect Populations: Patterns, Causes and Implications. Plenum Press, NY.

Fleischer, R.C., R.F. Johnston and W.J. Klitz. 1983. Allozymic heterozygosity and morphological variance in house sparrows. Nature 304:628-630.

Florida Natural Areas Inventory and Florida Department of Natural Resources. 1990. Guide to the natural communities of Florida. FNAI & DNR, Tallahassee, FL.

Flory, C.H. 1960. Hurricane Gracie damages timber stands in South Carolina. pp. 48-50 *In* Report of the state commission of forestry for the year July 1, 1959 to June 30, 1960. South Carolina Forestry Comm., Columbia.

Fons, W.L. 1940. Influence of forest cover on wind velocity. J. For. 38:481-486.

Forman, R.T.T. and M. Godron. 1986. Landscape ecology. John Wiley & Sons. New York, N. Y.

Foster, D.R. 1988. Species and stand response to catastropic wind in central New England, U.S.A. J. Ecology 76:135-151.

Foster, J.H. 1912. Forest conditions in Louisiana. Bull. 114. U. S. Dept. Agr., For. Serv., Washington, DC.

Foti, T.L. and S.M. Glenn. 1991. The Ouachita Mountain landscape at the time of settlement. pp. 49-66 *In* D. Henderson and L.D. Hedrick (eds.). Restoration of old growth forests in the Interior Highlands of Arkansas and Oklahoma, Proc. Conf. Winrock International, Morrilton, AR.

Franklin, I.R. 1980. Evolutionary changes in small populations. pp. 135-149 *In* M.E. Soulé and B.A. Wilcox (eds.). Conservation biology. An evolutionary-ecological perspective. Sinauer Associates, Sunderland, MA

Fraser, A.I. 1962. The soil and roots as factors in tree stability. Forestry 35:117-127.

Freeman, J.T. 1984. Woodsman, spare that woodpecker. Defenders 59:4-13.

Gaines, G.D., K.E. Franzreb, D.H. Allen, K. S. Laves and W.L. Jarvis. 1995. Red-cockaded woodpecker management on the Savannah River Site: a management/research success story. *In* D.L. Kulhavy, R.G. Hooper and R. Costa (eds.). Red-cockaded Woodpecker: Recovery, Ecology and Management, Center for Applied Studies, College of Forestry, Stephen F. Austin State Univ., Nacogdoches, TX.

Gara, R.I. 1967. Studies on the attack behavior of the southern pine beetle. I. The spreading and collapse of outbreaks. Contrib. Boyce Thompson Inst. 24:349-354.

Gara, R.I. and J.E. Coster. 1968. Studies on the attack behavior of the southern pine beetle. III Sequence of tree infestation within stands. Contrib. Boyce Thompson Inst. 24:77-86.

Garren, K. H. 1943. Effects of fire on vegetation of the southeastern United States. Bot. Review 9:617-654.

George C. Marshall Institute. 1992. Global Warming Update. The George C. Marshall Institute, Washington, D.C.

Gerstacker, F. 1854. Wild sports in the far west. E. L. Steeves and H. R. Steeves (eds.). Reprinted in 1968 from English translation of 1854. Duke Univ. Press, Durham, N.C.

Gibson, A. M. 1965. Oklahoma: a history of five centuries. Harlow Publishing Corp., Norman, Okla.

Gilbertson, R.L. and L. Ryvarden. 1987. North American Polypores, Vol. 1. Lubrecht and Cramer Publ., New York.

Gilmour, J.D. 1926. Clear-cutting of pulpwood lands. For. Chron. 2:1-2.

Gilpin, M.E. and M.E. Soulé. 1986. Minimum viable populations: Processes of species extinction. pp. 19-34 *In* M.E. Soulé (ed.). Conservation Biology: The Science of Scarcity and Diversity. Sinauer, Sunderland, MA.

Givens, L.S. 1971. Introduction to the symposium on the ecology and management of the Red-cockaded Woodpecker. pp. 1-3 In R.L. Thompson (ed.). The Ecology and Management of the Red-cockaded Woodpecker. Bureau of Sport Fisheries and Wildlife, U.S. Dept. Int., and Tall Timbers Res. Stat., Tallahassee FL.

Goggans, R., R.D. Dixon and L.C. Seminara. 1987. Habitat use by 3-toed and black-backed woodpeckers, Deschutes National Forest, Oregon. Oregon Dept. Fish Wildl. and U. S. Dept. Agr. Deschutes Nat. Forest.

Goldman, S.E. and R.T. Franklin. 1977. Development and feeding habits of southern pine beetle larvae. Ann. Entomol. Soc. Amer. 70:54-56.

Goodman, D. 1987. The demography of chance extinction. pp. 11-34 *In* M.E. Soulé (ed.) Viable Populations for Conservation. Cambridge Univ. Press, MA.

Gordon, D.T. 1973. Damage from wind and other causes in mixed white fir-red fir stands adjacent to clearcuttings. U.S. Dept. Agr., For. Serv., Res. Pap. PSW-90. Berkeley, CA.

Gould, C. N. 1928. Travels through Oklahoma. Harlow Pub. Co., Oklahoma City, OK.

Grano, C.X. 1953. Wind-firmness of shortleaf and loblolly pines. Southern Lumberman 187:116.

Grano, C.X. 1970. Eradicating understory hardwoods by repeat prescribed burning. U. S. Dept. Agr. For. Serv. Res. Pap. SO-56, South. For. Exp. Station, New Orleans, LA.

Gratkowski, H.J. 1956. Windthrow around staggered settings in old-growth Douglas-fir. Forest Sci. 2:60-74.

Gray, W.M. 1990. Strong association between West African rainfall and U.S. landfall of intense hurricanes. Sci. 249:1251-1256.

Green, P.E. 1978. Analyzing multivariate data. Dryden Press, Hinsdale, IL

Greiner, E.C., G.F. Bennett, E. M. White and R.F. Coombs. 1975. Distribution of the avian hematozoa of North America. Canadian J. Zoology 53: 1762-1787.

Gresham, C.A., T.M. Williams and D.J. Lipscomb. 1991. Hurricane Hugo wind damage to southeastern U.S. coastal forest tree species. Biotropica 23(4a): 420-426.

Grier, J.W. 1980a. Ecology: a simulation model for small populations of animals. Computing 6:116-121.

Grier, J.W. 1980b. Modeling approaches to bald eagle population dynamics. Wildl Soc. Bull. 8:316-322.

Griffin, L.H., R.N. Conner and C.S. Adkisson. 1975. Repairs for a wind damaged hairy woodpecker nest. Raven 46:26.

Grimes, T.L. 1977. Relationship of red-cockaded woodpecker (*Picoides borealis*) productivity to colony area characteristics. Master's Thesis. Dep. For. Res. Clemson Univ. , Clemson, S.C.

Grubb, T. C., Jr. 1978. Weather dependent foraging rates of wintering woodland birds. Auk 95: 370-376.

Grumbine, A.A. 1936. Management plan. Francis Marion National Forest. Unpublished report.

Gruschow, G.F. and K.B. Trousdell. 1958. Incidence of heartrot in mature loblolly pine in coastal North Carolina. J. For. 56:221-221.

Guldin, J.M. 1986. Ecology of shortleaf pine, pp. 25-40 *In* P. A. Murphy (ed.). Proc. Symposium on the Shortleaf Pine Ecosystem. U.S. Dept. Agr. For. Serv., Monticello, AR.

Guyette, R.P. and B.E. Cutter. 1991. Tree-ring analysis of fire history of a post oak savanna in the Missouri Ozarks. Nat. Areas J. 11:93-99.

Guyette, R. and E.A. McGinnes, Jr. 1982. Fire history of an Ozark glade in Missouri. Trans. Mo. Acad. Sci. 16:85-93.

Haig, S.M., J.R. Belthoff and D.H. Allen. 1993a. Examination of population structure in red-cockaded woodpeckers using DNA profiles. Evolution 47:185-194.

Haig, S.M., J.R. Belthoff and D.H. Allen. 1993b. Population viability analysis for a small population of red-cockaded woodpeckers and an evaluation of enhancement strategies. Conservation Bio. 7:289-301.

Handford, P. 1980. Heterozygosity at enzyme loci and morphological variation. Nature 286:261-262.

Hardesty, J.L. 1992. The red-cockaded woodpecker and ecosystem management on Eglin AFB, Florida: directions and priorities. Fish & Wildl. Coop. Res. Unit, Univ. Florida, Gainesville, FL.

Hardesty, J.L. 1992. The red-cockaded woodpecker and ecosystem management on Eglin Air Force Base, Florida: directions and priorities. Summary Report from the Workshop, 1-2 Feb. 1992, Eglin Air Force Base, Gainesville, Florida.

Hardesty, J.L., R.J. Smith, C.J. Petrick and B.W. Hagedorn. 1995. Status and distribution of the fourth largest population of red-cockaded woodpeckers: preliminary results from Eglin AFB, Florida. *In* D.L. Kulhavy, R.G. Hooper and R. Costa (eds.). Red-cockaded Woodpecker: Recovery, Ecology and Management, Center for Applied Studies, College of Forestry, Stephen F. Austin State Univ., Nacogdoches, TX.

Hardin, G. 1968. The tragedy of the commons. Science 162:1243- 1248.

Hardin, K.I. and K.E. Evans. 1977. Cavity nesting bird habitat in the oak-hickory forest...a review. U.S Dept. Agr. For. Serv., Gen. Tech. Rep. NC-30, St. Paul, MN.

Harestad, A.S. and F.L. Bunnell. 1979. Home range and body weight—a reevaluation. Ecology 60:389-402.

Harlow, R.F. 1983. Effects of fidelity to nest cavities on the reproductive success of the red-cockaded woodpecker in South Carolina. pp 94-97 *In* D. Wood (ed.). Red-cockaded Woodpecker Symposium II Proceedings. State of Florida Game and Fresh Water Fish Commission, Tallahassee, FL

Harlow, R.F. and A.T. Doyle. 1990. Food habits of southern flying squirrels (*Glaucomys volans*) collected from red-cockaded woodpecker (*Picoides borealis*) colonies in South Carolina. Amer. Midl. Nat. 124:187-191.

Harlow, R.F. and M.R. Lennartz. 1983. Interspecific competition for red-cockaded woodpecker cavities during the nesting season in South Carolina. pp. 41-43 *In* D.A. Wood (ed.) Red-cockaded Woodpecker symposium II. Florida Game and Fresh Water Fish Comm., Tallahassee.

Harlow, R.F., R.G. Hooper and M.R. Lennartz. 1983. Estimating numbers of red-cockaded woodpecker colonies. Wildl. Soc. Bull. 11:360-363.

Harris, L.D. 1984. The fragmented forest: island biogeographic theory and the preservation of biotic diversity. Univ. Chicago Press, Chicago, Ill.

Harris, L.D. and P.B. Gallagher. 1989. New initiatives for wildlife conservation: the need for movement corridors. pp. 11-34 *In* G. Mackintosh (ed.). Preserving communities and corridors. Defenders of Wildl., Washington, D.C.

Harris, L.D. and J. Scheck. 1991. From implications to applications: the dispersal corridor principle applied to the conservation of biological diversity. pp. 189-220 *In* Saunders, D.A. and R.J.Hobbs (eds.). Surrey Beatty & Sons.

Harris, R.B. and F.W. Allendorf. 1989. Genetically effective population size of large mammals: an assessment of estimators. Cons. Biol. 3(2):181-191.

Harvey, M.J. and R.W. Barbour. 1965. Home range of *Microtus ochrogaster* as determined by a modified minimum area method. J. Mammal. 46:398-402.

Hatch, S. L., K. N. Gandhi and L. E. Brown. 1990. Checklist of the vascular plants of Texas. Texas Agr. Exp. Sta. MP-1655.

Hawley, F. 1941. Tree-ring analysis and dating in the Mississippi drainage. Univ. Chicago Press, Chicago, IL.

Hebb, E.A. and A.F. Clewell 1976. A remnant stand of

old-growth slash pine in the Florida panhandle. Bull. of the Torrey Botanical Club. 103 (1):1 -9.

Heckel, D.G. and M.R. Lennartz. In review. Estimation of effective population size for the red-cockaded woodpecker (*Picoides borealis*). American Naturalist.

Hedden, R.L. 1978. The need for intensive forest management to reduce southern pine beetle activity in east Texas. South. J. Appl. For. 2:19-22.

Hedden, R.L. 1984. Piedmont risk. *In* G.N. Mason (comp.) Predicting Southern Pine Beetle Trends. U. S. Dept. Agr. Integrated Pest Management. RD & A Prog. SPB Prediction Handbook.

Hedden, R.L. 1985. Simulation of southern pine beetle-associated timber loss using CLEMBEETLE. pp. 288-291 *In* Branham, S.J. and R.C. Thatcher (eds.). Integrated Pest Manag. Res. Symp. Proc. U.S. Dept. Agr. For. Serv. Gen. Tech. Rep. SO-56, New Orleans, LA.

Hedden, R.L. and R.P. Belanger. 1985. Predicting Susceptibility to southern pine beetle attack in the Coastal Plain, Piedmont, and Southern Appalachians. pp. 233-238 *In* S.J. Branham and R.C. Thatcher (eds). Integrated Pest Management Research Symposium: The Proceedings. U.S. Dept. Agr. For. Serv. Gen. Tech. Rep. SO-56, New Orleans, LA.

Hedden, R.L. and R.F. Billings. 1977. Seasonal variations in fat content and size of the southern pine beetle in East Texas. Ann. Entomol. Soc. Amer. 70:876-880.

Hedden, R.L. and R.F. Billings. 1979. Southern pine beetle: factors influencing the growth and decline of summer infestations in east Texas. For. Sci. 25:547-556.

Hedrick, L. 1992. Tree 249-3. Alabama's Trees. For. 11(2):18-21.

Heidt, G.A. 1977. Utilization of nest boxes by the southern flying squirrel, *Glaucomys volans*, in central Arkansas. Proc. Ark. Acad. Sci. 31:55-57.

Heikkenen, H.J. 1977. Southern pine beetle: a hypothesis reregarding its primary attractant. J. For. 75:412-413.

Heinselman, M. L. 1973. Fire in the virgin forests of the Boundary Waters Canoe Area, Minnesota. Quaternary Res. 3:329-382.

Henry, A.J. 1896. Notes concerning the West India Hurricane of September 29-30, 1896. Monthly Weather Review 24:368-369.

Henry, V.G. 1989. Guidelines for preparation of biological assessments and evaluations for the red-cockaded woodpecker. U.S. Fish and Wildl. Serv., Southeastern Reg., Atlanta, GA.

Hepting, G.H. 1971. Diseases of forest and shade trees of the United States. U. S. Dept. Agr. For. Serv. Agr. Handbook No. 386, Washington, D.C.

Hester, F.E. and J. Dermid. 1973. The world of the wood duck. J.B. Lippincott Co. Philadelphia.

Hicks, R.R. Jr. 1980. Climatic, site and stand factors (affecting the southern pine beetle). pp. 55-68 *In* R.C. Thatcher, J.L. Searcy, J.E. Coster and G.D. Hertel (eds.). The Southern Pine Beetle. U. S. Dept. Agr., For. Serv. Tech. Bull. 1631.

Hicks, R.R. Jr., J.E. Coster and G.N. Mason. 1987. Forest insect hazard rating. J. For. 85(10):20-26.

Hicks, R.R. Jr., J.E. Howard, J.E. Coster and K.G. Watterston. 1978. The role of tree vigor in susceptibility of loblolly pine to southern pine beetle. pp. 177-181 *In* C.A. Hollis and A.E. Squillace (eds.), Proceedings fifth North American forest biology workshop. Univ. Florida Schl. For. Res., Gainesville.

Hirshleifer, J. 1976. Price theory and applications. Prentice-Hall, Inc., Englewood Cliffs, New Jersey.

Hodges, J.D. and L.S. Pickard. 1971. Lightning in the ecology of the southern pine beetle, *Dendroctonus frontalis* (Coleoptera: Scolytidae). Can. Entomol. 103: 44-51.

Hodges, J.D., W.W. Elam and W.F. Watson. 1977. Physical properties of the oleoresin system of the four major southern pines. Can. J. For. Res. 7:520-525.

Hodges, J.D., W.W. Elam, W.F. Watson and T.E. Nebeker. 1979. Oleoresin characteristics and susceptibility of four southern pines to southern pine beetle (Coleoptera: Scolytidae) attack. Can. Entomol. 111:889-896.

Honea, C.R., T.E. Nebeker and T.J. Straka. 1987. Economic evaluation of seven southern pine beetle hazard rating systems. J. Environ. Manage. 24:259-266.

Honess, C. W. 1923. Geology of the southern Ouachita Mountains of Oklahoma. Bull. 32, Parts I and II. Okla. Geological Surv., Norman, OK.

Hook, D.D., M.B. Buford and T.M. Williams. 1991. Impact of Hurricane Hugo on the South Carolina coastal plain forest. J. Coastal Res. SI No. 8, 291-300.

Hooper, R.G. 1982. Use of dead cavity trees by red-cockaded woodpeckers. Wildl. Soc. Bull. 10:163-164.

Hooper, R.G. 1983. Colony formation by red-cockaded woodpeckers: hypotheses and management implications. pp. 72-77 *In* D.A. Wood (ed.). Red-cockaded Woodpecker Symposium II Proceedings. State of Florida Game and Fresh Water Fish Commission, Tallahassee, FL

Hooper, R.G. 1988. Longleaf pines used for cavity trees by red-cockaded woodpeckers. J. Wildl. Manage. 52:392-398.

Hooper, R.G. and R.F. Harlow. 1986. Forest stands selected by foraging red-cockaded woodpeckers. U.S. Dept. Agr. For. Serv., Southeastern For. Exp. Sta., Res. Paper SE-259, Ashville, NC.

Hooper, R.G. and M.R. Lennartz. 1981. Foraging behavior of the red-cockaded woodpecker in South Carolina. Auk 98:321-334.

Hooper, R.G. and M.R. Lennartz. 1983. Roosting behavior of red-cockaded woodpecker with insufficient cavities. J. Field Ornithology 54:72-76.

Hooper, R.G. and M.R. Lennartz. 1995a. Short-term response of a high density red-cockaded woodpecker population to removal of foraging habitat. *In* D.L. Kulhavy, R.G. Hooper and R. Costa (eds.). Red-cockaded Woodpecker: Recovery, Ecology and Management, Center for Applied Studies, College of Forestry, Stephen F. Austin State Univ.,

Nacogdoches, TX.

Hooper, R.G. and C.J. McAdie. 1995b. Hurricanes as a factor in the long-term management of red-cockaded woodpeckers. *In* D.L. Kulhavy, R.G. Hooper and R. Costa (eds.). Red-cockaded Woodpecker: Recovery, Ecology and Management, Center for Applied Studies, College of Forestry, Stephen F. Austin State Univ., Nacogdoches, TX.

Hooper, R.G. and H.D. Muse. 1989. Sequentially observed periodic surveys of management compartments to monitor red-cockaded woodpecker populations. U. S. Dept. Agr. For. Serv., Southeastern For. Exp. Sta., Res. Pap. SE-276, Ashville, NC.

Hooper, R.G., A.F. Robinson, Jr. and J.A. Jackson. 1980. The red-cockaded woodpecker: notes on life history and management. U.S. Dept. Agr. For. Serv., Southeastern Area, State and Private Forestry. Gen. Rep. SA-GR 9. Atlanta, GA.

Hooper, R.G., L.J. Niles, R.F. Harlow and G.W. Wood. 1982. Home ranges of red-cockaded woodpeckers in coastal South Carolina. Auk 99:675-682.

Hooper, R.G., J.C. Watson and R.E.F. Escano. 1990. Hurricane Hugo's initial effects on red-cockaded woodpeckers in the Francis Marion National Forest. Trans. Fifty-fifth No. Amer. Wildl. Nat. Resour Conf. 55:220-224.

Hooper, R.G., D.L. Krusac and D.L. Carlson. 1991a. An increase in a population of red-cockaded woodpeckers. Wildl. Soc. Bull. 19:277-286.

Hooper, R.G., M.R. Lennartz and H.D. Muse. 1991b. Heart rot and cavity tree selection by red-cockaded woodpeckers. J. Wildl. Manage. 55:323-327.

Hooper, R.G., D.L. Carlson and J.C. Watson. In press. The red-cockaded woodpecker in Francis Marion National Forest before and after Hurricane Hugo. *In* Proceedings of the Nat. Council of the Paper Ind. for Air and Stream Imp., Southern Regional Meeting, Charleston, SC.

Hopkins, M.L. and T.E. Lynn. 1971. Some characteristics of red-cockaded woodpecker cavity trees and management implications in South Carolina. pp. 140-169 *In* R. L. Thompson (ed.). The ecology and management of the Red-cockaded Woodpecker. Bureau of Sport Fish and Wildlife and Tall Timbers Research Station, Tallahassee, FL

Hovis, J.A. 1982. Population biology and vegetation requirements of the red-cockaded woodpecker (*Picoides borealis*) in pine habitats of north Florida. M.S. thesis, Univ. Florida, Gainesville.

Hovis, J.A. and R.F. Labisky. 1985. Vegetative associations of red-cockaded woodpecker colonies in Florida. Wildl. Soc. Bull. 13:307-314.

Hoyt, S.F. 1957. The ecology of the pileated woodpecker. Ecology 38: 246-256.

Hsi, G. and J.H. Nath. 1970. Wind drag within simulated forest canopies. J. Appl. Met. 9:592-602.

Hudson, W.E. (ed.) 1991. Landscape linkages and biodiversity. Defenders of Wildl., Washington, D.C.

Hughes, R.M. and R.F. Noss. 1992. Biological diversity and biological integrity: current concerns for lakes and streams. Fisheries 17(3):11-19.

Hunter, M.L. Jr. 1990. Wildlife, forests, and forestry—principles of managing forests for biological diversity. Prentice Hall. Englewood Cliffs, NJ.

Hurst, C. 1965. Hurricane Hilda—an aftermath of tangled woodlands. Southern Lumberman 210:36,38.

Irvine, R.E. 1970. The significance of windthrow for *Pinus radiata* management in the Nelson District. New Zealand J. For. 15(1):57-68.

Jackson, J.A. 1970. Predation of a black rat snake on yellow-shafted flicker nestlings. Wilson Bull. 82: 329-330.

Jackson, J.A. 1971. The evolution, taxonomy, distribution, past populations and current status of the red-cockaded woodpecker. pp. 4-29 *In* R.L. Thompson (ed.). The Ecology and Management of the Red-cockaded Woodpecker. Bureau of Sport Fisheries and Wildlife, U.S. Dept. Int., and Tall Timbers Research Station, Tallahassee, FL.

Jackson, J.A. 1974. Gray rat snakes versus red-cockaded woodpeckers: predator-prey adaptations. Auk 91:342-47.

Jackson, J.A. 1976. How to determine the status of a woodpecker nest. Living Bird 15:205-221.

Jackson, J.A. 1977a. A device for capturing tree cavity roosting birds. North Amer. Bird Bander 2:14-15.

Jackson, J.A. 1977b. Determination of the status of red-cockaded woodpecker colonies. J. Wildl. Management. 41:448-452.

Jackson, J.A. 1977c. Red-cockaded woodpeckers and pine red heart disease. Auk 94: 160-63.

Jackson, J.A. 1978a. Analysis of the distribution and population status of the redcockaded woodpecker. pp. 101-110 *In* R.R. Odum and L. Landers (eds.). Proc. Rare and Endangered Wildlife Symposium. Georgia Dept. Natural Resources, Game and Fish Division. Tech. Bull. WL 4.

Jackson, J.A. 1978b. Competition for cavities and red-cockaded woodpecker management. pp. 103-112 *In* S.A. Temple (ed.). Endangered birds: management techniques for threatened species. Univ. Wisconsin Press, Madison.

Jackson, J.A. 1978c. Pine bark redness as an indicator of red-cockaded woodpecker activity. Wildl. Soc. Bull. 6:171-172.

Jackson, J.A. 1978d. Predation by a gray rat snake on red-cockaded woodpecker nestlings. Bird-Banding 49: 187-188.

Jackson, J.A. 1979a. Age characteristics of red-cockaded woodpeckers. Bird-Banding 50(1):23-29.

Jackson, J.A. 1979b. Tree surfaces as foraging substrate for insectivorous birds. pp. 69-93 *In* J.G. Dickson, R.N. Conner, R.R.. Fleet, J.A. Jackson and J.C. Kroll (eds.). The role of insectivorous birds in forest ecosystems. Academic Press, NY

Jackson, J.A. 1982. Capturing woodpecker nestlings with a noose—a technique and its limitations. North Amer. Bird Bander 7:90-92.

Jackson, J.A. 1985. Cavity tree killed by red-cockaded woodpeckers. Chat 49:72-75.

Jackson, J.A. 1986. Biopolitics, management of federal lands, and the conservation of the red-cockaded woodpecker. Amer. Birds 40(5): 1162-1168.

Jackson, J.A. 1987. The Red-cockaded woodpecker. pp. 479-494 *In* Silvester, R.L., W.J. Chandler, K. Barton and L. Labate (eds.). Audubon Wildlife Rep. Academic Press, NY.

Jackson, J.A. 1988. The southeastern pine forest ecosystem and its birds: past, present, and future. pp. 119-159 *In* J.A. Jackson (ed.), Bird Conservation 3. Univ. of Wisconsin Press, Madison, WI

Jackson, J.A. 1989. The southeastern pine forest ecosystem and its birds: past, present, and future. pp. 119-159 *In* J.A. Jackson (ed.). Bird conservation 3. Univ. Wisconsin Press, Madison, WI

Jackson, J.A. 1990. Intercolony movements of red-cockaded woodpeckers in South Carolina. J. Field Ornithol. 61: 149-55.

Jackson, J.A. 1995. The red-cockaded woodpecker: two centuries of knowledge, three decades under the Endangered Species Act. *In* D.L. Kulhavy, R.G. Hooper and R. Costa (eds.). Red-cockaded Woodpecker: Recovery, Ecology and Management, Center for Applied Studies, College of Forestry, Stephen F. Austin State Univ., Nacogdoches, TX.

Jackson, J.A. and B.J.S. Jackson. 1986. Why do red-cockaded woodpeckers need old trees? Wildl. Soc. Bull. 14: 318-322.

Jackson, J.A. and S.D. Parris. 1991. A simple, effective net for capturing cavity roosting birds. North Amer. Bird Bander 16:30-31.

Jackson, J.A. and R.L. Thompson. 1971. A glossary of terms used in association with the red-cockaded woodpecker. pp. 187-188 *In* R.L. Thompson (ed.), The Ecology and Management of the Red-cockaded Woodpecker, Bureau of Sport Fisheries and Wildlife, U.S. Dep. Interior, and Tall Timbers Research Station, Tallahassee, FL

Jackson, J.A., R. Weeks and P. Shindala. 1976. The present status and future of red-cockaded woodpeckers in Kentucky. Kentucky Warbler 52:75-80.

Jackson, J.A., B.J. Schardien and R. Weeks. 1978. An evaluation of the status of some red-cockaded woodpecker colonies in east Texas. Bull. Tex. Ornithol. Soc. 11:2-9.

Jackson, J.A., W.W. Baker, V. Carter, T. Cherry and M.L. Hopkins. 1979a. Recovery plan for the red-cockaded woodpecker. U.S. Fish and Wildlife Serv., Atlanta, GA

Jackson, J.A., M.R. Lennartz and R.G. Hooper. 1979b. Tree age and cavity initiation by red-cockaded woodpeckers. J. For. 77:102-103.

Jackson, J.A., B.J. Schardien and P.R. Miller. 1983. Moving red-cockaded woodpecker colonies: relocation or phased destruction? Wildl. Soc. Bull. 11:59-62.

Jackson, J.A., R.N. Conner and B.J.S. Jackson. 1986. The effects of wilderness on the endangered red-cockaded woodpecker. pp. 71-78 *In* D.L. Kulhavy, and R. N. Conner (eds.). Wilderness and Natural Areas in the Eastern United States: A Management Challenge, Center for Applied Studies, College of Forestry, Stephen F. Austin State Univ., Nacogdoches, TX.

Jacobs, D.F., D.W. Cole and J.R. McBride. 1985. Fire history and perpetuation of natural coast redwood ecosystems. J. For. 83:494-497.

James, D.A. and J.C. Neal. 1989. Update of the status of the red-cockaded woodpecker in Arkansas. Unpubl. Final Report, Univ. Arkansas and Arkansas Game and Fish Comm.

James, F.C. 1991. Signs of trouble in the largest remaining population of red-cockaded woodpeckers. Auk 108:419-423.

James, F.C. 1995. The status of the red-cockaded woodpecker in 1990 and the prospect for recovery. *In* D.L. Kulhavy, R.G. Hooper and R. Costa (eds.). Red-cockaded Woodpecker: Recovery, Ecology and Management, Center for Applied Studies, College of Forestry, Stephen F. Austin State Univ., Nacogdoches, TX.

James, F.C., C. Hess, G. Hagan and B. Kotrla. 1995. Population structure and annual turnover rates of cavities of the red-cockaded woodpecker in the Apalachicola National Forest. *In* D.L. Kulhavy, R.G. Hooper and R. Costa (eds.). Red-cockaded Woodpecker: Recovery, Ecology and Management, Center for Applied Studies, College of Forestry, Stephen F. Austin State Univ., Nacogdoches, TX.

Jansen, D.K. and G.A. Patterson. 1983. Late nesting and low nesting success of red-cockaded woodpeckers in the Big Cypress National Preserve, Florida, in 1983. pp. 99 *In*: Proceedings of the Red-cockaded Woodpecker Symposium II, (D.A. Wood, Ed.), Florida Game and Freshwater Fish Commission, Tallahassee, FL.

Jasumback, T. and D. Luepke. 1993. Evaluation of the Trimble Pathfinder Professional GPS receiver under a hardwood tree canopy. In press. U.S. For. Serv., Montana Tech. and Development Center, Missoula, MT.

Jerauld, A.E., R.S. DeLotelle and J.R. Newman. 1983. Restricted red-cockaded woodpecker clan movement. pp.97-99 *In* D.A. Wood (ed.). Red-cockaded Woodpecker Symposium II. Florida Game and Fresh Water Fish Comm.

Johnson, D.L. 1977. Inbreeding in populations with overlapping generations. Genetics 87:581-591.

Johnson, F.L. and G.D. Schnell. 1985. Wildland fire history and the effects of fire on vegetative communities at Hot Springs National Park, Arkansas. Final rept. to Nat. Park Serv., Santa Fe, NM, Okla. Biol. Surv., Univ. Oklahoma, Norman, OK.

Johnson, P.C. and J.E. Coster. 1978. Probability of attack by southern pine beetle in relation to distance from an attractive host tree. For. Sci. 24:574-580.

Jones, J.K. Jr., D.C. Carter, H.H. Genoways, R.S. Hoffman, D.W. Rice and C. Jones. 1986. Revised checklist of North American mammals north of

Mexico, 1986. Occasional Pap. Museum Texas Tech Univ., No. 107.

Juanes, F. 1986. Population density and body size in birds. Amer. Nat. 128:921-929.

Kale, H.W., II, B. Pranty, B. Stith and W. Biggs (eds.). 1992. The atlas of the breeding birds of Florida. Nongame Wildlife Program Final Report. Contract GFC-84-017. Florida Game and Fresh Water Fish Commission. Tallahassee, FL.

Kalisz, P.J. and S.E. Boettcher. 1991. Active and abandoned red-cockaded woodpecker habitat in Kentucky. J. Wildl. Manage. 55:146-154.

Kalkstein, L.S. 1976. Effects of climatic stress on outbreaks of the southern pine beetle. Environ. Entomol. 5:653-658.

Kappes, J. Jr. and L.D. Harris. 1995. Interspecific competition for red-cockaded woodpecker cavities in Apalachicola National Forest. *In* D.L. Kulhavy, R.G. Hooper and R. Costa (eds.). Red-cockaded Woodpecker: Recovery, Ecology and Management, Center for Applied Studies, College of Forestry, Stephen F. Austin State Univ., Nacogdoches, TX.

Kat, P.W. 1982. The relationship between heterozygosity for enzyme loci and developmental homeostasis in peripheral populations of aquatic bivalves (Unionidae). Amer. Nat. 119:824-832.

Kattan, G. 1988. Food habits and social organization of acorn woodpeckers in Colombia. Condor 90:100-106.

Kelly, J.F. 1991. The influence of habitat quality on the population decline of the red-cockaded woodpecker in the McCurtain County Wilderness Area, Oklahoma. M. S. Thesis, Oklahoma State Univ., Stillwater.

Kelly, W., R.P. Belanger, M.D. Connor, T.R. Dell, J.E. DeSteiguer, R. Kiester, R.N. Kitchens, P.L. Lorio, Jr. and K.M. Swain. 1986. Long-term strategies and research needs for managing Southern forests to reduce southern pine beetle impacts. U. S. Dept. Agr. For. Serv. Southern Region.

Kilham, L. 1976. Winter foraging and associated behavior of pileated woodpeckers in Georgia and Florida. Auk 93:15-24.

Kilham, L. 1983. Life history studies of woodpeckers of eastern North America. Nuttall Ornithological Club, Cambridge, MA.,Publ. No. 20.

King, E. 1972. Rainfall and epidemics of the southern pine beetle. Environ. Entomol. 1:279-285.

King, W.B., J.A. Jackson, H.W. Kale, II, H.F. Mayfield, R.L. Plunkett, Jr., J.M. Scott, P.F. Springer, S.A. Temple and S.R. Wilbur. 1977. The Recovery Team— Recovery plan approach to conservation of endangered species: a status summary and appraisal. Auk 94(4, Suppl.):1DD-19DD.

Kingston, R. 1991. Ron Kingston's snake/predator guard. Sialia 13: 56-57.

Kinzer, G.W., A.F. Fentiman, Jr., T.L. Page, R.L. Foltz, J.P. Vité and G.B. Pitman. 1969. Bark beetle attractants: identification, synthesis and field bioassay of a compound isolated from *Dendroctonus*. Nature 221:447- 448.

Kish, L. 1965. Survey sampling. John Wiley & Sons, Inc., New York.

Knight, F.B. and H.J. Heikkenen 1980. Principles of forest entomology. McGraw-Hill Co., New York, Ny.

Knight, H.A. 1978. The South is losing its pines. Forest Farmer. November/ December, pp. 10-12.

Kodric-Brown, A. and J.H. Brown. 1978. Influences of economics, interspecific competition, and sexual dimorphism on territoriality of migrant rufous hummingbirds. Ecology 59:285-296.

Koenig, W.D. and R.L. Mumme. 1987. Population ecology of the cooperatively breeding acorn woodpecker. Princeton Univ. Press. Princton, NJ.

Koenig, W.D. and F.A. Pitelka. 1981. Ecological factors and kin selection in the evolution of cooperative breeding in birds. pp. 261-280 *In* R.D. Alexander and D.W. Tinkle (eds.). Natural selection and social behavior. Chiron Press, New York, NY.

Komarek, E.V. 1968. Lightning and lightning fires as ecological forces. Proc. Tall Timbers Ecology Conf. 8:169-197.

Komarek, E.V. 1974. Effects of fire on temperate forests and related ecosystems: southeastern United States. pp. 251-277 *In* T.T. Kozlowski and C.E. Ahlgren (eds.). Fire and Ecosystems, Academic, New York.

Kozlowski, T.T., P.J. Kramer and S.G. Pallardy. 1991. The physiological ecology of woody plants. Academic Press, NY.

Kramer, P.J. and T.T. Kozlowski. 1979. Physiology of woody plants. Academic Press, New York, N.Y.

Kroll, J.C. and R.R. Fleet. 1979. Impact of woodpecker predation on overwintering within-tree populations of the southern pine beetle (*Dendroctonus frontalis*). pp. 269-281 *In* J.G. Dickson, R.N. Conner, R.R.. Fleet, J.A. Jackson and J.C. Kroll (eds.). The role of insectivorous birds in forest ecosystems. Academic Press, NY

Krueger, H. 1991. Harry Krueger's snake trap. Sialia 13: 63-67.

Krusac, D.C., J.M. Dabney and J.J. Petrick 1995. An ecological approach to recovering the red-cockaded woodpecker on southern national forests. *In* D.L. Kulhavy, R.G. Hooper and R. Costa (eds.). Red-cockaded Woodpecker: Recovery, Ecology and Management, Center for Applied Studies, College of Forestry, Stephen F. Austin State Univ., Nacogdoches, TX.

Ku, T.T., J.M. Sweeney and V.B. Shelburne. 1980. Gulf coastal plain (Arkansas). *In* J.E. Coster and J.L. Searcy (eds.)., Site, stand, and host characteristics of southern pine beetle infestations. U. S. Dept. Agr. For. Serv. Tech. Bull. 1612. Pineville, LA.

Kuiper, L.C. and M.P. Coutts. 1992. Spatial disposition and extension of the structural root system of Douglas-fir. Forest Ecology and Management 47:111-125.

Kulhavy, D.L., J.H. Mitchell and R.N. Conner. 1988. The southern pine beetle and the red-cockaded woodpecker: potential for interaction. pp.337-343 *In* T.L. Payne and H. Saarenmaa (eds.). Integrated

Control of Scolytid Bark Beetles. Virginia Polytechnic Institute and State Univ., Blacksburg.

Labisky, R.F. and M.L. Porter. 1984. Home range and foraging habitat of red-cockaded woodpeckers in north Florida. Department Wildl. and Range Sci., School of For. Res. and Cons., Univ. of Florida, Gainesville, FL.

LaBranche, M.S. 1988. Reproductive ecology of the red-cockaded woodpecker in the Sandhills of North Carolina. M.S. Thesis, No. Carolina State Univ., Raleigh, NC.

Lacis, A.A. and B.E. Carlson. 1992. Keeping the sun in proportion. Nature 360:297.

Lacy, R.C. 1991. VORTEX. v.4.1. Chicago Zoological Soc., Brookfield, Illinois

Lancia, R.A., J.P. Roise, D.A. Adams and M.R. Lennartz. 1989. Opportunity Costs of red-cockaded woodpecker foraging habitat. So. J. Appl. For. 13: 81-85.

Lande, R. and B.F. Barrowclough. 1987. Effective population size, genetic variation, and their use in population management. pp. 87-123 *In* M.E. Soulé (ed.). Viable Populations for Conservation. Cambridge Univ. Press, New York.

Landers, J.L. 1987. Prescribed burning for managing wildlife in southeastern pine forests. pp. 19-27 *In* J.G. Dickson and O.E. Maughan (eds.). Managing southern forests for wildlife and fish., U.S. Dept. Agr. For. Serv. Gen. Tech. Rep. SO-65, New Orleans, LA.

Landers, J.L., N.A. Byrd and R. Komarek 1990. A holistic approach to managing longleaf pine communities. pp. 135-167 *In* R.M. Farrar (ed.). Proc. of the symposium on the management of longleaf pine. U. S. Dept. Agr. For. Serv. Gen. Tech. Rep. SO-75.

Langdon, O.G. 1963. Growth patterns of *Pinus elliotii* var. *densa*. Ecology 44:82:5-827.

Langley, R.B. 1991. The mathematics of GPS. GPS World. July/August 1991:45-50.

Larson, P.R. 1963. Stem form and development of forest trees. Forest Sci. Monogr. 5.

Laves, K.S. 1992. Establishment of a viable population of red-cockaded woodpecker at the Savannah River Site—Ann. Rep. (Unpublished). U. S. Dept. Agr. For. Serv. Clemson, SC.

Lawrence, L.D.K. 1967. A comparative life-history study of 4 species of woodpeckers. Amer. Ornithol. Union, Ornithological Monographs No. 5.

Lawson, E.R. 1990. Shortleaf pine. pp. 316-326 *In* R.M. Burns and B.H. Honkala (eds.). Silvics of North America. Vol. 1, Conifers. Agr. Handb. 654. U. S. Dept. Agr., For. Serv., Washington, DC.

Lay, D.W. 1969. Destined for oblivion. Texas Parks Wildl. 27(2):12-15.

Lay, D.W. and D.N. Russell. 1970. Notes on the red-cockaded woodpecker in Texas. Auk 87:781-786.

Lay, D.W. and D.A. Swepston. 1973. Job No. 10: Red-cockaded woodpecker study. Texas Parks & Wildl. Dep. Fed. Aid in Wildl. Restor. Act Proj. W-80-R-16.

Leary, R.F. and F.W. Allendorf. 1989. Fluctuating asymmetry as an indicator of stress: implications for conservation biology. Trends Ecol. Evol. 4:214-217.

Leick, A. 1990. GPS satellite surveying. John Wiley & Sons, New York.

Lennartz, M.R. 1983. Sociality and cooperative breeding of red-cockaded woodpeckers, (*Picoides borealis*). Ph.D. Dissertation, Clemson Univ., Clemson, SC

Lennartz, M.R. 1988. The red-cockaded woodpecker: old growth species in a second growth landscape. Natural Areas J. 8(3):160-166.

Lennartz, M.R. and R.F. Harlow. 1979. The role of parent and helper red-cockaded woodpeckers at the nest. Wilson Bull. 91:331-335.

Lennartz, M.R. and D.G. Heckel. 1987. Population dynamics of a red-cockaded woodpecker population in Georgia piedmont loblolly pine habitat. pp. 48-55 *In* Odum R.R., K.A. Riddleberger and J.C. Ozier (eds.). Proc. of the Third Southeastern Nongame and Endangered Wildlife Symposium. Georgia Department Nat. Res., Game and Fish Div. Athens, GA

Lennartz, M.R. and D.G. Heckel. 1988. Population dynamics of a red-cockaded woodpecker population in Georgia piedmont loblolly pine habitat. pp. 48-55 *In* Proc. (D.A. Wood, ed.). Florida Game Fresh Water Fish Comm., Tallahassee, FL.

Lennartz, M.R. and V.G. Henry 1985. Red-cockaded woodpecker recovery plan. U.S. Fish and Wildlife Serv., Atlanta, GA.

Lennartz, M.R. and J.D. Metteauer. 1986. Test of a population estimation technique for red-cockaded woodpeckers. Proc. Ann. Conf. Southeastern. Assoc. Fish and Wildl. Agencies 40:320-324.

Lennartz, M.R., P.H. Geissler, R. F. Harlow, R.C. Long, K.M. Chitwood and J.A. Jackson. 1983a. Status of the red-cockaded woodpecker populations on federal lands in the south. pp 7-12 *In* D.A. Wood (ed.) Proceedings of Red-cockaded Woodpecker Symposium II, Florida Game and Freshwater Fish Commission. Tallahassee.

Lennartz, M.R., H.A. Knight, J.P. McClure and V.A. Rudis. 1983b. Status of red-cockaded woodpecker nesting habitat in the South. pp. 13-19 *In* D. Wood (ed.). Red-cockaded Woodpecker Symposium II. State of Florida Game and Fresh Water Fish Commission, Tallahassee.

Lennartz, M.R., R.G. Hooper and R.F. Harlow. 1987. Sociality and cooperative breeding in red-cockaded woodpeckers (*Picoides borealis*). Behav. Ecol. Sociobiol. 20:77-88.

Lenth, R. 1987. Power Pack User's Guide. Russ Lenth, Iowa City, IA.

Leopold, A. 1930. The American game policy in a nutshell. *In* The American Game Policy. Seventeenth Amer. Game Conf. (facsimile reprint) The American Game Policy And Its Development 1928-1930, Wildl. Manage. Instit., Washington,D.C.

Leopold, A. 1949. A Sand County Almanac with essays on conservation from Round River. Ballantine Books 1970.

Lerner, I.M. 1954. Genetic Homeostasis. Oliver and Boyd, London.

Leuschner, W.A., H.E. Burkhart, G.D. Spittle, I.R. Ragenovich and R.N. Coulson. 1976. A descriptive study of host site variables associated with occurrence of *Dendroctonus frontalis* Zimm. in east Texas. Southwest. Entomol. 1:141-149.

Levene, H. 1960. Robust tests for the equality of variances. pp. 278-292 *In* I. Olkin (ed.) Contributions to Probability and Statistics. Stanford Univ. Press, Palo Alto.

Lewis, A. 1924. La Harpe's first expedition in Oklahoma, 1718-1719. Chron. Okla. 2(4):331-349.

Lewontin, R.C. 1974. The genetic basis of evolutionary change. Columbia Univ. Press, New York, NY.

Light, D.L. 1993. The national aerial photography program as a geographic information system resource. Photogra. Eng. & Remote Sen. 59(1)61-65.

Ligon, J.D. 1968. Sexual differences in foraging behavior in two species of *Dendrocopos* woodpeckers. Auk 85:203-215.

Ligon, J.D. 1970. Behavior and breeding biology of the red-cockaded woodpecker. Auk 87:255-278.

Ligon, J.D. 1971. Some factors influencing numbers of the red-cockaded woodpecker. pp. 30-43 *In* R.L. Thompson (ed.). Proc. Symp The Ecology and Management of the Red-cockaded Woodpecker. Bureau of Sport Fisheries and Wildlife, U. S. Dept. Int., and Tall Timbers Res. Stat., Tallahassee FL.

Ligon, J.D., P.B. Stacy, R.N. Conner, C.E. Bock and C.S. Adkisson. 1986. Report of the American Ornithologists' Union Committee for the Conservation of the Red-cockaded Woodpecker. Auk 103:848-855.

Lih, M.P. and F.M. Stephen. 1986. The southern pine beetle integrated modeling system at the University of Arkansas: a user's guide. Department of Entomology, Univ. of Arkansas, Fayetteville, AR.

Lillard, R.G. 1947. The great forest. Alfred A. Knopf, New York.

Lipscomb, D.J. and T.M. Williams. 1988. pp. 219-224 *In* J.H. Miller (ed.). Proceedings of the Fifth Biennial Southern Silvicultural Research Conference. Micro-Computer Map Based Data Retrieval System for Forest Managers. U. S. Dept. Agr. For. Serv. Gen. Tech. Rep. SO-74:

Lipscomb, D.J. and T.M. Williams. 1991. Developing a GIS for forest management in the 1990's. pp. 551-560 *In* Resource Technology 90, Second International Symposium on Advanced Technology in Natural Resource Management. Amer. Soc. Phot. and Remote Sensing. Bethesda MD.

Lipscomb, D.J. and T.M. Williams. 1995. Impact of Hurricane Hugo on cavity trees of a red-cockaded woodpecker population and natural recovery after two and a half years. *In* D.L. Kulhavy, R.G. Hooper and R. Costa (eds.). Red-cockaded Woodpecker: Recovery, Ecology and Management, Center for Applied Studies, College of Forestry, Stephen F. Austin State Univ., Nacogdoches, TX.

Little, E. L. Jr. and E. E. Olmstead. 1931. An ecological study of the southeastern Oklahoma protective unit, Oklahoma For. Serv. (Manuscript edited by W. T. Penfound, Univ. Okla. Libr.).

Locke, B.A. and R.N. Conner. 1983. A statistical analysis of the orientation of entrances to red-cockaded woodpecker cavities. pp. 108-109 *In* D.A. Wood (ed.). Red-cockaded Woodpecker Symposium II proceedings. State of Florida Game and Fresh Water Fish Comm., Tallahassee, FL

Locke, B.A., R.N. Conner and J.C. Kroll. 1983. Factors effecting colony site selection by red-cockaded woodpeckers. pp. 46-50 *In* D.L. Wood (ed.) Red-cockaded Woodpecker symposium II. Proc. Florida Game and Fresh Water Fish Commission, Tallahassee, FL

Lockett, S.H. 1969. Louisiana as it is: a geographical and topographical description of the state. Louisiana State Univ. Press, Baton Rouge.

Loeb, S.C. 1993. Use and selection of red-cockaded woodpecker cavities by southern flying squirrels. J. Wildl. Manage. 57:329-335.

Loeb, S.C. and E.E. Stevens. 1995. Turnover of red-cockaded woodpecker nest cavities on the Piedmont Plateau. *In* D.L. Kulhavy, R.G. Hooper and R. Costa (eds.). Red-cockaded Woodpecker: Recovery, Ecology and Management, Center for Applied Studies, College of Forestry, Stephen F. Austin State Univ., Nacogdoches, TX.

Loeb, S.C., W.D. Pepper and A.T. Doyle. 1992. Habitat characteristics of active and abandoned red-cockaded woodpecker colonies. So. J. Appl. For. 16:120-125.

Lohrey, R.E. and S.V. Kossuth. 1990. Slash pine. pp. 338-347 *In* R.M. Burns and B.H. Honkala (eds.). Silvics of North America. Vol. 1, U. S. Dept. Agr., For. Serv., Conifers. Agr. Handb. 654. Washington, D.C.

Lorio, P.L. Jr. 1966. *Phytopathora cinnamomi* and *Pythium* species associated with loblolly pine decline in Louisiana. Plant Disease Rep. 50: 596-97.

Lorio, P.L. Jr. 1968. Soil, site, and stand conditions related to southern pine beetle activity in Hardin County, Texas. J. Econ. Entomol. 51: 565-66.

Lorio, P.L. Jr. 1980. Rating stands for susceptibility to southern pine beetle. pp. 71-105 *In* Thatcher, R.C., J.L. Searcy, J.E. Coster, and G.D. Hertel (eds.). 1980. Southern Pine Beetle. U.S. Dept. Agr. For. Serv. and Sci. and Educ. Admin. Tech. Bull. 1631.

Lorio, P.L. Jr. 1986. Growth-differentiation balance: a basis for understanding southern pine beetle-tree interactions. For. Ecol. and Manag. 14:259-273.

Lorio, P.L. Jr. and W.H. Bennet. 1974. Recurring southern pine beetle infestations near Oakdale, Louisiana. U. S. Dept. Agr. For. Serv. Res. Pap. SO-95, New Orleans, LA.

Lorio, P.L. Jr. and R.A. Sommers. 1981. Use of available resource data to rate stands for southern pine beetle risk. pp. 75-78 *In*: R.L. Hedden, S.J. Barras and J.E. Coster (eds.). Proc., Symp. on Hazard Rating Syst. in For. Insect Pest Manage. U. S. Dept. Agr. For. Serv.

Gen. Tech. Rep. WO-27, Washington, DC.
Lorio, P.L. Jr. and D.O. Yandle. 1978. Distribution of lightning-induced southern pine beetle infestations. Southern Lumberman, Jan. pp. 12-13.
Lorio, P.L. Jr., R.A. Sommers, C.A. Blanche, J.D. Hodges and T.E. Nebeker. 1990. Modeling pine resistance to bark beetles based on growth and differentiation balance principles, pp. 402-409. *In*: R.K. Dixon, R.S. Meldaho, G.A. Ruark and W.G. Warren (eds.). Process Modeling of Forest Growth Responses to Environmental Stress. Timber Press, Portland, OR.
Love, G.J., S.A. Wilkin and M.H. Goodwin, JR. 1953. Incidence of blood parasites in birds collected in southwestern Georgia. J. Parasitology 39: 52-57.
Lovelady, C.N., P.E. Pulley, R.N. Coulson and R.O. Flamm. 1991. Relation of lightning to herbivory by the southern pine bark beetle guild (Coleoptera: Scolytidae). Environ. Entomol. 20: 1279-84.
Lowery, G.H., Jr. 1960. Louisiana birds. Louisiana State Univ. Press, Baton Rouge.
Lubencho, J., et al. 1991. The sustainable biosphere initiative: an ecological research agenda. Ecology 72:371-412.
Ludlum, D.M. 1963. Early American hurricanes: 1492-1870. American Meteorological Soc. Lancaster Press. Lancaster, PA.
MacArthur, R.H. and E.O. Wilson. 1967. The theory of island biogeography. Princeton Univ. Press, Princeton. N.J.
Madansky, A. 1988. Prescriptions for working statisticians. Springer-Verlag, NY.
Mandelbrot, B. 1983. The fractal geometry of nature. W. H. Freeman and Co., New York, N.Y.
Mangini, A.C., M.C. Hicks and R.C. Kertz. 1988. Biological evaluation of southern pine beetle on the Noxubee National Wildlife, Mississippi. U. S. Dept. Agr., For. Serv. Alexandria Field Office Rep. No. 89-2-3.
Mannan, R.W. 1984. Summer area requirements of pileated woodpeckers in western Oregon. Wildl. Soc. Bull. 12:265-268.
Margules, C.R., A.O. Nicholls and R.L. Pressey. 1988. Selecting networks of reserves to maximize biological diversity. Biol. Conserv. 43:63-76.
Marx, L. 1964. The machine in the garden. Oxford Univ. Press. New York, NY.
Mason, G.N. and C.M. Bryant. 1984. Establishing southern pine beetle hazard from aerial stand data and historical records. For. Sci. 30(2): 375-382.
Mason, G.N., R.R. Hicks, C.M. Bryant, M.L. Matthews, D.L. Kulhavy and J.E. Howard. 1981 Rating southern pine beetles by aerial photography. pp. 75-78 *In*: R.L. Hedden, S.J. Barras and J.E. Coster (eds.). Proc., Symp. on Hazard Rating Syst. in For. Insect Pest Manage. U. S. Dept. Agr. For. Serv. Gen. Tech. Rep. WO-27.
Mason, G.N., P.L. Lorio, R.P. Belanger and W.A. Nettleton. 1985. Rating the susceptibility of stands to southern pine beetle attack. U. S. Dept. Agr. Cooperative State Res. Serv., Integrated Pest Manage. Handbook No. 645, Washington D.C.
Mason, R.R. 1971. Soil moisture and stand density affect oleoresin exudation pressure in a loblolly pine plantation. For. Sci. 17: 170-177.
Masters, R.E. 1991. Effects of timber harvest and prescribed fire on wildlife habitat and use in the Ouachita Mountains of eastern Oklahoma. Ph.D. Thesis, Oklahoma State Univ., Stillwater.
Masters, R.E., J.E. Skeen and J. A. Garner. 1989. Red-cockaded woodpecker in Oklahoma: an update of Wood's 1974-77 study. Proc. Okla. Acad. Sci. 69:27-31.
Maxwell, R.S. and R.D. Baker. 1983. Sawdust empire, the Texas lumber industry, 1830-1940. Texas A&M Univ. Press, College Station, TX.
Maynard, C.J. 1881. The birds of eastern North America. C.J. Maynard & Co., Newtonville, MA
McComb, W.C., S.A. Bonney, R.M. Sheffield and N.D. Cost. 1986. Snag resources in Florida—are they sufficient for average populations of primary cavity nesters? Wildl. Soc. Bull. 14:40-48.
McFarlane, R.W. 1992. A stillness in the pines: the ecology of the red-cockaded woodpecker. W.W. Norton.
McKinley, D. 1951. Another plea for ecological thinking. Bluebird 18(4):2-3.
McMahan, C. A., R. G. Frye and K. L. Brown. 1984. The vegetation types of Texas including cropland. Texas Parks & Wildl. Dept. PWD Bull. 7000-120.
McMahon, J.P. 1992. Forest industry's commitment to the public. J. For. 90(10):38-40.
McNab, B.K. 1963. Bioenergetics and the determination of home range size. Amer. Nat. 97:133-140.
McWilliams, W. H. and R. G. Lord. 1988. Forest resources of East Texas. U. S. Dept. Agr. For. Serv. Resour. Bull. SO-136.
Meine, C. 1988. Aldo Leopold: his life and work. Univ. Wisconsin Press, Madison, WI.
Mengel, R.M. and J.A. Jackson. 1977. Geographic variation of the red-cockaded woodpecker. Condor 79:349-355.
Mergen, F. 1954. Mechanical aspects of wind-breakage and windfirmness. J. For. 52:119-125.
Meroney, R.N. 1968. Characteristics of wind and turbulence in and above model forests. J. Appl. Met. 7:780-788.
MFC. 1979. Summary of forestry commission activities following Hurricane Frederick. Unpublished report by Mississippi Forestry Commission, Jackson, MS.
Miller, G.L. 1978. The population, habitat, behavioral and foraging ecology of the red-cockaded woodpecker (*Picoides borealis*) in southeastern Virginia. Unpublished M.A. thesis, College of William and Mary, Williamsburg, VA, .
Millsap, B.A., J.A. Gore, D.E. Runde and S.I. Cerulean. 1990. Setting priorities for the conservation of fish and wildlife species in Florida. Wildlife Monographs 111:1-57.
Mitchell, J.H. 1987. Hazard and risk rating of red-

cockaded woodpecker colony areas and relative susceptibility of cavity trees to the southern pine beetle. M.S. Forestry Thesis. Stephen F. Austin State Univ., Nacodoches, TX.

Mitchell, J.H., D.L. Kulhavy, R.N. Conner and C.M. Bryant V. 1991. Susceptibility of red-cockaded woodpecker colony areas to southern pine beetle infestation in East Texas. So. J. Appl. For. 15(3):158-162.

Mitchell, R.M. and R.T. Carson. 1989. Using surveys to value public goods. Resources for the Future, Washington, D. C.

Mitton, J.B. and M.C. Grant. 1984. Associations among protein heterozygosity, growth rate, and developmental homeostasis. pp. 479-499 *In* R.F. Johnston (ed.). Ann. Rev. Ecol. and Syst. Vol. 15. Ann. Rev., Palo Alto, CA

Mohr, C.O. 1947. Table of equivalent populations of North American small mammals. Amer. Midl. Natur. 37:223-249.

Mohr, C.T. 1897. The timber pines of the southern United States. U. S. Dept. Agr., For. Bull. 13.

Montague, W.G., J.C. Neal, J.E. Johnson and D.A. James 1995. Techniques for excluding southern flying squirrels from cavities of red-cockaded woodpeckers. *In* D.L. Kulhavy, R.G. Hooper and R. Costa (eds.). Red-cockaded Woodpecker: Recovery, Ecology and Management, Center for Applied Studies, College of Forestry, Stephen F. Austin State Univ. , Nacogdoches, TX.

Mumme, R.L., W.D. Koenig and F.A. Patelka. 1983. Are acorn woodpecker territories aggregated? Ecology 64:1305-1307.

Murphey, E.E. 1939. *Dryobates borealis* (Vieillot) red-cockaded woodpecker. pp. 72-79 *In* A.C. Bent (ed.). Life histories of North American woodpeckers. U.S. Natl. Mus. Bull. 174.

Muul, I. 1968. Behavior and physiological influences on the distribution of the flying squirrel, *Glaucomys volans*. Misc. Publ. Mus. Zool. Univ. Michigan 134:1-66.

Myers, N. 1993. The question of linkages in environment and development. BioScience. 43:302-310.

Myers, R.L. 1990. Scrub and high pine. pp. 151-193 *In* R.L. Myers and J.J. Ewel (eds.). Ecosystems of Florida. Univ. Press. Florida, Gainesville, FL.

Nash, R. 1987. Aldo Leopold's intellectual heritage. pp. 63-88 *In* J.B. Callicott (ed.) Companion To a Sand County Almanac. Univ. of Wisconsin Press, Madison, Wisc.

National Oceanic and Atmospheric Administration. 1989. Climatological data, South Carolina, Dec. 1989, Vol. 92, No. 12. National Climatic Data Center, Asheville, NC.

Neal, J.C. 1992. Factors affecting breeding success of red-cockaded woodpeckers in the Ouachita National Forest, Arkansas. M.S. Thesis, Univ. Arkansas, Fayetteville, AR.

Neal, J.C. and W.G. Montague. 1991. Past and present distribution of the red-cockaded woodpecker *Picoides borealis* and its habitat in the Ouachita Mountains, Arkansas. Proc. Ark. Acad. Sci. 45:71-75.

Neal, J.C., W.G. Montague and D.A. James. 1992. Sequential occupation of cavities by red-cockaded woodpeckers and red-bellied woodpeckers in the Ouachita National Forest. Proc. Ark. Acad, Sci. 46:106-108.

Nebeker, T.E. and J.D. Hodges. 1983. Influence of forestry practices on host-susceptibility to bark beetles. Zeit. ange. Entomol. 96(2):194-208.

Nebeker, T.E., J.D. Hodges, B.K. Karr and D.M. Moehring. 1985. Thinning practices in southern pines—with pest management recommendations. U. S. Dept. Agr. For. Serv. Tech. Bull. 1703.

Nebeker, T.E., J.D. Hodges, C.R. Honea and C.A. Blanche. 1988. Preformed defensive system in loblolly pine: variability and impact on management practices. pp. 147-162 *In* T.L. Payne and J. Saarenmaa (eds.). Integrated control of Scolytid bark beetles. Virginia Poly. Institute and State Univ, Blacksburg.

Nebeker, T.E., M. Pelligrine, R.A. Tisdale and J.D. Hodges. 1995. Site/stand factors associated with red-cockaded woodpecker colonies on the Noxubee National Wildlife Reffuge, Mississippi. *In* D.L. Kulhavy, R.G. Hooper and R. Costa (eds.). Red-cockaded Woodpecker: Recovery, Ecology and Management, Center for Applied Studies, College of Forestry, Stephen F. Austin State Univ., Nacogdoches, TX.

Neel, L. 1971. Some observations and comments on the red-cockaded woodpecker in the Thomasville-Tallahassee Game Preserve Region. pp. 137-139 *In* R.L. Thompson (ed.). The Ecology and Management of the Red-cockaded Woodpecker. Bureau of Sport Fisheries and Wildlife, U.S. Dept. Int., and Tall Timbers Res. Sta., Tallahassee, FL.

Neenan, M. and J.L. Spencer-Smith. 1975. An analysis of the problem of lodging with particular reference to wheat and barley. J. Agr. Sci., Camb. 85:495-507.

Nei, M., T. Maruyama and R. Chakraborty. 1975. The bottleneck effect and genetic variability of populations. Evolution 29:1-10.

Nelson, R.M. 1931. Decay in loblolly pine on the Atlantic Coastal Plain. Virginia For. Serv. Publ. 43:58-59.

Nelson, T. C. and W. M. Zillgitt. 1969. A forest atlas of the south. U. S. Dept. Agr. For. Serv. South. For. Exp. Station, New Orleans, LA., and Southeastern. For. Exp. Station, Asheville, N.C.

Nesbitt, S.A., D.T. Gilbert and D.B. Barbour. 1978. Red-cockaded woodpecker fall movements in a Florida flatwoods community. Auk 95:145-151.

Nesbitt, S.A., B.A. Harris, R.W. Repenning and C.B. Brownsmith. 1982. Notes on red-cockaded woodpecker study techniques. Wildl. Soc. Bull. 10(2):160-163.

Nesbitt, S.A., A.E. Jerauld and B.A. Harris. 1983. Red-cockaded woodpecker summer range size in southwest Florida. pp. 68-71 *In*: Proceedings of the

Red-cockaded Woodpecker Symposium II, (D.A. Wood, ed.), Florida Game and Freshwater Fish Commission, Tallahassee, FL.

Neumann, C.J. 1987. The National Hurricane Center risk analysis program. NOAA Technical Memorandum NWS NHC 38. NOAA, National Weather Serv., National Hurricane Center, Coral Gables, FL.

Neumann, C.J., B.R. Jarvinen, A.C. Pike and J.D. Elms. 1987. Tropical cyclones of the North Atlantic Ocean 1871-1986. Historical Climatology Series 6-2. NOAA, National Weather Serv., National Climatic Data Center, Asheville, N.C.

Neustein, S.A. 1965. Windthrow on the margins of various sizes of felling areas. pp.166-171 *In* Report of forest research for the year ended March 1964. Forestry Comm., London.

Newcomb, L.S. 1960. U.S. Forest Service protects red-cockaded woodpecker nest trees. M.O.S. Newsletter 5(4):6-7.

Nicholson, C.P. 1977. The red-cockaded woodpecker in Tennessee. Migrant 48:54-62.

Nix, L.E. and T.F. Ruckelshaus. 1990. Long-term effects of thinning on stem taper of old-field, plantation loblolly pine in the Piedmont. pp. 202-207 *In* S.S. Coleman and D.G. Neary (eds.). U. S. Dept. Agr., For. Serv., Gen. Tech. Rep. SE-70. Sixth Biennial Southern Silvicultural Research Conf. Asheville, N.C.

Nobles, M.K. 1965. Identification of cultures of wood-inhabiting hymenomycetes. Can. J. Bot. 43:1097-1139.

Nolan, V. Jr. 1959. Pileated woodpecker attacks pilot black snake at tree cavity. Wilson Bull. 71: 381-382.

Norton, B.G. 1988. The constancy of Leopold's land ethic. Conserv. Biol. 2(1):93-102.

Norton, D.A. 1989. Tree windthrow and forest soil turnover. Can. J. For. Res. 19:386-389.

Norusis, M.J. 1985. SPSS^X Introductory Statistics Guide. SPSS Inc., Chicago, Illinois.

Noss, R.F. 1987. Corridors in real landscapes: a reply to Simberloff and Cox. Conserv. Biol. 1:159-164.

Noss, R.F. 1989. Longleaf pine and wiregrass: keystone components of an endangered ecosystem. Nat. Areas. J. 9:211-213.

Nuttall, T. (S. Lottinville, ed.). 1980. A J. travels into the Arkansas territory during the year 1819. Univ. Oklahoma Press, Norman, OK.

O'Neill, R.V., J.R. Krummel, R.H. Gardner, G. Sugihara, B. Jackson, D.L. DeAngelis, B.T. Milne, M.G. Turner, B. Zygmunt, S.W. Christensen, V.H. Dale and R.L. Graham. 1988. Indices of landscape pattern. Landscape Ecol. 1(3):153-162.

Oberholser H.C. 1938. The bird life of Louisiana. State of Louisiana, Dept. Conservation, New Orleans.

Odom, R.R. 1983. Georgia's red-cockaded woodpecker relocation experiment: a 1983 update. pp 106-108 *In* D. A. Wood (ed.). Red-cockaded Woodpecker Symp. II Proc. FL. Game & Fresh Water Fish Comm., US FWS & US For. Serv., Tallanassee.

Odom, R.R., J. Rappole, J. Evans, D. Charbonneau and D. Palmer. 1982. Red-cockaded woodpecker relocation experiment in coastal Georgia. Wildl. Soc. Bull. 10:197-203.

Odum, E.P. and E.J. Kuenzler. 1955. Measurement of territory and home range size in birds. Auk 72:128-137.

Office of Endangered Species and International Activities, compiler. 1973. Threatened wildlife of the United States. Bureau of Sport Fisheries and Wildlife, U.S. Dept. Interior, Resource Publ. 114.

Oklahoma Department of Wildlife Conservation. 1991a. McCurtain County Wilderness Area Management Plan. Game Div., Oklahoma City, Okla. (unpubl.).

Oklahoma Department of Wildlife Conservation. 1991b. McCurtain County Wilderness Area Implementation Plan. Game Div., Oklahoma City, Okla. (unpubl.).

Oklahoma Department of Wildlife Conservation. 1991c. McCurtain County Wilderness Area Environmental Assessment. Game Div., Oklahoma City, Okla. (unpubl.).

Oklahoma Game and Fish Department. 1954. Biennial report—Oklahoma Game and Fish Department (1952-1954). Oklahoma Game and Fish Department., Oklahoma City, Okla.

Oklahoma State Game and Fish Commission. 1934. Biennial report—State Game and Fish Commission (1932-1934). State Game and Fish Comm., Oklahoma City, Okla.

Oliver, C.D. 1992. A landscape approach—achieving and maintaining biodiversity and economic productivity. J. For. 90 (9):20-25.

Oliver, C.D. and B.C. Larson, 1990. Forest stand dynamics. McGraw-Hill Co. NY. 467 p.

Ortego, B. and D. Lay. 1988. Status of red-cockaded woodpecker colonies on private lands in east Texas. Wildlife Soc. Bull. 16:403-405.

Ortego, B., R.N. Conner and C. Rudolph. 1988. Status of the red-cockaded woodpecker in Texas, 1985-1987. Bull. Texas Ornithological Soc. 21:22-25.

Orville, R. E. 1991. Lightning ground flash density in the contiguous United States-1989. Monthly Weather Rev. 119:573-577.

Paine, T.D., F.M. Stephen and R.G. Cates. 1985. Induced defenses against *Dendroctonus frontalis* and associated fungi: variation in loblolly pine resistance. pp. 167-176 *In* S.J. Branham and R.C. Thatcher (eds.). Integrated Pest Management Research Symp.: Proc. U. S. Dept. Agr. For. Serv. Gen. Tech. Rep. SO-56, New Orleans, LA.

Palmer, A.R. and C. Strobeck. 1986. Fluctuating asymmetry: measurement, analysis, patterns. Ann. Rev. Ecol. Syst. 17:391-421.

Pangburn, C.H. 1919. A three months' list of the birds of Pinellas County, Florida. Auk 36:393-405.

Parker, W.T. and M.K. Phillips. 1991. Application of the experimental population designation to recovery of endangered red wolves. Wildl. Soc. Bull. 19:73-79.

Patterson, G.A. and W.B. Robertson. 1981. Distribution and habitat of the red-cockaded woodpecker in Big Cypress National Preserve. U.S. Nat. Park Serv. South Florida Res. Center, Rept. T-613.

Patterson, G.A. and W.B. Robertson, Jr. 1983. An instance of red-cockaded woodpeckers nesting in a dead pine. pp. 99-100 *In* D.A. Wood (ed.), Red-cockaded Woodpecker Symposium II Proceedings. Florida Game and Fresh Water Fish Commission, Tallahassee, FL.

Payne, T.L. 1980. Life history and habits. pp 7-28 *In* R.C. Thatcher, J.L. Searcy, J.E. Coster and G.D. Hertel (eds.). The southern pine beetle. U. S. Dept. Agr. For. Serv. Expanded South Pine Beetle Res. Appl. Prog. For. Serv. Sci. and Education Tech. Bull. 1631.

Payne, T.L. and R.N. Coulson. 1985. Role of visual and olfactory stimuli in host selection and aggregation behavior by *Dendroctonus frontalis*. pp. 73-82 *In* Safranyik, L. (ed.). The role of the host in population dynamics of forest insects. Proc. IUFRO Symp., Banff, Canada.

Pessin, L.J. 1933. Forest associations in the uplands of the lower gulf coastal plain (longleaf pine belt). Ecology 14:1-14.

Peters, R.H. 1991. A critique for ecology. Cambridge Univ. Press.

Peters, R.H. and J.V. Raelson. 1984. Relations between individual size and mammalian population density. Amer. Nat. 124:498-517.

Peterson, G.L. and A. Randall, eds. 1984. Valuation of wildlife resource benefits, Westview Press, Boulder, CO.

Phelps, F.M. 1914. The resident bird life of the Big Cypress Swamp region. Wilson Bull. 26:86-101.

Platt, W.J., G.W. Evans and S.L. Rathbun 1988. The population dynamics of a long-lived conifer (*Pinus palustris*). Amer. Nat. 131:491-525.

Platt, W.J., J.S. Glitzenstein and K.R. Strent. 1989. Evaluating pyrogenicity and its effects on vegetation in longleaf pine savannas. pp. 143-191 *In* Proc. 17th Tall Timbers Fire Ecology Conf.

Porter, M.L. and R.F. Labisky. 1986. Home range and foraging habitat of red-cockaded woodpeckers in northern Florida. J. Wildl. Manag. 50:239-247.

Porterfield, R.L. and C.E. Rowell. 1980. Characteristics of southern pine beetle infestations southwide. *In* J.E. Coster and J.L. Searcy (eds.). Site, stand, and host characteristics of southern pine beetle infestations. U.S. Dept. Agr. For. Serv. Tech. Bull. 1612.

Post, W. and J.S. Greenlaw. 1989. Metal barriers protect near-ground nests from predators. J. Field Ornithol. 60: 102-103.

Powell, M.D., P.P. Dodge and M.L. Black. 1991. The landfall of Hurricane Hugo in the Carolinas: surface wind distribution. Weather and Forecasting 6:379-399.

Price, T.S. and C. Doggett (eds.). 1982. A history of southern pine beetle outbreaks in the southeastern United States. Ga. For. Comm., Macon, GA.

Pritchett, W.L.and R.F. Fisher 1987. Properties and management of forest soils, second edition. John Wiley & Sons. New York.

Purvis, J.C., S.F. Sidlow, D.J. Smith, W. Tyler and I. Turner. 1990. Hurricane Hugo. Climate Rep. G-37, South Carolina Water Res. Comm., Columbia.

Putz, F.E. and R.R. Sharitz. 1991. Hurricane damage to old-growth forest in Congaree Swamp National Monument, South Carolina, U.S.A. Can. J. Forest Res. 21:1765-1770.

Quarles,S.P., J.A. Macleod and T.R. Lundquist. 1991. The Endangered Species Act and its application to private lands. Amer. For. Resour. Alliance, Tech. Bull. No. 91-06.

Quattro, J.M. and R.C. Vrijenhoek. 1989. Fitness differences among remnant populations of the endangered Sonoran topminnow. Science 245:976-978.

Rafferty, J.E. 1979. Cultural resource reconnaissance and project-oriented survey, Noxubee National Wildlife Refuge, Mississippi. Mississippi State Univ., Dept. Anthro., Starkville, MS.

Ralls, K., J.D. Ballou and A. Templeton. 1988. Estimates of lethal equivalents and the cost of inbreeding in mammals. Conserv. Biol. 2:185-193.

Ramey, P. 1980. Seasonal, sexual, and geographical variation in the foraging ecology of red-cockaded woodpeckers (*Picoides borealis*). Unpublished M.S. thesis, Mississippi, Mississippi State Univ., Starkville.

Randall, A., J.P. Hoehn and C.S. Swanson. 1990. Estimating the recreational, visual, habitat, and quality of life benefits of Tongass National Forest. U.S. Dept. Agr. For. Serv. Gen. Tech. Rep. RM-192, Ft. Collins, CO.

Reap, R. M. and D. R. MacGorman. 1989. Cloud-to-ground lightning: climatological characteristics and relationships to model fields, radar observations, and severe local storms. Mon. Weather Review 117:518-535.

Reed, J.M., J.H. Carter, III, J.R. Walters and P.D. Doerr. 1988a. An evaluation of indices of red-cockaded woodpecker populations. Wildl. Soc. Bull. 16(4): 406-410.

Reed, J.M., P.D. Doerr and J.R. Walters. 1988b. Minimum viable population size of the red-cockaded woodpecker. J. Wildl. Manage. 52:385-391.

Reinman, J.P. 1985. A survey of the understory vegetation communities of the St. Marks National Wildlife Refuge pinelands. U.S. Fish and Wildl. Serv. Unpubl. Rep., St. Marks, FL.

Reinman, J.P. 1995. Population status and management of red-cockaded woodpeckers on St. Marks National Wildlife Refuge 1980-1992. *In* D.L. Kulhavy, R.G. Hooper and R. Costa (eds.). Red-cockaded Woodpecker: Recovery, Ecology and Management, Center for Applied Studies, College of Forestry, Stephen F. Austin State Univ. , Nacogdoches, TX.

Reller, A.W. 1972. Aspects of behavioral ecology of red-headed and red-bellied woodpeckers. Amer. Midl. Nat. 88:270-290.

Renken, R.B. and E.P. Wiggers. 1989. Forest characteristics related to pileated woodpecker territory size in Missouri. Condor 91:642-652.

Renwick, J.A.A. and J.P. Vité. 1969. Bark beetle attractants: mechanism of colonization by *Dendroctonus*. Nature 224:1222-1223.

Repasky, R.R. 1984. Home range and habitat utilization of the red-cockaded woodpecker. M.S. thesis, North Carolina State Univ. Raleigh.

Reynolds, R.T., J.M. Scott and R.A. Nussbaum. 1980. A variable circular plot method for estimating bird numbers. Condor 82:309-313.

Richardson, D.M. and D. Smith. 1992. Hardwood removal in red-cockaded woodpecker colonies using a shear V-blade. Wildl. Soc. Bull. 20:428-433.

Richardson, D.M. and J.M. Stockie. 1995. Response of a small red-cockaded woodpecker population to intensive management at Noxubee National Wildlife Refuge. *In* D. L. Kulhavy, R. G. Hooper and R. Costa (eds.). Red-cockaded Woodpecker: Recovery, Ecology and Management, Center for Applied Studies, College of Forestry, Stephen F. Austin State Univ. , Nacogdoches, TX.

Robbins, C., D. Bystrak and P. Geissler. 1986. The breeding bird survey: its first fifteen years, 1965-1979. U. S. Dept. Int., Fish and Wildl. Serv. Res. Publ. 157. Washington, D.C.

Roberts, R.C. 1979. Habitat and resource relationships in acorn woodpeckers. Condor 81:1-8.

Robertson, W.B. and J.A. Kushlan. 1974. The south Florida avifauna. pp. 414-452 *In* Gleason, P.J. (ed.). Environments of South Florida: Present and Past. Miami Geol. Soc. Miami, FL

Robinson, V.E. 1965. Big bad Betsy. In Oct. 15 issue, Southern Lumberman. pp. 20-22.

Rohlf, F.J. and R.R. Sokal. 1969. Statistical tables. W. H. Freeman and Co., San Francisco. CA

Roise, J.P., J. Chung, R. Lancia and M. Lennartz. 1990. Red-cockaded woodpecker habitat and timber management: production possibilities. So J. Appl For. 14: 6-11.

Ross, W.G, D.L. Kulhavy, R.N. Conner and J. Sun. 1991. Physiology of red-cockaded woodpecker cavity trees: implications for management. pp. 558-566 *In* S.S. Coleman and D.G. Neary (eds.)., Proc. Sixth Biennial Southern Silvicultural Res. Conf. U. S. Dept. Agr. For. Serv. Gen. Tech. Rep. 70, Asheville, NC.

Rudolph, D.C. and R.N. Conner. 1991. Cavity tree selection by red-cockaded woodpeckers in relation to tree age. Wilson Bull. 103:458-467.

Rudolph, D.C. and R.N. Conner. 1995. The impact of southern pine beetle induced mortality on red-cockaded woodpecker cavity trees. *In* D.L. Kulhavy, R.G. Hooper and R. Costa (eds.). Red-cockaded Woodpecker: Recovery, Ecology and Management, Center for Applied Studies, College of Forestry, Stephen F. Austin State Univ. , Nacogdoches, TX.

Rudolph, D.C., H. Kyle and R.N. Conner. 1990a. Competition for red-cockaded woodpecker roost and nest cavities: effects of resin age and entrance diameter. Wilson Bull. 102:23-36.

Rudolph, D.C., R.N. Conner and J. Turner. 1990b. Red-cockaded woodpeckers vs rat snakes: the effectiveness of the resin barrier. Wilson Bull. 102:14-22.

Rudolph, D.C., R.N. Conner, D.K. Carrie and R.R. Shaefer. 1992. Experimental reintroduction of red-cockaded woodpeckers. Auk 109:914-916.

Rudolph, D.C., R.N. Conner and R.R. Shaefer. 1995. Red-cockaded woodpecker detection of red heart infection. *In* D.L. Kulhavy, R.G. Hooper and R. Costa (eds.). Red-cockaded Woodpecker: Recovery, Ecology and Management, Center for Applied Studies, College of Forestry, Stephen F. Austin State Univ. , Nacogdoches, TX.

Ruggiero, L.F., K.B. Aubry, A.B. Carey and M.H. Huff. (eds.). 1991. Wildlife and vegetation of unmanaged Douglas-fir forests. U. S. Dept. Agr. For. Serv. Gen. Tech. Rep. PNW-285, Portland, OR.

Runkle, J. R. 1982. Guidelines and sample protocol for sampling forest gaps. U. S. Dept. Agr., For. Serv., Gen. Tech. Rep. PNW-GTR-283, Portland, OR.

Runkle, J.R. 1991. Natural disturbance regimes and the maintenance of stable regional floras. pp. 31-48 *In* D. Henderson and L.D. Hedrick (eds.). Proc. of Conf. Restoration of old growth forests in the interior highlands of Arkansas and Oklahoma.

Ruth, R.H. and R.A. Yoder. 1953. Reducing wind damage in the forests of the Oregon coast range. U. S. Dept. Agr., For. Serv., Res. Pap. No 7. Pacific Northwest Forest and Range Exp. Sta., Portland, OR.

Saffir, H.S. 1977. Design and construction requirements for hurricane resistant construction. Amer. Soc. Civil Engineers, NY.

SAS Institute Inc. 1985. SAS users' guide: statistics, version 5th ed. SAS Institute Inc., Cary, S.C. 956pp.

SAS. 1988. SAS/STAT user's guide, release 6.03 edition. SAS Institute Inc., SAS Circle, Cary, NC.

Saunders, D.A., R.J. Hobbs and C.R. Margules. 1991. Biological consequences of ecosystem fragmentation: a review. Conserv. Bio. 5:18-32.

Savill, P.S. 1983. Silviculture in windy climates. For. Abstracts 44:473-488.

Schenck, V. 1973. Little southern 'redhead' gets foresters to change harvest rules. Inland Bird Banding News 45:5-6.

Schoener, T.W. 1968. Sizes of feeding territories among birds. Ecology 49:123-141.

Scholander, P.F., H.T. Hammel, E.D. Bradstreet and E.A. Hemingsen. 1965. Sap pressure in vascular plants. Science 148: 339-346.

Schopmeyer, C.S. and P.R. Larson. 1955. Effects of diameter, crown ratio, and growth rates on gum yields of slash and longleaf pine. J. For. 53:822-826.

Schowalter, T. D., R. N. Coulson and D. A. Crossley, Jr. 1981. Role of southern pine beetle and fire in maintenance of structure and function of the southeastern coniferous forest. Environ. Entomol. 10: 821-825.

Schroeder, R.L. 1983a. Habitat suitability index models: downy woodpecker. U.S. Fish Wildl. Serv. FWS/OBS-82/10.38.

Schroeder, R.L. 1983b. Habitat suitability index models:

pileated woodpecker. U.S. Fish Wildl. Serv. FWS/OBS-82/10.39.

Schumacher F.X. and T.S. Coile. 1960. Growth and yields of natural stands of southern pines. T.S. Coile, Inc., Durham, NC.

Schwarz, G.F. 1907. The longleaf pine in virgin forest: a silvical study. John Wiley & Sons, New York.

Scott, J.M., F. Davis, B. Csuti, R. Noss, B. Butterfield, C. Groves, H. Anderson, S. Caicco, F. D'Erchia, T.C. Edwards, Jr., J. Ullman and R.G. Wright. 1993. Gap analysis: a geographic approach to protection of biological diversity. Wildl. Monogr. 123: 1-141.

Scuderi, L.A. 1993. A 2000-year tree ring record of annual temperatures in the Sierra Nevada Mountains. Sci. 259:1433-1436.

Seamon, L.H. 1954. Hurricane Hazel. Climatological data, national summary. 5:381-385.

Selander, R.K. 1964. Speciation in the wrens of the genus *Campylorhynchus*. Univ. Calif. Publ. Zool. 74:1-305.

Shaffer, M.L. 1981. Minimum population size for species conservation. BioScience 31:131-134.

Shaffer, M.L. 1987. Minimum viable populations: coping with uncertainty. pp. 69-86. *In* M.E. Soulé (ed.). Viable Populations for Conservation. Cambridge Univ. Press, New York.

Shannon, C.E. and W. Weaver. 1962. The mathematical theory of communication. Univ. Ill. Press, Urbana. IL.

Shaprio, A.E. 1983. Characteristics of red-cockaded woodpecker cavity trees and colony areas in southern Florida. Florida Sci. 2:89-95.

Shaw, W.W. 1984. Problems in Wildlife Valuation in Natural Resource Management. *In* G.L. Peterson and A. Randall (eds.). Valuation of Wildland Resource Benefits. eds. Westview Press: Boulder, CO.

Sheffield, R.M. and M.T. Thompson. 1992. Hurricane Hugo: effects on South Carolina's forest resource. U. S. Dept. Agr., For. Serv., Res. Pap. SE-284. Asheville, NC.

Sherrill, D. M. and V. M. Case. 1980. Winter home ranges of four clans of red-cockaded woodpeckers in the Carolina sandhills. Wilson Bull. 92: 369-375.

Short, L.L. 1982. Woodpeckers of the world. Monogr. Ser. 4, Delaware Museum of Natural History. Greenville, DE

Shumway, C. 1986. A Summary of Forest Service GIS Activities. pp. 49-52. Proc. Geographic Infor. Sys. Workshop. Amer. Soc. Photogram. & Remote Sensing.

Simberloff, D. and J. Cox. 1987. Consequences and costs of conservation corridors. Conserv. Bio. 1:63-71.

Simberloff, D., J. Farr, J. Cox and D. Mehlman. 1992. Movement corridors: conservation bargains or poor investments? Conserv. Bio. 6:492-503.

Simpson, R.H. and H. Riehl. 1981. The hurricane and its impacts. Louisiana State Univ. Press, Baton Rouge, LA.

Skelly, J.M. 1976. Levels of *Fomitopsis annosa* in root systems of southern pine beetle attacked versus non-attacked trees. *In* Proc. southwide forest diseases workshop. U. S. Dept. Agr. For. Serv., Southeastern. Area, State and Private Forestry, Atlanta, GA.

Skorupa, J.P. 1979. Foraging ecology of the red-cockaded woodpecker in South Carolina. M.S. thesis, Univ. Calif., Davis, Davis, CA.

Skorupa, J.P. and R.W. McFarlane. 1976. Seasonal variation in foraging territory of red-cockaded woodpeckers. Wilson Bull. 88:662-665.

Smith, B.H., S.M. Garr and P.E. Cole. 1982. Problems of sampling and inference in the study of fluctuating dental asymmetry. Amer. J. Phys. Anth. 58:281-289.

Smith, D.M. 1962. The practice of silviculture. John Wiley & Sons, Inc., New York.

Smith, D.M. 1986. The Practice of Silviculture. 8th ed. John Wiley & Sons, Inc., New York.

Smith, F.E. 1970. Analysis of ecosystems. pp. 7-18 *In* D. Reichle (ed.). Analysis of temperate forest ecosystems Springer-Verlag, New York.

Smith, K. L. 1986. Historical perspective. pp. 1-8 *In* P. A. Murphy (ed.). Proc. of a symposium on the shortleaf pine ecosystem. U.S. Dept. Agr. For. Serv. South. For. Exp. Station, Monticello, AR.

Smith, M.W. 1981. Comments on herbicide injection for habitat maintenance of red-cockaded woodpecker colonies. The Mississippi Kite. 11:52-53.

Smith, M.W., C.F. Aquadro, M.H. Smith, R.K. Chesser and W.J. Etges. 1982. Bibliography of electrophoretic studies of biochemical variation in natural vertebrate populations. Texas Tech Press, Lubbock, TX.

Snyder, J.R. 1978. Analysis of vegetation in Croatan National Forest. Ph.D. Dissertation, Univ. of North Carolina, Chapel Hill, NC.

Soil Conservation Service. 1969. U. S. Dept. Agr.

Sokal, R.R. and F.J. Rohlf. 1969. Biometry: the principles and practice of statistics in biological research. W. H. Freeman and Co., San Francisco. 776p.

Solomon, B., et al. 1991. Global Climate Change: Management Strategies For Wisconsin. Wisc. Department Nat. Resour., PUBL- AM-066-91.

Soulé, M.E. 1979. Heterozygosity and developmental stability: another look. Evolution 73:396-401.

Soulé, M.E. 1980. Thresholds for survival: maintaining evolutionary fitness and potential. pp. 151-169 *In* M.E. Soulé and B.A. Wilcox (eds.). Conservation Bio. Sinauer Associates, Sunderland, MA.

Soulé, M.E. (ed.). 1987. Viable populations for conservation. Cambridge Univ. Press.

Sousa, P.J. 1983a. Habitat suitability index models: Lewis' woodpecker. U.S. Fish Wildl. Serv. FWS/OBS-82/10.32.

Sousa, P.J. 1983b. Habitat suitability index models: Williamson's sapsucker. U.S. Fish Wildl. Serv. FWS/OBS-82/10.47.

Sousa, P.J. 1987. Habitat suitability index models: Hairy woodpecker. U.S. Fish Wildl. Serv. Biol. Rep. 82(10.146).

Spurr, S.H. and B.V. Barnes. 1980. Forest ecology.John Wiley & Sons. NY.

Squillace, A.E. 1966. Geographic variation in the slash

pine. For. Sci. Monogr. 10. 56 pp.
Stabb, M.A., M.E. Gartshore and P.L. Aird. 1989. Interactions of southern flying squirrels, *Glaucomys volans*, and cavity-nesting birds. Can. Field-Nat. 103:401-403.
Stacey, P.B. and J.D. Ligon. 1987. Territory quality and dispersal options in the Acorn woodpecker, and a challenge to the habitat-saturation model of cooperative breeding. Amer. Nat. 130:654-676.
Stacey, P.B. and J.D. Ligon. 1991. The benefits of philopatry hypothesis for the evolution of cooperative breeding: variation in territory quality and group size effects. Amer. Nat. 137:831-846.
Stahle, D.W., J.G. Hehr, G.G. Hawks, Jr., M.K. Cleaveland and J.R. Baldwin. 1985. Tree-ring chronologies for the southcentral United States. Dept. Geography, Univ. Arkansas, Fayetteville, AR.
Stangel, P.W. and M.R. Lennartz. 1988. Survival of red-cockaded woodpecker nestlings unaffected by sampling blood and feather pulp for genetic studies. J. Field Ornithology 59: 389-394.
Stangel, P.W., M.R. Lennartz and M.H. Smith. 1992. Genetic variation and population structure of red-cockaded woodpeckers. Cons. Bio. 6:283-292.
Statistical Sciences, Inc. 1991. S-plus User's Manual, Version 3.0. Statistical Sciences, Inc. Seattle, WA.
Steirly, C.C. 1949. A note on the red-cockaded woodpecker. Raven 20:6-7.
Steirly, C.C. 1950. Nest cavities of the red-cockaded woodpecker. Raven 21:2-3.
Steirly, C.C. 1957a. Nesting ecology of the red-cockaded woodpecker in Virginia. Atlantic Nat. 12:280-292.
Steirly, C.C. 1957b. Nesting ecology of the red-cockaded woodpecker in Virginia. Raven 28:24-36.
Stevens, E.E. In prep. A stable population of red-cockaded woodpeckers in the Georgia Piedmont.
Stickel, L.F., W.H. Stickel and F.C. Schmid. 1980. Ecology of a Maryland population of black rat snakes (*Elaphe o. obsoleta*). Amer. Midl. Nat. 103: 1-14.
Stokes, M.A. and T.L. Smiley. 1968. An introduction to tree ring dating. Univ. Chicago Press, Chicago.
Straney, D.O. 1978. Variance partitioning and nongeographic variation. J. Mammal. 59:1-11.
Swain, K.M. 1979. Minimizing timber damage from hurricanes. Southern Lumberman 239(2968):107-109.
Swain, K.M. and M.C. Remion. 1981. Direct control methods for the southern pine beetle. U. S. Dept. Agr, For. Serv. Agr. Handb. 575. Washington, D.C.
Swofford, D.L. and R.B. Selander. 1981. BIOSYS-1: a FORTRAN program for the comprehensive analysis of electrophoretic data in population genetics and systematics. J. Hered. 72:281-283.
Taha, H.A. and F.M. Stephen. 1984. Modeling with imperfect data: a case study simulating a biological system. Simulation. 42(3):109-115.
Tanner, J.T. 1942. The Ivory-billed woodpecker. Nat. Audubon Soc. Res. Rep. No. 1. Dover Pub., NY.
Tansey, J.B. and C.C. Hutchins, Jr. 1988. South Carolina's forests. U. S. Dept. Agr. For. Serv., Res. Bull. SE-103.
Taylor, M.W. 1977. A comparison of three edge indexes. Wildl. Soc. Bull. 5(4):192-193.
Taylor, W.E. and R.G. Hooper. 1991. A modification of Copeyon's drilling technique for making artificial red-cockaded woodpecker cavities. U. S. Dept. Agr., For. Serv., Gen. Tech. Rep. SE-72. Southeastern For. Exp. Sta. Asheville, NC: .
Telewski, F.W. and M.J. Jaffe. 1986. Thigmomorphogenesis: the role of ethylene in the response of *Pinus taeda* and *Abies fraseri* to mechanical perturbation. Physiol. Plant. 66:227-233.
Terborgh, J. and B. Winter. 1980. Some causes of extinction. pp. 119-133 *In* M.E. Soulé and B.A. Wilcox (eds.). Cons. Bio., Sinauer Associates, Sunderland, MA.
Texas Forest Service. 1978. Texas forest pest activity 1976-1977 and forest pest control section biennial report. Texas For. Serv., Publ. 117:1-18, College Station, TX.
Thatcher, R.C. and L.S. Pickard. 1964. Seasonal variations in activity of the southern pine beetle in east Texas. J. Econ. Emtomol. 57:840-842.
Thatcher, R.C. and L.S. Pickard. 1967. Seasonal development of the southern pine beetle in east Texas. J. Econ. Entomol. 60:656-658.
Thatcher, R.C., G.N. Mason and G.D. Hertel. 1986. Integrated pest management in southern pine forests. U. S. Dept. Agr., For. Serv., Agr. Handb. 650. Coop. State Res. Serv..
Thompson, M.T. 1991. Forest statistics for the Coastal Plain of Virginia, 1991. U. S. Dept. Agr. For. Serv. Res. Bull. SE-122, Asheville, NC.
Thompson, R.L. (ed.). 1971. The ecology and management of the red-cockaded woodpecker. Bureau of Sport Fisheries and Wildlife, U.S. Dept. Int., and Tall Timbers Research Station, Tallahassee, FL
Thompson, R.L. 1976. Change in status of red-cockaded woodpecker colonies. Wilson Bull. 88:491-492.Thompson, R.L. 1983. Red-cockaded Woodpecker Symposium II-an overview. pp. 3-4 In D.A. Wood (ed.). Red-cockaded Woodpecker Symposium II Proc., Florida Game and Fresh Water Fish Comm., Tallahassee, FL.
Thompson, R.L. and W.W. Baker. 1971. A survey of red-cockaded woodpecker requirements. pp. 170-186 *In* Proc. Symp. The ecology and management of the Red-cockaded Woodpecker (R.L. Thompson, ed.). U.S. Bur. Sport Fish and Wildl. and Tall Timbers Res. Station, Tallahassee, FL.
Times Mirror Magazines. 1992. Natural resource conservation: where environmentalism is headed in the 1990s. Times Mirror Magazines National Environmental Survey. Times Mirror Magazines, Washington, D.C.
Touliatos, P. and E. Roth. 1971. Hurricanes and trees: ten lessons from Camille. J. For. 69:285-289.
Townsend, J.F. 1990. Preliminary data report, Hurricane Hugo in the Charleston area. National Weather Serv., Charleston, SC.

Travis, J. 1977. Seasonal foraging ecology in a downy woodpecker population. Condor 79:371-375.
Trimble Navigation, Ltd. 1992. General reference for the GPS Pathfinder Professional System. Trimble Navigation, Sunnyvale, CA.
Trousdell, K.B. 1955. Hurricane damage to loblolly pine on Bigwoods Experimental Forest. Southern Lumberman 191:35-37.
Trousdell, K.B., W.C. Williams and T.C. Nelson. 1965. Damage to recently thinned loblolly pine stands by Hurricane Donna. J. For. 63:96-100.
Turchin, P., P.L. Lorio Jr., A.D. Taylor and R.F. Billings. 1991. Why do populations of southern pine beetles (Coleoptera: Scolytidae) fluctuate? Environ. Entomol. 20:401-409.
Turner, M.G. 1990. Landscape changes in nine rural counties in Georgia. Photogrammetric Engineering and Remote Sensing 56(3):379-386.
Turner, T. 1987. Trouble on the split estate. Defenders 62(1):28-37.
U.S. Army. 1984. Policy and management guidelines for red-cockaded woodpecker on Army installations. U.S. Army, Washington, DC.
U.S. Army. 1992. Revised Fort Bragg forest management plan. U.S. Army, Fort Bragg, N.C.U.S. Army. 1993. Draft management guidelines for the red-cockaded woodpecker on army installations. U.S. Army, Washington, D.C.
U.S. Department of Energy, Savannah River Field Office. 1991. The Savannah River Site Natural Resources Management Plan. New Ellenton, SC.
U.S. Department of Interior and U.S. Department of Agriculture. 1992. America's biodiversity strategy: actions to conserve species and habitats. U. S. Dept. Int. Office of Prog. Anal. and U. S. Dept. Agr. Natur. Resour. and Environ., Washington, D.C. (preprint 6/15/92).
U.S. Department of Interior, Fish and Wildlife Service. 1985. Red-cockaded Woodpecker Recovery Plan. Department of the Interior, Fish and Wildlife Serv., Region 4. Atlanta, GA.
U.S. Department of Interior. 1968a. List of endangered and extinct wildlife in America. Bureau of Sport Fisheries and Wildlife, Washington, D.C.
U.S. Department of Interior. 1968b. Rare and endangered fish and wildlife of the United States. U.S. Bur. Sport Fish. and Wildl., Res. Pub. 34
U.S. Fish and Wildlife Service. 1970. Listing of the red-cockaded woodpecker as endangered. Federal Register 35:16047, October 13, 1970.
U.S. Fish and Wildlife Service. 1979. Red-cockaded woodpecker recovery plan. U.S. Fish and Wildl. Serv., Atlanta, GA.
U.S. Fish and Wildlife Service. 1980. Biological opinion for the effects of woodland management at Fort Bragg. 4-2-80-F-31, U.S. Fish and Wildl. Serv., Atlanta, GA.
U.S. Fish and Wildlife Service. 1985a. Biological opinion for the effects of proposed 5-year range modernization program and main cantonment expansion at Fort Bragg. 4-2-84-897, U.S. Fish and Wildl. Serv., Asheville, NC.
U.S. Fish and Wildlife Service. 1985b. Red-cockaded Woodpecker Recovery Plan. U.S. Fish and Wildlife Service, Atlanta, GA. 88 pp.
U.S. Fish and Wildlife Service. 1987. Management guidelines for red-cockaded woodpeckers on national wildlife refuges. U.S. Fish and Wildl. Serv., Atlanta, GA. U.S. Fish and Wildlife Service. 1988. Endangered Species Act of 1973, as amended. U.S.D.I., Fish and Wild. Serv., Washington, D.C.
U.S. Fish and Wildlife Service. 1989. A wildlife management plan for the forested uplands of the St. Marks National Wildlife Refuge. U.S. Fish Wildl. Service, St. Marks FL.
U.S. Fish and Wildlife Service. 1990a. Biological opinion for the effects of military and associated activities at Fort Bragg, Camp Mackall and the Sandhills Gamelands. 4-0-90-001. U.S. Fish and Wildl. Serv., Atlanta, GA.
U.S. Fish and Wildlife Service. 1990b. Unpublished red-cockaded woodpecker data from the Fort Stewart Army Base, Savannah, GA
U.S. Fish and Wildlife Service. 1992a. Biological opinion for the effects of continued operation and use of the Coleman Danger Area, Fort Bragg. U.S. Fish and Wildl. Service, Raleigh, NC.
U.S. Fish and Wildlife Service. 1992b. Biological opinion for the effects of proposed construction of Installation Materials and Maintenance Division complex on Fort Bragg. U.S. Fish and Wildl. Serv., Raleigh, NC.
U.S. Forest Service. 1931. Management plan for Choctawhatchee N.F.: 1931 Revision. U.S. Dept. Agr., U.S. For. Serv., Atlanta, GA.
U.S. Forest Service. 1939. Management plan for Choctawhatchee N.F.: 1939 Revision. U.S. Dept. Agr., For. Serv., Atlanta, GA.
U.S. Forest Service. Revised 1976. Volume, yield, and stand tables for second-growth southern pines. U.S. Dept. Agr., For. Serv., Misc. Pub. No. 50, Washington. 202 p.
U.S. Forest Service. 1979. Wildlife habitat management handbook, chapter 420, red-cockaded woodpecker. U.S. Dept. Agr., For. Serv., Handbook 2609.23R, Atlanta, GA (unpub. administrative document).
U.S. Forest Service. 1985. Wildlife habitat management handbook, chapter 420 (420-420.4), red-cockaded woodpecker. U.S. Dept. Agr., For. Serv., Handbook 2609.23R. Southern Region Atlanta, GA.
U.S. Forest Service. 1986. Service foresters handbook. U. S. Dept. Agr. For. Serv., Southern Region, State and Private Forestry. Misc. Rep. R8-MR11., Atlanta, GA.
U.S. Forest Service. 1988. The South's fourth forest: alternatives for the future. U.S. Dept. Agr. For. Serv., For. Resour. Rep. No. 24, Washington DC.
U.S. Forest Service. 1991a. Draft long-term red-cockaded woodpecker management strategy. U.S. Dept. Agr., For. Serv., Atlanta, GA.

U.S. Forest Service. 1991b. The Savannah River Site red-cockaded woodpecker management plan. Savannah River Forest Station, New Ellenton, SC.

U.S. Forest Service. 1992. The Savannah River Site wildlife, fisheries, and botany operations plan. (In Draft). Savannah River Forest Station, New Ellenton, SC.

U.S. Forest Service. 1993. Draft Environmental Impact Statement for the Management of the red-cockaded woodpecker and its habitat on National Forests in the Southern Region. U.S. Dept. Agr., For. Serv., Atlanta, GA.

U.S. Forest Service. Statement of Policy.

Urban, E.K. (comp.). 1991. 91st Christmas Bird Count, St. Catherines Island, GA. Am Birds 45(4) 726.

Van Balen, J.B. and P.D. Doerr. 1978. The relationship of understory to red-cockaded woodpecker activity. Proc. Southeastern. Assoc. Fish and Wildl. Agencies 32:82-92.

Van Hooser, D.W. and A. Hedlund. 1969. Timber damaged by Hurricane Camille in Mississippi. U.S. For. Serv. Resources Note SO-96. New Orleans, LA.

Van Lear, D.H. 1985. Prescribed fire—its history, uses and effects in southern forest ecosystems. pp. 57-75 *In* D.D. Wade (comp.). Prescribed fire and smoke management in the south: conf. proc. U. S. Dept. Agr. For. Serv. Southern For. Exp. Sta., Asheville, N.C.

Van Sickle, C.C. and A. Hedlund. 1969. Timber losses from Hurricane Camille. For. Farmer 29:6-7.

Van Valen, L. 1962. A study of fluctuating asymmetry. Evolution 16:125-142.

Vieillot, L.J.P. 1807. Histoire naturelle des oiseaux de l'Amerique septrionale. 2:66. Chez Desray, Paris, France.

Vissage, J.S., P.E. Miller and A.J. Hartsell. 1992. Forest statistics for Louisiana Parishes. 1991. U.S. Dept. Agr., For. Serv. Resour. Bull. SO-168. New Orleans, LA.

Vrijenhoek, R.C. and S. Lerman. 1982. Heterozygosity and developmental stability under sexual and asexual breeding systems. Evolution 36:768-776.

Vrijenhoek, R.C., M.E. Douglas and G.K. Meffe. 1985. Conservation genetics of endangered fish populations in Arizona. Sci. 229:400-402.

Waddell, K.L., D.D. Oswald and D.S. Powell. 1989. Forest statistics of the United States, 1987. U. S. Dept. Agr. For. Serv. Resour. Bull. PNW-RB-168, Portland, OR.

Waddington, C.H. 1942. Canalization of development and the inheritance of acquired characters. Nature 150:563-565.

Waddington, C.H. 1948. Polygenes and oligogenes. Nature 156:394.

Wahlenberg, W.G. 1946. Longleaf pine—its use, ecology, regeneration, protection, growth, and management. C. Lathrop Pack Forestry Foundation, Washington, D.C.

Wahlenberg, W.G. 1960. Loblolly pine: its use, ecology, regeneration, protection, growth and management. Duke Univ. , School of Forestry, Durham, NC.

Waldrop, T.A., D.L. White and S.M. Jones. 1992. Fire regimes for pine-grassland communities in the southeastern United States. For. Ecol. Manage. 47:195-210.

Walker, L.C. 1991. The southern forest—a chronicle. Univ. Texas Press, Austin, TX.

Walker, L.R., D.J. Lodge and R.B. Waide. 1991. An introduction to hurricanes in the Caribbean. Biotropica 23:313-316.

Walsberg, G.E. 1978. Brood size and the use of time and energy by the Phainopepla. Ecology 59:147-153.

Walters, J.R. 1990. Red-cockaded woodpeckers: a "primitive" cooperative breeder. pp. 69-101 *In* P.B. Stacey and W.D. Koenig (eds.). Cooperative Breeding in Birds: Long-term Studies of Ecology and Behavior. Cambridge Univ. Press, Cambridge, England

Walters, J.R. 1991. Application of ecological principles to the management of endangered species: the case of the red-cockaded woodpecker. Ann. Rev. Ecol. Syst. 22:505-523.

Walters, J.R., P.D. Doerr and J.H. Carter, III. 1988. The cooperative breeding system of the red-cockaded woodpecker. Ethology 78:275-305.

Walters, J.R., C.K. Copeyon and J.H. Carter, III. 1992a. Test of the ecological basis of cooperative breeding in red-cockaded woodpeckers. Auk 109:90-97.

Walters, J.R., P.D. Doerr and J.H. Carter III. 1992b. Delayed dispersal and reproduction as a life-history tactic in cooperative breeders: fitness calculations from Red-cockaded woodpeckers. Amer. Nat. 139: 623-643.

Walters, J.R., J.H. Carter, III and P.D. Doerr. 1995a. Long-term response to drilled artificial cavities by red-cockaded woodpeckers in the North Carolina Sandhills. *In* D.L. Kulhavy, R.G. Hooper and R. Costa (eds.). Red-cockaded Woodpecker: Recovery, Ecology and Management, Center for Applied Studies, College of Forestry, Stephen F. Austin State Univ., Nacogdoches, TX.

Walters, J.R., P.P. Robinson, W. Starnes and J. Goodson. 1995b. The relative effectiveness of artificial cavity starts and artificial cavities in inducing the formation of new groups of Red-cockaded woodpeckers. *In* D.L. Kulhavy, R.G. Hooper and R. Costa (eds.). Red-cockaded woodpecker: Recovery, Ecology and Management, Center for Applied Studies, College of Forestry, Stephen F. Austin State Univ., Nacogdoches, TX.

Watson, J.C., R.G. Hooper, D.L. Carlson, W.E. Taylor and T.C. Milling. 1995. Restoration of the red-cockaded woodpecker population on the Francis Marion National Forest: three years post-Hugo. *In* D.L. Kulhavy, R.G. Hooper and R. Costa (eds.). Red-cockaded Woodpecker: Recovery, Ecology and Management, Center for Applied Studies, College of Forestry, Stephen F. Austin State Univ., Nacogdoches, TX.

Webb, T., III. 1987. The appearance and disappearance of major vegetational assemblages: long-term vegetational dynamics in eastern North America. Vegetatio. 69: 177-187.

Weidman, R.H. 1920. A study of windfall loss of western yellow pine in selection cuttings fifteen to thirty years old. J. For. 18:616-622.

Weir, M.J.C. 1973. Airflow patterns from photogrammetric measurements of windblown timber. Photogramm. Rec. 7:731-736.

Wells, B.W. and I.V. Shunk. 1931. The vegetation and habitat factors of the coarser sands of the North Carolina coastal plain: an ecological study. Ecol. Monogr. 1:467-520.

Wiens, J.A. 1981. Single-sample surveys of communities: are the revealed patterns real? American Naturalist 117:90-98

Wilcox, B.A. 1982. Insular ecology and conservation. pp. 95-117 *In* M.E. Soulé and B.A. Wilcox (eds.). Cons. Bio., Sinauer Associates, Sunderland, MA.

Wilkinson, L. 1990. SYSTAT: The system for statistics. Evanston, IL: SYSTAT, Inc.

Wilkinson, R.C., R.W. Britt, E.A. Spence and S.M. Seiber. 1978. Hurricane—tornado damage, mortality and insect infestations of slash pine. So. J. Appl. For. 2:132-134.

Williams, M. 1989. Americans and their forests. Cambridge Univ. Press, Cambridge.

Williams, T.M. and D.J. Lipscomb. 1983. A logging history of Hobcaw Barony. Clemson Dept. For. Res. Series. No. 38.

Wilson, A. and C.L. Bonaparte. 1831. American ornithology or the natural history of the birds of the United States. Vol. 1. Constable and Co., Edinburgh.

Wilson, B.F. 1984. The growing tree. Univ. Massachusetts Press, Amherst, MA.

Wilson, B.F. and R.R. Archer. 1979. Tree design: some biological solutions to mechanical problems. BioScience 29:293-298.

Withgott, J.H., J.C. Neal and W.G. Montague. 1995. A technique to deter climbing by rat snakes on cavity trees of red-cockaded woodpeckers. *In* D.L. Kulhavy, R.G. Hooper and R. Costa (eds.). Red-cockaded Woodpecker: Recovery, Ecology and Management, Center for Applied Studies, College of Forestry, Stephen F. Austin State Univ., Nacogdoches, TX.

Wood, D.A. 1977. Status, habitat, home range, and notes on the behavior of the red-cockaded woodpecker in Oklahoma. M.S. Thesis, Oklahoma State Univ., Stillwater.

Wood, D.A. 1983a. Foraging and colony habitat characteristics of the red-cockaded woodpecker in Oklahoma. pp.51-58 *In* D.A. Wood (ed.). Red-cockaded Woodpecker Symposium II Florida Game and Fresh Water Fish Comm. Tallahassee.

Wood, D.A. 1983b. Observations on the behavior and breeding biology of the red-cockaded woodpecker in Oklahoma. pp. 92-94 *In* D.A. Wood (ed.). Proc. Red-cockaded Woodpecker Symposium II Florida Game and Fresh Water Fish Comm., Tallahassee, Florida.

Wood, D.A. (ed.). 1983c. Red-cockaded Woodpecker Symposium II. State of Florida Game and Fresh Water Fish Commission, Tallahassee, FL

Wood, D.A. and J.C. Lewis. 1977. Status of the red-cockaded woodpecker in Oklahoma. Proc. Annual Conference Southeastern Association Fish and Wildlife Agencies 31:276-282.

Wood, D.A. and A.S. Wenner. 1983. Status of the red-cockaded woodpecker in Florida: 1983 update. pp. 89-91 *In* D.A. Wood (ed.). Red-cockaded Woodpecker Symposium II, Proceedings. Florida Game and Fresh Water Fish Commission, Tallahassee, FL.

Wood, G.W., L.J. Niles, R.M. Hendrick, J.R. Davis and T.L. Grimes. 1985a. Compatibility of even-aged management and red-cockaded woodpecker conservation. Wildl. Soc. Bull. 13:5-17.

Wood, G.W., L.J. Niles, R.M. Hendricks and T.L. Grimes. 1985b. Influences of clearcutting on red-cockaded woodpecker reproduction and nestling tending. For. Bull. 45., Department Forestry, Clemson Univ., Clemson, SC. 11p.

Wooten, M.C. and M.H. Smith. 1986. Fluctuating asymmetry and genetic variability in a natural population of *Mus musculus*. J. Mamm. 67:725-732.

Wright, S. 1931. Evolution in Mendelian populations. Genetics 16:97-159.

Wyckoff, D. G. and L. R. Fisher. 1985. Preliminary testing and evaluation of the Grobin Davis archeological site, 34Mc-253, McCurtain County, OK. Archaeological Res. Surv. Rep. 22. Oklahoma Arch. Surv., Univ. Oklahoma, Norman, OK.

Zahavi, A. 1974. Communal nesting by the Arabian Babbler. Ibis 116:84-87.

Zahner, R. and F.W. Whitmore. 1960. Early growth of radically thinned loblolly pine. J. For. 58:628-634.

Zar, J.H. 1984. Biostatistical Analysis. Prentice Hall, Englewood Cliffs, NJ.

Zedaker, S.M., H.E. Burkhart and A.R. Stage 1987. General principles and patterns of conifer growth and yield. pp. 203-241 *In* Walstad, J.D and P.J. Kuch (eds). Forest vegetation management for conifer production. John Wiley & Sons, New York.

Zink, R.M., M.F. Smith and J.L. Patton. 1985. Associations between heterozygosity and morphological variance. J. Hered. 76:415-420.

Subject Index

D

E

F

G

H

I

J

K

L

M

Q

R

S

T

U

V

W

X

Y

Index of Scientific Names

Contributors

David H. Allen
North Carolina Wildlife Resource Commission,
New Bern, NC 28560
pp. 81-88

W. Wilson Baker
Tall Timbers Research, Inc.,
Tallahassee, FL 32312
pp. 489-493

W. Wilson Baker
The Nature Conservancy, 625 N. Adams St.,
Tallahassee, FL 32301
pp. 457-469

Haven R. Barnhill
Southern Research Station
Department of Forest Resources
Clemson University, Clemson, SC 29634
pp. 323-331

Ed Barron
Texas Forest Service, College Station, TX
pp. 477-481

C. M. Bartlett
University College of Cape Breton
Sydney, Nova Scotia, Canada B1P 6L2.
pp. 332-333

Tim Beaty
Directorate of Engineering and Housing
Fish and Wildlife Section
Fort Stewart, GA 31320
pp. 270-276

Reed Bowman
Archbold Biological Station
Lake Placid, FL 33852
pp. 415-426

Dana Bradshaw
Center for Conservation Biology
The College of William and Mary
Williamsburg, VA 23185
pp. 482-488

Leonard A. Brennan
Department of Wildlife and Fisheries
Mississippi State University, MS.
pp. 309-319

Jacqueline J. Britcher
Endangered Species Branch
XVIII Airborne Corps and Fort Bragg HQ
Fort Bragg, NC 28307
pp. 89-97

R. R. Cahal, III,
College of Forestry
Stephen F. Austin State University
Nacogdoches, TX, 75962
pp. 427-430

Mark A. Cantrell
Endangered Species Branch
XVIII Airborne Corps and Fort Bragg HQ
Fort Bragg, NC 28307
pp. 89-97

Danny L. Carlson
USDA Forest Service
Francis Marion National Forest
McClellanville, SC 29458
pp. 172-182

Dawn Carrie
USDA Forest Service, Cleveland, TX 77327
pp. 320-322

J.H. Carter III
Department of Zoology
North Carolina State University
Box 7617, Raleigh, NC 27695-7617
pp. 248-258, 380-384

J.H. Carter III
Dr. J.H. Carter III & Associates
Southern Pines, NC 28388
pp. 372-379

J. E. Cely
Nongame-Heritage Trust Section,
South Carolina Wildlife and Marine Resources
Department
Columbia; SC 29202
pp. 470-476

G. M. Chrismer
College of Forestry
Stephen F. Austin State University
Nacogdoches, TX, 75962
pp. 214-218

Richard N. Conner
Wildlife Habitat and Silviculture Laboratory
Southern Research Station
USDA Forest Service
Nacogdoches, TX 75962
pp. 183-195, 208-213, 335-352, 410-414

Jeffery L. Cooper
Department of Wildlife and Fisheries
Mississippi State University, MS.
pp. 309-319

Carole K. Copeyon
Department of Zoology
North Carolina State University
Box 7617, Raleigh, NC 27695
pp. 380-384

Ralph Costa
USDI, Fish and Wildlife Service
Red-cockaded Woodpecker Field Office
College of Agriculture, Forestry and Life Sciences
Clemson University, Clemson, SC 29634
pp. 3-5, 22-27, 67-74, 385-388

Robert N. Coulson
Knowledge Engineering Laboratory
Department of Entomology
Texas A&M University
College Station, TX 77843
pp. 191-195

James Cox
Florida Game and Fresh Water Fish Commission
Tallahassee, FL 32399-1600
pp. 457-464

Joseph M. Dabney
USDA Forest Service, Atlanta, GA 30367
pp. 61-66

Roy S. DeLotelle
DeLotelle and Guthrie, Inc.
1220 SW 96th Street, Gainesville, FL 32607
pp. 259-269, 270-276

Greg DeMuth
Environment Division
Orlando Utilities Commission, Orlando, FL
pp. 259-269

Philip M. Dixon
Savannah River Ecology Laboratory
Drawer E, Aiken, SC 29802
pp. 239-247

Phillip D. Doerr
Department of Zoology
North Carolina State University
Box 7617, Raleigh, NC 27695-7617
pp. 248-258, 380-384

Lieutenant General Charles E. Dominy
Director of the Army Staff
pp. 16-21

R. Todd Engstrom
Tall Timbers Research, Inc.
Tallahassee, FL 32312
pp. 372-379, 489-493

Robert J. Epting
Department of Resource Management
St. Johns River Water Management District,
P.O. Box 1429, Palatka, FL 32178
pp. 259-269, 270-276

Ronald E. F. Escano
USDA Forest Service
Washington, DC 20090-6090
pp. 28-35

D. P. Ferral
Nongame-Heritage Trust Section
South Carolina Wildlife and
Marine Resources Department
Columbia, SC 29202
pp. 470-476

Jeffrey W. Fitzgerald
Knowledge Engineering Laboratory
Department of Entomology
Texas A&M University
College Station, TX 77843
pp. 191-195

Kathleen E. Franzreb
USDA Forest Service
Southern Research Station
Department of Forest Resources,
Clemson University, Clemson, SC 29634
pp. 81-88, 323-331

J. G. Gage
College of Forestry
Stephen F. Austin State University
Nacogdoches, TX, 75962
pp. 427-430

Glen D. Gaines
USDA Forest Service
Savannah River Forest Station
New Ellenton, SC 29809
pp. 81-88

Janice Goodson
Department of Zoology
North Carolina State University,
Box 7617, Raleigh, NC 27695-7617
pp. 367-372

Gregory Hagan
Department of Biological Science
Florida State University
Tallahassee, FL 32306
pp. 353-360

Bruce W. Hagedorn
Natural Resources Division
Eglin Air Force Base, FL. 32578
pp. 494-502

Jeffrey L. Hardesty
Florida Cooperative Fish and Wildlife Research Unit
Department of Wildlife, Ecology and Conservation
University of Florida, Gainesville, FL 32611
The Nature Conservancy
Gainesville, FL 32611
pp. 494-502

Larry D. Harris
University Florida,, Gainesville, FL 32611.
pp. 389-393

Charles A. Hess
Department of Biological Science
Florida State University
Tallahassee, FL 32306
pp. 353-361

Charles A. Hess
USDA Forest Service, P.O. Box 579
Bristol, FL, 32321
pp. 385-388

John D. Hodges
Department of Forestry
Mississippi State University
Mississippi State, MS 39762
pp. 196-207, 431-435

Erich L. Hoffman
Endangered Species Branch
XVIII Airborne Corps and Fort Bragg HQ
Fort Bragg, NC 28307
pp. 89-97

Thomas P. Holmes
USDA Forest Service,
Southern Research Station
Research Triangle Park, NC 27709
pp. 219-226

Robert G. Hooper
U.S.D.A., Forest Service
Southern Research Station
Charleston, SC 29414
pp. 59-60, 145-166, 172-182, 283-289

Christopher Huh
Archbold Biological Station
Lake Placid, FL 33852
pp. 415-426

George A. Hurst
Department of Wildlife and Fisheries
Mississippi State University, MS
pp. 309-319

Jerome A. Jackson
Department of Biological Sciences
Mississippi State University
Mississippi State, MS 39762
pp. 42-48, 277-282

Douglas A. James
Department of Biological Sciences
University of Arkansas
Fayetteville, AR 72701
pp. 401-409

Frances C. James
Department of Biological Science
Florida State University
Tallahassee,FL 32306
pp. 353-360, 437-451

William L. Jarvis
USDA Forest Service
Savannah River Forest Station
New Ellenton, SC 29809
pp. 81-88

James E. Johnson
U.S. Fish and Wildlife Service
Arkansas Cooperative Fish and Wildlife Research Unit
University of Arkansas,
Fayetteville, AR 72701
pp. 401-409

John J. Kappes, Jr.
University Florida, Gainesville, FL 32611
pp. 389-393

John Kleinhofs
Forest Manager, West-Central Region
Georgia-Pacific Corporation
Crossett, AR 71635
pp. 75-80

Bowie Kotrla
Department of Biological Science
Florida State University
Tallahassee, FL 32306
pp. 353-360

Randall A. Kramer
School of the Environment
Duke University, Durham , NC 27707
pp. 219-226

Mike Krueger
Texas Parks & Wildlife Department,
P.O. Box 207, Lampassas, TX
pp. 477-481

Dennis L. Krusac
USDA Forest Service, Atlanta, GA 30367
pp. 61-66

David L. Kulhavy
College of Forestry
Stephen F. Austin State University
Nacogdoches, TX 75962
pp. 131-136, 145-147, 214-218, 410-414, 427-430

Kevin S. Laves
USDA Forest Service
Southern Research Station
Clemson University, Clemson, SC 29634
pp. 81-88

Michael R. Lennartz
U.S.D.A. Forest Service
Washington, DC 20090-6090
pp. 283-289

Bruce D. Leopold
Department of Wildlife and Fisheries
Mississippi State University, MS
pp. 309-319

Donald J. Lipscomb
The Belle W. Baruch Forest Science Institute
Clemson University
Georgetown, SC 29442
pp. 137-143, 167-171

Susan C. Loeb
Southern Research Station
USDA Forest Service
Department of Forest Resources,
Clemson University,
Clemson, SC 29634-1003
pp. 361-366

Kathleen E. Lucas
Department of Wildlife and Fisheries
Mississippi State University, MS.
pp. 309-319

M. P. Luttrell
Southeastern Cooperative Wildlife Disease Study
College of Veterinary Medicine
The University of Georgia,
Athens, GA 30602
pp. 332-333

Richard Martin
The Nature Conservancy of Louisiana
P.O. Box 4125
Baton Rouge, LA 70821
pp. 452-456

Ronald E. Masters
Department of Forestry
Oklahoma State University
Stillwater, OK 74078
pp. 290-302

Colin J. McAdie
U.S.D.C., NOAA
National Weather Service
National Hurricane Center
Coral Gables, FL 33146
pp. 148-166

Robert W. McFarlane
McFarlane & Associates, Houston, TX 77036
pp. 303-308

Marvin C. Meier
USDA Forest Service, Atlanta, GA
pp. 11-15

Timothy E. Milling
USDA Forest Service
Southern Research Station
Charleston, SC 29414
pp. 172-182

Warren G. Montague
Poteau Ranger District, USDA Forest Service
P.O. Box 2255, Waldron, AR 72958
pp. 394-409

Kenneth Moore
USDA Soil Conservation Service
Leesville, LA 71446
pp. 320-322

Joseph C. Neal
Arkansas Cooperative Fish and Wildlife Research
Unit, University of Arkansas
Fayetteville, AR 72701
pp. 394-409

T. Evan Nebeker
Department of Entomology and Plant Pathology
Mississippi State University
Mississippi State, MS 39762
pp. 196-207, 431-435

Forrest L. Oliveria
USDA Forest Service
Forest Pest Management
Pineville, LA 71360
pp. 191-195

Brent Ortego
Texas Parks & Wildlife Department
2601 N. Azalea, Victoria, TX 77901
pp. 320-322, 477-481

Stephen D. Parris
Environmental Division
Directorate of Engineering and Housing
Fort Polk, LA 71459
pp. 277-282

Matthew Pelligrine
Department of Forestry
Mississippi State University
Mississippi State, MS 39762
pp. 196-207, 431-435

H. Franklin Percival
Florida Cooperative Fish and Wildlife Research Unit, Department of Wildlife
University of Florida
Gainesville, FL 32611
pp. 494-502

John J. Petrick
USDA Forest Service, Atlanta, GA 30367
pp. 61-66

Carl J. Petrick
Natural Resources Division
Eglin Air Force Base, FL. 32578
pp. 494-502

Patricia M. Purcell
Dr. J.H. Carter III & Associates
Southern Pines, NC 28388
pp. 372-379

Dixie Watts Reaves,
Department of Economics
Duke University, Durham 27707
pp. 219-226

Joseph P. Reinman
U. S. Fish and Wildlife Service
St. Marks National Wildlife Refuge
St. Marks, FL 32355
pp. 106-111

David M. Richardson
Noxubee National Wildlife Refuge
U.S. Fish & Wildlife Service
Route 1, Box 142
Brooksville, MS 39739
pp. 98-105

Pamela P. Robinson
Department of Zoology
North Carolina State University
Box 7617, Raleigh, NC 27695-7617
pp. 367-371

W. G. Ross
College of Forestry
Stephen F. Austin State University
Nacogdoches, TX, 75962
pp. 214-218, 410-414, 427-430

D. Craig Rudolph
Wildlife Habitat and Silviculture Laboratory
Southern Research Station
USDA Forest Service
Nacogdoches, TX 75962
pp. 183-195, 208-213, 338-352

R. Schaefer
Wildlife Habitat and Silviculture Laboratory
Southern Research Station
USDA Forest Service
Nacogdoches, TX 75962
pp. 338-342

John E. Skeen
Oklahoma Dept. of Wildlife Conservation,
1801 N. Lincoln,
Oklahoma City, OK 73105
pp. 290-302

Emlyn B. Smith
U.S. Forest Service
Catahoula Ranger District, 5325 Highway 8
Bently, LA 71407-9725
pp. 452-456

Ruthe J. Smith
Florida Cooperative Fish and Wildlife Research Unit, Department of Wildlife,
University of Florida,, Gainesville, FL 32611
pp. 494-502

Bruce A. Sneddon
United States Army
Office of the Deputy Chief of Staff for Operations and Plans,
Department of the Army
Washington, D.C. 20310
pp. 36-41

P. W. Stangel
National Fish and Wildlife Foundation
1120 Connecticut Ave., NW, Suite 900
Washington, DC 20036
pp. 239-247, 332-333

Warren Starnes
USDA Forest Service, Croatan Ranger District
141 E. Fisher Avenue, New Bern, NC 28560
pp. 367-371

Earnest E. Stevens
USDA Forest Service
Southern Research Station
Department of Forest Resources,
Clemson University, Clemson, SC 29634-1003
pp. 225-238, 361-366

James M. Stockie
Noxubee National Wildlife Refuge
U.S. Fish & Wildlife Service
Route 1, Box 142
Brooksville, MS 39739
pp. 98-105

Sheron W. Sweeney
National Council of the Paper Industry for Air and Stream Improvement, Inc.
Department of Aquaculture, Fisheries and Wildlife
Clemson University, Clemson, SC 29634-0362
pp. 503-509

William E. Taylor
USDA Forest Service
Francis Marion National Forest
Moncks Corner, SC, 29461
pp. 172-182

Robert A. Tisdale
Department of Entomology and Plant Pathology
Mississippi State University
Mississippi State, MS 39762
pp. 196-207, 431-435

John Turner
Director, Fish and Wildlife Service
pp. 6-10

Jimmy S. Walker
USDA Forest Service
Atlanta, GA 30367
pp. 112-130

Jeffrey R. Walters
Department of Zoology
North Carolina State University
Box 7617, Raleigh, NC 27695-7617
pp. 248-258, 367-371, 380-384

J. Craig Watson
USDA Forest Service
Francis Marion National Forest
Moncks Corner, SC 29461
pp. 172-182

James Whitehead
Oklahoma Department of Wildlife Conservation,
1801 N. Lincoln,, Oklahoma City, OK 73105
pp. 290-302

T. Bently Wigley
National Council of the Paper Industry for Air and Stream Improvement, Inc.
Department of Aquaculture, Fisheries and Wildlife
Clemson University
Clemson, SC 29634-0362
pp. 503-509

Thomas M. Williams
The Belle W. Baruch Forest Science Institute
Clemson University, Georgetown, SC 29442
pp. 137-147, 167-171

James H. Withgott
Department of Biological Sciences
University of Arkansas, Fayetteville, AR 72701
pp. 394-400

Don Wood
Florida Game and Fresh Water Fish Commission
Tallahassee, FL 32399-1600
pp. 457-464

Gene W. Wood
Consulting Forest Wildlife Ecologist
Georgia-Pacific Corporation, Seneca, SC 29678
pp. 75-80

Gene W. Wood
Department of Aquaculture, Fisheries and Wildlife,
Clemson University,
Clemson, SC 29634
pp. 49-60

David L. Kulhavy, currently Piper Professor of Forest Entomology and Landscape Ecology, research experience with red-cockaded woodpeckers includes site/stand factors as they affect stand health and long-term monitoring of red-cockaded woodpecker clusters and surrounding stands as part of a forest landscape ecology project. In 1995, he was selected for the University of Idaho Alumni Hall of Fame; and as Distinguished Professor by the Alumni Association at Stephen F. Austin State University. He received the Distinguished Achievement Award in Teaching from the Entomological Society of America in 1993; the regional award from the National Association for Interpretation and the Texas Forestry Association research award. Dr. Kulhavy received a bachelor degree in Zoology from San Diego State University; a master's degree in Forest Entomology and a Ph.D. in Forestry Sciences from the University of Idaho.

Since 1976, Robert G. Hooper, a research wildlife biologist with the Southern Research Station, USDA Forest Service, for 29 years, has conducted research on the ecology and management of the red-cockaded woodpecker. Focusing on foraging habitat requirements, prey-habitat relationships, cavity tree selection, cluster formation, population trends, population monitoring techniques, hurricane risk assessment, sociobiology, spatial behavior, land use history and artificial cavity technology, Hooper's current studies involve red-cockaded woodpecker populations in South Carolina, Georgia, Florida and Louisiana. Hooper received his bachelor of science degree in biology from Lynchburg College and a master of science degree in wildlife management from Virginia Polytechnic Institute and State University. He received the USDA Distinguished Service award for Hurricane Hugo recovery efforts.

Ralph Costa, currently Red-cockaded Woodpecker Recovery Coordinator, U. S. Fish and Wildlife Service, worked for the U. S. Forest Service with the red-cockaded woodpecker on the Daniel Boone National Forest in Kentucky (1978), the Kisatchie National Forest (1985-1988), and the Apalachicola National Forest (1988-1991); before transferring to the Fish and Wildlife Service in 1991. His current assignments include red-cockaded woodpecker recovery; developing conservation coalitions, partnerships and economic incentives for private lands; and public outreach and environmental education. Costa received a bachelor of science degree (1973) in wildlife biology and a masters of science degree in watershed management (1976) from the University of Arizona. Since joining the Fish and Wildlife Service, he has received 2 Special Achievement Awards, 2 Quality Performance Awards and the Regional Director's Honor Award for 1993.

Derrick Hamrick, photographer, grew up in the suburbs of Raleigh, NC, where he became interested in observing wildlife. Hamrick is a self-taught photographer with assignments in Australia, Kenya, Panama, Cuba, Israel, Belize and Italy. Credits include National Geographic World, Ranger Rick, American Forests, BBC Wildlife, National Wildlife, Nature Photographer and Bird Watcher's Digest. He uses Nikon equipment and high-speed flash to freeze motion of his subjects in startling detail. Currently he resides along the banks of the Neuse River in rural Wake County, NC.